An Illustrated Guide to Anatomical Eponyms

Hans J. ten Donkelaar • Marina Quartu
David Kachlík

An Illustrated Guide to Anatomical Eponyms

Springer

Hans J. ten Donkelaar
Department of Neurology
Radboud University Medical Centre
Nijmegen, The Netherlands

David Kachlík
Second Faculty of Medicine
Department of Anatomy
Charles University
Prague, Czech Republic

Marina Quartu
Department of Biomedical Science
University of Cagliari
Cagliari, Italy

ISBN 978-3-031-91663-2 ISBN 978-3-031-91664-9 (eBook)
https://doi.org/10.1007/978-3-031-91664-9

Cover Image: The fresco of a public dissection in the lunette of the Eastern wall of the Auditorium of the Jesuit College in Jindřichův Hradec, Southern Bohemia, Czech Republic, was made by a fresco-secco technique between 1628 and 1633. The picture of the fresco was kindly provided by the Director of the Museum of Photography, Ondřej Mašek.

This Springer imprint is published by the registered company Springer Nature Switzerland AG
The registered company address is: Gewerbestrasse 11, 6330 Cham, Switzerland

Preface

Introduction

The history of eponymous anatomical terminology was beautifully summarized by Jessie Dobson (1962). Apparently, Andreas Vesalius (1543) introduced personal names into the anatomical nomenclature. In the first edition of the Nomina Anatomica, the Basler Nomina Anatomica (BNA), Wilhelm His (1895) stated that personal names are met not only in Anatomy but also in other natural sciences and in particular in medicine. Others, such as Friedrich Gustav Jacob Henle, in fact declared war against an originally much more extended personal nomenclature, and later anatomists quite vigorously fought against the use of eponyms. In later editions of the Nomina Anatomica most eponyms were simply deleted, but in the Terminologia Anatomica (1998) a list of anatomical eponyms was provided. More recently, the Federal International Programme for Anatomical Terminology (FIPAT) of the International Federation of Associations of Anatomists (IFAA) presented the more common anatomical eponyms in their second edition of the Terminologia Embryologica (TE2 2017), the Terminologia Neuroanatomica (TNA 2017), and the second edition of the Terminologia Anatomica (TA2 2019).

Pros and Cons to the use of eponyms in anatomical terminology have been repeatedly published (Whitworth 2007; Woywood and Matteson 2007; Olry 2014a, b; Burdan et al. 2016). Updates of Dobson's work were presented by Donáth (1969), Olry (1995), Páč (2000), Bartolucci and Forbis (2004) and for various parts of anatomy such as otology (Lustig et al. 1998; Mudry 2001), fascial eponyms (Adstrum 2015), and eponyms in obstetrics and gynecology (Baskett 2019). More extensive sources such as Pagel (1901), Dumesmil and Bonnet-Roy (1947), Fischer (1962), Hubotter (1962), the third edition of Gurtl, Wernich, and Hirsch's (1883) original "Biographisches Lexikon," Haymaker and Schiller (1968), Clarke and O'Malley (1996), and Swanson (2014) were available and for histology Krstić (1984) and Dhom (2001). The website WhoNamed It? was also a reliable source for eponyms. For Italian eponyms, Mezzogiorno (1991), Rizzi et al. (2014), and Zaffe (2015) were of great value, and for the Dutch history of Medicine, Lindeboom (1981). An excellent source on pre-Vesalian anatomists was Vollmuth (2004). For data on early Czech anatomy, Grim et al. (2014) was of great help. Although eponyms may appear anachronistic, "Their elimination appears utopian on the weight of habits and their usage is occasionally easier than the official anatomical terminology" (Olry 1995). So, in line with FIPAT's current view on the use of eponyms, we prepared a concise illustrated dictionary of anatomical eponyms, including the more obvious anthropological, embryological, and histological eponyms. For the nervous system and the senses, most eponyms were published in ten Donkelaar et al. (2018). The illustrations from those who gave their names to eponyms come from Pagel (1901), others from the Wellcome Library in London, as well as from various Universities, Libraries, and Musea. Wikimedia Commons has been a great source. The illustrations used to explain eponyms are largely derived from our own collection (ten Donkelaar and Oostra 2014; ten Donkelaar 2020; ten Donkelaar et al. 2018, 2023).

What Are Eponyms?

According to the 28th edition of Stedman's Medical Dictionary (2008), an eponym is: "The name of a disease, structure, operation or procedure, usually derived from the name of the person who discovered or described it first." The word eponym derives from the Greek for "named after." Other dictionaries show comparable definitions. Today, eponyms are used in the nonpossessive form. Here, we restrict ourselves to anatomical, embryological, and histological eponyms. As will be shown in **Part I** *The Classical Eponyms*, in various countries, different eponyms are given for the same structure. Not always an eponym is given to who discovered or first described a certain structure. In **Part III**, we present anatomical eponyms used in various clinical disciplines. We present eponyms with their proper accents wherever possible but otherwise in the English spelling. In **Part II**, the alphabetical lists of eponyms, for German eponyms the preposition "von" is hardly used, so: Haller, Albrecht von, and Luschka, Hubert, von. Exceptions are von Baer, Karl Ernst and von Economo, Constantin, but they are listed under B and E, respectively, for consistency. For Dutch eponyms, the prepositions "de" and "van" are part of the family name, so: de Graaf, Reinier is found under D, and van Diemerbroeck, IJsbrand, van Helmont, Jean Baptiste, van Horne, Jan, van Leeuwenhoek, Antonie, and van den Spieghel, Adriaan under V. The same holds for French eponyms starting with "Le" as in Le Fort, to be found under L. For the place of birth and where they worked, the terms used at that time are given, so as examples: Dorpat (currently Tartu, Estland), Königsberg (currently Kaliningrad, Russia), and Breslau (currently Wróclaw, Poland).

Image Credits

Most of the engravings, portraits, or photographs are more than 100 years old, and so are in the Public Domain, and freely available on the internet. The following figures date back to **Pagel (1901):** Abernethy, Adamkiewicz, Artl, von Baer, Baudelocque, Bell (Charles), Bichat, Billroth, Blumenbach, Boerhaave, Broca, Brodmann, Brücke, Burdach, Charcot, Cooper, Cruveilhier, Dogiel, Donders, Dupuytren, Edinger, Exner, Falloppio, Flechsig, Forel, Guyon, Henle, Hensen, Heubner, Huxley, Hyrtl, Key, Kölliker, Kupffer, Lannelongue, Littré, Ludwig, Luschka, Magendie, Malgaigne, Malpighi, Merkel, Meynert, Morgagni, Müller, Nélaton, Owen, Pirogoff, Purkinje, Recklinghausen, Reichert, Remak, Retzius (Gustav), Riedel, Robin, Rokitansky, Roser, Sattler, Schwalbe, Schwann, Stilling, Testut, Traube, Troeltsch, Turner, Valentin, Vesalius, Virchow, Volkmann, Waldeyer, Westphal, Zaufal, Zinn, and Zuckerkandl. From the **Wellcome Collection CCBY**, we acknowledge the use of portraits of: Addison, Baillarger, Bartholin (Caspar), Campbell, Clarke, Cloquet, Cotugno, the Dejerines, Douglas, Farabeuf, Glisson, Gowers, Gratiolet, Hilton, Hochstetter (Anatomical Lesson), Houston, Hunter (John), Hunter (William), Ingrassia, Keith, Langhans, McBurney, Monakow, Monro, Paré, Petit (Pourfour du), Quain, Reil, Santorini, Sappey, Scarpa, Skene, Smellie, Valverde de Amusco, van Deventer, van Helmont, van Hoboken, van Leeuwenhoek, Vicq d'Azyr, Vieussens, Weber (EF), Wernicke, and Willis. The **Royal College of Surgeons** is thanked for the portraits of Colles and Cowper, and the **American College of Surgeons** for the photograph of Ochsner. From the **Dutch Iconographic Bureau** the engravings of de Graaf, Sylvius (François de le Boë), and van Horne are acknowledged.

From **Wikimedia Commons** we acknowledge the portraits of Albarrán, Aschoff, Aselli, Auerbach, Bartholin (Thomas the Elder), Botallo, Bowman, Cantlie, Civinini, Claudius, Coiter, Corti, Crampton, Cuvier, Czermak, Daubenton, Demours, Denonvilliers, Disse, Drummond, Duret, Ecker, Eustachi, Fabrici, Ferrein, Fontana, Froriep, Fuse, Ganser, Gaskell, Gerdy, Gerlach, Gerota, Giacomini, Gimbernat, Glaser, Golgi, Goll, Gudden, Haller, Harvey, Heister, Heschl, Hirschsprung, His Jr, His Sr, Hoche, Hoffa, Hovelacque, Howell, Huschke, Ihering, Jacobsohn, Jolly, Juvara, Kaes, Klümpke, Kobelt, Kocher, Lancisi, Langer, Langerhans,

Langley, Lauth (Ernest), Lauth (Thomas), Lecat, Leydig, Lobstein, Louis, Lugaro, Luys, MacLean, Mall, Marchi, Mascagni, Meckel JF (the Elder), Meckel JF (the Younger), Meibom, Meissner, Nissl, Nuel, Obersteiner, Oddi, Onufrowitz, Pacini, Paneth, Panizza, Pavlov, Pawlik, Petit (JL), Pick, Pitres, Prussak, Quénu, Ramón y Cajal, Ranke, Ranvier, Rathke, Reissner, Retzius (Anders), Riolan, Rivinus, Rolando, Rose, Rosenthal, Rouget, Rusconi, Schlemm, Schultze (Friedrich), Sertoli, Siebenmann, Soemmerring, Spemann, Staderini, Struthers, Swammerdam, Tawara, Tenon, Thebesius, Theile, Tiedemann, Todaro, Trolard, Türck, Valsalva, van Diemerbroeck, van den Spieghel, Varolio, Vater, Verga (Buste), Verheyen, Vitali, Wachendorff, Walther, Weber (WE), Winslow, Wolff, Worm, Wutzer, Zenker, and Ziehen, all are in the public domain.

From **Wikipedia** we acknowledge the portraits of Acrel, Aranzio, Bellini, Boettcher, Chiarugi, Chopart, Deiters, von Economo, Foville, Helvetius, Imlach, Jacobson, Jarjavay, Kiesselbach, Michels, Pacchioni, Palfijn (Statue), Pertick, Poirier (Wikimanche), Quatrefages de Bréau, Rouvière, Rüdinger, Ruffini, Sedillot, Sylvius (Jacques), Treitz, Tyson, Unna, Wenckebach, Winterhalter, and Wrisberg, all are in the public domain; and from **Wikidata**, the portraits of Descemet, Folli, Jackson, Lenhossék, Lisfranc, Lissauer, Nuck, Sommer, and Uchimura, also in the public domain.

Several portraits come for various **Universities**, **Libraries**, or **Musea**, all are in the public domain or free to use. The portraits of Albinus and Rau come from the University of Leiden (Collectiones Icones Leidense), the portraits of Arnold and Nuhn from the University Library of Heidelberg, the engraving of Bidloo stems from his atlas, the engraving of Blasius and the portrait of Doornik from the Municipal Archive Amsterdam, and the portrait of Jelgersma comes also from the University of Leiden. The photo of Brödel comes from the Alan Mason Chesney Archives of the Johns Hopkins Medical Institutions. The portrait of Chassaignac comes from the Bibliothèque de l'Academie nationale de médecine in Paris. The photos of Ebner and Fuchs come from the Archiv der Universität Wien. From the National Library of Medicine (Digital Collection), we used the portraits of Hering and Papez. The portrait of Kerckring hangs in the Hamburger Kunsthalle. The portrait of Larrey comes from Chemin de Mémoire, Ministères des Armées, Paris. The photograph of Larsell is from the Oregon Health & Sciences University Historical Collection and Archives, Digital Collections. The photograph of Lorente de Nó comes from the Real Academia de la Historia, Madrid. The portrait of Marshall is in the Victoria and Albert Museum in London.

For the Anatomical Lesson of Bolk in the Anatomisch Embryologisch Laboratorium, AMC, we thank Roelof-Jan Oostra (Amsterdam). For the Anatomical Lessons of Camper and Ruysch we acknowledge the Amsterdam Museum. The original of the portrait of Rosenmüller is in the Rijksmuseum, Amsterdam, and that of the portrait of Stensen is found in the Uffizi Museum in Florence. The Anatomical Lesson of Nicolaes Tulp is found in the Mauritshuis in The Hague. The Anatomical Lesson of Velpeau is in the Hôpital Musée d'Assistance Publique, Paris. The fresco showing Wirsung is at the University of Padua. All are in the public domain.

Several portraits or photos come from all kind of sources: The photo of Bechterew was taken by Karl Boella (in public domain). The photograph of Bochdalek is from Kachlik D, Čech P, Vincenz Alexander Bochdalek (1801–83). *J Med Biogr* 2011; 19:38–43. The photo of Dorello was kindly provided by Mariangela Fortunati (Narni, near Terni, Umbria). The photograph of Eberstaller comes from the Department of Anatomy in Graz (courtesy Michael Pretterklieber). The photograph of Foix comes from Caplan LR, Charles Foix (1882–1927). *J Neurol* 2010; 257:1941–1942. The eighteenth-century engraving of Galenus was made by Georg Paul Busch. The photograph of Hans Kuypers was kindly provided by Roger Lemon (London). The photograph of Landsmeer was kindly provided by John Kauer (Nijmegen). The photograph of Latarjet comes from Reveron RR, André Latarjet (1877–1947). Anatomist and surgeon specialized in sports medicine. *Ital J Anat Embryol* 2014; 119:250–254. Mackenrodt's photograph comes from Ludwig H, Alwin Mackenrodt (1859–1925). Die quere Beckenfaszie. *Der Gynäkologe* 2003; 36:165–166. The photograph of Hilde Mangold comes from Fässler PE, Sander K, Hilde Mangold (1898–1924) and Spemann's organizer: Achievements and tragedy. *Roux Arch Dev Biol* 1996; 205:323–332 (courtesy Klaus Sander, Freiburg). The photograph of Walle Nauta was kindly provided by the late Ton Lohman (Monnickendam). The photographs of Rudolf Nieuwenhuys and Luis Puelles were kindly provided by Loreta Medina (Lleda). We thank André Parent (Quebec) for the portrait of Berengario da Carpi and some other illustrations by Berengario. The portrait of Sudeck is from Teichmann W, Egger C, Schröder H-J, *Hundert Jahre Vereinigung Nordwestdeutscher Chirurgen.* Hamburg 2009, p 71 (free to use). The photo of Tsai comes from Wang Y, Li Z, Qian Y, Guo B, Tsai Chiao: The founder of physiology and aviation, aerospace and navigation medicine in China. *Protein & Cell* 2022; 13:779–784 (Open Access). The photograph of Uchimura is also in the public domain (Vintage baseball memorabilia of Japan; no copyright restrictions). The 1903 photograph of the Vogts comes from the Archives der Universität Düsseldorf. The photograph of Jan Voogd was kindly provided by Tom Ruigrok (Rotterdam).

Portraits of Casserio, Guidi, Pensa, Sterzi, and Tigri were kindly provided by Marina Quartu and Alessandro Riva (Cagliari). From Zaffe (2015), the portraits of Amici, Beccari, and Galeazzi were obtained (in public domain). The Anatomical Lesson of Gruber and the photos of Hrdlička, Kohn, and Toldt were kindly provided by Ondřej Naňka (Prague), who also provided several other figures from Grim et al. (2014).

The figures on the works of the various eponymists presented are acknowledged *separately* in the *legends* of these figures.

Cover Picture

The fresco of a public dissection in the lunette of the Eastern wall of the Auditorium of the Jesuit College in Jindřichův Hradec, Southern Bohemia, Czech Republic, was made by a fresco-secco technique between 1628 and 1633. It is a unique work for which no comparable piece can be found in contemporary painting in Central Europe. It was discovered in 2003 during the reconstruction of the Jesuit dormitory and restored from 2003 to 2006 by the academic painter Tomáš Švéda. The Jesuit College was founded by Adam II and his wife Kateřina of Montfort and built from 1595 to 1602, when five basic classes of the new gymnasium were formed. After the Jesuit Order was dissolved in 1773, the buildings were used for miltary purposes. In 1995, the entire area was reconstructed and now houses the Museum of Photography.

The central motif in the fresco shows a teacher and a student above a corpse lying on the dissection table. The act of dissection, framed by the mortal symbols of the scientific allegory of death in a muscular figure and a skeleton, is watched by other figures. The picture of the fresco was kindly provided by the Director of the Museum of Photography, Ondřej Mašek.

Nijmegen, The Netherlands — Hans J. ten Donkelaar
Monserrato, Italy — Marina Quartu
Prague, Czech Republic — David Kachlík

References

Adstrum S (2015) Fascial eponyms may help elucidate terminological and nomenclatural development. J Bodywork Mov Therapies 19:516–525

Baskett TF (ed) (2019) Eponyms and names in obstetrics and gynaecology, 3rd edn. Cambridge University Press, Cambridge

Bartolucci S, Forbis P (2004) Stedman's medical eponyms book. Lippincott, Williams & Wilkins, Baltimore

Brüel A, Christensen EI, Tranum-Jensen J, Qvortrup K, Geneser F (2012) Genesers histologi. Munksgaard, Copenhagen

Burdan F, Dworzański W, Cendrowska-Pinkosz M, Burdan M, Dworzańska A (2016) Anatomical eponyms – unloved names in medical terminology. Folia Morphol 75:413–438
Dhom G (2001) Geschichte der Histopathologie. Springer, Berlin/Heidelberg/New York
Dobson J (1962) Anatomical eponyms, 2nd edn. Livingstone, Edinburgh/London
Donáth T (1969) Anatomical dictionary with nomenclatures and explanatory notes. Akadémiai Kiadó, Budapest.
Dumesmil R, Bonnet-Roy F (1947) Les médecins célèbres. Masson, Paris
Fischer I (1962) Biographisches Lexikon der hervorragenden Ärzte der letzten 50 Jahre, 3rd edn, Bd 1 u. 2. Urban & Schwarzenberg, Munich-Berlin
Grim M, Naňka O, Černy K (2014) Anatomie od Vesalia po současnost. 1514–2014, Grada, Prague.
Haymaker W, Schiller F (eds) (1968) The founders of neurology. One hundred and forty-six biographical sketches by eighty-eight authors, 2nd edn. Thomas, Springfield
His W (1895) Die Anatomische Nomenklatur. Nomina Anatomica. Arch Anat Entwgesch (Suppl):1–180
Hubotter F (Hrsg) (1962) Gurtl, Wernich and Hirsch's Biographisches Lexikon der hervorragenden Ärzte aller Zeiten und Völker, 3rd edn. Bd 1–5 u. Ergänzungsbd. Urban & Schwarzenberg, Munich-Berlin
Krstić RV (1984) Illustrated encyclopedia of human histology. Springer, Berlin/Heidelberg/New York/Tokyo
Lindeboom GA (1981) Geschiedenis van de medische wetenschap in Nederland (in Dutch), 2nd edn. Fibula-Van Dishoeck, Haarlem
Lustig LR, Jackler RK, Mandelcorn R (1998) The history of otology through its eponyms I: Anatomy. Am J Otol 19:371–389
Mezzogiorno V (1991) Glossario Anatomico. Napoli
Mudry A (2001) The origin of eponyms used in cochlear anatomy. Otol Neurotol 22:258–263
Olry R (1995) Dictionary of anatomical eponyms. Fischer, Stuttgart/Jena/New York
Olry R (2014a) Anatomical eponyms, Part 1: To look on the bright side. Clin Anat 27:1142–1144
Olry R (2014b) Anatomical eponyms, Part 2: The other side of the coin. Clin Anat 27:1145–1148
Páč L (2000) Slovník Anatomických Eponym, Galén, Prague (2nd edn, 2010)
Pagel JC (1901) Biographisches Lexikon hervorragender Ärzte des neunzehnten Jahrhunderts. Urban & Schwarzenberg, Berlin/Vienna
Rizzi A, Benagiano V, Ribatti D (2014) Dizionario degli eponimi anatomici. Stilo Editrice, Modena
Stedman's Medical Dictionary (2008) 28th edition. Lippincott, Williams & Wilkins, Baltimore
Swanson LW (2014) Neuroanatomical terminology. Oxford University Press, New York
TA2 (2019) Terminologia Anatomica, 2nd edn. https://FIPAT.library.dal.ca. Federative International Programme for Anatomical Terminology
TE2 (2017) Terminologia Embryologica, 2nd edn. https://FIPAT.library.dal.ca. Federative International Programme for Anatomical Terminology
ten Donkelaar HJ (2020) Clinical neuroanatomy. Brain circuitry and its disorders, 2nd edn. Springer, Cham
ten Donkelaar HJ, Oostra RJ (eds) (2014) Klinische Anatomie en Embryologie, 4th edn. Springer Media, Houten
ten Donkelaar HJ, Kachlík D, Tubbs RS (2018) An illustrated Terminologia Neuroanatomica. A concise encyclopedia of human neuroanatomy. Springer, Cham
ten Donkelaar HJ, Lammens M, Hori A (2023) Clinical neuroembryology. development and developmental disorders of the human central nervous system, 3rd edn. Springer, Cham
Terminologia Anatomica (1998) International Anatomical Terminology. FCAT, Thieme, Stuttgart
TNA (2017) Terminologia Neuroanatomica. https://FIPAT.library.dal.ca. Federative International Programme for Anatomical Terminology; an updated digital version (2022) is available at https://ifaa.unifr.ch

Vesalius A (1543) De humani corporis fabrica. Oporinus, Basel
Vollmuth R (2004) Das anatomische Zeitalter. Die Anatomie der Renaissance von Leonardo da Vinci bis Andreas Vesal. Neuer Merkur, Munich.
Whitworth JA (2007) Should eponyms be abandoned? No. BMJ 335:425
WhoNamed It?: The dictionary of medical eponyms. https://www.whonamedit.com
Woywood A, Matteson E (2007) Should eponyms be abandoned? Yes. BMJ 335:424
Zaffe D (2015) On the great anatomists. A memory e-book, 3rd version. BMN, Modena

Acknowledgments

The help of Kohei Shiota (Kyoto) for the Japanese eponyms is greatly acknowledged. Several new drawings were prepared by Ad Gruter (Nieuwegein) and several illustrations for Chaps. 22, 23 and 31 by Anhelina Khadanovich (Prague). For their help with illustrations, we also thank Akira Hori (Hannover), Roelof-Jan Oostra and Bjarke Jensen (Amsterdam), Koos Jaap van Zwieten (Hasselt), Lucas Boer (Nijmegen), and Rafael Coello (Guayaquil, Ecuador). We also appreciate the reuse of several figures in various Chapters of Part III, provided by Alberto Caggiati (Rome), Hans-Jörg Mischinger (Graz), Richard Douard (Paris), Luciano A Favorito (Rio de Janeiro), Salah Mansour (Beirut), Yuki Watanabe (Miyagi, Japan), Harvey M Goldstein (San Antonio), Alexander J Towbin (Cincinnati), MS Levine (Philadelphia), Francesco Allessandrino (Miami), Linda Kelahan (Chicago), Rutger Jan Nievelstein (Utrecht), Francesca Maccioni (Rome), Ali Devrim Karaosmanoglu (Ankara), Mushtak Al-Atabi (Selangor, Malaysia), and David W Robinson (Winston-Salem). For the reuse of several figures from the Danish edition of Finn Geneser's Histology Textbook by Brüel et al. (*Genesers histologi* 2012, Munksgaard, Copenhagen), we greatly appreciate the help of Britta Østergaard of Munksgaard.

Competing Interests The authors have no competing interests to declare that are relevant to the content of this manuscript.

Contents

Contributors

Albert Caggiati Department of Anatomy, Universita di Roma, Rome, Italy

Joannes R. M. Cruysberg Molenhoek, The Netherlands

Ondřej Naňka Department of Anatomy, First Faculty of Medicine, Charles University, Prague, Czech Republic

Veronika Němcová Department of Anatomy, First Faculty of Medicine, Charles University, Prague, Czech Republic

Michael Pretterklieber Department of Macrosopic and Clinical Anatomy, University of Graz, Graz, Austria

Jiři Šedý Department of Anatomy, Second Faculty of Medicine, Charles University, Prague, Czech Republic

Vedat Topsakal Department of Otorhinolaryngology, Head and Neck Surgery, University Hospital Brussel, Vrije Universiteit Brussel, Brussels, Belgium

Medical Illustrator

Ad Gruter Ereprijs 3, 3434 CN Nieuwegein, The Netherlands

Part I

The Classical Eponyms

Vignette Part I Title page *Institutiones Anatomicae* by Caspar Bartholin, The Elder, revised by his son Thomas in 1641, published by Hackium in Leiden. The illustration comes from a French translation in 1697 by Henault in Paris, and shows many of the old anatomists

1.1 Introduction

During the period of intellectual revival in the sixteenth century, the custom of using eponymous terms arose, and the use of Latin as a common language enabled scholars throughout Europe to become familiar with each others writings and discoveries. Early medieval anatomists, such as Mondino de Luzzi and Berengario da Capri, introduced a systematic study of anatomy and dissection, and paved the way for Vesalius great work *De humani corporis fabrica* from 1543. After Vesalius, in particular Italian, Danish, Dutch and French anatomists studied human anatomy and their names are recognized in many eponyms. In the following centuries, German, English, French, Italian, Czech, Austrian, Hungarian and Russian anatomists greatly improved our knowledge of human anatomy. This resulted in the use of different names as *eponyms* in several languages for the same structure. More recently, also American, Japanese, Spanish and Swedish anatomists received their eponyms.

In Chap. 1 of this part, the major classical eponyms on arteries, bands, bodies, bundles, canals, corpuscles, ducts, fasciae, fibres, folds, foramina, fossae, ganglia, glands, ligaments, membranes, muscles, nerves, nodes, nuclei, plexuses, spaces, triangles, tubercles, valves and veins are summarized. Further data and illustrations on the various authors mentioned are provided in Part II *The Anatomical Eponyms from A to Z*. We did not restrict ourselves to anatomical eponyms but also included anthropological, embryological and histological terms. More recent, specific clinical eponyms are discussed in Part III *Anatomical Eponyms in Clinical Disciplines*.

The Classical Anatomical Eponyms

1

Hans J. ten Donkelaar and David Kachlík

1.1 Introduction

During the period of intellectual revival in the sixteenth century, the custom of using eponymous terms arose (Dobson 1962). The use of Latin as a common language enabled scholars throughout Europe to become familiar with each others writings and discoveries. But earlier anatomists such as Mondino de Luzzi and Berengario da Capri by introducing a systematic study of anatomy and dissection paved the way for Vesalius great work *De humani corporis fabrica* from 1543. After Vesalius, in particular Italian, Danish, Dutch and French anatomists studied human anatomy and their names are recognized in many eponyms. In the following centuries, German, English, French, Italian, Czech, Austrian, Hungarian and Russian anatomists greatly improved our knowledge of human anatomy. This resulted in the use of different names as eponyms in several languages for the same structure. More recently, also American, Japanese, Spanish and Swedish anatomists received their eponyms.

In this Chapter, the major classical eponyms on arteries, bands, bodies, bundles, canals, corpuscles, ducts, fasciae, fibres, folds, foramina, fossae, ganglia, glands, ligaments, membranes, muscles, nerves, nodes, nuclei, plexuses, spaces, triangles, tubercles, valves and veins are summarized. Further data and illustrations are provided in Part II *The Anatomical Eponyms from A to Z*. We did not restrict ourselves to anatomical eponyms but also included anthropological, embryological and histological terms. More recent, specific clinical eponyms are discussed in Part III *Anatomical Eponyms in Clinical Disciplines*.

1.2 Arteries

There are many eponyms for **arteries** but apart from those of Adamkiewicz, Charcot, Drummond, Guidi (Vidius) and Heubner, they are hardly used anymore.

Artery of Adamkiewicz: *Eponym* for the great radicular artery; *see* Adamkiewicz, Albert Wojciech.

Artery of Bernasconi and Cassinari: *Eponym* for the marginal tentorial branch; *see* Bernasconi, Vittorio, and Cassinari, Valentino.

Artery of Charcot: *Eponym* for any of the lenticulostriate arteries, known as 'artery of haemorrhage'; *see* Charcot, Jean Martin.

Artery of Chaussier, Great muscular: *Eponym* for a branch of the lateral circumflex femoral artery supplying the quadriceps femoris muscle; *see* Chaussier, François.

Arteries of Demel: *Eponym* for anterior and posterior oesophagotracheal arteries; *see* Demel, Rudolf.

Artery of Drummond, Marginal: *Eponym* for the marginal colic artery; *see* Drummond, David.

Arteries of Dubreuil-Chambardel: *Eponym* for several arteries, (a) an inconstant branch of the inferior epigastric artery, and (b) the recurrent interosseous artery; *see* Dubreuil-Chambardel, Louis.

Arteries of Duret: Duret described several arteries of the brain; *see* Duret, Henri.

Artery of Dwight, Parallel: *Eponym* for a comparable system to that of Drummond along the mesenteric margin of the jejunum and ileum.

Artery of Falloppio: *Eponym* for the uterine artery; *see* Falloppio, Gabriele.

Arteries of Flint: *Eponym* for the arcuate arteries of the kidney; *see* Flint, Austin.

Arteries of Foix: Foix introduced the terms thalamogeniculate, thalamoperforated and thalamotuberal arteries for the blood supply of the thalamus; *see* Foix, Charles.

Artery of Glaser: *Eponym* for the anterior tympanic artery; *see* Glaser, Johann Heinrich.

Artery of Guidi (Vidius): *Eponym* for the artery of the pterygoid canal (Vidian artery); *see* Guidi, Guido (Vidius, Vidus).

Artery of Haller: *Eponym* for (a) the dorsal pancreatic artery, (b) the superior thoracic artery, and (c) the supreme thoracic artery; *see* Haller, Albrecht von.

H. J. ten Donkelaar et al., *An Illustrated Guide to Anatomical Eponyms*, https://doi.org/10.1007/978-3-031-91664-9_1

Artery of Heubner, Recurrent: *Eponym* for the long striate artery; *see* Heubner, Otto.

Arteries of Kadyi, Central: *Eponym* for the sulcomarginal arteries that enter the spinal cord through the anterior median fissure; *see* Kadyi, H.

Artery of Konstantinovich: *Eponym* for the dorsal rectal artery, a branch of the superior rectal artery; *see* Konstantinovich, Vikentzi Bonifatiyevich.

Artery of Kugel, Anastomotic: *Eponym* for a rare variant, an atrial anastomotic branch between branches of the right coronary artery and the circumflex branch of the left coronary artery; *see* Kugel, Maurice.

Arteries of Lazorthes: *Eponym* for arteries supplying the cervical and lumbosacral enlargements of the spinal cord; *see* Lazorthes, Guy.

Arteries of Michels: *Eponym* for the retroduodenal arteries; *see* Michels, Nicholas Aloysius.

Artery of Moskowitz, Meandering mesenteric: *Eponym* for an inconstant intermesenteric artery, corresponding to the intermesenteric artery of Villemin; *see* Moskowitz, Myron.

Artery of Neubauer: *Eponym* for the thyroid ima artery; *see* Neubauer, Johann Ernst.

Artery of Percheron: *Eponym* for a variant of the posterior cerebral circulation, characterized by a solitary branch supplying bilaterally the paramedian thalami and the rostral midbrain; *see* Percheron, Gérard.

Arteries of Ranvier: *Eponym* for medium-calibre arteries with a tunica media rich in smooth muscle cells; *see* Ranvier, Louis Antoine.

Artery of Rauber: *Eponym* for the proper hepatic artery; *see* Rauber, August Antonius.

Artery of Rolando: *Eponym* for the middle cerebral artery; *see* Rolando, Luigi.

Artery of Salvi, Anastomotic tarsal: *Eponym* for a short branch from the dorsal tarsal artery to the posterior tibial artery; *see* Salvi, Giunio.

Arteries of Spalteholz: *Eponym* for arterioles with a wall of endothelial lining and sparse smooth muscle cells; *see* Spalteholz, Karl Werner.

Artery of Sylvius: *Eponym* for the middle cerebral artery; *see* Sylvius, François de le Boë.

Artery of Testut: *Eponym* for the inferior pancreatic artery; *see* Testut, Jean.

Arteries of Theile, Pericardiophrenic: *Eponym* for branches of the superior phrenic artery supplying the pericardium; *see* Theile, Friedrich Wilhelm.

Arteries of Uchimura, Long and short: *Eponym* for branches of the posterior cerebral artery supplying the hippocampal formation; *see* Uchimura, Yushi.

Artery of Villemin, Mesenteric: *Eponym* for a rare anastomosis between the stems of both mesenteric arteries; see Villemin, Fernand.

Artery of Wallenberg: *Eponym* for the posterior inferior cerebellar artery; *see* Wallenberg, Adolf.

Artery of Weber: *Eponym* for a tympanic branch of the external carotid artery; see *Weber*, Moritz Ignatz.

Artery of Wilkie: *Eponym* for the supraduodenal artery; *see* Wilkie, David Percival Dalbreck.

Artery of Wollschlaeger and Wollschlaeger: *Eponym* for an inconstant dural branch of the superior cerebellar artery; *see* Wollschlaeger, Paul, and Wollschlaeger, Gertraud.

Artery of Zinn: *Eponym* for the central retinal artery; *see* Zinn, Johann Ernst.

1.3 Bands

The term **band** is used for various structures from a ligament (Aranzio, Clado, Retzius) to a band of myelinated fibres (Broca, Gennari, Vicq d'Azyr). The moderator band already illustrated by Vesalius, also received the eponyms of King and Poirier. Sometimes, the term bandelette is used as synonym (bandelette of Giacomini, bandelette of Maissiat).

Band of Aranzio (Aranzius): *Eponym* for the ligamentum venosum of the liver; *see* Aranzio, Giulio Caesare.

Band of Broca, Diagonal: *Eponym* for a band of fibres extending into the amygdaloid body from the paraolfactory gyrus on the medial side of the frontal lobe ventrally and along the lateral margin of the optic tract, marking the caudal boundary of the anterior perforated substance or olfactory tubercle; also known as the olfactory radiation of Zuckerkandl; *see* Broca, Pierre Paul.

Band of Clado: *Eponym* for the suspensory ligament of the ovary; *see* Clado, Spiro.

Band of Gennari: *Eponym* for the occipital stripe, a conspicuous band of myelinated fibres marking the striate cortex of the occipital lobe; also known as band of Vicq d'Azyr; *see* Gennari, Francesco.

Band or **Bandelette of Giacomini:** *Eponym* for the band of the dentate gyrus; *see* Giacomini, Carlo.

Band of King, Moderator: *Eponym* for the septomarginal trabecula, the moderator band, already known for a long time since Leonardo da Vinci; *see* King, Thomas Wilkinson.

Band of Leonardo: *Eponym* for the moderator band; *see* Leonarda da Vinci.

Band or **Bandelette of Maissiat:** *Eponym* for the iliotibial tract of the fascia lata; *see* Maissiat, Jacques Henri.

Band of Poirier, Accessory: *Eponym* for the moderator band; *see* Poirier, Paul Julien.

Band of Pringle: *Eponym* for a peritoneal band from the mesocolon to the duodenojejunal flexure; *see* Pringle, Seton Sidney.

Band of Remak: *Eponym* for axon; *see* Remak, Robert.

Band of Retzius: *Eponym* for the fundiform ligament of the penis; *see* Retzius, Anders Adolf.

Band of Vicq d'Azyr: *Eponym* for the occipital stripe, a conspicuous band of myelinated fibres marking the striate cortex; also known as band of Gennari; *see* Vicq d'Azyr, Felix.

1.4 Bodies

The term **body** is used variously, for the mesonephros (Wolff, Oken) and the parathyroid glands (Sandström), for arachnoid granulations (Pacchioni), for fibrous structures (Savage, Symington), for paraganglia (Kohn, Zuckerkandl), for nodules (Aranzio) or corpuscles (Hassall), for intracellular structures (Barr, Call-Exner, Hensen, Herring, Langerhans and Nissl), and even for the subthalamic nucleus (Luys).

Body of Aranzio (Aranzius): *Eponym* for the nodules on the semilunar valvules of the aortic valve; *see* Aranzio, Giulio Caesare.

Body of Balbiani, Vitelline: *Eponym* for a cluster of organelles (mitochondria, flattened cisterns of Golgi apparatus, endoplasmatic reticulum and lysosomes) in primary oocytes; *see* Balbiani, Edouard Gérard.

Body of Barr: *Eponym* for the chromatin body (sexual chromatin) found in about 90% of female and 10% of male interphase nuclei; *see* Barr, Murray.

Body of Biondi: *Eponym* for an inclusion body (psammoma) in the cranial pia mater; *see* Biondi, Adolfo.

Bodies of Call-Exner: *Eponym* for small cavities found among the ovarian granulosa cells of growing and mature follicles; *see* Call, Emma Louise, and Exner, Siegmund.

Bodies of Deetjen: *Eponym* for the platelets; *see* Deetjen, Hermann.

Bodies of Hensen: *Eponym* for whorllike aggregates of smooth endoplasmatic reticulum found in outer hair cells of the spiral organ; *see* Hensen, Viktor.

Bodies of Herring: *Eponym* for spherical thickenings along neurosecretory axons of the hypothalamohypophysial tract, particularly abundant in the infundibulum and the posterior lobe of the pituitary gland; *see* Herring, Percy Theodore.

Body of Highmore: *Eponym* for the mediastinum testis; *see* Highmore, Nathaniel.

Bodies of Howell-Jolly: *Eponym* for remnants of nonextruded nuclei from orthochromatic erythroblasts; *see* Howell, William Henry, and Jolly, Justin Marie Jules.

Bodies of Kohn: *Eponym* for the paraganglia; *see* Kohn, Alfred.

Bodies of Langerhans: *Eponym* for platelike structures found in the cells of Langerhans; *see* Langerhans, Paul.

Body of Luys: *Eponym* for the subthalamic nucleus; *see* Luys, Jules Bernard.

Bodies of Nissl: *Eponym* for the basophilic granules in the cytoplasm of neurons, also known as tigroid substance; *see* Nissl, Franz.

Body of Oken: *Synonym* for the mesonephros of Wolff; *see* Oken, Lorenz.

Bodies of Pacchioni: *Eponym* for the arachnoid granulations in the superior sagittal sinus; *see* Pacchioni, Antonio.

Bodies of Sandström: *Eponym* for the parathyroid glands; *see* Sandström, Ivar.

Body of Savage, Perineal: *Eponym* for the perineal body; *see* Savage, Henry.

Body of Symington, Anococcygeal: *Eponym* for the anococcygeal ligament; *see* Symington, Johnson.

Body of Waldeyer: *Eponym* for the retrosternal adipose tissue, formed by thymic residues; *see* Waldeyer, Heinrich Wilhelm Gottfried.

Bodies of Weibel-Palade: *Eponym* for storage granules of endothelial and endocardial cells; *see* Weibel, Ewald Rudolf, and Palade, George Emil.

Body of Wolff: *Eponym* for the mesonephros; *see* Wolff, Caspar Friedrich.

Bodies of Zuckerkandl: *Eponym* for the paraaortic (chromaffin) bodies; *see* Zuckerkandl, Emil.

1.5 Bundles

The term **bundle** is mostly used for fibre tracts in the central nervous system, also described as fasciculi or tracts. Other uses are in the atrioventricular bundle of His, Kent and Paladino and their crura of Tawara, and for certain bundles of muscle fibres (Helie, Killian and Weissmann).

Bundle of Bachmann: *Eponym* for the interatrial bundle of the cardiac conducting system; *see* Bachmann, Jean George.

Bundle of Barnes: *Eponym* for corticospinal fibres leaving the pyramidal tract at the pontine level and descending lateral to the inferior olive; *see* Barnes, Charles.

Bundle of Bechterew: *Eponym* for (a) the central tegmental tract, and (b) the spinoolivary tract; *see* Bechterew, Vladimir Michailovich von.

Bundles of Dejerine: *Eponym* for the pes lemnisci bundles; *see* Dejerine, Joseph Jules.

Bundle of Edinger, Predorsal: *Eponym* for the tectospinal tract, also described by Held and Loewenthal; *see* Edinger, Ludwig.

Bundle of Flechsig, Oval: *Eponym* for the septomarginal fasciculus of the spinal cord; *see* Flechsig, Paul Emil.

Bundle of Foville: *Eponym* for the stria terminalis; *see* Foville, Achille Louis.

Bundle of Gierke, Respiratory: *Eponym* for the solitary tract, also described by Krause; *see* Gierke, Hans Paul Bernhard.

Bundle of Held: *Eponym* for the tectospinal tract, also described by Edinger and Loewenthal; *see* Held, Hans.

Bundle of Helie: *Eponym* for the vertical and ansiform bundles of the superficial musculature of the uterus; *see* Helie, Louis Theodore.

Bundle of His: *Eponym* for the atrioventricular bundle of the cardiac conducting system, also described by Kent and Paladino; *see* His, Wilhelm Jr.

Bundle of Hoche: *Eponym* for an aberrant bundle of the pyramidal tract leaving it at the level of the facial nerve and crossing the midline to join the contralateral pyramidal tract; *see* Hoche, Alfred Erich.

Bundle of James: *Eponym* for the anterior internodal tract of the cardiac conducting system; *see* James, Thomas Naum.

Bundle of Kent: *Eponym* for the atrioventricular bundle of the cardiac conducting system, also described by His Jr and Paladino; *see* Kent, Albert Frank Stanley.

Bundle of Killian: *Eponym* for the lowest, cricopharyngeal fibres of the inferior constrictor muscle of the pharynx; *see* Killian, Gustav.

Bundle of Krause, Respiratory: *Eponym* for the solitary tract, also described by Gierke; *see* Krause, Wilhelm Johann Friedrich.

Bundle of Loewenthal: *Eponym* for the tectospinal tract, also described by Edinger and Held; *see* Loewenthal, Wilhelm.

Bundle of Löwy: *Eponym* for the angular bundle or peduncle of the flocculus; *see* Löwy, R.

Bundle of Marie: *Eponym* for the sulcomarginal fasciculus of the spinal cord; *see* Marie, Pierre.

Bundle of Meynert: *Eponym* for the habenulointerpeduncular tract; *see* Meynert, Theodor.

Bundle of Monakow: *Eponym* for the rubrospinal tract; *see* Monakow, Constantin von.

Bundle of Munzer: *Eponym* for the tectopontine tract; *see* Munzer, Egmont.

Bundle of Muratoff: *Eponym* for the subcallosal bundle; *see* Muratoff, Wladimir Aleksandrovich.

Bundle of Paladino: *Eponym* for the atrioventricular bundle of the cardiac conducting system, also described by His Jr. and Kent; *see* Paladino, Giovanni.

Bundle of Pick: *Eponym* for an anomalous bundle of nerve fibres in the medulla oblongata, accompanying the pyramidal tract but then ascending to the nucleus ambiguus; *see* Pick, Arnold.

Bundle of Probst: *Eponym* for a bundle of longitudinal fibres (long callosal fasciculus) in the absence of the corpus callosum; *see* Probst, Moritz.

Bundle of Rasmussen: *Eponym* for the olivocochlear bundle; *see* Rasmussen, Grant Lister.

Bundle of Russell, Hook: *Eponym* for the uncinate fasciculus of the cerebellum; *see* Russell, James S. Risien.

Bundle of Schultze: *Eponym* for the interfascicular fasciculus or comma tract of the spinal cord; *see* Schultze, Friedrich.

Bundle of Schütz: *Eponym* for the posterior or dorsal longitudinal fasciculus; *see* Schütz, Hugo.

Bundle of Spitzka: *Eponym* for fibres of a posterior longitudinal bundle connecting the third and sixth nerve nuclei, now believed to be part of the medial longitudinal fasciculus; *see* Spitzka, Edward Charles.

Bundles of Tawara: *Eponym* for the right and left crus of the atrioventricular bundle of the cardiac conducting system; *see* Tawara, Sunao.

Bundle of Thorel: *Eponym* for the posterior internodal tract of the cardiac conducting system; *see* Thorel C.

Bundle of Vicq d'Azyr: *Eponym* for the mammillothalamic tract; *see* Vicq d'Azyr, Felix.

Bundle of Wenckebach: *Eponym* for the middle internodal tract of the cardiac conducting system; *see* Wenckebach, Karel Frederik.

Bundle of Weissmann: *Eponym* for the aggregation of striated muscular fibres of a neuromuscular spindle; *see* Weissmann, August.

1.6 Canals

The term **canal** is frequently used as synonym for ducts. Where evident, the term *synonym* is used. Here, it is also obvious that eponyms exist in various languages. This is particularly evident for the scleral venous sinus, named after Fontana, Lauth and Schlemm.

Canal of Alcock: *Eponym* for the pudendal canal in the ischioanal or ischiorectal fossa; *see* Alcock, Benjamin.

Canal of Aranzio (Arantius): *Synonym* for the embryonic ductus venosus; *see* Aranzio, Giulio Caesare.

Canal of Arnold: *Eponym* for (a) the hiatus of the lesser petrosal nerve, and (b) the mastoid canaliculus, containing the auricular branch of the vagus nerve; *see* Arnold, Friedrich.

Canal of Bernard: *Eponym* for the accessory pancreatic duct, earlier described by Santorini; *see* Bernard, Claude.

Canal of Bichat: *Eponym* for the cistern of the great cerebral vein; *see* Bichat, Marie François Xavier.

Canal of Blasius (Blaes): *Synonym* for the parotid duct, described in the same year by Stensen (*see* Duct of Stensen); *see* Blasius, Gerard.

Canal of Bochdalek: *Synonym* for the embryonic thyroglossal duct, also described by Vater and His Sr; *see* Bochdalek, Vincenc Alexander.

Canal of Braun: *Eponym* for the neurenteric canal, also described by Kowalewsky; *see* Braun, Max.

Canals of Breschet: *Eponym* for canals for the diploic veins in the skull, later also described by Dupuytren; *see* Breschet, Gilbert.

Canal of Calori, Hypophysial: *Eponym* for the craniopharyngeal canal, an inconstant canal traversing the sphenoidal bone, also described by Landzert; *see* Calori, Luigi.

Canal of Civinini: *Eponym* for the anterior canaliculus of the chorda tympani, also described by Glaser and Huguier; *see* Civinini, Filippo.

Canal of Cloquet: *Eponym* for the hyaloid canal of the developing eye, also described by Stilling; *see* Cloquet, Jules Germaine.

Canal of Corti: *Eponym* for the inner tunnel of the spiral organ of the cochlea; see Corti, Alfonso.

Canal of Cotugno: *Eponym* for (a) the endolymphatic duct, and (b) the vestibular canaliculus; *see* Cotugno, Domenico.

Canal of Cruveilhier, Brachial: *Eponym* for the brachial canal containing the brachial vessels and the median and ulnar nerves; *see* Cruveilhier, Jean Baptiste.

Canal of Cuvier: *Eponym* for the sinus venosus of the developing heart; *see* Cuvier, Georges Leopold Chrétien Frédéric Dagobert.

Canal of Dorello: *Eponym* for the abducens nerve canal, earlier described by Gruber; *see* Dorello, Primo.

Canals of Dupuytren: *Eponym* for canals for the diploic veins in the skull, earlier described by Breschet; *see* Dupuytren, Guillaume.

Canal of Falloppio: *Eponym* for the facial canal in the petrous part of the temporal bone; *see* Falloppio, Gabriele.

Canal of Ferrein: *Eponym* for the rivus lacrimalis of the eye; *see* Ferrein, Antoine.

Canal of Fontana: *Eponym* for the scleral venous sinus, later also described by Schlemm and Lauth; *see* Fontana, Abbada Felice Gaspar Ferdinand.

Canal of Gärtner: *Synonym* for the longitudinal duct of the epoophoron, earlier recognized by Malpighi; *see* Gärtner, Hermann Treschow.

Canal of Gruber: *Eponym* for (a) the canal of the abducens nerves, later also described by Dorello, and (b) the cruropopliteal canal; *see* Gruber, Wenzel.

Canal of Guidi (Vidius): *Eponym* for the bony canal of the pterygoid process of the sphenoidal bone (Vidian canal); *see* Guidi, Guido.

Canal of Guyon: *Eponym* for the ulnar canal in the hand; *see* Guyon, Jean Casimir Felix.

Canal of Hannover: *Eponym* for a potential retrozonular space between the ciliary zonule and the vitreous body, also described by Petit; *see* Hannover, Adolph.

Canals of Havers: *Eponym* for vascular canals in the compact tissue of bone, previously described by Antonie van Leeuwenhoek; *see* Havers, Clopton.

Canal of Henle: *Eponym* for the adductor canal in the thigh; *see* Henle, Friedrich Gustav Jacob.

Canal of Hensen: *Synonym* for the ductus reuniens of the membranous labyrinth of the inner ear; *see* Hensen, Viktor.

Canals of Hering: *Eponym* for biliary canals with epithelial lining in a hepatic lobule; *see* Hering, Heinrich Ewald.

Canal of Hirschfeld: *Eponym* for the tympanic canaliculus, also described by Jacobson and Sappey; *see* Hirschfeld, Ludwig Moritz.

Canals of Hoyer: *Eponym* for arteriovenous anastomoses, also described by Sucquet; *see* Hoyer, Heinrich F.

Canal of Huguier: *Eponym* for the anterior canaliculus of the chorda tympani, also described by Civinini; *see* Huguier, Pierre Charles.

Canal of Hunter: *Eponym* for the adductor canal; *see* Hunter, John.

Canal of Hyrtl: *Eponym* for the musculofibular or musculoperoneal canal, containing the fibular or peroneal artery; *see* Hyrtl, Joseph.

Canal of Jacobson: *Eponym* for the tympanic canaliculus, also described by Sappey and Hirschfeld; *see* Jacobson, Ludwig Levin.

Canal of Jesenský, Furcated: *Eponym* for an accessory interradicular root canal in molars; *see* Jesenský, Jan.

Canal of Kowalewsky: *Eponym* for the neurenteric canal, also described by Braun; *see* Kowalewsky, Nikolaus Osipovich.

Canal of Kupffer: *Eponym* for the ureteric diverticle, an embryonic outgrowth from the mesonephric duct to form the ureter; *see* Kupffer, Karl Wilhelm von.

Canals of Lambert: *Eponym* for miscroscopic collateral airways between the distal bronchiolar tree and adjacent pulmonary alveoli; *see* Lambert, Margaret Waugh.

Canal of Landzert: *Eponym* for the craniopharyngeal canal, an inconstant canal traversing the sphenoidal bone, also described by Calori; *see* Landzert, Theodor.

Canal of Lauth: *Eponym* for the scleral venous sinus, also described by Fontana and Schlemm; *see* Lauth, Ernest Alexandre.

Canal of Löwenberg: *Eponym* for the cochlear duct, also described by Reissner; *see* Löwenberg, Benjamin Benno.

Canal of Malpighi: *Eponym* for the longitudinal duct of the epoophoron, later also described by Gärtner; *see* Malpighi, Marcello.

Canal of Müller: *Synonym* for the embryonic paramesonephric duct; *see* Müller, Johannes Peter.

Canal of Nuck: *Eponym* for a patent vaginal process of the peritoneal cavity in females; *see* Nuck, Anton.

Canal of Petit: *Eponym* for a potential space between the ciliary zonule and the vitreous body, also described by Hannover; *see* Petit, François Pourfour du.

Canals of Recklinghausen: *Eponym* for lymphatic canaliculi, cell spaces in tissues; *see* Recklinghausen, Friedrich Daniel von.

Canal of Reissner: *Eponym* for the cochlear duct, also described by Löwenberg; *see* Reissner, Ernst.

Canal of Richet: *Eponym* for a canal for the umbilical vein on the anterior wall of the abdomen, made by the passage of some fibres of the transversus abdominis muscle; *see* Richet, Louis Alfred.

Canal of Rosenthal: *Eponym* for the spiral canal of the cochlea; *see* Rosenthal, Friedrich Christian.

Canal of Rüdinger: *Eponym* for the upper compartment of the auditory tube between the lateral and medial laminae of the cartilage; *see* Rüdinger, Nicolaus.

Canal of Santorini: *Synonym* for the accessory pancreatic duct, later also described by Charcot; *see* Santorini, Giovanni Domenico.

Canal of Sappey: *Eponym* for the tympanic canaliculus, also described by Hirschfeld and Jacobson; *see* Sappey, Marie Philibert Constant.

Canals of Scarpa: *Eponym* for canals of the nasopalatine nerves and blood vessels; *see* Scarpa, Antonio.

Canal of Schlemm: *Eponym* for the scleral venous sinus, also described by Fontana and Lauth; *see* Schlemm, Friedrich.

Canals of Siebenmann: *Eponym* for small vascular canals in the aqueduct of the cochlea; *see* Siebenmann, Friedrich.

Canals of Stensen: *Eponym* for the paired incisive canals in the maxilla; *see* Stensen, Niels.

Canal of Stilling: *Eponym* for the hyaloid canal, also described by Cloquet; *see* Stilling, Benedict.

Canals of Sucquet: *Eponym* for arteriovenous anastomoses, also described by Hoyer; *see* Sucquet, JP.

Canal of Theile: *Eponym* for the transverse pericardial sinus; *see* Theile, Friedrich Wilhelm.

Canal of Tourtual: *Eponym* for the major palatine canal; *see* Tourtual, Kaspar Theobald.

Canal of van Horne: *Synonym* for the thoracic duct; *see* van Horne, Jan.

Canals of van Leeuwenhoek: *Eponym* for vascular canals in the compact tissue of bone, observed by van Leeuwenhoek in 1675 and described by Havers in 1691; *see* van Leeuwenhoek, Antonie.

Canal of Velpeau: *Eponym* for the inguinal canal; *see* Velpeau, Alfred Armand Louis Marie.

Canals of Volkmann: *Eponym* for perforating canals in compact bone connecting those of Havers with the periosteum, containing blood vessels; *see* Volkmann, Alfred Wilhelm.

Canals of Walther: *Synonym* for the minor sublingual ducts; *see* Walther, Augustin Friedrich.

Canal of Wharton: *Synonym* for the submandibular duct; *see* Wharton, Thomas.

Canal of Wolff: *Synonym* for the embryonic mesonephric duct; *see* Wolff, Kaspar Friedrich.

1.7 Cells

The term **cell** is used for all kinds of tissue with hardly double eponyms. Most of the eponyms for cells are still in use.

Cells of Beale: *Eponym* for nerve cells in the cardiac ganglia; *see* Beale, Lionel Smith.

Cells of Bergmann, Glial: *Eponym* for certain glial cells in the cerebellum; *see* Bergmann, Gottlieb Heinrich.

Cell of Betz: *Eponym* for the giant pyramidal cell of the primary motor cortex; *see* Betz, Vladimir Alexsandroče.

Cells of Blanes: *Eponym* for interneurons in the internal glomerular layer of the olfactory bulb; *see* Blanes.

Cells of Boettcher, Outer glandular: *Eponym* for the basal external glandular cells of the spiral organ of the cochlea; *see* Boettcher, Arthur.

Cell of Cajal, Horizontal: *Eponym* for the horizontal cell of Cajal found in the molecular layer of the cerebral cortex; *see* Rámon y Cajal, Santiago.

Cell of Cajal, Interstitial: *Eponym* for the interstitial cell of Cajal or enteric pacemaker cell, a pacemaker cell of the intestine, found scattered among the smooth muscle cells; *see* Rámon y Cajal, Santiago.

Cell of Cajal-Retzius: *Eponym* for the horizontal cell of the developing cerebral cortex; *see* Rámon y Cajal, Santiago, and Retzius, Gustav.

Cells of Clara: *Eponym* for exocrine cells of the bronchial epithelium, currently known by the descriptive term club cell; *see* Clara, Max.

Cells of Claudius: *Eponym* for the cuboidal external supporting cells lining the external spiral sulcus of the cochlear duct; *see* Claudius, Friedrich Matthias.

Cells of Corti, Pillar: *Eponym* for the inner and outer pillar cells of the spiral organ of the cochlea; *see* Corti, Alfonso.

Cells of Davidoff: *Eponym* for cells located at the base of glands in the small intestine, also described by Paneth; *see* Davidoff, M von.

Cells of Deiters: *Eponym for* the outer or external phalangeal cells of the spiral organ of the cochlea; *see* Deiters, Otto Friedrich Karl.

Cell of Dogiel: *Eponym* for a multipolar nerve cell within the prevertebral ganglia; *see* Dogiel, Alexander Stanislavovic.

Cell of Fañanas: *Eponym* for the feathered astrocyte in the cerebellar cortex; *see* Fañanas, Jorge Rámon.

Cells of Feyrter: *Eponym* for respiratory neuroendocrine cells, also described by Kulchitsky; *see* Feyrter, Friedrich.

Cells of Gegenbauer: *Eponym* for osteoblasts; *see* Gegenbauer, Karl.

Cells, Golgi, Large: *Eponym* for large inhibitory interneurons in the upper part of the granular layer of the cerebellum; *see* Golgi, Camillo.

Cells, Golgi, Small: *Eponym* for small inhibitory interneurons in the lower half of the granular layer of the cerebellar cortex; *see* Golgi, Camillo.

Cells, Golgi type I: *Eponym* for interneurons with a long axon, consisting of projection neurons, commissural neurons and association neurons; *see* Golgi, Camillo.

Cells, Golgi type II: *Eponym* for interneurons with a short axon, the local circuit interneurons; *see* Golgi, Camillo.

Cells of Golgi, Mitral: *Eponym* for olfactory bulb cells that form complex synapses, called glomeruli, with the axons of olfactory receptor neurons; *see* Golgi, Camillo.

Cells of Goormaghtigh: *Eponym* for the extraglomerular mesangial or juxtaglomerular cells, also described by Zimmermann; *see* Goormaghtigh, Norbert.

Cell of Grünwald, Interlamellar: *Eponym* for the pneumatization of the middle nasal concha; *see* Grünwald, Ludwig.

Cells of Hamperl: *Eponym* for the uterine natural killer cells, a special population of lymphocytes in the endometrium; *see* Hamperl, Herwig.

Cells of Heidenhain: *Eponym* for parietal cells of gastric glands; *see* Heidenhain, Rudolf Peter Heinrich.

Cells of Held: *Eponym* for the inner border cells of the spiral organ of the cochlea; *see* Held, Hans.

Cells of Hensen: *Eponym* for the outer or external border cells of the spiral organ of the cochlea; *see* Hensen, Viktor.

Cells of Hofbauer: *Eponym* for macrophages in the placental villi contributing to the suppression of immune responses between mother and foetus; *see* Hofbauer, Isfred.

Cells of Höhl: *Eponym* for postmitotic and undifferentiated daughter cells of pre-odontoblasts; *see* Höhl, Erwin.

Cells of Hortega: *Eponym* for microglia; *see* Hortega, Pio del Rio.

Cells of Ito: *Eponym* for perisinusoidal (fat storing) cells in the perisinusoidal spaces of Disse; *see* Ito, Toshio.

Cells of Koganei: *Eponym* for clump cells in the iris; *see* Koganei, Yoshikiyo.

Cells of Kölliker: *Eponym* for (a) osteoclasts, and (b) spermatoblasts; *see* Kölliker, Rudolf Albert von.

Cell of Kolmer: *Eponym* for the epiplexus cell of the choroid plexus; *see* Kolmer, Walter.

Cells of Kulchitsky: *Eponym* for respiratory neuroendocrine cells, also described by Feyrter; *see* Kulchitsky, Nicholai.

Cells of Kupffer: *Eponym* for perisinusoidal cells in the lining of blood channels in the liver; *see* Kupffer, Karl Wilhelm von.

Cells of Langendorff: *Eponym* for the principal cells of the thyroid gland; *see* Langendorff, Oskar.

Cells of Langerhans: *Eponym* for clear or dendritic cells of the epidermis; *see* Langerhans, Paul.

Cells of Langhans: *Eponym* for the cells of the cytotrophoblast; *see* Langhans, Theodor.

Cells of Leydig: *Eponym* for the interstitial cells of the testis; *see* Leydig, Franz von.

Cells of Lugaro: *Eponym* for the fusiform horizontal cells in the cerebellar cortex; *see* Lugaro, Ernesto.

Cells of Martinotti: *Eponym* for the inverted pyramidal neuron in the cerebral cortex; *see* Martinotti, Giovanni.

Cells of Merkel: *Eponym* for oval or rounded cells located around the basal layer cells of the epithelial glandular ridges; *see* Merkel, Friedrich Sigmund.

Cells of Meynert: *Eponym* for solitary cells in the primary visual cortex (BA17); *see* Meynert, Theodor.

Cell(s) of Mosher: *Eponym* for (a) the ethmoidal bulla, and (b) extensions of the ethmoidal air cells beneath the ethmoidal bulla; *see* Mosher, Harris Peyton.

Cells of Mugnaini: *Eponym* for the unipolar brush cells of the cerebellar cortex; *see* Mugnaini, Enrico.

Cells of Müller: *Eponym* for the radial glial cells of the retina; *see* Müller, Heinrich.

Cells of Neumann: *Eponym* for nucleated red blood cells; *see* Neumann, Ernst Francis Christian.

Cells of Nussbaum: *Eponym* for cells in the pyloric glands of the stomach, also described by Stöhr; *see* Nussbaum, Moritz.

Cells of Onodi: *Eponym* for a sphenoethmoidal cell, a variant of a posterior ethmoidal air cell in intimate relationship with the optic nerve just distal to the optic chiasm; *see* Onodi, Adolf.

Cells of Paneth: *Eponym* for cells located at the base of glands in the small intestine that contain large acidophilic refractile granules, also described by Davidoff; *see* Paneth, Joseph.

Cells of Popescu: *Eponym* for telocytes, interstitial cells first described in the uterus and the uterine tubes; *see* Popescu, Laurentin Mircea.

Cells of Purkinje: *Eponym* for the large output cells of the cerebellar cortex, forming the layer of Purkinje cells; *see* Purkinje, Jan Evangelista.

Cells of Renshaw: *Eponym* for recurrent inhibitory interneurons in the spinal cord; *see* Renshaw, Birdsey.

Cells of Rouget: *Eponym* for pericytes; *see* Rouget, Charles Marie Benjamin.

Cells of Saxer: *Eponym* for 'primary emigrant corpuscles', a primitive form of leucocytes in embryonic mesenchymal tissues; *see* Saxer, Franz.

Cells of Schultze: *Eponym* for the olfactory cells of the nasal mucosa; *see* Schultze, Maximilian Johann Sigismund.

Cell of Schwann: *Eponym* for the schwannocytus; *see* Schwann, Theodor.

Cells of Sertoli: *Eponym* for elongated cells in the seminiferous tubules of the testis to which spermatids are attached during spermiogenesis; *see* Sertoli, Enrico.

Cells of Stöhr: *Eponym* for pyloric cells in the stomach, also described by Nussbaum; *see* Stöhr, Philipp.

Cell of von Economo: The von Economo neuron (VEN) described by von Economo and Koskinas (1925) in the frontoinsular cortex and the anterior limbic area in great apes and humans but not in other primates. VENs are also present in elephants and whales and may be a specialization related to large brain size; *see* Economo, Constantin von.

Cells of Vignal: *Eponym* for embryonic connective tissue cells lying upon the axis cylinders of foetal nerve fibres, forming a complete sheath; *see* Vignal, Guillaume.

Cells of Vitelli: *Eponym* for small air-filled cells in the sinudural angle in a well pneumatized ear; *see* Citelli, Salvatore.

Cells of Welsh: *Eponym* for oxyphil cells of the parathyroid gland; *see* Welsh, David Arthur.

Cells of Zimmermann: *Eponym* for the extraglomerular mesangial cells, also described by Goormaghtigh; *see* Zimmermann, Karl V.

1.8 Corpuscles

The term **corpuscle** is mainly used for sensory receptors in the skin for which usually eponyms are in use. In addition, the term is used for structures in the thymus (Gierke, Hassall, Virchow-Hassall) and in the kidney and spleen (Malpighi).

Corpuscles of Gierke: *Eponym* for thymic corpuscles, also described by Hassall and Virchow; *see* Gierke, Hans Paul Bernhard.

Corpuscles of Grandry: *Eponym* for sensory end-organs in the epidermis, particularly of the mouth and tongue; *see* Grandry, M.

Corpuscles of Hassall: *Eponym* for concentrically arranged epithelial cells in the thymus, also described by Gierke and Virchow; *see* Hassall, Arthur Hill.

Corpuscles of Malpighi: *Eponym* for (a) the renal corpuscles, and (b) splenic corpuscles (lymphatic nodules of the white pulp of the spleen); *see* Malpighi, Marcello.

Corpuscles of Meissner: *Eponym* for the tactile corpuscles of the skin, described together with Rudolf Wagner; *see* Meissner, Georg, and Wagner, Rudolf.

Corpuscle of Pacini: *Eponym* for the lamellar corpuscle of the skin, a large fast-adapting receptor sensitive to vibration, also known as corpuscle of Vater-Pacini; *see* Pacini, Filippo.

Corpuscles of Rainey: *Eponym* for calcium nodules occurring in the process of calcification of tissues; *see* Rainey, George.

Corpuscle of Ruffini: *Eponym* for the bulbous corpuscle of the skin; *see* Ruffini, Angelo.

Corpuscles of Troeltsch: *Eponym* for minute spaces, resembling corpuscles, between the radial fibres of the eardrum; *see* Troeltsch, Anton Friedrich von.

Corpuscle of Valentin: *Eponym* for small amyloid bodies described in nervous tissue; *see* Valentin, Gabriel Gustav.

Corpuscle of Vater-Pacini: *Eponym* for the lamellar corpuscle of the skin; *see* Vater, Abraham, and Pacini, Filippo.

Corpuscles of Virchow-Hassall: *Eponym* for concentrically arranged epithelial cells in the thymus, also described by Gierke; *see* Virchow, Rudolf Ludwig Karl.

Corpuscles of Wagner: *Eponym* for tactile nerve endings described together with Meissner; *see* Wagner, Rudolf.

1.9 Ducts

Frequently, the term **canal** is used as *synonym* for duct. Here, often, various eponyms for the same structure are in use in French, German, Italian and English.

Duct of Aranzio (Arantius): *Eponym* for (a) the embryonic ductus venosus, and (b) the posterior part of the fourth ventricle; *see* Aranzio (Arantius), Giulio Caesare.

Duct of Bartholin: *Eponym* for the major sublingual duct; *see* Bartholin, Caspar, The Younger.

Ducts of Bellini: *Eponym* for the papillary ducts of the kidney; *see* Bellini, Lorenzo.

Duct of Bernard: *Eponym* for the accessory pancreatic duct, earlier described by Santorini; *see* Bernard, Claude.

Duct of Blasius: *Eponym* for the parotid duct, in the same year described by Stensen; *see* Blasius, Gerard.

Duct of Bochdalek: *Eponym* for the embryonic thyroglossal duct, also described by Vater and His Sr; *see* Bochdalek, Vincenc Alexander.

Duct of Boettcher: *Eponym* for the utriculosaccular duct; *see* Boettcher, Arthur.

Duct of Botallo: *Eponym* for the embryonic ductus arteriosus; *see* Botallo, Leonardo.

Duct of Cuvier: *Eponym* for the embryonic common cardinal vein; *see* Cuvier, Georges.

Duct of Gärtner: *Eponym* for the longitudinal duct of the epoophoron, also described as canal by Malpighi; *see* Gärtner, Hermann Treschow.

Duct of Hensen: *Eponym* for the ductus reuniens of the membranous labyrinth of the inner ear; *see* Hensen, Viktor.

Duct of His: *Eponym* for the embryonic thyroglossal duct, also described by Bochdalek and Vater; *see* His Sr.

Duct of Hoffmann: *Eponym* for the pancreatic duct, also described by Wirsung; *see* Hoffmann, Moritz.

Ducts of Luschka: *Eponym* for tubular structures, which sometimes branch from the bile ducts in the wall of the gallbladder, probably aberrant bile ducts; *see* Luschka, Hubert von.

Duct of Müller: *Eponym* for the embryonic paramesonephric duct; *see* Müller, Johannes Peter.

Duct of Oken: *Synonym* for the embryonic mesonephric duct of Wolff; *see* Oken, Lorentz.

Duct of Pecquet: *Eponym* for the thoracic duct (described in animals); *see* Pecquet, Jean.

Duct of Reichel, Cloacal: *Eponym* for the embryonic communication between the hindgut and the anlage of the urinary bladder; *see* Reichel, Friedrich Paul.

Ducts of Rivinus: *Eponym* for the minor sublingual ducts, also described by Walther; *see* Rivinus, Augustus Quirinus.

Duct of Ruysch: *Eponym* for a small diverticulum in the fetal nasal septum associated with the vomeronasal organ; *see* Ruysch, Frederik.

Duct of Santorini: *Eponym* for the accessory pancreatic duct; *see* Santorini, Giovanni Domenico.

Ducts of Schueller: *Eponym* for the paraurethal ducts, also described by Skene; *see* Schueller, Karl Heinrich Maximilian.

Ducts of Skene: *Eponym* for the paraurethal ducts, also described by Schueller; *see* Skene, Alexander Johnston Chalmers.

Duct of Stensen: *Eponym* for the parotid duct, in the same year described by Blasius; *see* Stensen, Niels.

Duct of van Horne: *Eponym* for the thoracic duct (described in humans); *see* van Horne, Jan.

Duct of Vater: *Eponym* for the embryonic thyroglossal duct, also described by Bochdalek and His Sr; *see* Vater, Abraham.

Ducts of Walther: *Eponym* for the minor sublingual ducts, also described by Rivinus, *see* Walther, August Friedrich.

Duct of Wharton: *Eponym* for the submandibular duct; *see* Wharton, Thomas.

Duct of Wirsung: *Eponym* for the pancreatic duct, also described by Hoffmann; *see* Wirsung, Johann Georg.

Duct of Wolff: *Eponym* for the embryonic mesonephric duct, also described by Oken; *see* Wolff, Kaspar Friedrich.

1.10 Fasciae

For the **fasciae**, we largely follow Adstrum S. *Fascial eponyms may help elucidate terminological and nomenclatural development.* J Body Work Mov Ther 2015; 19:516–525. Several terms are still used as eponyms, especially for the fasciae of Camper, Cloquet, Colles, Cooper, Denonvilliers, Dupuytren, Gerota, Hesselbach, Hyrtl, Scarpa, Tenon, Treitz and Zuckerkandl.

Fascia of Abernethy: *Eponym* for the fascia covering the external iliac artery; *see* Abernethy, John.

Fascia of Buck: *Eponym* for (a) the deep fascia of the penis, and (b) the deep perineal fascia investing the perineal muscles, also described by Gallaudet; *see* Buck, Gurdon.

Fascia of Camper: *Eponym* for the adipose layer of the superficial abdominal fascia of the abdominal subcutaneous tissue; *see* Camper, Petrus.

Fascia of Charpy: *Eponym* for the prevesical fascia, forming the posterior wall of the prevesical space; *see* Charpy, Adrien.

Fascia of Cloquet: *Eponym* for the part of the fascia lata covering the hiatus saphenus; *see* Cloquet, Jules Germain.

Fascia of Colles: *Eponym* for (a) the membranous layer of the subcutaneous tissue of the perineum, (b) the subcutaneous tissue of the penis, and (c) the superficial fascia of the scrotum (fascia dartos); *see* Colles, Abraham.

Fascia of Cooper: *Epomym* for the cremasteric fascia; *see* Cooper, Astley Paston.

Fascia of Cruveilhier: *Eponym* for the superficial perineal fascia; *see* Cruveilhier, Jean Baptiste.

Fascia of Delbet: *Eponym* for the vesicoumbilical fascia; *see* Delbet, Paul.

Fascia of Denonvilliers: *Eponym* for the rectovesical septum or fascia in males, also described by Tyrrell, and the rectovaginal septum or fascia in females; *see* Denonvilliers, Charles Pierre.

Fascia of Dupuytren: *Eponym* for the palmar aponeurosis; *see* Dupuytren, Guillaume.

Fascia of Foville: *Eponym* for the stria terminalis; *see* Foville, Achille Louis.

Fascia of Fredet: *Eponym* for the preduodenopancreatic fascia; *see* Fredet, Pierre.

Fascia of Gallaudet: *Eponym* for (a) the innominate fascia under the subcutaneous tissue of the anterior abdominal wall, and (b) the deep perineal fascia covering the perineal muscles, also described by Buck; *see* Gallaudet, Bern.

Fascia of Gerota: *Eponym* for the renal fascia (also known as capsule of Gerota); *see* Gerota, Dumitru.

Fascia of Godman: *Eponym* for a continuation of the pretracheal layer of the cervical fascia to the thorax and pericardium; *see* Godman, John Davidson.

Fascia of Gruber: *Eponym* for the middle layer of the cervical fascia, also described by Paulet, Richet and Testut; *see* Gruber, Wenzel.

Fascia of Halban: *Eponym* for fascial strips between the trigone of the urinary bladder and the anterior part of the vaginal wall, also known as pubocervical fascia or vesicovaginal septum; *see* Halban, Josef.

Fascia of Hesselbach: *Eponym* for the cribriform fascia, part of the fascia lata; *see* Hesselbach, Franz Kaspar.

Fascia of Hyrtl: *Eponym* for the transversalis fascia, also described by Velpeau; *see* Hyrtl, Josef.

Fascia of Luschka: *Eponym* for the endothoracic or parietal thoracic fascia; *see* Luschka, Hubert von.

Fascia of Paulet: *Eponym* for the middle layer of the cervical fascia, also described by Gruber, Richet and Testut; *see* Paulet, Vincent.

Fascia of Porter: *Eponym* for the pretracheal layer of the cervical fascia; *see* Porter, William Henry.

Fascia of Richet: *Eponym* for (a) a fold of peritoneum enveloping the obliterated umbilical vein, also known as canal of Richet, (b) the middle layer of the cervical fascia, also described by Gruber, Paulet and Testut; *see* Richet, Louis Alfred.

Fascia of Rouvière: *Eponym* for the axillary fascia; *see* Rouvière, Henri.

Fascia of Scarpa: *Eponym* for the membranous layer of the subcutaneous tissue of the abdomen; *see* Scarpa, Antonio.

Fascia of Sibson: *Eponym* for the suprapleural membrane; *see* Sibson, Francis.

Fascia of Tarin: *Eponym* for the dentate gyrus; *see* Tarin, Pierre.

Fascia of Tenon: *Eponym* for the fascia bulbi; *see* Tenon, Jacques René.

Fascia of Testut: *Eponym* for the middle layer of the cervical fascia, also described by Gruber, Paulet and Richet; *see* Testut, Jean Léon.

Fascia of Thomson: *Eponym* for (a) the iliopectineal fascia, the most inferior fibres of the fascia of the external oblique muscle attached to the posterior surface of the inguinal ligament, and (b) the iliopubic tract; *see* Thomson, Alexander.

Fascia of Thompson: *Eponym* for the iliopubic tract, the thickened inferior margin of the transversalis fascia; *see* Thompson, Henry.

Fascia of Toldt: *Eponym* for (a) the retropancreatic fascia behind the body of the pancreas, and (b) the prerenal fascia, also known as membrane of Toldt; *see* Toldt, Karl.

Fascia of Treitz: *Eponym* for the retropancreatic fascia behind the head of the pancreas; *see* Treitz, Wenzel.

Fascia of Troeltsch: *Eponym* for the salpingopharyngeal fascia; *see* Troeltsch, Anton Friedrich von.

Fascia of Tyrrell: *Eponym* for the rectovesical fascia or septum, also described by Denonvilliers; *see* Tyrrell, Frederick.

Fascia of Velpeau: *Eponym* for the transversalis fascia, also described by Hyrtl; *see* Velpeau, Alfred.

Fascia of Waldeyer: *Eponym* for the retrorectal fascia; *see* Waldeyer, Heinrich Wilhelm Gottfried.

Fascia of Zuckerkandl (Emil): *Eponym* for the posterior layer of the renal fascia; *see* Zuckerkandl, Emil.

Fascia of Zuckerkandl (Otto): *Eponym* for the rectovaginal fascia or septum; *see* Zuckerkandl, Otto.

1.11 Fibres

The term **fibres** is used for fibres in the central nervous system, for (bundles of) fibres of muscles, for certain ligaments, and in the old days also for the reticular formation (Schroeder van der Kolk, Stilling).

Fibres of Arnold, Arcuate: *Eponym* for the short association fibres of the cerebral cortex, currently known as U-fibres of Meynert; *see* Arnold, Friedrich.

Fibres of Bergmann: *Eponym* for the ascending fibres of the Bergmann glial cells traversing the cerebellar cortex, but earlier by Crampton; *see* Bergmann, Gottlieb Heinrich.

Fibres of Bowman: *Eponym* for the meridional fibres of the ciliary muscle, also described by Brücke in the same year, but earlier by Crampton; *see* Bowman, William.

Fibres of Brücke: *Eponym* for the meridional fibres of the ciliary muscle, also described by Bowman in the same year; *see* Brücke, Ernst Wilhelm.

Fibres of Crampton: *Eponym* for the meridional fibres of the ciliary muscle, also described by Bowman and Brücke; *see* Crampton, Phillip.

Fibres of Gratiolet: *Eponym* for the optic radiation; *see* Gratiolet, Louis Pierre.

Fibres of Helie: *Eponym* for a vertical and ansiform bundle of the superficial musculature of the uterus; *see* Helie, Louis Theodore.

Fibres of Henle: *Eponym* for the layer of inner cone fibres in the central area of the retina; *see* Henle, Friedrich Gustav Jacob.

Fibres of Kaplan: *Eponym* for the attachment of the iliotibial tract to the distal femur; *see* Kaplan, Emanuel.

Fibres of Korff: *Eponym* for argyrophilic strands of ground substance with reticular microfibrils, originating in the pulp of the tooth; *see* Korff, Karl von.

Fibres of Müller: *Synonym* for the radial glial cells of the retina (cells of Müller); *see* Müller, Heinrich.

Fibres of Nansen: *Eponym* for the ascending and descending branches of the dorsal spinal nerve roots; *see* Nansen, Fridtjof.

Fibres of Prussak: *Eponym* for connective tissue fibres between the flaccid and tense part of the tympanic membrane; *see* Prussak, Alexander.

Fibres of Purkinje: *Eponym* for the terminal ramifications of the cardiac conducting system beneath the endocardium of the ventricles; *see* Purkinje, Jan Evangelista.

Fibres of Remak: *Eponym* for unmyelinated fibres; *see* Remak, Robert.

Fibres of Rouget: *Eponym* for the radial fibres of the ciliary muscle; *see* Rouget, Charles Marie Benjamin.

Fibres of Rouget-Javara: *Eponym* for phrenicooesophageal muscle fibres; *see* Rouget, Charles Marie Benjamin, and Juvara, Ernest.

Fibres of Sappey: *Eponym* for smooth muscle fibres in the check ligaments of the eyeball; *see* Sappey, Marie Philibert Constant.

Fibres of Schroeder van der Kolk: *Eponym* for the reticular formation of the spinal cord and hindbrain, also described as such by Stilling; *see* Schroeder van der Kolk, Jacob Ludlow Conrad.

Fibres of Sharpey: *Eponym* for connective tissue fibres between the periosteum and bone; see Sharpey, William.

Fibres of Stilling: *Eponym* for the reticular formation of the spinal cord and hindbrain, also described as such by Schroeder van der Kolk; *see* Stilling, Benedict.

Fibres of Tomes: *Eponym* for projections of odontoblasts, which run inside dentinal tubules perpendicular to the surface of a tooth; *see* Tomes, John.

Fibres of Weitbrecht: *Eponym* for retinacular fibres of the neck of the femur; *see* Weitbrecht, Josias.

1.12 Folds

For several **folds**, still eponyms are in use (folds of Douglas, Houston, Kohlrausch, Luschka, Nélaton, and Troeltsch).

Fold of Amantini, Pectineofoveal: *Eponym* for the pectineofoveal fold; *see* Amantini, Cesare.

Fold of Braune: *Eponym* for a perpendicular mucous fold at the ostium of the stomach, described as the cardiac sphincter between the oesophagus and the stomach, also described by Gubaroff; *see* Braune, Christian Wilhelm.

Fold of Broca, Falciform: *Eponym* for the lateral border of the anterior perforated substance of the brain; *see* Broca, Pierre Paul.

Fold of Douglas: *Eponym* for the rectouterine fold; *see* Douglas, James.

Folds of Duncan: *Eponym* for peritoneal folds of the uterus; *see* Duncan, James Matthews.

Fold of Falloppio-Kerckring: *Eponym* for the circular folds of the small intestines, especially distinct in the duodenum and jejunum; *see* Falloppio, Gabriele, and Kerckring, Dirk.

Fold of Gubaroff: *Eponym* for a perpendicular mucous fold at the ostium of the stomach, described as the cardiac sphincter between the oesophagus and the stomach, also described by Braune; *see* Gubaroff, Alexander Petrowitsch von.

Fold or **Valve of Hasner:** *Eponym* for the lacrimal fold or valve in the nasolacrimal duct, also described by Bianchi and Cruveilhier; *see* Hasner, Joseph.

Folds of Havers: *Eponym* for collections of adipose tissue in the major joints, covered by synovial membrane; *see* Havers, Clopton.

Fold of Houston: *Eponym* for the superior transverse rectal fold; *see* Houston, John,

Fold of Jackson: *Eponym* for the pericolic membrane, a peritoneal fold or adhesion between the caecum or ascending colon and the right posterior abdominal wall; *see* Jackson, Jabez North.

Fold of Kohlrausch: *Eponym* for the middle transverse rectal fold; *see* Kohlrausch, Otto Ludwig Bernhard.

Fold of Luschka: *Eponym* for the ventral vascular ileocaecal fold; *see* Luschka, Hubert von.

Fold of Marshall, Vestigial: *Eponym* for a pericardial fold containing the obliterated remains of the left superior vena cava; *see* Marshall, John.

Fold of Nélaton: *Eponym* for the inferior transverse rectal fold; *see* Nélaton, Auguste.

Folds of Rathke: *Eponym* for two transient lateral folds, assumed to be forming the urogenital septum dividing the rectum from the urinary bladder; *see* Rathke, Martin Heinrich.

Folds of Rindfleisch: *Eponym* for the semilunar folds of the serous surface of the pericardium, embracing the beginning of the aorta; *see* Rindfleisch, Eduard.

Folds of Rivinus, Mallear: *Eponym* for the anterior and posterior mallear folds of the tympanic membrane, usually attributed to Troeltsch; *see* Rivinus, August Quirinus.

Folds of Schultze: *Eponym* for amniotic folds enclosing the yolk stalk at the origin of the umbilical cord from the placenta, the remains normally are incorporated in the placenta; *see* Schultze, Bernhard Sigismund.

Fold of Treves, Bloodless: *Eponym* for the bloodless inferior ileocaecal fold of the vermiform appendix; see Treves, Frederick.

Folds of Troeltsch: *Eponym* for the anterior and posterior mallear folds of the tympanic membrane, also described by Rivinus; *see* Troeltsch, Anton Friedrich von.

Fold of Vater: *Eponym* for a mucous fold in the duodenum, just above the major duodenal papilla; *see* Vater, Abraham.

Fold of Veraguth: *Eponym* for a skin fold that runs obliquely from the lower outside to the upper inside of the upper eyelid, assumed to be characteristic for patients suffering from depression; *see* Veraguth, Otto.

Fold of Willis: *Eponym* for the inferior medullary velum of the fourth ventricle; *see* Willis, Thomas.

Fold of Zaufal: *Eponym* for the salpingopharyngeal folds; *see* Zaufal, Emanuel.

1.13 Foramina

The term **foramen** of **foramina** is used for large openings such as the foramen ovale (Botallo), the omental or epiploic foramen (Duverney, Winslow), for openings of veins (Lannelongue, Thebesius, Vieussens), but especially for openings for peripheral nerves (Civinini, Ferrein, Jacobson) and of the ventricles of the brain (Key-Retzius, Luschka, Magendie, Mierzejewski, Monro).

Foramen of Bichat: *Eponym* for the cistern of the great cerebral vein; see Bichat, Marie François Xavier.

Foramen of Bochdalek: *Eponym* for the embryonic pleuroperitoneal hiatus; *see* Bochdalek, Vincenc Alexander.

Foramen of Botallo: *Eponym* for the foramen ovale of the heart; *see* Botallo, Leonardo.

Foramen of Civinini: *Eponym* for the inconstant pterygospinous foramen; *see* Civinini, Filippo.

Foramen of Duverney: *Eponym* for the omental or epiploic foramen, later also described by Winslow; *see* Duverney, Joseph Guichard.

Foramen of Ferrein: *Eponym* for (a) the hiatus of the facial canal, earlier described by Falloppio as hiatus, and (b) the hiatus of the canal for the greater petrosal nerve; *see* Ferrein, Antoine.

Foramen of Huschke: *Eponym* for an inconstant opening in the floor of the bony part of the external acoustic meatus near the tympanic membrane; *see* Huschke, Emil.

Foramen of Hyrtl: *Eponym* for an occasional foramen in the sphenoidal bone for the motor portion of the trigeminal nerve (his foramen crotaphyticobuccinatorius); *see* Hyrtl, Joseph.

Foramen of Jacobson: *Eponym* for the opening of the tympanic canaliculus in the temporal bone in the ridge between the jugular foramen and the carotid canal, through which the tympanic branch of the glossopharyngeal nerve enters the tympanic cavity; *see* Jacobson, Ludwig Levin.

Foramen of Key-Retzius: *Eponym* for the lateral aperture of the fourth ventricle, also described by Luschka and Mierzejewsky; *see* Key, Ernst Axel Henrik, and Retzius, Gustav Magnus.

Foramen of Knox: *Eponym* for an inconstant supracondylar foramen of the humerus; *see* Knox, Robert.

Foramina of Lannelongue: *Eponym* for the venous openings of the smallest cardiac veins into the right atrium, already described by Vieussens and Thebesius; *see* Lannelongue, Odilon.

Foramen of Luschka: *Eponym* for the lateral aperture of the fourth ventricle, also described by Key-Retzius and Mierzejewsky; *see* Luschka, Hubert von.

Foramen of Magendie: *Eponym* for the median aperture of the fourth ventricle; *see* Magendie, François.

Foramen of Mierzejewsky: *Epnonym* for the lateral aperture of the fourth ventricle, also described by Luschka, Key and Retzius; *see* Mierzejewsky, Jan Lucian.

Foramen of Monro: *Eponym* for the interventricular foramen of the brain; *see* Monro, Alexander.

Foramen of Morand: *Eponym* for the foramen caecum of the skull, which is situated in the anterior cranial fossa between the crista galli and the frontal crest; *see* Morand, Sauveur François.

Foramina of Morgagni: *Eponym* for the foramina nervosa of the cochlea; *see* Morgagni, Giovanni Battista.

Foramen of Pacchioni: *Eponym* for (a) the tentorial notch, and (b) the foramen of the diaphragma sellae; *see* Pacchioni, Antonio.

Foramen of Rouvière: *Eponym* for a rare dehiscence in the capsule of the shoulder joint between the middle and inferior glenohumeral ligaments; *see* Rouvière, Henri.

Foramen of Sappey: *Eponym* for the helicotrema, also described by Scarpa; *see* Sappey, Marie Philibert Constant.

Foramen of Soemmerring: *Eponym* for the fovea centralis of the retina; *see* Soemmerring, Samuel Thomas von.

Foramen of Stensen: *Eponym* for the incisive foramen of the maxilla; *see* Stensen, Niels.

Foramina of Swan: *Eponym* for the foramina for the superior dental nerves; *see* Swan, Joseph English.

Foramina of Thebesius: *Eponym* for the openings of the smallest cardiac veins into the right atrium, also described by Vieussens and Lannelongue; *see* Thebesius, Adam Christian.

Foramen or **Space of Velpeau, Quadrangular:** *Eponym* for a space determined by the teres major muscle caudally, the teres minor muscle cranially, the humerus laterally and the long head of the triceps brachii muscle medially; *see* Velpeau, Alfred Armand Louis Marie.

Foramen of Vesalius: *Eponym* for a small inconstant foramen in the sphenoidal bone anteromedial to the oval foramen, through which an emissary vein from the cavernous sinus may pass to the pterygoid plexus; *see* Vesalius, Andreas.

Foramen of Vicq d'Azyr: *Eponym* for the foramen caecum of the medulla oblongata; *see* Vicq d'Azyr, Felix.

Foramina of Vieussens: *Eponym* for the openings of the smallest cardiac veins into the right atrium, also described by Thebesius and Lannelongue; *see* Vieussens, Raymond de.

Foramen of Weitbrecht: *Eponym* for a gap in the capsule of the shoulder joint between the superior and middle glenohumeral ligaments; *see* Weitbrecht, Josias.

Foramen of Winslow: *Eponym* for the omental or epiploic foramen, earlier described by Duverney; *see* Winslow, Jacob Benignus.

1.14 Fossae

A large group of **fossae** has received an eponym, many for parts of bones, peritoneal fossae (Brösike, Claudius, Gruber-Landzert, Ionesco, Luschka, Poisson, Treitz, Treves, Waldeyer), and for superficial structures (Gerdy, Malgaigne, Mohrenheim).

Fossa of Allen: *Eponym* for an inconstant fossa on the neck of the femur; *see* Allen, Harrison.

Fossa of Bichat: *Eponym* for the pterygopalatine fossa; *see* Bichat, Marie François Xavier.

Fossa of Brösike: *Eponym* for the intermesocolic fossa of the peritoneal cavity; *see* Brösike, Gustav.

Fossa of Charpy: *Eponym* for the retromuscular fossa behind the lower part of the rectus abdominis muscle; *see* Charpy, Adrien.

Fossa of Claudius: *Eponym* for a peritoneal fossa, housing the ovary in a multipara, located between the internal iliac artery and the sacrum; *see* Claudius, Friedrich Matthias.

Fossa of Cruveilhier: *Eponym* for the scaphoid fossa of the sphenoidal bone; *see* Cruveilhier, Jean Baptiste.

Fossa or **Triangle of Gerdy:** *Eponym* for the infraclavicular fossa, also described by Mohrenheim; *see* Gerdy, Pierre Nicholas.

Fossa of Gruber-Landzert: *Eponym* for the inferior duodenal recess of the peritoneal cavity; *see* Gruber, Wenzel Leopold, and Landzert, Theodor.

Fossa of Ionesco: *Eponym* for the duodenojejunal fossa of the peritoneal cavity, a term sometimes also applied to the retroduodenal fossa; *see* Ionesco, Thoma.

Fossa of Jobert de Lamballe: *Eponym* for the fossa in the popliteal region bounded proximally by the adductor magnus muscle and distally by the gracilis and sartorius muscles; *see* Jobert de Lamballe, Antoine Joseph.

Fossa of Krause: *Eponym* for the ovarian fossa, also described by Claudius and Waldeyer, *see* Krause, Karl Friedrich Theodor.

Fossa of Landzert: *Eponym* for the inferior duodenal recess in the peritoneal cavity, also described by Gruber, and usually described as fossa of Gruber-Landzert; *see* Landzert, Theodor.

Fossa of Luschka: *Eponym* for the superior ileocaecal fossa of the peritoneal cavity; *see* Luschka, Hubert von.

Fossa or **Triangle of Malgaigne:** *Eponym* for the carotid fossa or triangle; *see* Malgaigne, Joseph François.

Fossa or **Triangle of Marcille:** *Eponym* for the lumbosacral triangle, bounded by the medial border of the psoas major muscle, the lateral margin of the vertebral column and the iliolumbar ligament; *see* Marcille, Maurice.

Fossa of Merkel: *Eponym* for the central fossa between the two ventricles of the larynx; *see* Merkel, Karl Ludwig.

Fossa of Mohrenheim: *Eponym* for the infraclavicular fossa, also described by Gerdy; *see* Mohrenheim, Joseph Jacob.

Fossa of Morgagni: *Eponym* for the navicular fossa of the male urethra; *see* Morgagni, Giovanni Battista.

Fossa of Poisson: *Eponym* for the infraduodenal peritoneal recess of the peritoneal cavity; *see* Poisson, Francis.

Fossa of Rosenmüller: *Eponym* for the pharyngeal recess; *see* Rosenmüller, Johann Christian.

Fossa of Scarpa: *Eponym* for the fossa iliopectinea; *see* Scarpa, Antonio.

Fossa of Sylvius: *Eponym* for the lateral cerebral fossa (Sylvian fossa); *see* Sylvius, François de le Boë.

Fossa of Tarin: *Eponym* for the interpeduncular fossa of the brain stem; *see* Tarin, Pierre.

Fossa of Treitz: *Eponym* for (a) the superior and inferior duodenal recesses of the peritoneal cavity; (b) the subcaecal fossa of the peritoneal fossa; and (c) the intersigmoid recess; *see* Treitz, Wenzel.

Fossa of Treves: *Eponym* for the inferior ileocaecal fossa of the peritoneal cavity; *see* Treves, Frederick.

Fossa of Velpeau: *Eponym* for the ischiorectal fossa; *see* Velpeau, Alfred Armand Louis Marie.

Fossa of Waldeyer: *Eponym* for (a) the ovarian fossa, also described by Claudius and Krause, and (b) the mesentericoparietal fossa of the peritoneal cavity; *see* Waldeyer, Heinrich Wilhelm Gottfried.

1.15 Ganglia

The term **ganglion** is used in various ways, for peripheral nerve ganglia in particular but also for lymph nodes, especially in the French literature.

Ganglion of Acrel: *Eponym* for a pseudoganglion on the posterior interosseous nerve on the back of the wrist; *see* Acrel, Olof.

Ganglion of Andersch: *Eponym* for the inferior ganglion of the glossopharyngeal nerve; *see* Andersch, Carol Samuel.

Ganglion of Arnold: *Eponym* for (a) the otic ganglion, (b) the thoracic splanchnic ganglia, and (c) the glomus caroticum; *see* Arnold, Friedrich.

Ganglia of Auerbach: *Eponym* for collections of parasympathetic cells in the myenteric plexus of the digestive tract; *see* Auerbach, Leopold.

Ganglion of Axenfeld: *Eponym* for an accessory ciliary ganglion; *see* Axenfeld, Theodor.

Ganglion of Bendz: *Eponym* for the inferior ganglion of the vagus nerve, also described by Wutzer; *see* Bendz, Henrik Carl Bang.

Ganglion of Bezold: *Eponym* for an aggregate of nerve cells in the interatrial septum of the heart, also described by Bidder; *see* Bezold, Albert von.

Ganglia of Bidder: *Eponym* for an aggregate of nerve cells in the interatrial septum of the heart, also described by Bezold; *see* Bidder, Heinrich Friedrich.

Ganglion of Blandin: *Eponym* for (a) the submandibular ganglion, later also described by Meckel and Langley, and (b) the inconstant sublingual ganglion; *see* Blandin, Philippe.

Ganglion of Bochdalek: *Eponym* for a thickening in the superior dental plexus of the junction of the anterior and middle alveolar nerves. However, it is not a true ganglion since it consists of interlacing bundles of nerve fibres and does not contain nerve cells; *see* Bochdalek, Vincenc Alexander.

Ganglion of Bock: *Eponym* for the carotid ganglion; *see* Bock, August.

Ganglion of Boettcher: *Eponym* for a small ganglion on the cochlear nerve in the internal acoustic meatus, innervating that part of the cochlear duct located in the cavity of the vestibular ganglion, sometimes also used as synonym for the spiral ganglion of Corti; *see* Boettcher, Arthur.

Ganglion of Corti: *Eponym* for the spiral or cochlear ganglion; *see* Corti, Alfonso.

Ganglion of Ehrenritter: *Eponym* for the superior ganglion of the glossopharyngeal nerve; *see* Ehrenritter, Johannes.

Ganglion of Frankenhäuser: *Eponym* for the parasympathetic ganglion of the uterovaginal plexus, also described by Lee; *see* Frankenhäuser, Ferdinand.

Ganglion of Froriep: *Eponym* for a temporary collection of nerve cells on the dorsal aspect of the hypoglossal nerve in the embryo; *see* Froriep, August von.

Ganglion of Gasser: *Eponym* for the trigeminal ganglion; *see* Gasser, Johann Laurentius.

Ganglion of Gudden: *Eponym* for the interpeduncular nucleus of the brain stem; *see* Gudden, Johannes Bernhard Aloys von.

Ganglion of Haller, Thyroid: *Eponym* for the middle cervical ganglion of the sympathetic trunk; *see* Haller, Albrecht von.

Ganglion of Hirschfeld: *Eponym* for the superior cervical ganglion of the sympathetic trunk; *see* Hirschfeld, Ludwig Moritz.

Ganglion of Langley: *Eponym* for the submandibular ganglion, also described by Blandin and Meckel; *see* Langley, John Newport.

Ganglion of Laumonier: *Eponym* for the carotid ganglion in the cavernous plexus, previously described by Petit and Schmiedel; *see* Laumonier, Jean Baptiste Phillippe Nicolas Réné.

Ganglion of Lee: *Eponym* for a parasympathetic ganglion of the uterovaginal plexus, also described as ganglion of Frankenhäuser; *see* Lee, Robert.

Ganglion of Lobstein: *Eponym* for an accessory ganglion of the greater thoracic splanchnic nerve in connection with the coeliac plexus; *see* Lobstein, Johann Georg Christian Friedrich Martin; .

Ganglion of Ludwig: *Eponym* for a small collection of parasympathetic nerve cells in the interatrial septum of the heart; *see* Ludwig, Karl Friedrich Wilhelm.

Ganglion of Mayer: *Eponym* for the inconstant sublingual ganglion; *see* Mayer, Sigmund.

Ganglion of Meckel: *Eponym* for (a) the pterygopalatine ganglion, and (b) the submandibular ganglion; *see* Meckel, Johann Friedrich, the Elder.

Ganglion of Meynert: *Eponym* for the basal nucleus of the forebrain composed of large (cholinergic) cells; *see* Meynert, Theodor.

Ganglion of Petit: *Eponym* for the carotid ganglion in the cavernous plexus, also described by Schmiedel and Laumonier; see Petit, François Pourfour du.

Ganglion of Rauber: *Eponym* for the sublingual ganglion; *see* Rauber, August Antonius.

Ganglia of Remak: *Eponym* for (a) an autonomic ganglion in the opening of the inferior vena cava in the diaphragm, and (b) autonomic ganglia in the nerves of the stomach; *see* Remak, Robert.

Ganglion of Ribes: *Eponym* for a small sympathetic ganglion on the anterior communcating artery of the brain; *see* Ribes, François.

Ganglion of Scarpa: *Eponym* for the vestibular ganglion; *see* Scarpa, Antonio.

Ganglion of Schacher: *Eponym* for the ciliary ganglion; *see* Schacher, Polycarp Gottlieb.

Ganglion of Schmiedel: *Eponym* for the carotid ganglion in the cavernous plexus, also described by Petit and Laumonier; *see* Schmiedel, Kasimir Christoph.

Ganglion of Soemmerring: *Old term* for the substantia nigra; *see* Soemmerring, Samuel Thomas von.

Ganglion of Troisier: *Eponym* for a lymph node located above the left clavicula belonging to the supraclavicular lymph nodes, palpably enlarged in metastasis from a malignant tumor, also known as node of Virchow-Troisier; *see* Troisier, Charles-Emile.

Ganglion of Valentin: *Eponym* for a ganglion on the superior alveolar nerve; *see* Valentin, Gabriel Gustav.

Ganglion of Vieussens: *Eponym* for the coeliac ganglion; *see* Vieussens, Raymond de.

Ganglion of Walther: *Eponym* for the ganglion impar; *see* Walther, Johann Gottlieb.

Ganglion of Wrisberg: *Eponym* for a large cardiac ganglion located below the arch of the aorta to the right of the ligamentum arteriosum; *see* Wrisberg, Heinrich August.

Ganglion of Wutzer: *Eponym* for the inferior ganglion of the vagus nerve, also described by Bendz; *see* Wutzer, Karl Wilhelm.

Ganglion of Winterhalter: *Eponym* for an aggregation of sympathetic nerve cells forming the ovarian ganglion; *see* Winterhalter, Elisabeth Hermine.

1.16 Glands

The term **gland** was used in various ways, for secreting glands and for lymph nodes.

Gland(s) of Albarrán: *Eponym* for (a) the middle lobe of the prostate, and (b) infracervical glands, small accessory prostatic glands in the mucosa of the male urethra, halfway between the internal urethral opening and the seminal colliculus or verumontanum; *see* Albarrán, Joaquin Maria.

Gland of Aselli: *Eponym* for a single large lymph node ventral to the abdominal aorta, draining all lymph from the intestines in many small animals, also known as Aselli pancreas; *see* Aselli, Gaspar.

Gland of Bartholin: *Eponym* for (a) the greater vestibular gland, also described by Duverney and Tiedemann; and (b) the sublingual salivatory gland; *see* Bartholin, Caspar Secundus.

Glands of Bauhin: *Eponym* for the anterior (apical) lingual glands, also described by Blandin and Nuhn; *see* Bauhin, Caspar, The Younger.

Glands of Blandin: *Eponym* for the anterior (apical) lingual glands, also described by Bauhin and Nuhn; see Blandin, Philippe Frédéric.

Glands of Boerhaave: *Eponym* for the sweat glands; *see* Boerhaave, Herman.

Glands of Bowman: *Eponym* for mucous glands in the olfactory part of the nasal cavity, also described by Kölliker; *see* Bowman, William.

Glands of Bruch: *Eponym* for lymphatic nodules in the palpebral conjunctiva; *see* Bruch, Karl Wilhelm Ludwig.

Glands of Brunner: *Eponym* for duodenal glands, also described by Wepfer, his father-in-law, Galeazzi, Lieberkühn and Severino; *see* Brunner, Johann Konrad.

Glands of Ciaccio: *Eponym* for accessory lacrimal glands mostly in the upper eyelid, also described by Krause, Henle, Terson and Wolfring; *see* Ciaccio, Giuseppe Vincenzo.

Gland of Cloquet: *Eponym* for the proximal deep inguinal lymph node, also described by Rosenmüller; *see* Cloquet, Jules Germain.

Glands of Cowper: *Eponym* for the bulbourethral glands, earlier described by Méry; *see* Cowper, William.

Glands of Duverney: *Eponym* for vulvovaginal glands, discovered in the cow, comparable to the greater vestibular gland of Bartholin; *see* Duverney, Joseph Guichard.

Glands of Ebner: *Eponym* for serous glands in relation to the vallate or circumvallate papillae of the tongue; *see* Ebner, Victor von.

Glands of Galeazzi: *Eponym* for the intestinal glands, also described by Lieberkühn, Severino, Brunner and Wepfer; *see* Galeazzi, Domenico.

Glands of Gay: *Eponym* for circumanal glands; *see* Gay, Alexander Heinrich.

Glands of Gley: *Eponym* for the parathyroid glands; *see* Gley, Eugène.

Glands of Guérin: *Eponym* for female paraurethral glands, later also described by Skene; *see* Guérin, Alphonse François Marie.

Gland of Harder: *Eponym* for an accessory lacrimal gland, also described by Ciaccio, Krause and others; *see* Harder, Johann Jacob.

Glands of Havers: *Eponym* for collections of adipose tissue in the major joints, covered by synovial membrane; *see* Havers, Clopton.

Glands of Henle: *Eponym* for accessory lacrimal glands mostly in the upper eyelid, also described by Ciaccio, Krause, Terson and Wolfring; *see* Henle, Friedrich Gustav Jacob.

Glands of Knoll: *Eponym* for glands in the vestibular folds of the larynx; *see* Knoll, Philipp.

Glands of Kölliker: *Eponym* for the olfactory glands in the nasal cavity, also described by Bowman; *see* Kölliker, Rudolf Albert von.

Glands of Krause: *Eponym* for accessory lacrimal glands mostly in the upper eyelid, described both by Karl Friedrich Theodor Krause and his son Wilhelm, as well as by others.

Glands of Lieberkühn: *Eponym* for intestinal glands, also described by Brunner, Galeazzi and Severino; *see* Lieberkühn, Johann Nathaniel.

Glands of Littré: *Eponym* for the urethral glands in males, also described by Morgagni; *see* Littré, Alexis.

Gland of Luschka: *Eponym* for the coccygeal body; *see* Luschka, Hubert von.

Glands of Meibom: *Eponym* for tarsal glands of the eyelids; *see* Meibom, Heinrich.

Gland of Méry: *Eponym* for the bulbourethral gland, also described by Cowper; *see* Méry, Jean.

Glands of Meyer: *Eponym* for aggregations of glands beneath the tongue, found in the hyoglossus muscle; *see* Meyer, Georg Hermann von.

Glands of Moll: *Eponym* for the ciliary or palpebral sweat glands, later also described by Sattler; *see* Moll, Jacob Antonius.

Glands of Montgomery: *Eponym* for the areolar glands of the breast; *see* Montgomery, William Fetherston.

Glands of Morgagni: *Eponym* for the paraurethral glands; *see* Morgagni, Giovanni Battista.

Glands of Nuhn: *Eponym* for the anterior (apical) lingual glands, also described by Bauhin and Blandin; *see* Nuhn, Anton.

Glands of Pechlin: *Eponym* for the aggregated lymphatic nodules of the ileum, usually described as glands or patches of Peyer; *see* Pechlin, Johannes Nicolaas.

Glands of Peyer: *Eponym* for the aggregated lymphatic nodules of the ileum, earlier described by Pechlin but usually as plaques of Peyer; *see* Peyer, Johann Conrad.

Gland of Rivinus: *Eponym* for the sublingual gland; *see* Rivinus, Augustinus Quirinus.

Gland of Rosenmüller: *Eponym* for the proximal deep inguinal lymph node, also described by Cloquet (gland or node of Cloquet); *see* Rosenmüller, Johann Christian.

Glands of Sattler: *Eponym* for the ciliary or palpebral sweat glands, earlier described by Moll; *see* Sattler, Hubert.

Glands of Serres: *Eponym* for 'gingival glands', islands of epithelium present in the gums of infants; *see* Serres, Antoine Etienne Renaud Augustin.

Glands of Severino, Internal: *Eponym* for the tubular glands of the small intestine, also described by Brunner, Galeazzi and Lieberkühn; *see* Severino, Marco Aurelio.

Glands of Skene: *Eponym* for the paraurethral glands of females, earlier described by Guérin; *see* Skene, Alexander Johnston Chalmers.

Gland of Sorgius: *Eponym* for the sentinel lymph node of the mammary gland, the caudalmost of the pectoral lymph nodes on the second or third slip of the anterior serratus muscle; *see* Sorgius, Wilhelm.

Gland of Stahr: *Eponym* for a lymph node on the facial artery; *see* Stahr, Hermann.

Glands of Stensen: *Eponym* for the nasal glands; *see* Stensen, Niels.

Glands of Suzanne: *Eponym* for the mucous gland found in the floor of the mouth close to the midline; *see* Suzanne, Jean George.

Glands of Terson: *Eponym* for accessory lacrimal glands, already described by Ciaccio, Henle, the Krauses and Wolfring; *see* Terson, Albert.

Glands of Theile: *Eponym* for mucous glands in the wall of the cystic duct and the neck of the gallbladder; *see* Theile, Friedrich Wilhelm.

Gland of Tiedemann: *Eponym* for the greater vestibular gland, already described by Bartholin and Duverney; *see* Tiedemann, Friedrich.

Gland of Troisier: *Eponym* for a lymph node located above the left clavicle belonging to the supraclavicular lymph nodes, palpably enlarged as the result of a malignant neoplasm, also described by Virchow, and known as the gland of Virchow-Troisier; *see* Troisier, Charles Emile.

Glands of Tyson, Preputial: *Eponym* for peculiar sebaceous glands in the dermis of the corona of the glans penis and the inner surface of the prepuce; *see* Tyson, Edward.

Glands of Vesalius: *Eponym* for bronchial lymph nodes; *see* Vesalius, Andreas.

Gland of Virchow-Troisier: *Eponym* for the supraclavicular lymph node above the left clavicula, a signal node for malignancy; *see* Virchow, Rudolf Ludwig Karl, and Troisier, Charles Emile.

Glands of Vitali: *Eponym* for ungual sweat glands present on the anterior fourth of the nail bed; *see* Vitali, Giovanni.

Glands of Wasmann: *Eponym* for the gastric glands; *see* Wasmann, Adolphus.

Glands of Weber: *Eponym* for mucous glands on the posterior border of the tongue; *see* Weber, Ernst Heinrich.

Glands of Wepfer: *Eponym* for duodenal glands, also described by Brunner, his son-in-law; *see* Wepfer, Johann.

Glands of Wolfring: *Eponym* for accessory lacrimal glands mostly in the upper eyelid, also described by Ciaccio, Krause, Henle and Terson; see Wolfring, Ernst.

Gland of Woelfler: *Eponym* for an accessory thyroid gland; *see* Woelfler, Anton.

Glands of Zeis: *Eponym* for the ciliary glands of the eyelids; see Zeis, Eduard.

Glands of Zoja: *Eponym* for small accessory thyroid glands; *see* Zoja, Giovanni.

Gland of Zuckerkandl: *Eponym* for an ectopic thyroid gland located in front of the hyoid bone; *see* Zuckerkandl, Emil.

1.17 Ligaments

The term **ligament** is especially used for fibrous structures, many of which are still in use as eponyms, but also for rudiments of embryonic vascular structures such as the ligamentum arteriosum of Botallo and the ligamentum venosum of Aranzio (Arantius).

Ligament of Aranzio (Arantius): *Eponym* for the ligamentum venosum of the liver; *see* Aranzio, Julio Caesar.

Ligament of Arnold: *Eponym* for the superior ligament of the incus; *see* Arnold, Friedrich.

Ligament of Bardinet: *Eponym* for the olecranohumeral ligament, part of the ulnar collateral ligament of the elbow joint; *see* Bardinet, Bartholemy Alphonse..

Ligament of Barkow: *Eponym* for (a) the plantar calcaneonavicular ligament, (b) the oblique tibiofibular ligament, (c) the pisohamate ligament, (d) a thin innominate ligament connecting both occipital condyles, and (e) the anterior and posterior portions of the fibrous capsule of the elbow joint; *see* Barkow, Hans Karl Leopold.

Ligament of Bauhin: *Eponym* for the ischiofemoral ligament of the hip joint; *see* Bauhin, Caspar.

Ligament of Bellini: *Eponym* for the ischiofemoral ligament of the hip joint; *see* Bellini, Lorenzo..

Ligament of Bérard: *Eponym* for the suspensory ligament of the pericardium; *see* Bérard, Auguste.

Ligament of Berry: *Eponym* for the suspensory ligament of the thyroid gland; *see* Berry, James.

Ligament of Bertin: *Eponym* for the iliofemoral ligament, also described as Y-ligament by Bigelow; *see* Bertin, Exupère Joseph.

Ligament of Bichat: *Eponym:* for the inferior fascicle of the posterior sacroiliac ligament, attached to the third sacral transverse tubercle; *see* Bichat, Marie François Xavier.

Ligament of Bigelow, Y: *Eponym* for the iliofemoral ligament, earlier described by Bertin; *see* Bigelow, Henry Jacob.

Ligament of Botallo: *Eponym* for the ligamentum arteriosum; *see* Botallo, Leonardo.

Ligament of Bourgery: *Eponym* for (a) the oblique popliteal ligament, and (b) the intertransverse ligament of the lumbar part of the vertebral column; *see* Bourgery, Jean-Baptiste Marc.

Ligament of Brodie: *Eponym* for the transverse humeral ligament; *see* Brodie, Charles Gordon.

Ligament of Burns: *Eponym* for the superior horn of the falciform margin of the saphenous opening, three years earlier described by Hey; *see* Burns, Alan.

Ligament of Caldani: *Eponym* for the coracoclavicular ligament of the shoulder joint; *see* Caldani, Leopold Marco Antonio.

Ligament of Campbell: *Eponym* for the suspensory ligament of the axilla, previously described by Gerdy; *see* Campbell, William Francis.

Ligament of Camper: *Eponym* for the perineal membrane; *see* Camper, Petrus.

Ligament of Carcassone, Perineal: *Eponym* for the thickened anterior border of the perineal membrane; *see* Carcassone, Bernard Gauderic.

Ligament of Casseri: *Eponym* for the anterior ligament of the malleus; *see* Casseri, Giulio Caesare.

Ligament of Charpy: *Eponym* for the right phrenicocolic ligament, a peritoneal fold between the right lateral abdominal wall and the right colic flexure; *see* Charpy, Adrien.

Ligament of Chopart: *Eponym* for the bifurcate ligament of the foot; *see* Chopart, François.

Ligament of Civinini: *Eponym* for the pterygospinous ligament; *see* Civinini, Filippo.

Ligament of Clado: *Eponym* for the appendiculoovarian ligament, a mesenteric fold from the vermiform appendix to the right broad ligament of the uterus; *see* Clado, Spiro.

Ligaments of Cleland: *Eponym* for cutaneous ligaments of the digits; *see* Cleland, John.

Ligament of Cloquet: *Eponym* for the obliterated part of the vaginal process of the peritoneal cavity from the testis to the peritoneum at the deep inguinal ring; *see* Cloquet, Jules Germain.

Ligament of Colles: *Eponym* for the reflected inguinal ligament; see Colles, Abraham.

Ligament of Cooper: *Eponym* for: (a) the suspensory ligament of the breast, (b) the pectineal ligament, and (c) an oblique ligament from the olecranon to the coronoid process; *see* Cooper, Astley Paston.

Ligament of Cowper: *Eponym* for that part of the fascia lata, giving origin to the pectineus muscle; *see* Cowper, William.

Ligaments of Cruveilhier: *Eponym* for the plantar metatarsophalangeal ligaments; *see* Cruveilhier, Jean.

Ligament of Denonvilliers: *Eponym* for the rectovesical septum or fascia. The female counterpart is the rectovaginal septum or fascia; *see* Denonvilliers, Charles Pierre..

Ligament of Denucé, Arcuate: *Eponym* for the quadrate ligament of the proximal radioulnar joint; *see* Denucé, Jean Louis Paul.

Ligament of Falloppio: *Eponym* for the inguinal ligament, also described by Vesalius and Poupart; *see* Falloppio, Gabriele.

Ligaments of Farabeuf: *Eponym* for the glenohumeral ligaments; *see* Farabeuf, Louis Hubert.

Ligament of Ferrein: *Eponym* for (a) the lateral temporomandibular ligament, and (b) the vocal cord; *see* Ferrein, Antoine.

Ligament of Flood: *Eponym* for the superior glenohumeral ligament of the shoulder joint, also described by Farabeuf; *see* Flood, Valentine.

Ligament of Gerber: *Eponym* for an accessory band of the ascending band of the cruciform ligament attached onto the apex of the dens of the axis; *see* Gerber, Friedrich Andreas.

Ligament of Gerdy: *Eponym* for (a) the suspensory ligament of the axilla, also described by Campbell, and (b) the superficial transverse metacarpal ligament; *see* Gerdy, Pierre Nicolas.

Ligament of Gerlach: *Eponym* for the fibrous anulus of the tympanic membrane; *see* Gerlach, Joseph.

Ligament of Gillette, Suspensory: *Eponym* for the cricopharyngeal ligament, his suspensory ligament of the oesophagus, long fibres of the oesophagus attached to the posterior aspect of the cricoid cartilage; *see* Gillette, Eugène Paulin.

Ligament of Gimbernat: *Eponym* for the lacunar ligament; *see* Gimbernat, Manuel Louise Antonio.

Ligament of Grayson: *Eponym* for a ligament originating from the palmar aspect of the flexor tendon sheath. Its fibres project at right angles to this sheath, pass palmar to the digital nerves and vessels, and insert into the skin at the same level as their origin; *see* Grayson, J.

Ligament of Gruber: *Eponym* for the petroclinoid ligament; *see* Gruber, Wenzel.

Ligament of Günz: *Eponym* for part of the superficial layer of the obturator muscle; *see* Günz, Justus Gottfried.

Ligaments of Haller: *Eponym* for the medial and lateral arcuate ligaments of the diaphragm; *see* Haller, Albrecht von.

Ligament of Helvetius: *Eponym* for lateral muscle bands of the stomach; *see* Helvetius, Claude-Adrien.

Ligaments of Henle: *Eponyms* for (a) the suspensory ligament of the ovary, his infundibulopelvic ligament, also described by Testut as lumboovarian ligament, (b) the inferior transverse ligament of the scapula, (c) the lumbocostal ligament, (d) the transverse perineal ligament, also described by Krause and Waldeyer, and (e) medial and lateral accessory ligaments of the temporomandibular joint. The falx inguinalis was also referred to as ligament of Henle but Henle's description concerns a lateral extension of the rectus abdominis sheath (Anson BJ, McVay CB *Inguinal hernia: I. The anatomy of the region.* Surg Gynec Obst 1938; 66:186–191); *see* Henle, Friedrich Gustav Jacob.

Ligament of Hensing: *Eponym* for the phrenicocolic ligament; *see* Hensing, Friedrich Wilhelm.

Ligament of Hesselbach: *Eponym* for the interfoveolar ligament; *see* Hesselbach, Franz Kaspar.

Ligament of Hey: *Eponym* for the superior horn of the falciform margin of the saphenous opening, also described by Burns; *see* Hey, William.

Ligament of Holl: *Eponym* for the intracrural ligament joining the cavernous bodies of the clitoris in front of the external urethral opening; *see* Holl, Moritz.

Ligament of Hueck: *Eponym* for the trabecular meshwork of the sclera, also known as pectinate ligament, first described by Fontana; *see* Hueck, Alexander Friedrich.

Ligament of Humphry: *Eponym* for the anterior meniscofemoral ligament; *see* Humphry, George Murray.

Ligament of Hunter: *Eponym* for the round ligament of the uterus; *see* Hunter, William.

Ligament of Jarjavay: *Eponym* for the sacrouterine fold, also described by Petit; *see* Jarjavay, Jean François.

Ligament of Krause: *Eponym* for (a) the lateral costotransverse ligament, and (b) the transverse perineal ligament, also described by Henle and Waldeyer; *see* Krause, Karl Friedrich Theodor.

Ligament of Landsmeer, Oblique retinacular: *Eponym* for a ligament at the level of the proximal interphalangeal joint associated with the extensor mechanism, showing transverse and oblique components; *see* Landsmeer, Johan Matthijs Frederik.

Ligaments of Lannelongue: *Eponym* for the sternopericardial ligaments, also described by Luschka; *see* Lannelongue, Odillon Marc.

Ligament of Lauth: *Eponym* for (a) the inferior pubic ligament, (b) the transverse ligament of the atlas, and (c) the costoclavicular ligament (his rhomboid ligament); *see* Lauth, Thomas.

Ligament of Lisfranc: *Eponym* for the interosseous tarsometatarsal ligament between the second metatarsal bone and the medial cuneiform bone; *see* Lisfranc, Jacques.

Ligament of Lockwood, Suspensory: *Eponym* for the suspensory ligament of the eyeball; *see* Lockwood, Charles Barrett.

Ligament of Luschka: *Eponym* for (a) the sternopericardial ligaments (also described by Lannelongue), and (b) the xiphopericardial ligament, also known as the inferior sternopericardial ligament; *see* Luschka, Hubert von.

Ligament of Mackenrodt: *Eponym* for the cardinal or transverse cervical ligament of the uterus; *see* Mackenrodt, Alwin.

Ligament of Maissiat: *Eponym* for the iliotibial tract of the fascia lata; *see* Maissiat, Jacques Henri.

Ligament of Malgaigne, Accessory: *Eponym* for a supernumerary ligament of the shoulder between the acromion and the greater tubercle of the humerus; *see* Malgaigne, Joseph François.

Ligament of Mauchart: *Eponym* for the alar or check ligaments of the extraocular muscles; *see* Mauchart, Burkhard David.

Ligament of Meckel, JF, The Younger: *Eponym* for the interosseous tibiofibular liament; *see* Meckel, Johann Friedrich, The Younger.

Ligament of Meckel, PF: *Eponym* for the anterior mallear ligament; *see* Merkel, Philipp Friedrich.

Ligaments of Morris: *Eponym* for longitudinal and short internal lateral ligaments of the temporomandibular joint; *see* Morris, Henry.

Ligament of Osborne: *Eponym* for the fascia investing the ulnar nerve in the cubital tunnel; *see* Osborne, Geoffrey Vaughan.

Ligament of Petit: *Eponym* for the sacrouterine fold, also described by Jarjavay; *see* Petit, Antoine.

Ligament of Pétrequin: *Eponym* for a fibrous band connecting the membranous layer of the subcutaneous abdominal tissue to the fascia lata; *see* Pétrequin, Joseph Pierre Eléonor.

Ligament of Pinto: *Eponym* for the discomallear ligament of the temporomandibular joint, connecting the articular dis to the middle ear; *see* Pinto, Olympio Faissol.

Ligament of Poupart: *Eponym* for the inguinal ligament, already described by Vesalius and Falloppio; *see* Poupart, François. In the French literature, the term 'ligament externe de Henle' is often used.

Ligament of Proctor: *Eponym* for the fibrous insertion of the salpingopharyngeal fascia of Troeltsch to the skull base at the sphenopetrosal suture; *see* Proctor, Bruce.

Ligament of Rauber: *Eponym* for the salpingopharyngeal fold; *see* Rauber, August Antonius.

Ligament of Retzius: *Eponym* for (a) the deep attachment of the inferior expensor retinaculum in the tarsal sinus, and (b) the pelvoprostatic ligament; *see* Retzius, Anders Adolf.

Ligament of Santorini: *Eponym* for an inconstant cricopharyngeal ligament; *see* Santorini, Giovanni Domenico.

Ligament of Sappey: *Eponym* for (a) the anterior triangular ligament of the first sternochondral joint, and (b) the posterior, thickened part of the temporomandibular joint; *see* Sappey, Marie Philibert Constant.

Ligament of Scarpa: *Eponym* for the superior horn of the falciform margin of the saphenous opening; *see* Scarpa, Antonio.

Ligament of Sebileau, Suspensory: *Eponym* for the costopleural ligament; *see* Sebileau, Pierre.

Ligament of Soemmerring: *Eponym* for the suspensory ligament of the lacrimal gland; *see* Soemmerring, Samuel Thomas von.

Ligament of Spinner: *Eponym* for an inconstant fibrous layer between the flexor digitorum superficialis muscle and the humeral head of the flexor carpi ulnaris muscle, attached to the medial epicondyle of the humerus and the medial surface of the coronoid process of the ulna; *see* Spinner, Morton.

Ligaments of Stanley, Cervical: *Eponym* for fibres of the capsule of the hip joint reflected onto the neck of the femur; *see* Stanley, Edward.

Ligament of Struthers: *Eponym* for an inconstant fibrous band located on the medial side of the distal part of the humerus; this structure has been noted earlier by Tiedemann, Magendie and Knox; *see* Struthers, John.

Liament of Testut: *Eponym* for the radioscapholunate ligament of the wrist; *see* Testut, Jean Léon.

Ligament of Testut, Lumboovarian: *Synonym* for the suspensory ligament of the ovary, his lumboovarian ligament, described as infundibulopelvic ligament by Henle; *see* Testut, Jean Léon.

Ligament of Teutleben: *Eponym* for the pulmonary ligament of the pleura; *see* Teutleben, Friedrich Ernst Karl von.

Ligament of Thomson: *Eponym* for the iliopubic tract; *see* Thomson, Henry.

Ligament of Tourtual: *Eponym* for the anterior mallear ligament; *see* Tourtual, Kaspar Theobald.

Ligament of Treitz: *Eponym* for the suspensory ligament of the duodenum; *see* Treitz, Wenzel.

Ligament of Tuffier, Inferior: *Eponym* for the mesenterico-parietal fold, the part of the mesentery inserted into the iliac fossa; *see* Tuffier, Theodore.

Ligament of Valsalva: *Eponym* for the anterior ligament of the auricle; *see* Valsalva, Antonio Marie.

Ligament of Vesalius: *Eponym* for the inguinal ligament; *see* Vesalius, Andreas.

Ligament of Waldeyer: *Eponym* for the transverse perineal ligament, also described by Henle and Krause; *see* Waldeyer, Heinrich Wilhelm Gottfried.

Ligament of Weitbrecht: *Eponym* for (a) the oblique radio-ulnar ligament, and (b) the anterior ligament of the radiocarpal joint; *see* Weitbrecht, Josias.

Ligament of Whitnall, Superior transverse: *Eponym* for a condensation of fascial sheaths around the levator palpebrae superioris muscle; *see* Whitnall, Samuel Ernest.

Ligament of Wieger: *Eponym* for the ring-shaped hyalocapsular ligament, a thickening of the vitreous body, attached to the posterior surface of the lens; see Wieger, Germain.

Ligament of Winslow: *Eponym* for (a) the fibular collateral ligament, (b) the transverse ligament of the knee joint, and (c) the posterior meniscofemoral ligament of the knee joint, also described by Wrisberg; *see* Winslow, Jacob Benignus.

Ligament of Wrisberg: *Eponym* for the posterior meniscofemoral ligament, also described by Winslow; *see* Wrisberg, Heinrich August.

Ligament of Young: *Eponym* for the trapeziometacarpal ligament between the trapezium and the second and third metacarpals; *see* Young, Robert Bruce.

Ligament of Zaglas: *Eponym* for a short fibrous band of the posterior sacroiliac ligament extending from the posterior superior iliac spine to the second transverse tubercle of the sacrum; *see* Zaglas, John.

Ligament of Zinn: *Eponym* for the ciliary zonule or the suspensory ligament of the lens; *see* Zinn, Johann Gottfried.

1.18 Membranes

Eponyms for the term **membrane** are still mainly in use for structures of the eye (Bowman, Bruch, Demours, Descemet, Duddell, Ruysch, Verhoeff) and the ear (Corti) but also for embryonic structures (von Baer, Hunter, Reissner, Rivinus, Scarpa, Shrapnell).

Membrane of von Baer: *Eponym* for the chorionic membrane; *see* Baer, Karl Ernst von.

Membrane of Bertelli-Laimer: *Eponym* for fibrous tissue with which the oesophagus and stomach are attached to the diaphragm; it redraws the stomach after swallowing; *see* Bertelli, Dante, and Laimer, Edouard.

Membrane of Bowman: *Eponym* for the anterior limiting layer of the cornea; *see* Bowman, William.

Membrane of Bruch: *Eponym* for the basal complex of the choroid of the eyeball; *see* Bruch, Karl Wilhelm Ludwig.

Membrane of Brunn: *Eponym* for the olfactory epithelium of the nasal conchae; *see* Brunn, Albert von.

Membrane of Cloquet: *Eponym* for the part of the fascia lata covering the saphenous opening; *see* Cloquet, Jules Germain.

Membrane of Corti: *Eponym* for the tectorial membrane of the cochlear duct; *see* Corti, Alfonso.

Membrane of Demours: *Eponym* for the posterior limiting membrane of the cornea, also described by Descemet, but probably first by Duddell; *see* Demours, Pierre.

Membrane of Descemet: *Eponym* for the posterior limiting membrane of the cornea, also described by Demours and Duddell; *see* Descemet, Jean.

Membrane of Duddell: *Eponym* for the posterior limiting membrane of the cornea, usually described as membrane of Descemet. Duddell was apparently the first to describe this membrane; *see* Duddell, Benedict.

Membrane of Fielding: *Eponym* for the tapetum lucidum of the retina of some nocturnal animals; *see* Fielding, George H.

Membrane of Hannover: *Eponym* for the stratum intermedium of the enamel; *see* Hannover, Adolph.

Membrane of Hardesty: *Eponym* for the outer stripe of the tectorial membrane; *see* Hardesty, Irving.

Membrane of Henle: *Old term* for the basal lamina of the choroid, currently known as the membrane of Bruch; *see* Henle, Friedrich Gustav Jacob.

Membrane of Hunter: *Eponym* for the mucosa of the pregnant uterus (deciduous membrane); *see* Hunter, William.

Membrane of Jackson: *Eponym* for the pericolic membrane, a peritoneal fold or adhesion between the caecum or ascending colon and the right abdominal wall; *see* Jackson, Jabez North.

Membrane of Jacob: *Eponym* for the retinal layer containing the rods and cones; *see* Jacob, Arthur.

Membrane of Liliequist: *Eponym* for an arachnoid membrane separating the chiasmatic, interpeduncular and prepontine cisterns; *see* Liliequist, Bengt.

Membrane of Nassmyth: *Eponym* for the cuticle of the dental enamel; *see* Nasmyth, Alexander.

Membrane of Payr: *Eponym* for a peritoneal membrane passing over the left colic flexure; *see* Payr, Erwin.

Membrane of Pritchard, Reticulated: *Eponym* for the intercellular membrane in the ampulla of the semicircular duct; *see* Pritchard, Urban.

Membrane of Reissner: *Eponym* for the vestibular membrane of the cochlea; *see* Reissner, Ernst.

Membrane of Rivinus: *Eponym* for the flaccid part of the tympanic membrane, mostly known as membrane of Shrapnell; *see* Rivinus, Augustus Quirinus.

Membrane of Ruysch: *Eponym* for the choriocapillary layer of the choroid; *see* Ruysch, Frederik.

Membrane of Salzmann: *Eponym* for an external vitreous membrane, a basal membrane between the vitreous body and the choroid; *see* Salzmann, Maximilian,

Membrane of Scarpa: *Eponym* for the secondary tympanic membrane that closes the round window; *see* Scarpa, Antonio.

Membrane of Schneider: *Eponym* for the nonolfactory part of the nasal mucosa; *see* Schneider, Conrad Viktor.

Membrane of Schultze: *Eponym* for the olfactory mucous membrane, also described by Brunn; *see* Schultze, Maximilian Johann Sigismund.

Membrane of Shrapnell: *Eponym* for the flaccid part of the tympanic membrane, earlier described by Rivinus; *see* Shrapnell, Henry Jones.

Membrane of Slavjanski: *Eponym* for the basement membrane separating the granulosa from the inner theca of the ovarian follicle; *see* Slavjanski, Kronid Fedorovich von.

Membrane of Toldt: *Eponym* for the prerenal layer of the renal fascia; *see* Toldt, Karl.

Membrane of Tourtual: *Eponym* for the quadrangular membrane of the larynx; *see* Tourtual, Kaspar Theobald.

Membrane of Verhoeff: *Eponym* for the interdigitation zone of the retina; *see* Verhoeff.

Membrane of Wachendorff: *Eponym* for the embryonic pupillary membrane, earlier described by Albinus; *see* Wachendorff, Eberhard Jacob.

Membrane of Zeissl: *Eponym* for the stratum compactum of the subglandular layer of the gastric mucosa; *see* Zeissl, Maximilian von.

Membrane of Zinn: *Eponym* for the anterior layer of the iris; *see* Zinn, Johann Gottfried.

1.19 Muscles

There are many eponyms with the term **muscle** but they are all hardly used, apart from parts of the ciliary muscle (Bowman, Brücke, Crampton, Rouget).

Muscle of Aeby: *Eponym* for the depressor labii inferioris muscle, also described by Klein; *see* Aeby, Christopher Theodor.

Muscle of Albinus: *Eponym* for: (a) the risorius muscle, also described by Santorini, (b) the rhomboid muscles, (c) the scalenus minimus muscle (his lateral scalenus muscle), described later by Sibson as fourth scalenus muscle, and (d) the tensor tympani muscle, already described by Eustachi; *see* Albinus, Bernhard Friedrich.

Muscle of Bell: *Eponym* for muscular strands from the internal urethal openings to the uvula, bounding the vesical trigone; *see* Bell, John.

Muscle of Bochdalek: *Eponym* for the triticeoglossal muscle, an inconstant band of muscle fibres between the root of the tongue and the triticeal cartilage of the larynx; *see* Bochdalek, Vincenc Alexander.

Muscle of Bowman: *Eponym* for the meridional fibres of the ciliary muscle, described in the same year as Brücke, but early by Crampton; *see* Bowman, William.

Musle of Braune: *Eponym* for the puborectalis muscle; *see* Braune, Christian Wilhelm.

Muscle of Brücke: *Eponym* for the meridional fibres of the ciliary muscle, also described by Bowman, but earlier by Crampton; *see* Brücke, Ernst Wilhelm Ritter.

Muscle of Calori, Stylomaxillary: *Eponym* for an accessory muscle from the styloid process to the mandibular angle; *see* Calori, Luigi.

Muscle of Casseri: *Eponym* for the coracobrachial muscle; *see* Casseri, Giulio Caesare.

Muscle of Coiter: *Eponym* for the corrugator supercilii muscle; *see* Coiter, Volcher.

Muscle of Crampton: *Eponym* for the meridional fibres of the ciliary muscle, as also described by Bowman and Brücke; *see* Crampton, Phillip.

Muscle of Duverney: *Eponym* for the lacrimal (deep) part of the orbicularis oculi muscle, later also described by Horner; *see* Duverney, Joseph Guichard.

Muscle of Ellis: *Eponym* for the corrugator cutis ani muscle; *see* Ellis, George Vines.

Muscle of Eustachi: *Eponym* for the tensor tympani muscle; *see* Eustachi, Bartolomeo.

Muscle of Falloppio: *Eponym* for the pyramidal muscle; *see* Falloppio, Gabriele.

Muscle of Folli: *Eponym* for the anterior mallear ligament; *see* Folli, Cecilio.

Muscle of Gantzner: *Eponym* for an inconstant muscle extending from the superficial flexor of the digits to the deep flexor of the digits; *see* Gantzner, Carol Friedrich Ludovic.

Muscle of Gavard: *Eponym* for oblique fibres (the internal layer) of the muscular layer of the stomach wall; *see* Gavard, Hyacinthe.

Muscle of Grynfeltt: *Eponym* for the dilator pupillae muscle, previously described by Kölliker; *see* Grynfeltt, Joseph Casimir.

Muscle of Gruber: *Eponym* for an accessory fibular or peroneal muscle; *see* Gruber, Wenzel.

Muscle of Guthrie: *Eponym* for the external urethral sphincter; *see* Guthrie, George James.

Muscle of Henle: *Eponym* for (a) the anterior auricular muscle, and (b) the pubococcygeus muscle (Henle's musculus levator ani); *see* Henle, Friedrich Gustav Jacob.

Muscle of Hilton: *Eponym* for the (inferior) aryepiglotticus muscle; *see* Hilton, John.

Muscle of Horner: *Eponym* for the lacrimal (deep) part of the orbicularis oculi muscle, already described by Duverney; *see* Horner, William Edmonds.

Muscle of Houston: *Eponym* for a variant of the bulbospongiosus muscle in the penis (compressor venae dorsalis penis muscle); *see* Houston, John.

Muscle of Hyrtl: *Eponym* for the iliopsoas muscle; *see* Hyrtl, Joseph.

Muscle of Jung: *Eponym* for the pyramidal muscle of the auricle; *see* Jung, Karl Gustav.

Muscle of Killian: *Eponym* for the lowest, cricopharyngeal fibres of the inferior constrictor of the pharynx, immediately above the oesophageal musculature; *see* Killian, Gustav.

Muscle of Klein: *Eponym* for the compressor labii muscle, a marginal bundle of the lips or rectus labii proprius muscle, as also described by Aeby; *see* Klein, Eduard Emmanuel.

Muscle of Kohlrausch: *Eponym* for longitudinal muscle fibres of the rectal wall; *see* Kohlrausch, Otto Ludwig Bernhard.

Muscle of Kölliker: *Eponym* for dilator pupillae muscle; *see* Kölliker, Rudolf Albert.

Muscle of Krause: *Eponym* for the cutaneomucous muscle; *see* Krause, Karl Friedrich Theodor.

Muscle of Lesshaft: *Eponym* for the levator ani muscle; *see* Lesshaft, Pjotr Frantsovitsch.

Muscle of Luschka: *Eponym* for the rectouterine muscle; *see* Luschka, Hubert von.

Muscle of Macalister: *Eponym* for the inconstant cricotracheal muscle; *see* Macalister, Alexander.

Muscle of Marcacci: *Eponym* for a sheet of smooth muscle fibres underlying the areola and nipple of the mammary gland; *see* Marcacci, Arturo.

Muscle of Merkel: *Eponym* for the inconstant ceratocricoid muscle; *see* Merkel, Karl Ludwig.

Muscle of Müller: *Eponym* for two smooth muscles: (a) the smooth orbitalis muscle, and (b) the superior tarsal muscle in the upper eyelid, sometimes also used for the inferior tarsal muscle; *see* Müller, Heinrich.

Muscles of Oehl: *Eponym* for strands of muscle fibres in the tendinous chords of the left atrioventricular valve; *see* Oehl, Eusebio.

Muscle of Portal: *Eponym* for the articularis cubiti muscle, fibres from the shaft of the humerus inserted to the anterior aspect of the capsule of the elbow joint; *see* Portal, Antoine.

Muscle of Pozzi: *Eponym* for the inconstant extensor digitorum brevis muscle of the hand; *see* Pozzi, Samuel Jean.

Muscle of Reisseisen: *Eponym* for smooth muscle fibres of the smallest bronchial tubes; *see* Reisseisen, François Daniel.

Muscle of Riolan: *Eponym* for (a) the cremaster muscle, and (b) the palpebral part of the orbicularis oculi muscle; *see* Riolan, Jean.

Muscle of Rouget: *Eponym* for the radial fibres of the ciliary muscle; *see* Rouget, Charles Marie Benjamin.

Muscle of Roux: *Eponym* for the rectourethral muscle; *see* Roux, Caesar.

Muscle of Rüdinger: *Eponym* for a muscle layer internal to the circular fibres of the rectum (his musculus dilator internus); see Rüdinger, Nicolaus.

Muscle of Ruysch: *Eponym* for the muscular tissue of the fundus of the uterus; *see* Ruysch, Frederik.

Muscle of Santorini: *Eponym* for (a) the procerus muscle, (b) the risorius muscle, also described by Albinus, and (c) the incisurae helicis muscle; *see* Santorini, Giovanni Domenico.

Muscle of Sappey, Deep stylohyoid: *Eponym* for an accessory muscle bundle of the stylohyoid muscle; *see* Sappey, Marie Philibert Constant.

Muscle of Sappey, Transverse chin: *Eponym* for transverse inferior fibres of the depressor anguli oris muscle; *see* Sappey, Marie Philibert Constant.

Muscle of Sebileau: *Eponym* for the deep layer of the tunica dartos; *see* Sebileau, Pierre.

Muscle of Sibson: *Eponym* for the scalenus minimus muscle; *see* Sibson, Francis.

Muscle of Soemmerring: *Eponym* for the levator glandulae thyroideae muscle; *see* Soemmerring, Samuel Thomas von.

Muscle of Theile: *Eponym* for the superficial transverse perineal muscle; *see* Theile, Friedrich Wilhelm.

Muscle of Tod: *Eponym* for the oblique auricular muscle; *see* Tod, David.

Muscle of Toynbee: *Eponym* for (a) the tensor tympani muscle, already well known by Eustachi and given its name by Albinus, and (b) the rectococcygeus muscle; *see* Toynbee, Joseph.

Muscle of Treitz: *Eponym* for the suspensory ligament of the duodenum that may contain muscle fibres; *see* Treitz, Wenzel.

Muscle of Valsalva: *Eponym* for the tragicus muscle of the auricle; *see* Valsalva, Antonio Maria.

Muscle of van Diemerbroeck: *Eponym* for the adductor quartus muscle, a supernumerary muscle of the medial group of thigh muscles, formed by superior fibres of the adductor magnus muscle; *see* van Diemerbroeck, IJsbrand.

Muscle of Wilson: *Eponym* for the external urethral sphincter, also described by Guthrie; *see* Wilson, James.

1.20 Nerves

There are many eponyms with the term **nerve** but they are only occasionally used, apart from the nerves of Cruveilhier, Guidi (Vidius), Hering, Jacobson, Latarjet and Oort.

Nerve of Andersch: *Eponym* for the tympanic branch of the glossopharyngeal nerve, usually described as nerve of Jacobson; *see* Andersch, Carol Samuel.

Nerve of Arnold: *Eponym* for (a) the auricular branch of the vagus nerve, also known as Alderman's nerve, (b) the tentorial branch of the ophthalmic nerve (the recurrent nerve of Arnold), and (c) the greater occipital nerve; *see* Arnold, Friedrich.

Nerve of Bell, External: *Eponym* for the long thoracic nerve; *see* Bell, Charles.

Nerve of Bell, Internal: *Eponym* for the phrenic nerve; *see* Bell, Charles.

Nerve of Bock, Pharyngeal: *Eponym* for the pharyngeal branch of the maxillary nerve; *see* Bock, August.

Nerve of Casseri: *Eponym* for the musculocutaneous nerve (his perforating nerve); *see* Casseri, Giulio Caesare.

Nerve of Cotugno: *Eponym* for the nasopalatine nerve, also described by Scarpa; *see* Cotugno, Domenico.

Nerve of Cruveilhier: *Eponym* for the vertebral nerve; *see* Cruveilhier, Jean.

Nerve of Cyon: *Eponym* for the depressor nerve of the heart, a branch of the parasympathetic superior laryngeal nerve, also described by Hofer and Ludwig; *see* Cyon, Elie de.

Nerve of Estienne: *Eponym* for the recurrent laryngeal nerve; *see* Estienne, Charles.

Nerve of Exner: *Eponym* for an inconstant middle laryngeal nerve; *see* Exner, Siegmund.

Nerve of Galen: *Eponym* for the communicating branch of the superior laryngeal nerve with the recurrent laryngeal nerve; *see* Galen.

Nerves of Gaskell: *Eponym* for the accelerator nerves of the heart; *see* Gaskell, Walter Holbrook.

Nerve of Guidi (Vidius): *Eponym* for the nerve of the pterygoid canal, composed of the greater and deep petrosal nerves; *see* Guidi, Guido (Vidius, Vidus).

Nerve of Hering: *Eponym* for the carotid sinus nerve; *see* Hering, Heinrich Ewald.

Nerve of Hirschfeld: *Eponym* for the lingual branch of the facial nerve; *see* Hirschfeld, Ludwig Moritz.

Nerve of Hofer: *Eponym* for the depressor nerve, a branch of the superior laryngeal nerve to the cardiac plexus, also described by Cyon and Ludwig; *see* Hofer, Gustav.

Nerve of Jacobson: *Eponym* for (a) the tympanic nerve, and (b) the vomeronasal nerve; *see* Jacobson, Ludwig Levin.

Nerve of Langenbeck: *Eponym* for a lateral supraclavicular nerve (his nervus superficialis scapulae); *see* Langenbeck, Konrad Johann Martin.

Nerves of Latarjet: *Eponym* for the terminal branches of the anterior and posterior vagal trunk, running along the lesser curvature of the stomach; *see* Latarjet, André.

Nerves of Latarjet, Principal ureteric: *Eponym* for the inferior and superior principal ureteric nerves; see Latarjet, André.

Nerve of Ludwig, Depressor: *Eponym* for the aortic nerve, a branch of the vagus nerve, which ends in the aortic arch and the base of the heart, also described by Cyon and Hofer; *see* Ludwig, Karl Friedrich Wilhelm.

Nerve of Luschka: *Eponym* for (a) the posterior ethmoidal nerve, and (b) the meningeal branch of each spinal nerve, which supplies structures within the vertebral canal, also known as sinuvertebral nerve; *see* Luschka, Hubert von.

Nerve of Oort: *Eponym* for the cochlear communicating (vestibulocochlear) anastomosis, the efferent fibres of the cochlear nerve, which leave the brain stem with the vestibular nerve and join the cochlear nerve to end in the spiral organ of Corti; *see* Oort, Henricus.

Nerve of Panizza, Gustatory: *Eponym* for the glossophryngeal nerve; *see* Panizza, Bartolomeo.

Nerve of Pavlov: *Eponym* for the inferior cervical cardiac nerve; *see* Pavlov, Ivan Petrovich.

Nerves of Randacio: *Eponym* for branches arising from the pterygopalatine ganglion; *see* Randacio, Francesco.

Nerve of Sapolini: *Eponym* for the intermediate nerve, also known as nerve of Wrisberg; *see* Sapolini, Giuseppe.

Nerve of Scarpa: *Eponym* for the nasopalatine nerve, also described by Cotugno; *see* Scarpa, Antonio.

Nerve of Tiedemann: *Eponym* for a minute sympathetic nerve said to run with the central retinal artery; *see* Tiedemann, Friedrich.

Nerve of Valentin: *Eponym* for a nerve connecting the pterygopalatine ganglion with the abducens nerve; *see* Valentin, Gabriel Gustav.

Nerve, Vidian: *Eponym* for the nerve of the pterygoid canal; *see* Guidi, Guido (Vidius, Vidus).

Nerve of Voit: *Eponym* for a branch of the anterior ramus of the vestibular nerve, supplying the macula of the utriculus; *see* Voit, Max.

Nerve of Walter: *Eponym* for a branch of the least thoracic splanchnic nerve to the renal plexus; *see* Walter, Johann Gottlieb.

Nerve of Walther: *Eponym* for the greater thoracic splanchnic nerve; *see* Walther, Augustin Friedrich.

Nerve of Willis: *Eponym* for the accessory nerve; *see* Willis, Thomas.

Nerve of Wrisberg: *Eponym* for (a) the intermediate nerve, also descibed by Sapolini, and (b) the medial brachial cutaneous nerve; *see* Wrisberg, Heinrich August.

1.21 Nodes

The term **node** is used in various ways: for the atrioventricular (Aschoff-Tawara, Koch) and sinuatrial (Keith-Flack) nodes, for lymph nodes, for the node of Ranvier, and for the primitive node (Hensen).

Node of Aschoff-Tawara: *Eponym* for the atrioventricular node of the cardiac conducting system; *see* Aschoff, Karl Ludwig, and Tawara, Sunao. In the German literature, it is known as node of Koch.

Node of Bardeleben: *Eponym* for a popliteal lymph node, also described by Poirier and Cunéo; see Bardeleben, Karl von.

Node of Bayer: *Eponym* for an inconstant parauterine lymph node at the crossing of the uterine artery and the ureter; *see* Bayer, Heinrich.

Node of Botallo: *Eponym* for a lymph node on the ligamentum arteriosum; *see* Botallo, Leonardo.

Node of Cabanas: *Eponym* for the superomedial superficial inguinal node, a sentinel lymph node for carcinoma of the penis; *see* Cabanas, Ramon Maximilian.

Node of Calori: *Eponym* for the cystic lymph node, also described by Mascagni and Lund; *see* Calori, Luigi.

Node of Cloquet: *Eponym* for the proximal deep inguinal lymph node, also described by Rosenmüller; *see* Cloquet, Jules Germain.

Nodes of Gerota, Anorectal: *Eponym* for pararectal lymph nodes, also described by Poirier and Cunéo; see Gerota, Dumitru.

Node of Gillette-Henle: *Eponym* for a retropharyngeal lymph node; *see* Gillette, Eugène Paulin, and Henle, Friedrich Gustav Jacob.

Node of Hensen: *Eponym* for the primitive node, also described by Hubrecht as protocordal knot; *see* Hensen, Victor.

Node of Keith-Flack: *Eponym* for the sinuatrial node of the cardiac conducting system; *see* Keith, Arthur, and Flack, Martin.

Node of Koch: *Eponym* for the atrioventricular node of the cardiac conducting system in the German literature, usually known as node of Aschoff-Tawara; *see* Koch, Walter.

Node of Küttner: *Eponym* for the jugulodigastric lymph node; *see* Küttner, Arthur.

Node of Lund: *Eponym* for the cystic lymph node, also described by Mascagni and Calori; *see* Lund, Fred Bates.

Node of Mascagni: *Eponym* for the cystic lymph node, also described by Calori and Lund; *see* Mascagni, Paolo.

Nodes of Oelsner: *Eponym* for the superior and inferior axillary lymph nodes in the 2nd and 3rd, and 4th and 5th intercostal spaces, respectively; *see* Oelsner, L.

Nodes of Poirier and Cunéo: *Eponym* for various lymph nodes, including among others (a) abdominoaortic lymph nodes, (b) juxta-articular lymph nodes, (c) pararectal lymph nodes, also described by Gerota, and (d) popliteal lymph nodes, also described by Bardeleben. Reference: *Système lymphatique.* In: Poirier and Charpy (eds) *Traité d'anatomie humaine*, 2nd ed. Masson, Paris, 1902, pp. 1172–1180.

Nodes of Quénu: *Eponym* for the inguinal lymph nodes; *see* Quénu, Edouard André Victor Alfred.

Node of Ranvier: *Eponym* for gaps in the myelin sheath of a peripheral nerve fibre, where the myelin sheath is discontinuous and at which the axolemma is free from myelin; *see* Ranvier, Louis Antoine.

Node of Rosenmüller: *Eponym* for the proximal deep inguinal lymph node, also described by Rosenmüller, also described by Cloquet; *see* Rosenmüller, Johann Christian.

Nodes of Rotter: *Eponym* for the interpectoral axillary lymph nodes; *see* Rotter, Josef.

Nodes of Rouvière: *Eponym* for various lymph nodes, including (a) the prelaryngeal or cricoid lymph node, (b) the lateral retropharyngeal lymph node, and (c) the xyphoid or prepericardial lymph node also described by Sappey. Reference: *Anatomie des lymphatiques de l'homme.* Masson, Paris, 1932.

Node of Sorgius: *Eponym* for the sentinel lymph node of the mammary gland, the caudalmost pectoral lymph node, located at the second or third slip of the serratus anterior muscle; *see* Sorgius, Wilhelm.

Node of Tawara: *Eponym* for the atrioventricular node of Aschoff-Tawara; *see* Tawara, Sunao.

Node of Virchow-Troisier: *Eponym* for a supraclavicular lymph node above the left clavicula, a signal node; *see* Virchow, Rudolf Ludwig Karl, and Troisier, Charles Emile.

Node of Wood: *Eponym* for the tonsillar lymph node; *see* Wood, Alexander.

1.22 Nuclei

The term **nucleus** is mainly used for centres in the central nervous system, and here, eponyms are still in use.

Nucleus of Barrington: *Eponym* for the pontine micturition centre; *see* Barrington, James Fitzmaurice.

Nucleus of Bechterew: *Eponym* for (a) the superior vestibular nucleus, (b) the reticulotegmental nucleus, and (c) the posteromedial or dorsomedial part of the red nucleus; *see* Bechterew, Vladimir Michaelovich von.

Nucleus of Blumenau: *Eponym* for the lateral cuneate nucleus; *see* Blumenau, Leonid Wassiljewitsch.

Nucleus of Broca, Diagonal band: *Eponym* for a nucleus in the basal forebrain related to the diagonal band; *see* Broca, Pierre Paul.

Nucleus of Burdach: *Eponym* for the cuneate nucleus; *see* Burdach, Karl Friedrich.

Nucleus of Cajal, interstitial: *Eponym* for the interstitial nucleus of the medial longitudinal fasciculus; *see* Ramón y Cajal, Santiago.

Nucleus of Clarke: *Eponym* for the dorsal or posterior thoracic nucleus, also described by Stilling; *see* Clarke, Jacob Augustus Lockhart.

Nucleus of Darkschewitsch: *Eponym* for what is currently known as the elliptic nucleus in the prerubral tegmentum; *see* Darkschewitsch, Liverij Osipovich.

Nucleus of Deiters: *Eponym* for the lateral vestibular nucleus, also described by Laura; *see* Deiters, Otto Friedrich Karl.

Nucleus of Edinger-Westphal: *Eponym* for an accessory nucleus of the oculomotor nerve; *see* Edinger, Ludwig, and Westphal, Karl Friedrich Otto.

Nucleus of Ganser: *Eponym* for the basal nucleus of the forebrain composed of large (cholinergic) neurons, already described by Meynert; *see* Ganser, Siegbert.

Nucleus of Goll: *Eponym* for the gracile nucleus; *see* Goll, Friedrich.

Nuclei of Gudden: *Eponym* for the dorsal and ventral tegmental nuclei; *see* Gudden, Bernhard von.

Nucleus of Jacobsohn: *Eponym* for the posterior or dorsal paramedian nucleus of the brain stem; *see* Jacobsohn, Louis.

Nuclei of Kohnstamm: *Eponym* for the superior and inferior salivatory nuclei of the brain stem; *see* Kohnstamm, Oskar.

Nucleus of Kölliker: *Eponym* for the intermediate substance of the spinal grey matter; see Kölliker, Rudolf Albert von.

Nucleus of Kölliker-Fuse: *Eponym* for the subparabrachial nucleusn of the brain stem; *see* Kölliker, Rudolf Albert, and Fuse, Gennosuke.

Nucleus of Laura: *Eponym* for the lateral vestibular nucleus, earlier described by Deiters; *see* Laura, Giovanni Battista.

Nucleus of Luys: *Eponym* for the subthalamic nucleus; *see* Luys, Jules Bernard.

Nucleus of Marburg: *Eponym* for the prepositus nucleus of the brain stem; *see* Marburg, Otto.

Nucleus of Massazza: *Eponym* for the intermediolateral nucleus of the spinal cord; *see* Massazza, Adolfo.

Nucleus of Meessen-Olszewski: *Eponym* for the pars alpha of the gigantocellular nucleus of the reticular formation; *see* Meessen, Hubert, and Olszewski, Jerzy.

Nucleus of Meynert: *Eponym* for the basal nucleus of the forebrain composed of large (cholinergic) neurons; *see* Meynert, Theodor.

Nucleus of Monakow: *Eponym* for the lateral cuneate nucleus of the brain stem; *see* Monakow, Constantin von.

Nucleus of Onuf: *Eponym* for the nucleus of the pudendal nerve in the spinal cord; *see* Onuf (Onufrowitz), Bronislaw.

Nucleus of Perlia: *Eponym* for the interoculomotor nucleus; *see* Perlia, Richard.

Nucleus of Roller: *Eponym* for the subhypoglossal nucleus; sometimes also used as eponym for the inferior vestibular nucleus; *see* Roller, Christian Friedrich Wilhelm.

Nucleus of Schwalbe: *Eponym* for the medial vestibular nucleus; *see* Schwalbe, Gustav Albert.

Nucleus of Staderini: *Eponym* for the intercalated nucleus of the brain stem; *see* Staderini, Rutilio.

Nucleus of Stilling: *Eponym* for the dorsal or posterior thoracic nucleus, also described by Clarke; *see* Stilling, Benedikt.

Nucleus of Tsuchida: *Eponym* for the central caudal nucleus; *see* Tsuchida, Usabro.

Nucleus of Voit: *Eponym* for an accessory dentate nucleus of the cerebellum; *see* Voit, Max.

Nucleus of Ziehen: *Eponym* for the sagulum nucleus; *see* Ziehen, Theodor.

1.23 Plexuses

The term **plexus** is mainly used for peripheral nerve plexuses and choroid plexuses. Particularly for the venous plexuses, eponyms are also still in use (Batson, Hyrtl, Kobelt, Lazorthes, Leber and Santorini).

Plexus of Andersch: *Eponym* for the neural tonsillar plexus; *see* Andersch, Carol Samuel.

Plexus of Arnold: *Eponym* for the cavernous plexus on the cavernous part (C4) of the internal carotid artery in the cavernous sinus; *see* Arnold, Friedrich.

Plexus of Auerbach: *Eponym* for the myenteric plexus; *see* Auerbach, Leopold.

Plexuses of Batson: *Eponym* for the vertebral venous plexuses; *see* Batson, Oscar Vivian,

Plexus of Bochdalek: *Eponym* for the choroid plexus of the fourth ventricle; *see* Bochdalek, Vincenc Alexander.

Plexus of Bradlaw: *Eponym* for the predentinal plexus, a marginal nerve plexus of free nerve endings between the odontoblast layer and the dentin in permanent teeth; *see* Bradlaw, Robert.

Plexus of Cruveilhier: *Eponym* for the vertebral plexus formed by the dorsal or posterior rami of the first three cervical nerves; *see* Cruveilhier, Jean..

Plexus of Exner: *Eponym* for superficial tangential fibres of the molecular layer of the isocortex; *see* Exner, Siegmund.

Plexus of Frankenhäuser: *Eponym* for the neural uterovaginal plexus, also described by Lee; *see* Frankenhäuser, Ferdinand.

Plexus of Haller: *Eponym* for the laryngeal plexus, formed by fibres from the superior laryngeal nerve and sympathetic fibres; *see* Haller, Albrecht von.

Plexus of Heller: *Eponym* for an arterial plexus in the wall of the small intestine; *see* Heller, Arnold Ludwig Gotthilf.

Plexus of Hovelacque: *Eponym* for the superior hypogastric plexus; *see* Hovelacque, André Edouard Emile.

Plexus of Hyrtl: *Eponym* for the venous plexus beneath the isthmus of the thyroid gland; *see* Hyrtl, Joseph.

Plexus of Jacobson: *Eponym* for the tympanic plexus on the promontory of the labyrinthine wall of the tympanic cavity; *see* Jacobson, Ludwig Levin.

Plexus of Jacques: *Eponym* for a nerve plexus within the muscular layer of the uterine tube; *see* Jacques, Paul.

Plexus of Kobelt: *Eponym* for the venous plexus of the clitoris; *see* Kobelt, Georg Ludwig.

Plexuses of Lazorthes: *Eponym* for the anterior and posterior tibial plexuses, branches of the anterior and posterior tibial nerves that supply the distal segment of the anterior and posterior tibial arteries, respectively; *see* Lazorthes, Guy.

Plexus of Leber: *Eponym* for a small venous plexus between the canal of Schlemm and the spaces of Fontana, formed by plexiform veins of the iris associated with the ciliary processes of the ciliary body; *see* Leber, Theodor.

Plexus of Lee: *Eponym* for the uterovaginal plexus, also described by Frankenhäuser; *see* Lee, Robert.

Plexus of Mascagni: *Eponym* for the internal (hypogastric) and external iliac lymphatic plexuses; *see* Mascagni, Giovanni Paolo.

Plexus of Meissner: *Eponym* for the internal or inner submucosal plexus of the intestines, also described by Remak; *see* Meissner, Georg.

Plexus of Panizza: *Eponym* for a plexus of lymphatic vessels in the lateral fossae of the prepuce; *see* Panizza, Bartolomeo.

Plexus of Patey: *Eponym* for a neurovenous plexus around the facial nerve within the parotid gland; *see* Patey, David Howard.

Plexus of Quénu, Haemorrhoidal: *Eponym* for a series of lymphatic plexuses in the mucosa and skin of the anus; *see* Quénu, Edouard André Victor Albert.

Plexus of Raschkow: *Eponym* for the subodontoblastic plexus, a plexus of nerve fibres located in the subodontoblastic layer; *see* Raschkow, Isaac..

Plexus of Remak: *Eponym* for the submucosal plexus of the intestines, also described by Meissner; *see* Remak, Robert.

Plexus of Santorini, Venous: *Eponym* for the prostatic venous plexus in males or the vesical venous plexus in females, a venous plexus at the bottom of the retropubic space, draining among others the dorsal vein of the penis and the dorsal vein of the clitoris, respectively; *see* Santorini, Giovanni Domenico.

Plexus of Sappey, Subareolar: *Eponym* for a lymphatic plexus under the areola of the mamma; *see* Sappey, Marie Philibert Constant.

Plexus of Schabadasch: *Eponym* for the external or outer submucosal neural plexus located in the outer part of the submucosa of the intestines; *see* Schabadasch, Arnold Leontowitsch.

Plexus of Teichmann: *Eponym* for superficial and deep lymphatic plexuses in the wall of the stomach; *see* Teichmann, Ludwig Karl.

Plexus of Trolard: *Eponym* for the hypoglossal venous plexus at the extracranial aspect of the hypoglossal canal connecting veins around the occipital condyle with the jugular bulb; *see* Trolard, Paulin.

Plexus of Walther: *Eponym* for the intracavernous sympathetic plexus; *see* Walther, Augustin Friedrich.

1.24 Spaces

The term **space** is usually applied to spaces between certain structures, but also as a *synonym* for a sac.

Space of Berger: *Eponym* for a potential retrolental space, a central hyalocapsular space between the posterior aspect of the lens and the anterior aspect of the vitreous body; *see* Berger, Emil.

Space of Boettcher: *Synonym* for the endolymphatic sac, also described by Cotugno; *see* Boettcher, Arthur.

Space of Bogros: *Eponym* for retroinguinal space; *see* Bogros, Annet-Jean.

Space of Burns: *Eponym* for the suprasternal space; *see* Burns, Allan.

Space of Chassaignac: *Eponym* for a potential space between the pectoralis major muscle and the mammary gland; *see* Chassaignac, Charles Marie Edouard.

Space of Cloquet: *Eponym* for the retrozonular space between the ciliary zonule and the vitreous body; *see* Cloquet, Jules Germain.

Space of Colles: *Eponym* for the superficial perineal space; *see* Colles, Abraham.

Space of Cotugno: *Eponym* for the endolymphatic sac, later also described by Boettcher; *see* Cotugno, Domenico.

Spaces of Czermak: *Eponym* for the interglobular lacunae or interglobular dentin gaps in the continuity of dentin; *see* Czermak, Johann Nepomuk.

Space of Disse: *Eponym* for the perisinusoidal space between the sinusoids and the hepatic cells; *see* Disse, Josef.

Space of Donders: *Eponym* for a space between the dorsum of the tongue and the hard palate when the mandibula is in rest position following expiration; *see* Donders, Franciscus.

Space of Erggelet: *Eponym* for a funnel-shaped dilatation of the anterior end of the hyaloid canal behind the lens; *see* Erggelet, Heinrich.

Spaces of Fontana: *Eponym* for the trabecular meshwork or pectineal ligament, spaces of the iridocorneal angle; *see* Fontana, Abbada Felix.

Space of Grynfeltt, Superior lumbar: *Eponym* for a space bounded by the serratus posterior inferior muscle and the 12th rib cranially, the erector spinae muscle medially and the internal oblique abdominal muscle laterally; *see* Grynfeltt, Joseph Casimir.

Spaces of Havers: *Eponym* for spaces in bone formed by the enlargement of the canals of Havers; *see* Havers, Clopton.

Space of Henke: *Eponym* for the retropharyngeal space; *see* Henke, Wilhelm.

Space of His, Perivascular: *Synonym* for the Virchow-Robin space; *see* His, Wilhelm Sr.

Space of Kiernan: *Eponym* for the interlobular space in the liver; *see* Kiernan, Francis.

Space of Kretschmann: *Eponym* for a slight depression in the epitympanic recess below the superior recess of the tympanic cavity of Prussak; *see* Kretschmann, Friedrich.

Spaces of Kuhnt: *Eponym* for shallow recesses, openings into the posterior chamber of the eye between the ciliary body and the ciliary zonules; *see* Kuhnt, Hermann.

Spaces of Magendie: *Eponym* for subarachnoid spaces between the pia mater and the arachnoid at the level of the sulci of the brain; *see* Magendie, François.

Space of Malacarne: *Eponym* for the posterior perforated substance; *see* Malacarne, Michele Vincenzo Giacinto.

Space of Mall: *Eponym* for the periportal space between the limiting membrane and the portal canal of the liver continuous with the space of Disse; *see* Mall, Franklin Paine.

Space of Meckel: *Eponym* for the dural space in the middle cranial fossa containing the trigeminal ganglion; *see* Meckel, Johann Friedrich, the Elder.

Space of Mohrenheim: *Eponym* for the clavipectoral or deltopectoral trigone; *see* Mohrenheim, Joseph Jacob.

Space of Nuel: *Eponym* for the intermediate tunnel of the spiral organ, located between the outer pillar and the outer hair cells; *see* Nuel, Jean Pierre.

Space of Parona: *Eponym* for a space between the pronator quadratus muscle and the overlying flexor tendons of the forearm; *see* Parona, Francesco.

Space of Poirier: *Eponym* for the space overlying the volar aspect of the proximal capitate between the volar radiocapitate and volar radiotriquetral ligaments; *see* Poirier, Paul Julien.

Space of Proust: *Eponym* for the rectovesical pouch of the peritoneal cavity; *see* Proust, P.T.

Space of Prussak: *Eponym* for the superior recess of the tympanic membrane; *see* Prussak, Alexander.

Space of Reinke: *Eponym* for loose connective tissue under the epithelium of the vocal cords.

Space of Retzius: *Eponym* for the retropubic space; *see* Retzius, Anders Adolf.

Spaces of Robin: *See* spaces of Virchow-Robin.

Space of Schwalbe: *Eponym* for the episcleral space of the optic nerve, the space within the internal sheath of the optic nerve, filled with cerebrospinal fluid and continuous with the subarachnoid space; *see* Schwalbe, Gustav Albert.

Space of Tarin: *Eponym* for the interpeduncular fossa; *see* Tarin, Pierre.

Space of Tenon: *Eponym* for the episcleral space, the space between the fascial sheath of the eyeball and the sclera.

Space of Traube: *Eponym* for the semilunar area on the thoracic wall over which resonance of the stomach is obtained, comparable to the triangle of Labbé; *see* Traube, Ludwig.

Space or **Foramen of Velpeau, Quadrangular:** *Eponym* for a space determined by the teres major muscle caudally, the teres minor muscle cranially, the humerus laterally, and the long head of the triceps brachii muscle medially; *see* Velpeau, Alfred Armand Louis Marie.

Space(s) of Virchow-Robin: *Eponym* for the perivascular spaces, a tunnel-like extension of the subarachnoid space surrounding around blood vessels entering the CNS in fixed tissue sections; *see* Virchow, Rudolf Ludwig Karl, and Robin, Charles Philippe.

Space of Waldeyer: *Eponym* for the tubular space between the bladder wall and the intramural part of the ureter; *see* Waldeyer, Heinrich Wilhelm Gottfried.

Space of Zang: *Eponym* for the supraclavicular fossa; *see* Zang, Christoph Bonifacius.

1.25 Triangles (Trigones)

The term **triangle** or **trigone** is mostly used to indicate certain areas of the skin, for muscular triangles, and sometimes in the nervous system.

Triangle or **Trigone of Arnold:** *Eponym* for the vagal trigone; *see* Arnold, Friedrich.

Triangle of Assézat: *Eponym* for a triangle formed by a line connecting the nasion with the alveolar and nasal points, used to determine the degree of prognathism; *see* Assézat, Jules.

Triangle of Babcock: *Eponym* for a relatively radiolucent triangular area seen in the subcapital region of the neck of the femur on anteroposterior radiographs of the hip, caused by loosely arranged trabeculae; *see* Babcock, William Wayne.

Triangle of Béclard: *Eponym* for the triangle bounded by the posterior border of the hyoglossus muscle, the posterior belly of the digastric muscle and the greater horn of the hyoid bone; *see* Béclard, Pierre Augustin.

Triangle or **Gap of Bochdalek:** *Eponym* for the lumbocostal or vertebrocostal triangle, a triangular area between the costal and lumbar parts of the diaphragm devoid of muscle fibres that may lead to a posterior diaphragmatic hernia; *see* Bochdalek, Vincenc Alexander.

Triangle of Bryant: *Eponym* for the iliofemoral triangle, a surgical marking for diagnosing fractures of the neck of the femur; *see* Bryant, Thomas.

Triangle of Budde-Rocko: *Eponym* for the hepatocystic triangle, bounded by the inferior border of the liver, laterally by the gallbladder and the cystic duct, and medially by the common hepatic duct; *see* Budde-Rocko triangle.

Triangle or **Trigone of Calot:** *Eponym* for the cystohepatic trigone, a triangle bounded by the cystic artery, the cystic duct and the common hepatic duct; *see* Calot, Jean-François.

Triangle of Farabeuf: *Eponym* for a triangle in the upper part of the neck bounded dorsally by the internal jugular vein, ventrally by the facial vein and caudally by the hypoglossal nerve; *see* Farabeuf, Louis Hubert.

Triangle or **Fossa of Gerdy:** *Eponym* for the infraclavicular fossa, also described by Mohrenheim; *see* Gerdy, Pierre Nicholas.

Triangle of Grynfeltt, Superior lumbar: *Eponym* for the superior lumbar triangle bounded by the serratus posterior inferior muscle and the 12th rib cranially, the erector spinae muscle medially and the internal oblique abdominal muscle laterally, also described by Lesshaft and Krause; *see* Grynfeltt, Joseph Casimir.

Triangle of Guillain-Mollaret: *Eponym* for the dentatorubroolivary circuit, an important feedback circuit for the dentate nucleus and the brain stem; *see* Guillain, Georges Charles, and Mollaret, Pierre.

Triangle of Guyon: *Eponym* for a triangle determined by the inferior jugular vein, the thyro-linguofacial vein and the posterior belly of the digastric muscle; *see* Guyon, Jean Casimir Felix.

Triangle of Henke: *Eponym* for a triangular area between the descending part of the inguinal fold, the lateral part of the fold and the lateral border of the rectus abdominis muscle; *see* Henke, Wilhelm.

Triangle of Henle: *Synonym* for the lumbocostal triangle of the diaphragm of Bochdalek; *see* Henle, Friedrich Gustav Jecob.

Triangle of Hesselbach: *Eponym* for a triangular area in the lower abdominal wall bounded cranially by the falx inguinalis, the inferior epigastric vessels laterally, the lateral border of the rectus abdominis muscle medially and the inguinal ligament caudally; *see* Hesselbach, Franz Kaspar.

Triangle of Hueter: *Eponym* for the equilateral triangle formed by the medial and lateral epicondyli of the humerus and the olecranon of the ulna in a 90° flexed elbow; *see* Hueter, Carl.

Triangle of Kager: *Eponym* for a sharply marginated radiolucent triangle seen posteriorly on lateral radiographs of the ankle. It represents the Kager fat pad; *see* Kager, Hans.

Triangles of Kiernan: *Eponym* for the interlobular triangles of the liver; *see* Kiernan, Francis.

Triangle of Killian-Jamieson: *Eponym* for the triangular area of the pharynx wall delimited by the oblique fibres of the thyropharyngeal muscle and the transverse fibres of the cricopharyngeal muscle; *see* Killian, Gustav.

Triangle of Koch: *Eponym* for a triangular area of the wall of the right atrium of the heart marking the position of the atrioventricular node; *see* Koch, Walter.

Triangle of Krause: *Eponym* for the superior lumbar triangle bounded by the serratus posterior inferior muscle and the 12th rib cranially, the erector spinae muscle medially and the internal obliue abdominal muscle laterally, also described by Grynfeltt and Lesshaft; *see* Krause, Karl Friedrich Theodor.

Triangle of Labbé: *Eponym* for an area where the stomach is normally in contact with the abdominal wall, comparable to the space of Traube; *see* Labbé, Léon.

Triangle of Langenbeck: *Eponym* for a triangular area formed by lines drawn from the anterior superior iliac spine to the greater trochanter of the femur and to the surgical neck of the femur; *see* Langenbeck, Bernhard Rudolf Konrad von.

Triangle or **Fissure of Larrey:** *Eponym* for the sternocostal triangle of the diaphragm, the place of a parasternal hernia, also noted by Morgagni and usually called hernia of Morgagni; *see* Larrey, Dominique Jean.

Triangle of Lesser: *Eponym* for the lingual triangle bounded by the anterior and posterior bellies of the digastricus muscle and the hypoglossal nerve; beneath its floor formed by the hyoglossus muscle, the lingual artery is found; *see* Lesser, Ladislaus Leon von.

Triangle of Lesshaft: *Eponym* for the superior lumbar triangle bounded by the serratus posterior inferior muscle and the 12th rib cranially, the erector spinae muscle medially and the inferior oblique abdominal muscle laterally, also described by Grynfeltt and Krause; *see* Lesshaft, Pjotr Frantsowitch.

Triangle or **Trigone of Lieutaud:** *Eponym* for the vesical trigone; *see* Lieutaud, Joseph.

Triangle of Macewen: *Eponym* for the suprameatal triangle or fossa on the squamous part of the temporal bone; *see* Macewen, William.

Triangle of Malgaigne: *Eponym* for the carotid fossa or triangle; *see* Malgaigne, Joseph François.

Triangle of Marcille: *Eponym* for the lumbosacral triangle, a triangular area bounded by the medial border of the psoas major muscle, the lateral margin of the vertebral column and the iliolumbar ligament; in it, the obturator nerve and lymph nodes are found; *see* Marcille, Maurice.

Triangle of Morgagni-Larrey: *Eponym* for the sternocostal triangle, also described by Larrey, through which a parasternal hernia may occur; *see* Morgagni, Giovanni Battista.

Triangle of Pawlik: *Eponym* for a triangular area of the anterior wall of the vagina in contact with thefundus of the bladder, and distinguished by the absence of vaginal rugae; *see* Pawlik, Karel.

Triangle of Petit, Inferior lumbar: *Eponym* for the inferior lumbar triangle between the iliac crest caudally, the margin of the external oblique abdominal muscle laterally, and the margin of the latissimi dorsi muscle medially; *see* Petit, Jean.

Triangle of Philippe-Gombault: *Eponym* for the sacral continuation of the septomarginal tract in the spinal cord, described by Philippe and by Gombault, his teacher; *see* Philippe, Claudien, and Gombault, François Alexis Albert.

Triangle of Pirogoff: *Eponym* for a triangle bounded by the intermediate tendon of the digastric muscle, the posterior border of the mylohyoid muscle and the hypoglossal muscle, containing the lingual artery; *see* Pirogoff, Nicolai Ivanovitch.

Triangle of Scarpa: *Eponym* for the femoral triangle; *see* Scarpa, Antonio.

Triangle of Sedillot: *Eponym* for the lesser supraclavicular fossa between the two origins of the sternocleidomastoideus muscle and the clavicula, also described by Zang; *see* Sedillot, Charles Emmanuel.

Triangle of Trautmann: *Eponym* for an area of the temporal bone bounded by the sigmoid sinus, the superior petrosal sinus and the vestibulocochlear nerve; *see* Trautmann, Moritz Ferdinand.

Triangle of Volkmann: *Eponym* for the posterolateral corner of the tibia; *see* Volkmann, Richard von.

Triangle of Ward: *Eponym* for an area of diminished density in the trabecular pattern of the neck of the femur; *see* Ward, Frederick Oldfield. Ward also described what is currently known as the triangle of Babcock: a relatively translucent triangular area seen in the subscapular region of the femoral neck.

Triangle of Weber: *Eponym* for an area on the sole of the foot formed by the heads of the first and fifth metatarsals and the midpoint of the plantar surface of the heel; *see* Weber, Wilhelm Eduard.

Triangle of Wilde: *Eponym* for the pyramid of light, the area at the anterior inferior part of the tympanic membrane, also known as cone of light of Wilde and Politzer; *see* Wilde, William Robert Wills.

Triangle of Zang: *Eponym* for the lesser supraclavicular fossa between the two origins of the sternocleidomastoideus muscle and the claviula, also described by Sedillot; *see* Zang, Christoph Bonifacius.

1.26 Tubercles

The term **tubercle** is mostly used for bony or cartilaginous protuberances.

Tubercle of Assaki: *Eponym* for an inconstant intraglenoid tubercle located just below the centre of the glenoid fossa of the scapula; *see* Assaki, George.

Tubercle of Carabelli: *Eponym* for an occasional fifth tubercle on the lingual surfac of the first upper upper molar; *see* Carabelli, George.

Tubercle of Chaput: *Eponym* for the anterior tibial tubercle, a small prominence on the distal end of the tibia, to which the anterior tibiofibular ligament inserts, also described by Tillaux; *see* Chaput, Henri.

Tubercle of Chassaignac: *Eponym* for the carotid tubercle on the sixth cervical vertebra; *see* Chassaignac, Charles Marie Edouard.

Tubercle of Czermak: *Eponym* for the tubercle of the epiglottis; *see* Czermak, Johann Nepomuk.

Tubercle of Darwin: *Eponym* for a small prominence occasionally present on the helix of the auricle, first noticed by Woolner, the sculptor who called Darwin's attention to it; *see* Darwin, Charles Robert.

Tubercle of Gerdy: *Eponym* for the tubercle on the lateral condyle of the tibia for the attachment of the iliotibial tract; *see* Gerdy, Pierre Nicolas.

Tubercle of Hochstetter: *Eponym* for the embryonic cerebellar tubercle; *see* Hochstetter, Ferdinand.

Tubercle of Le Fort: *Eponym* for the anterior fibular tubercle, a small prominence on the lateral malleolus to which the anterior tibiofibular ligament inserts, also described by Wagstaffe; *see* Le Fort, Léon Clément.

Tubercle of Lisfranc: *Eponym* for the scalene tubercle on the first rib for the attachment of the scalenus anterior muscle; *see* Lisfranc, Jacques.

Tubercle of Lister: *Eponym* for the dorsal tubercle of the radius, a prominence on the posterior surface of the distal end of the radius between the groove for the tendon of the extensor pollicis longus muscle medially, and the extensor carpi radialis brevis muscle laterally; *see* Lister, Joseph.

Tubercle of Lower: *Eponym* for the intervenous tubercle, a prominence on the wall of the right atrium between the openings of the superior and inferior venae cavae; *see* Lower, Richard.

Tubercle of Luschka: *Eponym* for (a) the urethral carina of the vagina, and (b) an inconstant tubercle on the superior angle of the scapula; *see* Luschka, Hubert von.

Tubercle of Morgagni: *Eponym* for the cuneiform cartilage of the larynx, also described by Wrisberg; see Morgagni, Giovanni Battista.

Tubercle of Müller: *Eponym* for the sinus tubercle, the first evidence of the embryonic uterus and vagina. It remains as the hymen; *see* Müller, Johannes Peter.

Tubercle of Princeteau: *Eponym* for the part of the temporal bone at the apex, where the superior petrosal sinus commences; *see* Princeteau, Laurent Raphael.

Tubercle of Rolando: *Eponym* for the tuber cinereum, a longitudinal prominence on the dorsolateral surface of the medulla oblongata along the lateral border of the cuneate tubercle; *see* Rolando, Luigi.

Tubercle of Santorini: *Eponym* for the corniculate tubercle in the aryepiglottic fold, elevated by the corniculate cartilage of the larynx; *see* Santorini, Giovanni Domenico.

Tubercle of Stieda: *Synonym* for the process of Stieda; *see* Stieda, Christian Hermann Ludwig.

Tubercle of Tillaux: *Eponym* for the anterior tibial tubercle, also described by Chaput; *see* Tillaux, Paul Jules.

Tubercle of Vater: *Eponym* of the major duodenal papilla, also known as papilla of Santorini; *see* Vater, Abraham.

Tubercle of Volkmann: *Eponym* for the posterior tibial tubercle, a small prominence on the distal end of the tibia to which the posterior tibiofbular lagament inserts; *see* Volkmann, Richard von.

Tubercle of Wagstaffe: *Eponym* for the anterior fibular tubercle, a small prominence on the lateral malleolus to which the anterior tibiofibular ligament inserts, also described by Le Fort; *see* Walstaffe, William Warwick.

Tubercle of Whitnall: *Eponym* for the orbital tubercle of the zygomatic bone; *see* Whitnall, Samuel Ernest.

Tubercle of Wrisberg: *Eponym* for the cuneiform cartilage of the larynx, also described by Morgagni; *see* Wrisberg, Heinrich August.

Tubercle of Zuckerkandl: *Eponym* for an inconstant tubercle of the thyroid gland, his processus posterior glandulae thyroideae; *see* Zuckerkandl, Emil.

1.27 Valves

The term **valve** or **valvula** appears in many eponyms but only a few remain to be used (Bauhin, Hasner, Morgagni). In many 'valves' it concerns only a fold. Valves of the nasolacrimal duct were described at various positions by Béraud, Bianchi, Bochdalek, Cruveilhier, Foltz, Hasner, Huschke,

Krause, Rosenmüller and Taillefer, but only the valve of Hasner, also described by Bianchi and Cruveilhier, deserves the term valve. The ileocaecal valve was described for the first time by Varolio, followed by Falloppio, Bauhin and Tulp, but it is also a fold.

Valves of Amussat: *Eponym* for spiral folds of the cystic duct, a series of crescentic folds of mucosa in the upper part of the cystic duct, described earlier by Heister; *see* Amussat, Jean Zulema.

Valvula of Amussat: *Eponym* for (a) posterior urethral valves, anomalous mucous folds at the level of the seminal colliculus, and (b) the uvula of the urinary bladder, an eminence of the mucosa, due to the middle lobe of the prostate; *see* Amussat, Jean Zulema.

Valves of Ball: *Eponym* for the anal valves, previously described by Morgagni; *see* Ball, Sir Charles Bent.

Valve of Bauhin: *Eponym* for the ileocaecal fold, first described by Varolio, and also described by Falloppio and Tulp; *see* Bauhin, Caspar.

Valve of Béraud: *Eponym* for the a small fold on the inner side of the lacrimal sac, where it passes into the nasolacrimal duct, also described by Krause; *see* Béraud, Bruno Jacques.

Valve of Bianchi: *Eponym* for the lacrimal valve in the nasolacrimal duct, as also described by Hasner and Cruveilhier; *see* Bianchi, Giovanni Battista.

Valve of Bochdalek: *Eponym* for a mucous fold at the puncta lacrimalia of the lacrimal canaliculi, also described by Foltz; *see* Bochdalek, Vincenc Alexander.

Valve of Braune: *Eponym* for a perpendicular mucous fold at the ostium of the stomach, his cardiac sphincter between the oesophagus and the stomach, also described by Gubaroff; *see* Braune, Christian Wilhelm.

Valve of Cruveilhier: *Eponym* for the lacrimal valve in the nasolacrimal duct, also described by Bianchi and Hasner; *see* Cruveilhier, Jean Baptiste.

Valve of Eustachi: *Eponym* for the valve of the inferior vena cava in the right atrium, also described by Sylvius; *see* Eustachi, Bartolomeo.

Valve of Falloppio: *Eponym* for the ileocaecal fold, also described by Varolio, Bauhin and Tulp; *see* Falloppio, Gabriele.

Valves of Foltz: *Eponym* for 'valvules' or constrictions, in fact mucous folds, within the lacrimal canaliculi, also described by Bochdalek; *see* Foltz, Jean Charles Eugène.

Valve of Gerlach: *Eponym* for an occasional mucous fold at the orifice of the vermiform appendix; *see* Gerlach, Joseph.

Valve of Gubaroff: *Eponym* for a perpendicular mucous fold at the ostium of the stomach, his cardiac sphincter between the oesophagus and the stomach, also described by Braune; *see* Gubaroff, Alexander Petrowitsch von.

Valve of Guérin: *Eponym* for the valve of the navicular fossa of the male urethra; *see* Guérin, Alphonse François Marie.

Valve of Hasner: *Eponym* for the lacrimal valve in the nasolacrimal duct, also described by Bianchi and Cruveilhier; *see* Hasner, Joseph, Ritter von Artha.

Valve of Heister: *Eponym* for a spiral fold of the cystic duct, also described by Amussat; *see* Heister, Lorenz.

Valve of Houston: *Eponym* for the superior transverse rectal fold; *see* Houston, John.

Valve of Huschke: *Eponym* for the prominent lower margin of the opening of the lacrimal canaliculi into the lacrimal sac, opposite the valve of Rosenmüller; *see* Huschke, Emil.

Valve of Hyrtl, Spiral: *Eponym* for a fold of the lower part of the nasolacrimal duct, also described by Taillefer; *see* Hyrtl, Joseph.

Valves of Kerckring: *Eponym* for the circular folds of the small intestine, already described by Falloppio; *see* Kerckring, Dirk.

Valve of Kohlrausch: *Eponym* for the largest, middle transverse rectal fold, earlier described by Houston; see Kohlrausch, Otto Ludwig Bernhard.

Valve of Krause: *Eponym* for a fold between the lacrimal sac and the nasolacrimal duct, also described by Béraud; *see* Krause, Karl Friedrich Theodor.

Valve of Mercier: *Eponym* for an occasional mucous fold of the urinary bladder partially occluding the internal urethral opening; *see* Mercier, Louis Auguste.

Valves of Morgagni: *Eponym* for the anal valves, also descibed by Haller and Ball; *see* Morgagni, Giovanni Battista.

Valve of Parchappe de Vinay: *Eponym* for the valve of the foramen ovale in the left atrium of the heart; *see* Parchappe de Vinay, Jean Baptiste Maximilien.

Valve of Rosenmüller: *Eponym* for the prominent upper margin of the opening of the lacrimal canaliculi into the lacrimal sac opposite the valve of Huschke; *see* Rosenmüller, Johann Christian.

Valve of Sylvius: *Eponym* for the valve of the inferior vena cava, also described by Eustachi; *see* Sylvius, François de le Boë.

Valve of Taillefer: *Eponym* for a fold of the lower part of the nasolacrimal duct, also described by Hyrtl; see Taillefer, Louis Auguste Horace Sydney Timoléon.

Valve of Tarin: *Eponym* for the inferior or posterior medullary velum of the roof of the fourth ventricle; *see* Tarin, Pierre.

Valve of Terrier: *Eponym* for the 'proximal valve' of the gallbladder, a fold between the gallbladder and the cystic duct; *see* Terrier, Louis Felix.

Valve of Thebesius: *Eponym* for the valve of the coronary sinus; *see* Thebesius, Adam Christian.

Valve of Tulp: *Eponym* for the ileocaecal fold, earlier described by Varolio, Falloppio and Bauhin; *see* Tulp, Nicolaes.

Valve of van Hoboken: *Eponym* for secondary windings of the vessels of the umbilical cord that form grooves externally and valve-like projections internally; *see* van Hoboken, Nicolaas.

Valve of Varolio: *Eponym* for the ileocaecal fold, described for the first time by Varolio, but later also by Falloppio, Bauhin and Tulp; *see* Varolio, Costanzio.

Valve of Vieussens: *Eponym* for (a) the superior medullary velum of the fourth ventricle, and (b) an inconstant valve-like structure at the transition of the great cardiac vein into the coronary sinus; *see* Vieussens, Raymond de.

1.28 Veins

The term **vein** is used in many eponyms but only a few remain in use (veins of Ascher, Breschet, Cuvier, Galen, Labbé, Rosenthal, Thebesius, Trolard, and in particular for perforating veins: Bassi, Boyd, Cockett, Dodd, Hach, Hunter, May, Sherman and Thierry).

Vein or **System of Albanese:** *Eponym* for an independent superficial venous system on the lateral aspect of the knee; *see* Albanese, Alfonso.

Veins of Ascher, Aqueous: *Eponym* for the aqueous veins, the efferent communications between the scleral venous sinus and the episcleral veins; *see* Ascher, Karl.

Veins of Baumgarten: *Eponym* for the paraumbilical veins along the round ligament of the liver, also described by Sappey and Burow; *see* Baumgarten, Paul Clemens von.

Vein of Bassi, Perforating: *Eponym* for the inferior paratendinous or para-Achillean perforating vein, also known as perforator of Bassi; *see* Bassi, Glauco.

Veins of Billroth: *Eponym* for the sinusoids of the spleen; *see* Billroth, Christian Albert Theodor.

Vein of Boyd, Perforating: *Eponym* for the superior paratibial perforating vein at the level of the knee connecting to the great saphenous vein; *see* Boyd AM.

Veins of Braune: *Eponym* for (a) an accessory superior intercostal vein, (b) the anterior saphenotibial vein, (c) posterior saphenotibial veins, and (d) saphenofibular veins; *see* Braune, Christian Wilhelm.

Veins of Breschet: *Eponym* for the diploic veins; *see* Breschet, Gilbert.

Vein of Browning: *Eponym* for the 'middle cerebral vein', the upper part of the superior anastomotic vein of Trolard; *see* Browning, William.

Veins of Burow: *Eponym* for (a) the renal veins, and (b) the paraumbilical veins along the round ligament of the liver, also described by Baumgarten and Sappey; *see* Burow, Karl August von.

Vein of Citelli: *Eponym* for the mastoid emissary vein connecting the sigmoid sinus and the occipital vein; *see* Citelli, Salvatore.

Veins of Cockett, Perforating: *Eponym* for the posterior tibial perforating veins (superior, middle and inferior) connecting the posterior accessory saphenous vein with the posterior tibial veins; *see* Cockett, Frank Bernard.

Vein of Cruveilhier, Accessory saphenous: *Eponym* for an accessory common venous trunk for the superficial posterior veins of the thigh; *see* Cruveilhier, Jean Baptiste.

Vein of Dandy: *Eponym* for the petrosal vein; *see* Dandy, Walter Edward.

Vein of Dodd, Perforating: *Eponym* for one of the medial femoral perforating veins; *see* Dodd, William Harold Alfred.

Vein of Galen, Great: *Eponym* for the great cerebral vein. The internal cerebral veins are sometimes also ascribed to Galen; *see* Galen.

Vein of Giacomini: *Eponym* for the femoral intersaphenous vein connecting the cranial extension of the small saphenous vein and the great saphenous vein, corresponding to the posterior circumflex femoral vein; *see* Giacomini, Carlo.

Vein of Gillot, Perforating: *Eponym* for the medial leg perforator; *see* Gillot, Claude.

Vein of Guidi (Vidius): *Eponym* for the vein of the pterygoid canal; *see* Guidi, Guido.

Vein of Hach, Perforating: *Eponym* for the posterolateral femoral perforating vein; *see* Hach, Wolfgang.

Vein of Henle: *Eponym* for the middle cardiac vein; *see* Henle, Friedrich Gustav Jacob.

Vein of Hunter: *Eponym* for a perforating vein in the adductor canal; *see* Hunter, John.

Vein of Hyrtl: *Eponym* for the femoropopliteal vein; *see* Hyrtl, Joseph.

Vein of Kiernan: *Eponym* for the intralobular vein of the liver; *see* Kiernan, Francis.

Vein of Kocher: *Eponym* for the middle thyroid vein, also described by Lichačeva; *see* Kocher, Theodor.

Vein of Konstantinovich: *Eponym* for the marginal vein of the anus; *see* Konstantinovich, Vikentzi Bonifatiyevich.

Vein of Krukenberg: *Eponym* for the central vein of the hepatic lobule; *see* Krukenberg, Adolph.

Vein of Labbé: *Eponym* for the inferior anastomotic vein of the cerebral cortex; *see* Labbé, Charles.

Vein of Latarjet: *Eponym* for the prepyloric vein, also described by Mayo; *see* Latarjet, André.

Veins of Le Fort: *Eponym* for the pleuropulmonary veins; *see* Le Fort, Léon Clément.

Vein of Leonardo: *Eponym* for the posterior accessory great saphenous vein of the leg, previously known as the posterior arcuate vein of the leg; *see* Leonardo da Vinci.

Vein of Lichačeva: *Eponym* for the middle thyroid vein, also described by Kocher; *see* Lichačeva, Natalia Borisovna.

Vein of Markowski: *Eponym* for the embryonic median prosencephalic vein; see Markowski, Józef.

Vein of Marshall: *Eponym* for the oblique vein of the left atrium; *see* Marshall, John.

Vein of May, Perforating: *Eponym* for the midcalf perforant vein; *see* May, Robert.

Vein of Mayo: *Eponym* for the prepyloric vein, also described by Latarjet; *see* Mayo, Charles Horace, and William.

Veins of Quain: *Eponym* for the colicospermatic veins, his colicoepigastric veins, collecting blood from the descending colon and draining into the left testicular vein; *see* Quain, Jones.

Veins of Retzius: *Eponym* for (a) the retroperitoneal portocaval anastomoses, also described by Ruysch, and (b) the vorticose veins of the eyeball, also described by Stensen and Ruysch; *see* Retzius, Anders Adolf.

Vein of Rolando: *Eponym* for the central vein, one of the superior central veins of the cerebral cortex, found in the central sulcus; *see* Rolando, Luigi.

Vein of Rosenthal, Basal: *Eponym* for the basal vein; *see* Rosenthal, Friedrich Christian.

Veins of Ruysch: *Eponym* for (a) the retroperitoneal portocaval anastomoses, later also described by Anders Retzius, and (b) the vorticose veins of the eyeball, also described by Stensen and later by Anders Retzius; *see* Ruysch, Frederik.

Vein(s) of Santorini: *Eponym* for (a) the parietal emissary vein, and (b) the prostatic (in males) or vesical (in females) venous plexus; *see* Santorini, Giovanni Domenico.

Veins of Sappey: *Eponym* for the paraumbilical veins along the round ligament of the liver, also described by Baumgarten and Burow; *see* Sappey, Marie Philibert Constant.

Veins of Sherman, Perforating: *Eponym* for the inferior and middle paratibial perforating veins; *see* Sherman R Shanton.

Veins of Stensen: *Eponym* for the vorticose veins of the eyeball, also described by Ruysch and Anders Retzius; *see* Stensen, Niels.

Vein of Sylvius: *Eponym* for the middle cerebral vein of the cerebral cortex, also known as central vein (Rhoton); *see* Sylvius, François de le Boë.

Vein of Testut, Central: *Eponym* for the principal suprarenal vein; *see* Testut, Jean Léon.

Veins of Thebesius: *Eponym* for the smallest cardiac veins (Thebesian veins), also described by Vieussens; *see* Thebesius, Adam Christian.

Vein of Thierry, Perforating: *Eponym* for the popliteal perforating vein; *see* Thierry, L.

Vein of Trolard: *Eponym* for (a) the superior anastomotic vein of the cerebral cortex, and (b) the inferior preoccipital vein connecting the venous plexus of Rektorzik and the cavernous sinus with the jugular bulb; *see* Trolard, Paulin.

Veins of Verheyen: *Eponym* for the stellate veins of the kidney; *see* Verheyen, Philippe.

Vein of Vesalius: *Eponym* for the sphenoidal emissary vein between the cavernous sinus and the pterygoid plexus; *see* Vesalius, Andreas.

Veins of Vieussens: *Eponym* for the smallest cardiac veins, also described by Thebesius; *see* Vieussens, Raymond de.

Part II

Anatomical Eponyms from A to Z

Hans J. ten Donkelaar, Marina Quartu, and David Kachlík

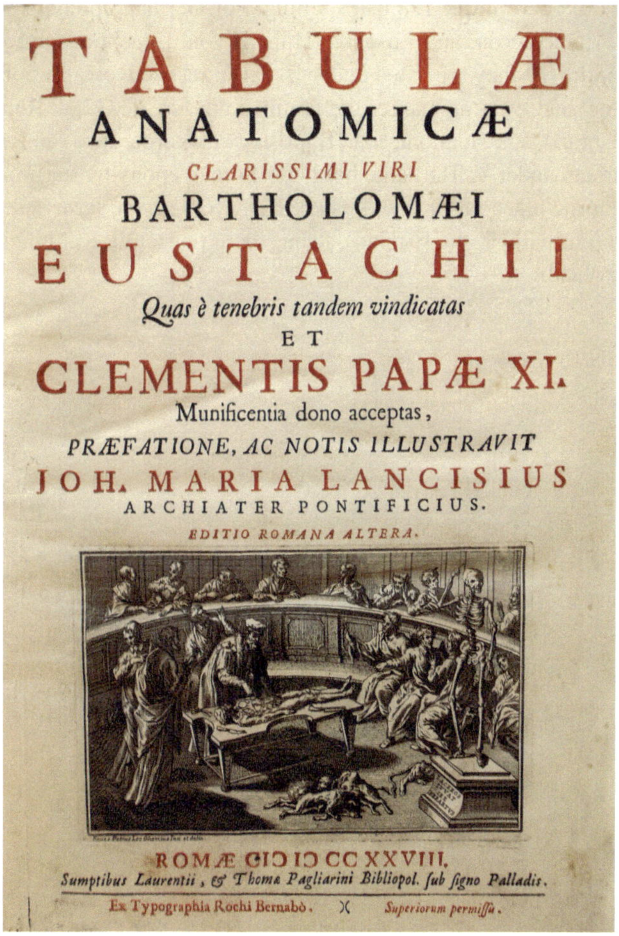
TABULÆ
ANATOMICÆ
CLARISSIMI VIRI
BARTHOLOMÆI
EUSTACHII
Quas è tenebris tandem vindicatas
ET
CLEMENTIS PAPÆ XI.
Munificentia dono acceptas,
PRÆFATIONE, AC NOTIS ILLUSTRAVIT
JOH. MARIA LANCISIUS
ARCHIATER PONTIFICIUS.
EDITIO ROMANA ALTERA.
ROMÆ CIϽ IϽ CC XXVIII.
Sumptibus Laurentii, & Thomæ Pagliarini Bibliopol. sub signo Palladis.
Ex Typographia Rochi Bernabò. *Superiorum permissu.*

Vignette Part II Title page *Tabulae Anatomicae Bartholomaei Eustachii*, published 1742 by JM Lancisi in Rome

1.1 Introduction

The history of eponymous anatomical terminology was beautifully summarized by Jessie Dobson (1962). Apparently, Andreas Vesalius (1543) introduced personal names into the anatomical nomenclature. In the first edition of the Nomina Anatomica, the Basler Nomina Anatomica (BNA), Wilhelm His (1895) stated that personal names are met not only in Anatomy but also in other natural sciences and in particular in medicine. Others, such as Friedrich Gustav Jacob Henle in fact declared war against an originally much more extended personal nomenclature, and later anatomists quite vigorously fought against the use of eponyms. In later editions of the Nomina Anatomica most eponyms were simply deleted, but in the Terminologia Anatomica (1998) a list of anatomical eponyms was provided. More recently, the Federal International Programme for Anatomical Terminology (FIPAT) of the International Federation of Associations of Anatomists (IFAA) presented the more common anatomical eponyms in their second edition of the Terminologia Embryologica (TE2 2017), the Terminologia Neuroanatomica (TNA 2017) and the second edition of the Terminologia Anatomica (TA2 2019).

According to the 28th edition of Stedman's Medical Dictionary (2008), an eponym is: "The name of a disease, structure, operation or procedure, usually derived from the name of the person who discovered or described it first". The word eponym derives from the Greek for 'named after'. Other dictionaries show comparable definitions. Today, eponyms are used in the nonpossessive form. Here, we restrict ourselves to anatomical, embryological and histological eponyms. We present eponyms with their proper accents wherever possible but otherwise in the English spelling. For German eponyms the preposition 'von' is hardly used, so: Haller, Albrecht von, and Luschka, Hubert, von. Exceptions are von Baer, Karl Ernst and von Economo, Constantin, but they are listed under B and E, respectively, for consistency. For Dutch eponyms, the prepositions 'de' and 'van' are part of the family name, so: de Graaf, Reinier is found under D, and van Diemerbroeck, IJsbrand, van Helmont, Jean Baptiste, van Horne, Jan, van Leeuwenhoek, Antonie, and van den Spieghel, Adriaan under V. The same holds for French eponyms starting with 'Le' as in Le Fort, to be found under L. For the place of birth and where they worked, the terms used at that time are given, so as examples: Dorpat (currently Tartu, Estland), Königsberg (currently Kaliningrad, Russia), and Breslau (currently Wróclaw, Poland).

Letter A

2

Abernethy, John (1764–1831): English surgeon born and working in London (Fig. A1a; for biography, *see* Thornton JL 1953 *John Abernethy*. London).

Fascia of Abernethy: *Eponym* for the fascia covering the external iliac artery. Reference: *Lectures on Anatomy, Surgery and Pathology*. Bulcock, London, 1828 (Fig. A1b).

Aberrant ductules; *Eponym* for aberrant ductules of the testis; *see* Haller, Albrecht von.

***Achilles tendon*:** *Eponym* for the tendo calcaneus. Achilles, Greek hero in the *Iliad* of Homer, was invulnerable except at the heel.

Ackermann, Jakob Fidelis (1765–1815): German physician, botanist and anatomist born in Rüdesheim. He was the successor of Soemmerring in Frankfurt am Main and worked later in Jena and Heidelberg.

Angle of Ackermann: *Eponym* for the angle formed by the base of the skull and the horizontal plane.

Plane of Ackermann: *Eponym* for the plane passing through the base of the tragus and the infranasal point.

Acrel, Olof (1717–1806): Swedish surgeon born and working in Stockholm and later in Uppsala who contributed greatly to the development of surgery in Sweden, including operative ophthalmology (Fig. A2).

Ganglion of Acrel: *Eponym* for a pseudoganglion on the posterior interosseous nerve on the back of the wrist. Reference: *Historia tumorum rariorum circa carpeum et in vola manus obvermentum.* Comment Soc Reg Scient Goetting 1779, Vol II: 112–138.

Adamkiewicz, Albert Wojciech (1850–1921): Polish pathologist born in Zerkow and working in Krakau, currently Kraków (Fig. A3a) who is known for the first extensive studies on the blood vessels of the spinal cord (1881, 1882). References: *Die Blutgefässe des menschlichen Rückenmarks. I. Teil: Die Gefässe der Rückenmarkssubstanz.* Sitzungsber Kais Akad Wiss Wien 1881, 84. Bd; *II. Teil: Die Gefässe der Rückenmarksoberfläche.* Ibid. 1882, 85. Bd.

Artery of Adamkiewicz: *Eponym* for the great radicular artery (Fig. A3b).

Adams, Robert (1791–1875): Irish surgeon and anatomist born and working in Dublin.

Arch of Adams: *Eponym* for a bony arch on the medial side of the neck of the femur, containing thick cortical bone. The German surgeon CW Streubel introduced the term 'Anatomische Knochenbogen' in 1847. In the English literature, it was wrongly translated as femoral calcar (calcar femorale; *see* Bartoníček J, Alt J, Naňka O *Internal architecture of the proximal femur: Calcar femorale or Adams' arch?* Int Orthopedics 2023; 47:1871–1877). Reference: *Hip-joint, abnormal conditions.* In: Todd RB, ed., *The Cyclopedia of Anatomy and Physiology of Man,* Vol II. Longman, London, 1836–1839, pp 780–825.

***Adam's apple*:** *Eponym* for the protrusion in the neck formed by the laryngeal prominence of the thyroid cartilage surrounding the larynx, typically visible in males, less frequently in females. Adam, the name given in the Bible to the first human.

Addison, Christopher, Viscount Addison of Stallingborough (1869–1951): English anatomist born in Hogsthorpe, Lincolnshire, and working in Sheffield and later in London, Member of Parliament and Minister of Health from 1918 to 1921 (Fig. A4).

Transpyloric line or **Plane of Addison:** *Eponym* for the upper transverse abdominal line or plane

H. J. ten Donkelaar et al., *An Illustrated Guide to Anatomical Eponyms*, https://doi.org/10.1007/978-3-031-91664-9_2

LECTURES
ON
ANATOMY, SURGERY,
AND
PATHOLOGY;
INCLUDING
OBSERVATIONS ON THE NATURE AND TREATMENT
OF
LOCAL DISEASES.
DELIVERED AT ST. BARTHOLOMEW'S HOSPITAL.

BY JOHN ABERNETHY, F.R.S.
SURGEON TO ST. BARTHOLOMEW'S AND CHRIST'S HOSPITALS,
ETC. ETC. ETC.

LONDON:
PRINTED BY C. SMITH, ONE BELL YARD, STRAND; AND SOLD BY
JAMES BULCOCK, 163, STRAND.
AND ALL MEDICAL AND OTHER BOOKSELLERS.
1828.

Fig. A1 (**a**) Portrait of Abernethy. (From Pagel 1901); (**b**) Title page *Lectures on Anatomy, Surgery and Pathology*. Bulcock, London, 1828

Fig. A2 Portrait of Acrel. (From Wikipedia; in public domain)

through the level of the first lumbar vertebra, found halfway the xiphoid process of the sternum and the umbilicus. On this plane, the gallbladder, the pylorus, the neck of the pancreas, the kidneys and the aortic hiatus of the diaphragm are located. This plane forms part of five transverse planes: (1) through the jugular notch of the sternum, (2) transthoracic, (3) transpyloric, (4) through the superior border of the pubic symphysis, and (5) transtubercular.

***Adonis' belt, Apollo's belt*:** *Eponym* for a part of the human anatomy referring to the two shallow grooves of the human abdomen running from the iliac crest to the os pubis. Adonis, Apollo, Greek mythological characters.

Aeby, Christopher Theodor (1835–1885): Swiss anatomist and anthropologist born in Guttenbrunn and working in Bern and later in Prague.

Epiarterial bronchus: *Eponym* for the superior bronchus of the right lung. Reference: *Der Bronchialbaum der Säugethiere und des Menschen.* Engelmann, Leipzig, 1880.

Muscle of Aeby: *Eponym* for the depressor labii inferioris muscle, also described by Klein. Reference: *Die Muskulatur der menschlichen Mundspalte.* Arch Mikr Anat 1878; 16:651–664.

Fig. A3 (**a**) Photo of Adamkiewicz. (From Pagel 1901); (**b**) The great radicular artery of Adamkiewicz shown in a postmortem spinal cord specimen. The artery of Adamkiewicz (*white arrow*) and the anterior spinal artery (*white arrow heads*) are shown on the right and the anterior median vein (*black arrow heads*) and the great anterior radicular vein (*black arrow*) are shown on the left. (From Bowen BC, et al. 1995 AJNR Am J Neuroradiol 17:483–494; with permission from the American Society of Neuroradiology, and ten Donkelaar HJ, Kachlík D, Tubbs RS *An Illustrated Terminologia Neuroanatomica*. Springer, Cham, 2018)

Fig. A4 Photo of Addison. (From the Wellcome Collection CCBY; in public domain)

Plane of Aeby: *Eponym* for a craniometric plane perpendicular to the median plane of the cranium, cutting the nasion and the basion. Reference: *Eine neue Methode zur Bestimmung der Schädelform von Menschen und Säugethieren.* Westermann, Braunschweig, 1862.

Albanese, Alfonso R (twentieth century): Argentinian surgeon working in Buenos Aires.

System of Albanese: *Eponym* for a lateral venous system (his subdermal lateral system) of the lower limb, an independent third superficial venous system on the lateral aspect of the knee. Reference: *Lateral subdermic varicose vein system of the legs. Its superficial treatment by the chiseling tube method* (with AM and EF Albanese). Vasc Surg 1969; 3:81–89.

Albarrán, Joaquin Maria (1860–1912): Cuban urologist born in Sagua-la-Grande who, after studying in Spain, worked as urogenital surgeon in Paris (Fig. A5).

Fig. A5 Photo of Albarrán. (From Wikimedia Commons, in public domain)

Gland of Albarrán: *Eponym* for the middle lobe of the prostate. Reference: *Médecine opératoire des voies urinaire.* Paris, 1909.

Infracervical glands of Albarrán: *Eponym* for small accessory prostatic glands in the mucosa of the urethra, halfway between the internal urethral opening and the seminal colliculus or verumontanum. Reference: *Les tumeurs de vessie.* Steinheil, Paris, 1891.

Albini, Giuseppe (1827–1911): Italian physiologist born in Milan and, after studying in Pavia, Lombardy, under Panizza and Brücke in Vienna, working in Krakau and later in Parma and Naples.

Nodules of Albini: *Eponym* for minute fibrous nodules on the margins of the mitral and tricuspid valves, representing rests of foetal tissue. Reference: *Ueber die Noduli an den Atrio-ventrikularklappen des Menschen.* Wochenblatt d.k.u.k. Ges d Aezrte in Wien 1856, 2, No 26, June 23.

Albinus (Weiss), Bernhard Siegfried (1697–1770): German anatomist and surgeon born in Frankfurt an der Oder who worked most of his life in Leiden (Fig. A6). References: *Historia musculorum hominis.* Haak and Luchtmans, Leiden, 1734; *Tabulae sceleti et musculorum corpus humani.* Verbeek, Leiden, 1747 (illustrated by Jan Wandelaar; *see* Fig. A7).

Fig. A6 Engraving of Albinus by J Houbraken, based on a painting by Carel Isaak de Moor. (From Collectiones Icones Leidenses; http://socratesleidenuniv.nl; in public domain)

Muscle of Albinus: *Eponym* for several muscles described by him: (a) the risorius muscle, also described by Santorini; (b) the rhomboid muscles; (c) the scalenus minimus muscle (his lateral scalenus muscle), described later by Sibson as fourth scalenus muscle; and (d) the tensor tympani muscle, already described by Eustachi.

Tendo sublimis truncatus of Albinus: In his *Historia musculorum hominis* (1734), Albinus illustrated what became known as the **chiasma tendinum of Camper**; *see* Camper, Petrus.

Albrecht, Karl Martin Paul (1851–1894): German anatomist and physician born and working in Hamburg.

Bone of Albrecht: *Eponym* for the basiotic bone, a transient bone between the basisphenoid and basioccipital bones during development.

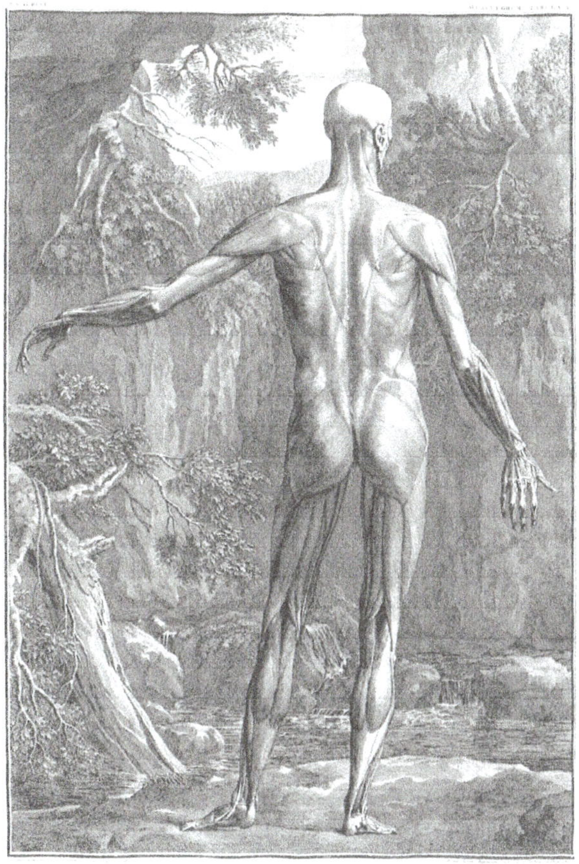

Fig. A7 Example of *Tabulae sceleti et musculorum corpus humani* of Albinus. (Verbeek, Leiden, 1747; illustrated by Jan Wandelaar; in public domain)

Alcock, Benjamin (1801–1859): Irish surgeon born in Kilkenny and working in Dublin. Later, he left for the USA (for biography, *see* O'Rahilly R 1948 *Benjamin Alcock*. Cork University Press).

Canal of Alcock: *Eponym* for the pudendal canal in the ischioanal or ischiorectal fossa (*see* **Fig. H16**). Reference: *Iliac Arteries,* in: *Todd's Cyclopedia of Anatomy and Physiology.* 1836–1839, Vol II:835.

***Alderman's nerve*:** The auricular branch of the vagus nerve is also known as *Alderman's nerve* because traditionally, Aldermans (Municipal magistrates in UK and US) during a gargantuan Alderman's dinner used to stimulate their jaded appetites by dropping cold water behind the ear. This practice reflexly encouraged gastric peristalsis. The nerve was also described by Arnold.

Allen, Harrison (1841–1897): American anatomist and zoologist born and working in Philadelphia, and among

Fig. A8 Portrait of Amici. (Courtesy Marina Quartu, Cagliari)

others as Director of the Wistar Institute. Reference: *A System of Human Anatomy.* Philadelphia, 1882.

Fossa of Allen: *Eponym* for an inconstant fossa on the neck of the femur.

Amantini, Cesare (nineteenth century): Italian anatomist in Perugia.

Pectineofoveal fold of Amantini: *Eponym* for the most prominent medial fold of the fibres of Weitbrecht. Reference: *Di una men nova ripiegatura sinoviale dell'articolazione dell'anca (Piega pettineofoveale); contributo anatomico allo studio delle fratture intra-articolari del collo del femore.* Tipografia di Vincenzo Santucci, Perugia, 1889.

Amici, Giovanni Battista (1786–1863): Italian mathematician and physicist born and working in Modena, where he improved the lenses for microscopes (Fig. A8).

Stria of Amici: *Eponym* for the dark line, generally known as Z line, which runs transversally through bands of myofibrils and marks the limits of a sarcomere, also described by Dobie and Krause.

***Ammon*:** Egyptian God, representing a ram's head with horn. The term is used in cornu Ammonis, the horn of Ammon, part of the hippocampal formation, and was proposed by de Garengeot in 1742. Reference: *Splanchnologie ou l'anatomie des viscères,* Tome 2, 2nd ed. Paris, 1742.

Ampulla of Vater: *Eponym* for the hepatopancreatic ampulla, the common termination of the bile duct and the pancreatic duct; *see* Vater, Abraham.

Amussat, Jean Zuléma (1796–1856): French pathologist and surgeon born in Saint-Maixent and working in Paris.

Valves of Amussat: *Eponym* for spiral folds of the cystic duct, a series of crescentic folds of mucosa in the upper part of the cystic duct, described earlier by Heister.

Valvula of Amussat: *Eponym* for (a) the posterior urethral valves that are anomalous mucous folds at the level of the seminal colliculus, and (b) the uvula of the urinary bladder, an eminence of its mucosa due to the middle lobe of the prostate. Reference: *Leçons sur les rétentions d'urine.* Paris, 1832.

Amyand, Claudius (1660–1740): French surgeon born in Mornac of a Huguenot family who emigrated to England and worked in London.

Hernia of Amyand: *Eponym* for an inguinal hernia in which the vermiform appendix of the caecum is involved. Reference: *Of an inguinal rupture, with a pin in the appendix caeci, incrusted with stone; and some observations on wounds in the gut.* Phil Trans Roy Soc London 1735; 39;329–355.

Amygdala of Eustachi: *Eponym* for the tubal tonsil, later also described by Gerlach; *see* Eustachi, Bartolomeo.

Anastomoses: Many anastomoses between cranial nerves were described by Bischoff (Bischoff EPE *Mikroskopische Analyse der Anastomosen der Kopfnerven.* Leutner, Munich 1865; English translation by Sachs E Jr. and Valtin EW *Microscopic Analysis of the Anastomoses between the Cranial Nerves.* University Press of New England, Hanover, New Hampshire, 1977) and by Hovelacque (Hovelacque A *Anatomie des nerfs crâniens et rachidiens et du système grand sympathique chez l'homme.* Doin, Paris, 1927). Sometimes, the term ansa or loop is used for anastomoses such as for the ansa of Galen, Haller and Hyrtl, but the proper name should be communicating branch. A few anastomoses between arteries have been described (Clado, Haller, Riolan and Villemin).

Anastomosis of Arnold: *Eponym* for a communicating branch between the facial nerve and the greater occipital nerve; *see* Arnold, Friedrich.

Anastomosis of Berettini: *Eponym* for a communicating branch between the median and ulnar nerves in the ulnar part of the palm; *see* Berettini, Pietro.

Anastomosis of Bichat: *Eponym* for a communicating branch between the glossopharyngeal and vagus nerves; *see* Bichat, Marie, François Xavier.

Anastomosis of Clado: *Eponym* for the anastomosis between the appendicular and ovarian arteries within the appendiculoovarian ligament; *see* Clado, Spiro.

Anastomosis of Cruveilhier: *Eponym* for a communicating branch between a branch of the medial antebrachial cutaneous nerve and a branch of the ulnar nerve; *see* Cruveilhier, Jean.

Anastomosis of Galen: *Synonym* for the ansa or nerve of Galen, a communicating branch between the superior laryngeal and the recurrent laryngeal nerves; *see* Galen.

Anastomosis of Haller: *Synonym* for the ansa of Haller: (a) a communicating branch between the facial nerve and the glossopharyngeal nerve, and (b) a communicating branch between the glossopharyngeal nerve and the auricular branch of the vagus nerve; *see* Haller, Albrecht von.

Anastomosis maxima or **magna of Haller:** *Eponym* for the anastomosis between the superior and inferior mesenteric arteries, attributed to Riolan, whom Haller gave the credit for this structure *see* Haller, Albrecht von.

Anastomosis of Henle: *Eponym* for a communicating branch between the trigeminal and facial nerves near the pons; *see* Henle, Jakob.

Anastomosis of His: *Eponym* for a communicating loop between the right and left hypoglossal nerves, also described by Hyrtl; *see* His, Wilhelm Sr.

Anastomoses of Hoyer: *Eponym* for arteriovenous anastomoses, also described by Sucquet; *see* Hoyer, Heinrich.

Anastomosis of Huguier: *Eponym* for an anastomosis around the isthmus of the uterus between both uterine arteries; *see* Huguier, Pierre Charles.

Anastomosis of Hyrtl, Intercostobrachial: *Eponym* for a communicating loop between the lateral cutaneous branch o the second or third intercostal nerve and the medial brachial cutaneous nerve; *see* Hyrtl, Joseph.

Anastomosis of Hyrtl, Umbilical: *Eponym* for an anastomosis between the umbilical arteries; *see* Hyrtl, Joseph.

Anastomosis or **Loop of Hyrtl:** *Eponym* for a communicating loop between the right and left hypoglossal nerves, also described by His Sr; *see* Hyrtl, Joseph.

Anastomosis of Jacobson: *Eponym* for a communicatiing branch between the inferior ganglion of the glossopharyngeal nerve via the tympanic nerve and plexus with the lesser petrosal nerve; *see* Jacobson, Ludwig Levin.

Anastomosis of Kaplan: *Eponym* for an inconstant communicating branch between the superficial and dorsal branches of the ulnar nerve, dorsal to the ulnar canal; *see* Kaplan, Emanuel.

Anastomosis of Marinacci: *Eponym* for a communicating branch of the ulnar nerve to the median nerve in the forearm; *see* Marinacci, Albert A.

Anastomosis of Martin-Gruber: *Eponym* for a communicating branch of the median nerve to the ulnar nerve in the forearm; *see* Martin, Roland, and Gruber, Wenzel.

Anastomosis of Riché-Cannieu: *Eponym* for a deep branch of the ulnar nerve and the recurrent branch of the thenar region; *see* Riché, P, and Cannieu, A.

Anastomosis of Riolan: *Eponym* for a rare anastomosis between the superior and inferior mesenteric arteries, the name proposed by Haller; *see* Riolan, Jean.

Anastomoses of Sucquet: *Eponym* for arteriovenous anastomoses, *see* Sucquet, JP.

Anastomosis of Tandler: *Eponym* for an hypothetical ventral longitudinal anastomosis of the embryonic blood vessels for the intestines; *see* Tandler, Julius.

Anastomosis of Villemin: *Eponym* for an inconstant anastomosis between the superior and inferior mesenteric arteries, also known as intermesenteric artery of Villemin, also described by Moskowitz and Gonzales; *see* Villemin, Fernand.

Anastomosis of Wrisberg: *Eponym* for a communicating branch between the medial brachial cutaneous nerve and the intercostobrachial nerve; *see* Wrisberg, Heinrich August.

Andernach, Johann Winther von (1487–1574): German physician born in Andernach, and after studying in Utrecht, working in Leuven, Strassburg and Paris. He is known for his studies on wormian bones. Reference: *Institutionem anatomicarum.* Paris, 1536.

Ossicles of Andernach: *Eponym* for sutural bones, later also described by Worm, his son-in-law.

Andersch, Carol Samuel (1732–1777): German anatomist in Göttingen, a student of Albrecht von Haller who described various parts of the glossopharyngeal nerve, published posthumously in 1797 by his nephew (*Tractatio anatomico-physiologica de nervis humani corporis aliquibus, quam edidit Ernst Ph. Andersch.* Fasch, Königsberg, 1797; Fig. A9a).

Ganglion of Andersch: *Eponym* for the inferior ganglion of the glossopharyngeal nerve (Fig. A9b).

Nerve of Andersch: *Eponym* for the tympanic branch of the glossopharyngeal nerve, mostly described as nerve of Jacobson (Fig. A9b).

Plexus of Andersch: *Eponym* for the nervous tonsillar plexus.

Angles: Angles were defined particularly for anthropology starting with Daubenton and Camper. In particular, the facial angle of Camper received much attention.

Angle of Ackermann: *Eponym* for the angle formed by the base of the skull and the horizontal plane; *see* Ackermann, Jakob Fidelis.

Angles of Broca: Broca defined various angles for anthropological research, basilar, facial, frontal, occipital parietal, suboccipital and supraoccipital; *see* Broca, Pierre Paul.

Angle of Camper, Facial: *Eponym* for the angle formed by the profile of the face and the horizontal plane determined by two lines, the first passing though the alveolar margin of the maxilla and the anterior surface of the frontal bone, and the second from the external acoustic meatus to the anterior nasal spine; *see* **Fig. C5** and Camper, Petrus.

Angle of Daubenton: *Eponym* for his occipital angle; *see* Daubenton, Louis Jean Marie.

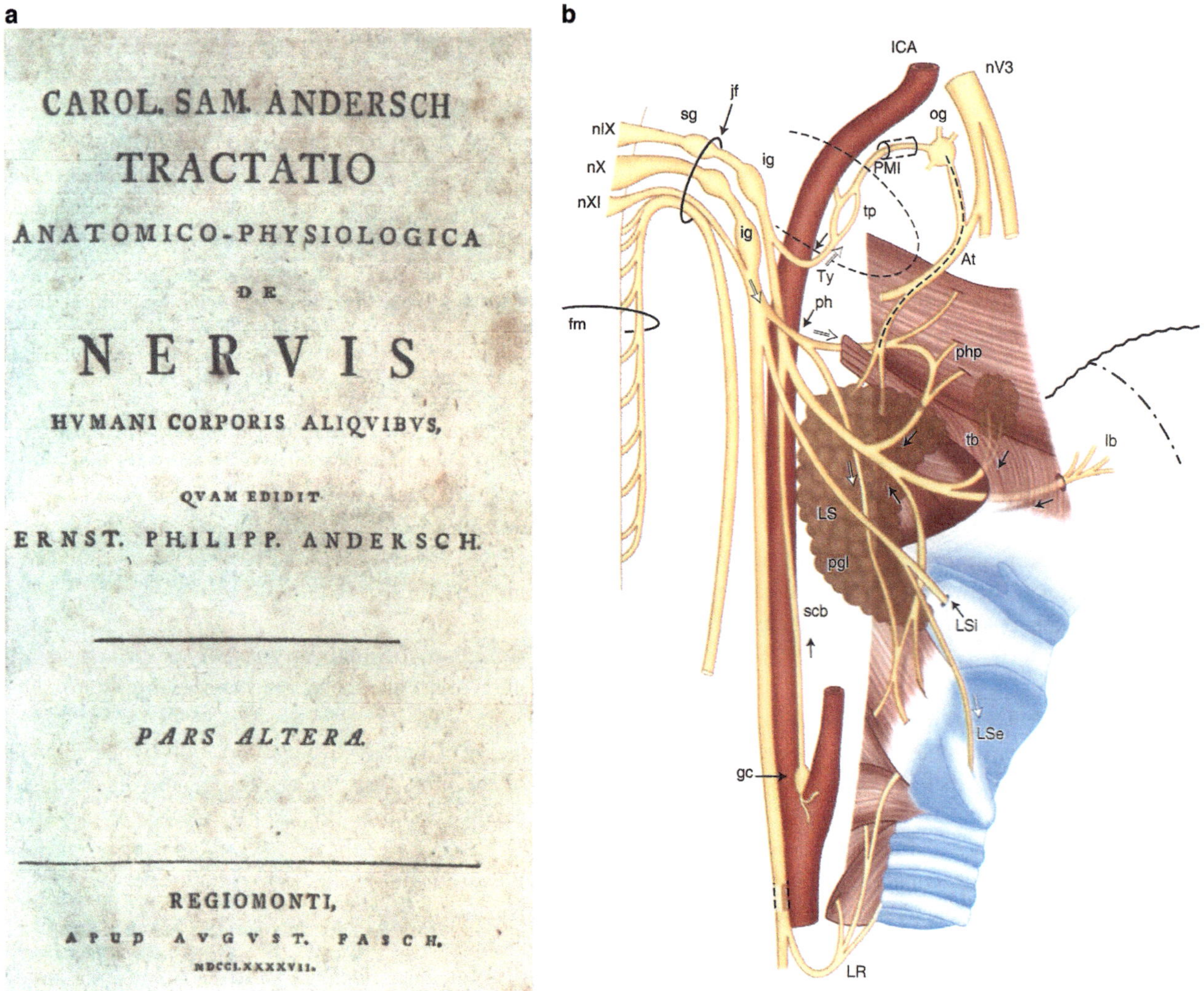
CAROL. SAM. ANDERSCH
TRACTATIO
ANATOMICO-PHYSIOLOGICA
DE
NERVIS
HVMANI CORPORIS ALIQVIBVS,
QVAM EDIDIT
ERNST. PHILIPP. ANDERSCH.

PARS ALTERA.

REGIOMONTI,
APVD AVGVST. FASCH.
MDCCLXXXXVII.

Fig. A9 (**a**) Andersch, Title page *De nervis humani corporis*. Fasch, Königsberg, 1797; (**b**) ganglion of Andersch shown in a figure on the glossopharyngeal, vagus and accessory nerves. *At* auriculotemporal nerve, *fm* foramen magnum, *gc* glomus caroticum, *ICA* internal carotid artery, *if* inferior ganglion, *jf* jugular foramen, *lb* lingual branch, *LR* recurrent laryngeal nerve, *LS*, *LSe*, *LSi* superior laryngeal nerve with external and internal branches, *nV3* mandibular nerve, *nIX*, *nX*, *nXI* glossopharyngeal, vagus and accessory nerves, *og* otic ganglion, *pgl* parotid gland, *ph* pharyngeal branches, *php* pharyngeal plexus, *PMI* minor petrous nerve, *scb* carotid sinus branch, *sg* superior ganglion, *tb* tonsillar branch, *tp* tympanic plexus, *Ty* tympanic nerve. (From ten Donkelaar HJ, Kachlík D, Tubbs RS *An Illustrated Terminologia Neuroanatomica*. Springer, Cham, 2018)

Angle of His: *Eponym* for the cardiac notch of the stomach; *see* His, Wilhelm Jr.

Angle of Louis (Ludovicus): *Eponym* for the angle between the manubrium and the body of the sternum; *see* Louis, Pierre Charles Alexandre.

Angle of Pirogoff: *Eponym* for the venous angle between the internal jugular and subclavian veins; see Pirogoff, Nicolai Ivanovitch.

Angle of Quatrefages: *Eponym* for the parietal angle used in craniometry; *see* Quatrefages de Bréau, Jean Louis Armand.

Angle of Ranke: *Eponym* for a cephalic angle used in craniometry; *see* Ranke, Johannes.

Angle of Theile: *Eponym* for the acute angle at the confluence of the superior mesenteric and splenic veins, which represents the origin of the portal vein; *see* Theile, Friedrich Wilhelm.

Angle of Topinard, Facial: *Modification* of the facial angle of Camper, formed between a line joining the glabella and the anterior nasal spine, and one joining the anterior nasal spine and the external acoustic meatus.

Angle of Welcker, Sphenoidal: *Eponym* for the angle of the basicranial axis; *see* Welcker Hermann.

Ansae: The term **ansa** is used for a loop of a nerve but sometimes also as a *synonym* for an anastomosis.

Ansa of Galen: *Eponym* for a communicating branch between the superior laryngeal nerve and the recurrent laryngeal nerve; *see* Galen.

Ansa of Haller: *Eponym* for (a) a communicating branch between the facial and the glossopharyngeal nerves, and (b) a communicating branch between the glossopharyngeal nerve and the auricular branch of the vagus nerve, also described by Sappey and known as ansa of Haller-Sappey; *see* Haller, Albrecht von.

Ansa of Hyrtl: *Synonym* for the anastomosis of Hyrtl; *see* Hyrtl, Joseph.

Ansa of Laignel-Lavastine: *Eponym* for a communicating branch between a branch of the right vagus nerve and the left coeliac ganglion; see Laignel-Lavastine, M.

Ansa of Rusconi: *Eponym* for the blastopore; *see* Rusconi, Mauro.

Ansa of Sappey: *Eponym* for a communicating branch between the glossopharyngeal nerve and the auricular branch of the vagus nerve, earlier described by Haller and also known as ansa of Haller-Sappey; *see* Sappey, Marie Philibert Constant.

Ansa of Vieussens: *Eponym* for the ansa subclavia between the middle cervical ganglion and the cervicothoracic or stellate ganglion; *see* Vieussens, Raymond de.

Ansa of Wrisberg: *Eponym* for a loop formed by a branch of the right vagus nerve, the coeliac ganglion and the greater thoracic splanchnic nerve; *see* Wrisberg, Heinrich August.

Anson, Barry Joseph (1894–1974): American anatomist known for his accurate description of the falx inguinalis of Henle and other parts of the musculo-aponeurotic layer of the abdominal wall. Reference: *The anatomy of the inguinal and hypogastric regions of the abdominal wall.* Anat Rec 1938; 70:211–225 (with CB McVay).

Antrum:

Antrum of Highmore: *Eponym* for the maxillary sinus that was already figured by Leonardo da Vinci; *see* Highmore, Nathaniel.

Antrum of Valsalva: *Eponym* for the mastoid antrum of the middle ear cavity; *see* Valsalva, Antonio Maria.

Antrum of Willis: *Eponym* for the pyloric antrum of the stomach; see Willis, Thomas.

Anular tendon of Gerlach: *Eponym* for the fibrocartilaginous ring of the tympanic membrane; *see* Gerlach, Joseph.

Anulus:

Anulus of Haller: *Synonym* for the insula of Haller, a doubling of the thoracic duct for part of its course through the thorax; *see* Haller, Albrecht von.

Anulus of Vieussens: *Eponym* for the limbus of the fossa ovalis in the right atrium of the heart; *see* Vieussens, Raymond de.

Anulus of Waldeyer: *Eponym* for the lymphatic ring formed by the lingual, palatine, tubal and pharyngeal tonsils, also described by Bickel as ring of Bickel; *see* Waldeyer, Heinrich Wilhelm Gottfried.

Anulus of Zinn: *Eponym* for the common tendinous ring of the external eye muscles; *see* Zinn, Johann Gottfried.

Anus (in the past used for opening):

Anus of Bartholin: *Eponym* for the opening of the mesencephalic or cerebral aqueduct into the third ventricle, also described by Vieussens; *see* Bartholin, Thomas, The Elder.

Anus of Eustachi: *Eponym* for the opening of the auditory tube into the pharynx; *see* Eustachi, Bartolomeo.

Anus of Vieussens: *Eponym* for the opening of the mesencephalic or cerebral aqueduct into the third ventricle, also described by Bartholin (Thomas, the Elder); *see* Vieussens, Raymond de.

Appendages of Morgagni: *Eponym* for (a) the appendix testis, a small sessile body on the superior pole of the testis, a remnant of the paramesonephric duct; and (b) the vesicular appendage of the epoophoron, a small peduncular vesicle attached to the fimbria of the uterine tube; *see* Morgagni, Giovanni Battista.

Appendix of Morgagni: *Eponym* for the pyramidal lobe of the thyroid gland, also described by Lallouette; *see* Morgagni, Giovanni Battista.

a

b

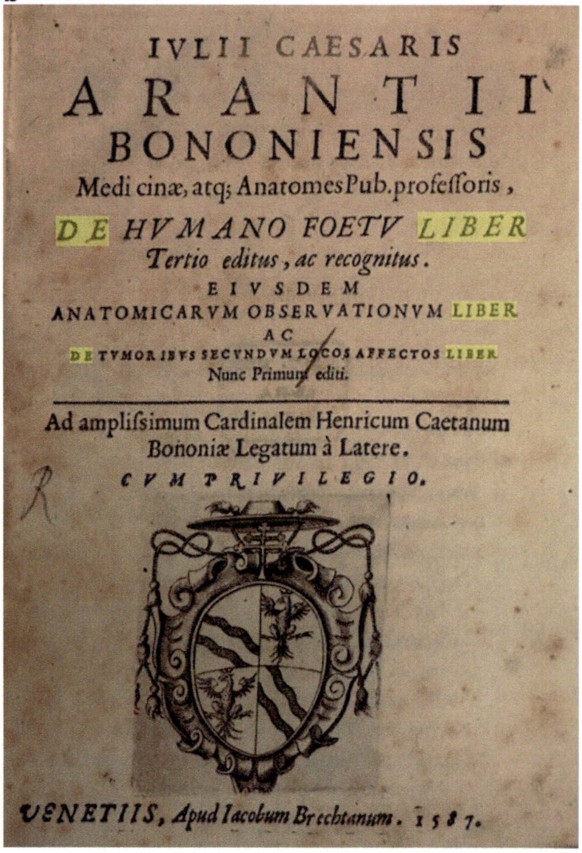
IVLII CAESARIS
ARANTII
BONONIENSIS
Medi cinæ, atq; Anatomes Pub. professoris,
DE HVMANO FOETV LIBER
Tertio editus, ac recognitus.
EIVSDEM
ANATOMICARVM OBSERVATIONVM LIBER
AC
DE TVMORIBVS SECVNDVM LOCOS AFFECTOS LIBER
Nunc Primum editi.

Ad amplissimum Cardinalem Henricum Caetanum
Bononiæ Legatum à Latere.
CVM PRIVILEGIO.

VENETIIS, Apud Iacobum Brechtanum. 1587.

Fig. A10 (**a**) Engraving of Aranzio. (Wikipedia; in public domain); (**b**) Title page *De humano foetu opisculum.* Brechtanus, Venice, 1587

Aqueducts:

Aqueduct of Falloppio: *Eponym* for the canal of the facial nerve; *see* Falloppio, Gabriele.

Aqueduct of Sylvius: *Eponym* for the mesencephalic or cerebral aqueduct; *see* Sylvius, François de le Boë.

Aranzio (Arantius), Giulio Caesare (1530–1589): Italian anatomist and surgeon born in Bologna and holding the first Chair of anatomy at the University of Bologna (1570). He was a pupil of Vesalius and is known for describing various structures (Fig. A10a). References: *De humano foetu opusculum.* Bologna, 1564, and Brechtanus, Venice, 1587 (Fig. A10b); *Observationes Anatomicae.* Basel, 1579. In 1587, Aranzio gave the first description of the hippocampus and compared the protrusion on the floor of the temporal horn of the lateral ventricle to a hippocampus or sea horse. In 1729, JG Duvernoy first illustrated the hippocampus (Duvernoy HM 1998 *The Human Hippocampus.* Springer, Berlin-Heidelberg-New York; *see* Duvernoy, Johann Georg and **Fig. D26**).

Bodies or **Nodules of Aranzio:** *Eponym* for the nodules of the semilunar valvules of the aortic valve.

Ligament or **Band of Aranzio:** *Eponym* for the ligamentum venosum of the liver (*see* **Fig. C29**).

Canal or **Duct of Aranzio:** *Eponym* for the embryonic ductus venosus (*see* **Fig. C29**).

Ventricle or **Duct of Aranzio:** *Eponym* for the posterior part of the fourth ventricle leading to the central canal of the spinal cord.

Arcades:

Arcade of Braune: *Eponym* for a superficial vein at the root of the toes, opening into the superficial plantar venous plexus by intercapitular veins; *see* Braune, Christian Wilhelm.

Arcade of Bühler: *Eponym* for a rare interconnection between the coeliac trunk and the superior mesenteric artery; *see* Chap. 22.

Arcade of Mondor: *Eponym* for an arterial segment connecting the descending branch of the left colic artery and the ascending branch of the first sigmoid artery; *see* Mondor, Henri Jean Justin.

Arcade of Riolan: *Eponym* for rare arterial anastomoses of the vessels of the large intestine; *see* Riolan, Jean.

Arcade of Treitz: *Eponym* for a vascular arcade formed by the inferior mesenteric artery and the left superior colic artery in the left mesocolon; *see* Treitz, Václav.

Arcade of Villemin: *Eponym* for the intermesenteric artery, an inconstant connection between both mesenteric arteries; *see* Villemin, Fernand.

Archambault, LaSalle (1879–1940): American neurologist.

Loop of Archambault: *Eponym* for the anterior or temporal loop of the optic radiation, mostly known as the Meyer loop, but also described by Cushing and Flechsig. Reference: *Le faisceau longitudinal inférieur et le faisceau optique central.* Rev Neurol 1905; 13:1053–1066.

Arches:

Arch of Adams: *Eponym* for a bony arch on the medial side of the neck of the femur, containing thick cortical bone; *see* Adams, Robert.

Arch of Barkow: *Eponym* for the anastomosis between the right, middle and left omental or epiploic branches of the gastroomental or gastroepiploic arteries along the free margin of the greater omentum, also described by Haller; *see* Barkow, Hans Karl Leopold.

Arch of Haller: *Eponym* for the medial and lateral arcuate ligaments; *see* Haller, Albrecht von.

Arch of Haller-Barkow: *Eponym* for the anastomosis between the right, middle and left omental or epiploic branches of the gastroomental or gastroepiploic arteries along the free margin of the greater omentum; *see* Barkow, Hans Karl Leopold, and Haller, Albrecht von.

Arch of Hyrtl: *Eponym* for the anastomosis between the right and left gastroomental or gastroepiploic arteries along the greater curvature of the stomach; *see* Hyrtl, Joseph.

Arch of Kirk: *Eponym* for an anastomosis between the dorsal pancreatic artery and the superior pancreaticoduodenal artery; *see* Chap. 22.

Arch of Rio Branco: *Eponym* for an anastomosis between the superior and inferior pancreaticoduodenal arteries; *see* Chap. 22.

Arch of Riolan: *Synonym* for the arcade of Riolan.

Areas: The term **area** is mostly used to describe a certain area of the cerebral cortex.

Area of Arnold: *Eponym* for the vagal trigone or ala cinerea of the rhomboid fossa of the brain stem; *see* Arnold, Friedrich.

Area of Broca: *Eponym* for (a) the triangular and opercular parts of the inferior frontal gyrus, and (b) the paraolfactory area; *see* Broca, Pierre Paul.

Areas of Brodmann: Based on the analysis of Nissl-stained preparations, Brodmann (1909) subdivided the human cerebral cortex into 52 areas (BA1-BA52); *see* Brodmann, Korbinian.

Area of Campbell: *Eponym* for the precentral area of the cerebral cortex; *see* Campbell, Alfred Walter.

Area of Pitres: *Eponym* for the prefrontal area of the cerebral cortex; *see* Pitres Jean-Albert.

Area of Rexed: *Eponym* for the central part of the spinal grey matter or spinal area X surrounding the central canal; *see* Rexed, Bror.

Area of Rolando: *Eponym* for the precentral and postcentral gyri of the cerebral cortex adjacent to the central sulcus; *see* Rolando, Luigi.

Areas of von Economo: *Eponym* for another cytoarchitectonic subdivision of the cerebral cortex, numbering the areas after the lobes of the cerebral cortex; *see* Economo, Constantin von.

Area of Wernicke: *Eponym* for the sensory speech area on the posterior third of the superior temporal gyrus; *see* Wernicke, Carl.

Arlt, Carl Ferdinand Ritter von (1812–1887): Bohemian ophthalmologist (Fig. A11) born in Teplice and working in Prague (1849) and later in Vienna (1856–1883).

Line of Artl: *Eponym* for a vertical line corresponding to the projection of the nasolacrimal duct and extending from the medial palpebral ligament to the beginning of the nasolabial fold.

Sinus of Arlt: *Eponym* for a dilatation of the lacrimal sac. Reference: *Ueber den Thränenschlauch.* Von Graefes Arch Ophthalmol 1854–1855, 135–160.

Fig. A11 Portrait of Arlt. (From Pagel 1901)

Fig. A12 Photo of Arnold. (From the University Library of Heidelberg; in public domain)

Arnold, Friedrich (1803–1890): German anatomist born near Landau and studying in Heidelberg (Fig. A12). Subsequently, he worked in Zürich, Freiburg, Tübingen and finally in Heidelberg. He described many structures in the peripheral and central nervous system (Fig. A13). References: *Der Kopftheil des vegetativen Nervensystems beim Menschen in anatomischer und physiologischer Hinsicht.* Heidelberg and Leipzig, 1831); *Tabulae anatomicae, quas ad naturam accurate descriptas in lucem edidit.* Höhr, Zürich, 1838–1843; *Handbuch der Anatomie des Menschen, Bd.I-III.* Herder, Freiburg, 1843–1851; *Icones Nervorum Capitis*, Heidelberg, 1860.

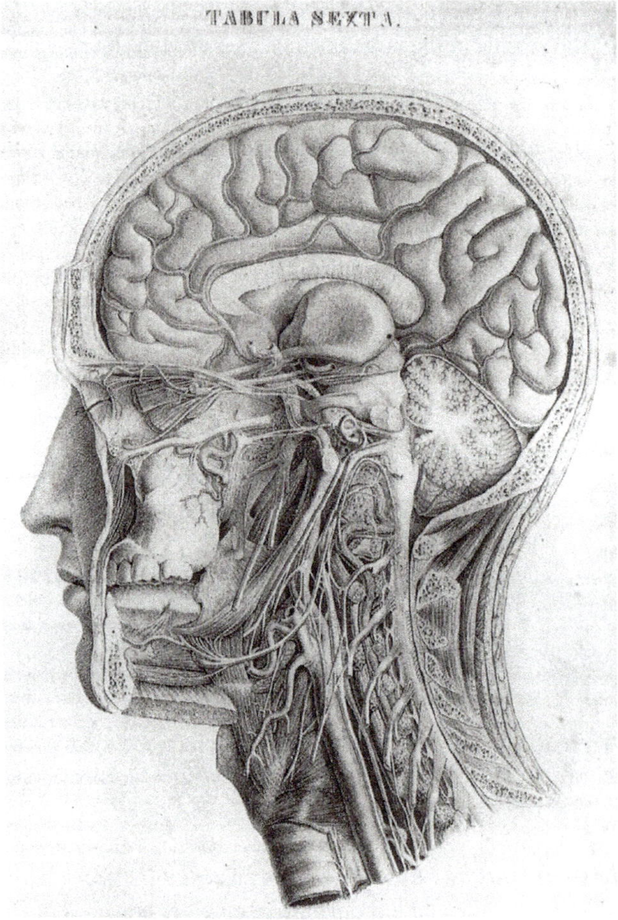

Fig. A13 Dissection of the cranial nerves by Arnold. (Tabula Sexta, from *Icones Nervorum Capitis.* Heidelberg, 1860)

Anastomosis of Arnold: *Eponym* for a communicating branch between the facial nerve and the greater occipital nerve; *see* Arnold, Friedrich.

Area or **Trigone of Arnold:** *Eponym* for the vagal trigone or ala cinerea of the rhomboid fossa of the brain stem.

Arcuate fibres of Arnold: *Eponym* for the short association fibres of the cerebral cortex (Fig. A14), currently known as U-fibres (of Meynert).

Canal of Arnold: *Eponym* for (a) the hiatus of the lesser petrosal nerve, b) the mastoid canaliculus, containing the auricular branch of the vagus nerve; and (c) an inconstant foramen petrosum or innominate canaliculus, containing the lesser petrosal nerve.

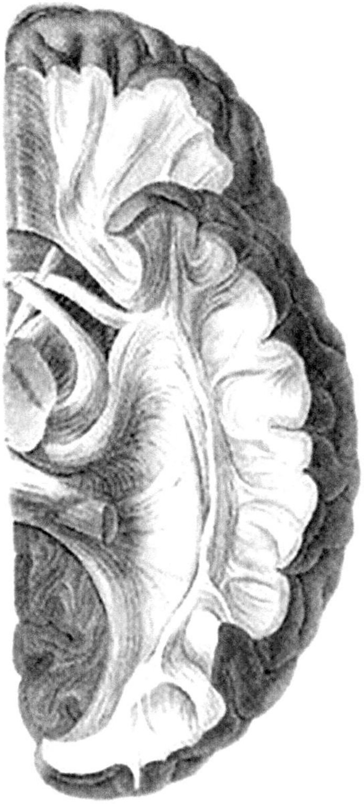

Fig. A14 Horizontal section of the brain with the arcuate fibres of Arnold. (From *Tabulae Anatomicae*, Hohr, Zürich, 1838; courtesy Jeremy Schmahmann, Boston). This gross dissection of the white matter revealed the arcuate fibres, the short association fibres of the cerebral cortex that were named after him, but currently are known as U-fibres of Meynert

Ganglion of Arnold: *Eponym* for (a) the otic ganglion, (b) the thoracic splanchnic ganglia and (c) the glomus caroticum.

Intrapleural lymph nodes of Arnold: *Eponym* for small lymph nodes along the suprapleural lymph vessels. Reference: *Ueber das Vorkommen lymphatischen Gewebes in den Lungen.* Arch Pathol Anat Physiol Klin Med 1880, 80:315–336.

Ligament of Arnold: *Eponym* for the superior ligament of the incus.

Nerve of Arnold: *Eponym* for (a) the auricular branch of the vagus nerve (also known as Alderman's nerve), (b) the tentorial branch of the ophthalmic nerve (the recurrent nerve of Arnold) and (c) the greater occipital nerve. Many other nerves are sometimes also named after Arnold such as the supratrochlear and infratrochlear nerves.

Fig. A15 Photo of Aschoff. (From Wikimedia Commons; in public domain)

Plexus of Arnold: *Eponym* for the cavernous plexus on the cavernous segment (C4) of the internal carotid artery within the cavernous sinus.

Reticular substance of Arnold: *Eponym* for the white reticular substance of the entorhinal cortex.

Tract of Arnold: *Eponym* for the frontal corticopontine tract, named so by Theodor Meynert.

Arteries: *See* **The Classical Anatomical Eponyms.**

Ascher, Karl (1887–1971): American ophthalmologist.

Aqueous veins of Ascher: *Eponym* for the aqueous veins, the efferent communications between the scleral venous sinus and the episcleral veins. Reference: *Aqueous veins.* Amer J Ophth 1942; 25:31–38.

Aschoff, Karl Albert Ludwig (1866–1942): German pathologist born in Berlin (Fig. A15) working in Göttingen, London and Liverpool, Marburg and Freiburg known for the atrioventricular node.

Node of Aschoff-Tawara: *Eponym* for the atrioventricular node of the conducting system of the heart described together with Suano Tawara (*see* **Fig. B1**). References: *Die Herzstörungen in ihren Beziehungen*

Fig. A16 Engraving of Aselli. (From Wikimedia Commons; in public domain; courtesy Marina Quartu, Cagliari)

Fig. A17 Photo of Auerbach. (From Wikimedia Commons; in public domain)

zu den spezifischen Muskelsystemen des Herzens. Verh Dtsch Path Ges 1910; 14:3–35; *Ueber den feineren Aufbau des Reizleitungs-Systems des Herzens.* Verh Dtsch Pharm Ges Arch Exp Path 1930; 157:94–95.

Sinus of Aschoff: *Eponym* for mucous crypts in the wall of the gallbladder, also described by Rokitansky, probably pathological; generally known as sinus of Aschoff-Rokitansky.

Aselli (Asellio), Gasparo (1581–1626): Italian anatomist and surgeon born in Cremona and working in Pavia, Lombardy (Fig. A16) who described 'milk vessels' in a dog as the first indication for the lymphatic system. Later, the thoracic duct was discovered by Pecquet in animal studies and by van Horne in humans. Reference: *De lactibus sive lacteis venis.* Milan, 1627.

Gland or **Pancreas of Aselli:** *Eponym* for a single large lymph node ventral to the abdominal aorta that receives all the lymph from the intestines in many small animals.

Vasa of Aselli: *Eponym* for the chylus vessels, he described as 'milk vessels' in his animal studies.

Assaki, George (1855–1899): Romanian physician born in Iaşi and working in Paris, Lille and Bucharest.

Tubercle of Assaki: *Eponym* for an inconstant intraglenoid tubercle located just below the centre of the glenoid fossa of the scapula. Reference: *Contribution à l' anatomie et à la physiologie de la cavité glénoide de l'omoplate.* CR Hebdom Séan Mém Soc Biol, Paris 1885; pp 371–373.

Assézat, Jules (1832–1876): French journalist and anthropologist.

Triangle of Assézat: *Eponym* for a triangle formed by a line connecting the nasion with the alveolar and nasal points, used to determine the degree of prognathism. Reference: *Recherches sur les proportions de la face.* Communic Assoc Franç Avanc Sciences 1874.

Auerbach, Leopold (1828–1897): German neuropathologist born and working in Breslau, currently Wróclaw, Poland (Fig. A17) who is known for his studies on the enteric nervous system (Fig. A18). Reference: *Ueber einen Plexus myentericus, einen bisher unbekannten ganglionervösen Apparat im Darm der Wirbelthiere.* Breslau, 1862.

Ganglia of Auerbach: *Eponym* for collections of parasympathetic cells in the myenteric plexus of the digestive tube.

Plexus of Auerbach: *Eponym* for the myenteric plexus, the plexus of autonomic nerve fibres between the longitudinal and circular muscular layers of the gut.

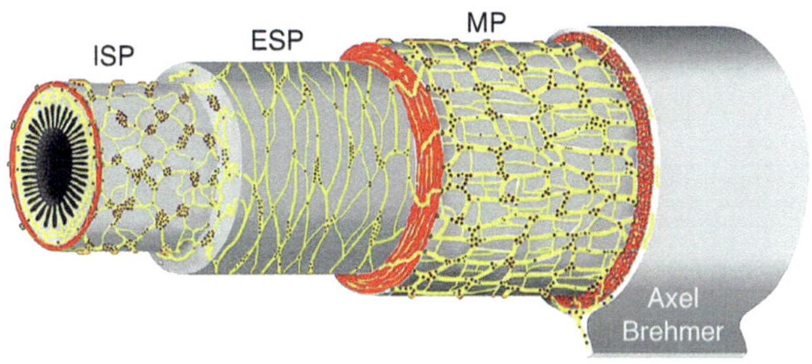

Fig. A18 The enteric plexuses, overview. *ESP* external submucosal plexus of Schabadasch, *ISP* internal submucosal plexus of Meissner, *MP* myenteric plexus of Auerbach. (From ten Donkelaar HJ, Kachlík D, Tubbs RS *An Illustrated Terminologia Neuroanatomica*. Springer, Cham, 2018; courtesy Axel Brehmer, Erlangen)

Axenfeld, Karl Theodor Paul Polycarpos (1867–1930): German ophthalmologist born in Smyrna and working in Marburg, Breslau and finally Rostock.

Ganglion of Axenfeld: *Eponym* for an accessory ciliary ganglion.

Loop of Axenfeld: *Eponym* for an inconstant extension of the posterior ciliary nerve making a loop through the sclera in the area of the anterior ciliary arteries or between the rectus muscle insertions posterior to the junction of the ciliary muscle at the pars plana of the ciliary body. Reference: *Ueber intrasklerale Nervenschleifen.* Ber Dtsch Ophthalmol Ges 1902; 30:134–137.

Axes:

Axis of Broca, Orbitofrontal: *Eponym* for the axis passing through the optic canal and the middle of the orbital base; *see* Broca, Pierre Paul.

Axis of Forel: The longitudinal axis of the cerebrum; for the cerebrum usually a series of coronal or frontal sections oriented perpendicular to the Forel axis (from the frontal to the occipital pole) is used; *see* Forel, Auguste Henri.

Axis of Meynert: Along the floor of the fourth ventricle; for the brain stem usually a series of sections perpendicular to the Meynert axis is used; *see* Meynert, Theodor.

Letter B

3

Babcock, William Wayne (1872–1963): American surgeon born in New York and working in Salt Lake City and Philadelphia.

Triangle of Babcock: *Eponym* for a relatively radiolucent triangular area seen in the subcapital region of the neck of the femur on anteroposterior radiographs of the hip, caused by loosely arranged trabeculae.

Bachmann, Jean George (1877–1959): American physiologist.

Bundle of Bachmann: *Eponym* for the interatrial bundle of the cardiac conducting system, a division of the anterior internodal tract, which continues into the left atrium (Fig. B1). It is believed to be a specialized pathway for interatrial conduction. Reference: *The interauricular time interval.* Am J Physiol 1916; 41:309–320.

Baer, Karl Ernst von (1792–1876): German embryologist born in Pilbe, currently Välka-Maarja (Estonia), then a Baltic province of Russia (Fig. B2). He specialized in embryology and zoology and worked in Dorpat, Königsberg and St. Petersburg (1834–1867). He may be viewed as the founder of modern embryology and discovered the mammalian ovum in 1827. He proposed that the ovum forms four layers of tissue from which the organs and other tissues arise. References: *De ovi mammalium atque hominis genesi.* Voss, Leipzig, 1827 (Fig. B3a, b); *Entwickelungsgeschichte der Thiere*, 3 Vols. Bornträger, Königsberg, 1820–1837.

Cavity of von Baer: *Eponym* for the cavity in the blastocoele.

Line of von Baer: *Eponym* for the Göttingen horizontal line through the superior border of the zygomatic arch.

Membrane of von Baer: *Eponym* for the chorionic membrane of the placenta.

Plane of von Baer: *Eponym* for a plane, which corresponds to the superior border of both zygomatic arches.

Baillarger, Jules Gabriel François (1809–1890): French neurologist born in Monbazon and working in Paris known for his studies on the microscopy of the isocortex (Fig. B4). Reference: *Recherches sur la structure de la couche corticale des circonvolutions du cerveau.* Mém Acad Méd Paris 1840; 8:149–183.

Inner stripe of Baillarger: *Eponym* for the white line or stripe of nerve fibres in layer V of the isocortex (Fig. B5a, b).

Outer stripe of Baillarger: *Eponym* for the white line or stripe of nerve fibres in layer IV of the isocortex (Fig. B5a, b).

Balbiani, Edouard Gérard (1823–1899): French embryologist born in Port-au-Prince, Haiti, educated in Frankfurt am Main and Paris, and working at the Collège de France in Paris.

Body of Balbiani: *Eponym* for a cluster of organelles (mitochondria, flattened cisterns of Golgi apparatus, endoplasmatic reticulum and lysosomes) in primary oocytes. Reference: *Sur la constitution du germe dans l'oeuf animal avant la fecondation.* Séances Acad Sci D 1864; 58:584.

Ball, Sir Charles Bent (1851–1916): Irish surgeon born and working in Dublin known for his studies on diseases of the rectum. Reference: *The Rectum and Anus, their diseases and treatment.* London, 1887.

H. J. ten Donkelaar et al., *An Illustrated Guide to Anatomical Eponyms*, https://doi.org/10.1007/978-3-031-91664-9_3

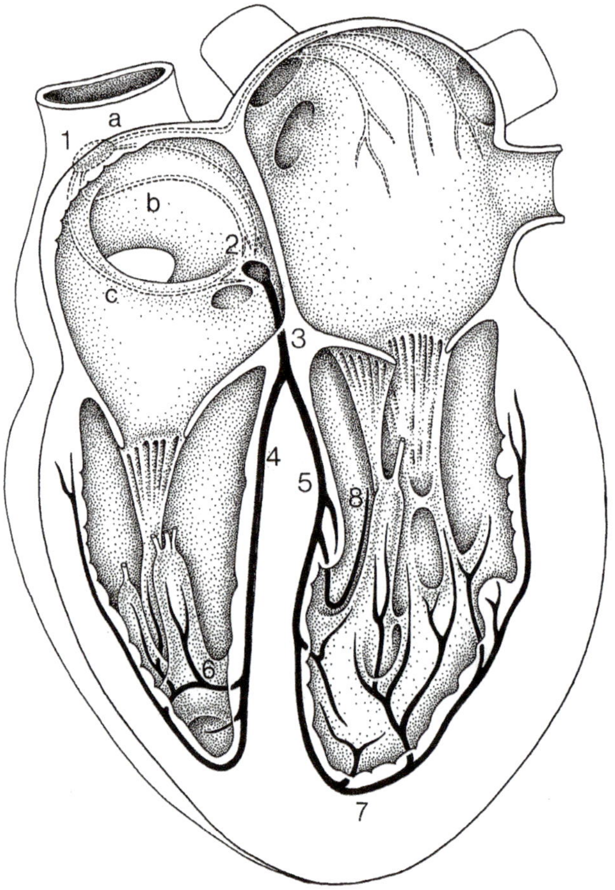

Fig. B1 The interatrial bundle. From the sinuatrial node, three internodal bundles pass to the atrioventricular node, the anterior of James (*a*) gives off the interatrial bundle of Bachmann. *1* sinuatrial node, *2* atrioventricular node, *3* bundle of His, *4* crus dextrum, *5* crus sinistrum, *6* moderator branch of crus dextrum, *7*, *8* anterior and posterior branches of crus sinistrum, *a-c* internodal bundles: *a* anterior of James, *b* middle of Wenckebach, *c* posterior of Thorel. (After Agur AM 1991 *Grant's Atlas of Anatomy*, 9th ed. Williams & Wilkins, Baltimore; and ten Donkelaar HJ, Kachlík D, Tubbs RS *An Illustrated Terminologia Neuroanatomica.* Spriner, Cham, 2018)

Valves of Ball: *Eponym* for the anal valves, mucous folds between the aboral ends of neighbouring anal columns, previously described by Morgagni.

Balli, Ruggero (1877–1958): Italian anatomist and radiologist born in Rovereto di Carpi (Modena) and working in Modena. He was a pioneer of radiobiology in Italy.

Sphincter of Balli: *Eponym* for the constriction between the descending colon and the sigmoid colon, which is highlighted on radiological examination of the large intestine. Reference: *The sphincters of the colon.* Radiology 1939; 33:372–378.

Bandl, Ludwig (1842–1892): Austrian obstetrician and gynaecologist born in Himberg and working in Vienna and Prague.

Fig. B2 Portrait of von Baer. (From Pagel 1901)

Contraction ring of Bandl: *Eponym* for a prominent muscular ring at the junction of the upper and lower segments of the uterus, resulting from obstructed labour. Reference: *Über Ruptur der Gebärmutter und ihre Mechanik.* Czermak, Vienna, 1876.

Bands: *See* **The Classical Anatomical Eponyms.**

Bardeleben, Karl von (1849–1919): German anatomist.

Articular lymph node of von Bardeleben: *Eponym* for a lymph node close to the popliteal artery, also described by Poirier and Cunéo. Reference: *Handbuch der Anatomie des Menschen*, 8 Bd. Fischer, Jena, 1896.

Bardinet, Bartholemy Alphonse (1809–1874): French physician in Limoges.

Ligament of Bardinet: *Eponym* for the olecranohumeral ligament, a part of the ulnar collateral ligament of the elbow joint. Reference: *Cas intéressant de fracture de l'olécrâne.* Gaz Méd Chir 1869; 6:110.

Barkow, Hans Karl Leopold (1798–1873): German anatomist born on the island of Rügen and working in Breslau who studied the ligaments of the hand and foot. Some are often known by his name. Reference: *Syndesmologie oder die Lehre von den Bändern, durch welche die Knochen des menschlichen Körpers zum Gerippe vereint werden.* Breslau, 1841.

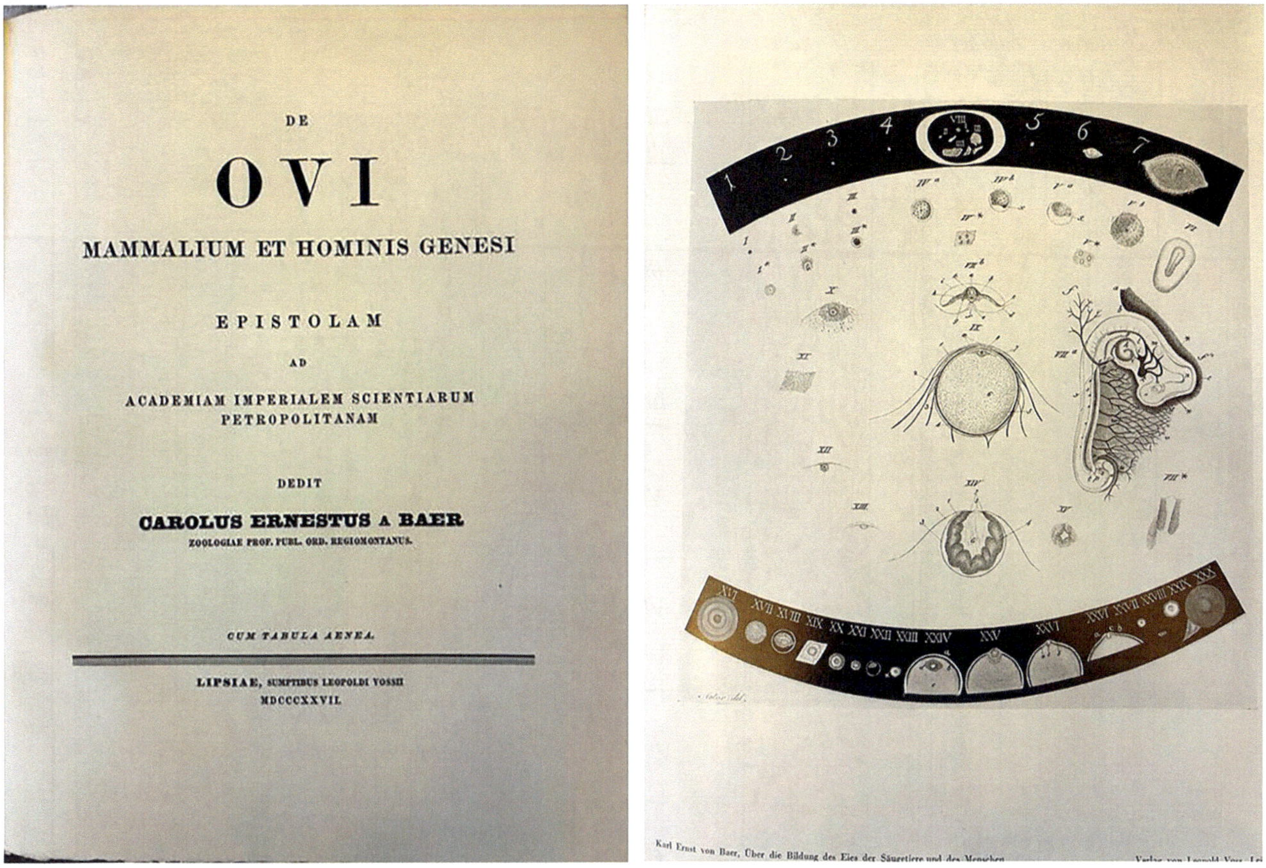
DE

OVI

MAMMALIUM ET HOMINIS GENESI

EPISTOLAM

AD

ACADEMIAM IMPERIALEM SCIENTIARUM
PETROPOLITANAM

DEDIT

CAROLUS ERNESTUS A BAER

ZOOLOGIAE PROF. PUBL. ORD. REGIOMONTANUS.

CUM TABULA AENEA.

LIPSIAE, SUMPTIBUS LEOPOLDI VOSSII
MDCCCXXVII.

Fig. B3 (**a**) von Baer's Title page *De Ovi mammalium et hominis genesi* (Voss, Leipzig, 1827); (**b**) Example of his illustrations, showing the development of the ovum. (Courtesy Ondřej Naňka, Prague)

Fig. B4 Portrait of Baillarger. (From the Wellcome Collection CCBY; in public domain)

Arch of Barkow: *Eponym* for the anastomosis between the right, middle and left omental or epiploic branches of the gastroomental or gastroepiploic arteries along the free margin of the greater omentum, also known as arch of Haller-Barkow.

Ligament of Barkow: *Eponym* for (a) the plantar calcaneonavicular ligament, (b) the pisohamate ligament; (c) the oblique tibiofibular ligament; and (d) a thin innominate horizontal ligament connecting both occipital condyles running in front of the alar ligaments and the dens of the axis..

Ligaments of Barkow: *Eponym* for the anterior and posterior portions of the fibrous capsule of the elbow joint.

Barnes, Stanley (late nineteenth, early twentieth century): English physician in London.

Bundle of Barnes: *Eponym* for corticospinal fibres leaving the pyramidal tract at the pontine level and

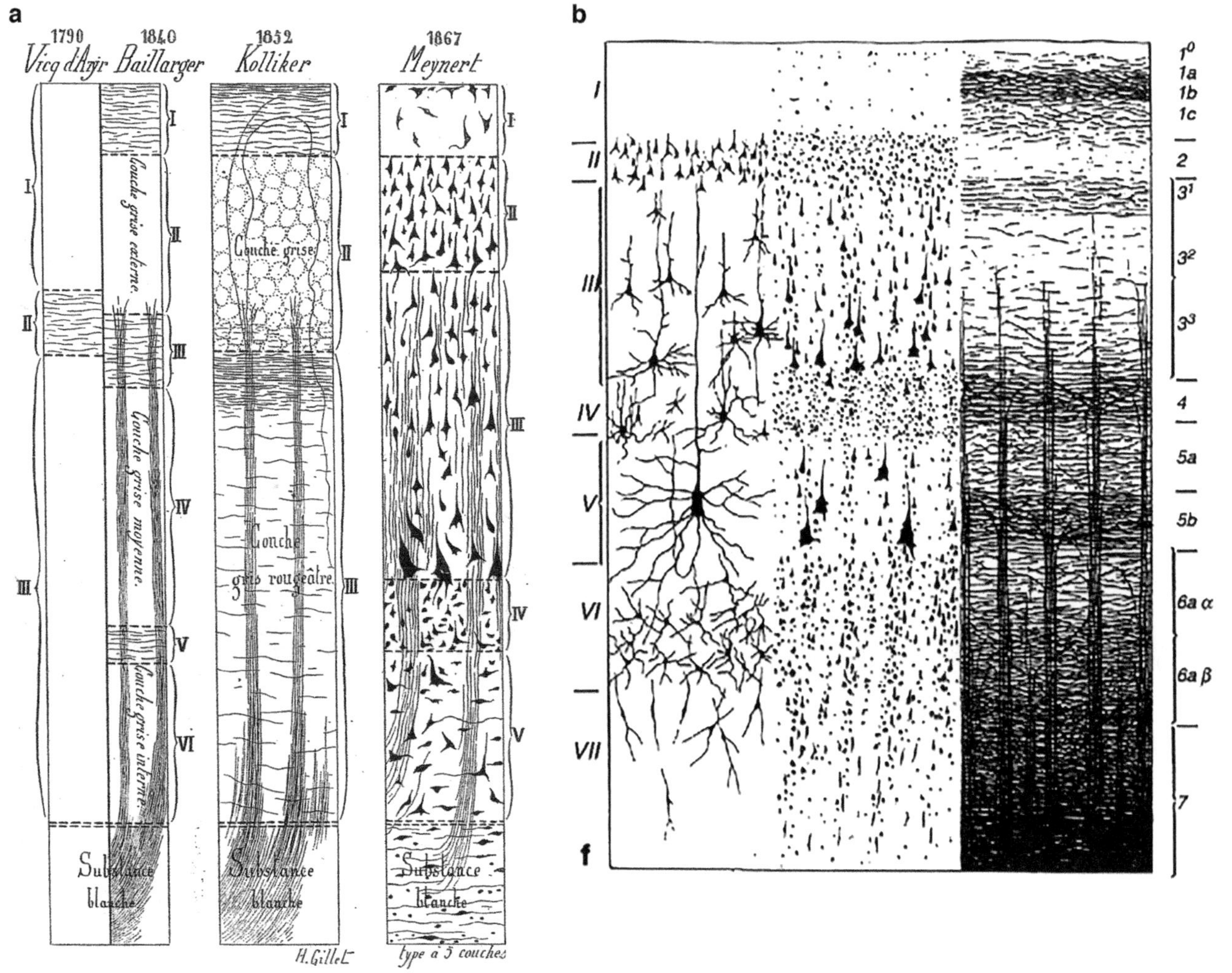

Fig. B5 (**a**) Stripes from Vicq d'Azyr till Meynert. (After Dejerine 1895); (**b**) stripes of Baillarger as present in isocortical layers: the left column shows Golgi-stained neurons, the middle column Nissl-stained cells and the right column myelinated fibres. (After Rose M *Cytoarchitektonik und Myeloarchitektonik der Grosshirnrinde*. In: Bumke O, Foerster O, eds, *Handbuch der Neurologie,* Vol I*: Allgemeine Neurologie I. Anatomie.* Springer, Heidelberg, 1935, pp 588–778; from ten Donkelaar HJ, Kachlík D, Tubbs RS *An Illustrated Terminologia Neuroanatomica.* Springer, Cham, 2018)

descending lateral to the inferior olive. Reference: *Degenerations in hemiplegia: with special reference to a ventrolateral pyramidal tract, the accessory fillet and Pick's bundle.* Brain 1901; 24:463–501.

Barr, Murray Llewellyn (1908–1995): Canadian pathologist born in Belmont, Ontario and working at the University of Western Ontario.

Body of Barr: *Eponym* for the chromatin body (sexual chromatin), which occurs in 90% of female and 10% of male interphase nuclei. Reference: *A morphological distinction between neurones of the male and female, and the behaviour of the nucleolar satellite during accelerated nucleoprotein synthesis.* Nature 1949; 163:676–677 (with EG Bertram).

Barrington, Frederick James Fitzmaurice (1884–1956): British surgeon and physiologist known for his experimental studies on micturition. Reference: *The effect of lesion of the hind- and midbrain on micturition in the cat.* Q J Exp Physiol 1925; 15:81–102.

Nucleus of Barrington: *Eponym* for the pontine micturition centre.

Bartholin (Bertelsen), Caspar, The Elder (1585–1629): Danish anatomist in Copenhagen who published a treatise

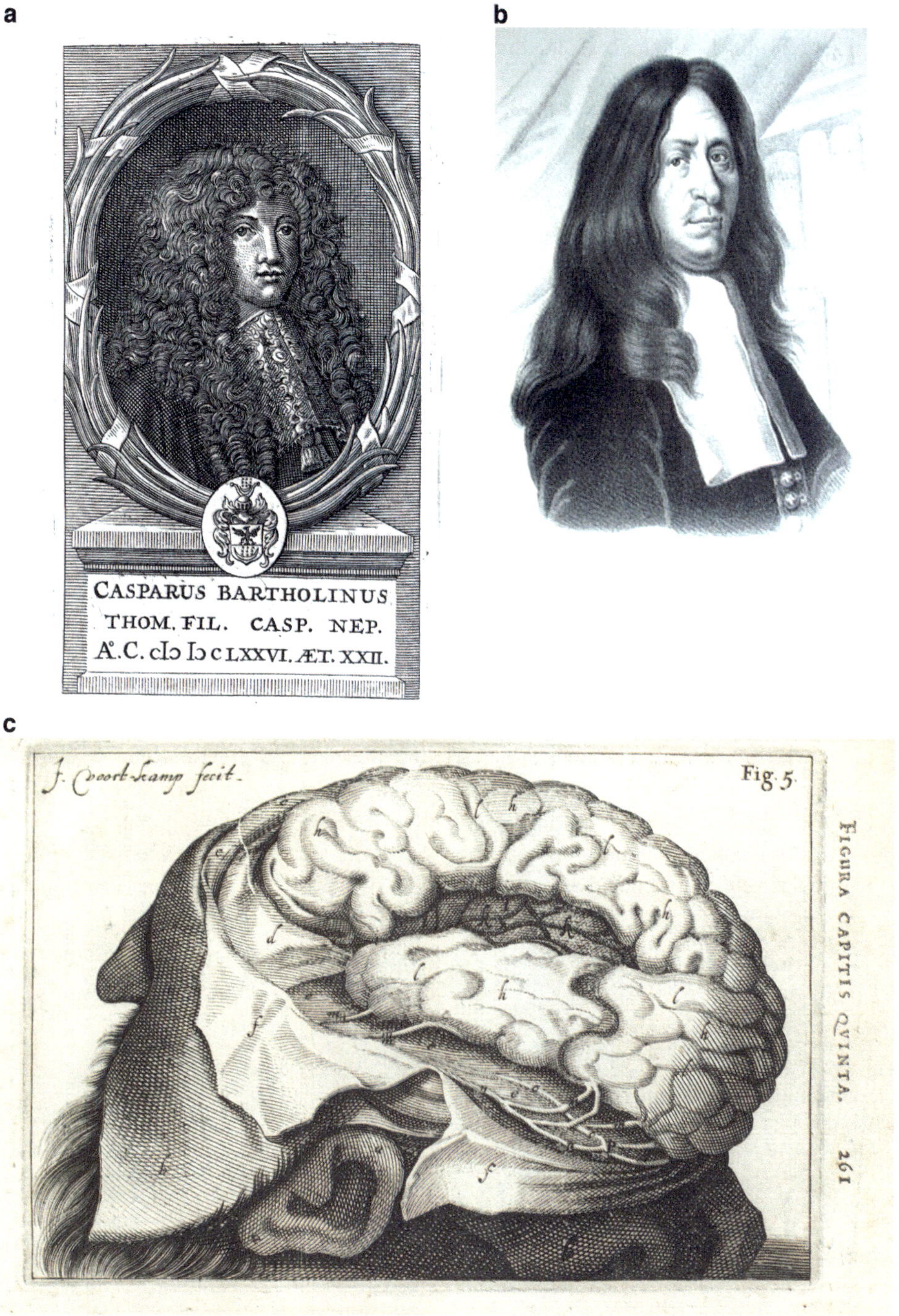

Fig. B6 (**a**) Portrait of Bartholin, Caspar, the Younger. (From the Wellcome Collection CCBY); (**b**) portrait of Bartholin, Thomas, the Elder. (From Wikimedia Commons; in public domain); (**c**) the first illustration of the lateral fissure of Sylvius. (From *Institutiones Anatomica.* Hackium, Leiden, 1641; in public domain)

on the brain (*Institutiones Anatomicae* 1611). His son Thomas revised the text and added illustrations including the one on the Sylvian fissure (Hackium, Leiden, 1641; Fig. B6c).

Bartholin (Bertelsen), Caspar, The Younger (1655–1738): Danish anatomist in Copenhagen (Fig. B6a), son of Thomas Bartholin (the Elder).

Duct of Bartholin: *Eponym* for the major sublingual duct. Reference: *De ductu salivati.* Utrecht, 1685.

Gland of Bartholin: *Eponym* for (a) the greater vestibular gland and (b) the sublingual salivatory gland, earlier described by Duverney (1676) and later by Tiedemann and Huguier. Reference: *De ovariis mulierum et generationis historia.* Zieger, Rome, 1677.

Bartholin (Bertelsen), Thomas, The Elder (1616–1680): Danish anatomist (Fig. B6b) who is said to have discovered the human thoracic duct and the lymphatic system (1652–1654). For the thoracic duct, however, credit should go to Pecquet (1651) and van Horne (1652).

Anus of Bartholin: *Eponym* for the opening of the mesencephalic or cerebral aqueduct into the third ventricle, also described by Vieussens; *synonym*: anus cerebri.

Basal nucleus of Meynert: *Eponym* for large, hyperchromatic neurons in the basal forebrain; *see* Meynert, Theodor.

Basal vein of Rosenthal: *Eponym* for the basal cerebral vein; *see* Rosenthal, Friedrich Christian.

Bassi, Glauco (1914–1987): Italian phlebologist born in Feitre and educated in Bologna and Padua, Veneto. During the Second World War, he was interned in a Nazi concentration camp. After the war, he went to Paris where he founded the first scientific society for the study of veins and their diseases. Reference: *The Varices of the Lower Limbs.* Edizione Minerva Medica, Torino, 1962 (French translation: *Les varices des membres inférieurs.* Doin, Paris, 1967).

Vein of Bassi, Perforating or Perforator: *Eponym* for the inferior paratendinous or para-Achillean perforator, located about 5 cm above the level of the calcaneal tuberosity next to the calcaneal tendon (Achilles tendon), connecting the small saphenous vein with the fibular veins, also known as perforator of Bassi.

Batson, Oscar Vivian (1894–1979): American otolaryngologist born in Sedalia, Pettis, Missouri and working in Philadelphia, Pennsylvania.

Plexuses of Batson: *Eponym* for the internal and external vertebral venous plexuses, a highly anastomotic network of valveless veins along the entire length of the vertebral column (Fig. B7). It connects pelvic veins and the azygos system to the external vertebral venous plexus, allowing the spread of tumour cells from the pelvic organs to the vertebral column. Reference: *The function of the vertebral veins and their role in the spread of metastasis.* Ann Surg 1940; 112:138–149.

Baudelocque, Jean-Louis (1746–1810): French obstetrician born in Heilly, Picardie, and working in Paris (Fig. B8a).

Diameter or **Line of Baudelocque:** *Eponym* for the external conjugate diameter of the pelvis, the distance between the superior margin of the symphysis pubis and the groove on the surface of the trunk below the spinous process of the fifth lumbar vertebra (Fig. B8b). Reference: *L'art des accouchements.* Méquignon, Paris, 1781.

Bauhin, Caspar (1560–1624): Swiss anatomist in Basel. Reference: *Institutiones anatomicae.* Basel, 1590.

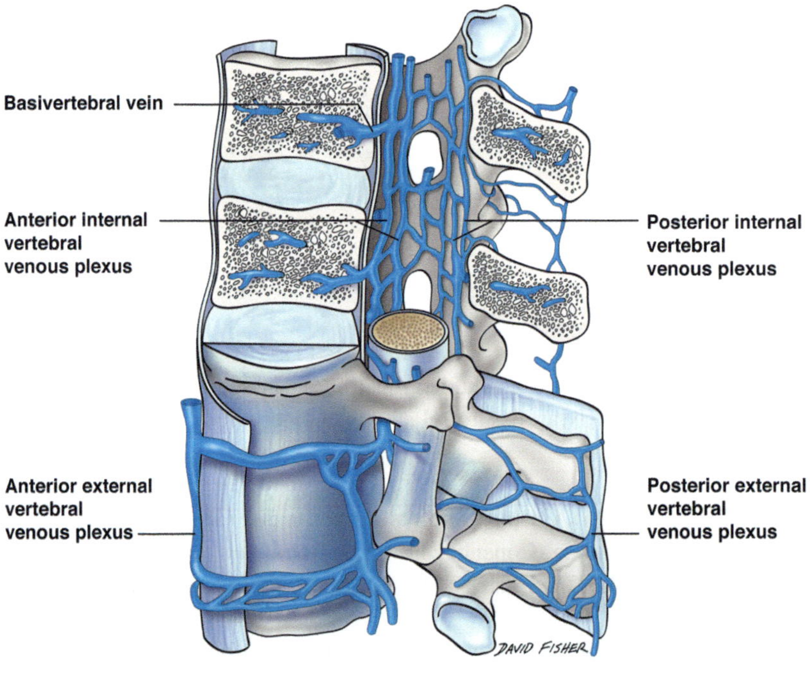

Fig. B7 Vertebral venous plexuses of Batson. (From ten Donkelaar HJ, Kachlík D, Tubbs RS *An Illustrated Terminologia Neuroanatomica.* Springer, Cham, 2018)

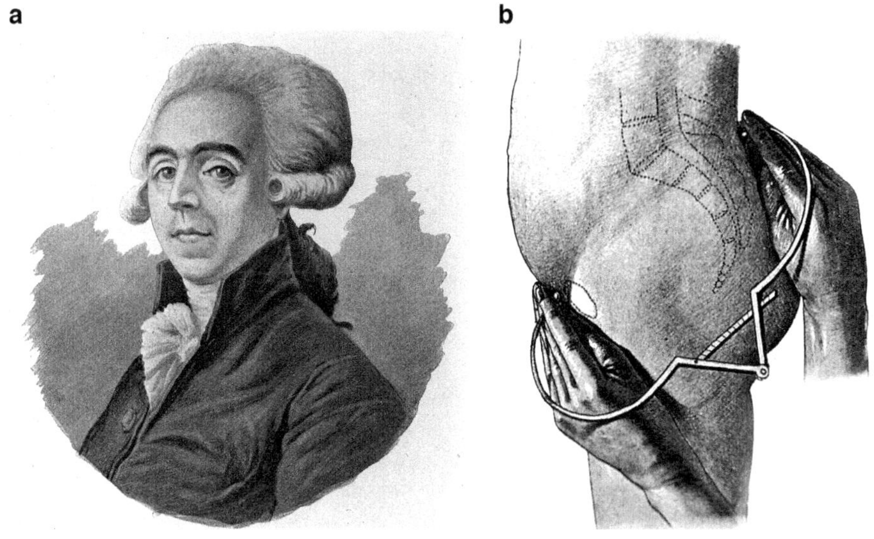

Fig. B8 (**a**) Portrait of Baudelocque. (From Pagel 1901); (**b**) external conjugate of pelvis. (From the Wellcome Collection CCBY; in public domain)

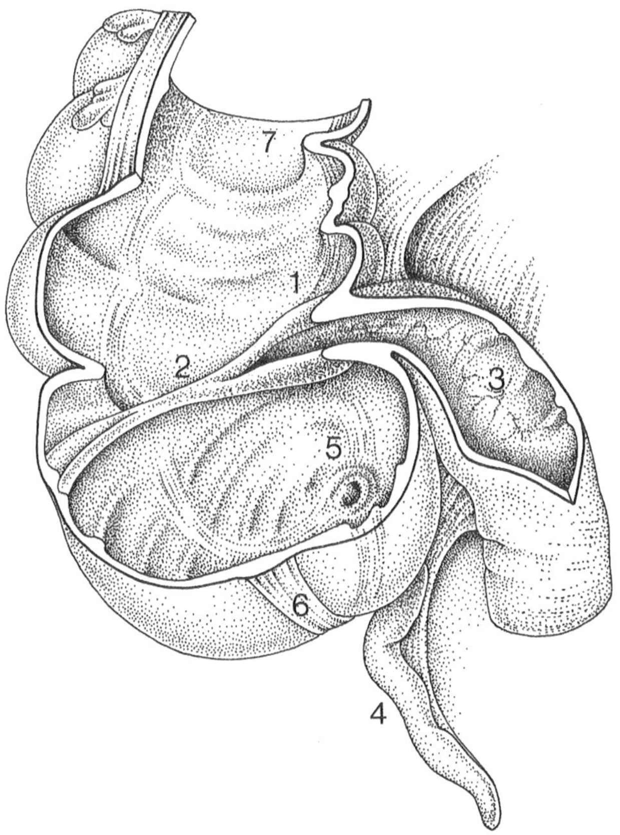

Fig. B9 The valve of Bauhin: *1* ileocaecal valve and ileocaecal ostium, *2* frenulum of the ileocaecal valve, *3* ileum, *4* vermiform appendix, *5* ostium of appendix, *6* taenia coli, *7* ascending colon. (After ten Donkelaar HJ, Oostra R-J *Klinische Anatomie en Embryologie,* 4th ed. Springer Media, Houten, 2014)

Glands of Bauhin: *Eponym* for the anterior (apical) lingual glands, also described by Blandin and Nuhn.

Ligament of Bauhin: *Eponym* for the ischiofemoral ligament of the hip joint, also described by Bellini.

Valve of Bauhin: *Eponym* for the ileocaecal fold, previously considered as a valve (Fig. B9), first described by Falloppio and later also by Tulp and Varolio.

Baumgarten, Paul Clemens (1848–1887): German pathologist.

Veins of Baumgarten: *Eponym* for the paraumbilical veins along the round ligament of the liver, also described by Sappey and Burow. Reference: *Ueber die Nabelvene des Menschen und ihre Bedeutung für die Circulationsstörung bei Lebercirrhose.* Arb Geb Path Anat Bakt, Path Anat Inst Tübingen, 1891; 1:1–46.

Bayer, Heinrich (1879–1926): German obstetrician and gynaecologist working in Strassburg.

Node of Bayer: *Eponym* for an inconstant parauterine lymph node at the crossing of the uterine artery and the ureter, also described by Poirier.

Beale, Lionel Smith (1828–1906): English physician and microscopist born and working in London known for his book on the microscope. Reference: *The Microscope and Its Applications to Clinical Medicine.* Highley, London, 1854.

Cells of Beale: *Eponym* for nerve cells in the cardiac ganglia. *On the anatomy of nerve fibres and cells and the ultimate distribution of nerve fibres.* Quart J Micr Sci 1863; 3:97–100.

a

b

NELLO BECCARI

PROFESSORE DI ANATOMIA COMPARATA
NELLA R. UNIVERSITÀ DI FIRENZE

NEUROLOGIA
COMPARATA

ANATOMO-FUNZIONALE
DEI VERTEBRATI, COMPRESO L'UOMO

SANSONI EDIZIONI SCIENTIFICHE
FIRENZE 1943-XXI

Fig. B10 (**a**) Photograph of Beccari. (Courtesy Marina Quartu, Cagliari); (**b**) Title page *Neurologia Comparata* (Sansoni, Florence, 1943)

Beccari, Nello (1883–1957): Italian anatomist, born in Bagno a Ripoli, Florence, working in Catania and Florence (Fig. B10a). Reference: *Neurologia Comparata.* Sansoni, Florence, 1943; Fig. B10b).

Frontal-olfacto-limbic fold of Beccari: *Eponym* for the fold on the medial surface of the cerebral hemisphere that unites the superior frontal gyrus to the cingulate and olfactory gyri.

Tegmental centre of Beccari: *Eponym* for the reticular formation of the medulla oblongata.

Bechterew, Vladimir Michailovich von (1857–1927): Russian psychiatrist born in Vjatka and working in Kasan and later in St. Petersburg (Fig. B11a) known for his studies on fibre pathways in the human brain. References: *Die Leitungsbahnen im Gehirn und Rückenmark.* In Russian, St. Petersburg 1894, in German: Georgi, Leipzig, 1899, and translated into French as *Les voies de conduction du cerveau et de la moëlle.* Storch, Lyon, 1900.

Bundle or **Tract of Bechterew:** *Eponym* for (a) the central tegmental tract and (b) the spinoolivary tract.

Nucleus of Bechterew: *Eponym* used for (a) the superior vestibular nucleus, (b) the reticulotegmental nucleus and (c) the posteromedial or dorsomedial part of the red nucleus.

Stripe of Bechterew-Kaes: *Eponym* for a stripe of nerve fibres in layer II of the isocortex, also described by Kaes (Fig. B11b).

Béclard, Pierre Augustin (1785–1866): French anatomist born in Angers and working in Paris.

Angle of Béclard: *Eponym* for an acute angle bounded cranially by the posterior belly of the digastric muscle and cranially by the greater horn of the hyoid bone, containing the lingual artery.

Triangle of Béclard: *Eponym* for a triangle bounded by the posterior border of the hyoglossus muscle, the

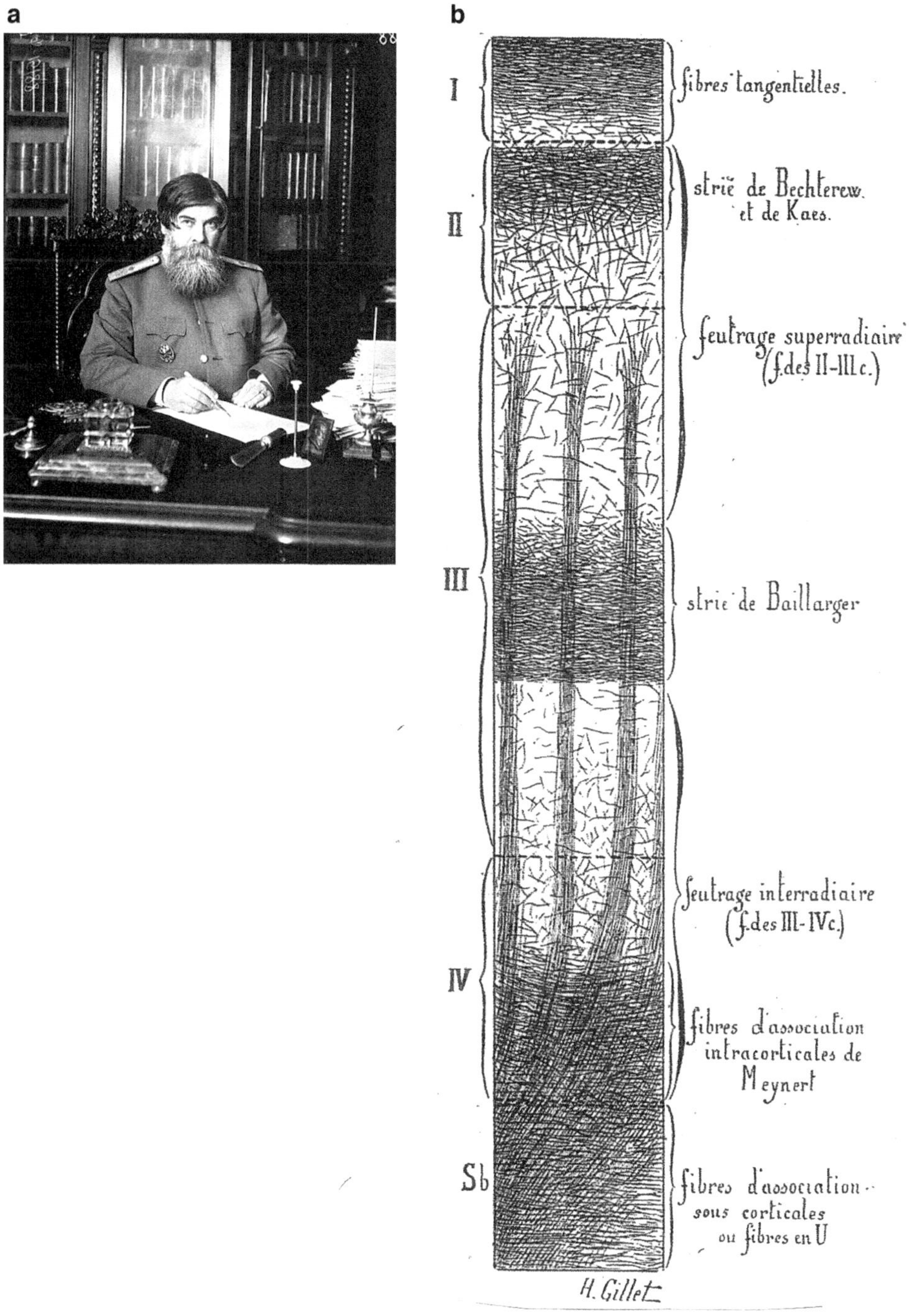

Fig. B11 (**a**) Photo of Bechterew by Karl Boella. (From Wikipedia; in public domain); (**b**) stripe of Bechterew-Kaes. (After a figure from Dejerine J *Anatomie des centres nerveux,* Tome Premier. Rueff, Paris, 1895)

posterior belly of the digastric muscle and the greater horn of the hyoid bone. Reference: *Extirpation de la parotide.* Arch Gén Méd 1824; 4:60–66.

Bell, John (1763–1820): Scottish surgeon and anatomist born and working in Edinburgh, brother of Sir Charles Bell. Reference: *Anatomy of the Human Body* (with Charles Bell). Longmans, London, 1816, 7th ed. 1829.

Muscle of Bell: *Eponym* for muscular strands from the internal openings of the urethra to the uvula, bounding the vesical trigone.

Bell, Sir Charles (1774–1842): Scottish surgeon and anatomist born and working in Edinburgh and later in London (Fig. B12) who demonstrated the motor function of the anterior or ventral roots of the spinal nerves. Reference: *On the nerves of respiration.* Edinburgh, 1822.

External respiratory nerve of Bell: *Eponym* for the long thoracic nerve.

Internal respiratory nerve of Bell: *Eponym* for the phrenic nerve.

Fig. B12 Portrait of Charles Bell. (From Pagel 1901)

Fig. B13 Portrait of Bellini. (From Wikipedia; in public domain)

Bellini, Lorenzo (1643–1704): Italian anatomist and physician born in Florence, working in Pisa and known for his studies on the kidney (Fig. B13). Reference: *Exercitationes anatomicae duae de structura et usu renum ut et de gustus organo novissime deprehenso.* Florence, 1662.

Ducts of Bellini: *Eponym* for the papillary ducts of the kidney.

Ligament of Bellini: *Eponym* for the ischiofemoral ligament of the hip joint, also described by Bauhin.

Tubuli of Bellini: *Eponym* for the straight part of the renal tubuli.

Bendavid, Robert (1940–2019): Canadian surgeon of Moroccan origin born in Casablanca, Morocco. In 1957, he emigrated to Canada where he studied medicine in Toronto and worked mainly in Toronto.

Venous circle of Bendavid: *Eponym* for a network of deep inferior epigastric, suprapubic and retropubic veins in the subinguinal space of Bogros. Reference: *The space of Bogros and the deep inguinal venous circulation.* Surg Gynaec Obst 1992; 174:356–357.

Bendz, Henrik Carl Bang (1816–1882): Danish physician and anatomist born in Odense.

Ganglion of Bendz: *Eponym* for the inferior or nodose ganglion of the vagus nerve, also described by Wuther. Reference: *Tractatus de connexu inter nervum vagum et accessorium Willissii.* Thesis, 1836.

Bérard, Auguste (1802–1846): French anatomist and surgeon born in Varrains and working in Paris. Reference: *Texture et développement des poumons.* Paris, 1836.

Ligament of Bérard: *Eponym* for the suspensory ligament of the pericardium.

Béraud, Bruno Jacques (1823–1865): French surgeon and anatomist born in Montreux and working in Paris. Reference: *Essai sur le cathéterisme du canal nasal, suivant la méthode de Laforest, nouveau procédé.* Paris, 1854.

Valve of Béraud: *Eponym* for a fold between the lacrimal sac and the nasolacrimal duct, already described by Krause.

Berengario da Carpi, Jacopo (around 1460-around 1530): Italian born in Carpi near Modena and working in Bologna (Fig. B14a) who made the first anatomical text with illustrations: *Anatomia Carpi. Isagoge breves perlucide ac uberime, in Anatomium humani corporis* 1522. For biography, see Parent A. *Berengario da Carpi and the renaissance of brain anatomy.* Front Neuroanat 2019; 13:11 (Fig. B14b). His work challenged Galenic physiology. Berengario wrote that he could not demonstrate Galen's rete mirabile, where the vital spirit was supposed to be transformed into animal spirit stored in the ventricles. He also showed that in contrast to Galen's view nerves are not hollow but solid structures.

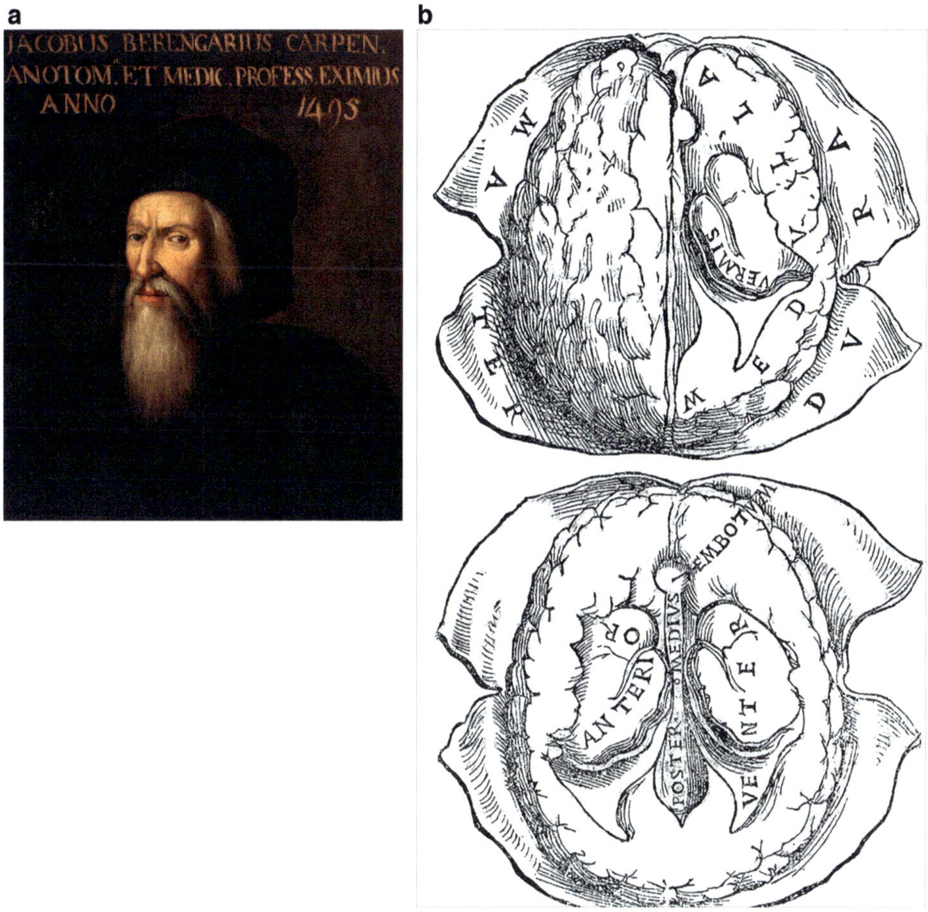

Fig. B14 (**a**) Portrait of Berengario da Carpi; (**b**) Drawings brain. (From Parent A 2019 *Berengario da Carpi and the renaissance of brain anatomy.* Front Neuroanat 13:11, Open Access; courtesy André Parent, Quebec)

Berettini, Pietro, also known as Pietro da Cortoni (1597–1669): Italian painter and architect born in Cortona. He drew highly expressive anatomical tables, which were printed only a century after his death. Reference: *Tabulae anatomicae.* Amadei, Rome, 1741.

Anastomosis of Berettini: *Eponym* for a communicating branch between the median and ulnar nerves in the ulnar portion of the palm. *See*: Shin EK, Meals RA *The historical importance of the hand in advancing the study of human anatomy.* J Hand Surg Am 2005; 30:209–221.

Berger, Emil (1855–1926): Austrian ophthalmologist born in Vienna and working in Paris.

Space of Berger: *Eponym* for the retrolental space, a central hyalocapsular space between the posterior aspect of the lens capsule and the anterior aspect of the vitreous lamina, being a potential space in the young, limited peripherally by the circular hyalocapsular ligament of Wieger. Reference: *Beiträge zur Anatomie des Auges in normalen und pathologischen Zustände.* Bergmann, Wiesbaden, 1887.

Bergmann, Gottlieb Heinrich (1781–1861): German psychiatrist working at the Asylum in Hildesheim. References: *Neue Untersuchungen über die innere Organization des Gehirns.* Helwing, Hannover, 1831; *Untersuchungen über die Struktur der Mark- und Rindensubstanz des grossen und kleinen Gehirns.* Arch Anat Physiol (Lpz) 1841, pp 126–175.

Cords of Bergmann: *Eponym* for the acoustic striae in the floor of the IVth ventricle of the brain, also described by Piccolomini.

Fibres of Bergmann: *Eponym* for the ascending fibres of the Bergmann glial cells of the cerebellum.

Glial cells of Bergmann: *Eponym* for radial astrocytes, certain glial cells in the cerebellum.

Bergmeister, Otto (1845–1918): Austrian ophthalmologist born in Silz, Tirol, and working in Vienna.

Papilla of Bergmeister: *Eponym* for a small mass of glial tissue that forms in the hyaloid artery during foetal life. It may persist as a prepapillary membrane.

Reference: *Beiträge zur Entwicklungsgeschichte des Säugethierauges.* Wien Embryol Inst Mitteilungen 1877; 1:63–84.

Bernard, Claude (1813–1878): French physiologist and physician born in Saint-Julien and working in Paris as successor of Magendie.

Canal or **Duct of Bernard:** *Eponym* for the accessory pancreatic duct, earlier described by Santorini. Reference: *Mémoire sur le pancréas et sur le rôle du suc pancréatique dans les phénomènes digestifs, particulièrement dans la digestion des matières grasses neutres.* Paris, 1856.

Bernasconi, Vittorio (1921–2016): Italian neuroradiologist working in Milan.

Artery of Bernasconi-Cassinari: *Eponym* for the marginal tentorial artery described together with Cassinari. Reference: *Un segno carotidografico tipico di meningioma del tentorio.* Chirurgia (Milan) 1956; 11:586–588.

Berry, Sir James (1860–1946): Canadian surgeon who specialized in surgery of the cleft palate and the thyroid gland. References: *Suspensory ligaments of the thyroid gland.* J Anat (Lond) 1888, 22; *Diseases of the Thyroid Gland and Their Surgical Treatment.* London, 1901.

Ligament of Berry: *Eponym* for the suspensory ligament of the thyroid gland.

Bertelli, Dante (1858–1946): Italian anatomist born in Caldana, Grosseto, and working in Padua, Veneto. Reference: *Atlante di embriologia e di anatomia topografica dell'uomo.* Vallardi, Milan, 1890.

Diaphragmatic bundle of Bertelli: *Eponym* for a muscular band that stands out from the aponeurosis of the transversus abdominis muscle at the level of the seventh costal cartilage.

Membrane of Bertelli-Laimer: *Eponym* for the phrenicooesophageal connective tissue that joins the diaphragmatic fascia and the adventitia of the oesophagus, also described by Laimer.

Supraglottic rhyme of Bertelli: *Eponym* for the vestibule of the larynx.

Bertin, Exupère Joseph (1712–1781): French anatomist born in Tremblay, Bretagne, working in Paris and known for his studies on the kidney and bones. References: *Sur les reins.* Paris, 1744; *Sur les cornets sphénoidales.* Paris, 1744; *Traité de osteologie*, 4 Vols. Paris, 1754.

Bones or **Ossicles of Bertin:** *Eponym* for the sphenoidal conchae, paired replacement bones, each of which arises from several ossification centres. After their separation from the nasal capsule, they form the anterior border of the sphenoidal sinus.

Columns of Bertin: *Eponym* for the renal columns, extensions of the cortical tissue of the kidney into the medullary substance between the pyramids.

Ligament of Bertin: *Eponym* for the iliofemoral ligament of the hip joint, also described as Y-ligament by Bigelow.

Betz, Vladimir Alexsandroče (1834–1894): Ukrainian anatomist in Kiev. Reference: *Anatomischer Nachweis zweier Gehirncentra.* Zbl Med Wiss 1874; 12:578–580, and 595–599.

Betz cells: *Eponym* for the giant pyramidal cells of the primary motor cortex.

Bezold, Albert von (1838–1868): German physiologist working in Jena and later in Würzburg.

Ganglia of Bezold: *Eponym* for an aggregate of nerve cells in the interatrial septum of the heart, also described by Bidder. Reference: *Untersuchungen ueber die Innervation des Herzens.* Leipzig, 1863.

Bianchi, Giovanni Battista (1681–1761): Italian anatomist born in Turin and working in Turin and later in Bologna.

Nodules of Bianchi: *Eponym* for the nodules of the semilunar valves, earlier described by Aranzio.

Valve of Bianchi: *Eponym* for the lacrimal valve in the nasolacrimal duct, also described by Hasner and Cruveilhier. Reference: *Ductuum lacrymalium novorum epistolaris dissertatio.* Leiden, 1723.

Bichat, Marie François Xavier (1771–1802): French anatomist and physician born in Thiorette en Bresse and working in Paris (Fig. B15a). Many structures are named after him in the French literature. Reference: *Traité*

Fig. B15 (**a**) Portrait of Bichat. (From Pagel 1901); (**b**) Title page from his *Traité d'anatomie descriptive* (Brisson and Gabon, Paris, 1801–1803).

a

b

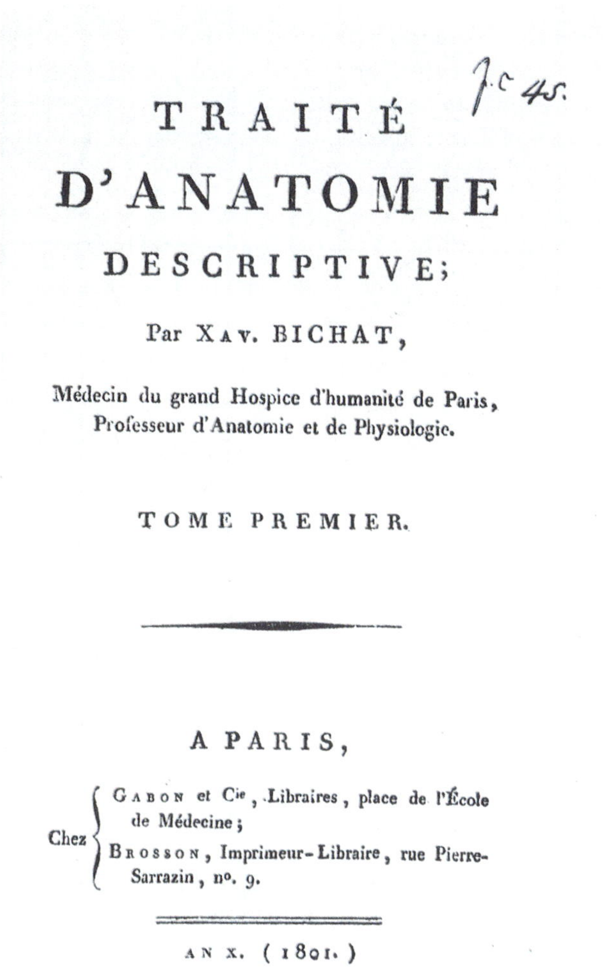
TRAITÉ
D'ANATOMIE
DESCRIPTIVE;
Par XAV. BICHAT,
Médecin du grand Hospice d'humanité de Paris,
Professeur d'Anatomie et de Physiologie.
TOME PREMIER.
A PARIS,
Chez GABON et Cie, Libraires, place de l'École de Médecine;
BROSSON, Imprimeur-Libraire, rue Pierre-Sarrazin, no. 9.
AN X. (1801.)

d'anatomie descriptive (with Buisson MFR and Roux PJ, 5 Vols. Brisson and Gabon, Paris, 1801–1803 (Fig. B15b).

Anastomosis of Bichat: *Eponym* for a communicating branch between the glossopharyngeal and vagus nerves.

Canal or **Foramen of Bichat:** *Eponym* for the subarachnoid quadrigeminal cistern or cistern of the great cerebral vein.

Fat pad or **Corpus of Bichat:** *Eponym* for the buccal fat pad.

Fissure of Bichat: *Eponym* for the transverse cerebral fissure, a fissure below the splenium of the corpus callosum.

Fossa of Bichat: *Eponym* for the pterygopalatine fossa.

Ligament of Bichat: *Eponym* for the inferior fascicle of the posterior sacroiliac ligament, attached to the third sacral transverse tubercle.

Tunic of Bichat: *Eponym* or the tunica intima of blood vessels.

Bickel, Gustav (nineteenth century): German physician in Wiesbaden.

Ring of Bickel: *Eponym* for a ring of lymphatic tissue formed by the lingual, palatine, tubal and pharyngeal tonsils, also known as ring of Waldeyer. Reference: *Ueber die Ausdehnung und den Zusammenhang des lymphatischen Gewebes in der Rachengegend.* Virchows Arch 1884, 97:340.

Bidder, Georg Friedrich Karl Heinrich (1810–1894): German-Estonian anatomist and physiologist from Courland (Estonia), working in Dorpat (currently: Tartu, Estonia).

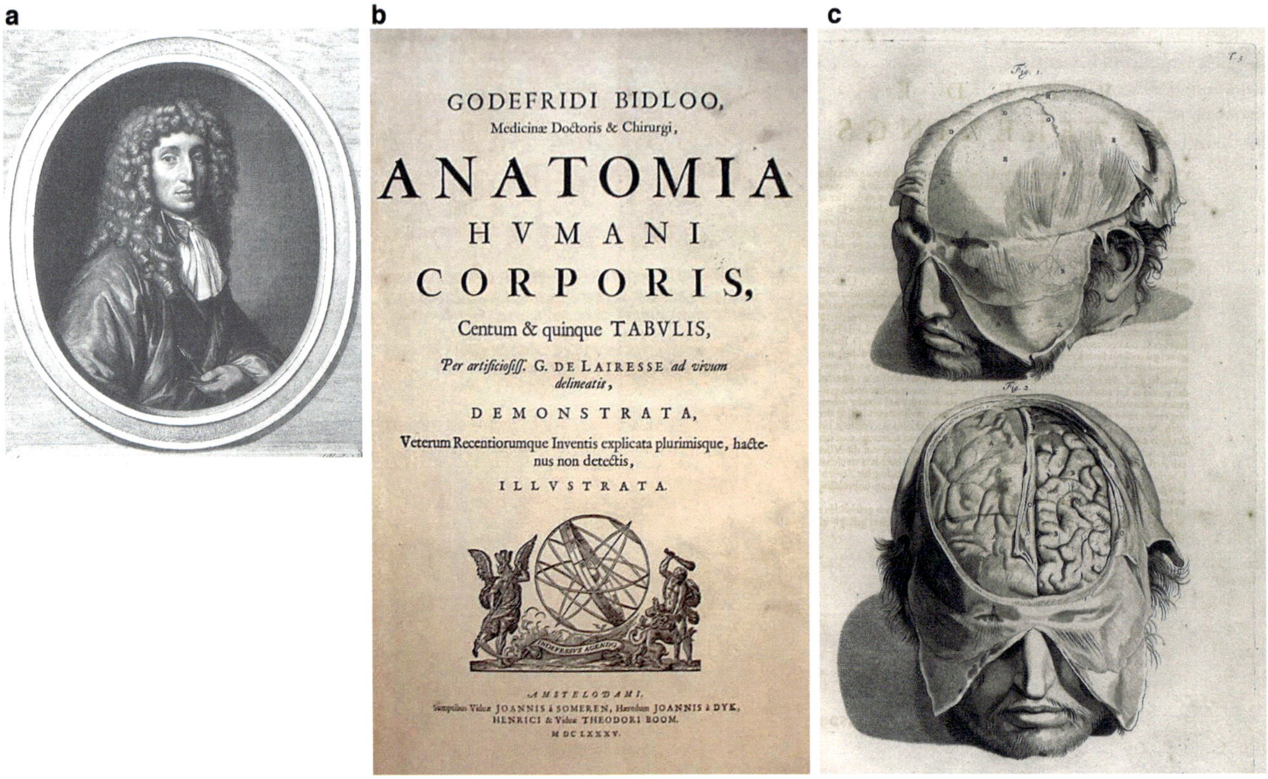

Fig. B16 (**a**) Engraving of Bidloo. (By Abraham Blotelingh after portrait by Gérard de Lairesse for Bidloo's anatomical atlas); (**b**) title page from his *Anatomia Humani Corporis* (Amsterdam, 1685); (**c**) dissection of the brain, in which the central sulcus may be noted

Ganglion of Bidder: *Eponym* for an aggregate of nerve cells in the interatrial septum of the heart, also described by Bezold. Reference: *Neurologische Beobachtungen.* Dorpat, 1836.

Bidloo, Govert (1649–1713): Dutch anatomist and surgeon born in Amsterdam and working in Leiden (Fig. B16a). Reference: *Anatomia humani corporis.* Leiden, 1685 (Fig. B16b; illustrated by Gérard de Lairesse). In his atlas, the central sulcus of the cerebral cortex was already illustrated (Fig. B16c). Bidloo's atlas was used by Cowper for an English edition with Cowper as author without crediting Bidloo.

Bigelow, Henry Jacob (1818–1890): American surgeon in Boston. Reference: *The Mechanism of Dislocation and Fracture of the Hip.* Lea, Philadelphia, 1869.

Ligament of Bigelow, Y: *Eponym* for the iliofemoral ligament of the hip joint, earlier described by Bertin.

Bill's bar*:** The vertical crest in the fundus of the internal acoustic meatus is known as ***'Bill's bar' after William House, a famous American ear-nose-throat surgeon. It separates the superior portion of the fundus of the internal acoustic meatus into an anterior channel for the facial nerve (the facial area) and a posterior channel for the superior part of the vestibular nerve, the utriculoampullary nerve (the superior vestibular area; Fig. B17).

Billroth, Christian Albert Theodor (1829–1894): German surgeon from the island of Rügen, and after working in Berlin, subsequently holding positions in Zürich and Vienna (Fig. B18). He is known for new techniques in the surgery of the stomach.

Cords of Billroth, Splenic: *Eponym* for radicles of splenic veins in the splenic pulpa. Reference: *Untersuchingen ueber die Entwickelung der Blutgefässe, nebst Beobachtungen aus der königlichen Universität-Klinik zu Berlin.* Reimer, Berlin, 1856.

Veins of Billroth: *Eponym* for the sinusoids of the spleen.

Biondi, Adolfo (1846–1917): Italian pathologist working in Naples.

Body of Biondi: *Eponym* for an inclusion body (psammoma) in the cranial pia mater.

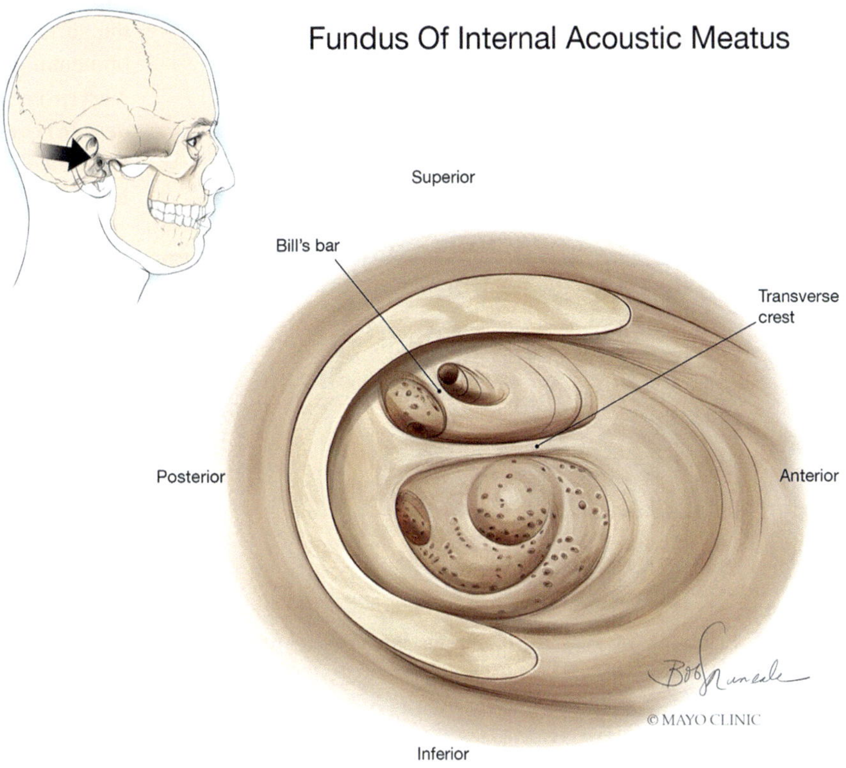

Fig. B17 The fundus of the internal acoustic meatus showing Bill's bar. (From ten Donkelaar HJ, Kachlík D, Tubbs RS *An Illustrated Terminologia Neuroanatomica*. Springer, Cham, 2018; courtesy Bob Morreale, Mayo Clinic, Rochester, MN)

Fig. B18 Photo of Billroth. (From Pagel 1901)

Birbeck, Michael Stanley, Clive (1925–2005): English scientist and electron microscopist working mainly in London.

Granules of Birbeck: *Eponym* for platelike structures found in stellate cells of the skin, earlier described by Langerhans. Reference: *An electron microscopic study of basal melanocytes and high-level clear cells (Langerhans cells) in vitiligo.* J Invest Dermat 1961; 37:51–64 (with AS Breathnach and JD Everall).

Bischoff, Theodor Ludwig Wilhelm (1807–1882): German anatomist and embryologist born in Hannover and working in Heidelberg, Giessen and Munich. His son, Ernst Philipp Eduard, working in his father's Anatomical Institute in Munich, performed an extensive microscopic analysis of the anastomoses between the cranial nerves, entitled *Mikroskopische Analyse der Anastomosen der Kopfnerven.* Leutner, Munich, 1865, which won the Award Winning Answer to the special problem proposed by competition by the Royal Medical Faculty of Munich in 1863 (English translation by Sachs E Jr. and Valtin EW *Microscopic Analysis of the Anastomoses between the Cranial Nerves.* University Press of New England, Hanover, New Hampshire, 1977).

Corona of Bischoff: *Eponym* for the corona radiata of the mammalian ovum. Reference: *Beiträge zur Lehre von den Eihüllen des menschlichen Fötus.* Bonn, 1834.

Blandin, Philippe Frédéric (1798–1849): French surgeon and anatomist in Paris. Reference: *Mémoire sur la struc-*

Fig. B19 Engraving of Blasius. (From the Municipal Archive, Amsterdam; in public domain)

ture et les mouvements de la langue dans l'homme. Arch Gén Méd 1823, I:457–470.

Ganglion of Blandin: *Eponym* for (a) the submandibular ganglion, later also described by Langley, and (b) the inconstant sublingual ganglion.

Glands of Blandin: *Eponym* for the anterior (apical) lingual glands, earlier described by Bauhin and later by Nuhn.

Blanes Viale, Tomas (1878–1900): Medical student born in Mercedes, Uruguay, and living in Madrid. He died prematurely of typhoid fever while in his fifth year of medicine. Among his first medical subjects, histology and pathology were taught by Ramón y Cajal. His skills as a cartoonist contributed to his incorporation in the Cajal's laboratory in Madrid.

Cells of Blanes: *Eponym* for interneurons in the internal glomerular layer of the olfactory bulb. Reference: *Sobre algunos puntos dudosos de la estructura del bulbo olfactorio.* Rev Trim Micrograf 1898; 3:99–127.

Blaschko, Alfred (1858–1922): German dermatologist born in Freienwalde an der Oder and working in Berlin and Stettin (currently Sczeczyn, Poland).

Lines of Blaschko: *Eponym* for a system of lines on the surface of the human body, representing a developmental growth pattern of the skin. Reference: *Die Nerven-verteilung in der Haut in ihrer Beziehung zu den Erkrankungen der Haut.* Beilage Verhandl Deutsch Dermatol Gesellschaft. VII. Congress, Breslau, 1911.

Blasius (Blaes), Gerard (1626–1682): Flemish comparative anatomist born in Oostiliet, near Bruges, and working in Leiden and Amsterdam (Fig. B19). He described the duct of the parotid gland (duct of Blasius, claimed in 1662), also described by Stensen in the same year. Reference: *Miscellanea anatomica hominis brutorumque fabricum diversam magna parte exhibentia.* Amsterdam, 1673.

Duct or **Canal of Blasius:** *Eponym* for the parotid duct, described in the same year as Stensen.

Blessig, Robert (1830–1878): Russian ophthalmologist born in St. Petersburg and working in Dorpat and later in St. Petersburg.

Lacunae of Blessig: *Eponym* for spaces in the anterior part of the retina near the ora serrata. Reference: *De retinae textura disquisitiones microscopicae.* Thesis, Dorpat, 1845.

Bloodless fold of Treves: *Eponym* for the inferior ileocaecal fold of the appendix; *see* Treves, Frederick.

Bloodless line of Brödel: *Eponym* for a relatively avascular zone between the segments supplied by the anterior and posterior branches of the renal artery (*see* Fig. B39b), also described by Hyrtl; *see* Brödel, Max.

Blumenau, Leonid Wassiljewitsch (1862–1932): Russian neurologist in St. Petersburg. Reference: *Ueber den äusseren Kern des Keilsstranges im verlängertem Mark.* Neurol Zbl 1891; 10:226–232, 589–590.

Nucleus of Blumenau: *Eponym* for the lateral cuneate nucleus of the caudal hindbrain.

Blumenbach, Johann Friedrich (1752–1840): German anthropologist, zoologist, physiologist and physician born in Gotha and working in Göttingen (Fig. B20). References: *Geschichte und Beschreibung der Knochen des menschlichen Körpers.* Göttingen, 1786; *De generis humani varietate natura,* 3rd ed of Thesis, Göttingen, 1795.

Fig. B20 Portrait of Blumenbach. (From Pagel 1901)

Clivus of Blumenbach: *Eponym* for the clivus, the sloping surface from the foramen magnum leading up to the sella turcica, formed by the bodies of the occipital and sphenoidal bones.

Line and **Plane of Blumenbach:** *Eponym* for the cranium in norma verticalis, lying on its base over a horizontal plane. The plane passes through the points of contact at the maxillary teeth anteriorly and the mastoid processes posteriorly (*see* Fig. P18).

Bochdalek, Vincenz (Vincenc) Alexander (1801–1883): Bohemian anatomist and pathologist born in Skřipov, near Opava, working in Prague (Fig. B21a). Reference: *Anleitung zur praktischen Zergliederung des menschli-*

Fig. B21 (**a**) Portrait of Bochdalek. (From Kachlik D, Čech P *Vincenz Alexander Bochdalek (1801–83)*. J Med Biogr 2011; 19:38–43; courtesy David Kachlík); (**b**) title page from his *Anleitung zur praktischen Gliederung des menschlichen Gehirns nebst einer anatomischen Beschreibung deszelben mit besonderer Rücksicht auf das kleine Gehirn* (Gottlieb Haase Söhne, Prague, 1833)

a

b

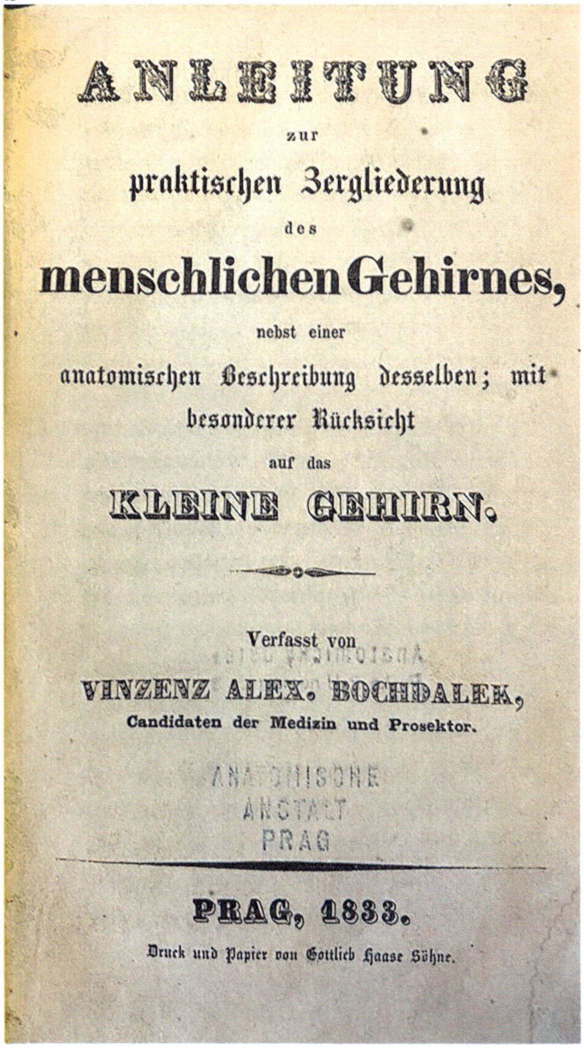

ANLEITUNG
zur
praktischen Zergliederung
des
menschlichen Gehirnes,
nebst einer
anatomischen Beschreibung desselben; mit
besonderer Rücksicht
auf das
KLEINE GEHIRN.

Verfasst von
VINZENZ ALEX. BOCHDALEK,
Candidaten der Medizin und Prosektor.

ANATOMISCHE
ANSTALT
PRAG

PRAG, 1833.
Druck und Papier von Gottlieb Haase Söhne.

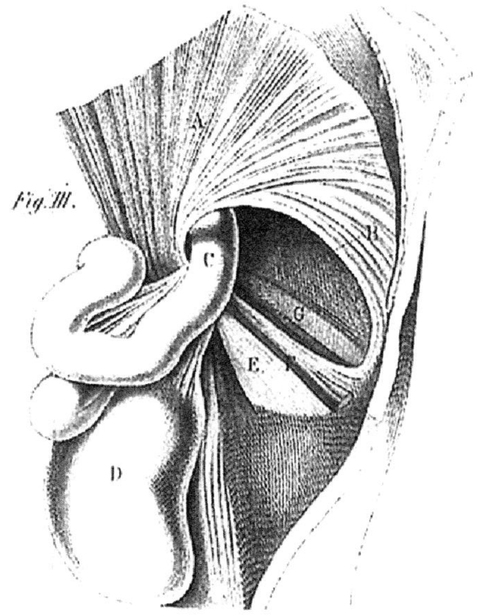

Fig. B22 Gap of Bochdalek. (From Bochdalek VA *Einige Betrachtungen über die Entstehung des angeborenen ZwerchfellBruches. Als Beitrag zur Anatomie der Hernien.* Vjschr Prakt Heilk 1848; 5:89–97)

chen Gehirns nebst einer anatomischen Beschreibung deszelben mit besonderer Rücksicht auf das kleine Gehirn. Gottlieb Haase Söhne, Prague, 1833 (Fig. B21b). For a biography, see Kachlík D, Čech P. *Vincenz Alexander Bochdalek (1801–83).* J Med Biogr 2011; 19:38–43.

Duct or **Canal of Bochdalek:** *Eponym* for the embryonic thyroglossal duct, also described by Vater and His Sr.

Flower basket or **Bouquet of Bochdalek:** *Eponym* for the fruticulus, that part of the choroid plexus of the fourth ventricle that extends through its lateral aperture, originally described as 'Blumenkörbchen'.

Foramen of Bochdalek: *Eponym* for the embryonic pleuroperitoneal hiatus.

Ganglion of Bochdalek: *Eponym* for a pseudoganglion in the dental plexus. Reference: *Neue Untersuchungen der Nerven des Ober- und Unterkiefers.* Prague, 1855.

Muscle of Bochdalek: *Eponym* for the triticeoglossal muscle, an inconstant band of muscle fibres between the root of the tongue and the triticeal cartilage.

Plexus of Bochdalek: *Eponym* for the choroid plexus of the fourth ventricle.

Fig. B23 Portrait of Boerhaave by Jan Wandelaar. (From Pagel 1901)

Triangle or **Gap of Bochdalek:** *Eponym* for the lumbocostal or vertebrocostal triangle, a triangular space between the costal and lumbar parts of the diaphragm devoid of muscle fibres that may lead to a posterior diaphragmatic hernia (Fig. B22). Reference: *Einige Betrachtungen über die Entstehung des angeborenen Zwerchfellbruches. Als Beitrag zur Anatomie der Hernien.* Vjschr Prakt Heilk 1848; 5:89–97.

Valve of Bochdalek: *Eponym* for a mucous fold at the puncta lacrimalia of the lacrimal canaliculi, earlier described by Foltz. Reference: *Beitrag zur Anatomie der Thränenorgane.* Vjschr Prakt Heilk 1866; 90:121.

Bock, August Carl (1782–1833): German anatomist and surgeon born in Magdeburg and working in Leipzig.

Ganglion of Bock: *Eponym* for the carotid ganglion.

Nerve of Bock, Pharyngeal: *Eponym* for the pharyngeal branch of the maxillary nerve. Reference: *Beschreibung des fünften Nervenpaares und seiner Verbindungen mit anderen Nerven, vorzüglich mit den Gangliensysteme.* Leipzig, 1817.

Bodies: *See* **The Classical Anatomical Eponyms.**

Boerhaave, Herman (1668–1738): Dutch physician born in Voorhout, near Leiden, and working in Leiden (Fig. B23). He described the sweat glands, first noted in a dissertation in 1693, and elaborated in an *Epistle to Ruysch* in 1722.

Fig. B24 Portrait of Boettcher. (From Wikipedia; in public domain)

Glands of Boerhaave: *Eponym* for the sweat glands.

Boettcher, Arthur (1831–1889): German anatomist and pathologist born in Bauske and working in Dorpat, who described many anatomical features of the internal ear (Fig. B24). Reference: *Observationes microscopicae de ratione qua nervus cochleae mammalium terminatur.* Schünmann and Mathiesen, Dorpat (currently: Tartu, Estonia), 1856.

Cells of Boettcher, Outer glandular: *Eponym* for the basal external glandular cells, small cells interposed between the basilar membrane and the cells of Claudius (*see* Fig. C21). Reference: *Weitere Beiträge zur Anatomie der Schnecke.* Arch Pathol Anat Physiol Klin Med 1859; 17:243–281.

Duct of Boettcher: *Eponym* for the utriculosaccular duct.

Ganglion of Boettcher: *Eponym* for a small ganglion on the cochlear nerve in the internal acoustic meatus, innervating that part of the cochlear duct located in the cavity of the vestibular ganglion; usually but incorrectly seen as a *synonym* for the spiral ganglion of Corti.

Sac or **Space of Boettcher:** *Eponym* for the endolymphatic sac, also described by Cotugno; so, the *eponym* sac of Boettcher-Cotugno would be appropriate.

Bogros, Annet Jean (1786–1823): French anatomist in Paris. Reference: *Essai sur l'anatomie chirurgical de la région iliaque et description d'un nouveau procédé pour faire la ligature des artères épigastrique et iliaque externe.* Didot Le Jeune, Paris, 1823.

Space of Bogros: *Eponym* for the retroinguinal or preperitoneal space between the peritoneum and the fascia transversalis containing the external iliac vessels.

Bolk, Lodewijk (1866–1930): Dutch anatomist born in Overschie (Rotterdam) and working in Amsterdam. He is known for his subdivision of the cerebellum (Fig. B25). Reference: *Das Cerebellum der Säugetiere.* Bohn, Haarlem, and Fischer, Jena, 1906.

Ansiform lobe of Bolk: *Eponym* for the semilunar lobules of the cerebellum (Fig. B26).

Complicated lobule of Bolk: *Eponym* for the posterior lobe of the cerebellum.

Median lobule of Bolk: *Eponym* for the posterior part and inerior parts of the vermis.

Simple lobule of Bolk: *Eponym* for the anterior part of the complicated lobule of the cerebellum.

Bolton, Joseph Shaw (1867–1946): British physician born in Whitby and working in Leeds.

Plane of Bolton: *Eponym* for a roentgenographic cephalometric plane connecting the point of Bolton to the nasion.

Point of Bolton: *Eponym* for a point at the intersection of the occipital condyle and the foramen magnum at the highest notch posterior to the occipital condyle.

Bones: The term **bones** is mostly used for sutural bones (Andernach, Eysson, Riolan and Worm) but also for accessory bones (Breschet, Rousseau) or sesamoid bones (Gruber, Vesalius).

Bone of Albrecht: *Eponym* for the basiotic bone, a transient bone between the basisphenoid and basioccipital bones during development; *see* Albrecht, Karl Martin Paul.

Fig. B25 The Anatomical Lesson of Bolk by Martin Monnickendam. (From the Anatomisch Embryologisch Laboratorium, AMC, Amsterdam; courtesy Roelof-Jan Oostra, Amsterdam). Bolk dissected an orang utan; at the left Boeke, at the right van den Broek and behind Barge. (Baljet B. *Het schilderij De anatomische les van professor Bolk door Martin Monnickendam.* Ned Tijdschr Geneeskd 1993; 137:2638–2644)

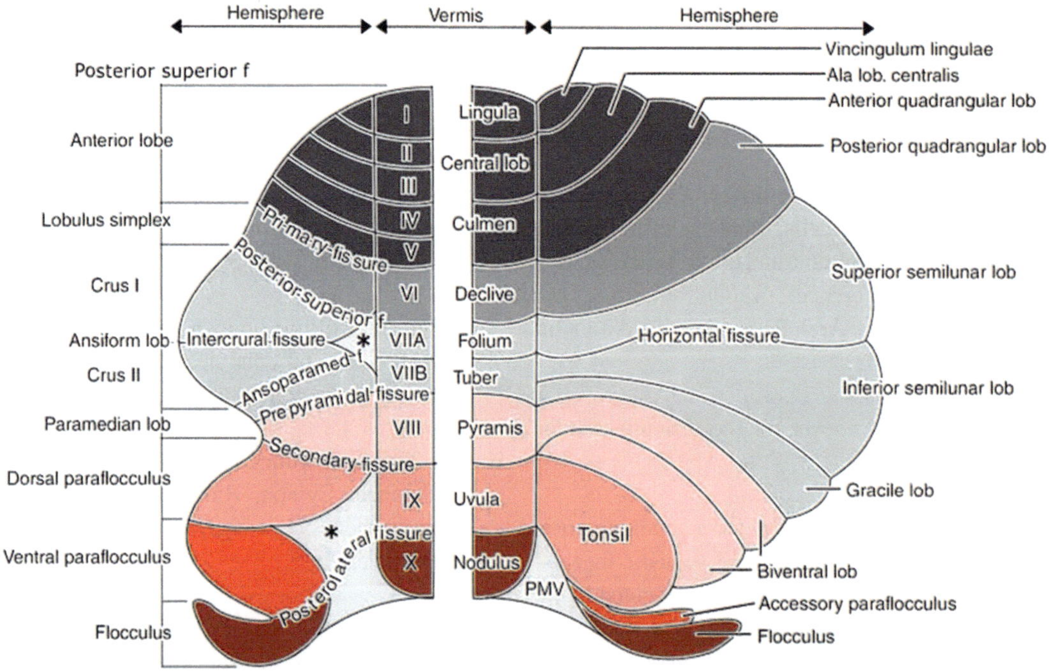

Fig. B26 Subdivision of the human cerebellum. (From ten Donkelaar HJ, Kachlik D, Tubbs RS *An Illustrated Terminologia Neuroanatomica.* Springer, Cham, 2018; courtesy Jan Voogd, Oegstgeest)

Bones or **Ossicles of Andernach:** *Eponym* for sutural bones, later also described by Worm, his son-in-law; *see* Andernach, Johann Winther von.

Bones or **Ossicles of Bertin:** *Eponym* for the sphenoidal conchae; *see* Bertin, Exupère Joseph.

Bones of Breschet: *Eponym* for inconstant suprasternal bones found in ligaments of the sternoclavicular joint; *see* Breschet, Gilbert.

Bones of Eysson: *Eponym* for inconstant mental bones in the mandibular syndesmosis; *see* Eysson, Henrique.

Bone of Goethe: *Eponym* sometimes used for the premaxillary bone (os incisivum); *see* Goethe, Johann Wolfgang von.

Bone of Gruber: *Eponym* for an accessory bone between the hamate, the capitate and the base of the third and fourth metacarpal bones; *see* Gruber, Wenzel Leopold.

Bone of Kerckring: *Eponym* or an isolated bone often found in the middle of the posterior margin of the foramen magnum; *see* Kerckring, Dirk.

Bone of Pirie: *Eponym* for the dorsal astragaloscaphoid bone; *see* Pirie, Geoege Alexander.

Bones of Riolan: *Eponym* for small sutural bones that may be present in the petrooccipital suture; *see* Riolan, Jean.

Bones of Rousseau: *Eponym* for an accessory lacrimal bone anterior to the proper lacrimal bone, also described by Rosenmüller and Weber; *see* Rousseau, Louis François Emanuel.

Bone of Sylvius: *Eponym* for the lenticular process of the malleus; *see* Sylvius, Jacobus.

Bone of Vesalius: *Eponym* for (a) the fabella, a sesamoid bone that may be present near the origin of the lateral head of the gastrocnemius muscle, and (b) the tuberosity of the fifth metatarsal bone when it occurs as a separate bone. *Eponyms* are also in use for accessory bones of the foot and hand, and are known as **Os vesalium pedis** and **Os vesalium manus**; *see* Vesalius, Andreas.

Bones of Worm: *Eponym* for sutural bones, also known as Wormian bones; *see* Worm, Ole.

Bonnet, Amédée (1802–1852): French surgeon and anatomist born in Ambérieux, Ain, and working in Lyon. Reference: *Traité des sections tendineuses et musculaires dans le strabisme, la myopie, la disposition à le fatique des yeux, le bégaiement, les pieds bots, etc, suivi d'un mémoire sur le névrotomie sous-cutanée.* Lyon, 1841.

Fig. B27 Plaquette of Botallo. (From Wikimedia Commons; in public domain)

Capsule of Bonnet: *Eponym* for the anterior part of the bulbar fascia or capsule of the eyeball (capsule of Tenon).

Borges, Albert F (1930–1990): Cuban plastic and reconstructive surgeon born in Havana and working in Havana and later in Falls Church, USA.

Lines of Borges: *Eponym* for the resting skin tension lines, rather similar to the lines of Kraissl. Reference: *Relaxed skin tension lines vs. other skin lines.* Plast Reconstr Surg 1984; 73:144–150.

Botallo, Leonardo (1530–1587?): Italian physician born in Asti, Piedmont, and physician to King Charles IX of France, subsequently holding positions in Pavia and Paris (Fig. B27). Reference: *Opera omnia medica et chirurgica.* Leiden, 1660, to which van Horne added some figures (Fig. B28). For biography, *see* Tozzo P, Zanatta A, D'Anguiolla G, Caenazzo L, Zampieri F. *Leonardo Botallo (1530–1587) and his pioneering contribution to traumatology, cardiology and dentology.* J Med Biogr 2022; 30:50–56.

Duct of Botallo: *Eponym* for the ductus arteriosus.

Foramen of Botallo: *Eponym* for the foramen ovale of the heart.

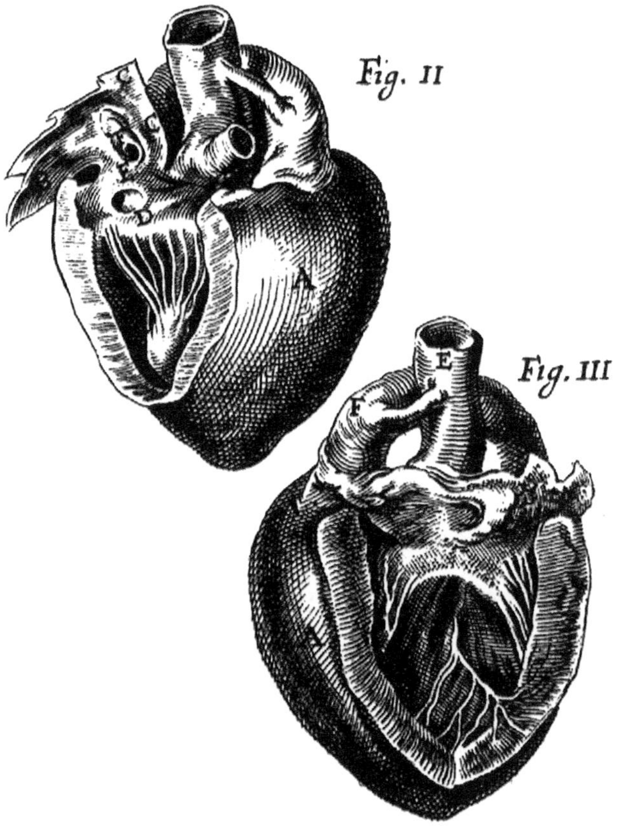

Fig. B28 Illustration by van Horne on duct and foramen of Botallo in Botallo's *Opera Omnia*. (From Wikimedia Commons; in public domain)

Ligament of Botallo: *Eponym* for the ligamentum arteriosum (Fig. B28).

Node of Botallo: *Eponym* for a lymph node at the ligamentum arteriosum.

Bourgery, Jean-Baptiste Marc (1797–1849): French surgeon and anatomist born in Orléans and working in Paris. Reference: *Traité complet de l'anatomie de l'homme comprenant la médecine opératoire*. Paris, 1837–1840.

Ligament of Bourgery: *Eponym* for (a) the oblique popliteal ligament, and (b) the intertransverse ligaments of the lumbar vertebral column.

Preisthmic lymph nodes of Bourgery: *Eponym* for accessory lymph nodes in front of the isthmus of the thyroid gland.

Tibiopopliteal lymph node of Bourgery: *Eponym* for an inconstant lymph node located on the tendinous arch of the soleus muscle.

Fig. B29 Photo of Bowman. (From Wikimedia Commons; in public domain)

Bowman, Sir William (1816–1892): English surgeon and anatomist born in Nantwick, Cheshire, and working in London. He is noted for his studies of the anatomy and physiology of the eye and of the kidney (Fig. B29). References: *On the structure and use of the Malpighian bodies of the kidney with observations on the circulation through that gland.* Phil Trans 1842; 132:57–80; *Lectures on the parts of the eye concerned in the operations, and on the structure of the retina.* Longmans, London, 1849; *see* also Jamison RL *Resolving an 80-year-old controversy: The beginning of the modern era of renal physiology.* Adv Physiol Edu 2014; 38:286–295.

Capsule of Bowman: *Eponym* for the glomerular capsule of the nephron in the kidney (Fig. B30).

Fibres or **Muscle of Bowman:** *Eponym* for the meridional fibres of the ciliary muscle, published in the same year as Brücke, but already described in 1805 by Crampton.

Glands of Bowman: *Eponym* for the olfactory glands, mucous glands in the olfactory part of the nasal cavity.

Membrane of Bowman: *Eponym* for the anterior limiting membrane of the cornea, also described by Reichert.

Boyd, Alexander Michael (1905–1973): English surgeon and phlebologist working in Manchester.

Vein of Boyd, Perforating: *Eponym* for the superior paratibial perforating vein some 2.5 cm below the level of the knee connecting the great saphenous vein and the posterior tibial veins. Reference: *Treatment of varicose veins: Possible changes of injection of sclerosing fluids.* Br Med J 1947 2452–454 (with DJ Robertson).

Boyden, Edward (1886–1977): American anatomist, known in particular for his studies on the segmental anatomy of the lungs. Reference: *Segmental Anatomy of the Lungs: A study of the patterns of the segmental bronchi and related pulmonary vessels.* McGraw-Hill, New York, 1953.

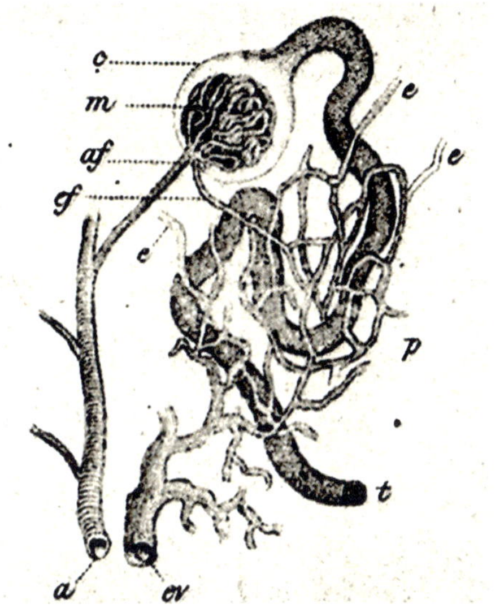

Fig. B30 Capsule of Bowman. (From Bowman W *On the structure and use of the Malpighian bodies of the kidney with observations on the circulation through the gland.* Phil Trans 1842; 132:57–80). The artery (*a*) gives a terminal twig (*af*) to a Malpighian tuft (*m*) from which emerges the efferent vessel (*ef*). Other efferent vessels (*e*) are also shown entering the capillary plexus (*p*) that surrounds the uriniferous tube (*t*). From this plexus, the renal vein (*ev*) arises

Four part sphincter of Boyden: *Eponym* for a four-part sphincter system for the bile duct, the pancreatic duct and the hepatopancreatic duct: (a) superior and (b) inferior (preampullary) sphincters of the bile duct, (c) the sphincter of the pancreatic duct, and (d) the sphincter of the hepatopancreatic or biliaropancreatic ampulla of Oddi (Fig. B31). References: *The sphincter of Oddi.* Surgery 1941; 9:443–445; *Hypertrophy of the sphincter choledochus: A cause of internal biliary fistula.* Surgery 1941; 10:567–571; *The anatomy of the choledochoduodenal junction in man.* Surg Gynecol Obst 1957; 104:641–652.

Boyer, Alexis (1757–1833): French surgeon and anatomist. Reference: *Traité complet d'anatomie ou description de toutes les parties du corps humain.* Paris, 1797–1799.

Bursa of Boyer: *Eponym* for the retrohyoid bursa, a tissue space anterior to the thyrohyoid membrane.

Bradlaw, Sir Robert Vivian (1905–1992): English dentist.

Plexus of Bradlaw, Predentinal: *Eponym* for the marginal nerve plexus of free nerve endings between the odontoblast layer and the dentin in permanent teeth (*see* Fig. 23.2). Reference: *The histology and histopathology of dental innervation.* Proc Roy Soc Med 1939; 32:1040–1053.

Fig. B31 Boyden's subdivision of sphincters of the ductus choledochus and of the hepatopancreatic ampulla. (After Boyden EA *Hypertrophy of the sphincter choledochus: A cause of internal biliary fistula.* Surgery 1941; 10:567–571)

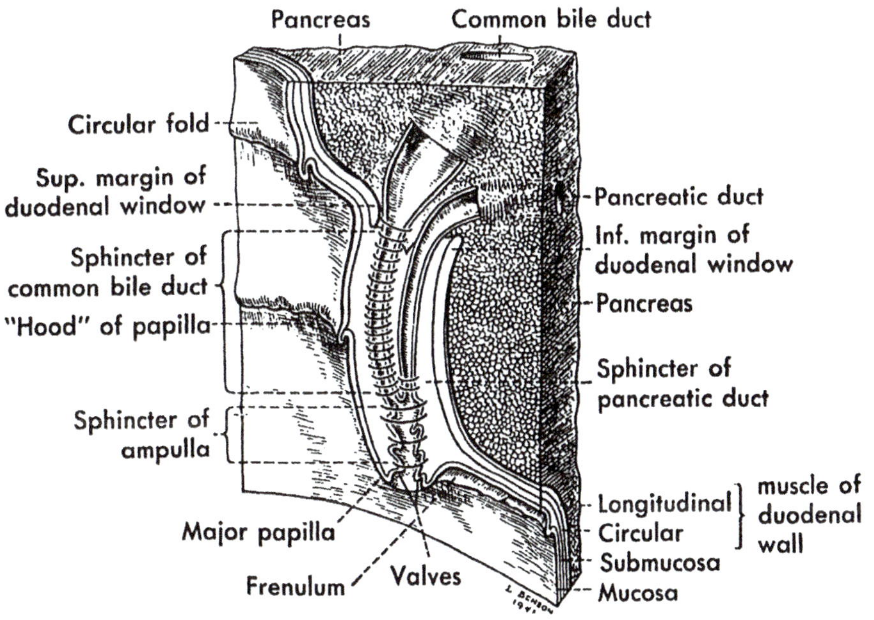

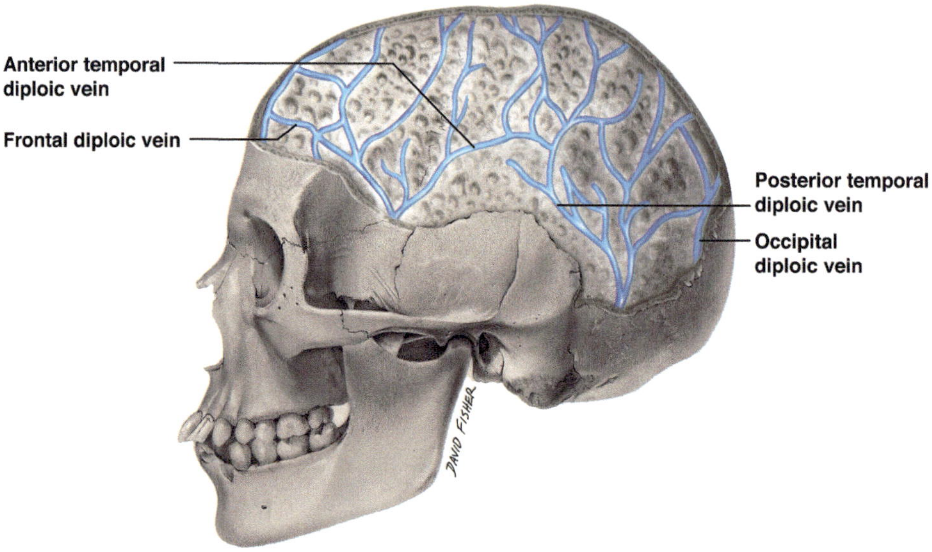

Fig. B32 The diploic veins of Breschet. (From ten Donkelaar HJ, Kachlík D, Tubbs RS *An Illustrated Terminologia Neuroanatomica*. Springer, Cham, 2018; courtesy David Fisher, Birmingham, AL)

Braun, Max (1850–1920): German zoologist and comparative anatomist born in Myslowitz, Silesia, working in Dorpat, Rostock and later in Königsberg.

Canal of Braun: *Eponym* for the embryonic neurenteric canal, also described by Kowalewsky.

Braune, Christian Wilhelm (1831–1892): German surgeon and anatomist born and working in Leipzig. Reference: *Topographisch-anatomischer Atlas*. Leipzig, 1875.

Arcade of Braune: *Eponym* for a superficial vein at the root of the toes, opening into the superficial plantar venous plexus by intercapitular veins.

Fold or **Valve of Braune:** *Eponym* for a perpendicular mucous fold at the cardiac ostium of the stomach, described as the cardiac sphincter between the oesophagus and the stomach.

Muscle of Braune: *Eponym* for the puborectalis muscle, part of the levator ani muscle.

Veins of Braune: *Eponym* for (a) an accessory superior intercostal vein, (b) anterior saphenotibial veins, (c) posterior saphenotibial veins, and (d) saphenofibular veins. Reference: *Die Venen des Fusses und Unterschenkels*. Leipzig, 1890.

Breschet, Gilbert (1784–1845): French anatomist born in Clermont and working in Paris as successor of Cruveilhier. Reference: *Traité d'anatomie humaine* (with Alexandre Brière de Boismont). Paris, 1834.

Bones of Breschet: *Eponym* for the inconstant suprasternal bones found in ligaments of the sternoclavicular joint.

Canals of Breschet: *Eponym* for canals for the diploic veins in the skull, later also described by Dupuytren. Reference: *Essai sur les veins du rachis*. Paris, 1819

Columella of Breschet: *Eponym* for the modiolus of the cochlea.

Hiatus of Breschet: *Eponym* for the helicotrema of the cochlea. Reference: *Etudes sur l'ouie*. Paris, 1834

Sinus of Breschet: *Eponym* for the sphenoparietal sinus along the margin of the lesser wing o the sphenoidal bone.

Veins of Breschet: *Eponym* for the diploic veins (Fig. B32). Reference: *Recherches anatomiques, physiologiques et pathologiques sur le système veineux*. Villaret, Paris, 1829.

Brissaud, Edouard (1852–1909): French physician and pathologist born in Besançon and working in Paris. Reference: *Anatomie du cerveau de l'homme; morphologie des hémisphères cérébraux, ou cerveau proprement dit*. Masson, Paris, 1893.

Sulcus of Brissaud: *Eponym* for the transverse parietal sulcus of the cerebral cortex.

Fig. B33 Photo of Broca. (From Pagel 1901)

Broca, Pierre Paul (1824–1880): French surgeon, anatomist and anthropologist in Paris (Fig. B33), born in Sainte Foy la Grande, who described aphasia as a manifestation of injury to the left inferior frontal gyrus and introduced the concept of cerebral localization (*see* Schiller, F 1979 *Paul Broca. Explorer of the Brain*. University of California Press; Paperback 1992 by Oxford University Press, New York-Oxford). He is also known for his studies of the cerebral hemisphere and for defining the *'Grand lobe limbique'*. References: *Remarques sur le siège de la faculté du language articulé, suivies d'une observation d'aphémie (perte de la parole).* Bull Soc Anat (Paris) 1861, 2e Série, 6:330–357; *Anatomie descriptive du corps humain.* Paris, 1865; *Nomenclature cérébrale: Dénomination et subdivision des hémisphères et des anfractuosités sur la surface.* Rev Anthropol 1878, 2:193–236; *Anatomie comparée des circonvolutions cérébrales. Le grand lobe limbique et la scissure limbique dans la série des mammifères.* Rev Anthropol 1878, 2:385–498. Broca's name is also used for several angles, lines, planes and points used in physical anthropology.

Angles of Broca: Broca defined various angles for anthropological research, basilar, facial, frontal, occipital, parietal, suboccipital and supraoccipital.

Area of Broca: *Eponym* for (a) the triangular and opercular parts of the inferior frontal gyrus and (b) the paraolfactory area of the cerebral cortex.

Axis of Broca, Orbital or **Visual:** *Eponym* for the axis passing through the optic canal and the middle of the orbital base.

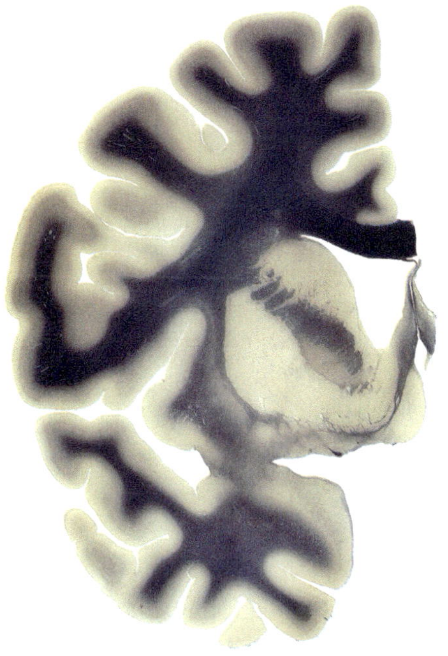

Fig. B34 Diagonal band of Broca shown in a Weigert-stained frontal section of the cerebrum from the Jelgersma Collection. (From ten Donkelaar HJ, Kachlík D, Tubbs RS *An Illustrated Terminologia Neuroanatomica.* Springer, Cham, 2018)

Band, Diagonal of Broca: *Eponym* for a band of fibres extending into the amygdaloid body from the paraolfactory gyrus on the medial side of the frontal lobe ventrally and along the lateral margin of the optic tract, marking the caudal boundary of the anterior perforated substance or olfactory tubercle (Fig. B34); also known as the olfactory radiation of Zuckerkandl.

Carrefour of the hemisphere of Broca: Broca's term for what now is known as the paraolfactory area, where limbic and extralimbic parts of the brain come together.

Centre of Broca: *Eponym* for the motor speech centre of Broca.

Convolution or **Gyrus of Broca:** *Eponym* for the left inferior frontal gyrus of the cerebral cortex, the so-called motor speech area.

Falciform fold of Broca: *Eponym* for the lateral border of the anterior perforated substance.

Great limbic lobe of Broca: *Eponym* for the cingulate and parahippocampal gyri, his 'grand lobe limbique' (Fig. B35).

Horizontal plane of head of Broca: *Eponym* for the alveolar-condyle plane of the skull passing from the alveolar point at the inferior border of the alveolar ridge to the most inferior part of the occipital condyles.

Limen of Broca: *Eponym* for the corpus callosum, the fornix and the anterior commissure.

Lines and Planes of Broca: Broca distinguished several reference lines and planes:

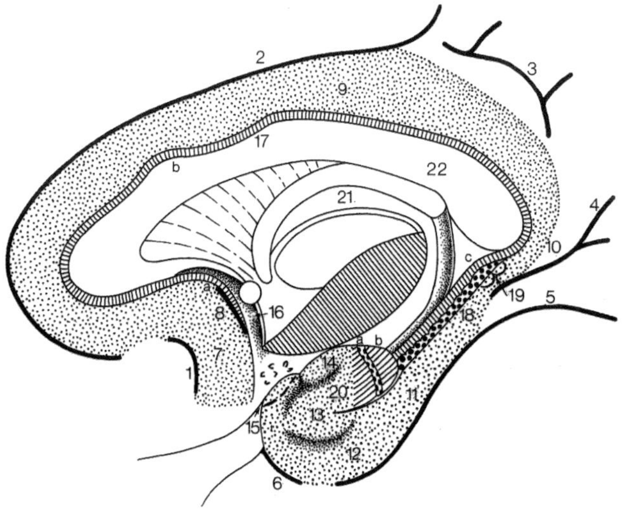

Fig. B35 The great limbic lobe of Broca. (After Duvernoy HM *The Hippocampus*, 2nd ed. Springer, Berlin-Heidelberg-New York, 1998; from ten Donkelaar HJ, Kachlík D, Tubbs RS *An Illustrated Terminologia Neuroanatomica.* Springer, Cham, 2018). *1* anterior paraolfactory sulcus, *2* cingulate sulcus, *3* subparietal sulcus, *4* rostral part of calcarine sulcus, *5* collateral sulcus, *6* rhinal sulcus, *7* subcallosal gyrus, *8* posterior paraolfactory sulcus, *9* cingulate gyrus, *10* isthmus, *11* parahippocampal gyrus, *12* entorhinal cortex, *13* ambient gyrus, *14* semilunar gyrus, *15* piriform cortex, *16* paraterminal gyrus, *17* indusium griseum, *18* dentate gyrus, *19* gyri of Andreas Retzius, *20* uncinate gyrus, *21* fornix, *22* corpus callosum, *a* bandelette of Giacomini, *b* apex of uncus

Horizontal plane of head (1862): *Eponym* for the alveolar-condylar plane from the alveolar point at the inferior border of the alveolar ridge to the inferior aspect of both occipital condyles.

Plane of horizontal vision (1862) or **Visual plane (1873):** *Eponym* for the natural attitude of the head that permits the eyes to reach the horizon without muscular contraction (Fig. B36). This plane was accepted by British anthropologists as well. German anthropologists preferred the plane of Virchow and Hoelder that came to be known as the Frankfurt-Virchow plane or 'Frankfurter Horizontale'. In 1877, Broca renamed this plane as **bi-orbital plane** (Fig. B36), corresponding to the **neuro-ocular plane** of Cabanis et al. (J Comput Assist Tomogr 1978; 2:150–155).

References: Broca P *Sur les projections de la tête, et sur le nouveau procédé céphalométrie.* Bull Soc Anthropol Paris, 1862; 3:514–544; Broca P *Nouvelles recherches sur le plan horizontal de la tête et sur le degré d'inclination des divers plans crâniens.* Bull Soc Anthropol Paris, 1873; 8:542–563; Broca P *Sur l'angle orbito-occipital.* Bull Soc Anthropol Paris, 1877; 12:325–333

Metopic lobule: *Eponym* for part of the internal frontal gyrus underneath the rostral sulcus of Eberstaller.

Metopic sulcus of Broca: *Eponym* for a sulcus of the inferior part of the inferior frontal gyrus, parallel to the cingulate gyrus.

Nucleus of the diagonal band of Broca: *Eponym* for a nucleus in the basal forebrain related to the diagonal band.

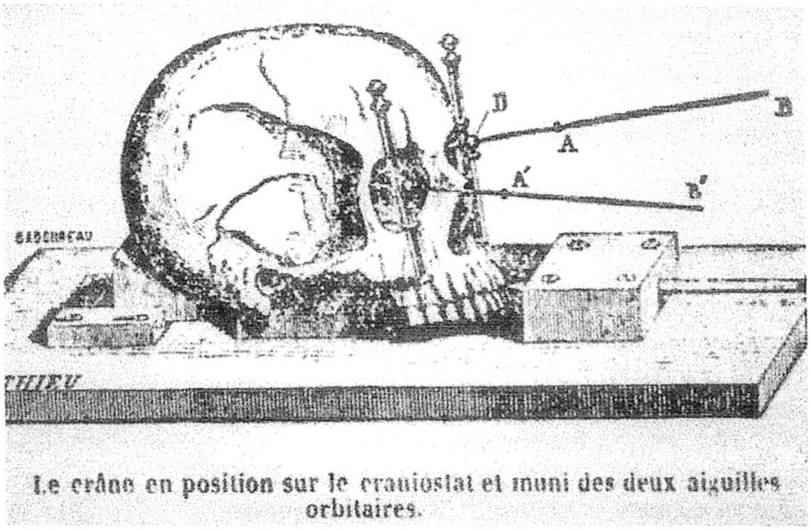

Fig. B36 Broca's biorbital plane. (From Broca P *Nouvelles recherches sur le plan horizontal de la tête et sur le degré d'inclination des divers plan crâniens.* Bull Soc Anthropol Paris, 1873; 8:542–563)

Fig. B37 Photo of Brodmann. (From Pagel 1901)

Pouch of Broca: *Eponym* for the pudendal sac, a pear-shaped encapsulated collection of connective and adipose tissue in the labium majus.

'Scissure limbique' of Broca: *Eponym* for the sulcus separating the limbic lobe from the rest of the brain, usually composed of the collateral and rhinal sulci (*see* Fig. B35).

Brodie, Sir Benjamin Collins (1783–1862): English anatomist and surgeon born in Winterslow, Wiltshire, and working in London, known for his description of various bursae.

Bursa of Brodie: *Eponym* for (a) the medial gastrocnemial subtendinosus bursa, and (b) the semimembranosus bursa. Reference: *Pathological and Surgical Observations on the Diseases of Joints,* 5th ed. London, 1850.

Brodie, Charles Gordon (1860–1933): English anatomist and surgeon in London.

Ligament of Brodie: *Eponym* for the transverse humeral ligament, but described by many anatomists before 1889 (Dobson 1962). Reference: *Note on the transverse humeral, coraco-acromial and coraco-humeral ligaments.* J Anat (Lond) 1889, 24:247.

Brodmann, Korbinian (1868–1918): German neurologist born in Liggersdorf and working in Tübingen, Halle and finally in Munich (Fig. B37) noted for his studies of the cytoarchitecture of the mammalian cerebral cortex, especially for his maps of the human cortex, which started to be published in 1908.

Areas of Brodmann: *Eponym* for the cortical areas distinguished cytoarchitectonically by Brodmann (Fig. B38). References: *Beiträge zur histologischen Lokalisation der Grosshirnrinde. VI. Mitteilung: Die Cortexgliederung des Menschen.* J Psychol Neurol (Lpz) 1908, 10:231–246; *Vergleichende Lokalisationslehre der Großhirnrinde in ihren Prinzipien dargestellt auf Grund des Zellenbaues.* Barth, Leipzig, 1909; translated by Lawrence J. Garey as *Brodmann's 'Localisation in the Cerebral Cortex'.* Imperial College Press, London, 1999.

Brödel, Max (1870–1941): German medical illustrator born in Leipzig who in 1893, invited by Franklin Mall, went to the USA, where he became one of the founders of medical illustration in the USA (Fig. B39a). At the Johns Hopkins Hospital, he illustrated a textbook on gynaecology by Howard Kelly, and collaborated with eminent surgeons such as Cushing, Halsted and Dandy. In 1911, he was appointed Director of the Department of Art as Applied to Medicine at Johns Hopkins. His three 'unpublished' drawings of the human ear are still famous (Brödel M *Three Unpublished Drawings of the Anatomy of the Human Ear.* Saunders, Philadelphia, PA, 1938).

Avascular, **Bloodless** or **White line of Brödel:** *Eponym* for the line of division of the kidney between the segments supplied by the anterior and posterior branches of the renal artery (Fig. B39b), also described by Hyrtl. Reference: *The intrinsic blood vessels of the kidney and their significance in nephrotomy.* Johns Hopkins Hosp Bull 1901; 12:10–13.

Brösike, Gustav (1853–1923): German surgeon and anatomist born in Puppen, then in East Prussia (currently Spychowo, Poland), and working in Berlin.

Fossa of Brösike: *Eponym* for the intermesocolic fossa of the peritoneal cavity (Fig. B40). Reference: *Ueber intra-abdominale Hernien und Bauchfelltaschen, nebst einer Darstellung der Entwicklung peritonealer Formationen.* Berlin, 1891.

Browne, Sir Denis John Wolko (1892–1967): English paediatric surgeon of Australian origin born in Melbourne and working in London.

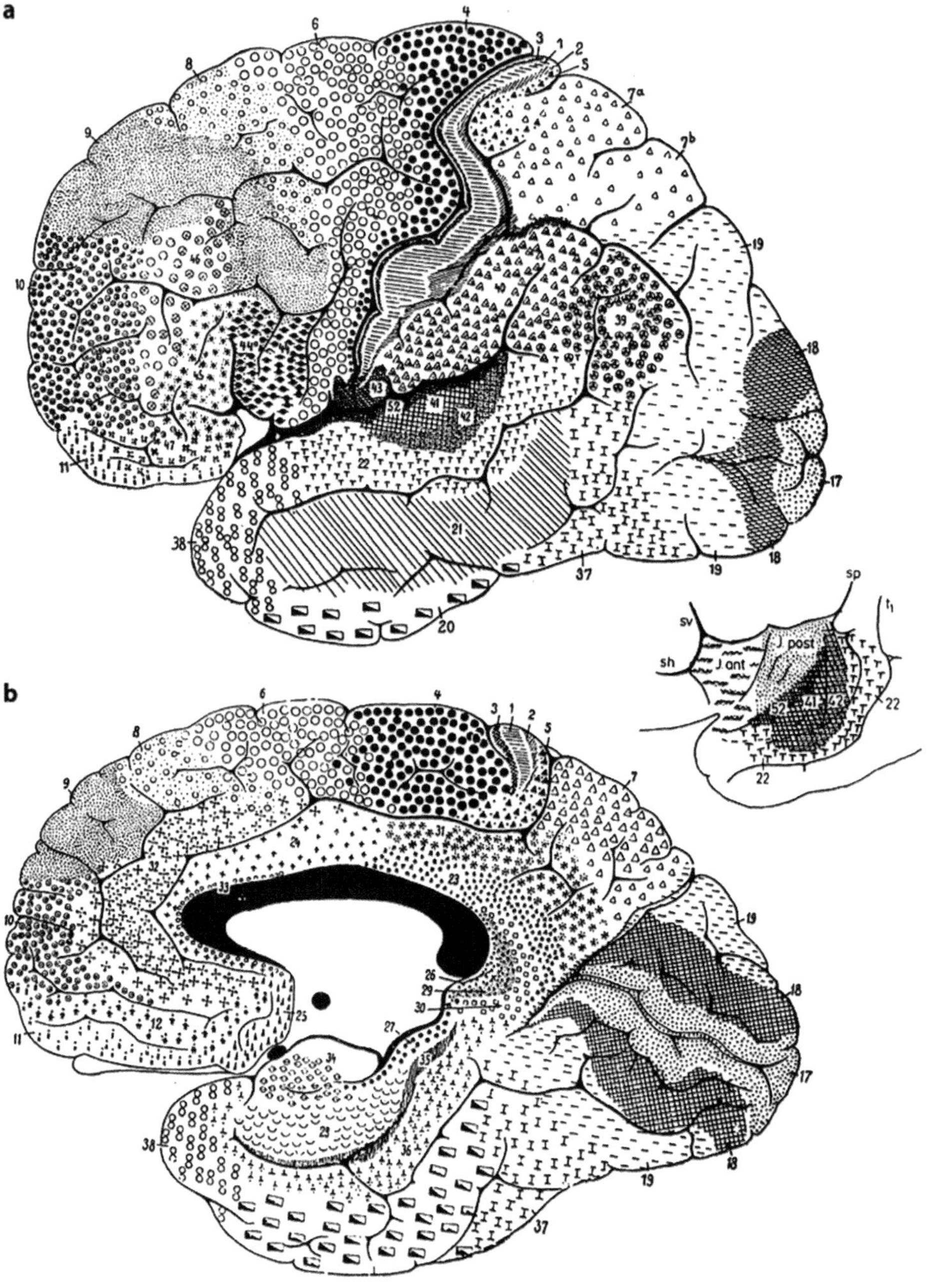

Fig. B38 Brodmann's subdivision of the human cerebral cortex in 52 areas; **a** lateral surface, **b** medial surface; the *insert* shows the insular cortex. (*J.ant* and *J.post*; from ten Donkelaar HJ, Lammens M, Hori A *Clinical Neuroembryology,* 3rd ed. Springer, Cham, 2023)

Pouch of Browne: *Eponym* for the superficial inguinal pouch, a pocket formed between the membranous layer of the subcutaneous tissue and the superficial abdominal fascia adjacent to the supericial inguinal ring, a common lodging site for an undescended or retractile testis. Reference: *Diagnosis of undescended testicle.* Br Med J 1938; 2:168–173.

Browning, William (1855–1941): American anatomist and neurologist in New York. His main studies concerned the normal and pathological circulation of the nervous system (Dobson 1962). Reference: *The Veins of the Brain and Its Envelopes: Their anatomy and bearing on the intracranial circulation.* Brooklyn, 1884.

Vein of Browning: *Eponym* for one of the superficial veins of the brain, his 'middle cerebral vein', the upper part of the superior anastomotic vein of Trolard.

Bruce, Alexander (1854–1911): Scottish neurologist born in Ardiffery and working in Edinburgh.

Tract of Bruce: *Eponym* for the septomarginal tract of the spinal cord, also described by Flechsig and Muir,

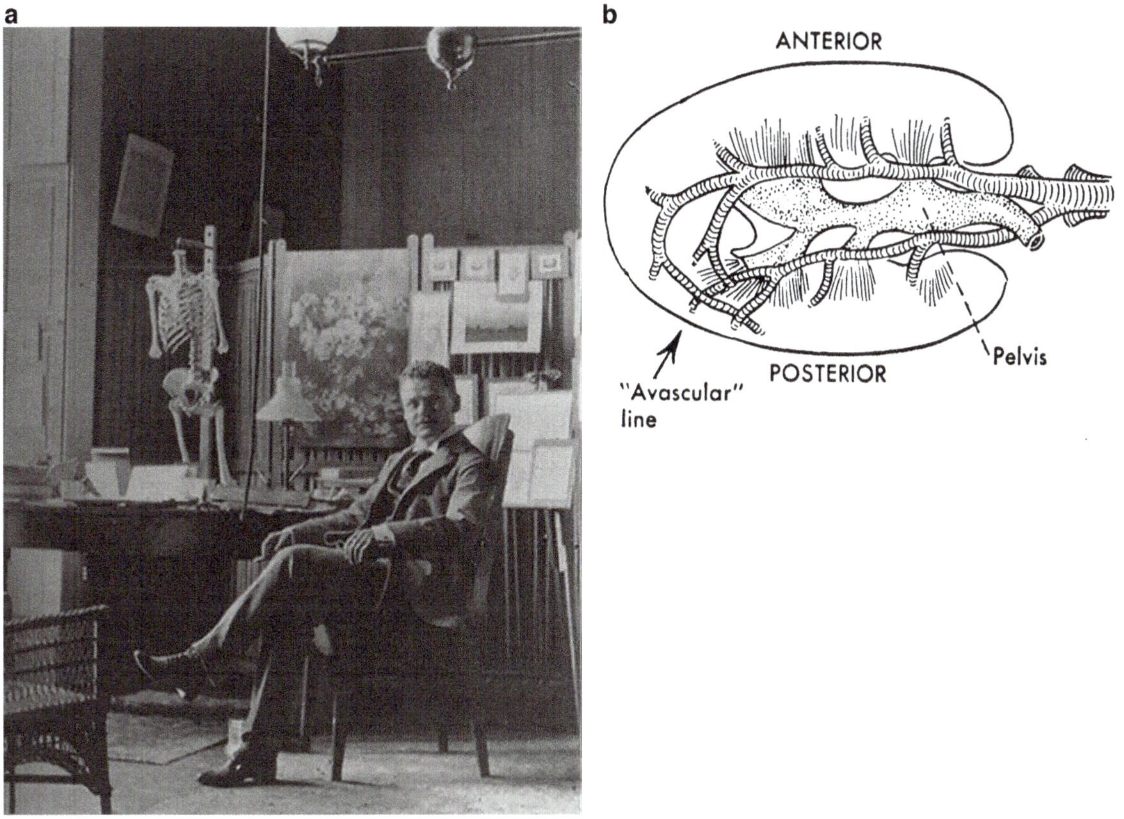

Fig. B39 (**a**) Photo of Brödel. (From the Alan Mason Chesney Medical Archives of the Johns Hopkins Medical Institutions; in public domain); (**b**) The bloodless or avascular line of Brödel, the region of overlap between the anterior and posterior branches. (After Hollinshead WH *Anatomy or Surgeons, Vol 2: The Thorax, Abdomen and Pelvis,* 3nd ed. Harper & Row, New York, 1971; and ten Donkelaar HJ, Oostra R-J *Klinische Anatomie en Embryologie,* 4th ed. Springer Media, Houten, 2014)

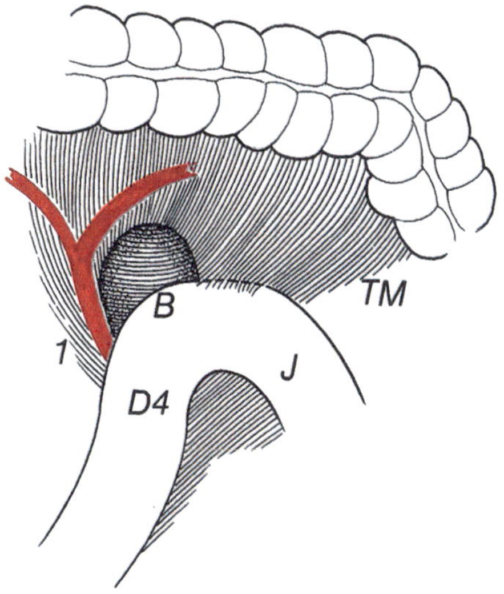

Fig. B40 Intermesocolic fossa of Brösike. (After Gray SW, Skandalakis JE *Embryology for Surgeons.* Saunders, Philadelphia, PA, 1972). *B* fossa of Brösike, *D4* fourth part of duodenum, *J* jejunum, *TM* transverse mesocolon, *1* middle colic artery (in *red*)

Fig. B41 The membrane of Bruch shown in a photomicrograph of the choroid. (From Brüel A, Christensen EI, Tranum-Jensen J, Qvortrup K, Geneser F 2012 *Genesers histologi*. Munksgaard, Copenhagen; with permission, and ten Donkelaar HJ, Kachlík D, Tubbs RS *An Illustrated Terminologia Neuroanatomica*. Springer, Cham, 2018). *1* basal complex (membrane of Bruch), *2* choroidocapillary layer (membrane of Ruysch), *3* substantia propria, *4* suprachoroid layer, *5* sclera

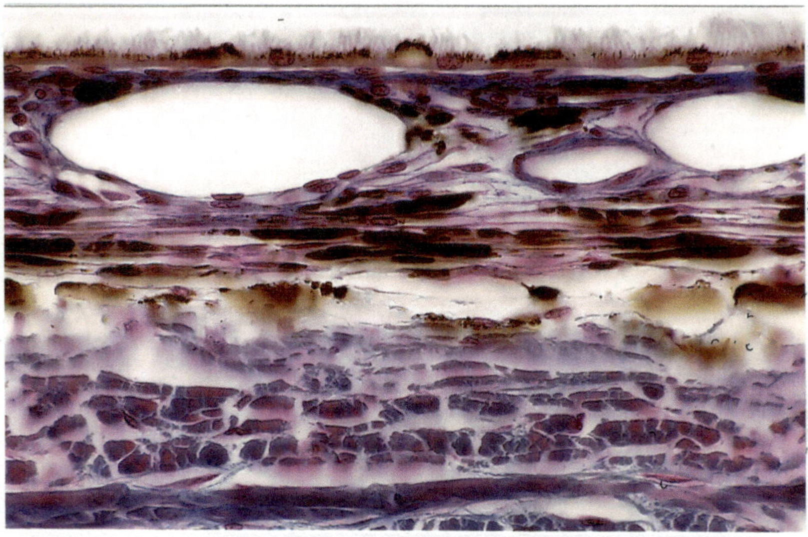

Fig. B42 Photo of Brücke. (From Pagel 1901)

and its sacral continuation as the triangle of Philippe-Gombault. References: *On a descending degeneration in the posterior columns in the lumbosacral region of the spinal cord*. Brain 1896, 19:333–345; *A Topographical Atlas of the Spinal Cord*. London, 1901.

Bruch, Carl Wilhelm Ludwig (1819–1884): German anatomist born in Mainz, working in Basel and later in Giessen. Reference: *Untersuchungen zur Kenntniss des körnigen Pigments der Wirbelthiere in physiologischer und pathologischer Hinsicht*. Meyer und Zeller, Zürich, 1844.

Glands of Bruch: *Eponym* for lymphatic nodules in the palpebral conjunctiva.

Membrane of Bruch: *Eponym* for the basal complex of the choroid of the eyeball (Fig. B41).

Brücke, Ernst Wilhelm, Ritter von (1819–1892): German physiologist and histologist born in Berlin and mainly working in Vienna (Fig. B42).

Fibres or **Muscle of Brücke:** *Eponym* for the meridional fibres of the ciliary muscle, also described by Bowman in the same year, but earlier by Crampton (1805). Reference: *Anatomische Beschreibung des menschlichen Augapfels*. Berlin, 1847.

Brunn, Albert von (1849–1895): German anatomist born in Zschorno and working in Göttingen, a student of Waldeyer.

Cell nests of Brunn: *Eponym* for epithelial cell masses within the lamina propria of the mucosa, mainly in the urinary bladder, and a reactive change. Reference: *Ueber drüsenähnliche Bildungen in der Schleimhaut des Nierenbeckens, des Ureters und der Harnblase beim Menschen*. Arch Mikr Anat 1893; 41:294–302.

Membrane of Brunn: *Eponym* for the olfactory epithelium of the nasal conchae. Reference: *Weitere Untersuchungen über das Riechepithel*. Arch Mikr Anat 1869; 17:141.

Brunner, Johann Konrad (1653–1727): Swiss anatomist born in Diessenhofer who studied under Duverney in Paris, and worked in Heidelberg and Strassburg.

Glands of Brunner: *Eponym* for duodenal glands, previously described in 1679 by Wepfer, his father-in-law. Reference: *Descriptio de glandulis in duodeno intestino detectis.* Heidelberg, 1687.

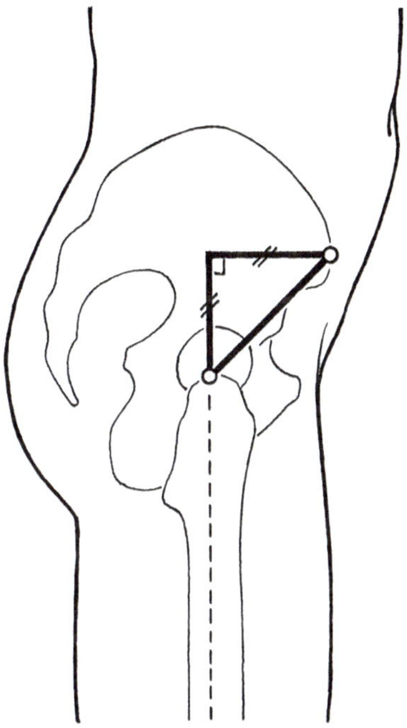

Fig. B43 Triangle of Bryant

Bryant, Thomas (1828–1914): English surgeon born in Kennington and working in London, known for his studies on diagnosing fractures of the neck of the femur. Reference: *The Diseases and Injuries of the Joints, Clinical and pathological observations.* London, 1861.

Triangle of Bryant: *Eponym* for the iliofemoral triangle, surgical marking for diagnosing fractures of the neck of the femur (Fig. B43).

Buck, Gordon (1807–1877): American surgeon born and working in New York.

Fascia of Buck: *Eponym* for (a) the deep fascia of the penis, and (b) the deep perineal fascia investing the perineal muscles, also described by Gallaudet. Reference: *A new feature in the anatomical structure of the genito-urinary organs not hitherto described.* Trans Amer Med Ass 1848; 1:367.

Budde-Rocko triangle: *Eponym* for the cystohepatic triangle bounded by the inferior border of the liver, laterally by the gall bladder and the cystic duct, and medially by the common hepatic duct (Fig. B44). Reference: Rocko JM, Swan KG, Di Gioia JM. *Calot's triangle revisited.* Surg Gynec Obstet 1981; 153:40–44. Its lower part is known as the triangle of Calot.

Budge, Julius Ludwig (1811–1888): German anatomist and physiologist born in Wetzlar and working in Bonn and later in Greifswald.

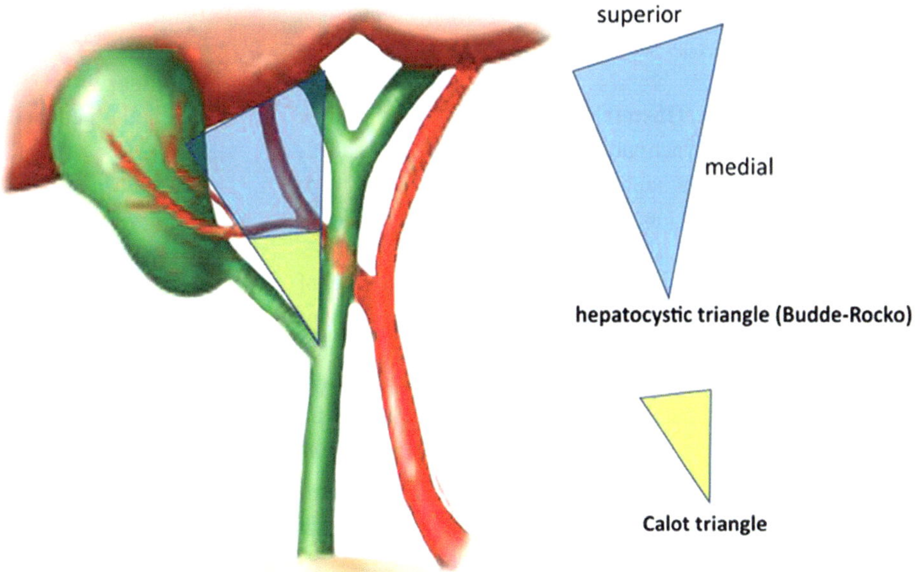

Fig. B44 Budde-Rocko triangle. (From Mischinger H-J, Wagner D, Korprat P, Becker H, Werkgartner G. Eur Surg *The "critical view of safety (CVS)" cannot be applied – What to do? Strategies to avoid bile duct injuries.* 2020; https://doi.org/10.1007/s10353-020-00660-1; CC-BY 4.0, courtesy Hans-Jörg Mischinger, Graz)

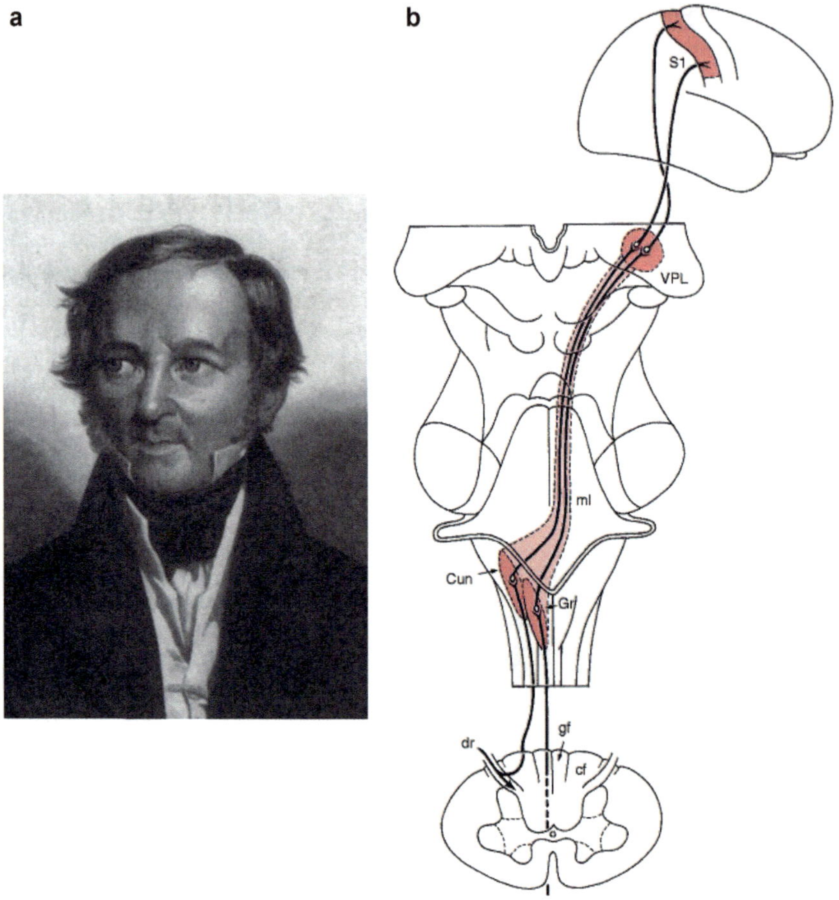

Fig. B45 (**a**) Portrait of Burdach. (From Pagel 1901); (**b**) The dorsal column-medial lemniscus pathway: Overview. *cf* cuneate fascicle, *Cun* cuneate nucleus, dr dorsal root, *gf* gracile fascicle, *Gr* gracile nucleus, *ml* medial lemniscus, *S1* primary somatosensory cortex, *VPL* ventral posterolateral nucleus. (From ten Donkelaar HJ, Kachlík D, Tubbs RS *An Illustrated Terminologia Neuroanatomica*. Springer, Cham, 2018)

Centre of Budge: *Eponym* for the ciliospinal centre in the cervical part of the spinal cord. References: *Untersuchungen ueber das Nervensystem.* Frankfurt, 1841–1842; *Bewegung der Iris.* Braunschweig, 1855.

Budin, Pierre-Constant (1846–1907): French obstetrician born in Enencourt-le-Sec and working in Paris.

Joint of Budin, Obstetrical: *Eponym* for the posterior intraoccipital synchondrosis, a cartilaginous connection between the squamous and lateral parts of the occipital bone in the newborn.

Bulbous corpuscle of Ruffini: The bulbous corpuscle of Ruffini is a slowly-adapting, spindle-shaped receptor sensitive to sustained pressure. It is located mainly in the gums, glans and joint capsules but it is ubiquitous. It forms an encapsulated unmyelinated multiply-branched nerve fibre situated deep in the dermis and the hypodermis as well; *see* Ruffini, Angelo.

Bundles: *See* **The Classical Anatomical Eponyms.**

Burdach, Karl Friedrich (1776–1847): German anatomist and physiologist (Fig. B45a) born in Leipzig and working in Dorpat (currently Tartu, Estonia) and Königsberg (currently Kaliningrad, Russia). Reference: *Vom Baue und Leben des Gehirns,* 3 Vols. Dyk, Leipzig, 1819–1826.

'Bindearm' of Burdach: *Eponym* for the superior cerebellar peduncle or brachium conjunctivum.

Nucleus of Burdach: *Eponym* for the cuneate nucleus of the caudal hindbrain (Fig. B45b).

Tract or **Fasciculus of Burdach:** *Eponym* for the cuneate fasciculus of the spinal cord (Fig. B45b).

Burns, Allan (1781–1813): Scottish anatomist born and working in Glasgow. References: *Observations on the structure of the parts concerned in crural hernia.* Edinb Med Surg J 1806; 2:265–274; *Observations on the Surgical Anatomy of the Head and Neck.* Bryce, Glasgow, 1811.

Ligament of Burns: *Eponym* for the superior horn of the falciform margin of the saphenous opening, although described three years after Hey.

Space of Burns: *Eponym* for the suprasternal space between the superficial and middle layers of the cervical fascia.

Burow, Karl August von (1809–1874): German anatomist and military surgeon born in Elbing, East Prussia (currently Elblag, Poland), and working in Königsberg (currently Kaliningrad, Russia). He is known for inventing new techniques in plastic surgery and wound healing.

Veins of Burow: *Eponym* for (a) the renal veins (Reference: *De vasibus sanguiferis renarum.* Königsberg, 1835), and (b) the paraumbilical veins, also described by Sappey and Baumgarten. Reference: *Beitrag zur Gefässlehre des Fötus.* Arch Anat Physiol 1838; pp 44–45.

Bursae:

Bursa of Boyer: *Eponym* for the retrohyoid bursa, a tissue space the hyoid bone and the thyrohyoid membrane; *see* Boyer, Alexis.

Bursa of Brodie: *Eponym* for (a) the medial gastrocnemial subtendinosus bursa, and (b) the semimembranosus bursa; *see* Brodie, Benjamin Collins.

Bursa of Calori: *Eponym* for an inconstant bursa between the trachea and the arcus aortae; *see* Calori, Luigi.

Bursa of Chassaignac: *Eponym* for a potential space between the pectoralis major muscle and the mammary gland; *see* Chassaignac, Charles Marie Edouard.

Bursa of Fabrici: *Eponym* for the site of hematopoiesis in birds; the bone marrow is the equivalent organ in mammals; *see* Fabrici, Girolamo.

Bursa of Poirier, Accessory: *Eponym* for (a) the bursa of the conoid ligament, (b) the bursa of the trapezoid ligament, and (c) the internal supracondylar bursa; *see* Poirier, Paul Julien.

Busi, Aristide (1874–1939): Italian radiologist working in Bologna and later in Rome.

Sphincter of Busi: *Eponym* for a group of circular fibres in the wall of the large intestine that mark the passage from the caecum to the ascending colon. Referemce: *Tecnica e diagnostica radiologica delle malattie chirurgicale.* Utet, Torino, 1933.

Letter C

4

Cabanas, Ramon Maximilian: American surgeon in Brooklyn. He recognized the sentinel lymph node for the penis.

Lymph node of Cabanas: *Eponym* for the superomedial superficial inguinal lymph node, the sentinel node for the penis. References: *An approach for the treatment of penile carcinoma.* Cancer 1977; 39:456–466; *Anatomy and biopsy of sentinel lymph nodes.* Urol Clin North Am 1992; 19:267–276.

Cabot, Richard Clarke (1868–1939): American physician.

Cabot rings: *Eponym* for remnants for nonextruded nuclei from orthochromatic erythroblasts, described in pernicious anemia. They are comparable to the Howell-Jolly bodies.

Cajal, Santiago Rámon y, *see* Rámon y Cajal, Santiago

Caldani, Leopold Marco Antonio (1725–1813): Italian anatomist in Padua, Veneto, as successor of Morgagni. He created the anatomical museum of the University of Padua. References: *Institutiones anatomicae.* Venice, 1787; *Icones anatomicae.* Venice 1801–1813 published with his nephew Floriano (1772–1836; Fig. C1).

Ligament of Caldani: *Eponym* for the coracoclavicular ligament of the shoulder.

Caldwell, William Edgar (1880–1943) and Moloy, Howard Carman (1903–1953): American obstetricians and gynaecologists in New York and New Yersey, respectively. They subdivided the female pelvis into android, anthropoid, gynecoid and platypelloid, using radiological pelvimetry but, by 1960, its clinical uselessness became apparent. Reference: Caldwell WE, and Moloy HC *Anatomical variations in the female pelvis and their effect on labor, with a suggested classification.* Amer J Obst Gyn 1933; 26:479–505.

Call, Emma Louise (1847–1937): American physiologist born in Newburyport, Massachusetts. She worked with Exner in Vienna as a postgraduate, specializing on the ovarium.

Call-Exner bodies: *Eponym* for small cavities found among the ovarian granulosa cells of growing and mature ovarian follicles. Their functional significance is unknown. Reference: *Zur Kenntniss des Graafschen Follikels und des Corpus luteum beim Kaninchen.* SB Akad Wiss Wien, Math-Naturw 1875; 71:321–328 (Call EM, Exner S).

Calleja y Sanchez, Julián (1836–1913): Spanish anatomist born and working in Madrid.

Islets of Calleja: *Eponym* for cell islands in the basal forebrain (*see* Fig. I6). Reference: *La region olfactoria del cerebro.* Moya, Madrid, 1893.

Calori, Luigi (1807–1896): Italian anatomist born in San Pietro in Casale and working in Bologna. Reference: *Tavole anatomiche rappresentanti la struttura del corpo umano.* Tipografia Sassi nelle Spaderie. Bologna, 1850–1855.

Bursa of Calori: *Eponym* for an inconstant bursa between the trachea and the aortic arch. Reference: *Di alcune borse mucose corrispondenti alla trachea e alla laringe e ad alcune delle parti adiacenti.* Mem R Acad Bologna 1874; 4:265.

Hepatic sinus of Calori: *Eponym* for a small enlargement of the inferior vena cava just above its junction with the hepatic veins.

H. J. ten Donkelaar et al., *An Illustrated Guide to Anatomical Eponyms*, https://doi.org/10.1007/978-3-031-91664-9_4

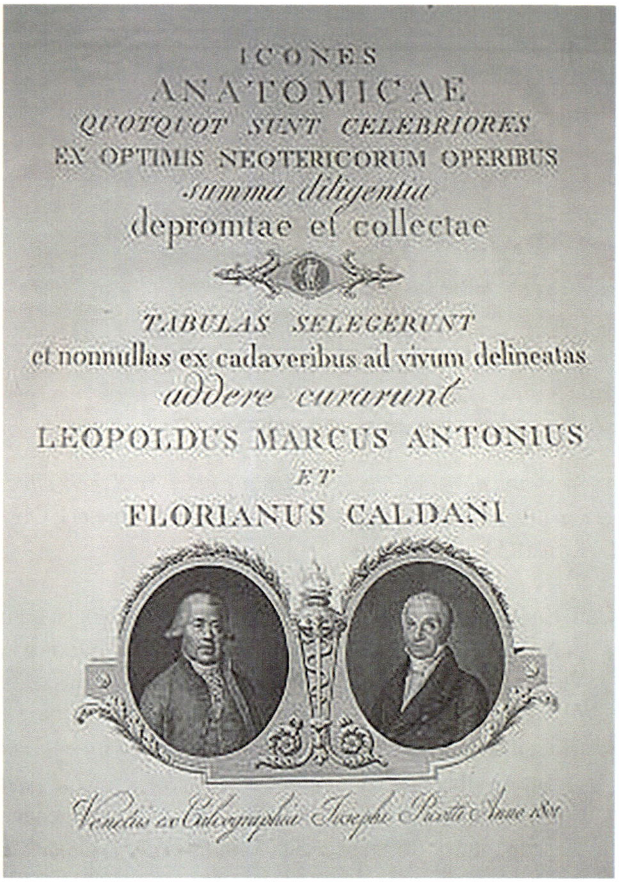
ICONES
ANATOMICAE
QUOTQUOT SUNT CELEBRIORES
EX OPTIMIS NEOTERICORUM OPERIBUS
summa diligentia
depromtae et collectae
TABULAS SELEGERUNT
et nonnullas ex cadaveribus ad vivum delineatas
addere curarunt
LEOPOLDUS MARCUS ANTONIUS
ET
FLORIANUS CALDANI
Venetiis ex Calcographia Josephi Picotti Anno 1801

Fig. C1 Title page of *Icones anatomicae* by Leopold and Floriano Caldani. (Venice, 1801–1813; courtesy Marina Quartu, Cagliari)

Hypophysial canal of Calori: *Eponym* for the cranio-pharyngeal canal, an inconstant canal traversing the sphenoidal bone, also described by Landzert.

Lymph node of Calori: *Eponym* for an inconstant gallbladder lymph node in the cystohepatic triangle, also described by Mascagni and Lund.

Renal sinus of Calori: *Eponym* for a small enlargement of the inferior vena cava just above its junction with the renal veins.

Stylomaxillary muscle of Calori: *Eponym* for an accessory muscle from the styloid process of the temporal bone to the angle of the mandibula. Reference: *Varietà des Muscoli del Tronco.* Bologna, 1868.

Calot, Jean-François (1861–1944): French surgeon from Arrens.

Trigone of Calot: *Eponym* for the cystohepatic trigone, a triangle bounded by the cystic artery, the cystic duct and the common hepatic duct (Fig. C2). Reference: *De la cholécystectomie (Ablation de la vésicule biliaire).* Thesis, Paris; Steinheil, Paris, 1890). It forms part of the Budde-Rocko triangle (*see* Fig. B44).

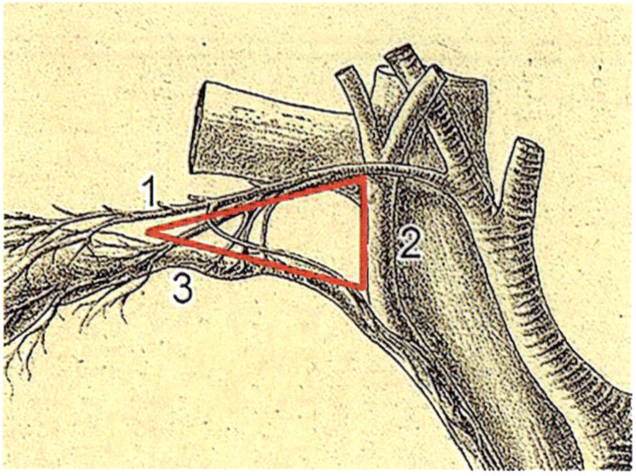

Fig. C2 Calot's drawing of his cystohepatic triangle. (From his Thesis, Steinheil, Paris, 1890; in public domain). *1* Cystic artery, *2* common hepatic duct, *3* cystic duct

Campbell, Alfred Walter (1868–1938): Australian pathologist born in Cunningham Plains, New South Wales. He studied in Vienna, Prague and Edinburgh and worked in various English asylums before returning to Australia in 1905, where he worked as a physician in Sydney (Fig. C3a). He is known for his histological studies of the cerebral cortex (Fig. C3b). Reference: *Histological studies of the Localisation of Cerebral Function.* Cambridge University Press, Cambridge, 1905.

Area of Campbell: *Eponym* for the precentral area of the frontal cortex.

Campbell, William Francis (1867–1926): American surgeon born and working in Brooklyn. Reference: *Textbook of Surgical Anatomy.* Philadelphia, 1908.

Ligament of Campbell: *Eponym* for the suspensory ligament of the axilla, previously described in 1851 by Gerdy.

Camper, Petrus (1722–1789): Dutch physician and anatomist born in Leiden and working in Franeker, Amsterdam and later in Groningen (Fig. C4). References: *Demonstrationum Anatomico-Pathologicarum, Liber Primus. Continens Brachii Humani Fabricam et Morbos.* Schreuder & Mortier, Amsterdam, 1760; Ibid., *Liber Secundus. Contineus Pelvis Humanae Fabricam et Morbos.* Schreuder & Mortier, Amsterdam. He also defined several anthropological angles, lines and planes.

Angle of Camper, Facial: *Eponym* for the angle formed by the profile of the face and the horizontal plane (Fig. C5a) determined by two lines, the first passing through

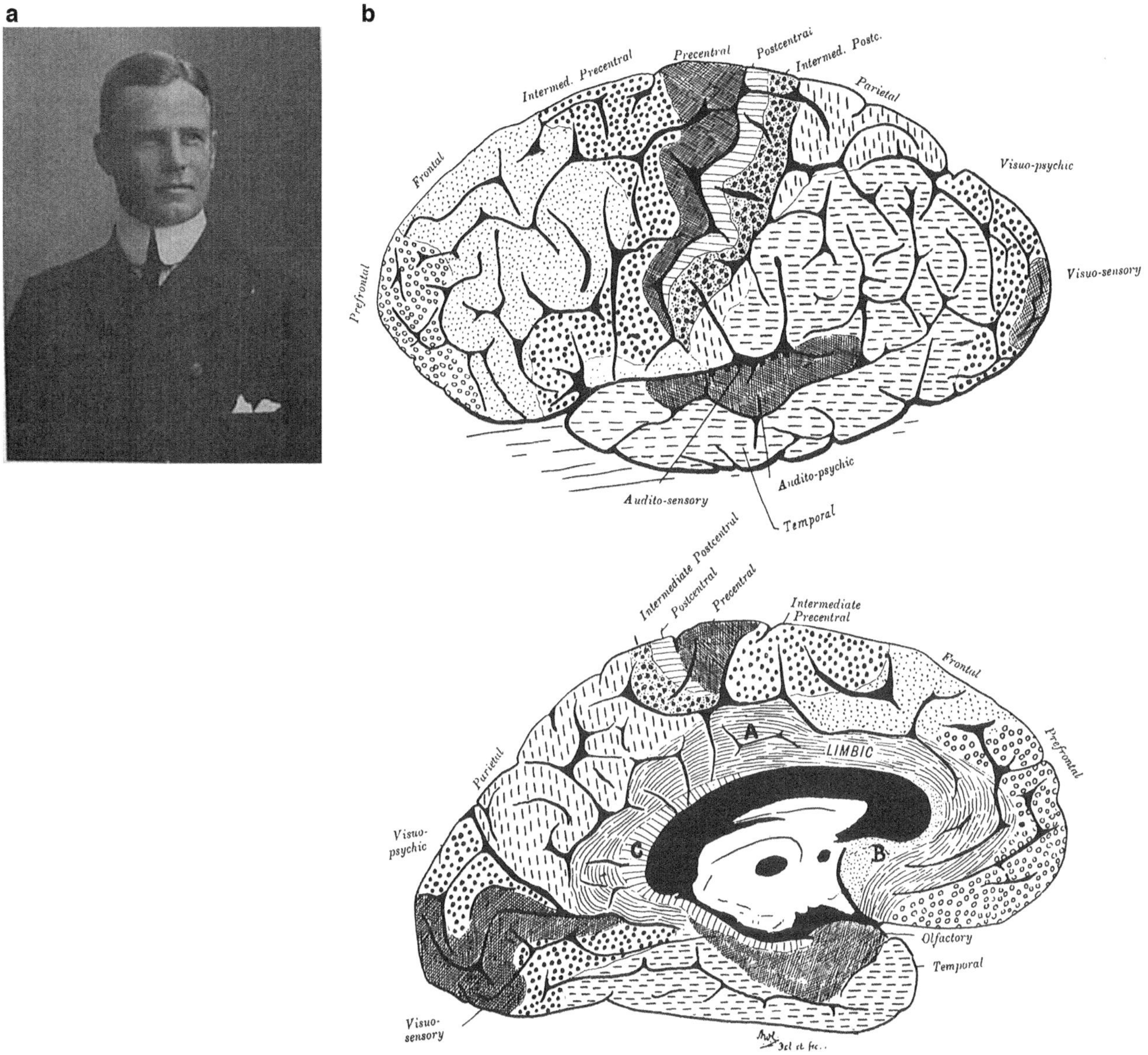

Fig. C3 (**a**) Photo of Campbell. (From the Wellcome Collection CCBY; in public domain); (**b**) his subdivision of cerebral cortex. (From Campbell AW *Histological Studies of the Localisation of Cerebral Function*. Cambridge University Press, Cambridge, 1905)

the alveolar margin of the maxilla and the anterior surface of the frontal bone, and the second through the external acoustic meatus to the anterior nasal spine. References: *Dissertation sur les variétés naturelles qui caractérisent la physionomie des Hommes des divers climats et les different âges*. Jansen, Paris, 1791; *Über den natürlichen Unterschied der Gesichtszüge in Menschen verschiedener Gegenden und verschiedenen Alters*. Vossische Buchhandlung, Berlin, 1792.

Chiasma tendinum of Camper: *Eponym* for the chiasma of the tendons of the superficial flexor muscles of the fingers near the proximal interphalangeal joint (Fig. C5b), a pattern similar to the chiasma opticum as Camper noted in 1760. He referred to an illustration in Albinus *Historia Musculorum Hominis* (Haak and Luchtmans, Leiden, 1734). Reference: IJpma FFA, van de Graaf RC, van Gulik TM. *Petrus Camper's work on the anatomy and pathology of the arm and hand in the 18th century*. J Hand Surg 2010; 35A:1382–1387.

Diaphragma of Camper: *Eponym* for the urogenital diaphragm.

Fascia of Camper: *Eponym* for the superficial adipose layer of the abdominal subcutaneous tissue.

Fig. C4 The Anatomical Lesson of Camper by Tibout Regiers commemorating his installment as 'praelector' of the Surgeon's Guild in 1755 in Amsterdam. (In Amsterdam Museum; in public domain)

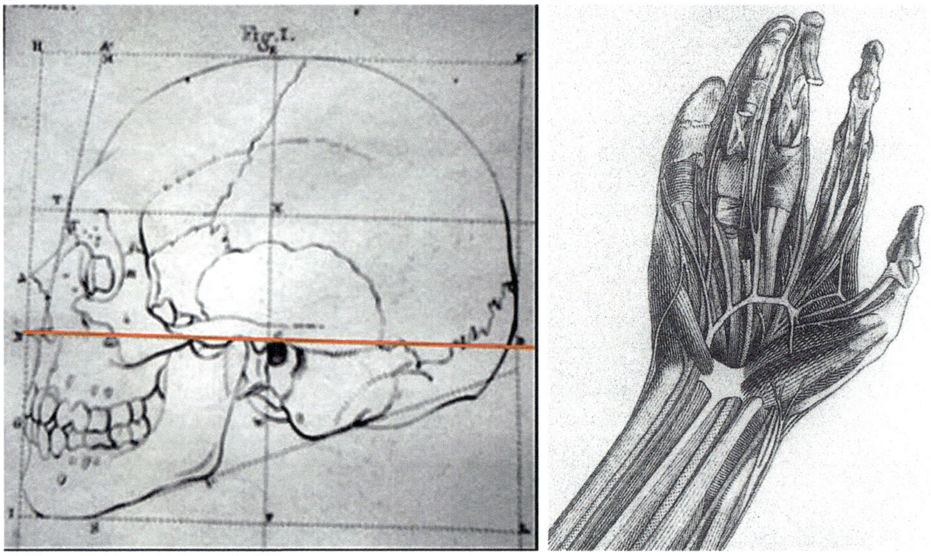

Fig. C5 (**a**) Camper's facial angle. (Based on an engraving in Camper's *Űber den natürlichen Unterschied der Gesichtszüge in Menschen verschiedener Gegenden und verschiedenen Alters.* Vossische Buchhandlung, Berlin, 1792; from Wikipedia; in public domain); (**b**) Camper's chiasma tendinum. (From *Demonstrationum Anatomico-Pathologicarum, Liber Primus, Continens Brachii Humani Fabricam et Morbos.* Schrueder & Mortier, Amsterdam, 1760)

Ligament of Camper: *Eponym* for the perineal membrane.

Line and **Plane of Camper:** *Eponym* for a reference line and plane in anthropology, a horizontal plane from the anterior nasal spine to the centre of the external acoustic meatus (*see* Fig. L24 and Fig. P18).

Canals: *See* **The Classical Anatomical Eponyms.**

Cannieu, A (nineteenth century): French physician.

Anastomosis of Riché-Cannieu: *Eponym* for a palmar communicating branch between the recurrent branch of the median nerve and the deep branch of the ulnar nerve, described in the same year as Riché, and therefore known as the anastomosis of Riché-Cannieu. Reference: *Note sur une anastomose entire le branche profundo de cubital et le médian.* Bull Soc Anat Physiol, Bordeaux 1897; 18:339–340.

Cannon, Walter Bradford (1871–1945): American physiologist born at Prairie du Chien, Wisconsin, and working

Fig. C6 Portrait of Cantley by H Salomon. (From Wikimedia Commons; in public domain)

at Harvard Medical School. He developed the theory of homeostasis and coined the term 'fight or flight response'. Reference: *The Wisdom of the Body.* Norton, New York, 1942.

Point or **Ring of Cannon:** *Eponym* for a sphincter of the transverse colon at the junction of the proximal and middle thirds, also known as **sphincter of Cannon-Böhm.** At this point, the innervation of the colon between the superior and inferior mesenteric plexuses overlaps. References: *The movements of the intestines studied by means of the Röntgen rays.* Am J Physiol 1902; 6:251–277; *The Mechanical Factors of Digestion.* Edward Arnold, London, 1911.

Cantlie, Sir James (1851–1926): Scottish physician born in Banffshire, Scotland, and working in London and later in Hong Kong (Fig. C6).

Line of Cantlie: *Eponym* for the anatomical midline of the liver connecting the fundus of the gallbladder with the inferior vena cava, described also by Rex and known as the line of Rex-Cantlie. Reference: *On a new arrangement of the right and left lobes of the liver.* Proc Anat Soc Great Britain and Ireland, 1897; 32:4–9.

Capsules:

Capsule of Bonnet: *Eponym* for the anterior part of the bulbar fascia of the eyeball, mostly known as capsule of Tenon; *see* Bonnet, Amédée.

Capsule of Bowman: *Eponym* for the glomerular capsule of the nephron in the kidney; *see* Bowman, William.

Capsule of Gerota: *Eponym* for the renal fascia; *see* Gerota, Dumitru.

Capsule of Glisson: *Eponym* for the fibrous capsule of the liver; *see* Glisson, Francis.

Capsule of Malpighi: *Eponym* for the hilar part of the splenic capsule; *see* Malpighi, Marcello.

Capsule of Tenon: *Eponym* for the bulbar fascia of the eyeball; *see* Tenon, Jacques René.

Carabelli, George, Edler von Lunkaszprie (1787–1842): Austrian dentist born in Pest (Hungary) and working in Vienna. Reference: *Systematisches Handbuch der Zahnheilkunde.* Vienna, 1831.

Tubercle or **Formation of Carabelli:** *Eponym* for an occasional fifth tubercle on the lingual surface of the first upper molar.

Carcassonne, Bernard Gauderic (eighteenth to nineteenth century): French anatomist and surgeon born and working in Montpellier where he received his M.D. in 1821 (Dobson 1962).

Ligament of Carcassonne, Perineal: *Eponym* for the thickened anterior border of the perineal membrane. Reference: *Recherches sur le manuel de l'opération de la taille souspubienne chez l'homme.* Montpellier, Thesis No 45, 1821.

Cartilages:

Cartilage of Huschke: *Eponym* for the vomeronasal cartilage, also described by Jacobson and Sappey; *see* Huschke, Emil.

Cartilage of Jacobson: *Eponym* for the vomeronasal cartilage, also described by Huschke and Sappey; *see* Jacobson, Ludwig Levin.

Cartilage of Luschka: *Eponym* for interarytenoid and other accessory cartilages connected with the arytenoid cartilages; *see* Luschka, Hubert von.

Cartilage of Meckel: *Eponym* for the cartilage of the first pharyngeal arch; *see* Meckel, Johann Friedrich, the Younger.

Cartilages of Meyer: *Eponym* for anterior sesamoid cartilages found in the anterior parts of the inferior thyroarytenoid ligaments; *see* Meyer, Edmund Victor.

Cartilage of Morgagni: *Eponym* for the cuneiform cartilage of the larynx, also described by Wrisberg; *see* Morgagni, Giovanni.

Cartilage of Reichert: *Eponym* for the cartilage of the second pharyngeal arch; see Reichert, Karl Bogislaus.

Cartilage of Santorini: *Eponym* for the corniculate cartilage of the larynx; *see* Santorini, Giovanni Domenico.

Cartilage of Sappey, Anterior accessory: *Eponym* for the vomeronasal cartilage, also described by Jacobson and Huschke; see Sappey, Marie Philibert Constant.

Cartilage of Seiler: *Eponym* for a small cartilaginous rod attached to the vocal process of the arytenoid cartilage; *see* Seiler, Carl.

Cartilage of Weitbrecht: *Eponym* for the articular disc of the acromioclavicular joint; *see* Weitbrecht, Josias.

Cartilage of Wrisberg: *Eponym* for (a) the epiglottis and (b) the cuneiform cartilage of the larynx; *see* Wrisberg, Heinrich August.

Caruncle of Santorini: *Eponym* for the opening of the accessory pancreatic duct, the minor duodenal papilla, into the descending part of the duodenum; *see* Santorini, Giovanni Domenico.

Carus, Carl Gustav (1789–1868): German zoologist, anatomist and gynaecologist born in Leipzig and working in Dresden.

Circle or **Curve of Carus:** *Eponym* for a curve having its centre at the pubic symphysis and embracing the pelvis outlet. Reference: *Lehrbuch der Gynaekologie, oder systematische Darstellung der Lehren von Erkenntniss und Behandlung eigentümlicher, gesunder und krankhafter Zustände, sowohl der Nichtschwangern, Schwangern und neugeborenen Kinder.* Fleischer, Leipzig, 1820.

Casseri (Casserius), Giulio Caesare (Julius; 1552–1616): Italian anatomist born in Piacenza and working in Padua, Veneto, as the great rival of Fabrici (Fig. C7a). References: *De vocis auditusque organis historia anatomica.* Ferrara, 1600–1601 (Fig. C7b); *Tabulae anatomicae*. Bucretius, Venice, 1627. In addition to these Tabulae, there are nine embryological Tabulae by Casseri, illustrating *De formatio foetu*, published posthumously in 1626 by van den Spieghel's son-in-law Liberale Crema (*see* Fig. V12). Casseri may also be credited for discovering, among others (Riva et al. 2001): (a) the arachnoid granulations, later described by Pacchioni; (b) the interventricular foramen, ascribed to Monro, and (c) the cerebral arterial circle, later described by Willis. For biography, *see* Riva A, Orrù B, Pirino A, Testa Riva F *Julius Casserius (1552–1616): The self-made anatomist of Padua's Golden Age.* Anat Rec 2001; 265:168–175.

Fontanelle of Casseri: *Eponym* for the mastoid fontanelle, a small fontanelle located at the mastoid angle of the parietal bone. It closes during the second year of life.

Ligament of Casseri: *Eponym* for the anterior ligament of the malleus, also described as muscle of Folli.

Muscle of Casseri, Perforated: *Eponym* for the coracobrachial muscle.

Nerve of Casseri: *Eponym* for the musculocutaneous nerve, Casseri's nervus perforans.

Cassinari, Valentino (1926–2014): Italian neurosurgeon working in Milan with Vittorio Bernasconi on the first description of the marginal tentorial artery; *see* Bernasconi, Vittorio.

Caves:

Cave of Douglas: *Eponym* for the rectouterine pouch of the peritoneal cavity; *see* Douglas, James.

Cave of Meckel: *Eponym* for the cavum trigeminale, a dural space for the trigeminal ganglion; *see* Meckel, Johann Friedrich, the Elder.

Cells: *See* **The Classical Eponyms.**

Chaput, Henri (1857–1919): French pathologist.

Tubercle of Chaput: *Eponym* for the anterior tibial tubercle onto which the anterior tibiofibular ligament inserts, also described by Tillaux and known as Tubercle of Chaput-Tillaux.

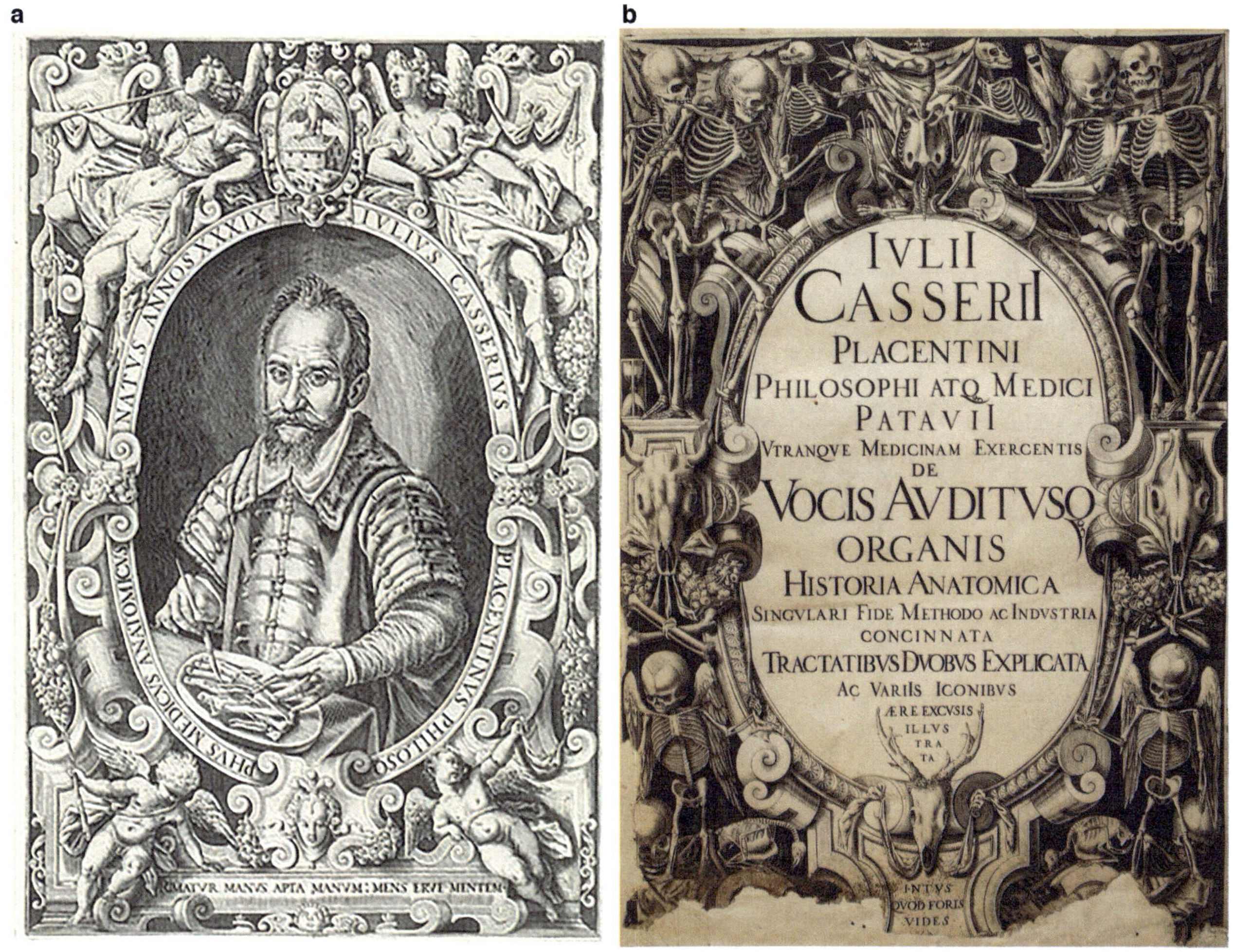

Fig. C7 (**a**) Engraving of Casseri. (Courtesy Alessandro Riva, Cagliari); (**b**) Title page *De vocis auditusque organis historia anatomica*. Ferrara, 1600–1601. (Courtesy Ondřej Naňka, Prague)

Fig. C8 Photo of Charcot. (From Pagel 1901)

Charcot, Jean Martin (1825–1893): French physician, anatomist and neurologist born and working in Paris known for his studies on nervous diseases (Fig. C8). Reference: *Leçons sur les localisations dans les maladies du cerveau faites à la Faculté de Médecine de Paris (1875).* Delahaye, Paris, 1876–1880.

Artery of Charcot: *Eponym* for any of the lenticulo-striate arteries, known as 'artery of haemorrhage'.

Portobiliary septa and **Spaces of Charcot:** *Synonyms* for the Septa and Spaces of Kiernan. Reference: *Leçons sur les maladies du foie, des voies biliares et des reins.* Prog Méd, Paris, 1877.

Charpy, Adrien (1848–1911): French anatomist born in Caluire (Rhône) working in Toulouse. He published *Traité d'anatomie humaine* with Poirier (Masson, Paris, 1899).

Fig. C9 Portrait of Chassaignac. (From the Bibliothèque de l'Anatomie nationale de Médecine; in public domain)

Fig. C10 Photo of Chiarugi. (From Wikipedia; in public domain)

Fascia of Charpy: *Eponym* for the prevesical fascia, forming the posterior wall of the prevesical space.

Fossa of Charpy: *Eponym* for the retromuscular fossa, a fat-filled submuscular space between the posterior aspect of the rectus abdominis muscle inferiorly and the transversalis fascia, bordered medially by the fusion of the transversalis fascia and the linea alba.

Ligament of Charpy: *Eponym* for the right phrenicocolic ligament (his abdominocolic ligament), a peritoneal fold between the right lateral abdominal wall and the right colic flexure.

Septum sagittale of Charpy: *Eponym* for the alar fascia of the neck (his 'cloison sagittale'), a paired sagittal fibrous layer between the carotid sheath and the prevertebral layer of the cervical fascia. Reference: *Aponèvroses du cou.* In: Poirier P, Charpy A *Traité d'anatomie humaine.* Masson, Paris, 1912, pp 258–280.

Chassaignac, Charles Marie Edouard (1805–1879): French surgeon born in Nantes and working in Paris (Fig. C9). Reference: *Etudes d'anatomie et de pathologie chirurgicale.* Paris, 1836–1851.

Space or **Bursa of Chassaignac:** *Eponym* for a potential space between the pectoralis major muscle and the mammary gland.

Tubercle of Chassaignac: *Eponym* for the carotid tubercle on the sixth cervical vertebra, serving clinically for compression of the common carotid artery.

Chaussier, François (1747–1828): French surgeon and anatomist born in Dijon and working in Paris and Dijon.

Artery of Chaussier, Great muscular: *Eponym* for a branch of the lateral circumflex femoral artery supplying the quadriceps femoris muscle. Reference: *Exposition sommaire des muscles du corps humain.* Dijon, 1789.

Chiari, Hans (1851–1916): Austrian pathologist born and working in Vienna. He is especially known for his work on the Chiari malformations. In 1891, he defined three types of cerebellar deformity associated with hydrocephalus, and later added a fourth type. Reference: *Ueber Veränderungen des Kleinhirns infolge von Hydrocephalie des Grosshirns.* Dtsch Med Wochenschr 1891; 17:1172–1175.

Network of Chiari: *Eponym* for an inconstant remnant of the right valve of the sinus venosus in the form of a fenestrated membrane consisting of threads and strands in the right atrium. Reference: *Ueber Netzbildungen im rechten Vorhofe des Menschen.* Beitr Pathol Anat 1897; 22:1–10.

Chiarugi, Giulio (1859–1944): Italian anatomist and embryologist born in Piacenza and working in Siena and Florence (Fig. C10). References: *Istituzioni di anatomia dell'uomo,* 1901–1920; *Trattato di embriologia,* 4 Vols, 1929–1944.

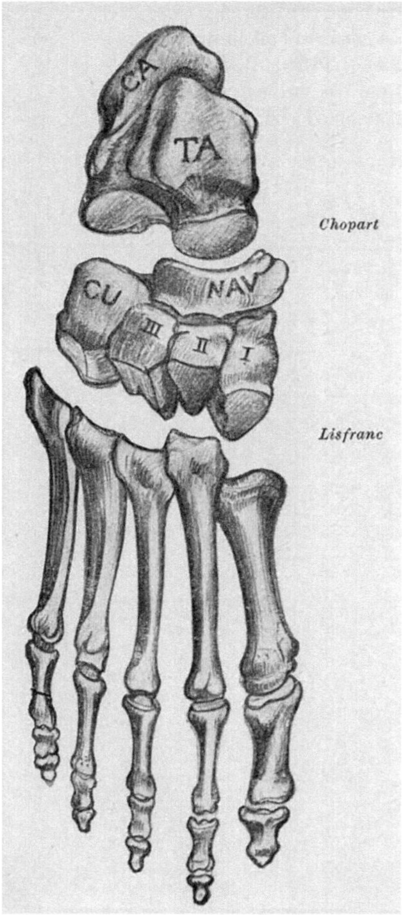

Fig. C11 (**a**) Portrait of Chopart; (**b**) joint of Chopart vs. joint of Lisfranc. (Both from Wikipedia; in public domain)

Dolichocephalic and **Brachiocephalic skulls:** *Eponyms* for skulls with an index, respectively, lower or higher than 79.

Chievitz, Johan Henrik (1850–1901): Danish anatomist born in Svendborg and working in Copenhagen.

Layer of Chievitz: *Eponym* for a transient anuclear layer of the developing retina. Reference: *Die Area und Fovea centralis retinae beim menschlichen Fetus.* Int Monatsschr Anat Physiol 1887; IV:201.

Organ of Chievitz: *Eponym* for the juxtaoral organ, an epithelial band on the inner side of the buccinator muscle, originally described by Chievitz as a rudimentary structure. Zenker, however, showed that it persists and contains numerous sensory corpuscles densely innervated by the buccal nerve. Reference: Zenker W *Juxtaoral Organ (Chievitz Organ).* Urban & Schwarzenberg, Munich-Baltimore, 1982.

Chopart, François (1743–1795): French surgeon in Paris, especially known for his studies on the foot (Fig. C11a).

Joint of Chopart: *Eponym* for the transverse tarsal joint (Fig. C11b).

Ligament of Chopart: *Eponym* for the bifurcate ligament of the foot.

Ciaccio, Giuseppe Vincenzo (1824–1901): Italian anatomist and histologist born in Catanzano, Calabria, and working in Bologna.

Glands of Ciaccio: *Eponym* for accessory lacrimal glands mostly in the upper eyelid, also described by Krause, Henle, Terson and Wolfring. Reference: *Osservazioni intorno alla struttura della congiuntiva umana.* Bologna, 1874.

Circles:

Circle of Carus: *Eponym* for a circle having its centre at the symphysis pubica and embracing the pelvic outlet; *see* Carus, Carl Gustav.

Circle of Haller: *Eponym* for (a) the vascular circle of the optic nerve, a network of branches of the short ciliary

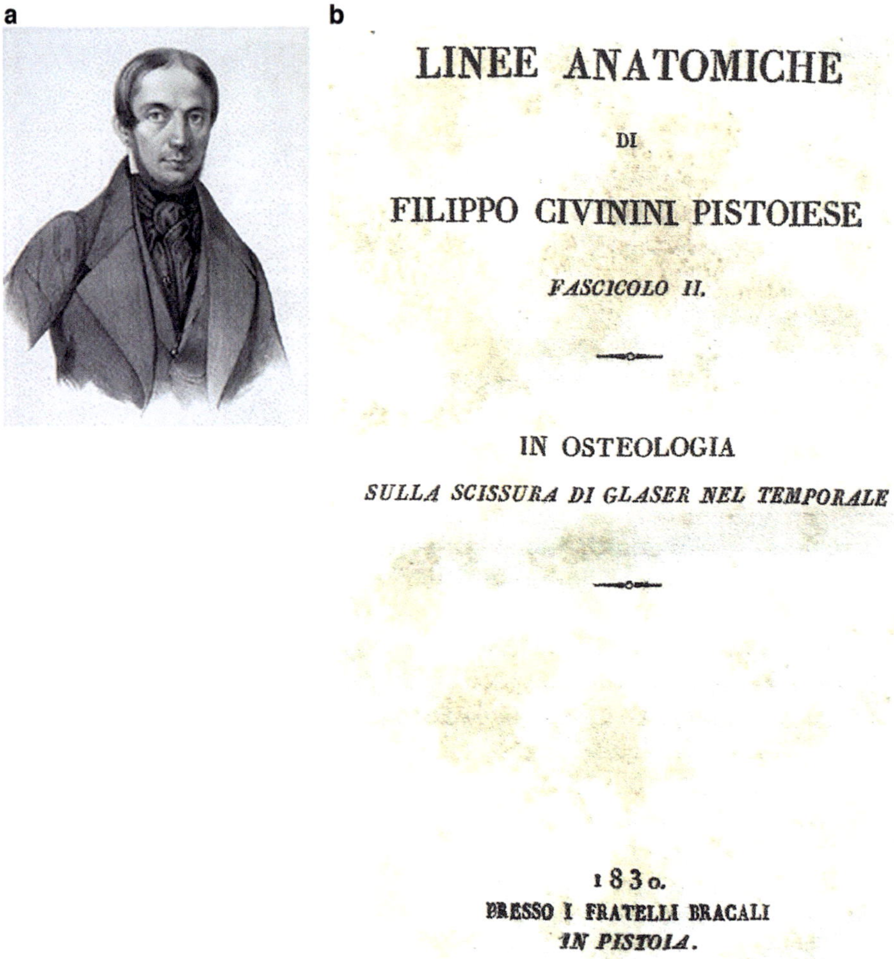
LINEE ANATOMICHE
DI
FILIPPO CIVININI PISTOIESE
FASCICOLO II.
IN OSTEOLOGIA
SULLA SCISSURA DI GLASER NEL TEMPORALE
1830.
PRESSO I FRATELLI BRACALI
IN PISTOIA.

Fig. C12 (**a**) Portrait of Civinini; (**b**) Title page *Linee anatomiche*. Fratelli Bracali, Fasc II, Pistoia, 1830. (Both from Wikimedia Commons; in public domain)

arteries on the sclera around the exit of the optic nerve, and (b) the areolar venous plexus, a venous plexus in the areola surrounding the nipple; *see* Haller, Albrecht von.

Circle of Huguier: *Eponym* for an anastomosis around the isthmus of the uterus between both uterine arteries; *see* Huguier, Pierre Charles.

Circle of Willis: *Eponym* for the cerebral arterial cicle or circulus arteriosus cerebri; *see* Willis,Thomas.

Cistern of Pecquet: *Eponym* for the cisterna chyli; *see* Pecquet, Jean.

Citelli, Salvatore (1875–1947): Italian otolaryngologist born in Regaliuto and working in Catania.

Angle of Citelli: *Eponym* for the sinudural angle formed between the dural plates of the middle cranial fossa superiorly, of the posterior cranially fossa anteriorly and the simoid sinus posteriorly, used to identify the superior petrosal sinus during surgery.

Cells of Citelli: *Eponym* for small air-filled cells in the sinudural angle in a well pneumatized ear.

Vein of Citelli: *Eponym* for the mastoid emissary vein connecting the sigmoid sinus and the occipital vein.

Civinini, Filippo (1805–1844): Italian anatomist, surgeon and pathologist born and working in Pistoia known for his studies of the sphenoid bone (Fig. C12a). Reference: *Linee anatomiche.* Fasc 1 and 2. Fratelli Bracali, Pistoia, 1829–1830 (Fig. C12b).

Canal of Civinini: *Eponym* for the anterior canaliculus of the chorda tympani, also described by Glaser and Huguier; but see Mudry (Otol Neurotol 2014; 36:1115–1120) for discussion whether this is an appropriate eponym.

Foramen or **Orifice of Civinini:** *Eponym* for the inconstant pterygospinous foramen of the sphenoidal bone.

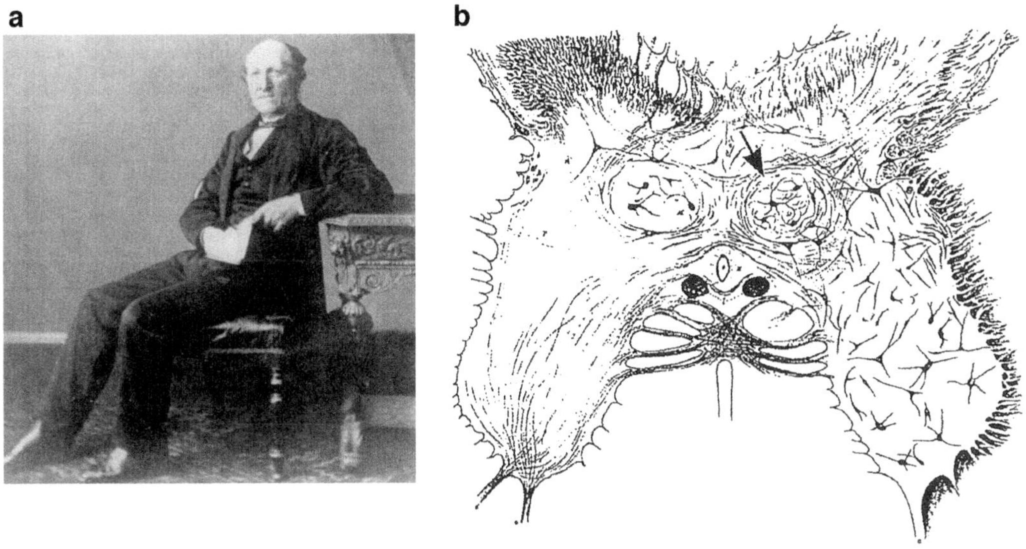

Fig. C13 (a) Photo of Clarke. (From the Wellcome Collection CCBY); (b) Clarke's nucleus as described in the spinal cord of the ox. (From *Recherches into the structure of the spinal cord.* Phil Trans 1851; 141:607–621)

Ligament of Civinini: *Eponym* for the pterygospinous ligament.

Process or **Spine of Civinini:** *Eponym* for the inconstant pterygospinous process of the sphenoidal bone.

Clado, Spiro (1856–1920): French gynaecologist in Paris. Reference: *Appendice caecal; anatomie, embryologie.* CR Soc Biol Paris (1892), 9th Ser, IV, Mem, pp 133–172.

Anastomosis of Clado: *Eponym* for the anastomosis between the appendicular and ovarian arteries in the appendiculoovarian ligament.

Band of Clado: *Eponym* for the suspensory ligament of the ovary.

Ligament of Clado: *Eponym* for the appendiculoovarian ligament, a mesenteric fold from the vermiform appendix to the right broad ligament of the uterus.

Point of Clado: *Eponym* for a point on the lateral margin of the right rectus abdominis muscle on the spinoumbilical line to evoke pain and appreciate tenderness on palpation.

Clara, Max (1899–1966): Austrian anatomist born in Schlern, South Tyrol, known for his Nazi sympathies.

Cells of Clara: *Eponym* for exocrine cells of the bronchial epithelium, currently known by the descriptive term club cell; for discussion, *see* Winkelmann A, Noack T *The Clara cell: A Third Reich Eponym.* Eur Resp J 2010; 36:722–727.

Clarke, Jacob Augustus Lockhart (1817–1880): English physician and neurologist born and working in London (Fig. C13a). Reference: *Researches into the structure of the spinal cord.* Phil Trans 1851; 141:607–621.

Column of Clarke: *Eponym* for the column of large spinal cells, giving rise to the posterior or dorsal spinocerebellar tract, also described by Stilling; therefore, also known as Stilling-Clarke nucleus.

Nucleus of Clarke: *Eponym* for the dorsal or posterior thoracic nucleus of the spinal cord, giving rise to the posterior or dorsal spinocerebellar tract of Flechsig (Fig. C13b).

Claudius (Cladius), Friedrich Matthias (1822–1869): German anatomist born in Lübeck and working in Kiel and later in Marburg (Fig. C14a).

Cells of Claudius: *Eponym* for the cuboidal external supporting cells lining the external spiral sulcus of the cochlear duct (*see* Fig. C21). Reference: *Physiologische Bemerkungen ueber das Gehörorgan.* Schwer, Kiel, 1858.

Fossa of Claudius: *Eponym* for a peritoneal fossa, housing the ovary in multipara, between the internal iliac artery and the sacrum (Fig. C14b). Reference: *Über die Lage des Uterus.* Ztschr Rat Med 1865; 22:248.

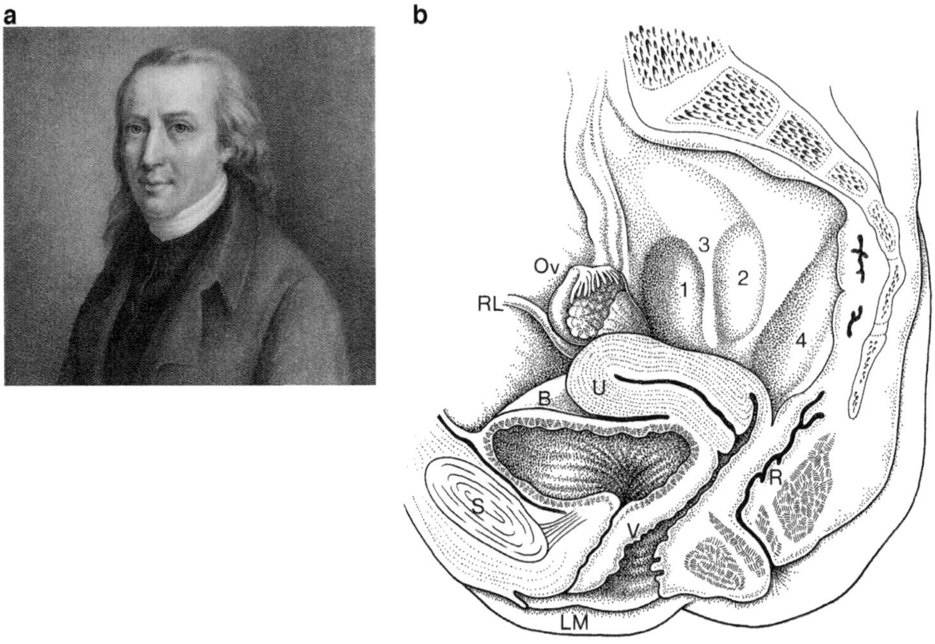

Fig. C14 (**a**) Portrait of Claudius. (From Wikimedia Commons; in public domain); (**b**) the fossa of Claudius. (After Rouvière H *Anatomie humain: Descriptive et topographique*. Tôme II, Tronc. 8th ed, revised by G Cordier. Masson, Paris, 1959). *B* bladder, *LM* labium majus, *Ov* ovarium with uterine tube, *R* rectum, *RL* round ligament of uterus, *S* symphysis pubica, *U* uterus, *1* ovarian fossa, *2* fossa of Claudius, *3* prominence of ureter, *4* pouch of Douglas

Fig. C15 The ligament of Cleland. (From Cleland J *On the cutaneous ligaments of the phalanges*. J Anat Physiol 1878; 12:526–572; in public domain)

Cleft of Schmidt-Lantermann: *Eponym* for the myelin cleft or incisure, a funnel-shaped structure within the myelin sheath of a peripheral nerve fibre; *see* Schmidt, Henry D. and Lantermann, A.J.

Cleland, John (1835–1925): Scottish anatomist born in Perth (Scotland) and working in Glasgow.

Ligaments of Cleland: *Eponym* for cutaneous ligaments of the digits (Fig. C15). These skin ligaments were originally described in 1742 by Weitbrecht. Reference: *On the cutaneous ligaments of the phalanges.* J Anat Physiol 1878; 12:526–527.

Clevenger, Shobal Veil Jr (1843–1920): American neurologist after whom sometimes the inferior temporal sulcus of the cerebral cortex is named.

Fissure of Clevenger: *Eponym* for the inferior temporal sulcus.

Clivus of Blumenbach: *Eponym* for the middle, ascending part of the posterior cranial fossa towards the sella turcica formed by the bodies of the occipital and sphenoidal bones; *see* Blumenbach, Johann Friedrich.

Cloquet, Jules Germain (1790–1883): French anatomist and surgeon born and working in Paris (Fig. C16a). References: *Recherches anatomiques sur les hernies de l'abdomen.* Méquignon-Marvis, Paris, 1817; *Recherches sur les causes et l'anatomie des hernies abdominales.* Thesis, Paris, 1819; *Manuel d'anatomie descriptive.* Paris, 1825 (Fig. C16b).

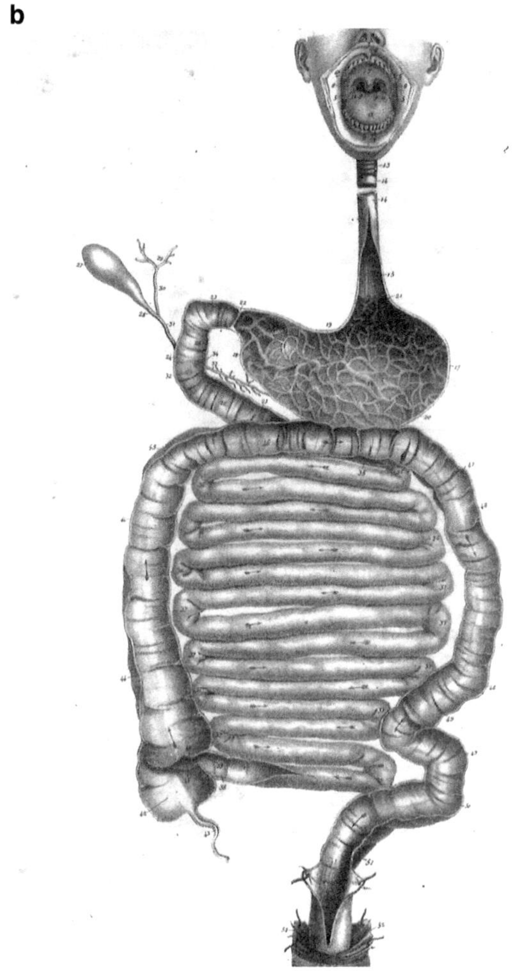

Fig. C16 (**a**) Portrait of Cloquet. (From the Wellcome Collection CCBY; in public domain); (**b**) digestive system. (From Cloquet JG 1825 *Manuel d'anatomie descriptive*, Paris, 1825)

Canal of Cloquet: *Eponym* for the hyaloid canal of the vitreous body, also described by Stilling.

Fascia or **Membrane of Cloquet:** *Eponym* for part of the fascia lata covering the saphenous opening.

Gland or **Node of Cloquet:** *Eponym* for the proximal deep inguinal lymph node located in the vascular lacuna adjacent to the inguinal canal, also described by Rosenmüller.

Hernia of Cloquet: *Eponym* for a femoral hernia, in which the intestine is found behind the blood vessels.

Ligament of Cloquet: *Eponym* for the obliterated part of the vaginal process of the peritoneal cavity, extending from the testis to the peritoneum at the deep inguinal ring.

Septum of Cloquet: *Eponym* for the hiatus saphenus of the fascia lata.

Space of Cloquet: *Eponym* for the retrozonular space between the ciliary zonule and the vitreous body. Reference: *Mémoire sur la membrane pupillaire et sur la formation du petit cercle artériel de l'iris.* Paris, 1818.

Cockett, Frank Bernard (1916–2014): Australia-born (Rockhampton) surgeon who came as a child to England where he became one of the leading experts on varices and their treatment. References: *The ankle blow-out syndrome: A new approach to the varicose ulcer.* Lancet 1953 (6749), pp 17–23; *The Pathology and Surgery of the Veins of the Lower Limb.* Churchill Livingstone, London, 1956 (with WHA Dodd; 2nd ed, 1976).

Veins of Cockett, Perforating: *Eponym* for the posterior tibial perforators (superior, middle and inferior) connecting the posterior accessory great saphenous vein with the posterior tibial veins.

Cohnheim, Julius Friedrich (1839–1884): German pathologist born in Demmin, Pomerania, and working in Kiel, Breslau and finally Leipzig.

Fields of Cohnheim: *Eponym* for polygonal areas seen in transverse sections of striated muscle fibres, representing bundles of myofibrils surrounded by myofibril-free sarcoplasm.

Coiter, Volcher (1534-around 1600): Dutch anatomist and surgeon born in Groningen who lived mostly in Bologna, Italy (Fig. C17). Reference: *De ossibus et cartilaginibus corporis partium Tabulae.* Rossium, Bologna, 1566.

Fig. C17 Engraving of Coiter. (From Wikimedia Commons; in public domain)

Muscle of Coiter: *Eponym* for the corrugator supercilii muscle.

Colles, Abraham (1773–1843): Irish anatomist and surgeon born in Milmount near Kilkenny and working mainly in Dublin known for his studies of the abdominal wall and perineum and for his description of the fracture of the distal portion of the forearm (Fig. C18a). Reference: *A Treatise on Surgical Anatomy.* Edinburgh, 1811.

Fascia of Colles: *Eponym* for (a) the membranous layer of the subcutaneous tissue of the perineum, also described by Cruveilhier; (b) the subcutaneous tissue of the penis, and (c) the superficial fascia of the scrotum (fascia dartos).

Ligament of Colles: *Eponym* for the reflected inguinal ligament (*ligamentum reflexum*), a small band detached from the pubic attachment of the lateral crus of the superficial inguinal ring, extending upwards and medially, behind the medial crus and blending with the opposite aponeurosis of the external oblique abdominal muscle (Fig. C18b).

Space of Colles: *Eponym* for the superficial perineal space.

Columns:

Columns of Bertin: *Eponym* for the renal columns, extensions of the cortical tissue of the kidney into the medul-

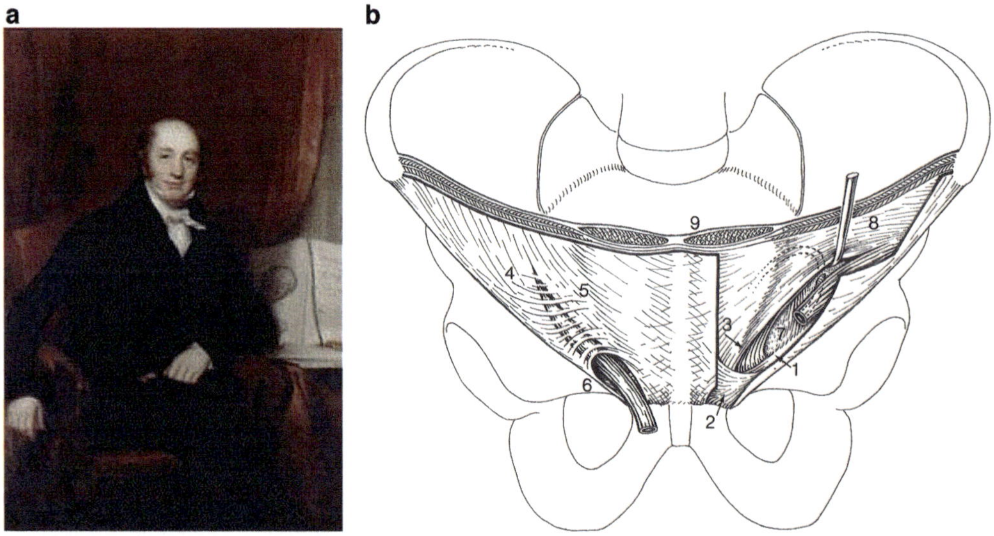

Fig. C18 (**a**) Portrait of Colles. (From the Royal College of Surgeons, London; in public domain); (**b**) the ligamentum reflexum of Colles. *1* inguinal ligament of Poupart, *2* reflected ligament of Colles, *3* inguinal falx, *4, 5* lateral and medial crus of external abdominal oblique muscle, *6* superficial inguinal anulus with funiculus spermaticus, *7* fascia transversalis, *8* internal oblique abdominal muscle, *9* rectus abdominus muscle with sheath. (After ten Donkelaar HJ, Oostra R-J *Klinische Anatomie en Embryologie,* 4th ed. Springer Media, Houten, 2014)

Fig. C19 Portrait of Cooper. (From Pagel 1901)

lary substance between the pyramids; *see* Bertin, Exupère Joseph.

Columns of Cotugno: *Eponym* for columns in the spiral lamina of the cochlea; *see* Cotugno, Domenico.

Columns of Morgagni: *Eponym* for the anal columns; *see* Morgagni, Giovanni Battista.

Commissures:

Commissure of Forel: *Eponym* for the posterior hypothalamic commissure; *see* Forel, Auguste Henri.

Commissure of Ganser: *Eponym* for the supreme supraoptic decussation; *see* Ganser, Sigbert Joseph Marie.

Commissure of Gudden: *Eponym* for the ventral supraoptic decussation; *see* Gudden, Bernhard Aloys von.

Commissure of Haller: *Eponym* for the commissura infima; *see* Haller, Albrecht von.

Commissure of Held: *Eponym* for the secondary auditory fibres from the anterior or ventral cochlear nucleus, which cross the midline dorsal to the trapezoid body to join the contralateral lateral lemniscus; *see* Held, Hans.

Commissure of Meynert: *Eponym* for the dorsal supraoptic decussation; *see* Meynert, Theodor.

Commissure of Probst: *Eponym* for the commissure of the dorsal nuclei of the lateral lemnisci; *see* Probst, Moritz.

Convolutions of Heschl: *See* Gyri of Heschl.

Cooper, Sir Astley Paston (1768–1841): English surgeon and anatomist born in Brooke, Norfolk and working in London as physician to King George IV, especially known for his studies of fasciae and ligaments (Fig. C19). References: *The Anatomy and Surgical Treatment of Abdominal Hernia.* Longman, London 1804–1807; *Treatise on Dislocation and on Fractures of the Joints.* London, 1825; *Observations on the Structure and Diseases of the Testis.* London, 1830; *Anatomy of Diseases of the Breast.* London, 1845.

Fascia of Cooper: *Eponym* for the cremasteric fascia.

Ligaments of Cooper: *Eponym* for (a) the suspensory ligaments of the breast, (b) the pectineal ligament, a lateral periosteal extension of the lacunar ligament along the pecten pubis; and (c) for an oblique ligament from the olecranon to the coronoid process of the ulna, part of the ulnar collateral ligament of the elbow joint.

Protrusion of Cooper-Hey: *Eponym* for a certain femoral hernia with two sacs, also known as bilocular femoral hernia.

Cords:

Cords of Bergmann: *Eponym* for the acoustic striae in the floor of the fourth ventricle, also described by Piccolomini; *see* Bergmann, Gottlieb Heinrich.

Cord of Leonardo (da Vinci): *Eponym* for the moderator band of the right ventricle of the heart; *see* Leonardo da Vinci.

Corpuscles: *See* **The Classical Eponyms.**

Corti, Alfonso, Marchese (1822–1888): Italian histologist born in Gambarana, Pavia (Lombardy) who worked in Vienna, Berlin, Utrecht and Turin (Fig. C20), noted for his investigations of the cochlea (Fig. C21). Reference: *Recherches sur l'organe de l'ouie des mammifères.* Z Wiss Zool 1851; 3:109–172.

Canal or **Tunnel of Corti:** *Eponym* for the inner tunnel of the spiral organ of the cochlea.

Ganglion of Corti: *Eponym* for the spiral or cochlear ganglion, sometimes also ascribed to Boettcher.

Fig. C20 Portrait of Corti. (From Wikimedia Commons; in public domain)

Membrane of Corti: *Eponym* for the tectorial membrane of the cochlear duct.

Organ of Corti: *Eponym* for the spiral organ.

Pillar cells or **Rods of Corti:** *Eponym* for the inner and outer pillar cells of the spiral organ.

Cotugno (Cotunnius), Domenico (1736–1822): Italian anatomist born in Ruvo di Puglia (Bari) and working in Naples known for his studies of the labyrinth (Fig. C22a). References: *De acquaeductibus auris humanae internae anatomica dissertatio* (Figs. C22b, C23). Ex typographica Sancti Thomae Aquinatis, Napoli, 1761; *De ischiade nervosa commentarius.* Fratres Simonis, Napoli, 1764.

Aqueduct or **Canal of Cotugno:** *Eponym* for (a) the endolymphatic duct and (b) the vestibular canaliculus in the petrous part of the temporal bone.

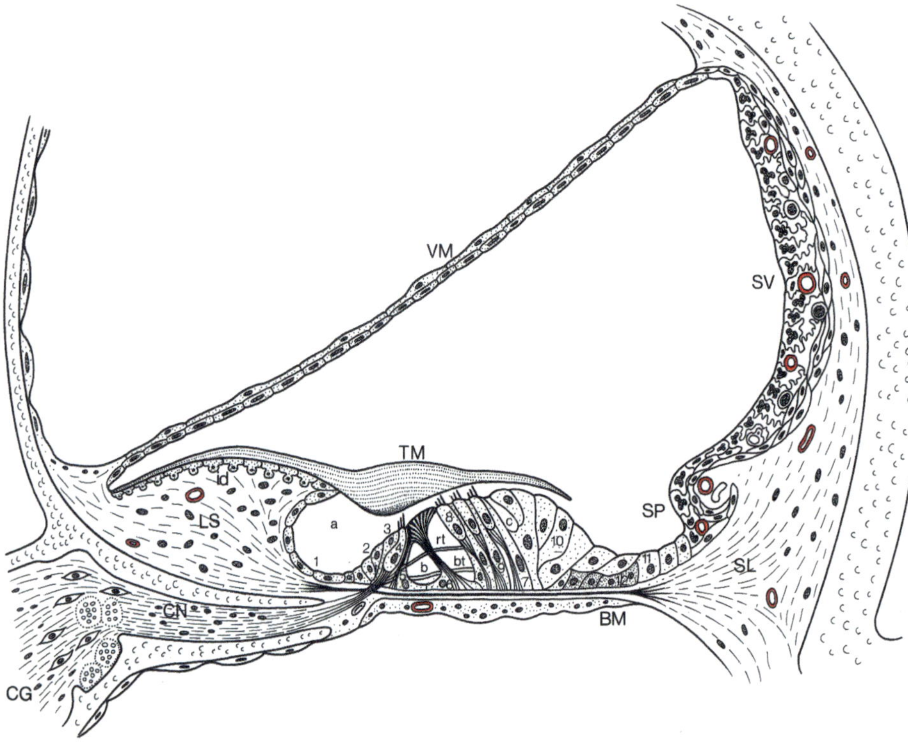

Fig. C21 The spiral organ of Corti. (After Schuknecht HF 1993 *Pathology of the Ear*, 2nd edn. Lea and Febiger, Philadelphia, PA; from ten Donkelaar HJ, Kachlik D, Tubbs RS *An Illustrated Terminologia Neuroanatomica.* Springer, Cham, 2018). *a* inner spiral tunnel, *b* inner tunnel (of Corti), *BM* basal lamina, *bt* basal tunnel fiber (afferent), *c* outer tunnel, *CG* cochlear ganglion, *CN* cochlear nerve, *id* interdental cells, *LS* limbus spiralis, *rt* radial tunnel fiber (efferent), *SL* spiral ligament, *SP* spiral prominence, *SV* stria vascularis, *TM* tectorial membrane, *VM* vestibular membrane, *1* inner border cell, *2* internal phalangeal cell, *3* inner hair cell, *4* inner pilar cell, *5* outer pilar cell, *6*, *8* external phalangeal cells (of Deiters), *7* outer hair cell, *9* outer border cell (of Hensen), *10* cuboidal external supporting cell (of Claudius), *11* basal external glandular cell (of Boettcher)

Fig. C22 (**a**) Portrait of Cotugno; (**b**) Title page *De acquaeductibus auris humanae internae anatomica dissertatio.* Sancti Thomae Aquinatis, Napoli, 1761; both from the Wellcome Collection CCBY; in public domain)

a

b

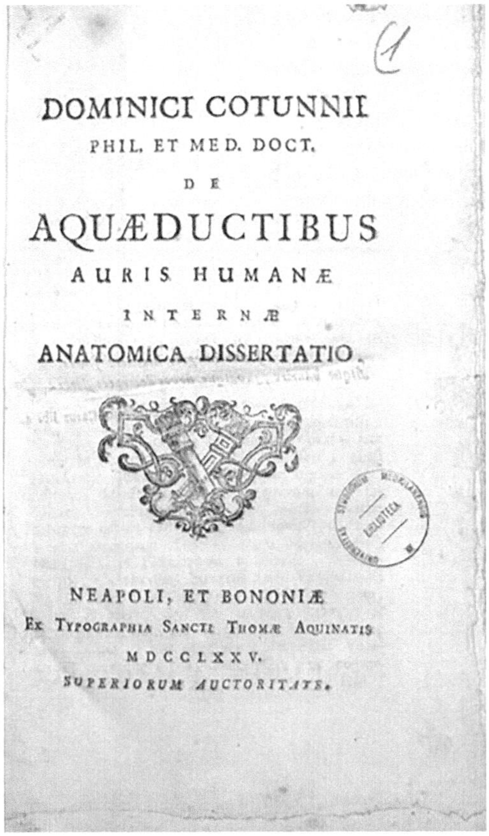

DOMINICI COTUNNII
PHIL. ET MED. DOCT.
DE
AQUÆDUCTIBUS
AURIS HUMANÆ
INTERNÆ
ANATOMICA DISSERTATIO.
NEAPOLI, ET BONONIÆ
Ex Typographia Sancti Thomæ Aquinatis
MDCCLXXV.
Superiorum auctoritate.

Columns of Cotugno: *Eponyms* for columns in the spiral lamina of the cochlea.

Liquor of Cotugno: *Eponym* for the perilymph in the osseous labyrinth of the inner ear.

Nerve of Cotugno: *Eponym* for the nasopalatine nerve.

Space of Cotugno: *Eponym* for the endolymphatic sac, later also described by Boettcher and known as sac of Boettcher-Cotugno.

Coufal, Emanuel: *See* Zaufal, Emanuel.

Couinaud, Claude Maurice (1922–2008): French surgeon born in Neuilly-sur-Seine, working in Paris and known for his extensive studies on the segmentation of the liver. Reference: Couinaud C *Le Foie. Études anatomiques et chirurgicales.* Masson, Paris, 1957.

Segments of Couinaud: *Eponym* for the segments of the liver (Fig. C24).

Courtney, Harold (twentieth century): American surgeon and proctologist in Syracuse, New York.

Space of Courtney, Subsphincteric: *Eponym* for the deep postanal space located posterior to the deep part of the external anal sphincter muscle between the levator ani muscle and the superficial part of the external anal sphincter muscle. Reference: *The posterior subsphincteric space; its relation to posterior horseshoe fistula.* Surg Gynec Obst 1949; 89:222–226.

Cowper, William (1666–1709): English surgeon born near Petersfield (Hampshire) and working in London (Fig. C25). He relied heavily on the work of Govert Bidloo and used the illustrations of Gérard de Lairesse in Bidloo's Atlas, adding an English text but without crediting Bidloo. Reference: *The Anatomy of Human Bodies.* Smith and Walford, London, 1698.

Glands of Cowper: *Eponym* for the bulbourethral glands, earlier described by Méry. Reference: *An account of two glands and their excretory ducts lately discovered in human bodies.* Phil Trans 1697; 21:364–369.

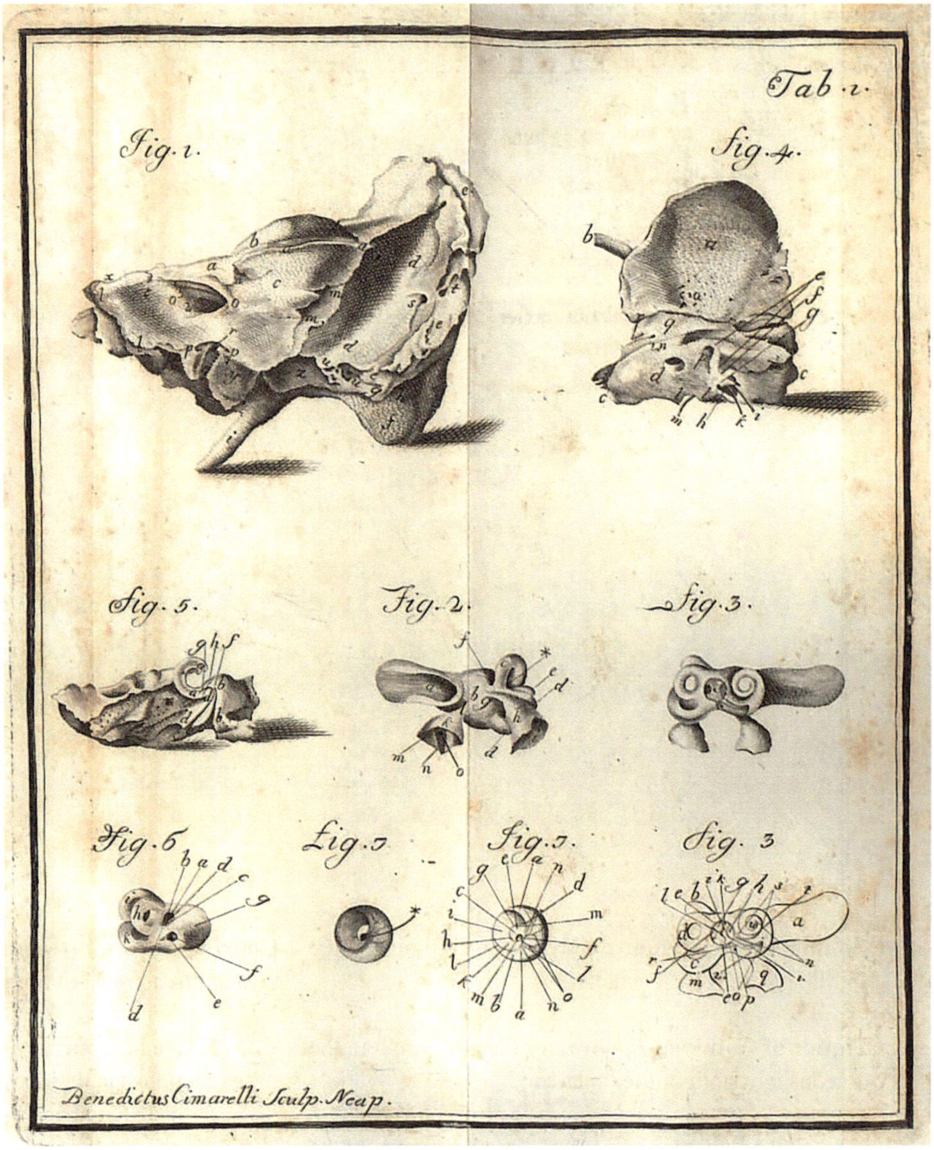

Fig. C23 Table 1 from his *De acquaeductibus auris humanae anatomica dissertatio*. Sancti Thomae Aquinatis, Napoli, 1761, showing Cotugno's knowledge of the ear. (From the Wellcome Collection CCBY; in public domain)

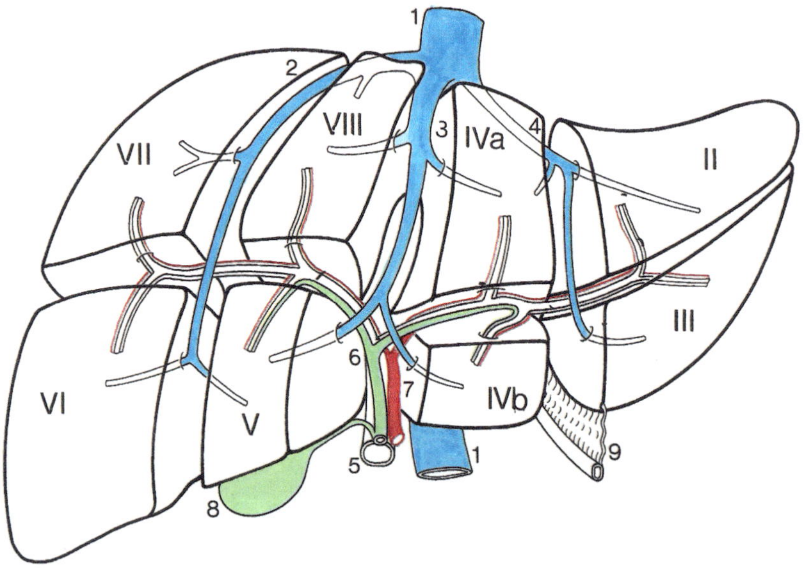

Fig. C24 The Couinaud segments of the liver. *1* inferior caval vein, *2* right hepatic vein, *3* intermediate hepatic vein, *4* left hepatic vein, *5* portal vein, *6* ductus choledochus, *7* proper hepatic artery, *8* gallbladder, *9* falciform ligament and ligamentum teres hepatis, *I-VIII* hepatic segments. (After ten Donkelaar HJ, Oostra R-J *Klinische Anatomie en Embryologie*, 4th ed. Springer, Houten, 2014)

Fig. C25 Portrait of Cowper. (From the Royal College of Surgeons, London; in public domain)

Fig. C26 Photo of Crampton. (From Wikimedia Commons; in public domain)

Ligament of Cowper: *Eponym* for part of the fascia lata, giving origin to the pectineus muscle.

Crampton, Sir Philip (1777–1858): English surgeon born in Dublin and working in London (Fig. C26).

Muscle of Crampton: *Eponym* for the meridional fibres of the ciliary muscle as also described later by Bowman and Brücke in 1847. Reference: *An Essay on the Entropion or Inversion of the Eyelids.* London, 1805.

Cruveilhier, Jean Baptiste (1791–1874): French physician, anatomist, surgeon and pathologist born in Limoges and working in Paris (Fig. C27a). He is known for his description of various anatomical structures as exemplified for the spinal cord (Fig. C27b). References: *Anatomie pathologique du corps humain.* Baillière, Paris, 1829; *Anatomie descriptive,* 2 Vols. Bechet Jeune, Paris, 1834; *Traité d'anatomie pathologique générale.* Paris, 1849–1865; *Traité d'anatomie descriptive.* Asselin, Paris, 1871.

Accessory saphenous vein of Cruveilhier: *Eponym* for an accessory common venous trunk for the superficial posterior veins of the thigh.

Alveolar plexus of Cruveilhier: *Eponym* for a plexiform anastomosis between the facial vein and the pharyngeal plexus.

Anastomosis of Cruveilhier: *Eponym* for a communicating branch between a branch of the medial cutaneous nerve of the forearm and a branch of the ulnar nerve.

Aponeurosis of Cruveilhier: *Eponym* for the pharyngobasilar fascia (his cephalopharyngeal aponeurosis).

Canal of Cruveilhier, Brachial: *Eponym* for the brachial canal, bounded ventrally by the coracobrachialis and biceps brachii muscles, dorsally by the medial brachial intermuscular septum, medially by the brachial fascia and laterally by the coracobrachialis muscle (proximally) and a space between the biceps brachii and brachialis muscles (distally), containing the brachial vessels and the median and ulnar nerves.

Fascia of Cruveilhier: *Eponym* for the membranous layer of the subcutaneous tissue of the perineum, i.e. the superficial perineal fascia, also described by Colles.

Fossa of Cruveilhier: *Eponym* for the scaphoid fossa of the sphenoidal bone.

Joint of Cruveilhier: *Eponym* for the median atlanto-axial joint.

Ligaments of Cruveilhier: *Eponym* for the plantar metatarsophalangeal ligaments.

Meningeal ramus of Cruveilhier: *Eponym* for the posterior meningeal artery.

Nerve of Cruveilhier: *Eponym* for the vertebral nerve.

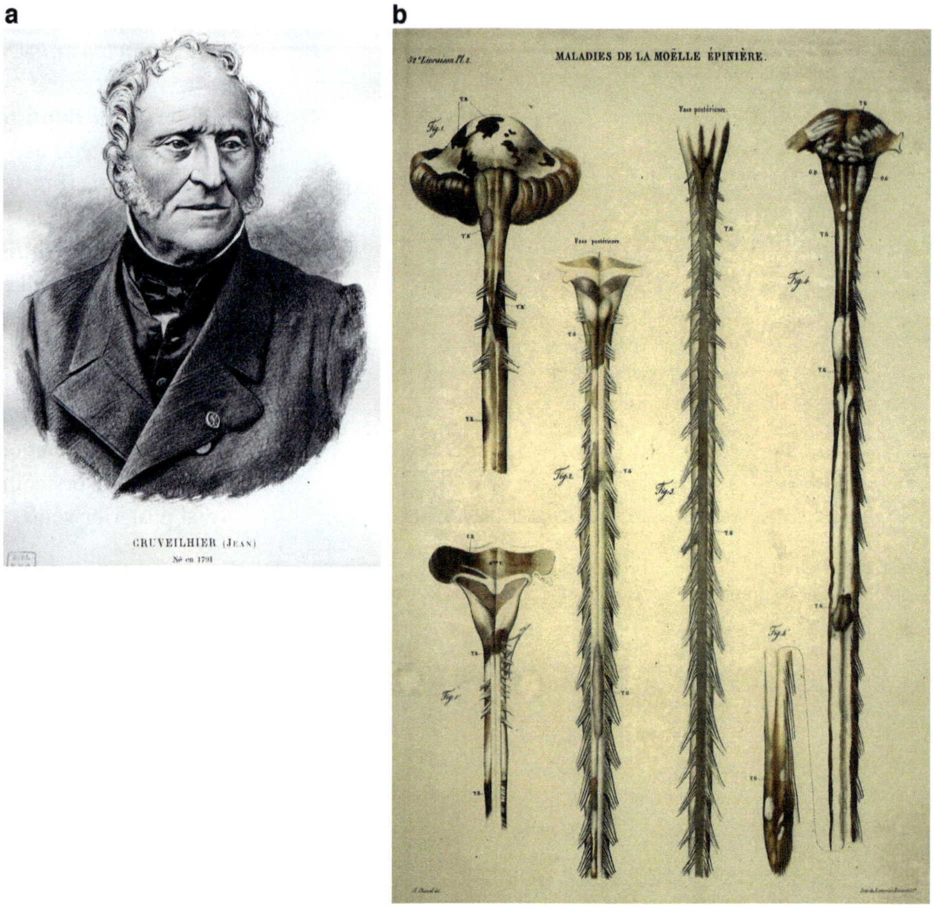

Fig. C27 (**a**) Portrait of Cruveilhier. (From Pagel 1901); (**b**) Cruveilhier's figure on the spinal cord in *Traité pathologique du corpus humain*. Baillière, Paris, 1829. (From Wikimedia Commons; in public domain)

Peduncular or **Pyramidal collar of Cruveilhier:** *Eponym* for a constriction of the pyramidal tract as it enters the pons.

Plexus of Cruveilhier: *Eponym* for the vertebral plexus formed by the dorsal or posterior rami of the first three cervical nerves.

Valve of Cruveilhier: *Eponym* for the lacrimal valve within the nasolacrimal duct, earlier described by Bianchi and Hasner.

Crypts of Lieberkühn: *Eponym* for the intestinal glands; *see* Lieberkühn, Johann Nathaniel.

Cunéo: *See* Poirier, Paul Julien.

***Cupid bow*:** *Eponym* for a facial feature when the double curve of the human upper lip resembles the bow of Cupid, outlined by the vermilion border of the lip. The peaks of the bow coincide with the columns of the philtrum giving a prominent bow appearance to the upper lip. Cupid was the Roman God of erotic love.

Curve of Spee: *Eponym* for a line touching the tops of the buccal cusps of the teeth from the first bicuspid to the third molar; *see* Spee, Ferdinand, Graf von.

Cushing, Harvey Williams (1869–1939): American neurosurgeon.

Loop of Cushing: *Eponym* for the anterior or temporal loop of the optic radiation, described also by Archambault, Flechsig and Meyer, usually named loop of Meyer.

Cuvier, Georges Leopold Chrétien Frédéric Dagobert, Baron (1769–1832): French zoologist and paleontologist born in Montbeliard and working in Paris (Fig. C28). Reference: *Leçons d'anatomie comparée.* Baudouin, Paris, 1799–1805.

Canal of Cuvier: *Eponym* for the sinus venosus of the developing heart (Fig. C29).

Duct of Cuvier: *Eponym* for the embryonic common cardinal vein (Fig. C29).

Fig. C28 Portrait of Cuvier. (From Wikimedia Commons; in public domain)

Line and **Plane of Cuvier:** Modifications of the horizontal line and plane of Camper.

Cyon, Elie de (1843–1912): Russian physiologist in St. Petersburg. Reference: *Die Reflexe eines der sensiblen Nerven des Herzens auf die motorischen der Blutgefässe.* Arb Physiol Anat (Lpz) 1866, 307–328 (with Ludwig).

Nerve of Cyon: *Eponym* for the depressor nerve of the heart, a branch of the parasympathetic superior laryngeal nerve, also described by Hofer and Ludwig.

Czermak (Čermák), Johann Nepomuk or Jan (1828–1873): Bohemian zoologist and comparative anatomist born in Prague, working in Graz and later as physiologist in Krakau, Jena and Leipzig (Fig. C30). He is noted for his studies on the microscopic anatomy of the teeth.

Lacunae or **Spaces of Czermak:** *Eponym* for the interglobular lacunae or interglobular dentin, gaps in the continuity of dentin. Reference: *Beiträge zur mikroskopischen Anatomie der menschlichen Zähne.* Z Wiss Zool 1850, pp 61–64.

Tubercle of Czermak: *Eponym* for the tubercle of the epiglottis. Reference: *Über die Inspektion des Cavum pharyngonasale und der Nasenhöhle durch Choanen vermittelst kleiner Spiegel.* Wien Med Wochenschr 1859; 9:518–520.

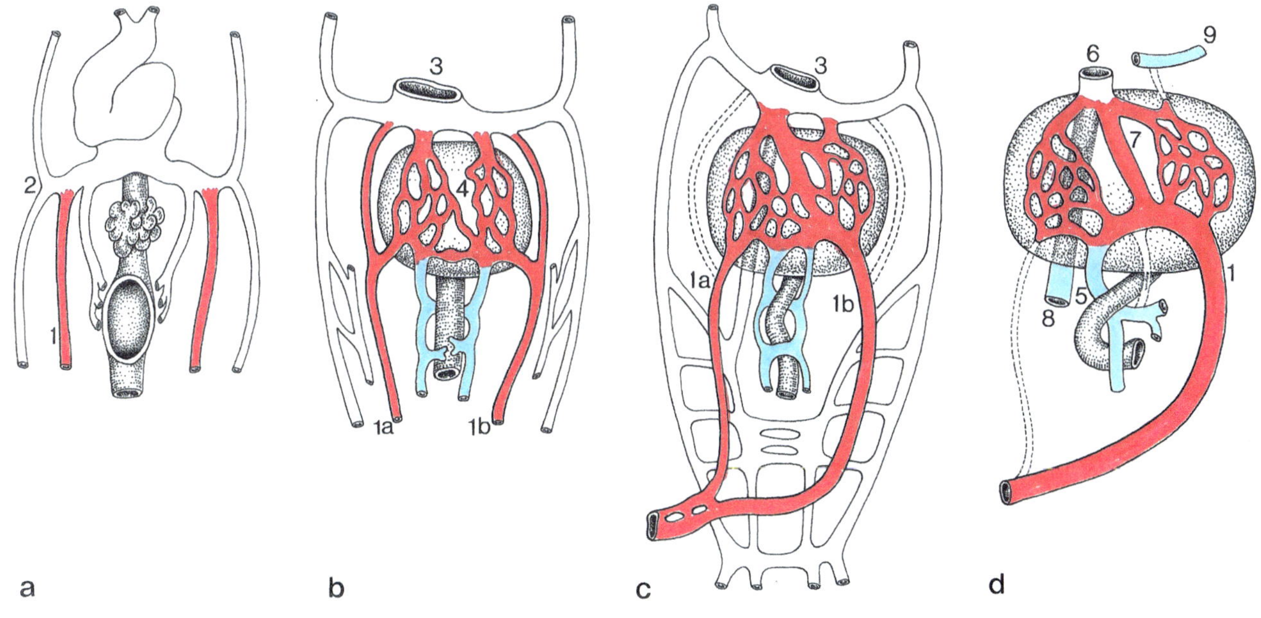

Fig. C29 The development of the vitelline and umbilical veins and the ducts and canal of Cuvier. *1*, *1a* right umbilical vein, *1b* left umbilical vein, *2* common cardinal vein (duct of Cuvier), *3* sinus venosus (canal of Cuvier), *4* liver sinusoids, *5* portal vein, *6* suparahepatic part of inferior caval vein, *7* ductus venosus, *8* infrahepatic part of inferior caval vein, *9* coronary sinus. (After ten Donkelaar HJ, Oostra R-J *Klinische Anatomie en Embryologie,* 4th ed. Springer, Houten, 2014)

Fig. C30 Photo of Czermak. (From Wikimedia Commons; in public domain)

Letter D

5

Dandy, Walter Edward (1866–1946): American neurosurgeon especially remembered for his introduction of ventriculography in 1918, and air encephalography in 1919, and for his surgical treatment of trigeminal neuralgia. References: *An operation for the cure of tic douloureux: Partial resection of the sensory root at the pons.* Arch Surg 1929; 18:687–734; *Treatment of trigeminal neuralgia by cerebellar route.* Ann Surg 1932; 96:787–795.

Vein of Dandy: *Eponym* for the petrosal vein.

Darkschewitsch, Liverij Osipovich (1858–1925): Russian neurologist in Moscow known for the nucleus named after him.

Nucleus of Darkschewitsch: *Eponym* for what is currently known as the elliptic nucleus in the prerubral tegmentum of the brain stem. Reference: *Ueber den oberen Kern des N. oculomotorius.* Arch Anat Physiol (Lpz) 1889, pp 107–116.

Davidoff, M von (? – 1904): German histologist.

Cells of Davidoff: *Synonym* for the cells of Paneth.

Darwin, Charles Robert (1809–1882): English naturalist born in Shrewsbury.

Tubercle of Darwin: *Eponym* for a small prominence occasionally present on the helix of the auricle (Fig. D1). It was first noticed by Thomas Woolner (1825–1892, sculptor), who called Darwin's attention to it.

Daubenton, Louis Jean Marie (1716–1799): French comparative anatomist, born in Mombard (Burgundy) and working as Conservator of the Museum of National History in Paris (Fig. D2). He is remembered for various anthropological indices of the skull. Reference: *Mémoire sur les différences de la situation du grand trou occipital dans l'homme et dans les animaux.* Mém Acad Sci Paris 1764, pp 568–575 (with JM Daele).

Angle of Daubenton: *Eponym* for his occipital angle.

Line of Daubenton and Daele: *Eponym* for the plane of the foramen magnum, the first search for a horizontal plane.

Decussations:

Decussation of Forel: *Eponym* for the ventral or anterior tegmental decussation; *see* Forel, Auguste Henri.

Decussation of Ganser: *Eponym* for the dorsal or posterior supraoptic decussation; *see* Ganser, Sigbert Joseph Marie.

Decussation of Held: *Eponym* for the crossing of part of the fibres from the cochlear nuclei forming the lateral lemniscus; *see* Held, Hans.

Decussation of Meynert: *Eponym* for the dorsal or posterior tegmental decussation; *see* Meynert, Theodor.

Decussation of Gudden: *Eponym* for the ventral or anterior supraoptic decussation; *see* Gudden, Bernhard von.

Decussation of Spitzka: *Eponym* for the decussation of the medial lemnisci in the medulla oblongata; *see* Spitzka, Edward Charles.

Decussation of Wernekinck: *Eponym* for the decussation of the superior cerebellar peduncles; *see* Wernekinck, Friedrich Christian Gregor.

Deetjen, Hermann (1867–1915): German physician.

Bodies of Deetjen: *Eponym* for platelets. Reference: *Eine Methode zur Fixierung der Bewegungszustände*

H. J. ten Donkelaar et al., *An Illustrated Guide to Anatomical Eponyms*, https://doi.org/10.1007/978-3-031-91664-9_5

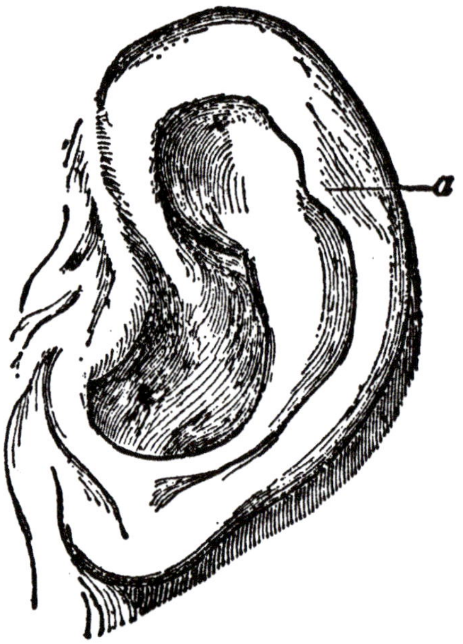

Fig. D1 Tubercle of Darwin. (From his *Descent of Man,* 2nd ed, 1882; illustrated by Woolner; Wikimedia Commons; in public domain)

Fig. D2 Portrait of Daubenton by Alexander Roslin. (From Wikimedia Commons; in public domain)

von Leukocyten und Blutplättchen. Münch Med Wochenschrift 1897; p 1192.

de Graaf, Reinier (1641–1673): Dutch anatomist and surgeon born in Schoonhoven and working in Delft (Fig. D3a). Reference: *Opera omnia.* Leiden, 1677; London, 1678.

Follicle of de Graaf: *Eponym* for the mature or vesicular ovarian follicle. Reference: *De mulierum organis generatione.* Hackiana, Leiden, 1672 (Fig. D3b). He also described the corpus luteum as substantia glandulosa. For biography, *see* Lindeboom GA *Reinier de Graaf (1641–1673).* Ned Tijdschr Geneeskd 1974; 118:789–795.

Deiters, Otto Friedrich Karl (1834–1863): German anatomist born and working in Bonn noted for his studies of the ear and the nervous system (Fig. D4). Reference: *Untersuchungen über Gehirn und Rückenmark des Menschen und der Säugetiere.* Vieweg, Braunschweig, 1865 (published posthumously by Max Schultze).

Ascending tract of Deiters: *Eponym* for an ascending component of the medial longitudinal fasciculus from the lateral vestibular nucleus to the ipsilateral oculomotor nucleus (Fig. D5).

Cells of Deiters: *Eponym* for the outer or external phalangeal cells of the spiral organ of the cochlea (*see* Fig. C21).

Nucleus of Deiters: *Eponym* for the lateral vestibular nucleus of the hindbrain, also described by Laura (Fig. D5).

Tract of Deiters: *Eponym* for the lateral vestibulospinal tract (Fig. D5).

Dejerine, Joseph Jules (1849–1917): French neurologist born to French parents in Geneva and studying and working in Paris as successor of Charcot at La Salpétrière. He is especially known for his extensive studies of the human nervous system. Reference: *Anatomie des centres nerveux,* 2 Vols. Rueff, Paris, 1895, 1901, written in collaboration with his American wife Augusta Marie Klumpke (Fig. D6).

Language zone of Dejerine: Lesions within this zone (Fig. D7) cause aphasia by affecting the Broca area (*B* as centre for motor images of articulation), the Wernicke area (*A* as centre for auditory images of words) and the parietooccipital cortex (*C* as centre for visual images of words).

Pes lemnisci bundles of Dejerine: *Eponym* for two aberrant bundles, which descend partly within the cerebral peduncle and partly in the area of the medial lemniscus (Fig. D8).

Retrolenticular segment of Dejerine: *Eponym* for the posterior part of the internal capsule of the brain (Fig. D9).

Sulci of Dejerine: *Eponym* for the anterior and posterior subcentral sulci of the cerebral cortex.

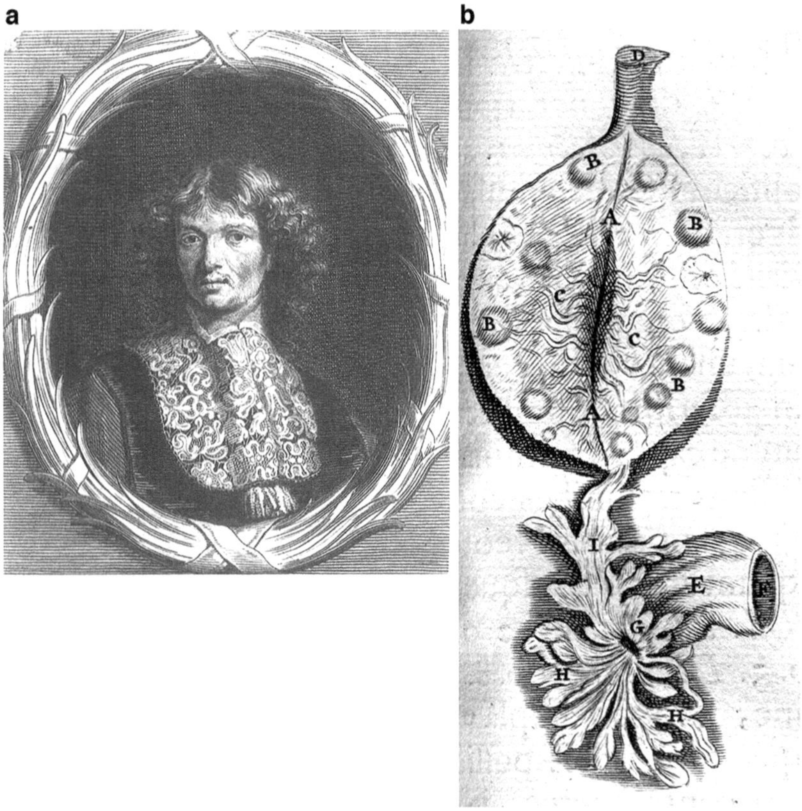

Fig. D3 (**a**) Engraving of de Graaf by Gerard Edelinck. (From the Dutch Iconographic Bureau; in public domain); (**b**) drawing of the ovarium from *De mulierum organis generatione.* Hackiana, Leiden, 1672. The eggs are marked by *B*, the uterine tube by *E*. (Courtesy Ondřej Naňka, Prague)

Fig. D4 Photo of Deiters. (From Wikipedia; in public domain)

Tract of Dejerine: *Eponym* for the anterior or ventral spinothalamic tract.

Delbet, Paul (1861–1924): French surgeon.

Fascia of Delbet: *Eponym* for the vesicoumbilical fascia. Reference: *Vessie, Urètre, Prostate, Verge, Périnée.* In: Poirier PJ, Charpy A *Traité d'anatomie humaine,* Tome V. Masson, Paris, 1907.

Demel, Rudolf (1891–1952): German surgeon.

Anterior and **Posterior oesophagotracheal arteries of Demel:** *Eponym* for the branches of the thoracic aorta supplying the oesophagus and the trachea. Reference: *Die Gefässversorgung der Speiseröhre.* Arch Klin Chir 1924; 128:453–504.

Point of Demel: *Eponym* for the area at the greater curvature of the stomach where the right and left gastroomental (gastroepiploic) arteries anastomose.

Demours, Pierre (1702–1795): French ophthalmologist and anatomist born in Marseille and working in Paris (Fig. D10). He is known for his work on the anatomy and pathology of the eye.

Membrane of Demours: *Eponym* for the posterior limiting membrane of the cornea, also described by Descemet (*see* Fig. D14), but probably first by Duddell (1729). Reference: *Lettre de Demours à M. Petit en réponse à sa critique d'un rapport sur une maladie de l'oeil survenu après inoculation de la petite vérole, contenant de nouvelles observations sur la structure de l'oeil, et quelques remarques générales de pratique relatives aux maladies de cet organe.* Paris, 1767.

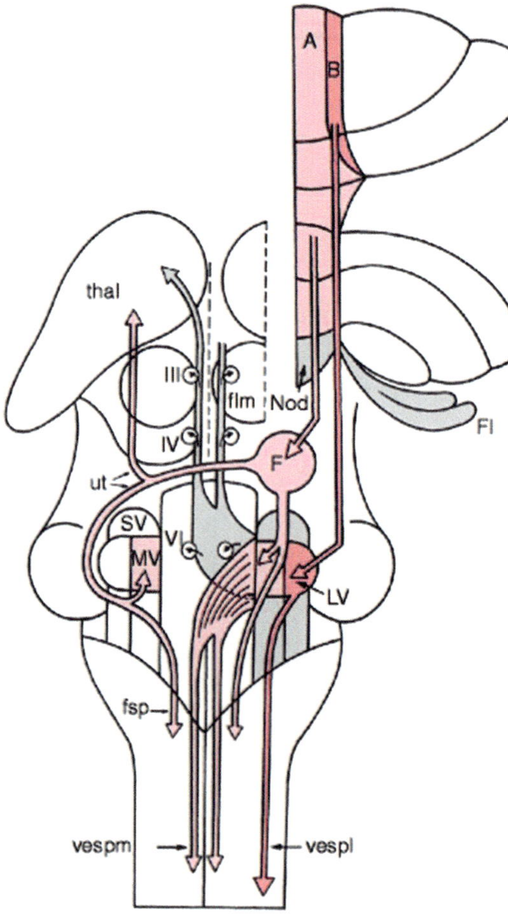

Fig. D5 Vestibulooculomotor and vestibulospinal projections. *A*, *B* longitudinal cortical cerebellar projection zones, *F* fastigial nucleus, *Fl* flocculus, *flm* fasciculus longitudinalis medialis, *fsp* fastigiospinal fibers, *LV*, *MV*, *SV* lateral, medial and superior vestibular nuclei, *Nod* nodulus, *thal* thalamus, *ut* uncinate tract, *vespl*, *vespm* lateral and medial vestibulospinal tracts, *III*, *IV*, *VI* oculomotor, trochlear and abducens nuclei. (From ten Donkelaar HJ *Clinical Neuroanatomy,* 2nd ed. Springer, Cham, 2020)

Fig. D6 The Dejerines. (From the Wellcome Collection CCBY, in public domain)

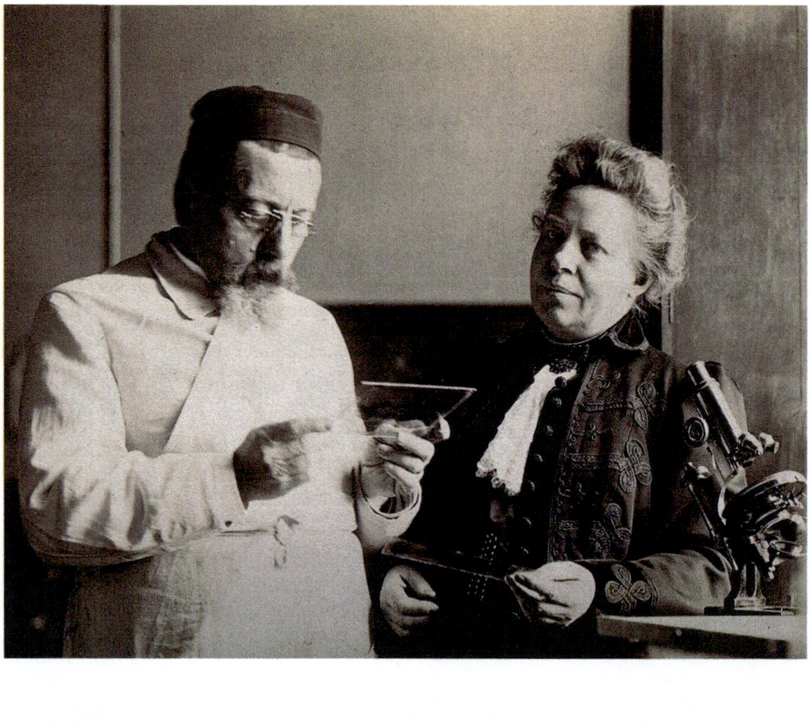

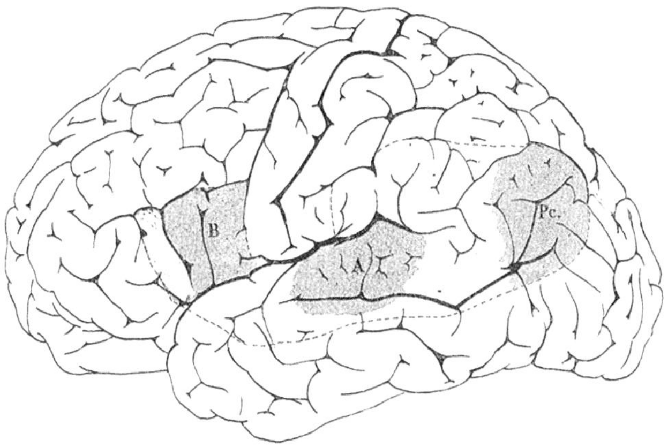

Fig. D7 Dejerine's "Zone of language". (From Dejerine J *Anatomie des centres nerveux, Tome Deux.* Rueff, Paris, 1901). Lesions within this zone, indicated by dashed lines, cause aphasia by affecting Broca's zone (*B*, "centre for motor images of articulation"), Wernicke's area (*A*, "centre for auditory images of words") or the parieto-occipital cortex (*Pc*, centre for visual images of words")

Denonvilliers, Charles-Pierre (1808–1872): French anatomist and surgeon born and working in Paris (Fig. D11).

Fascia or **Ligament of Denonvilliers:** *Eponym* for the rectovesical septum or fascia (Fig. D12). The female counterpart is the rectovaginal septum or fascia. Reference: *Anatomie du périnée.* Bull Soc Anat Paris 1836, pp 105–107. According to Barrett (2019) it was previously described by Frederick Tyrrell (1797–1843) and became also known as fascia of Tyrrell. However, in 1820 Pattison described a similar structure as prostatic fascia.

Denucé, Jean Louis Paul (1824–1889): French surgeon born at Ambarès (Gironde) and working in Bordeaux.

Ligament of Denucé, Arcuate: *Eponym* for the quadrate ligament of the proximal radioulnar joint. Reference: *Mémoire sur les luxations du coude.* Thesis No 69, Paris, 1854.

Descemet, Jean (1732–1810): French anatomist and surgeon born and working in Paris (Fig. D13).

Membrane of Descemet: *Eponym* for the posterior limiting membrane of the cornea, also described by Demours but before them in 1729 by Duddell (Fig. D14). References: *An sola lens crystalline cataractae sedes?* Paris, 1758; *Lettre contenant quelques remarques sur une lettre de M. Demours.* J Med Chir Pharm Paris 1769; 30:333–341; 33:40–70; 35:228–254.

Desjardins, Abel (1814–1866): French surgeon working in Paris.

Point of Desjardins: *Eponym* for a point on the right axilloumbilical line, marking the projection of the major duodenal papilla on the anterior abdominal wall.

'Deutsche Horizontale': *Synonym* for the Frankfurter line, which connects the lower border of the orbit with the upper border of the external acoustic meatus, also known as orbitomeatal plane.

Deventer, Hendrik van: *See* van Deventer, Hendrik.

Diameter of Baudelocque: *Eponym* for the external conjugate diameter of the pelvic inlet; *see* Baudelocque, Jean-Louis.

Dickinson, Robert Latou (1861–1950): American obstetrician and gynaecologist born in Jersey City, New Jersey, and after training in Germany and Switzerland, working in Brooklyn.

Sign of Dickinson: *Eponym* for the phenomenon that in early pregnancy, the uterus may be asymmetrical, since the implantation site can enlarge one horn before the other, also known as sign of Piskaçek. Reference: *The diagnosis of early pregnancy between the second and seventh weeks by bimanual examination.* NY J Gynaecol Obstet 1892; 10:544–555.

Diemerbroeck, IJsbrand van: *See* van Diemerbroeck, IJsbrand.

Disse, Joseph Hugo Vincenz (1852–1912): German anatomist and histologist born in Brakel and working in Strassburg as assistant to Waldeyer, then in Tokyo, Göttingen and finally in Marburg (Fig. D15a).

Space of Disse: *Eponym* for the perisinusoidal space between the sinusoids and the hepatic cells, visible only electronmicroscopically (Fig. D15b). Reference: *Über die Lymphbanen der Säugethierleber.* Arch Mikr Anat 1890; 36:203–224.

Sphincter of Disse: *Eponym* for a circular thickened muscular layer in the area of the renal papilla.

Dobie, William Murray (1828–1915): English physician and surgeon born in Liverpool and working in Chester.

Line of Dobie: *Eponym* for the dark line (isotropic band), generally known as Z-line, which runs transversally through bands of myofibrils and marks the limits of a sarcomere, also described by Amici and Krause.

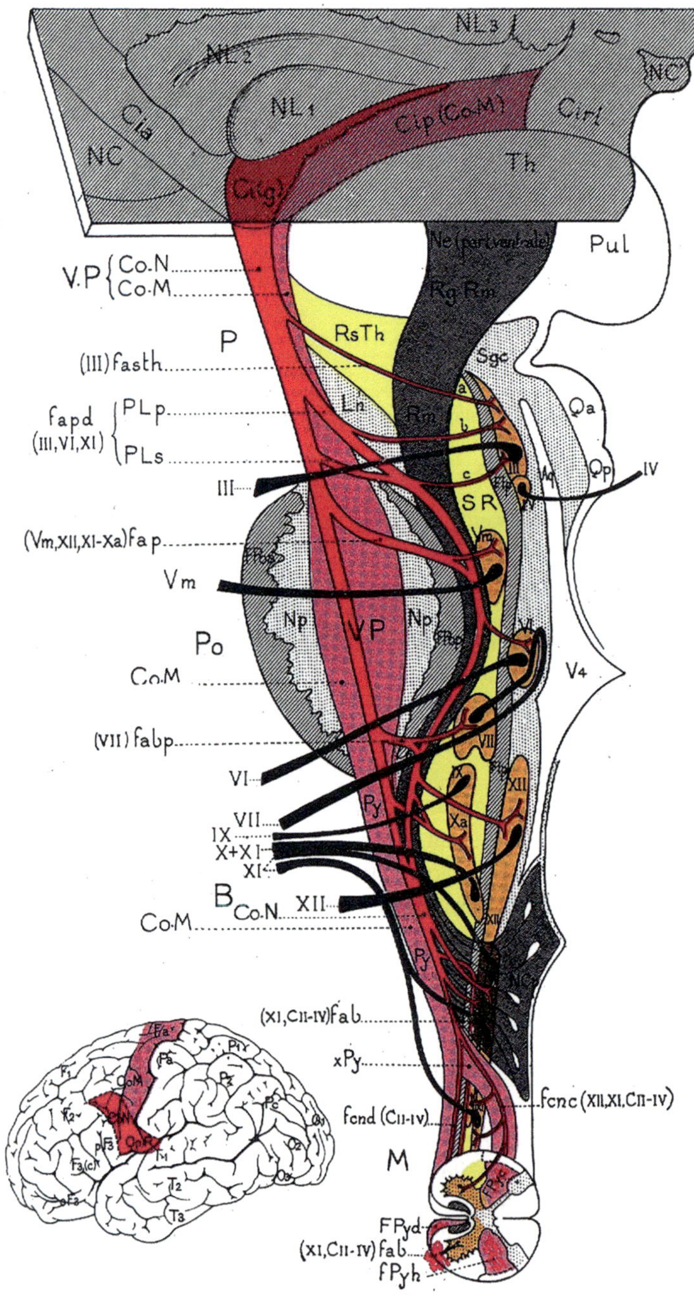

Fig. 43.

Fig. D8 Dejerine's view of the pyramidal tract with the pes lemnisci bundles joining the medial lemniscus. (After Dejerine J *Anatomie des centres nerveux, Tome Premier.* Rueff, Paris, 1895; from ten Donkelaar HJ *Clinical Neuroanatomy,* 2nd ed. Springer, Cham, 2020)

Dodd, William Harold Alfred (1899–1987): English surgeon born in Crewe, Cheshire, and working in Liverpool and later in London.

Popliteal vein of Dodd: *Eponym* for the vein of the popliteal fossa, an inconstant superficial vein draining as a perforating vein deep into the popliteal vein.

Veins of Dodd, Perforating or **Perforators:** *Eponym* for the medial femoral perforating veins of the lower limb (Fig. D16). References: *The varicose tributaries of the superficial femoral vein passing into Hunter's canal.* Postgraduate Med J 1959; 35:18–23; *The varicose tributaries of the popliteal vein.* Br J Surg 1965; 52:350–354; *The Pathology and Surgery of the Veins of the Lower Limb.* Churchill Livingstone, London, 1956 (with FB Cockett; 2nd ed, 1976).

Dogiel, Alexander Stanislavovic (1852–1922): Russian histologist in Tomsk and later in St. Petersburg (Fig. D17) known for the Dogiel end bulbs and cell types of

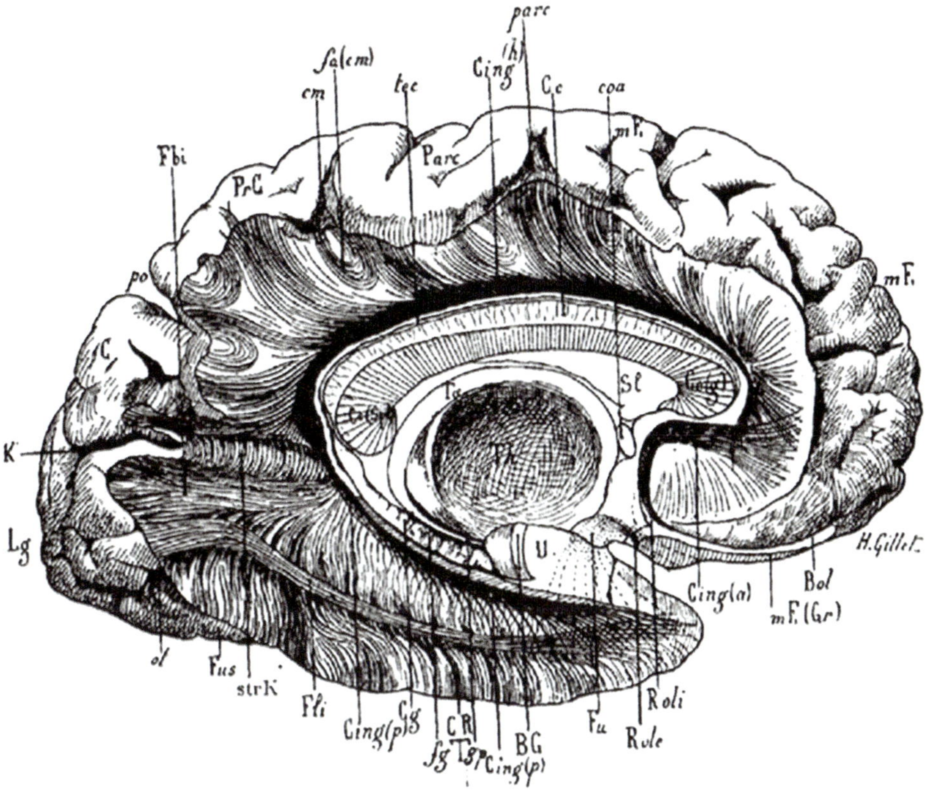

Fig. D9 Dissection of fibre tracts of the cerebral cortex, showing the cingulum (*Cing*), the inferior longitudinal fasciculus (*Fli*), the internal basal fasciculus of Burdach (*Fbi*) and the calcarine stratum (*strK*). Some other abbreviations: *BG* bandelette of Giacomini, Bol, olfactory bulb, *C* cuneus, *Cc* trunk of corpus callosum, *coa*, anterior commissure, *Fu* uncinate fasciculus, *K* calcarine sulcus, *Lg* lingual lobe, *Sl* septum lucidum. (From Dejerine J *Anatomie des centres nerveux, Tome Premier.* Rueff, Paris, 1895)

Fig. D10 Portrait of Demours, pastel by Maurice Quentin de la Tour. (From Wikimedia Commons; in public domain)

Fig. D11 Portrait of Denonvilliers. (From Wikimedia Commons; in public domain)

enteric neurons. Reference: *Ueber die Nervenendapparate in der Haut des Menschen,* Z Wiss Zool 1903; 75:46–111.

Cell of Dogiel: *Eponym* for a type of multipolar nerve cell within the prevertebral ganglia. Reference: *Ueber den Bau der Ganglien in den Geflechten des Darmes und der Gallenblase des Menschen und der Säugethiere.* Arch Anat Physiol, Anat Abt (Leipzig) 1899; pp 130–158.

Dogiel end bulbs or **Corpuscles:** *Eponym* for receptors with asymmetrical Schwann-cell lamination, found in mucocutaneous junctional areas, such as lips

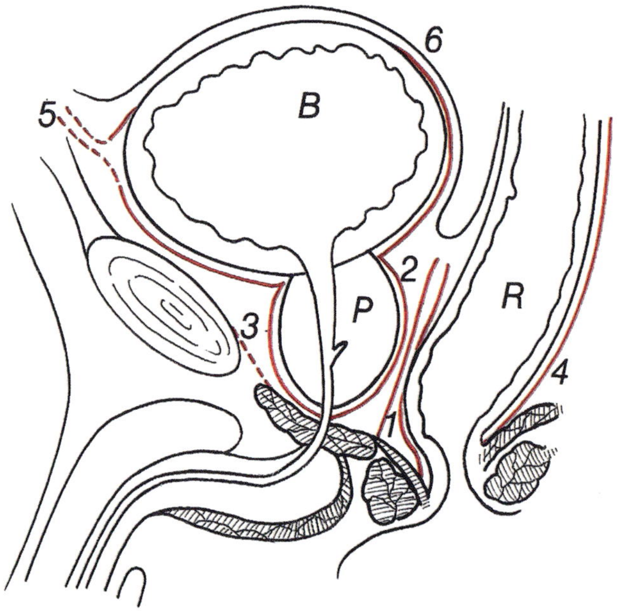

Fig. D12 Fascia of Denonvilliers. (After Hollinshead WH *Anatomy for Surgeons*, Vol 2. Harper & Row, New York, 1971). *B* bladder, *P* prostate, *R* rectum, *1* rectovesical septum, *2* fascial sheath of prostate, *3* puboprostatic ligament, *4* rectal fascia, *5* umbilicovesical fascia

Fig. D13 Portrait of Descemet. (From Wikidata; in public domain)

and external genitalia, also described by Krause (Chouchkov C *Cutaneous Receptors*. Adv Anat Embryol Cell Biol 1978; 54(5):1–63).

Dollinger, Johann Ignaz Josef (1770–1841): German physician and anatomist born in Bamberg, Bavaria, and working in Bamberg, Würzburg and Munich.

Tendinous ring of Dollinger: *Eponym* for an elastic ring around the periphery of the cornea, formed by a thickening of the posterior limiting membrane. Reference: *Illustratio iconographica fabricae oculi humani.* Würzburg, 1817.

Donaldson, James A (1930–2016): American ear surgeon born in St.Cloud, Minnesota, and working in Washington.

Line of Donaldson: *Eponym* for an imaginary line bisecting the lateral semicircular canal and transecting the posterior semicircular canal to locate the endolymphatic sac, situated inferior to the line. Reference: *Surgical Anatomy of Temporal Bone and Ear* (with BJ Anson). Saunders, Philadelphia, PA, 1967.

Donders, Franciscus (Frans) Cornelis (1818–1889): Durch ophthalmologist and founder of modern ophthalmology in the Netherlands, born in Tilburg and working in Utrecht (Fig. D18).

Space of Donders: *Eponym* for the space betwen the dorsum of the tongue and the hard palate when the mandible is in rest position following the expiratory cycle of respiration.

Doornik, Jacob Elisa (1777–1837): Dutch anatomist born in Leiden and working in Amsterdam (Fig. D19).

Line of Doornik: *Eponym* for a horizontal line from the incisors to the most prominent point of the occiput.

Plane of Doornik: *Eponym* for a horizontal plane from the incisors on both sides to the most prominent point of the occiput. Reference: *Wysgeerig natuurkundig onderzoek aangaande den oorspronkelyken mensch en de oorspronkelyke stammen van deszelfs geslachts.* Amsterdam, 1808.

Dorello, Primo (1872–1963): Italian anatomist (Fig. D20) known for his description of the abducens nerve canal (canal of Dorello), although it had already been described by Gruber (1859; *see* Gruber). Reference: *Considerazioni sopra la causa della paralisi transitoria dell'abducente nelle flogosi dell'orecchio medio.* In: Ferreri G (ed) *Atti della Clinica Oto-Rino-Laringoiatrica.* Tipografia del Campidoglio, Roma, 1905, pp 209–217.

Canal of Dorello: *Eponym* for the abducens nerve canal, earlier described by Gruber.

Douglas, James (1675–1742): Scottish anatomist and man-midwife born in Baads, near Edinburgh, and working in London (Fig. D21a). His main contribution was a monograph on the peritoneum. Reference: *A Description of the Peritoneum and of That Part of the Membrana cellularis Which Lies on Its Outside.* Roberts, London, 1730.

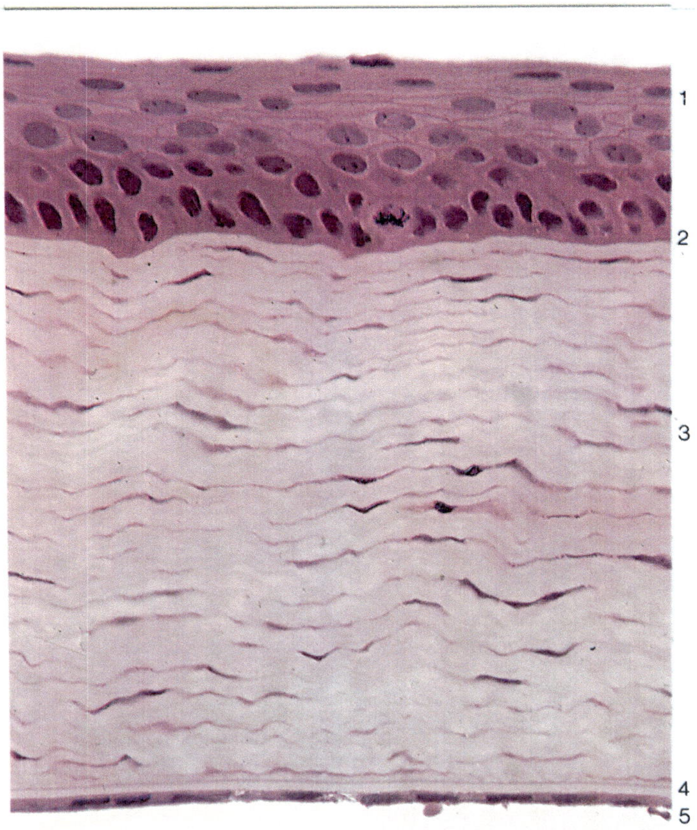

Fig. D14 The posterior limiting membrane of the cornea of Descemet, shown in a photomicrograph of a hematoxylin-eosin-stained section of part of the cornea from a guinea pig. (From Brüel A, Christensen EI, Tranum-Jensen J, Qvortrup K, Geneser F 2012 *Genesers histologi*. Munksgaard, Copenhagen; with permission). *1* anterior epithelium of cornea, *2* anterior limiting membrane of cornea (membrane of Bowman), *3* substantia propria of cornea, *4* posterior limiting membrane of cornea (membrane of Descemet), *5* posterior epithelium of cornea. (From ten Donkelaar HJ, Kachlík D, Tubbs RS *An Illustrated Terminologia Neuroanatomica*. Springer, Cham, 2018)

a

b

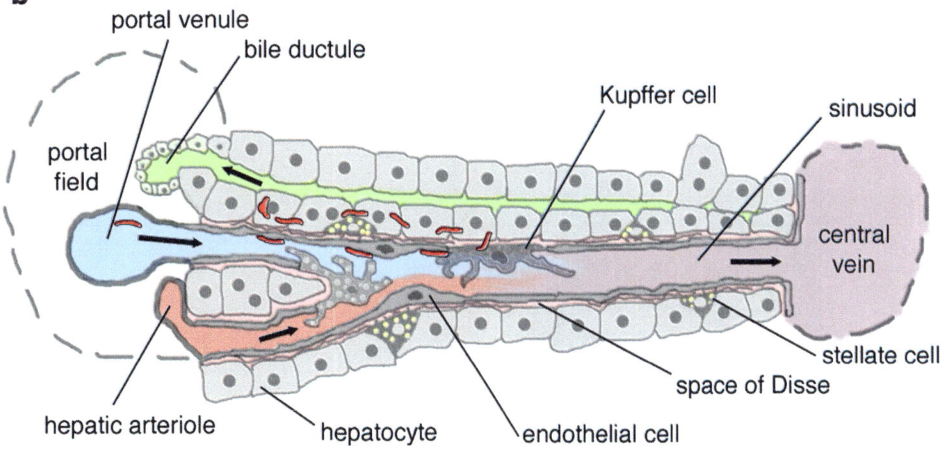

Fig. D15 (**a**) Portrait of Disse. (From Wikimedia Commons; in public domain); (**b**) space of Disse. (From Frevert U, Engelmann S, Zougbédé S, Stange J, Ng B et al. *Intravital observation of Plasmodium Berghei sporozoite infection of the liver.* PLoS Biology https://doi.org/10.1371/journal.pbio.0030192.9011; Open Access)

Fold of Douglas: *Eponym* for the rectouterine fold of the peritoneal cavity.

Line of Douglas, Arcuate: *Eponym* for the arcuate line representing the caudal boundary of the posterior layer of the rectus abdominis sheath (*see* **Fig. H12**).

Pouch or **Cul-de-sac of Douglas:** *Eponym* for the rectouterine pouch of the peritoneal cavity (Fig. D21b).

Septum of Douglas: *Eponym* for the embryonic cloacal membrane.

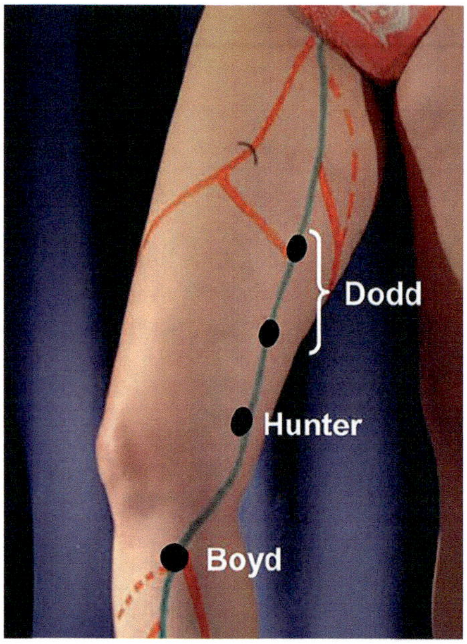

Fig. D16 Veins of Dodd, two of the perforating veins of the lower limb. (From Caggiati A, Mendoza E, Murena-Schmidt R, Lattimer CR *Anatomy of the Superficial Veins.* In: Mendoza E et al, eds *Duplex Ultrasound of Superficial Leg Veins.* Springer, Berlin-Heidelberg, 2014, pp 19–47; courtesy Alberto Caggiati, Rome)

Fig. D18 Photo of Donders. (From Pagel 1901)

Fig. D17 Photo of Dogiel. (From Pagel 1901)

Fig. D19 Portrait of Doornik. (From the Municipal Archive Amsterdam; in public domain)

Drummond, Sir David (1852–1932): Irish physician born in Rathgar, Dublin, and working in Newcastle and Durham (Fig. D22a).

Artery of Drummond, Marginal: *Eponym* for the marginal colic artery (Fig. D22b). Reference: *The arterial supply of the rectum and pelvic colon.* Brit J Surg 1914; 1:677–685.

Dua, Harminder S (twenty-first century): Indian-English ophthalmologist working in Nottingham.

Layer of Dua: *Eponym* for the preposterior limiting layer, anterior to the posterior limiting layer of Descemet in the cornea, composed of collagen with

Fig. D20 Photo of Dorello. (Courtesy Mariangela Fortunati, Narni, near Terni, Umbria)

abundant elastin and almost acellular. Reference: *Human corneal anatomy redefined: A novel pre-Descemet layer (Dua's layer).* Ophthalmology 2013; 120:1778–1785 (with Faraj LA, Said DG and others).

Dubreuil-Chambardel, Louis (1879–1927): French anatomist born in Marseille and working in Paris, known for his studies on the arteries of the limbs.

Accessory external pudendal artery of Dubreuil-Chambardel: *Epnym* for an inconstant branch of the inferior epigastric artery. Reference: *Traité des variations du système artériel. Variations des artères du pelvis et du membre inférieur.* Masson, Paris, 1925.

Anterior tibioperoneal trunk of Dubreuil-Chambardel: *Eponym* for an inconstant tibiofibular trunk, a common trunk for the origin of the fibular (peroneal) and anterior tibial arteries. Reference: *L'artère poplitée et ses branches terminales. Variations anatomiques et morphogénie.* Thesis, Paris, 1905.

Artery of Dubreuil-Chambardel: *Eponym* for the recurrent interosseous artery of the forearm. Reference: *Traité des variations du système artèriel. Variations des artères du membre supérieur.* Masson, Paris, 1926.

Ducts: *See* **The Classical Anatomical Eponyms.**

Duddell, Benedict (1695–1759): English surgeon and oculist who was the first to describe what is currently mostly known as membrane of Descemet (or Demours).

Membrane of Duddell: *Eponym* for the posterior limiting membrane of the cornea. Reference: *Treatise on the diseases of the horny coat of the eye and the various kinds of cataracts; to which is prefixed a method entirely new of scarifying the eyes for several disorders. With remarks on the practice of some oculists both at home and abroad.* Clark, London, 1729.

Duncan, Daniel (1649–1735): French physician born in Montauban (Garonne) who, forced by the Revocation of the Edict of Nantes, went to Berlin, The Hague and finally to London.

Ventricle of Duncan: *Eponym* for the cavum of the septum pellucidum, also described by Sylvius, Verga and Wenzel; Reference: *Explication nouvelle et méchanique des actions animales, ou il est traité des fonctions de l'âme, avec une méthode façile pour démonstrer exactement toutes les parties du cerveau sans couper sa propre substance et une discours sur sa formation.* Paris, 1678.

Duncan, James Matthews (1826–1890): Scottish obstetrician born in Aberdeen and working in Edinburgh.

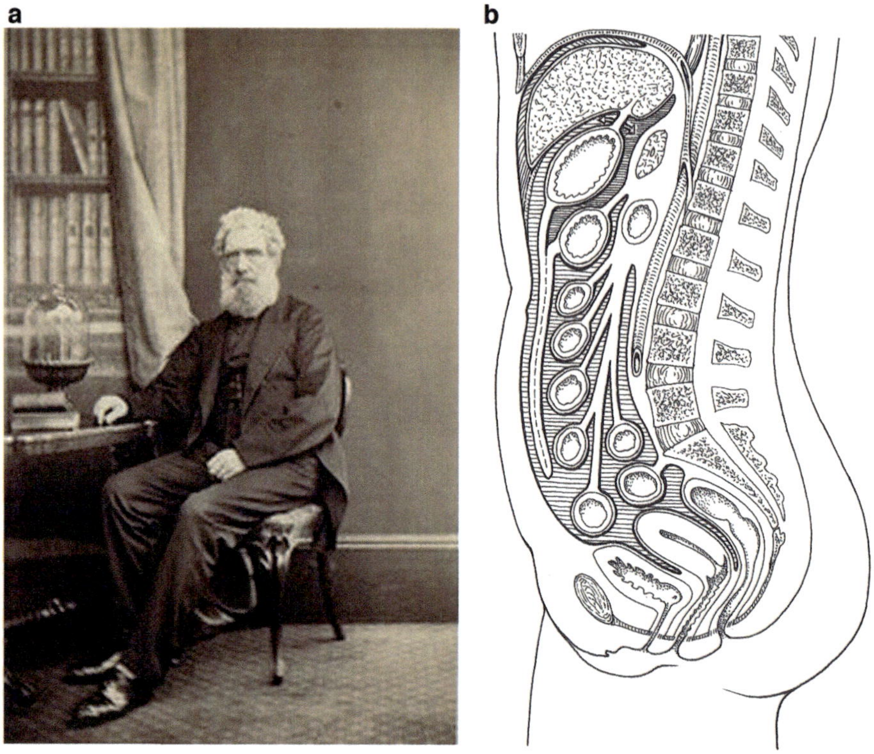

Fig. D21 (**a**) Photo of Douglas. (From the Wellcome Collection CCBY; in public domain); (**b**) pouch of Douglas, shown in a median section of the trunk; the *arrow* indicates the epiploic foramen. (From an old figure by the late Chris van Huijzen)

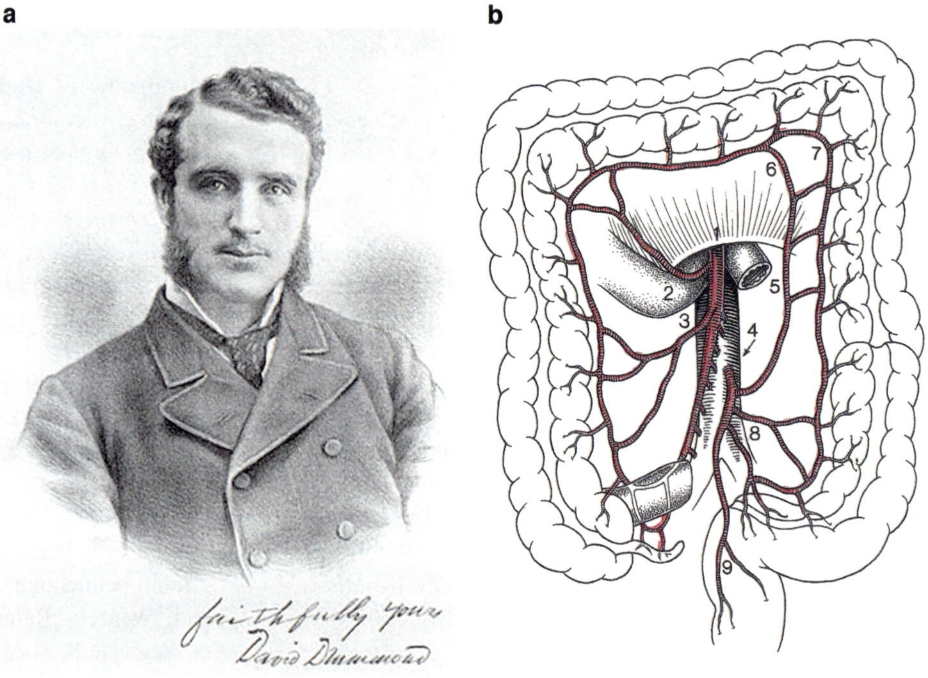

Fig. D22 (**a**) Portrait of Drummond. (From Wikimedia Commons; in public domain); (**b**) Marginal artery of Drummond. *1* superior mesenteric artery, *2* middle colic artery, *3* right colic artery, *4* inferior mesenteric artery, *5* left colic artery, *6* arc of Riolan, *7* marginal artery of Drummond. (After ten Donkelaar HJ, Oostra R-J *Klinische Anatomie en Embryologie,* 4th ed. Springer Media, Houten, 2014)

Folds of Duncan: *Eponym* for peritoneal folds of the uterus. Reference: *On the displacements of the uterus.* Edinburgh, 1854.

Dupuytren, Guillaume (1777–1835): French surgeon born in Pierre-Buffière and working in Paris (Fig. D23).

Canals of Dupuytren: *Eponym* for canals for the diploic veins in the skull, earlier described by Breschet.

Fascia of Dupuytren: *Eponym* for the palmar aponeurosis. Reference: *De le rétraction des doigts par suite d'une affection de l'aponévrose palmaire, opération* *chirurgicale qui convient en ce cas.* J Univ Hebd Méd Chir Prat 1831; 5:352–365.

Duret, Henri (1849–1921): French surgeon born in Conde-sur-Noireau and working in Paris and later in Lille known for his studies on the blood vessels of the brain (Fig. D24). Reference: *Recherches anatomiques sur le circulation de l'encéphale.* Arch Physiol Norm Pathol 1874; 6:60–91, 316–354, 664–693, 919–957.

Anterior internal frontal artery of Duret: *Eponym* for the frontopolar or prefrontal artery.

Anterior striated arteries of Duret: *Eponym* for the central branches of the anterior cerebral artery.

External striated arteries of Duret: *Eponym* for branches of the middle cerebral artery for the putamen (the putaminocaudate arteries of Foix and Lévy).

Internal optic arteries of Duret: *Eponym* for the thalamostriatal arteries of Foix and Hillemand.

Internal striatal arteries of Duret: *Eponym* for the striatal branches of the middle cerebral artery.

Middle internal frontal artery of Duret: *Eponym* for a branch of the anterior cerebral artery to the posterior part of the internal frontal gyrus.

Posterior internal frontal artery of Duret: *Eponym* for a branch of the anterior cerebral artery to the paracentral lobule and the precuneus.

Duverney, Joseph Guichard (1648–1730): French anatomist and otologist born in Feurs and working in Paris, and known as a pioneer in otology. Reference: *Traité de l'organe de l'ouie contenant la structure, les usages et les maladies de toutes les parties de l'oreille.* Michallet, Paris, 1683; new French edition by Langerak, Leiden, 1731; Latin version: *Tractatus de organo auditis continens structuram, usam, et morbos omnium auris partium.* Langerak, Leiden, 1730; English translation published 1737 by Baker, London. As an example of his work with the aid of a microscope, his Planche X from the 1731 edition, is shown (Fig. D25).

Canalis medius of Duverney: *Eponym* for the lateral semicircular duct of the inner ear.

Canalis semicircularis superior of Duverney: *Eponym* for the anterior semicircular duct of the inner ear.

Fig. D23 Portrait of Dupuytren. (From Pagel 1901)

Fig. D24 Photo of Duret. (From Wikimedia Commons; in public domain)

Fissures or **Incisures of Duverney:** *Eponym* for the notches in the cartilaginous part of the external acoustic meatus, also described by Santorini.

Foramen of Duverney: *Eponym* for the omental or epiploic foramen connecting the main peritoneal cavity with the omental bursa, also described by Winslow.

Glands of Duverney: *Eponym* for the vulvovaginal glands, which he discovered in cows, comparable to the greater vestibular glands of Bartholin.

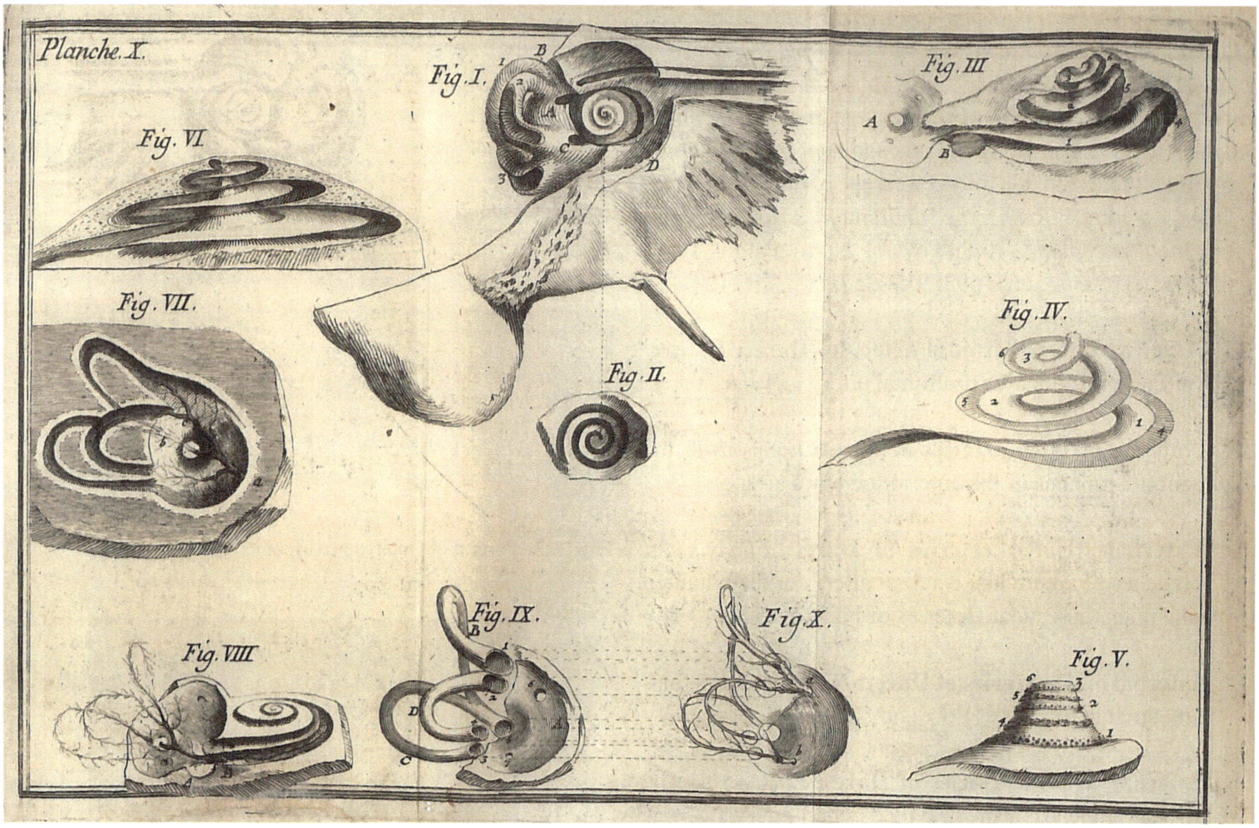

Fig. D25 The ear by Duverney. (Planche X from Duverney JG *Traité de l'organe de l'ouie contenant la structure, les usages et les maladies de toutes les parties de l'oreille.* Langerak, Leiden, 1731; from the Wellcome Collection CCBY)

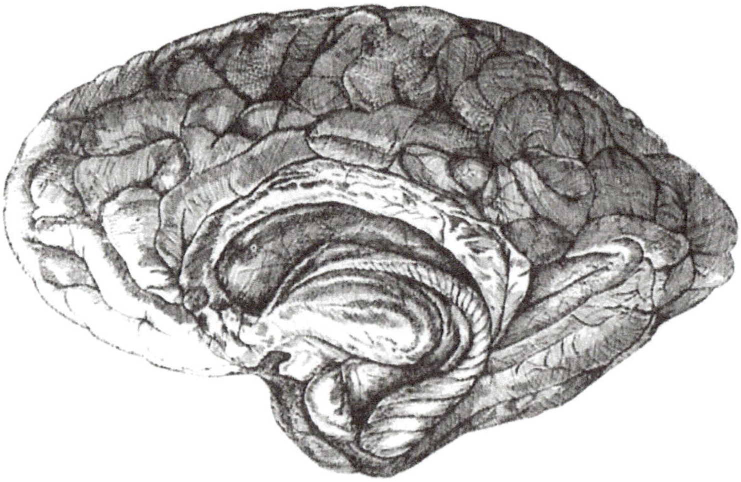

Fig. D26 The hippocampus as illustrated by Duvernoy JG *De sinibus cerebri.* Comm Acad Sci Imp Petropolitana, Tomus IV, St. Petersburg, 1729. (Redrawn by Lewis FT *The significance of the term hippocampus.* J Comp Neurol 1923; 35:213–230; in public domain)

Muscle of Duverney: *Eponym* for the lacrimal (deep) part of the orbicularis oculi muscle, later also described by Horner (1824).

Duvernoy, Johann Georg (1691–1759): German anatomist born in Mömpelgard, and working in Tübingen, where Haller was his student, and from 1725–1741 in St. Petersburg. After the first description of the structure by Arantius, he was the first to adequately illustrate the hippocampus (Fig. D26). Reference: *De sinibus cerebri.* Comm Acad Sci Imp Petropolitana, Tomus IV. St. Petersburg, 1729; *see* also Lewis FT *The significance of the term hippocampus.* J Comp Neurol 1923; 35:213–230).

Letter E

6

Eberstaller, Oscar (1851–1939): Austrian anatomist in Graz known for his studies on the gyri and sulci of the cerebral cortex (**Fig. E1**). References: *Zur Oberflächenanatomie der Grosshirnhemisphären. Wien Med Bl 1884; 7:479-482, 542-582, 644-646; Das Stirnhirn. Ein Beitrag zur Anatomie der Oberfläche des Großhirns.* Urban & Schwarzenberg, Vienna, 1890.

Sulcus of Eberstaller: *Eponym* for (a) the diagonal sulcus of the frontal lobe, a variable branch from the lateral sulcus, dividing the opercular part into two triangular parts, and (b) the second intermediate sulcus of the parietal lobe. The triangular part may be indented by an inconstant radiate sulcus of Eberstaller (**Fig. E2**). Eberstaller's transverse insular sulcus, separating the short and long gyri of the insula, is a *synonym* for the central sulcus of the insula of Reil.

Ebner, Victor, Ritter von Rofenstein (1842–1925): Austrian histologist and embryo**logis**t born in Bregenz and working in Innsbrück and later in Vienna (**Fig. E3**).

Glands of Ebner: *Eponym* for serous glands in relation to the vallate or circumvallate papillae of the tongue. Reference: *Die acinösen Drüsen der Zunge und ihre Beziehungen zu den Geschmacksorganen.* Graz, 1871.

Lines of Ebner: *Eponym* for cementing lines, distinct refractile borders of collagen-poor bone matrix forming the outer limit of secondary osteons (*see* **Fig. 23.1**).

Reticulum of Ebner: *Eponym* for a network of Sertoli cell processes in the seminiferous tubules of the testis. Reference: *Untersuchungen über den Bau der Samencanälchen.* Leipzig, 1871.

Ecker, Alexander (1816–1887): German anatomist and physiologist born in Freiburg and working in Basel and later in Freiburg known for his studies on the subdivision of the cerebral cortex (**Fig. E4**). Reference: *Die Hirnwindungen des Menschen nach eigenen Untersuchungen insbesondere über die Entwicklung derselben beim Foetus und mit Rücksicht auf das Bedürfniss der Aerzte dargestellt.* Vieweg, Braunschweig, 1869.

Gyrus of Ecker: *Eponym* for the descending occipital gyrus of the cerebral cortex.

Sulcus of Ecker: *Eponym* for the anterior or transverse occipital sulcus of the cerebral cortex.

Economo, Constantin Freiherr von (1876–1931): Romanian-born (Braila), Austrian neurologist of Greek origin (**Fig. E5**) known for his cytoarchitectonic studies of the brain as well as for his early description of *encephalitis lethargica.* Reference: *Die Cytoarchitektonik der Hirnrinde des erwachsenen Menschen* (with Georg N Koskinas, 1885–1975). Springer, Berlin, 1925; English translation by Triarhou LC: *Cytoarchitectonics of the Adult Human Cerebral Cortex.* Karger, Basel, 2008. For biography, *see* Lesky E *Constantin von Economo (1876–1931): Life and Work.* In: Brazier MAB, Petsche H (eds) *Architectonics of the Cerebral Cortex.* Raven Press, New York, pp 1–7.

Areas of von Economo: *Eponym* for his cytoarchitectonic subdivision of lobes of the cerebral cortex (**Fig. E6**).

Cell of von Economo: *Eponym* for the von Economo neuron (VEN), a large bipolar neuron, described by von Economo and Koskinas (1925) in the frontoinsular

H. J. ten Donkelaar et al., *An Illustrated Guide to Anatomical Eponyms*, https://doi.org/10.1007/978-3-031-91664-9_6

Fig. E1 Portrait of Eberstaller (from the Anatomical Institute Graz; courtesy Michael Pretterklieber, Graz)

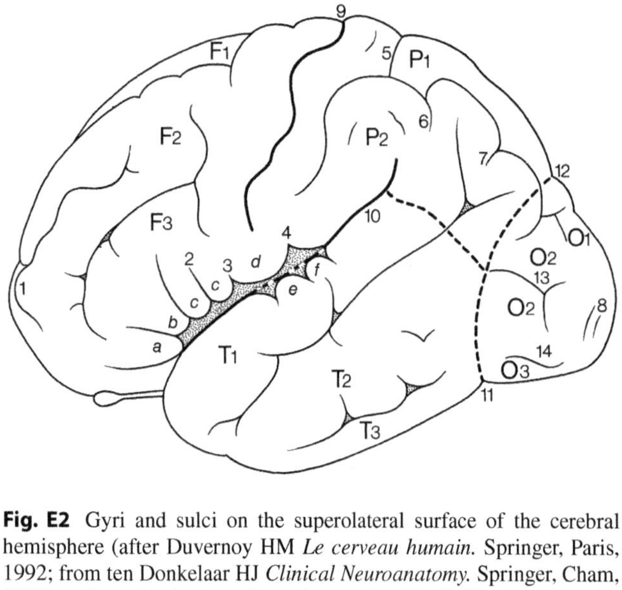

Fig. E2 Gyri and sulci on the superolateral surface of the cerebral hemisphere (after Duvernoy HM *Le cerveau humain.* Springer, Paris, 1992; from ten Donkelaar HJ *Clinical Neuroanatomy.* Springer, Cham, 2020). *a*, *b*, *c* orbital, triangular and opercular (divided into two parts by 2) parts of the inferior frontal gyrus, *d* subcentral gyrus, *e*, *f* transverse temporal gyri of Heschl, *F1*, *F2*, *F3* superior, middle and inferior frontal gyri, *O1*, *O2*, *O3* superior, middle and inferior occipital gyri, *P1*, *P2* superior and inferior parietal lobules, *T1*, *T2*, *T3* superior, middle and inferior temporal gyri, *1* frontomarginal sulcus of Wernicke, *2* diagonal sulcus of Eberstaller, *3* anterior subcentral sulcus, *4* posterior subcentral sulcus, *5* transverse parietal sulcus of Brissaud, *6* first intermediate sulcus of Jensen, *7* second intermediate sulcus of Eberstaller, *8* descending occipital gyrus of Ecker, *9* central sulcus, *10* lateral sulcus, *11* preoccipital notch of Meynert, *12* parieto-occipital sulcus, *13* lunate sulcus. *14* inferior occipital sulcus

Fig. E3 Photo of Ebner (from the Archiv der Uiversität Wien; in public domain)

cortex and the anterior limbic area in great apes and humans but not in other primates. VENs are also present in elephants and whales and may be a specialization related to large brain size (Allman JM, et al. Ann NY Acad Sci 2011; 1225:59–71).

Edinger, Ludwig (1855–1918): German anatomist born in Worms and working in Frankfurt am Main who described various structures in the brain (**Fig. E7**). His *Kammsystem des Fusses* (comb system) consists of pallidosubthalamic and striatonigral fibres. References: *Ueber den Verlauf der centralen Hirnnervenbahnen mit Demonstrationen von Präparaten.* Arch Psychiatr Nervenkr 1885; 16:858–859; *Vorlesungen über den Bau der nervösen Centralorgane des Menschen und der Thiere.* Vogel, Leipzig, 1893; *Vorlesungen über den Bau der nervösen Zentralorgane*, 8th ed. Vogel, Leipzig, 1911.

Bundle of Edinger, Predorsal: *Eponym* for the tectospinal tract, also described by Loewenthal.

Comb system of Edinger: *Eponym* for pallidosubthalamic and striatonigral fibres.

Nucleus of Edinger-Westphal: *Eponym* for the accessory nucleus of the oculomotor nerve. This nucleus

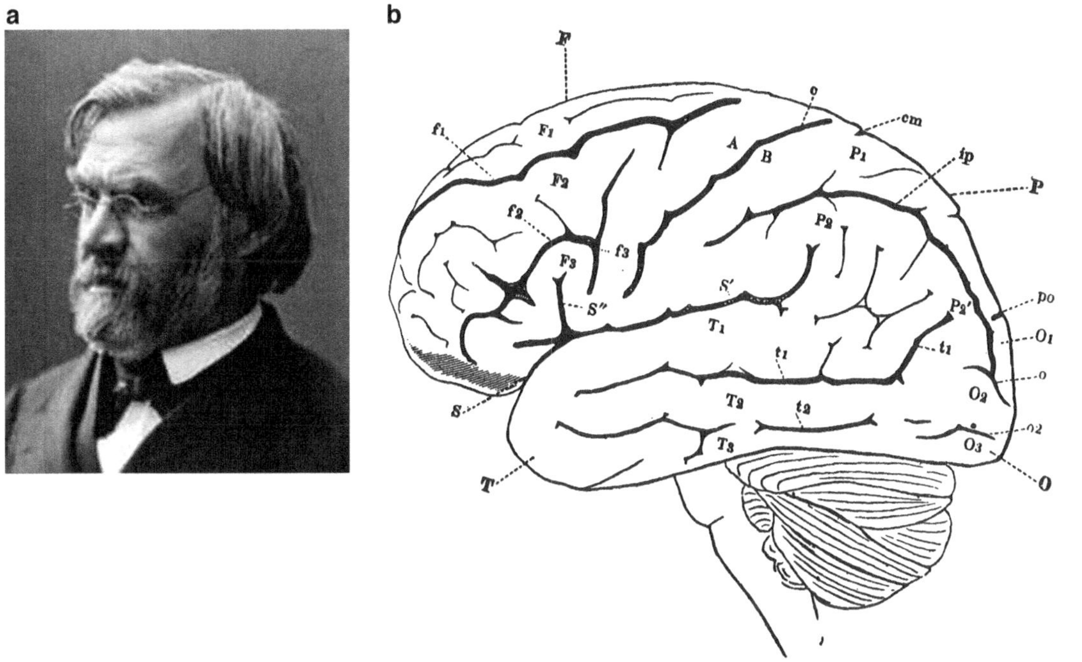

Fig. E4 (**a**) Photo of Ecker (from Wikimedia Commons; in public domain); (**b**) lateral view of the brain (from Ecker V *Die Hirnwindungen des Menschen nach eigenen Untersuchungen.* Vieweg, Braunschweig, 1869)

Fig. E5 Photo of von Economo by Georg Fayer (from Wikipedia; in public domain)

was identified by Edinger in human fetuses in 1885, and in adults by Carl Friedrich Otto Westphal in 1887 (*see* Westphal, Carl Friedrich Otto). Actually, the nucleus of Edinger-Westphal consists of several parts: (a) a preganglionic part, a group of parasympathetic neurons, and part of the oculomotor complex; (b) a non-preganglionic, centrally projecting part; and (c) an anterior medial nucleus.

Tract of Edinger: *Eponym* for the lateral spinothalamic tract (**Fig. E8**; 'Untere Schleife'). Reference: *Ueber die Verbindungen der sensiblen Nerven mit dem Zwischenhirn.* Anat Anz 1887; 2:145–153.

Egger, Fritz (1863–1938): Swiss internist born in Basel and working in Arosa and Halle.

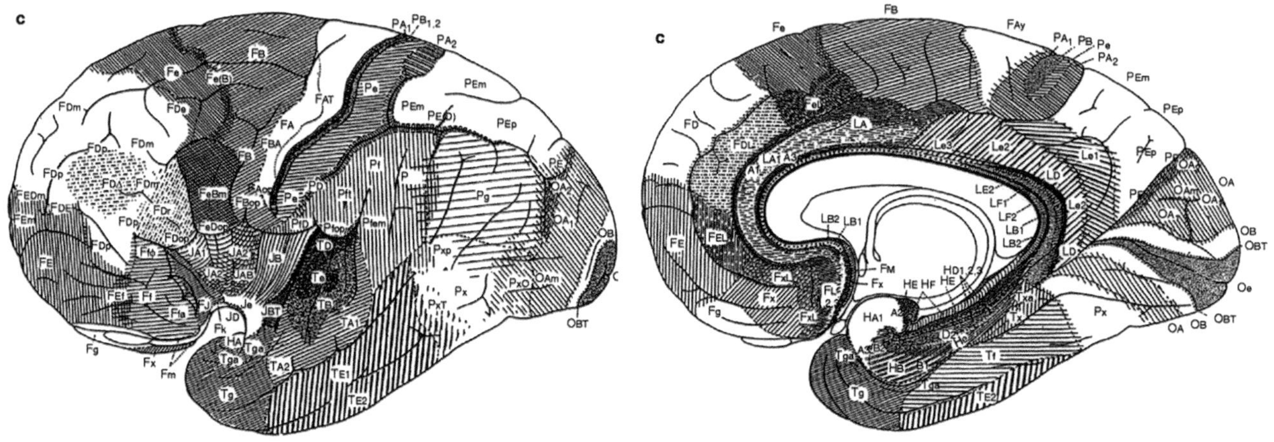

Fig. E6 Cortical maps of the lateral (**a**) and medial (**b**) surface of the brain showing the subdivision of the cerebral cortex by von Economo and Koskinas (*Die Cytoarchitektonik der Hirnrinde des erwachsenen Menschen.* Springer, Berlin, Heidelberg, New York, 1925; from ten Donkelaar HJ *Clinical Neuroanatomy,* 2nd ed. Springer, Cham, 2020)

Fig. E7 Photo of Edinger (from Pagel 1901)

Line of Egger: *Eponym* for a circular line, the outer border of the hyalocapsular ligament of Wieger. Reference: *Über Keratitis parenchymatosa circumscripta.* Thesis, University of Basel, 1888.

Ehrenritter, Johann (?-1790): Austrian anatomist born and working in Vienna.

Ganglion of Ehrenritter: *Eponym* for the superior ganglion of the glossopharyngeal nerve (*see* **Fig. A8b**). Reference: *Ueber den Nervus glossopharyngeus.* Med Chir Zeit Salzburg, 1790; 4:319 (Preliminary communication).

Ellis, George Viner (1812–1900): English physician and anatomist in London.

Muscle of Ellis: *Eponym* for the corrugator cutis ani muscle, longitudinal muscle fibes between the rectal muscles and the anal sphincter, also described by Kohlrausch.

Elschnig, Anton (1863–1939): Austrian ophthalmologist born in Leibnitz and working in Graz, Vienna and Prague. Reference: *Der normale Sehnerveneintritt des menschlichen Auges.* Denkschr Kais Akad Wiss Wien, Math-Naturw Kl 1901; 70:219–310.

Border tissue of Elschnig: *Eponym* for peripapillary border tissue, a ring of scleral collagenous tissue lying external to the glial sheaths, separating the lamina cribrosa of the sclera from the peripapillary scleral flange.

Membrane of Elschnig: *Eponym* for the internal limiting membrane covering the head of the optic nerve and isolating the ganglion cells of the retina from the peripapillary area.

Scleral ring of Elschnig: *Eponym* for a white circular band visible on the fundus of the eye, an anterior extension of the sclera between the choroid and the optic nerve, separating the intrapapillary area from the peripapillary area.

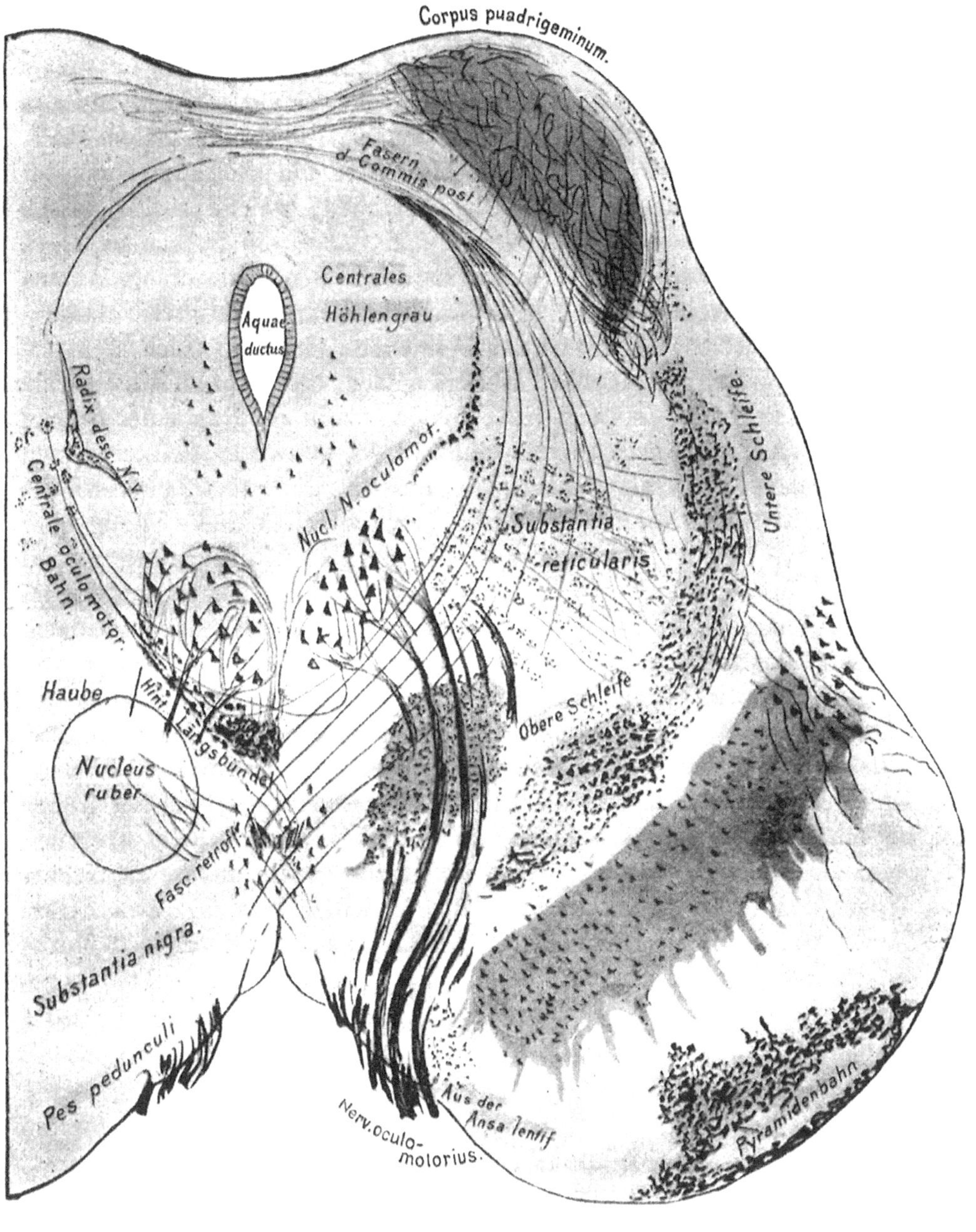

Fig. E8 Drawing of the midbrain by Edinger (from Edinger L *Vorlesungen über den Bau der nervösen Centralorgane des Menschen und der Thiere.* Vogel, Leipzig, 1893)

End bulbs:

End bulbs of Dogiel: *Eponym* for receptors with asymmetrical Schwann-cell lamination found in mucocutaneous junctional areas (the lips and external genitalia),also described by Krause; *see* Dogiel, Alexander Stanislavovic.

End bulbs of Krause: *Eponym* for receptors with asymmetrical Schwann-cell lamination found in mucocutaneous junctional areas (the lips and external genitalia), also described by Dogiel; *see* Krause, Wilhelm Johann Friedrich.

Englisch, Josef (1835–1915): Austrian physician born in Freudenthal (Silesia), now Bruntál, Czech Republic, and working in Vienna.

Sinus of Englisch: *Eponym* for an inconstant petrooccipital venous sinus. Reference: *Ueber eine constante Verbindung der Sinus cavernosus mit dem hinteren Ende des Petrosus inferior ausserhalb des Schädels.* Sitz Kais Akad Wiss Wien 1863; 58:27–29.

Erb, Wilhelm (1840–1921): German physician and pathologist born in Winnweiler and working in Leipzig and later in Heidelberg.

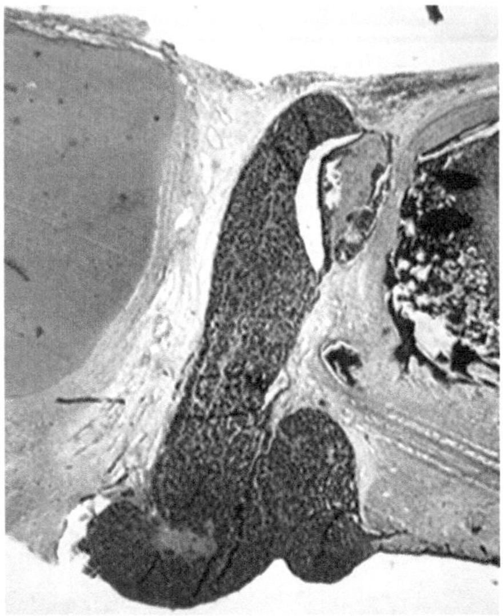

Fig. E9 Pharyngeal pituitary of Erdheim (from ten Donkelaar HJ, Lammens M, Hori A *Clinical Neuroembryology*, 3rd ed. Springer, Cham, 2023; courtesy Akira Hori, Hannover)

Point of Erb: *Eponym* for (a) the punctum nervosum of the upper part of the brachial plexus at the posterior border of the sternocleidomastoid muscle about 3 cm above the clavicle and in front of the transverse process of the sixth cervical vertebra; and (b) the auscultation point in the third intercostal space at the left sternal border, where sounds of pathologically altered heart valves can be heard. Reference: *Über eine eigentümliche Localisation von Lähmungen im Plexus brachialis.* Verh Naturhist Med Verein Heidelb 1874–1877; 1:130–136.

Erdheim, Jakob (1874–1937): Austrian pathologist born in Boryslaw (Galicia, currently Ukraine) and working in Vienna.

Pharyngeal hypophysis of Erdheim (Fig. E9): *Eponym* for ectopic pituitary tissue, earlier described by Killian (1888). Reference: *Über Hypophysenganggeschwülste und Hirncholesteatome.* Sitz Kais Akad Wiss Wien 1904, 62.

Erggelet, Heinrich (1883–1969): German ophthalmologist working in Jena and Göttingen.

Space of Erggelet: *Eponym* for a funnel-shaped dilatation of the anterior end of the hyaloid canal behind the lens. Reference: *Klinische Befunde bei fokaler Behandlung mit den Gullstrandschen Nernst Spaltlampe.* Klin Monatsbl Augenheilk 1914; 53:449–470.

Essick, Charles R (1883–1954): American anatomist working in Baltimore.

Corpus pontobulbare of Essick: *Eponym* for a transient structure in the brain stem through which neurons of the precerebellar systems reach their ultimate position. Reference: *The Corpus ponto-bulbare. A hitherto undescribed nuclear mass in the human hind brain.* Amer J Anat 1907; 7:119–135.

Estienne, Charles (1504–1564): French anatomist, like Jacques Dubois (Sylvius) largely following Galenic anatomy. He first observed the central canal of the spinal cord and the recurrent laryngeal nerve (**Fig. E10**). Reference: *De dissectione partium corporis humani libri tres.* Paris 1545.

Nerve of Estienne: *Eponym* for the recurrent laryngeal nerve.

Eustachi (Eustachio, Eustachius), Bartolomeo (c. 1510–1574): Italian anatomist born in San Severino (Marche), working in Rome and physician to the Pope (**Fig. E11a**). He prepared his accurate anatomical plates, the first on copper, in 1552, but these were overlooked and remained unpublished until 1714. Then, they were rediscovered in the Vatican Library and Pope Clement XI handed them over to Giovanni Maria Lancisi, his physician at that time. Lancisi edited and published Eustachi's 47 anatomical plates in 1714. Eustachi's name is still in use for the auditory tube (Eustachian tube) and many related structures. Reference: *Tabulae anatomicae* (edited by Lancisi). Gonzago, Rome, 1714 (**Fig. E11b**). Two examples of these tables are shown in **Figs. E12** and **E13**.

Amygdala or **Tonsil of Eustachi:** *Eponym* for the tubal tonsil.

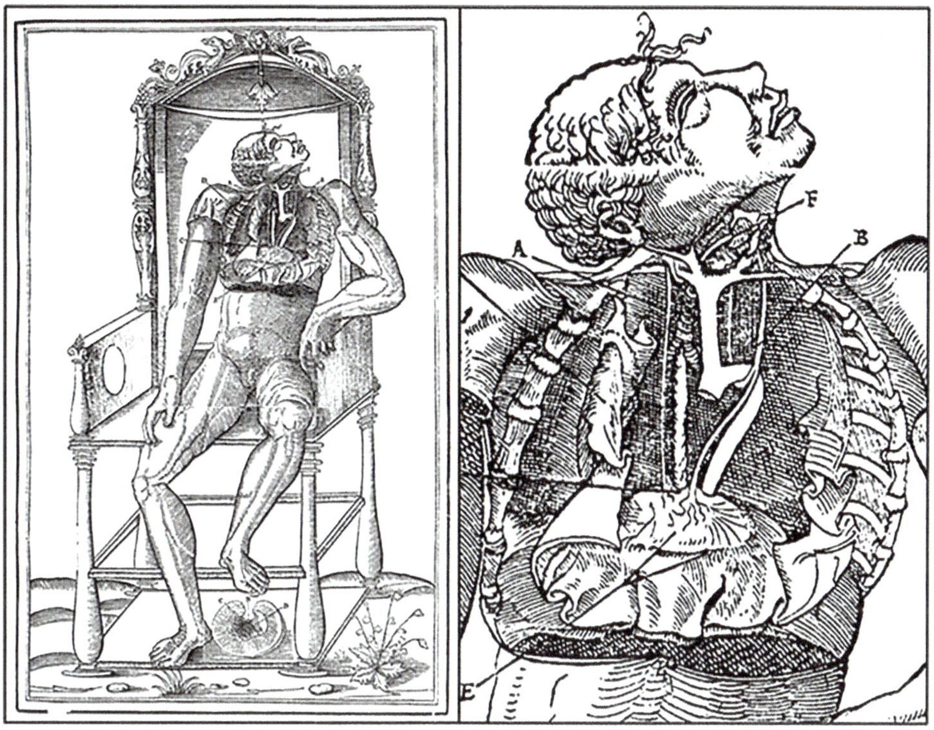

Fig. E10 Figure by Estienne on the recurrent laryngeal nerve (from Wikipedia; in public domain)

a b

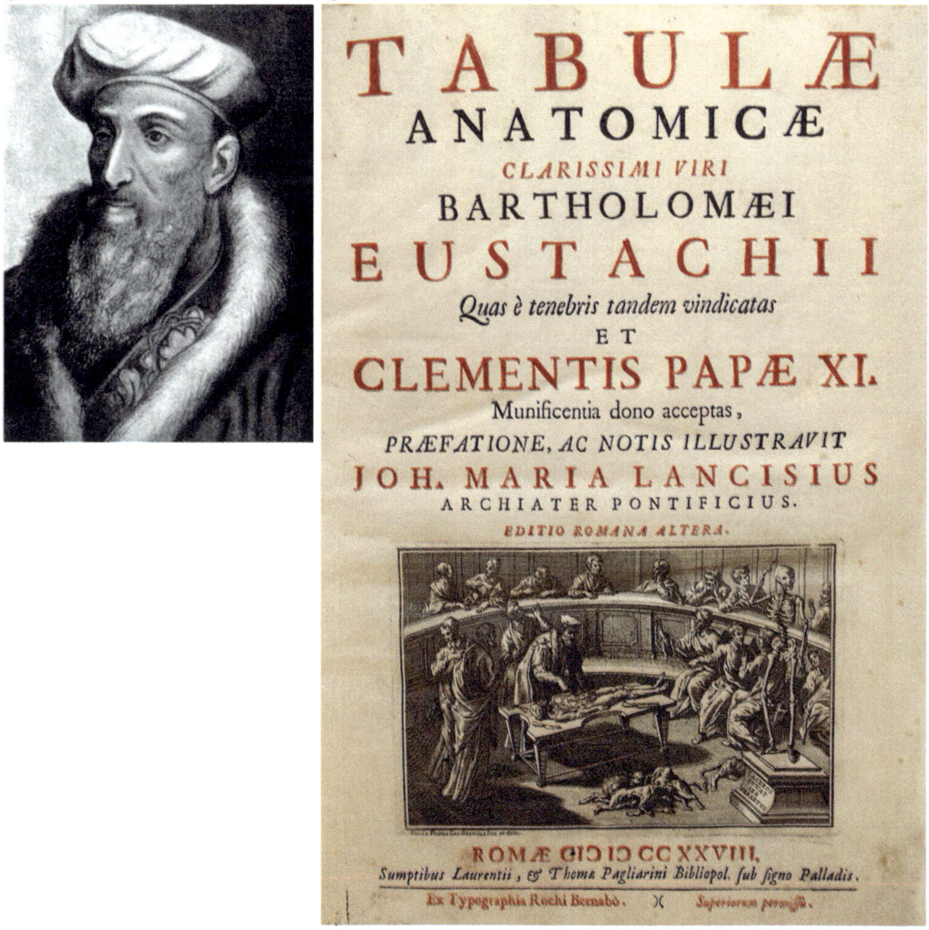

Fig. E11 **a** Portrait of Eustachi (from Wikimedia Commons; in public domain); **b** Title page *Tabulae Anatomicae Bartholomaei Eustachii* (Roma 1742 by JM Lancisi; courtesy Ondřej Naňka, Prague)

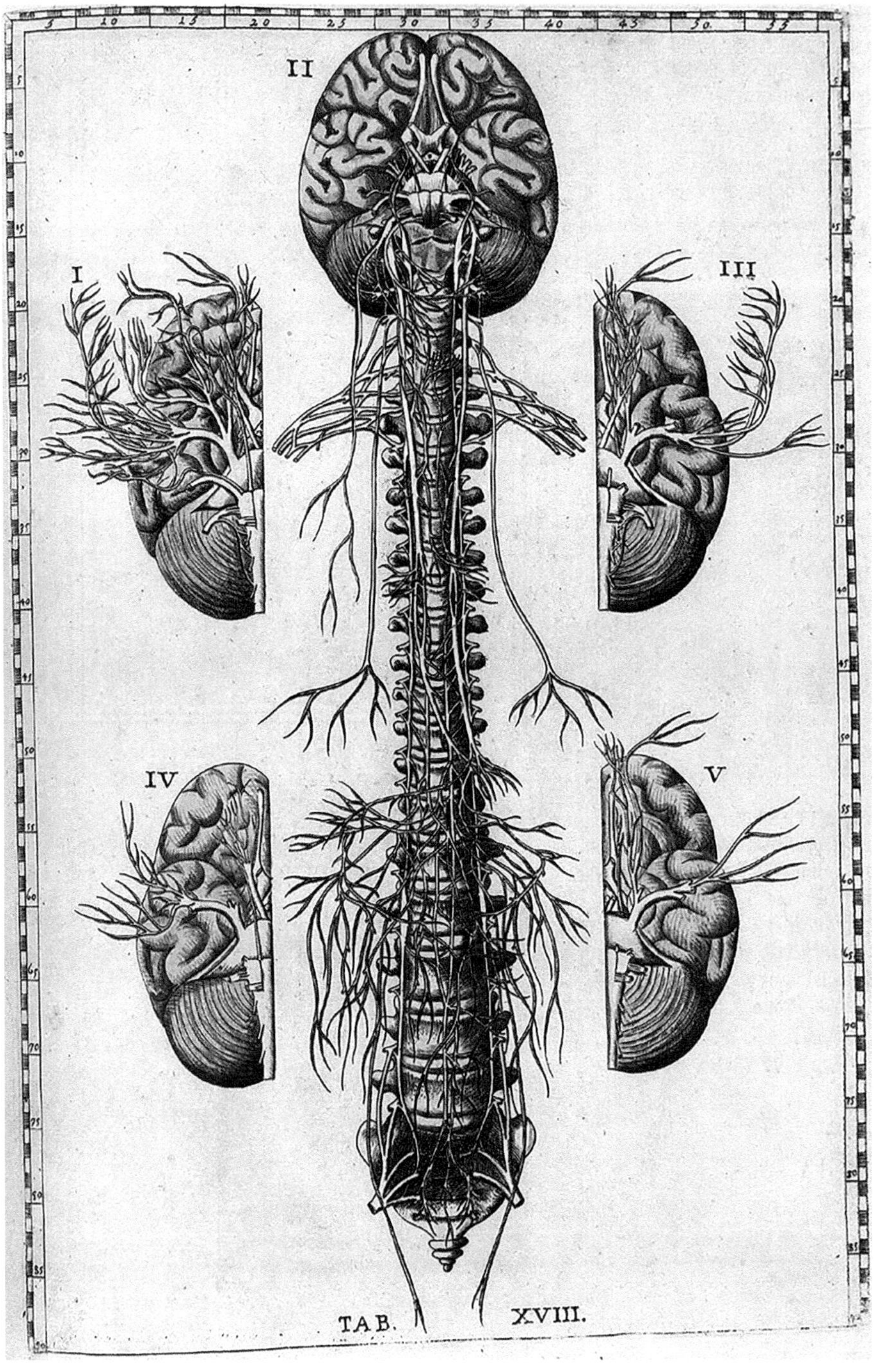

Fig. E12 The brain according to Eustachi

Anus of Eustachi: *Eponym* for the opening of the auditory tube into the pharynx.

Cushion of Eustachi: *Eponym* for the torus tubarius of the pharynx.

Muscle of Eustachi: *Eponym* for the tensor tympani muscle.

Ridge of Eustachi: *Eponym* for a rim between the fossa ovalis and the ostium of the coronary sinus into the right atrium of the heart (*see* **Fig. T13b**).

Tube of Eustachi: *Eponym* for the auditory tube.

Tuber of Eustachi: *Eponym* for a slight protrusion from the labyrinthine wall of the tympanic cavity below the oval window.

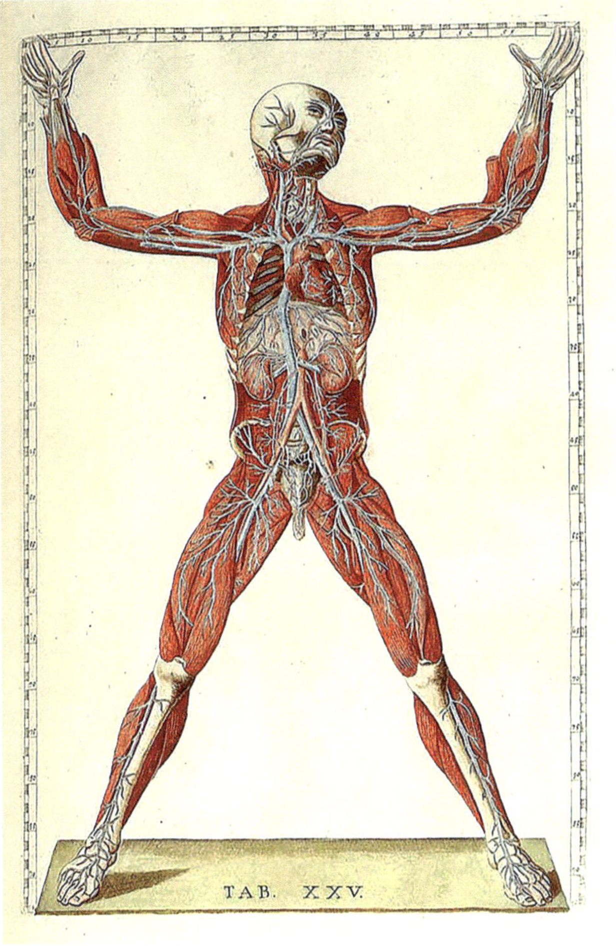

Fig. E13 Eustachi's 'blood vessel' man

Valve of Eustachi: *Eponym* for the valve of the inferior vena cava in the right atrium of the heart.

Eysson, Henrique (1620–1690): Dutch anatomist born and working in Groningen remembered for his work on the osteology of the foetus.

Bones of Eysson: *Eponym* for inconstant mental bones in the mandibular syndesmosis. Reference: *Tractatus anatomicus et medicus de ossibus infantis cognoscendis conservandis et curandis; accessit Volcheri Coiteri corundum ossium historia.* Groningen, 1659

Fig. E14 Photo of Exner (from Pagel 1901)

Exner, Siegmund (1846–1926): Austrian physiologist born and working in Vienna known for his studies on the function of the cerebral cortex (**Fig. E14**). The Exner writing centre is located in the posterior portion of the left middle frontal gyrus. Reference: *Die Lokalisation der Funktionen in der Grosshirnrinde des Menschen.* Braumüller, Vienna, 1881. With Emma Call, he described the bodies of Call-Exner.

Bodies of Call-Exner: *Eponym* for small cavities found among ovarian granulosa cells of growing and mature ovarian ollicles. Their functional significance is unknown. Reference: *Zur Kenntnis des Graafschen Follikels und des Corpus luteum beim Kaninchen.* SB Akad Wiss Wien Math-Naturw 1975; 71:321–328 (Call EM, Exner S).

Nerve of Exner: *Eponym* for an inconstant middle laryngeal nerve. Reference: *Die Innervation des Kehlkopfes.* Sitz Kais Akad Wiss Wien 1884, 89:63:118.

Plexus of Exner: *Eponym* for superficial tangential fibres of the molecular layer of the isocortex.

Letter F

7

Fabrici (Fabrizio) d'Acquapendente, Girolamo or Fabricius, Hieronymus (1533 or 1537–1619): Italian anatomist and embryologist born in Acquapendente who succeeded Falloppio in Padua (**Fig. F1**). He observed the valves in veins in 1574 and taught among many others Adriaan van den Spieghel and William Harvey (**Fig. F2**). Reference: *Opera omnia*. Padua, 1625. He is aso known for the first illustration of the lateral fissure or sulcus of the cerebral cortex (**Fig. F3**) as shown by Collice M, Collice R and Riva A. *Who discovered the Sylvian fissure?* Neurosurgery 2008; 63:623-628. It was Sterzi who found the famous *Tabulae Anatomicae (Pictae)* of Fabrici in the Marciana Library in Venice (*see* Sterzi, Giuseppe).

Bursa of Fabrici: *Eponym* for the cloacal bursa, the site of hematopoiesis in birds; the bone marrow is the equivalent organ in mammals.

Ship of Fabrici: *Eponym* for the fanciful resemblance of the contours of the occipital, sphenoidal and frontal bones to the outline of a ship.

Falloppio, Gabriele (1523–1563): Italian anatomist and surgeon born in Modena and working in Padua (Fig. F4). He made many important contributions to the anatomy of the ear and of the reproductive system. Reference: *Gabrielis Falloppii medici Mutinensis Observationes Anatomicae.* Ulmus, Venice, 1561 (Anatomical Observations of the Modena Physician Gabriele Falloppio; not illustrated).

Aqueduct or **Canal of Falloppio:** *Eponym* for the facial canal in the petrous part of the temporal bone.

Artery of Falloppio: *Eponym* for the uterine artery.

Folds of Falloppio-Kerckring: *Eponym* for the circular folds of the small intestine, especially distinct in the duodenum and jejunum, later also described by Kerckring.

Hiatus of Falloppio: *Eponym* for the hiatus of the facial canal, the opening in the cranial base leading to the facial canal, later also described by Ferrein.

Ligament of Falloppio: *Eponym* for the inguinal ligament, described also by Vesalius and Poupart.

Muscle of Falloppio: *Eponym* for the pyramidalis muscle.

Tube of Falloppio: *Eponym* for the uterine tube.

Valve of Falloppio: *Eponym* for the ileocaecal fold, also described by Varolio, Bauhin and Tulp.

Fañanas, Jorge Rámon (1886–1937): Spanish physician, second son of Santiago Rámon y Cajal, who used his mother's name.

Cell of Fañanas: *Eponym* for the feathered astrocyte in the cerebellar cortex, now known as a subtype of epithelial astrocytes. Reference: *Contribution al estudio de la neuroglia del cerebelo.* Trab Lab Invest Biol Madrid 1916; 14:163–179.

Farabeuf, Louis Hubert (1841–1910): French anatomist born in Bannost, Seine-et-Marne, and working in Paris (Fig. F5). He is known for his studies on joints and their ligaments (Fig. F6).

Ischio-supracervical ligament of Farabeuf: *Eponym* for the ischiofemoral ligament of the hip joint.

Line of Farabeuf: *Eponym* for the iliotrochanteric line between the posterior superior iliac spine and the

H. J. ten Donkelaar et al., *An Illustrated Guide to Anatomical Eponyms*, https://doi.org/10.1007/978-3-031-91664-9_7

Fig. F1 Engraving of Fabrici (from Wikimedia Commons; in public domain)

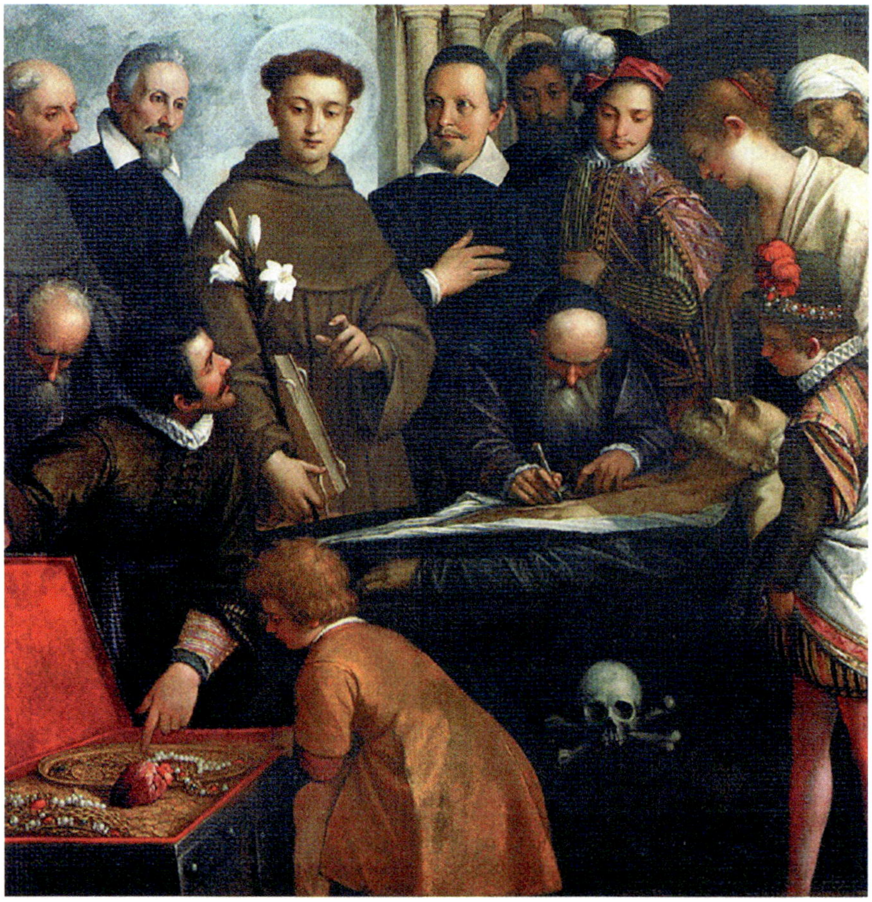

Fig. F2 The Anatomical lesson of Fabrici (detail of the 'Miracle of the miser's heart', painted around 1621 by Pietro Damini in the church of San Canziano in Padua; from Wikimedia Commons; in public domain). Fabrizi is shown dissecting the body of a miser whose heart is seen in the jewel case

most cranial point of the greater trochanter of the femur.

Pregleno-infrahumeral ligament of Farabeuf: *Eponym* for the inferior glenohumeral ligament. Reference: *Sur l'articulation scapulo-humérale.* Bull Soc Chir 1882.

Supragleno-prehumeral ligament of Farabeuf: *Eponym* for the middle glenohumeral ligament.

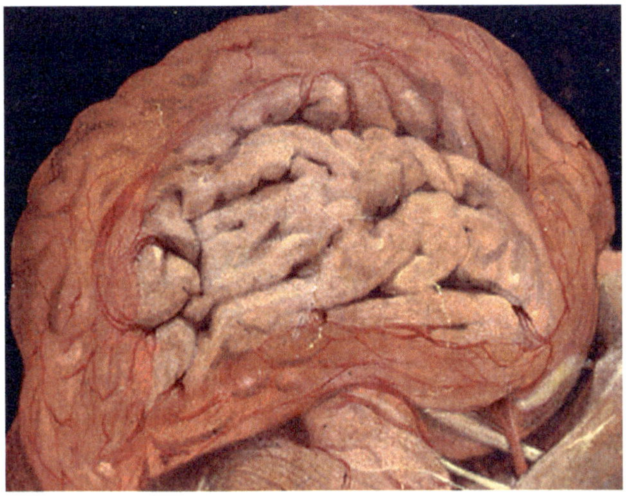

Fig. F3 Brain by Fabrici (1600, Detail of Tabula Picta 112.10, rotated 45 degrees), showing the lateral fissure and the course of the internal carotid artery and the optic nerve (courtesy Biblioteca Nazionale Marciana San Marco, Venice; from Wikimedia Commons; in public domain)

Fig. F4 Engraving of Falloppio (from Pagel 1901)

Supragleno-suprahumeral ligament of Farabeuf: *Eponym* for the superior glenohumeral ligament, also described by Flood.

Triangle of Farabeuf: *Eponym* for a triangle in the upper part of the neck bounded dorsally by the internal jugular vein, ventrally by the facial vein and caudally by the hypoglossal nerve. Reference: *Précis de manuel opératoire.* Paris, 1872.

Zygomaticomandibular band of Farabeuf: *Eponym* for the deep layer of the lateral ligament of the temporomandibular joint.

Zygomaticomandibular cord of Farabeuf: *Eponym* for the superficial layer of the lateral ligament of the temporomandibular joint.

Farre, Arthur (1811–1887): English obstetrician born and working in London.

Line of Farre, White: *Eponym* for the line of attachment of the peritoneum of the mesovarium to the substance of the ovary, also described by Romiti and Waldeyer. Reference: *The Uterus and Its Appendages.* In: Todd RE (ed) *Cyclopedia of Anatomy and Physiology.* 1859, Suppl Vol, p 548.

Fasciae: *See* **The Classical Anatomical Eponyms.**

Fasciculi:

Fasciculus of Burdach: *Eponym* for the cuneate fasciculus of the spinal cord; *see* Burdach, Karl Friedrich.

Fig. F5 Photo of Farabeuf (from the Wellcome Collection CCBY; in public domain)

Fasciculus of Flechsig: *Eponym* for the posterior or dorsal spinocerebellar tract; *see* Flechsig, Paul Emil. The term fasciculi of Flechsig is sometimes used for the propriospinal fibres surrounding the grey matter of the spinal cord.

Fasciculus of Foville: *Eponym* for the stria terminalis; *see* Foville, Achille Louis.

Fasciculus of Goll: *Eponym* for the gracile fasciculus of the spinal cord; *see* Goll, Friedrich.

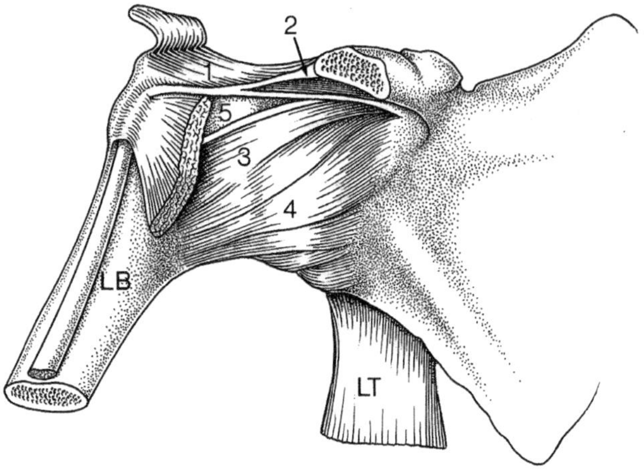

Fig. F6 Farabeuf, Illustration glenohumeral ligaments (after Rouvière H *Anatomie humaine. Descriptive et topographique, Tome III: Membres, système nerveux central.* tenth revised edition by A Delmas. Masson, Paris, 1967). *1* Coracohumeral ligament, *2,3,4* superior, middle and inferior glenohumeral ligaments, *5* foramen of Weitbrecht, *LB* long tendon of biceps brachii muscle, *LT* long tendon of triceps brachii muscle

Fasciculus of Gowers: *Eponym* for the anterior or ventral spinocerebellar tract; *see* Gowers, Wiliam Richard.

Fasciculus of Lissauer: *Eponym* for the posterolateral or dorsolateral fasciculus of the spinal cord; *see* Lissauer, Heinrich.

Fasciculus of Meynert, Retroflexus: *Eponym* for the habenulointerpeduncular tract; *see* Meynert, Theodor.

Fasciculus of Muskens: *Eponym* for the interstitiospinal tract; *see* Muskens, Louis Jacob Joseph.

Fasciculus of Russell: *Eponym* for the uncinate fasciculus of the cerebellum; *see* Russell, James S. Risien.

Fasciculus of Schultze: *Eponym* for the interfascicular fasciculus of the spinal cord (comma tract of Schultze); *see* Schultze, Friedrich.

Fasciculus of Schütz: *Eponym* for the posterior longitudinal fasciculus of the brain stem; see Schütz, Hugo.

Fasciculus of von Monakow: *Eponym* for the rubrospinal tract; *see* Monakow, Constantin von.

Fat pads:

Fat pad of Bichat: *Eponym* for the buccal fat pad; *see* Bichat, Marie François Xavier.

Fat pad of Hoffa: *Eponym* for the infrapatellar fat pad; *see* Hoffa, Albert.

Fig. F7 Engraving of Ferrein (from Wikimedia Commons; in public domain)

Fat pad of Imlach: *Eponym* for fat surrounding the round ligament of the uterus in the inguinal canal; *see* Imlach, Francis.

Fat pad of Kager: *Eponym* for fat within the Kager triangle, which normally appears lucent on radiographs; *see* Kager, Hans.

Fat pads of Ostmann: *Eponym* for lateral and medial fat pads around the pharyngeal opening of the auditoty tube; *see* Ostmann, Paul.

Ferrein, Antoine (1693–1769): French surgeon and anatomist born in Frespech and working in Paris as successor of Winslow (Fig. F7).

Canal of Ferrein: *Eponym* for the rivus lacrimalis of the eye.

Cords of Ferrein: *Eponym* for the vocal cords used by French anatomists.

Foramen of Ferrein: *Eponym* for the (a) hiatus of the facial canal, earlier described by Falloppio, and (b) the hiatus of the canal for the greater petrosal nerve.

Ligament of Ferrein: *Eponym* for (a) the lateral temporomandibular ligament and (b) the vocal cord.

Pyramid of Ferrein: *Eponym* for the cortical pyramids of the kidney, corresponding to the medullary pyramids of Malpighi. Reference: *Sur la structure des viscères glanduleux et en particulier sur celle des reins et du foie.* Hist Acad Roy Sci Paris, 1749.

Tube of Ferrein: *Eponym* for the convoluted part of the proximal tubule of the kidney, also described by Schachowa.

Feyrter, Friedrich (1895–1973): Austrian pathologist born in Vienna and working in Danzig (now Gdánsk, Poland), Graz and Göttingen. He developed the concept of the diffuse endocrine system. Reference: *Űber diffuse endokrine epitheliale Organe.* Barth, Leipzig, 1938.

Cells of Feyrter: *Eponym* for respiratory neuroendocrine cells, also described by Kulchitsky.

Fibres: *See* **The Classical Eponyms.**

Fielding, George Hunsley (1801–1871): British anatomist and ophthalmologist born in Hull and practizing in Tunbridge.

Membrane of Fielding: *Eponym* for the tapetum lucidum of the retina, a layer of highly differentiated cells interposed between the vascular layer and the choroidocapillary layer of the choroid in some nocturnal animals. Reference: *On a new membrane in the eye.* Philos Mag Series 1832; 3:113–116.

Filiminoff, IN: Russian anatomist known for his introduction of the allocortex. He introduced the terms archicortex, paleocortex, periarchicortex and peripaleocortex (Fig. F8). Reference: Filiminoff IN *A rational subdivision of the cerebral cortex.* Arch Neurol Psychiatry 1947; 58:296–311.

Fissures:

Fissure of Bichat: *Eponym* for the space below the splenium of the corpus callosum, also known as transverse cerebral fissure; *see* Bichat, Marie François Xavier,

Fissure of Clevenger: *Eponym* for the inferior temporal sulcus of the cerebral cortex; *see* Clevenger, Shobal.

Fissures of Duverney: *Eponym* for notches in the cartilaginous part of the external acoustic meatus, also described by Santorini; *see* Duverney, Joseph Guichard.

Fissure of Ecker: *Eponym* for the petrooccipital fissure of the skull; *see* Ecker, Alexander.

Fissure of Glaser: *Eponym* for the petrotympanic fissure of the skull; *see* Glaser, Johann Heinrich.

Fissure of Larrey: *Eponym* for the sternocostal triangle, a triangular space between the sternal and costal parts of

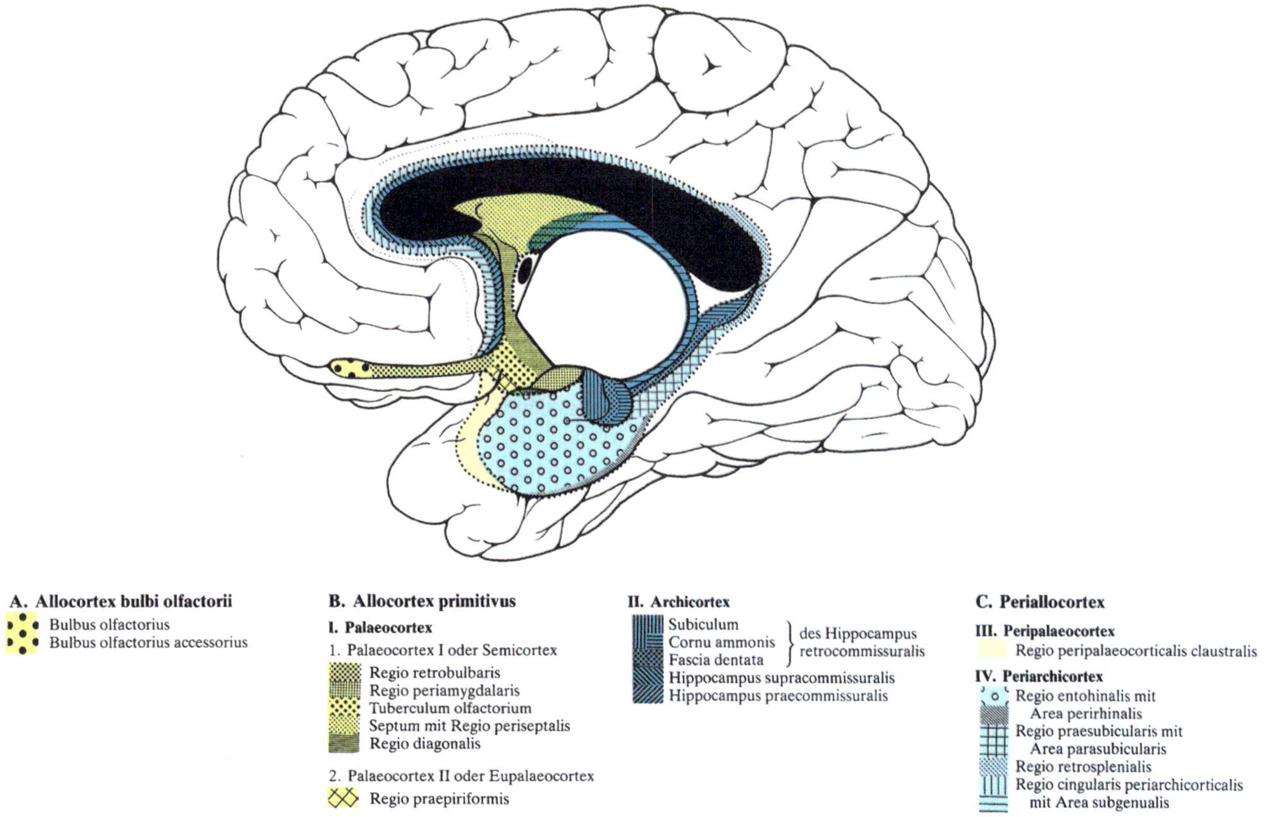

Fig. F8 Allocortex to illustrate the Filiminoff subdivision (from ten Donkelaar HJ, Kachlík D, Tubbs RS *An Illustrated Terminologia Neuroanatomica.* Springer, Cham, 2018)

Fig. F9 Photo of Flechsig (from Pagel 1901)

the diaphragm, also described by Morgagni; *see* Larrey, Dominique Jean.

Fissure of Pansch: *Eponym* for the intraparietal sulcus of the cerebral cortex; Pansch described this sulcus in the human brain in the same year as Turner in the brain of a chimpanzee; *see* Pansch, Adolph.

Fissure of Rouvière, Horizontal: *Eponym* for the horizontal fissure of the liver; *see* Rouvière, Henri.

Fissures of Santorini: *Eponym* for notches in the cartilaginous part of the external acoustic meatus, also described by Duverney; *see* Santorini, Giovanni Domenico.

Fissure of Schwalbe: *Eponym* for the choroid or optic fissure of the developing eye; *see* Schwalbe, Gustav Albert.

Fissure of Sylvius: *Eponym* for the lateral sulcus of the cerebral cortex (= Sylvian fissure); *see* Sylvius, François de le Boë.

Flack, Martin W. (1882–1931): English physiologist born in Borden, Kent, known for the sinuatrial node (node of Keith-Flack). Reference: *An investigation of the sino-auricular node of the mammalian heart.* J Physiol (Lond) 1910; 41:64.

Node of Keith-Flack: *Eponym* for the sinuatrial node of the cardiac conducting system (*see* **Fig. I5**).

Flechsig, Paul Emil (1847–1929): German neurologist and psychiatrist born in Zwickau and working in Leipzig (Fig. F9). He is known for his studies of CNS tracts and of their myelinization in fetuses (Fig. F10). Reference: *Die Leitungsbahnen im Gehirn und Rückenmark des Menschen auf Grund entwicklungsgeschichtlicher Untersuchungen.* Engelmann, Leipzig, 1876; *Anatomie des menschlichen Gehirns und Rückenmarks auf myelogenetischer Grundlage.* Thieme, Leipzig, 1920.

Fasciculi of Flechsig: *Eponym* sometimes used for the propriospinal fibres surrounding the grey matter of the spinal cord.

Loop of Flechsig: *Eponym* for the anterior or temporal loop of the optic radiation, also described by Archambault, Cushing and Meyer.

Oval bundle of Flechsig: *Eponym* for the septomarginal fasciculus of the spinal cord, also described by Bruce and Muir, and its sacral continuation as triangle of Philippe-Gombault.

Tract or **Fasciculus of Flechsig:** *Eponym* for the posterior or dorsal spinocerebellar tract.

Flint, Austin (1836–1915): American physiologist born in Northampton, Massachusetts, and working in New York.

Arteries of Flint: *Eponym* for the arcuate arteries of the kidney.

Flood, Valentine (1800–1847): Irish anatomist born and working in Dublin.

Ligament of Flood: *Eponym* for the superior glenohumeral ligament of the shoulder joint, also described by Farabeuf. Reference: *Discovery of a new ligament of the shoulder joint.* Lancet 1829–1830, p 672.

Foix, Charles (1882–1927): French neurologist known for his studies on the vascularization of the brain (Fig. F11). For biography, see Caplan LR *Charles Foix—The first modern stroke neurologist.* Stroke 1990; 21:348–356. Reference: *Les artères de l'axe encéphalique jusqu'au diencéphale inclusivement.* Rev. Neurol 1925; 2:705–739; with P Hillemand). For the thalamus, they introduced the terms thalamogeniculate, thalamoperforated and thalamotuberal arteries, and for the brain stem, the terms paramedian, lateral (short circumferential) and posterior (long circumferential) arteries. He is also recognized for his analysis of the fibre connections in the brain (Fig. F12).

Folds: *See* **The Classical Anatomical Eponyms.**

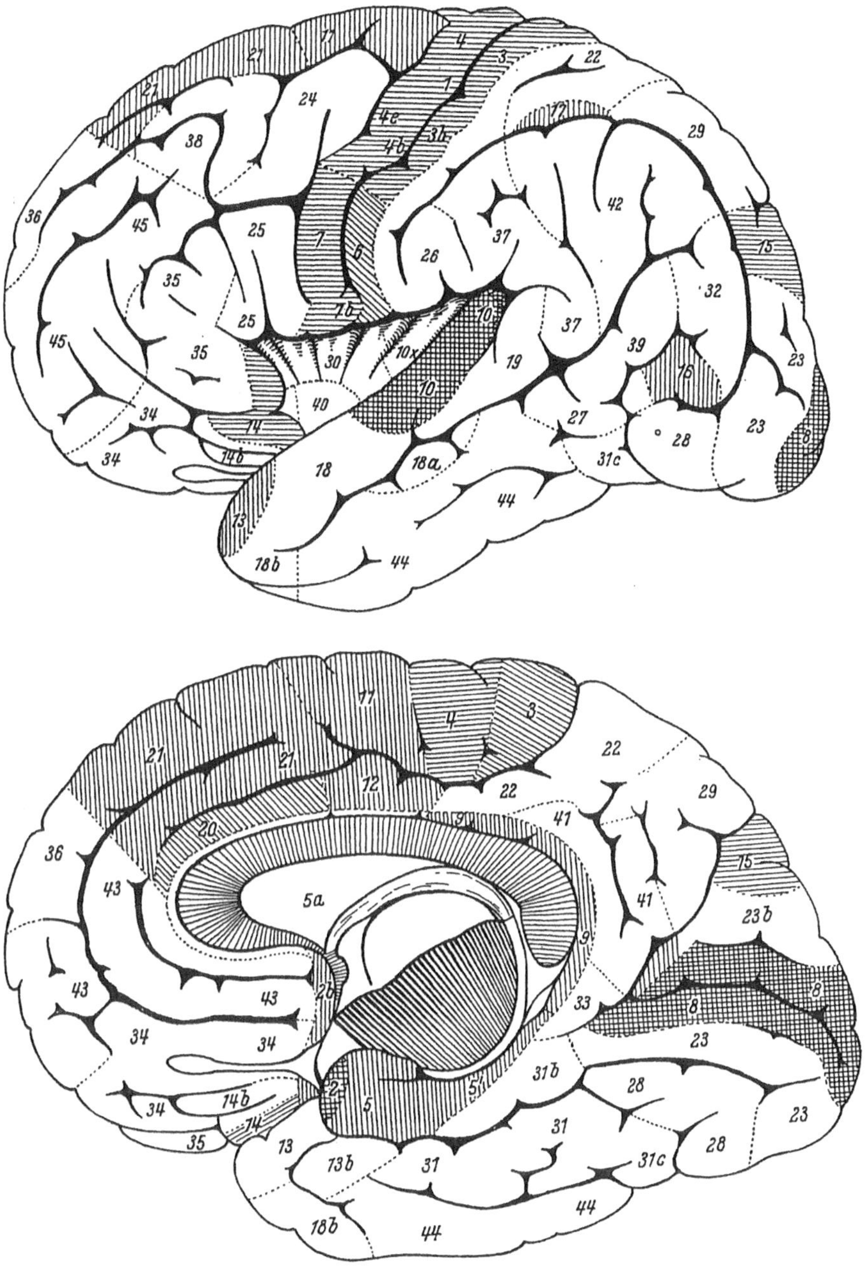

Fig. F10 Myelogenetic fields of Flechsig (from Flechsig PE *Anatomie des menschlichen Gehirns und Rückenmarks auf myelogenetischer Grundlage.* Thieme, Leipzig, 1920). Primordial fields corresponding to the primary motor and sensory areas show stainable myelin before or at term, intermediate fields corresponding to associative areas around the primary areas at 6–12 weeks, and terminal fields corresponding to higher-order association areas after the fourth postnatal month

Folli, Cecilio (1615–1660): Italian anatomist born in Modena and working in Venice (Fig. F13).

Muscle of Folli: *Eponym* for the anterior ligament of the malleus, also described by Casseri.

Process of Folli: *Eponym* for the anterior process of the malleus, also described by Rau. Reference: *Nova auris internae delineatio.* Venice, 1645.

Follicle of de Graaf: *Eponym* for the mature or vesicular ovarian follicle (Graafian follicle); *see* de Graaf, Reinier.

Foltz, Jean Charles Eugène (1822–1876): French anatomist and physiologist born in Nancy and working in Lyon.

Valvules of Foltz: *Eponym* for 'valvules' or constrictions, in fact mucous folds, within the opening of the lacrimal canaliculi, also described by Bochdalek. Reference: *Anatomie et physiologie des conduits lacrymeur.* Ann Soc Méd Lyon, 1862.

Fontana, Abbada (Abbot) Felice Gaspar Ferdinand (1730–1805): Italian anatomist born in Pomarola (South Tyrol), working in Pisa and later as Director of La Specola, the Natural History Museum of Florence (Fig. F14). He is

Fig. F11 Photo of Foix (from Caplan LR *Charles Foix (1882–1927)*. J Neurol 2010; 257:1941–1942; with permission)

especially known for his studies on the eye (Fig. F15). Reference: *Dei moti dell'iride.* Lucca, 1765.

Canal or **Sinus of Fontana:** *Eponym* for the scleral venous sinus, later also described by Lauth and Schlemm.

Spaces of Fontana: *Eponym* for the trabecular meshwork or pectinate ligament, spaces of the iridocorneal angle.

Trabecular meshwork of Fontana: *Eponym* for a network of anastomosing lamellae and trabeculae in the iridocorneal angle.

Foramina: *See* **The Classical Anatomical Eponyms.**

Fordyce, John Addison (1858–1925): American dermatologist born in Guernsey County, Ohio, who studied in Europe (on dermatology with Kaposi in Vienna), and after returning from Europe, working in New York.

Spots or **Granules of Fordyce:** *Eponym* for visible sebaceous glands around the edges of the lips (vermillion border) and inside the cheeks, originally described as a disease but soon found to be normal sebaceous glands. They may appear around the genital area as well. Reference: *A peculiar affection of the mucous membrane of the lips and oral cavity.* Arch Dermatol 1896; 132:1285.

Forel, Auguste Henri (1848–1931): Swiss neurologist born in Morges and working in Zürich (Fig. F16) known for several decussations and the H fields (H fields of Forel; H stands for 'Haube' or tegmentum; Fig. F17). Reference: *Beiträge zur Kenntnis des Thalamus opticus und der ihn umgebenden Gebilde bei den Säugethieren.* Zürich, 1872; *Untersuchungen über die Haubenregion und ihre Verknüpfungen im Gehirne des Menschen und einiger Säugethiere mit Beiträgen zu den Methoden der Gehirnuntersuchung.* Arch Psychiatr Nervenkr 1877; 7:393–495.

Campus of Forel: *Eponym* for the H fields of Forel in the prethalamic tegmentum.

Commissure of Forel: *Eponym* for the posterior hypothalamic commissure.

Decussation of Forel: *Eponym* for the anterior or ventral tegmental decussation, the decussation of the rubrospinal tract in the midbrain.

Fossae: *See* **The Classical Anatomical Eponyms.**

Foville, Achille Louis (1799–1878): French neurologist born in Pontoise and associated with the Asylums of Rouen and Charenton (Fig. F18). He is known for his studies of the cerebral cortex (Fig. F19). Reference: *Traité complet de l'anatomie, de la physiologie et de la pathologie du système nerveux cérébrospinal.* Fontin, Masson, Paris, 1844.

Fascia, Bundle or **Fasciculus of Foville:** *Eponym* for the stria terminalis of the brain.

Frankenhäuser, Ferdinand (1832–1894): German gynaecologist in Jena and later in Zürich. Reference: *Die Nerven der Gebärmutter und ihre Endigung in den glatten Muskelfasern: Ein Beitrag zur Anatomie und Gynäkologie.* Mauke, Jena, 1867.

Ganglion of Frankenhäuser: *Eponym* for the parasympathetic ganglion of the uterovaginal plexus (ganglion cervicale uteri). He completed the description of Robert Lee (1842).

Plexus of Frankenhäuser: *Eponym* for the uterovaginal plexus, also described by Lee (Fig. F20).

Frankfurter line: *Eponym* for the 'Deutsche Horizontale', which connects the lower border of the orbit with the upper border of the external acoustic meatus, also known as the orbitomeatal plane.

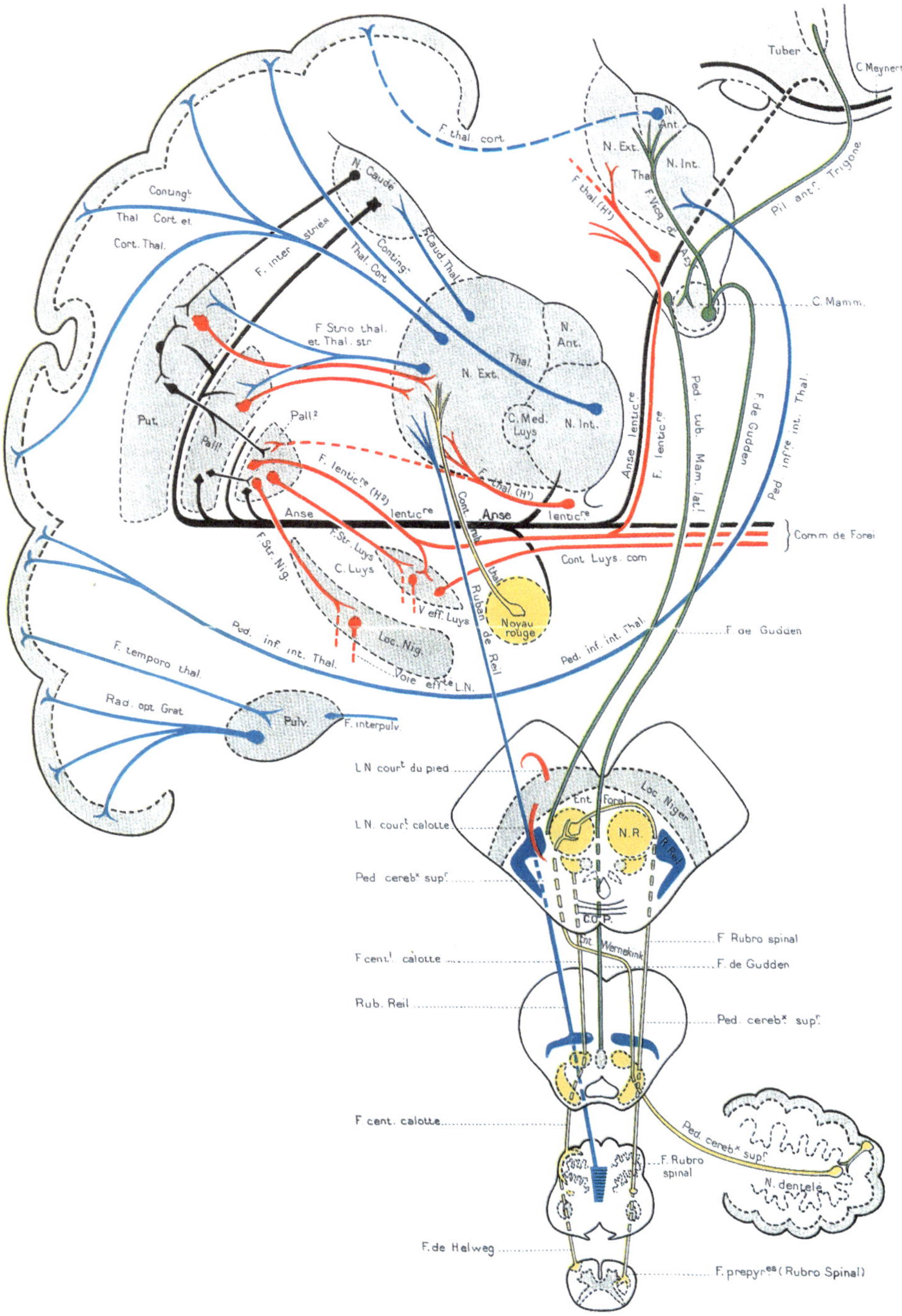

Fig. F12 Schema fibre connections brain by Foix C, Nicolesco J *Les noyaux gris centraux et la région mésencéphalo-sous-optique.* (Masson, Paris, 1925)

Fredet, Pierre (1870–1946): French surgeon born in Clermont Ferrand and best known for the Fredet-Ramstedt technique for treatment of hypertrophic pylorus stenosis.

Fascia of Fredet: *Eponym* for the preduodenopancreatic fascia.

Froriep, August von (1849–1917): German anatomist and physiologist born in Weimar and working in Tübingen (Fig. F21).

Ganglion of Froriep: *Eponym* for a temporary collection of nerve cells on the dorsal aspect of the hypoglossal nerve in the embryo. Reference: *Ueber ein Ganglion des Hypoglossus und Wirbelanlagen in der Occipitalregion.* Arch Anat Physiol (Lpz) 1882, 279–302.

Fruchaud, Henri René (1894–1960): French anatomist and surgeon born in Angers and working in Paris.

Myopectineal orifice of Fruchaud: *Eponym* or a large oval, funnel-like, potential orifice in a weak area where all inguinal and femoral hernias originate. It is bounded cranially by the internal oblique abdominis and transverse abdominis muscles, caudally by the superior pubic ramus and medially by the sheath of the

Fig. F13 Engraving of Folli (from Wikidata; in public domain)

Fig. F14 Portrait of Fontana (from Wikimedia Commons; in public domain)

rectus abdominis muscle. Reference: *L'anatomie chirurgicale des hernies de l'aine.* Doin, Paris, 1956.

Fuchs, Ernst (1851–1930): Austrian ophthalmologist born in Vienna working in Liège and later in Vienna (Fig. F22). Reference: *Beiträge zur normalen Anatomie der menschlichen Iris.* Von Graefe's Arch Ophthalmol 1885; 31:39–96.

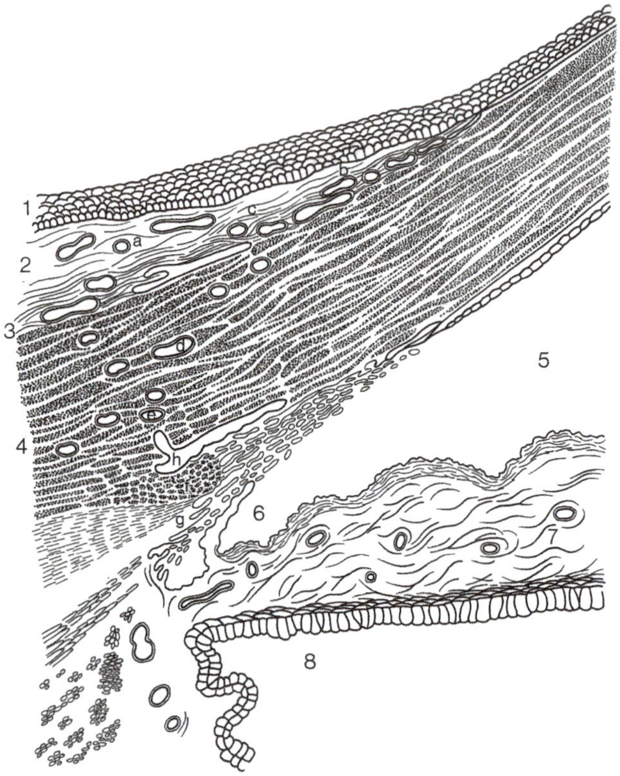

Fig. F15 The scleral venous sinus shown in a schematic drawing of the organization of the limbus and the iridocorneal angle (after Hogan MJ, Alvarado JA, Weddell JE *Histology of the Human Eye*. Saunders, Philadelphia, PA, 1971; from ten Donkelaar HJ, Kachlík D, Tubbs RS *An Illustrated Terminologia Neuroanatomica.* Springer, Cham, 2018). *1* conjunctival epithelium, *2* conjunctival connective tissue with: *a* conjunctival vessels, *b* peripheral corneal arcades of vessels, *c* episcleral vessels, *d* intrascleral vessels, *e* deep scleral vessels, *f* scleral spur, *g* ciliary muscle, *h* scleral venous sinus (canal of Schlemm), *3* episclera, *4* limbal (corneoscleral) stroma, *5* anterior chamber, *6* iridocorneal angle, *7* iris, 8 posterior chamber

Crypts or **Stomata of Fuchs:** *Eponym* for the crypts of the iris, depressions on the anterior surface of the iris near its pupillary border.

Roll of Fuchs: *Eponym* for the peripheral anterior margin of the iris, marking the iridocorneal angle from posterior.

Fuse, Gennosuke (1880–1946): Japanese anatomist (Fig. F23) who after graduating in Tokyo, worked for many years in Zürich with von Monakow.

Nucleus of Kölliker-Fuse: *Eponym* for the subparabrachial nucleus of the brain stem, described together with Kölliker.

Fig. F16 Photo of Forel (from Pagel 1901)

Fig. F18 Portrait of Foville (from Wikipedia; in public domain)

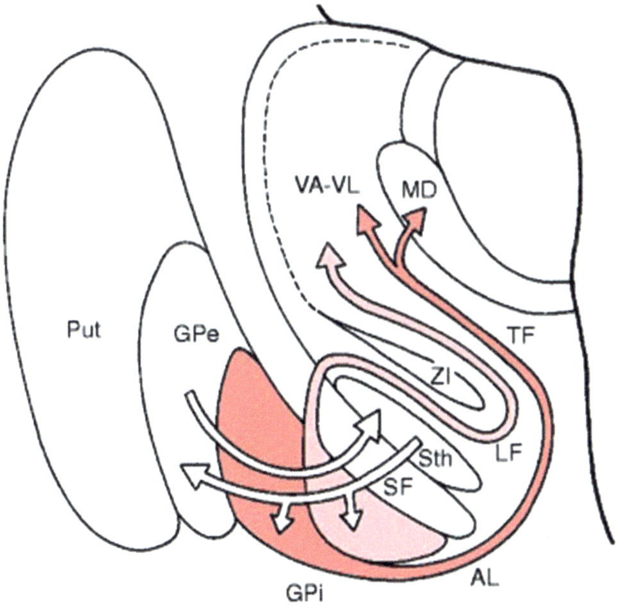

Fig. F17 The H fields of Forel: Pallidofugal projection systems shown in a coronal section. *AL* ansa lenticularis; *GPe, GPi* external and internal parts of globus pallidus; *IC* internal capsule; *LF* lenticular fascicle (**H2 field** of Forel); *MD* mediodorsal nucleus; *Put* putamen; *SF* subthalamic fascicle; *SN* substantia nigra; *Sth* subthalamic nucleus; *TF* thalamic fascicle (**H1 field** of Forel); *VA-VL* ventral anterior and ventral lateral nuclei; *ZI* zona incerta (after Parent A, Cossette M, Lévesque M *Anatomical considerations in basal ganglia surgery*. In: Lozano AM, ed., *Movement Disorder Surgery*, Vol 15. Prog Neurol Surg. Karger, Basel, 2000, pp. 21–30; from ten Donkelaar HJ, Kachlík D, Tubbs RS *An Illustrated Terminologia Neuroanatomica*. Springer, Cham, 2018)

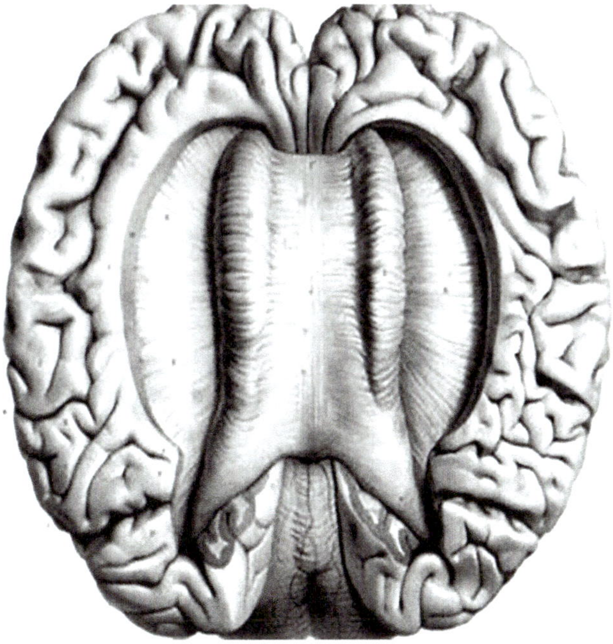

Fig. F19 Dorsal view of the brain showing the dissected corpus callosum (from Foville AL *Traité complet d'anatomie*. Fontin, Masson, Paris 1844)

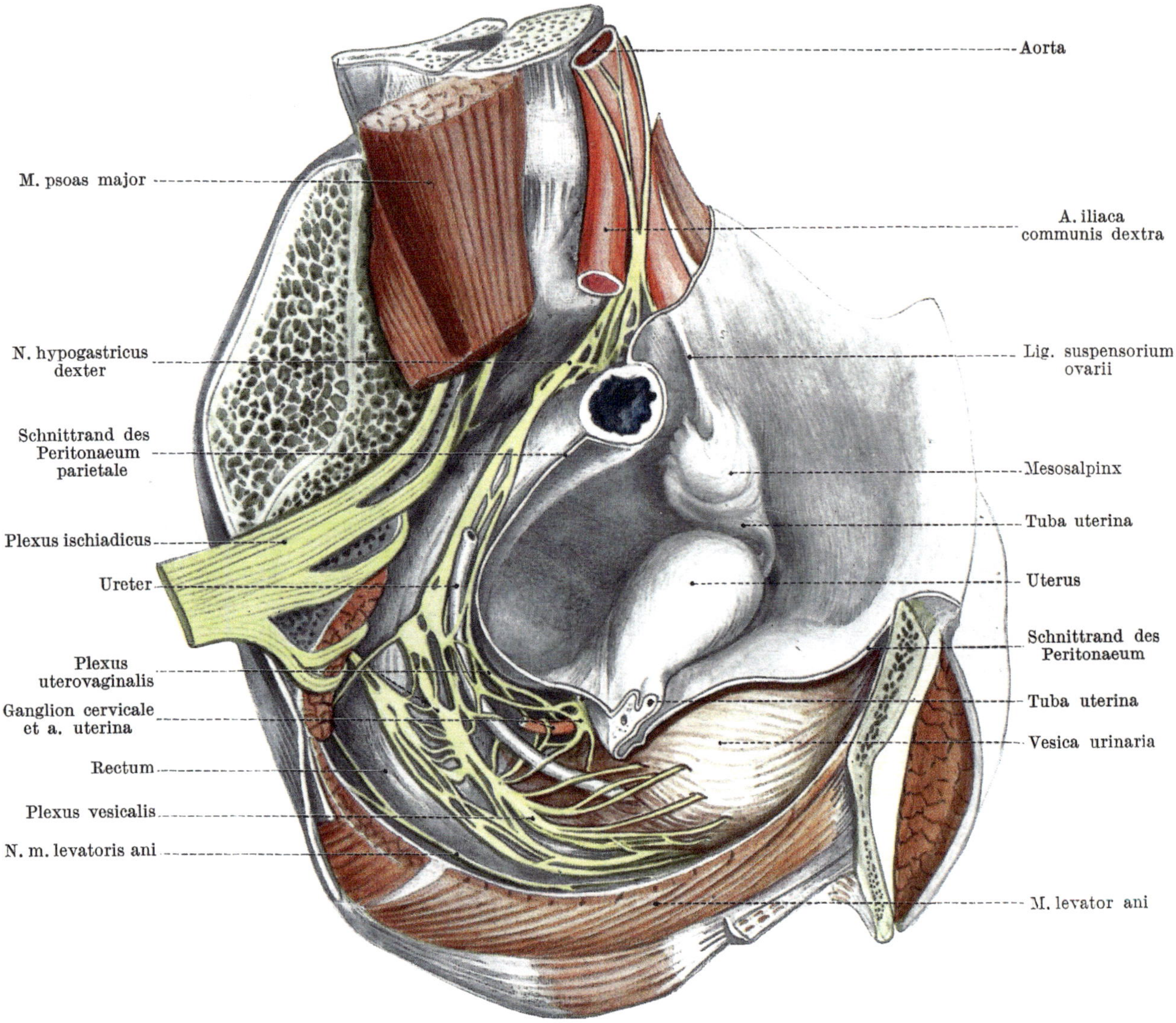

Fig. F20 The uterovaginal plexus (from Hafferl A *Lehrbuch der topographischen Anatomie,* second ed. Springer, Berlin, Göttingen, Heidelberg, 1957)

Fig. F21 Photo of Froriep (from Wikimedia Commons; in public domain)

Fig. F22 Portrait of Fuchs (from the Archiv der Unibersität Wien; in public domain)

Fig. F23 Photo of Fuse (from Wikimedia Commons; in public domain)

Letter G

8

Galeazzi (Galeati), Domenico Maria Gusmano (1716–1775): Italian physician born and working in Bologna (Fig. G1). In particular, he studied the intestines.

Glands of Galeazzi: *Eponym* for intestinal glands, also described by Lieberkühn. Reference: *De cribriforme intestinorum tunica.* Comm Bonon 1731, 1:359.

Galen (Galenus), Claudius (130–200): Greek physician and philosopher born in Pergamum in Asia Minor. After studying in Pergamum, Smyrna, Corinth and Alexandria, in 162 he went to Rome (Fig. G2a). Galen was the leading medical authority of the Christian world for 1400 years. He described many parts of the brain.

Ansa, Nerve or **Anastomosis of Galen:** *Eponym* for the communicating branch of the superior laryngeal nerve with the recurrent laryngeal nerve.

Great cerebral vein of Galen: *Eponym* for the great cerebral vein (Fig. G2b). The internal cerebral veins are sometimes also attributed to Galen.

Gallaudet, Bern (1860–1934): American anatomist and surgeon.

Fascia of Gallaudet: *Eponym* for (a) the innominate fascia under the subcutaneous tissue of the anterior abdominal wall, and (b) the deep perineal fascia covering the perineal muscles, also described by Buck. Reference: *A Description of the Planes of Fascia of the Human Body with Special Reference to the Fascia of the Abdomen, Pelvis and Perineum.* Columbia University Press, New York, 1931.

Ganglia: *See* **The Classical Eponyms.**

Ganser, Sigbert Joseph Marie (1853–1931): German psychiatrist born in Rhaunen and working in Dresden (Fig. G3). Reference: *Vergleichend-anatomische Studien über das Gehirn des Maulwurfs.* Morphol Jahrb 1882; 7:591–725.

Commissure of Ganser: *Eponym* for the supreme supraoptic decussation.

Nucleus of Ganser: *Eponym* for the basal nucleus of the forebrain composed of large neurons, also described by Meynert.

Gantzer, Carol Friedrich Ludovic (eighteenth–nineteenth century): German anatomist.

Muscle of Gantzer: *Eponym* for an inconstant muscle extending from the superficial flexor of the digits to the deep flexor of the digits. Reference: *Dissertatio anatomica musculorum varietates sistens.* Stark, Berlin, 1813.

Gärtner, Hermann Treschow (1785–1827): Danish surgeon born in the West Indies and working in the Norwegian Army.

Duct of Gärtner: *Eponym* for the longitudinal duct of the epoophoron (Fig. G4). This duct may have been already recognized by Malpighi in 1681 (Dobson 1962). Reference: *An anatomical description of a glandular organ observed in the uterus of some lower animals.* Edinb Med Surg 1822; 21:460–463.

Gaskell, Walter Holbrook (1847–1914): English physiologist born in Naples working in Cambridge (Fig. G5). Reference: *The Origin of the Vertebrates.* London, 1906.

Bridge of Gaskell: *Synonym* for the atrioventricular bundle of the cardiac conducting system.

Nerves of Gaskell: *Eponym* for the accelerator nerves of the heart.

H. J. ten Donkelaar et al., *An Illustrated Guide to Anatomical Eponyms*, https://doi.org/10.1007/978-3-031-91664-9_8

Gasser, Johann Laurentius (1723–1765): Austrian anatomist born and working in Vienna after whom the sensory ganglion of the trigeminal nerve was named. Reference: *Paris quinti nervorum encephali disquisitio anatomica.* Vienna, 1765 (In: Ludwig CF 1791 *Scriptores neurologici minores selecti*, p. 244).

Ganglion of Gasser: *Eponym* for the trigeminal ganglion (Fig. G6).

Fig. G1 Portrait of Galeazzi. (Courtesy Marina Quartu, Cagliari)

Gavard, Hyacinthe (1753–1802): French anatomist.

Muscle of Gavard: *Eponym* for oblique fibres (the internal layer) of the muscular layer of the stomach wall.

Gay, Alexander Heinrich (1842–1907): Russian anatomist born and working in Kasan.

Glands of Gay: *Eponym* for the circumanal glands. Reference: *Die Circumanaldrüsen des Menschen.* Sitz Akad Wiss Wien 1871, 63 (2): 329–332.

Gegenbauer, Karl (1826–1903): German comparative anatomist born in Würzburg and working in Würzburg, Jena and for 30 years in Heidelberg.

Cells of Gegenbauer: *Eponym* for osteoblasts.

Gennari, Francesco (1752–1797): Italian anatomist born and working in Parma who recognized the conspicuous

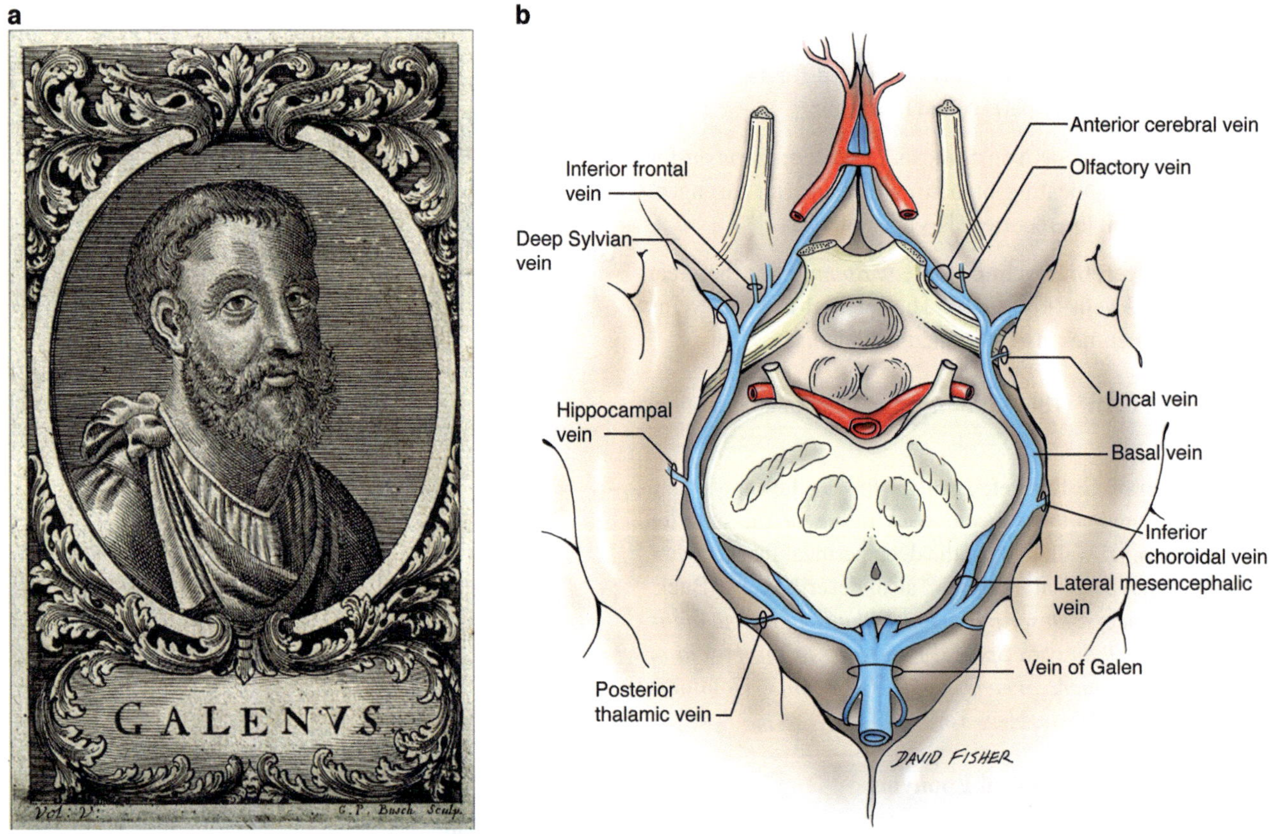

Fig. G2 (**a**) Galenus (eighteenth century engraving by George Paul Busch); and (**b**) the deep cerebral veins shown in a ventral view of the brain (from ten Donkelaar HJ *Clinical Neuroanatomy,* second ed. Springer, Cham, 2020; courtesy David Fisher, Birmingham, Alabama)

Fig. G3 Photo of Ganser (from Wikimedia Commons; in public domain)

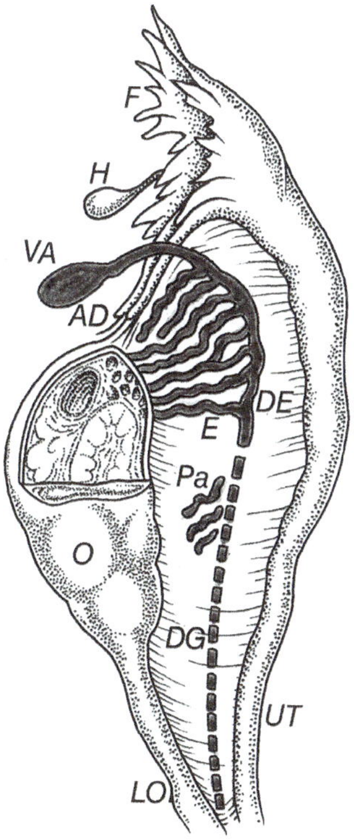

Fig. G4 Duct of Gärtner and other remnants of the mesonephric duct of Wolff in the female (in *grey*). *AD* aberrant ductule, *DE* duct of epoophoron, *DG* duct of Gärtner, *E* epoophoron, *F* fimbriae, *H* pedunculated hydatid of Morgagni, *LO* ligament of ovarium, *O* ovarium, *Pa* paroophoron, *UT* uterine tube, *VA* vesicular appendage (after Arey LB *Developmental Anatomy,* seventh ed. Saunders, Philadelphia, PA, 1965)

Fig. G5 Photo of Gaskell (from Wikimedia Commons; in public domain)

Fig. G6 Lateral view of the trigeminal ganglion of Gasser and the trigeminal nerve (from ten Donkelaar HJ, Kachlík D, Tubbs RS *An Illustrated Terminologia Neuroanatomica.* Springer, Cham, 2018; courtesy Ondřej Naňka, Prague)

myelinated band in the visual cortex, also described by Vicq d'Azyr. Reference: *De peculiari structura cerebri nonnullisque ejus morbis.* Parma, 1782 (Fig. G7a).

Band of Gennari: *Eponym* for the occipital stripe, a conspicuous band of myelinated fibres marking the striate cortex of the occipital lobe (Fig. G7b).

Gerber, Friedich Andreas (1797–1872): Swiss physician and anatomist working in Bern.

a

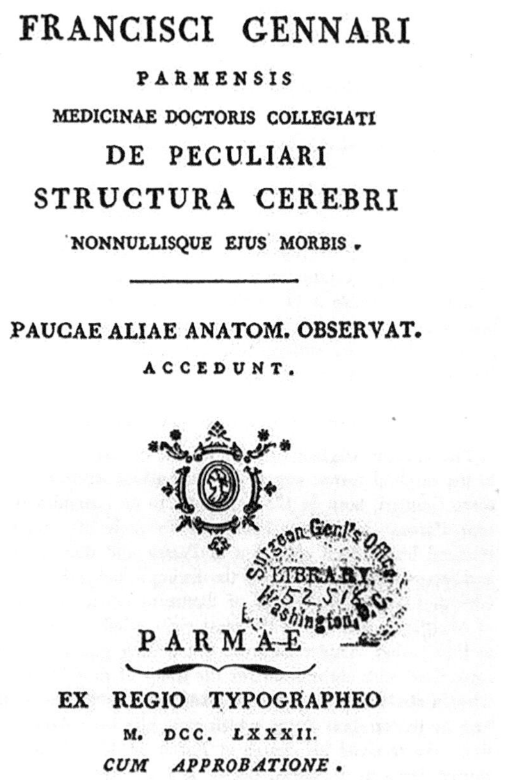
FRANCISCI GENNARI
PARMENSIS
MEDICINAE DOCTORIS COLLEGIATI
DE PECULIARI
STRUCTURA CEREBRI
NONNULLISQUE EJUS MORBIS.
PAUCAE ALIAE ANATOM. OBSERVAT.
ACCEDUNT.

PARMAE
EX REGIO TYPOGRAPHEO
M. DCC. LXXXII.
CUM APPROBATIONE.

b

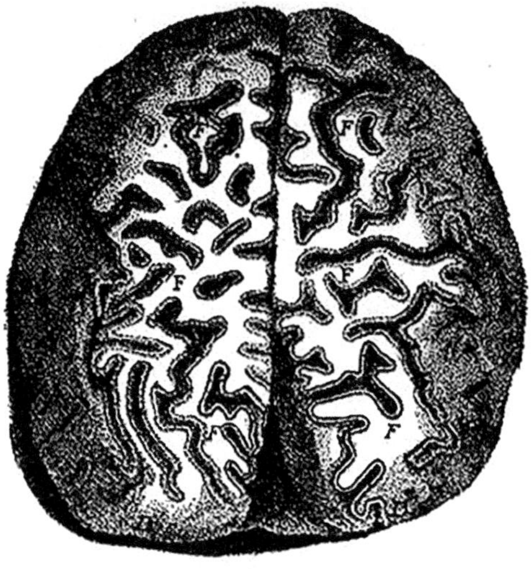

Fig. G7 (**a**) Title page of *De peculiari structura cerebri nonnullisque ejus morbis*. Parma, 1782; and (**b**) band of Gennari

Fig. G8 Portrait of Gerdy (from Wikimedia Commons; in public domain)

Ligament of Gerber: *Eponym* for an accessory band of the ascending band of the cruciform ligament attached onto the apex of the dens of the axis.

Gerdy, Pierre Nicholas (1797–1856): French surgeon and pathologist born in Loches and working in Paris (Fig. G8). Reference: *Chirurgie pratique complète.* Breslau, 1851.

Fontanel of Gerdy: *Eponym* for the sagittal fontanelle, an accessory fontanelle located within the sagittal suture in front of the point lambda. Reference: *Recherches et propositions d'anatomie, de pathologie et de toxicologie.* Thesis, Paris, 1823.

Fossa or Triangle of Gerdy: *Eponym* for the infraclavicular fossa, also described by Mohrenheim (*see* **Fig. M27**). Reference: *Anatomie des formes extérieures du corpsa humain, appliquée â la peinture, â la sculpture et â la chirurgie.* Béchet Jeune, Paris, 1829.

Ligament of Gerdy: *Eponym* for (a) the suspensory ligament of the axilla, later also described by Campbell; and (b) the superficial transverse metacarpal ligament.

Tubercle of Gerdy: *Eponym* for the tubercle on the lateral condyle of the tibia for the attachment of the iliotibial tract.

Fig. G9 Photo of Gerlach (from Wikimedia Commons; in public domain)

Gerlach, Joseph (1820–1896): German anatomist and physiologist born in Mainz and working in Erlangen (Fig. G9). Reference: *Handbuch der allgemeinen und speziellen Gewebelehre.* Mainz, 1848.

Anular tendon of Gerlach: *Eponym* for the fibrous anulus of the tympanic membrane.

Tonsil of Gerlach: *Eponym* the tubal tonsil, a cluster of lymphatic nodules in the mucosa of the pharyngeal ostium of the auditory tube, already noted by Eustachi. Reference: *Zur Morphologie der Tuba Eustachii.* Beilage zur Mschr Ohrenheilk 1875; 9:49.

Valve of Gerlach: *Eponym* for an occasional mucous fold at the opening of the vermiform appendix of the caecum. Reference: *Beobachtung einer tödtlichen Peritonitis, als Folge einer Perforation des Wormfortsatzes.* Z Rat Med Heidelberg 1847; 6:20.

Gerota, Dumitru (1867–1939): Rumanian surgeon born in Craiova and working in Bucharest (Fig. G10).

Anorectal lymph nodes of Gerota: *Eponym* for the pararectal lymph nodes, also described by Poirier and Cunéo. Reference: *Der anorectale Lymphapparat.* Sitz Preus Akad Wiss 1895, pp. 253–256.

Capsule or **Fascia of Gerota:** *Eponym* for the renal fascia (Fig. G11). Reference: *Beiträge zur Kenntniss des Befestigungsapparats der Niere.* Arch Anat Physiol (Lpz) 1895, Anat Abt, pp. 265–282.

Fig. G10 Photo of Gerota (from Wikimedia Commons; in public domain)

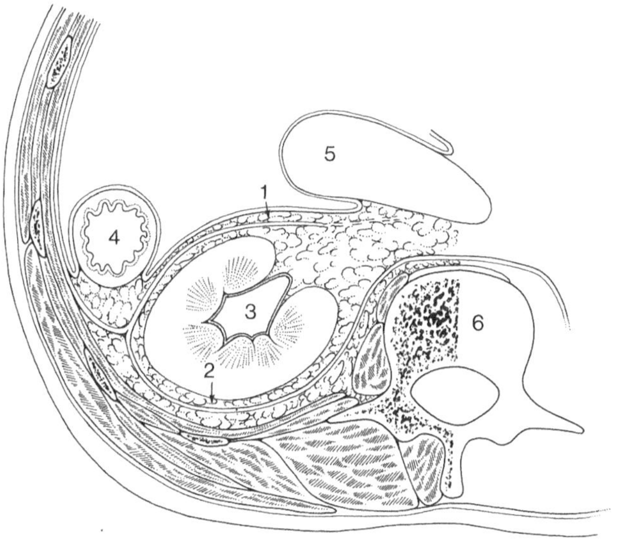

Fig. G11 Renal capsules (after ten Donkelaar HJ, Oostra R-J *Klinische Anatomie en Embryologie,* fourth ed. Springer Media, Houten, 2014). *1* prerenal fascia, *2* retrorenal fascia, *3* renal pelvis, *4* ascending colon, *5* duodenal flexure, *6* vertebra L2

Geschwind, Norman (1926–1984): American neurologist.

Area or **Territory of Geschwind:** *Eponym* for part of the inferior parietal lobule, involved in the perisylvian language networks. Reference: *Disconnection syndromes in animals and man. I and II.* Brain 1965; 88:237–294, 585–644.

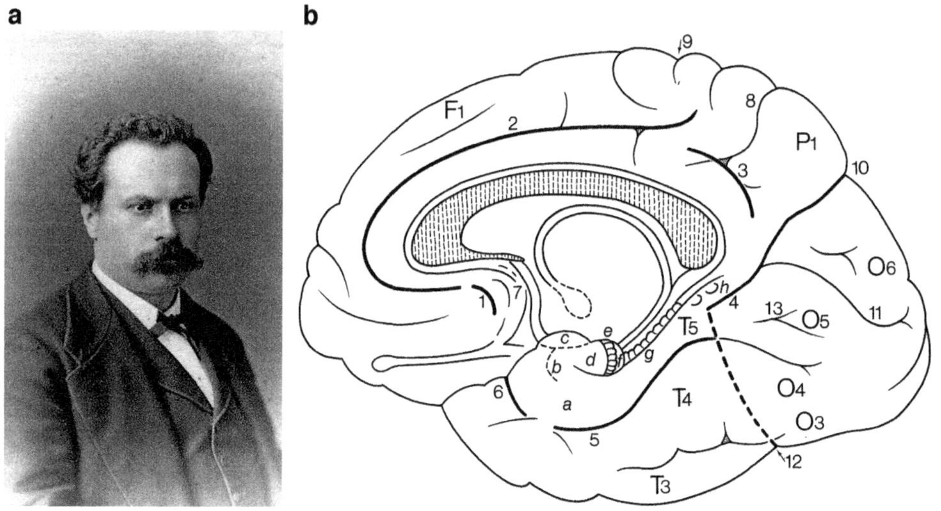

Fig. G12 (**a**) Photo of Giacomini (from Wikimedia Commons; in public domain); (**b**) gyri and sulci on the inferomedial surface of the cerebral hemisphere (after Duvernoy HM *Le cerveau humain.* Springer, Paris, 1992, from ten Donkelaar HJ *Clinical Neuroanatomy,* second ed. Springer, Cham, 2020). *a* entorhinal cortex, *b* ambient gyrus, *c* semilunar gyrus, *d* uncinate gyrus, *e* band of the dentate gyrus of Giacomini, *f* intralimbic gyrus or uncal apex, *g* dentate gyrus, *h* gyri of Andreas Retzius or subsplenial gyri, *F1* superior frontal gyrus, *P1* precuneus, *O3* inferior occipital gyrus, *O4* fusiform gyrus (occipital part), *O5* lingual gyrus, *O6* cuneus, *T3* inferior temporal gyrus, *T4* fusiform gyrus (temporal part), *T5* parahippocampal gyrus, *1–6* parts of the "limbic sulcus" or "scissure limbique": *1* anterior paraolfactory sulcus, *2* cingulate sulcus, *3* subparietal sulcus, *4* anterior part of parieto-occipital sulcus, *5* collateral sulcus, *6* rhinal sulcus, *7* posterior paraolfactory sulcus, *8* transverse parietal sulcus of Brissaud, *9* central sulcus, *10* parieto-occipital sulcus, *11* calcarine sulcus, *12* pre-occipital notch of Meynert, *13* lingual sulcus

Giacomini, Carlo (1840–1898): Italian anatomist born in Sale di Tortona and working in Turin (Fig. G12a). He invented a method for lasting hardening of the brain.

Band or **Bandelette of Giacomini:** *Eponym* for the band of the dentate gyrus of the cerebral cortex (Fig. G12b). Reference: *Fascia dentata del grande ippocampo nel cervello umano.* G Accad Med Torino 1883, and Arch Ital Biol 1898; 30:153–158.

Cardiac vertebrae of Giacomini: *Eponym* for the fourth to seventh thoracic vertebrae, corresponding to the topography of the heart.

Vein of Giacomini: *Eponym* for the femoral intersaphenous vein, connecting the cranial extension of the small saphenous vein and the great saphenous veins, corresponding to the posterior femoral circumflex vein. Reference: *Osservazioni anatomiche per servire allo studio della circolazione venosa, della estremita inferiori fatte all'istituto anatomico di Torino.* Vercellino, Turin, 1873.

Giannuzzi, Giuseppe Oronzio (1883–1976): Italian physiologist born in Altamura and, after studying in Paris and Berlin, working in Siena.

Demilunes or **Crescents of Giannuzzi:** *Eponym* for darkly stained crescentlike groups of serous cells at the end of mixed tubules of mixed salivary glands, also described by Heidenhain. They appeared to be artifacts.

Gierke, Hans Paul Bernhard (1847–1886): German anatomist born in Stettin and working in Tokyo and Breslau known for his studies on respiration. Reference: *Der Theile der Medulla oblongata deren Verletzung Athembewegungen hemmt und das Athemzentrum.* Arch Ges Physiol (Bonn) 1873; 7:583–600.

Bundle of Gierke, Respiratory: *Eponym* for the solitary tract of the brain stem.

Corpuscle of Gierke: *Eponym* for the thymic corpuscles, also described by Hassall and Virchow.

Gillette, Eugène Paulin (1836–1886): French surgeon born and working in Paris.

Ligament of Gillette, Suspensory: *Eponym* for the cricopharyngeal ligament, his suspensory ligament of the oesophagus, longitudinal fibres of the oesophagus attached to the posterior aspect of the cricoid cartilage.

Lymph node of Gillette-Henle: *Eponym* or the retropharyngeal lymph node, also described by Henle.

Space of Gillette: *Eponym* for the retropharyngeal space.

Gillot, Claude (1932–2024?): French anatomist working in Paris.

Vein of Gillot, Perforating: *Eponym* for the medial leg perforator. He also described transosseous perfora-

Fig. G13 Portrait of Gimbernat (from Wikimedia Commons; in public domain)

tors of the knee (see Uhl J-F, Valsamis M, Gillot G *The transosseous perforators of the knee.* Phlebology 2021; 28:61–66.

Gimbernat, Don Manuel Louise Antonio (1734–1816): Spanish surgeon and anatomist born in Cambrils, Catalunia, and working in Barcelona (Fig. G13). Later, he was appointed Director of the Royal College of Surgeons in Spain and Surgeon to King Carlos III.

Ligament of Gimbernat: *Eponym* for the lacunar ligament (Fig. G14). The triangular lacunar ligament is formed by the lowermost fibres of the aponeurosis of the external oblique muscle or inguinal ligament, turning and continuing as the pectineal ligament (of Cooper), inserting on the pecten pubis. Reference: *Nuevo metodo de operar en la hernia crural.* Madrid, 1793.

Giraldès, Joachim Albin Cardazo Cazado (1808–1875): Portuguese surgeon born in Porto, educated and working in Paris.

Organ of Giraldès: *Eponym* for the paradidymis, later also described by Waldeyer (*see* **Fig. H2**). Reference: *Note sur un nouvel organe glanduleux situé dans le cordon spermatique et pouvant donner naissance des kystes.* CR Soc Biol Paris 1859, pp. 123–145.

Glands: *See* **The Classical Anatomical Eponyms.**

Glaser, Johann Heinrich (1629–1675): Swiss anatomist and botanist born and working in Basel (Fig. G15a).

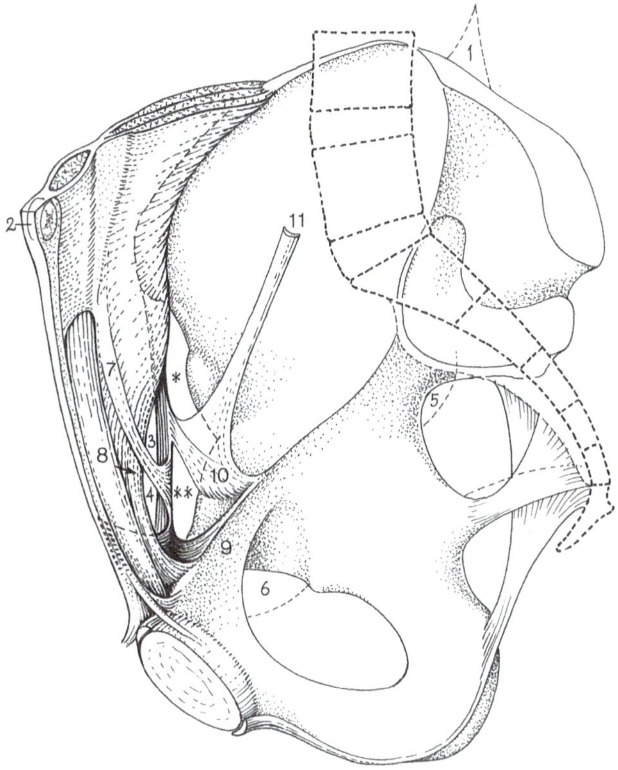

Fig. G14 The lacunar ligament of Gimbernat (based on a drawing by the late Chris van Huijzen, Nijmegen). *1* lumbar triangle, *2* umbilical anulus, *3* deep inguinal anulus, *4* superficial inguinal anulus, *5* major ischiadic foramen, *6* obturator foramen, *7* interfoveolar ligament, *8* falx inguinalis, *9* pectineal ligament joining the **lacunar ligament**, *10* iliopectineal arcus, *11* tendon of minor psoas muscle

Reference: *Tractatus posthumus de cerebro* (edited by Johann Jacob Stehelin, his son-in-law. Bertschi, Basel, 1680; Fig. G15b).

Artery of Glaser: *Eponym* for the anterior tympanic artery.

Fissure of Glaser: *Eponym* for the petrotympanic fissure, also described by Huguier and Civinini. *See* Mudry A *Glaser fissure, Huguier canal, and Civinini canal. A confused eponymical imbroglio.* Otol Neurotol 2014; 36:1115–1120.

Gley, Eugène (1857–1930): French physiologist born in Epinal and working in Paris.

Glands of Gley: *Eponym* for the parathyroid glands. Reference: *Les secretions internes; principes physiologiques; applications à le pathologie.* Paris, 1914. Dobson (1962) noted that the parathyroid glands were first described by Sandström in 1880 (Uppsala Lakarf Forhandl 15:441–471).

Glisson, Francis (1597–1677): English physician and anatomist born in Rampisham, Dorset, and working in

a

b

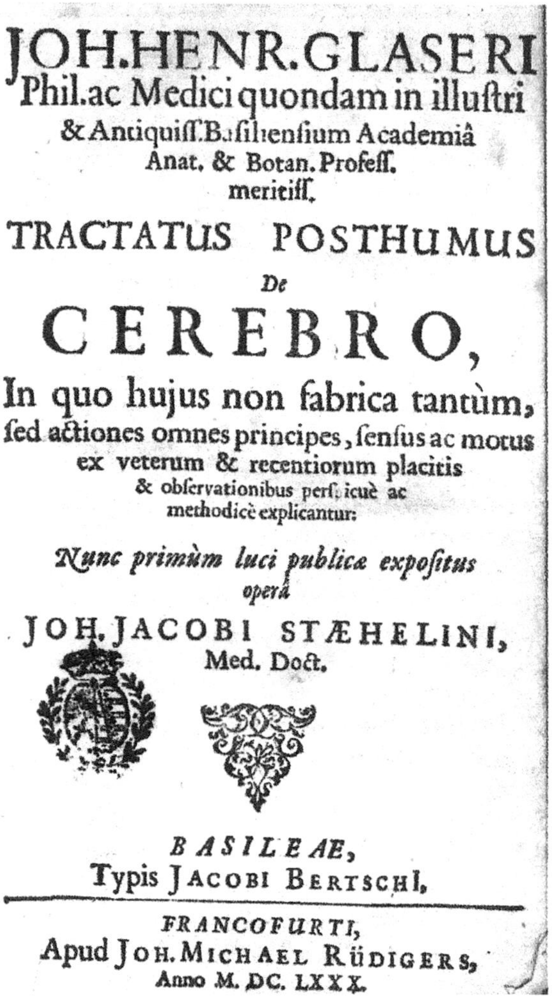

JOH.HENR.GLASERI
Phil.ac Medici quondam in illuſtri
& Antiquiſſ.Baſilienſium Academiâ
Anat. & Botan. Profeſſ.
meritiſſ.
TRACTATUS POSTHUMUS
De
CEREBRO,
In quo hujus non fabrica tantùm,
ſed actiones omnes principes, ſenſus ac motus
ex veterum & recentiorum placitis
& obſervationibus perſpicuè ac
methodicè explicantur:
Nunc primùm luci publicæ expoſitus
operâ
JOH. JACOBI STÆHELINI,
Med. Doct.
BASILEAE,
Typis JACOBI BERTSCHI,
FRANCOFURTI,
Apud JOH. MICHAEL RÜDIGERS,
Anno M. DC. LXXX.

Fig. G15 (**a**) Engraving of Glaser (from Wikimedia Commons; in public domain); (**b**) Title page of *Tractatus posthumus de cerebro.* Basel, 1680

Fig. G16 Engraving of Glisson (from the Wellcome Collection CCBY; in public domain)

London (Fig. G16), known for his studies of the liver. Reference: *Anatomia hepatis cui praemittuntur quaedem ad rem anatomicam universe spectantia, et nonnulla de Lymphae Ductibus subjiciuntur.* London, 1654.

Capsule of Glisson: *Eponym* for the fibrous capsule of the liver.

Portal triad (Trias hepatica) or **System of Glisson** (Fig. G17)**:** *Eponym* for the contents of the porta hepatis, composed of the proper hepatic artery, the portal vein and the common hepatic duct. Their further subdivisions around liver lobules are named after Kiernan

Sphincter of Glisson: *Eponym* for the inferior sphincter of the bile duct, just above the hepatopancreatic ampulla, also described in 1887 by Oddi (*see* **Fig. O4**).

Godman, John Davidson (1794–1830): American physician born in Annapolis, Maryland, and working in Philadelphia.

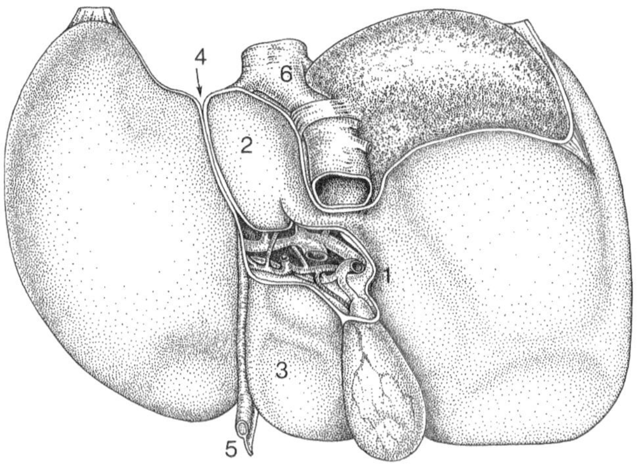

Fig. G17 Portral triad of Glisson (after ten Donkelaar HJ, Oostra R-J *Klinische Anatomie en Embryologie,* fourth ed. Springer Media, Houten, 2014). *1* porta hepatis, *2* caudate lobe, *3* quadrate lobe, *4* fissure of ligamentum venosum, *5* falciform ligament with ligamentum teres hepatis, *6* inferior vena cava

Fascia of Godman: *Eponym* for a continuation of the pretracheal layer of the cervical fascia to the thorax and the pericardium. Reference: *Anatomical Investigations Comprising Descriptions of Various Fasciae of the Human Body.* Philadelphia, 1824.

Goethe, Johann Wolfgang von (1749–1832): German poet, philosopher and scientist, born in Frankfurt am Main.

Bone of Goethe: *Eponym* sometimes used for the premaxillary bone (os incisivum), but also used for Os incae or interparietal bone. Reference: *Ueber den Zwischenkiefer des Menschen und der Thiere.* Jena, 1786.

Golgi, Camillo (1843–1926): Italian anatomist and pathologist born in Corteno and working in Parma and later in Sienna (Fig. G18). He introduced the silver chromate staining technique for nerve cells (1873) and described many features of silver-impregnated material including the Golgi tendon organ (1878), Golgi type I and type II neurons (1883, 1884), and the large and small Golgi cells of the cerebellar cortex. With Santiago Ramón y Cajal, he received the Nobel Price for Physiology and Medicine in 1906. References: *Sulla struttura della sostanza grigia del cervello.* Gazz Med Ital Lomb 1873; 33:244–246; *Recherches sur l'histologie des centres nerveux.* Arch Ital Biol 1883; 3:285–317; 1884; 4:92–123 (Edited version Arch Ital Biol 1959; 97:279–299); *Sulla fina anatomia degli organi centrali del sistema nervoso.* Hoepli, Milano, 1886.

Golgi apparatus: *Eponym* for the cytoplasmic organelle that, originally described in nerve cells, is present in all cells except erythrocytes.

Fig. G18 Photo of Golgi (from Wikimedia Commons; in public domain)

Golgi cells of cerebellum: *Eponym* for large and small cells in the granular layer of the cerebellar cortex.

Golgi type I and type II neurons: *Eponym* for interneurons with a long axon and a small axon, respectively.

Golgi-Mazzoni corpuscles: *Eponym* for a type of receptors with asymmetrical Schwann-cell lamination, found in the subcutaneous tissue of human fingers (Halata Z Adv Anat Embryol Cell Biol 1975; 50(5):1–77; Chouchkov C Adv Anat Embryol Cell Biol 1978; 54(5);1–63).

Golgi tendon organ: *Eponym* for the musculotendinous end-organ (Fig. G19). Reference: *Della terminazione del nervi nei tendini e di un nuovo apparato nervoso terminale musculotendineo.* Atti Soc Ital Sci Nat 1878; 272–275.

Mitral cells of Golgi: *Eponym* for olfactory bulb cells that form complex synapses, called glomeruli, with the axons of olfactory receptor neurons.

Goll, Friedrich (1829–1903): Swiss anatomist born in Zofingen and working in Zürich (Fig. G20). Reference: *Beiträge zur feineren Anatomie des menschlichen Rückenmarks.* Denkschr Med Chir Ges Zürich 1860, p. 130; *Zur feinere Anatomie der Medulla spinalis.* 1868.

Nucleus of Goll: *Eponym* for the gracile nucleus in the caudal hindbrain.

Tract or **Fasciculus of Goll:** *Eponym* for the gracile fasciculus in the spinal cord.

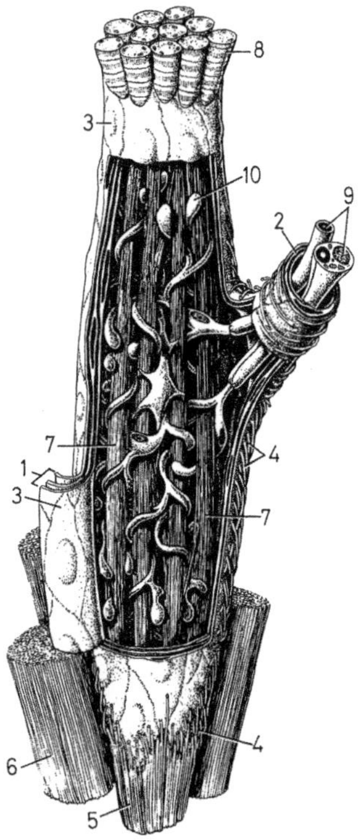

Fig. G19 The Golgi tendon organ (from Krstić R 1988 *Die Gewebe des Menschen und der Säugetiere*, second ed. Springer, Berlin-Heidelberg-New York; with permission; in ten Donkelaar HJ, Kachlík D, Tubbs RS *An Illustrated Terminologia Neuroanatomica*. Springer, Cham, 2018). *1* and *3* perineural capsule, *2* perineurium of peripheral nerve, *4* collagen microfibrils of the endomysium, *5–7* tendon fibers, *8* muscle fibers, *9* myelinated nerve fibers (type Ib), *10* terminal enlargement of axon

Fig. G20 Photo of Goll (from Wikimedia Commons; in public domain)

Gombault, François Alexis Albert (1844–1904): French neurologist and pathologist in Paris from Orléans.

Triangle of Philippe-Gombault: *Eponym* for the sacral continuation of the septomarginal fasciculus of the spinal cord, also described by Philippe, and known as triangle of Philippe-Gombault. References: *Contributions à l'étude des lésions systématisées des cordons blancs de la moelle epinière.* Arch Med Exp Anat Pathol 1894; 6:365–424; *Etat actuel des connaissances sur la systématisation des cordon postérieurs de la moelle epinière.* Sem Med 1895; 15:161–166.

Goormaghtigh, Norbert (1890–1960): Belgian pathologist born in Oostende and working in Gent.

Apparatus of Goormaghtigh: *Eponym* for the juxtaglomerular apparatus of the kidney.

Cells of Goormaghtigh: *Eponym* for the extraglomerular mesangial or juxtaglomerular cells, in 1933 also described by Zimmermann. Reference: *Les segments neuro-myo-artériele juxtaglomérulaires du rein.* Arch Biol 1932; 43:575–591.

Gottlieb, Bernhard (1885–1950): Austrian dentist working in Vienna.

Epithelial attachment of Gottlieb: *Eponym* for a collar of epithelial cells attached to the tooth surface and the gingival connective tissue. Today, this protective barrier is considered only a functional part, not histologically definable, of the gingivodental closure, dynamically changing over time, based also on dental hygiene.

Gowers, Sir Willam Richard (1845–1915): English physician born and working in London (Fig. G21) who published notable descriptions of many neurological disorders. Reference: *The Diagnosis of Diseases of the Spinal Cord.* London, 1880.

Tract of Gowers: *Eponym* for the anterior or ventral spinocerebellar tract.

Graafian follicle: *Eponym* for the mature or vesicular ovarian follicle; *see* de Graaf, Reinier.

Grandry, M. (nineteenth century): Belgian physician in Liège.

Corpuscles of Grandry: *Eponym* for sensory corpuscles, end-organs in the epidermis, particularly of the

Fig. G21 Photo of Gowers (from the Wellcome Collection CCBY; in public domain)

mouth and tongue. References: *Recherches sur le terminaison des nerfs cutanés chez l'homme.* J Anat (Paris) 1869; 6:390–395; *Recherches sur les corpuscles de Pacini.* J Anat (Paris) 1869; 6:395–398.

Gratiolet, Louis Pierre (1815–1865): French anatomist born in St. Foy la Grande and working in Paris (Fig. G22a) where he later succeeded Geoffroy St. Hilaire. He described the optic radiation, various convolutions of the cerebral cortex and suggested a subdivision of the cerebral cortex (Fig. G22b). References: *Mémoires sur les plis cérébraux de l'homme et des primates.* Bertrand, Paris, 1854; *Anatomie comparée du système nerveux considéré dans ses rapports avec l'intelligence.* Vol 2. Didet, Paris, 1857 (for Vol 1, *see* Leuret). *See* also: Parent A *Louis Pierre Gratiolet (1815–1865) and his contributions to the study of cerebral convolutionsin primates.* Neuroscience & Medicine 2014; 5:1, 43,511.

Fibres or **Radiation of Gratiolet:** *Eponym* for the optic radiation.

Gray, Edward George (1924–1999): English pioneer electron microscopist who decribed the Gray types I (symmetric) and II (asymmetric) synapses. Reference: *Axo-somatic and axo-dendritic synapses of the cerebral cortex: An electron-microscopic study.* J Anat (Lond) 1959; 93:420–433.

Grayson, J (twentieth century): English anatomist in Manchester.

Ligament of Grayson: *Eponym* for a ligament originating from the palmar aspect of the flexor tendon sheath. Its fibres project at right angles to this sheath, pass palmar to the digital nerves and vessels, and insert into the skin at the same level as their origin (Fig. G23). Reference: *The cutaneous ligaments of the digits.* J Anat (Lond) 1941; 75:164–165; see also Zwanenburg RL, McGrouther DA, Werker P *Grayson ligament: A revised description of its anatomy and function.* J Hand Surg 2019; 44:341.e1–341.e6.

Griffiths, John Daniel (1926–2001): Welsh surgeon born in Llanel and working in London.

Point of Griffiths: *Eponym* for the point at the left colic flexure where the marginal colic artery can be deficient. Reference: *Surgical anatomy of the blood supply of the distal colon.* Ann Roy Coll Surg 1956; 19:241–256.

Gruber, Wenzel (Václav) Leopold (1814–1890): Bohemian anatomist born in Krukanice and working in Prague as Prosector under Hyrtl (1842–1847), and from 1855 in St. Petersburg (Fig. G24). He described many anomalous forms of muscles. References: *Beiträge zur Anatomie des Keilbeins und Schläfenbeins.* In: Richter HE, Winter A (eds) *Schmidt's Jahrbücher der in- und ausländischen gesammten Medicin. 11. Anatomie und Physiologie.* Wigand, Leipzig 1859; *Über die Verbindung des Nervus medianus mit dem Nervus ulnaris am Unterarme des Menschen und der Säugethiere.* Arch Anat Physiol Klin Med (Lpz) 1870; 37:501–522.

Anastomosis of Martin-Gruber: *Eponym* for a communicating branch of the median nerve to the ulnar nerve.

Bone of Gruber: *Eponym* for an accessory bone between the hamate, the capitate and the base of the third and fourth metacarpal bones.

Canal of Gruber: *Eponym* for (a) the abducens nerve canal, usually described as canal of Dorello, and (b) the cruropopliteal canal.

Fascia of Gruber: *Eponym* for the middle layer of the cervical fascia, also described by Paulet, Richet and Testut.

Fossa of Gruber-Landzert: *Eponym* for the inferior duodenal recess of the peritoneal cavity, also described by Landzert (Fig. G25). Reference: *Über einen Fall nicht incarcerierter aber mit Incarceration complicierter Hernia interna mesogastrica.* Oesterr Zsch Prakt Heilk 1863; 9: 325–330, 341–345.

Ligament of Gruber: *Eponym* for the petroclinoid ligament.

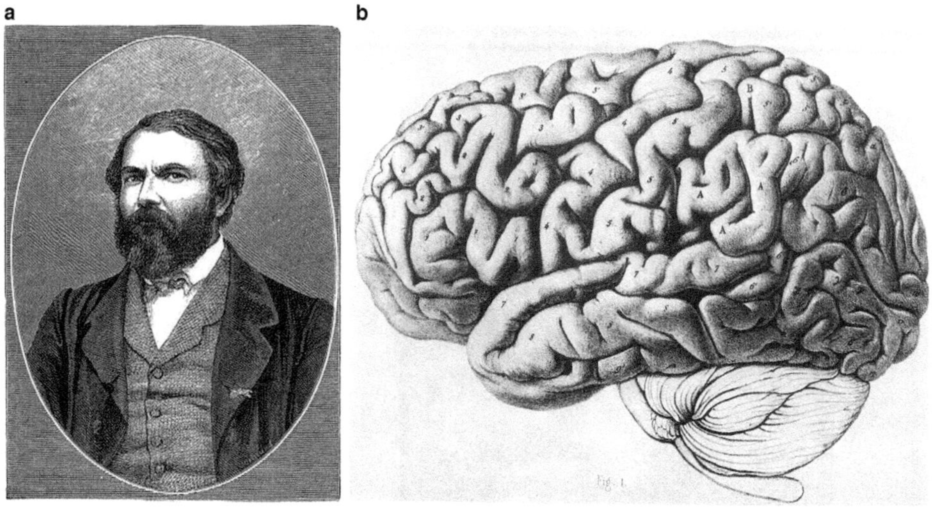

Fig. G22 (**a**) Portrait of Gratiolet (from the Wellcome Collection CCBY; in public domain); (**b**) Gratiolet's subdivision of the cerebral cortex (from Gratiolet LP *natomie comparé du système nerveux considéré dans ses rapports avec l'intelligence,* Vol 2. Didet, Paris, 1857)

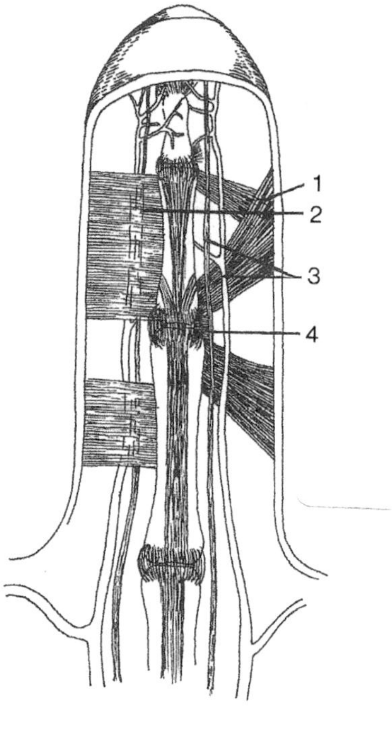

Fig. G23 Ligament of Grayson (after Milford LW *Retaining Ligaments of the Digits of the Hand.* Saunders, Philadelphia, PA, 1968). *1* Ligament of Cleland, *2* ligament of Grayson, *3* artery and nerve, *4* transverse reticular ligament

Muscle of Gruber: *Eponym* for an accessory fibular or peroneal muscle.

Suture of Gruber: *Eponym* for the petrosphenooccipital fissure.

Grünwald, Ludwig (1863–1927): Austrian otorhinolaryngologist born in Vienna and working in Munich. Reference: *Die Lehre von den Naseneiterungen mit besonderer Rücksicht auf die Erkrankungen des Sieb- und Keilbeins und deren chirurgische Behandlung.* Lehmann, Munich, 1893.

Cell of Grünwald, Interlamellar: *Eponym* for the pneumatization of the middle nasal concha restricted to the vertical lamella of the turbinate.

Sinus of Grünwald, Lateral: *Eponym* for the lateral sinus of the ethmoidal cells, located craniodorsal to the ethmoidal bulla.

Grynfeltt, Joseph Casimir (1840–1913): French obstetrician and gynaecologist born in Servian and working in Montpellier.

Dilator muscle of Grynfeltt: *Eponym* for the dilator pupillae muscle, previously described by von Kölliker (1852). Reference: *Le muscle dilatateur de la pupille chez les mammifères.* Firmin and Montaine, Montpellier, 1899.

Superior lumbar triangle or **Space of Grynfeltt:** *Eponym* for a space bounded by the serratus posterior inferior muscle and the 12th rib cranially, the erector spinae muscle medially and the internal oblique abdominal muscle laterally (Fig. G26), also described by Lesshaft and Krause.

Gubaroff, Alexander Petrowitsch von (1855–1931): Russian gynaecologist born and working in Moscow.

Semilunar fold or **Valve of Gubaroff:** *Eponym* for a perpendicular fold at the cardiac ostium of the stomach, desrbibed as cardiac sphincter between the oesophagus and the stomach, also described by Braune. Reference: *Ueber den Verschluss des memschlichen Magens an der Cardia.* Anat Physiol (Lpz) 1886–1887, Anat Abt, pp. 397–402.

Gudden, Johann Bernhard Aloys von (1823–1886): German psychiatrist born in Kleve, working in Zürich and later in Munich where he had a large psychiatric practice

Fig. G24 The Anatomical Lesson of Gruber, teaching a group of female students (courtesy Ondřej Naňka, Prague)

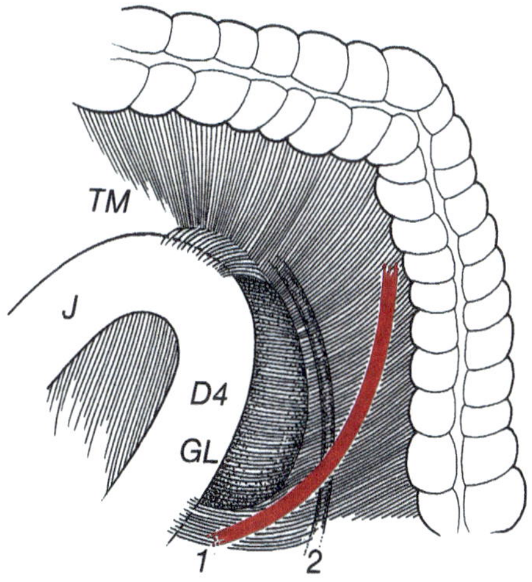

Fig. G25 Fossa of Gruber-Landzert (after Gray SW, Skandalakis JE *Embryology for Surgeons*. Saunders, Philadelphia, PA, 1972). *D4* fourth part of duodenum, *GL* fossa of Gruber-Landzert, *J* jejunum, *TM* transverse mesocolon, *1* left colic artery, *2* inferior mesenteric vein

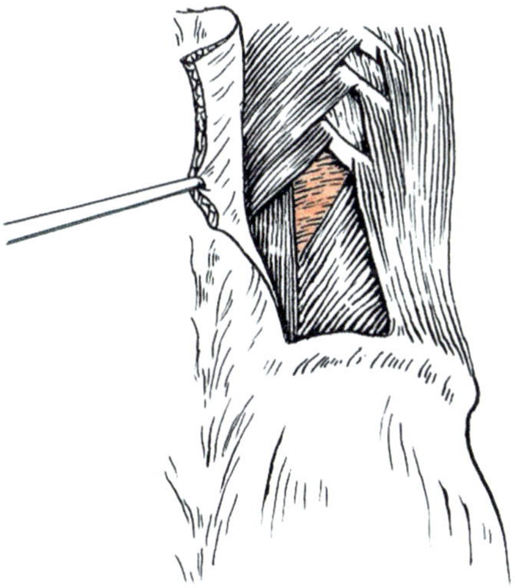

Fig. G26 Superior lumbar triangle of Grynfeltt (after Čihak R *Anatomie 1*. Grada, Prague, 2011; courtesy Ondřej Naňka, Prague)

(Fig. G27), which included King Ludwig II from Bavaria. He described retrograde degeneration of the thalamus after cortical lesions (Gudden atrophy). References: *Experimentaluntersuchungen über das peripherische und centrale Nervensystem.* Arch Psychiatr Nervenkr 1870; 2:693–723; *Ueber die Kreuzung der Fasern im Chiasma nervorum opticorum.* v. Graefe's Arch Ophthalmol 1874; 25, Abth 1:1–56.

Commissure of Gudden: *Eponym* for the most ventral supraoptic decussation.

Ganglion of Gudden: *Eponym* for the interpeduncular nucleus of the brain stem.

Nuclei of Gudden: *Eponym* for the dorsal and ventral tegmental nuclei.

Tract of Gudden: *Eponym* for the mammillotegmental tract.

Fig. G27 Photo of Gudden (from Wikimedia Commons; in public domain)

Guérin, Alphonse François Marie (1817–1895): French anatomist and surgeon born in Ploermel and working in Paris.

Glands of Guérin: *Eponym* for the paraurethral glands in females, later also described by Skene. Reference: *Maladies des organes génitaux externe de la femme.* Paris, 1864.

Sinus of Guérin: *Eponym* for the recess of the navicular fossa of the male urethra.

Valve of Guérin: *Eponym* for the valve of the navicular fossa of the male urethra. Reference: *Note sur une valvule existent normalement à le partie postérieure de la fosse naviculaire.* Gaz Méd Paris 1849, pp. 581 and 679.

Guidi, Guido (Vidius, Vidus; 1500–1569): Italian physician born in Florence, Tuscany, working first in Paris as physician to Francis I, where Vesalius was his student, and later in Pisa (Fig. G28). Reference: *De anatome corporis humani lib. Vii.* Iuntas, Venice, 1611.

Artery of Guidi (Vidius) or **Vidian artery:** *Eponym* for the artery of the pterygoid canal.

Canal of Guidi (Vidius): *Eponym* for the pterygoid canal of the sphenoidal bone.

Nerve of Guidi (Vidius) or **Vidian nerve:** *Eponym* for the nerve of the pterygoid canal, composed of the greater and deep petrosal nerves.

Fig. G28 Engraving of Guido Guidi by G Vascellini. (Courtesy Marina Quartu, Cagliari)

Vein of Guidi (Vidius) or **Vidian nerve:** *Eponym* for the vein of the pterygoid canal.

Guillain, Georges Charles (1876–1961): French neurologist born in Rouen and working in Paris.

Guillian-Mollaret Triangle: *Eponym* for the dentatorubroolivary circuit, an important feedback circuit for the dentate nucleus and the brain stem, described with Pierre Mollaret (1898–1987).

Günz, Justus Gottfried (1714–1755): German anatomist, physiologist and surgeon born in Königstein (Saxony) and working in Leipzig.

Ligament of Günz: *Eponym* for part of the superficial layer of the obturator membrane. Reference: *Observationum anatomico-chirurgicarum de Herniis Libellus.* Leipzig, 1744.

Guthrie, George James (1785–1856): English surgeon born in Wakefield and working in London.

Muscle of Guthrie: *Eponym* for the sphincter urethrae muscle. References: *On two new muscles of the membranous portion of the urethra.* Lond Med Surg J 1833; 3:491–496; *On the Anatomy and Diseases of the Neck of the Bladder.* London, 1834.

Fig. G29 Photo of Guyon (from Pagel 1901)

Guyon, Jean Casimir Félix (1831–1920): French surgeon and pathologist born on the island of Réunion and working in Paris (Fig. G29).

Canal of Guyon: *Eponym* for the ulnar canal of the wrist for the ulnar vessels and nerve (Fig. G30). Reference: *Note sur une disposition anatomique propre à la face antérieure de la région du poignet et non encore décrire.* Bull Soc Anat Paris 1861; 6:184–186.

Isthmus of Guyon: *Eponym* for the isthmus of the uterus. Reference: *Sur les cavités del'uterus à l'état de vacuité.* Thesis, Paris, 1858.

Triangle of Guyon: *Eponym* for a triangle determined by the inferior jugular vein, the thyro-linguofacial vein and the posterior belly of the digastric muscle.

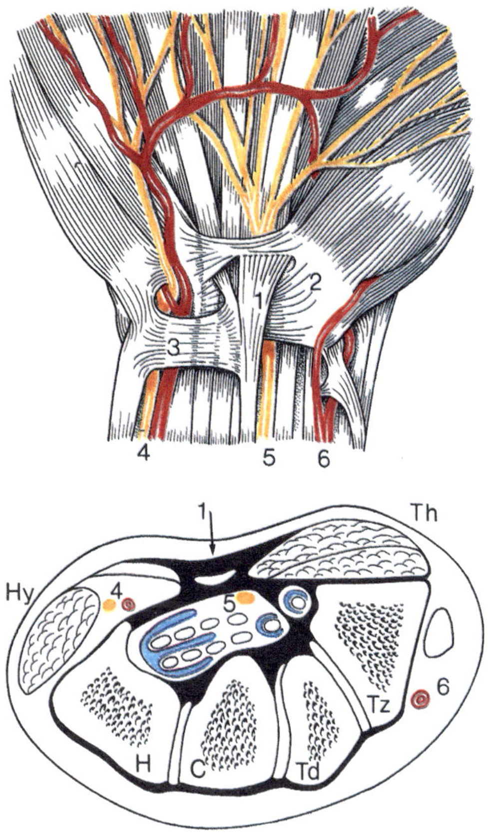

Fig. G30 (**a**) Canal of Guyon (based on a figure in the Prometheus Atlas; courtesy Ondřej Naňka, Prague); (**b**) section of the hand (after O'Rahilly R *Gardner-Gray-O'Rahilly Anatomy. A regional study of human struture,* 5th ed. Saunders, Philadelphia, PA, 1986). Arteries in *red*, nerves in *yellow*, in *blue* fascial sheaths of finger muscles. *C* Capitate, *H* hamate, *Hy* hypothenar muscles, *Th* thenar muscles (above: abductor pollicis brevis, below: opponens pollicis), *Td* trapezoid, *Tz* trapezium, *1* palmar aponeurosis, *2* flexor retinaculum, *3* roof of canal of Guyon, *4* ulnar artery and nerve, *5* median nerve, *6* radial artery

9 Letter H

Hach, Wolfgang (1930-today): German surgeon and phlebologist born in Berlin and working in Berlin, Pirmasens and Frankfurt am Main.

Vein of Hach, Perforating: *Eponym* for the posterolateral femoral perforating vein.

Haeckermann, Karl (nineteenth century): German anatomist working in Bremen.

Area of Haeckermann: *Eponym* for a triangular area located below the cricopharyngeal fibres in the posterior oesophageal wall, also known as area of Laimer-Haekermann. Reference: *Beitrag zur Lehre von der Entstehung der Divertikel des Oesophagus.* Thesis, Göttingen, 1891.

Halban, Josef (1870–1937): Austrian gynaecologist.

Fascia of Halban: *Eponym* for fascial strips between the trigone of the urinary bladder and the anterior part of the vaginal wall, also known as pubocervical fascia and vesicovaginal septum.

Haller, Albrecht von (1708–1777): Swiss anatomist and physiologist born in Bern and working in Göttingen and Bern (Fig. H1a). He was a student of Boerhaave, Albinus and Duverney. Haller's name is given eponymously to many anatomical structures, some are listed below. Reference: *Disputationes anatomicae selectae.* Vandenhoeck, Göttingen, 1746–1752.

Aberrant ductules of Haller: *Eponym* for aberrant ductules lying both cranial and caudal to the functional efferent ductuli of the testis (Fig. H1b). The cranial group is usually connected to the rete testis, the caudal ductule (also known as organ of Haller) is a blind diverticulum of the epididymal duct. These ductules represent mesonephric tubules that have not completed the connection between the rete testis and the epididymal duct, nor have they completely regressed (Gray SW, Skandalakis JE: *Embryology for Surgeons.* Saunders, Philadelphia, 1972).

Ansa or **Anastomosis of Haller:** *Eponym* for (a) an inconstant communicating branch of the facial nerve with the glossopharyngeal nerve and (b) a communicating branch between the glossopharyngeal nerve and the auricular branch of the vagus nerve, also described by Sappey, and known as the ansa of Haller-Sappey.

Anastomosis maxima or **magna of Haller:** *Eponym* for the anastomosis between the superior and inferior mesenteric arteries, attributed to Riolan, whom Haller gave the credit for this structure.

Anulus or **Insula of Haller:** *Eponym* for a doubling of the thoracic duct for part of its thoracic course; also known as anulus of Haller. Reference: *De ductus thoracica.* Göttingen, 1741.

Arch of Haller: *Eponym* for the crus of the diaphragm formed by the medial and lateral arcuate ligaments. Reference: *De musculis diaphragmatis dissertationem anatomicum.* Bern, 1733.

Arch of Haller, Epiploic: *Eponym* for the anastomosis between the omental or epiploic branches of the gastroomental arteries along the free margin of the greater omentum, also described by Barkow (Arch of Haller-Barkow).

Artery of Haller: *Eponym* for (a) the dorsal pancreatic artery, (b) the superior thoracic artery, and (c) the supreme thoracic artery.

Circle of Haller: *Eponym* for (a) the vascular circle of the optic nerve, a network of branches of the short cili-

H. J. ten Donkelaar et al., *An Illustrated Guide to Anatomical Eponyms*, https://doi.org/10.1007/978-3-031-91664-9_9

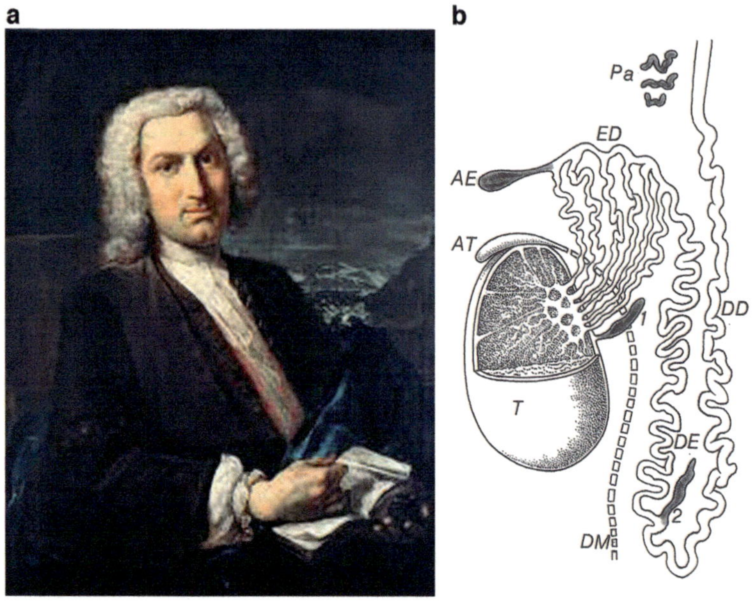

Fig. H1 (**a**) Portrait von Haller (from Wikimedia Commons; in public domain); (**b**) aberrant ductules of Haller and other mesonephric and paramesonephric vestigial structures (in *grey*). *AE* appendage of epididymis, *AT* appendage of testis, *DD* ductus deferens, *DE* duct of epididymis, *DM* obliterated duct of Müller, *ED* efferent ductules, *Pa* paradidymis, *T* testis, *1* cranial aberrant ductule, *2* caudal aberrant ductule (after Arey LB *Developmental Anatomy,* 7th ed. Saunders, Philadelphia, PA, 1965)

ary arteries on the sclera around the exit of the optic nerve, also described by Zinn, and (b) the areolar venous circle around the areola of the breast.

Commissure of Haller: *Eponym* for the commissura infima.

Cones of Haller: *Eponym* for the vascular cones formed by the convolutions of the efferent ductules and their blood vessels in the head of the epididymis.

Ganglion of Haller, Thyroid: *Eponym* for the middle cervical ganglion of the sympathetic trunk.

Gubernaculum vaginae of Haller: *Eponym* for the gubernaculum ovarii. Reference: *De viis feminis.* Göttingen, 1748.

Layer of Haller: *Eponym* for the vascular layer of the choroid of the eyeball.

Ligaments of Haller: *Eponym* for the medial and lateral arcuate ligaments of the diaphragm.

Organ of Haller: *Synonym* for the caudal aberrant ductule.

Plexus of Haller: *Eponym* for the laryngeal plexus, formed by fibres from the superior laryngeal nerve and sympathetic fibres.

Pouch or **Cul-de-sac of Haller:** *Eponym* for the oblique sinus of the pericardial cavity.

Tripus Halleri: *Eponym* for the coeliac trunk.

Unguis of Haller: *Old term* for the calcar avis, a protrusion into the occipital horn of the lateral ventricle of the brain, also described by Morand (1741).

Hamperl, Herwig (1899–1976): German pathologist working in Bonn.

Cells of Herwig: *Eponym* for the uterine natural killer cells (his 'fluorescierende Körnchenzellen'), a special population of lymphocytes in the endometrium. Reference: *Die fluorescierende Körnchenzellen ('Fluorocytes').* Virchows Arch Pathol Anat Physiol Klin Med 1950; 318:32–47.

Hamy, D (nineteenth century): French anthropologist.

Line of Hamy: *Eponym* for the glabella-lambda line, roughly parallel to the alveolar-condylar plane of Broca. Reference: *Discussion sur le plan horizontal de la tête par P. Broca.* Bull Soc Anthropol Paris 1873; 8:93.

Fontanelle of Hamy: *Eponym* for an accessory fontanelle behind the foramen magnum. Reference: *Fontanelles anomales du crâne humain.* 1871.

Hannover, Adolph (1814–1894): Danish anatomist born and working in Copenhagen well known for his studies on the nervous system and the eye. References: *Recherches microscopiques sur le système nerveux.* Copenhagen, 1844; *On the Anatomy, Physiology and Pathology of the*

Eye (in Danish). Copenhagen, 1845; *Das Auge*. Leipzig, 1852.

Canal of Hannover: *Eponym* for a potential retrozonular space between the ciliary zonule and the vitreous body, also described by Petit.

Membrane of Hannover, Intermediate enamel: *Eponym* for the stratum intermedium of the enamel, the inner layer of cells within the enamel organ of the dental germ in the foetus. Reference: *On the development and structure of the teeth of Mammifera* (with TW Jones). Br Foreign Med Chir Rev. 1857; 19:229–240.

Harder, Johann Jacob (1656–1711): Swiss anatomist born and working in Basel.

Gland of Harder: *Eponym* for an accessory lacrimal gland in the upper eyelid, also described by Krause and others. Reference: *Glandula nova lachrymalis una cum ductu excretorio in cervis et damis.* Acta Eruditorum, Leipzig 1694, pp. 49–52.

Hardesty, Irving (1866–1944): American anatomist and embryologist working in Berkeley and New Orleans.

Membrane of Hardesty: *Eponym* for the outer stripe of the tectorial membrane, also described as stripe by Kimura. Reference: *On the nature of the tectorial membrane and its probable role in the anatomy of hearing.* Am J Anat 1908; 8:109–179.

Hartmann, Arthur (1849–1931): German otologist born in Heidenheim and working in Vienna and later in Berlin.

Septum of Hartmann: *Eponym* for the petrosquamous septum, dividing the squamous and petrous parts of the mastoid cavity, also described by Körner. Reference: *Die Krankheiten des Ohres und deren Behandlung.* Fischer, Berlin, 1889.

Hartmann, Henri Albert Charles Antoine (1860–1952): French surgeon born and working in Paris who specialized in abdominal surgery.

Critical point of Hartmann: *Eponym* for the site on the large intestine where the last sigmoid artery meets the superior rectal artery, also described by Sudeck. Reference: *Some considerations upon high amputation of the rectum.* Ann Surg (Philadelphia) 1909; 50:1091.

Pouch of Hartmann: *Eponym* for a dilatation of the neck of the gallbladder. Reference: *Quelques points de l'anatomie et de la chirurgie des voies biliaires.* Bull Soc Anat Paris 1891; 5:480–500.

Recess of Hartmann: *Eponym* for an inconstant recess of the peritoneal cavity located ventral to the mesoappendix. Reference: *Les fossettes iléocaecales et la hernie de Rieux.* Bull Soc Anat, 1888.

Harvey, William (1578–1657): English physiologist born in Folkestone (Kent). He studied under Fabrici and Casseri, and then worked in London (Fig. H2a). One of his experiments on the blood circulation he demonstrated is shown in Fig. H2b. Reference: *Exercitatio anatomica de motu cordis et sanguinis in animalibus.* London, 1628.

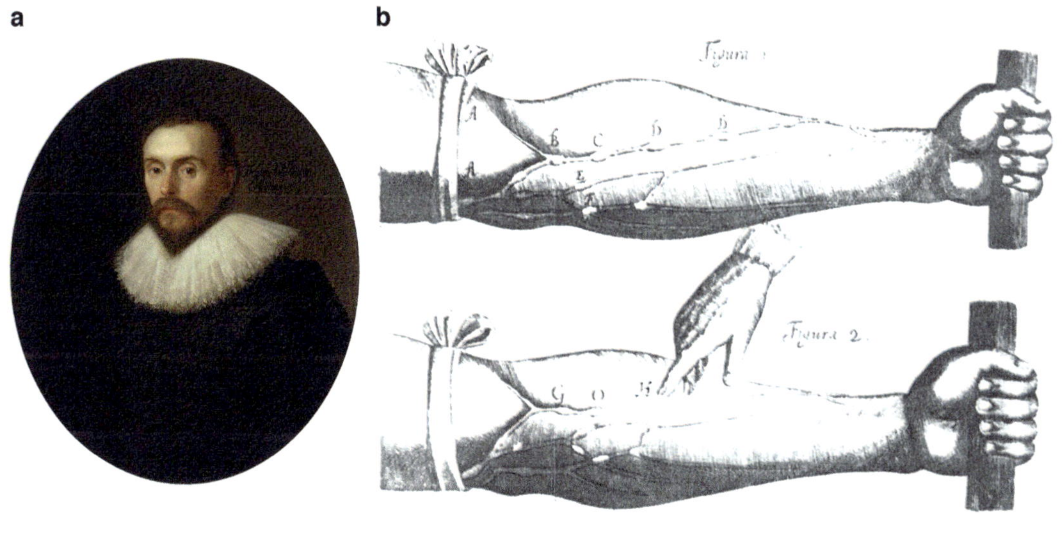

Fig. H2 (**a**) Portrait of Harvey; (**b**) experiment by Harvey demonstrating blood flow (both from Wikipedia Commons; in public domain)

Fig. H3 Portrait of Hassall (from Wikimedia Commons; in public domain)

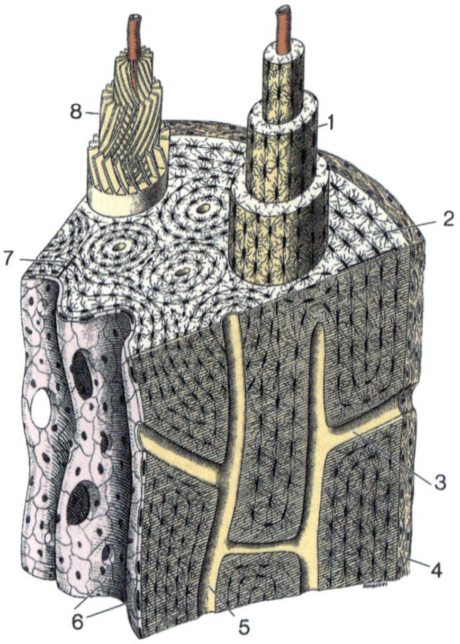

Fig. H4 Osteon (after ten Donkelaar HJ, Oostra R-J *Klinische Anatomie en Embryologie,* 4th ed. Springer Media, Houten, 2014). *1* system of Havers, *2* outer lamellae, *3* canal of Volkmann, *4* periost, *5* canal of Havers, *6* endost, *7* inner lamellae, *8* screwlike orientation of collagen fibres

Hasner, Joseph, Ritter von Artha (1819–1892): Bohemian ophthalmologist born and working in Prague.

Fold or **Valve of Hasner:** *Eponym* for the lacrimal valve within the nasolacrimal duct, also described by Bianchi and Cruveilhier. Reference: *Physiologie und Pathologie des Thränenableitungsapparates.* Prague, 1850.

Hassall, Arthur Hill (1817–1894): English physician born in Teddington, Middlesex, and working in London and later on the Isle of Wight (Fig. H3). He wrote the first English textbook on microscopic anatomy. Reference: *The Microscopic Anatomy of the Human Body in Health and Disease.* London, 1846.

Corpuscles of Hassall: *Eponym* for concentrically arranged epithelial reticular cells in the cortex of the thymus, also described by Gierke and Virchow.

Havers, Clopton (1657–1702): English physician born in London who qualified in Utrecht. He is known for his studies on the structure of bones (Fig. H4). Reference: *Osteologia Nova, or some new observations of the Bones and the Parts belonging to them, with the manner of their accretion and nutrition.* London, 1691.

Canals of Havers: *Eponym* for osteonic vascular canals in the compact tissue of bone, previously described by van Leeuwenhoek (1686).

Folds or **Glands of Havers:** *Eponym* for collections of adipose tissue in the major joints, covered by synovial membrane.

Lamellae of Havers: *Eponym* for bony septa surrounding the canals of Havers.

Spaces of Havers: *Eponym* for spaces in bone formed by the enlargement of canals of Havers.

System of Havers: *Eponym* for an osteon.

Heidenhain, Rudolf Peter Heinrich (1834–1897): German histologist and physiologist born in Marienwerder, Prussia (now Kwidzyn, Poland), and working in Berlin, Halle and Breslau (now Wróclaw).

Cells of Heidenhain: *Eponym* for parietal cells of the gastric glands. Reference: *Beiträge zur Histologie und Physiologie des Dünndarmschleimhaut.* Arch Ges Physiol 1888; 43, Suppl, pp. 1–103.

Demilunes or **Crescents of Heidenhain:** *Eponym* for darkly stained crescent-like groups of serous cells at the end of mixed tubules of mixed salivary glands, also described by Giannuzzi. They appeared to be artifacts.

Heimer, Lennart (1930–2007): Swedish-American neuroscientist born in Östersund, Sweden, and working at the Massachusetts Institute of Technology and the University

a

b

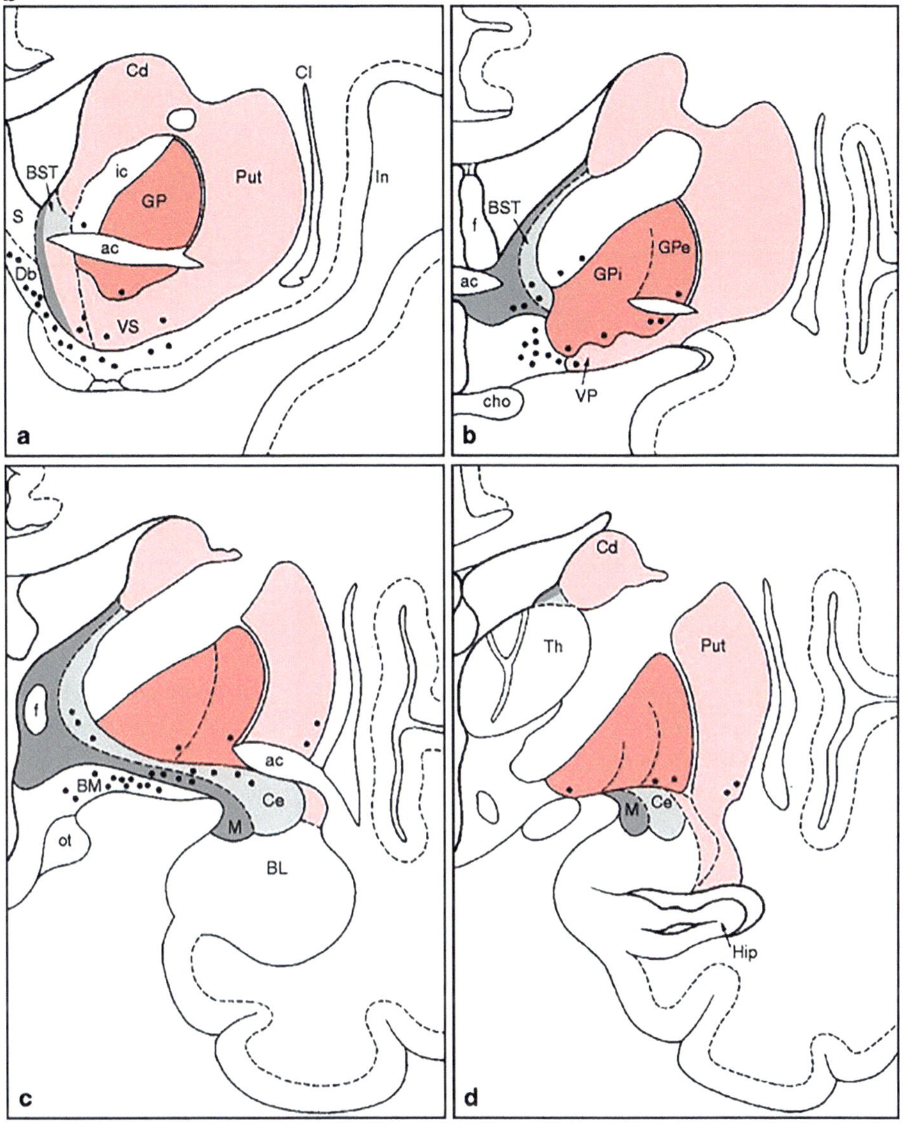

Fig. H5 (**a**) Photo of Heimer (from Haber S, McGinty JF, Mugnaini E, Zaborsky L *Lennart Heimer: In memoriam (1930–2007).* Brain Struct Funct 2008; 213:3–10; with permission); (**b**) Organization of the human basal forebrain with the current subdivision of the basal nuclei and the amygdaloid complex. Comparable structures are indicated in various colors. Transition areas are indicated as parts of the striatum, but do show some amygdaloid features. The *large dots* indicate the large, cholinergic cells of the basal nuleus of Meynert (*BM*). *ac* anterior commissure, *BL* basolateral amygdala, *BST* bed nucleus of the stria terminalis, *Cd* caudate nucleus, *Ce* central amygdaloid nucleus, *cho* chiasma opticum, *Cl* claustrum, *Db* diagonal band of Broca, *f* fornix, *GP* globus pallidus, *GPe*, *GPi* external and internal parts of globus pallidus, *Hip* hippocampus, *In* insula, *M* medial amygdaloid nucleus, *ot* optic tract, *Put* putamen, *S* septum, *Th* thalamus, *VP* ventral pallidum, *VS* ventral striatum (after Heimer L, et al. 1991 Prog Brain Res 87:109–165; from ten Donkelaar HJ, Kachlík D, Tubbs RS *An Illustrated Terminologia Neuroanatomica.* Springer, Cham, 2018)

of Virginia (Fig. H5a). He introduced the concepts of the ventral striatum and the ventral pallidum to emphasize that the ventral parts of the striatum and pallidum are primarily related to the limbic lobe and not to the isocortex (Fig. H5b). References: *The Human Brain and Spinal Cord,* 2nd ed. Springer, New York, 1995; *Anatomy of Neuropsychiatry.* Elsevier, Amsterdam, 2008 (with GW Van Hoesen, M Trimble and DS Zahm).

Heister, Lorenz (1683–1758): German anatomist, surgeon and botanist from Frankfurt am Main working in Altdorf and later in Helmstädt (Fig. H6). Reference: *Compendium anatomicum veterum recentiorumque observationes brevissime complectens.* Altdorf, 1717.

Diverticulum of Heister: *Eponym* for the superior bulb of the internal jugular vein.

Fig. H6 Engraving of Heister (from the Universitätsbibliothek Erlangen-Nürnberg in Wikimedia Commons; in public domain)

Valve of Heister: *Eponym* for the spiral fold of the cystic duct, later also described by Amussat.

Held, Hans (1866–1942): German anatomist and neurologist born in Neukloster, Mecklenburg, and working in Leipzig known for his studies on the acoustic system. References: *Die centralen Bahnen des Nervus acusticus bei der Katze.* Arch Anat Physiol (Lpz) 1891, pp. 271–290; *Die centrale Gehörleitung.* Ibid. 1893, pp. 201–248.

Bundle of Held: *Eponym* for the tectospinal tract, also described by Edinger and Loewenthal.

Calyx of Held: *Eponym* for a long, excitatory synapse in the mammalian auditory system, resembling the calyx of a flower, found in the trapezoid body and the lateral superior olivary nucleus.

Cells of Held: *Eponym* for the inner border cells of the spiral organ of the cochlea (*see* **Fig. C21**). References: *Untersuchungen über den feineren Bau des Ohrlabyrinthes der Wirbelthiere.* Teubner, Leipzig, 1902; *Die Cochlea der Säuger und der Vögel, ihre Entwicklung und ihr Bau, Receptionsorgane 1.* Springer, Berlin, 1926.

Commissure of Held: *Eponym* for the secondary auditory fibres from the anterior or ventral cochlear nucleus, which cross the midline dorsal to the trapezoid body to join the contralateral lateral lemniscus.

Decussation of Held: *Eponym* for the crossing of some of the fibres arising from the cochlear nuclei forming the lateral lemniscus.

End bulbs of Held: *Eponym* for another synapse found in the anteroventral cochlear nucleus.

Helie, Louis Theodore (1804–1867): French gynaecologist and anatomist born in Nantes and working in Paris.

Bundle or **Fibres of Helie:** *Eponym* for the vertical and ansiform bundles of the superficial muscle layer of the uterus. Reference: *Recherches sur la disposition des fibres musculaires développés par la grossesse.* Paris, 1865.

Heller, Arnold Ludwig Gotthilf (1840–1913): German pathologist born in Kleinhambach, Bavaria, and working in Kiel.

Plexus of Heller: *Eponym* for an arterial plexus in the wall of the small intestine. Reference: *Ueber die Blutgefässe der Dünndarmes.* Arb Physiol Anat (Lpz) 1873; 7:3–10.

Helvetius, Claude-Adrien (1715–1771): French physician and philosopher born and working in Paris (Fig. H7).

Collar of Helvetius: *Eponym* for the transition of the circular muscle fibres of the oesophagus to the oblique muscles fibres of the stomach at the cardiac notch of the stomach, also described by Willis.

Ligament of Helvetius: *Eponym* for lateral muscle bands of the stomach.

Helweg, Hans Christian Saxtroph (1847–1901): Danish psychiatrist born and working in Odense. He described the olivospinal tract (tract of Helweg), which is more likely to be a reticulospinal tract. Reference: *Studien über den centralen Verlauf der vasomotorischen Nervenbahnen.* Arch Psychiatr Nervenkr 1887; 19:104–183.

Fig. H7 Portrait of Helvetius (from Wkimideia Commons; in public domain)

Tract of Helweg, Triangular: *Eponym*, incorrectly used for the olivospinal tract. It stands out in Weigert-Pal preparations at the anterolateral border of the lower medulla oblongata and the first five cervical segments, but not when stained with the Klüver-Barrera technique. It consists mainly of extremely fine, myelinated fibres and, therefore, stands out as an area of pallor in the anterior columns of the cervical spinal cord and lower medulla oblongata. It may be reticulospinal, but certainly not an olivospinal tract (*see* Smith ML, Deacon P Brain 1981; 104:249–277 for further discussion).

Henke, Wilhelm (1834–1896): German artist, art critic and anatomist born in Jena and working in Rostock, Prague and finally in Tübingen.

Space of Henke: *Eponym* for the retropharyngeal space. Reference: *Topographische Anatomie des Menschen in Abbildung und Beschreibung.* Berlin, 1879–1883.

Triangle of Henke: *Eponym* for a triangular area between the descending part of the inguinal fold, the lateral part of the fold and the lateral border of the rectus abdominis muscle.

Henle, Friedrich Gustav Jacob (1809–1885): German anatomist, histologist and pathologist born in Fürth, working in Berlin, Zürich, Heidelberg and finally in Göttingen (Fig. H8a). He made many contributions to microscopic anatomy. Henle's name is given eponymously to many structures, although he was strictly against eponyms. References: *Handbuch der rationellen Pathologie,* 2 Vols. Vieweg, Braunschweig, 1846–1853; *Handbuch der systematischen Anatomie des Menschen,* 3 Vols. Vieweg, Braunschweig, 1855–1872.

Ampulla of Henle: *Eponym* for (a) the ampulla of the ductus deferens and (b) the ampulla of the uterine tube.

Anastomosis of Henle: *Eponym* for a communicating branch between the trigeminal and facial nerves at the pontine level.

Canal of Henle: *Eponym* for the adductor canal in the thigh.

Crypts of Henle: *Eponym* for infoldings of the conjunctiva.

Fibre layer or **Fibres of Henle:** *Eponym* for the layer of inner cone fibres in the central area of the retina.

Glands of Henle: *Eponym* for certain accessory lacrimal glands also described by Ciaccio, Krause and Wolfring.

Layer of Henle: *Eponym* for the external epithelial layer of cells in the internal sheath of the hair root.

Ligaments of Henle: *Eponyms* for (a) the suspensory ligament of the ovary, Henle's infundibulopelvic ligament, also described by Testut as lumboovarian ligament, (b) the inferior transverse ligament of the scapula, (c) the lumbocostal ligament between the transverse process of vertebra L1 and the caudal border of the 12th rib close to its medial end, (d) the transverse perineal ligament, also described by Krause and Waldeyer, and (e) lateral and medial accessory ligaments of the temporomandibular joint, comparable to the temporomandibular ligament and the sphenomandibular ligament, respectively. The falx inguinalis was also referred to as ligament of Henle but Henle's description concerns a lateral extension of the rectus abdominis sheath (Anson BJ, McVay CB *Inguinal hernia: I. The anatomy of the region.* Surg Gynec Obst 1938; 66:186–191).

Loop of Henle: *Eponym* for the nephronic loop of the kidney ('Schleife' of Henle; Fig. H8b, c).

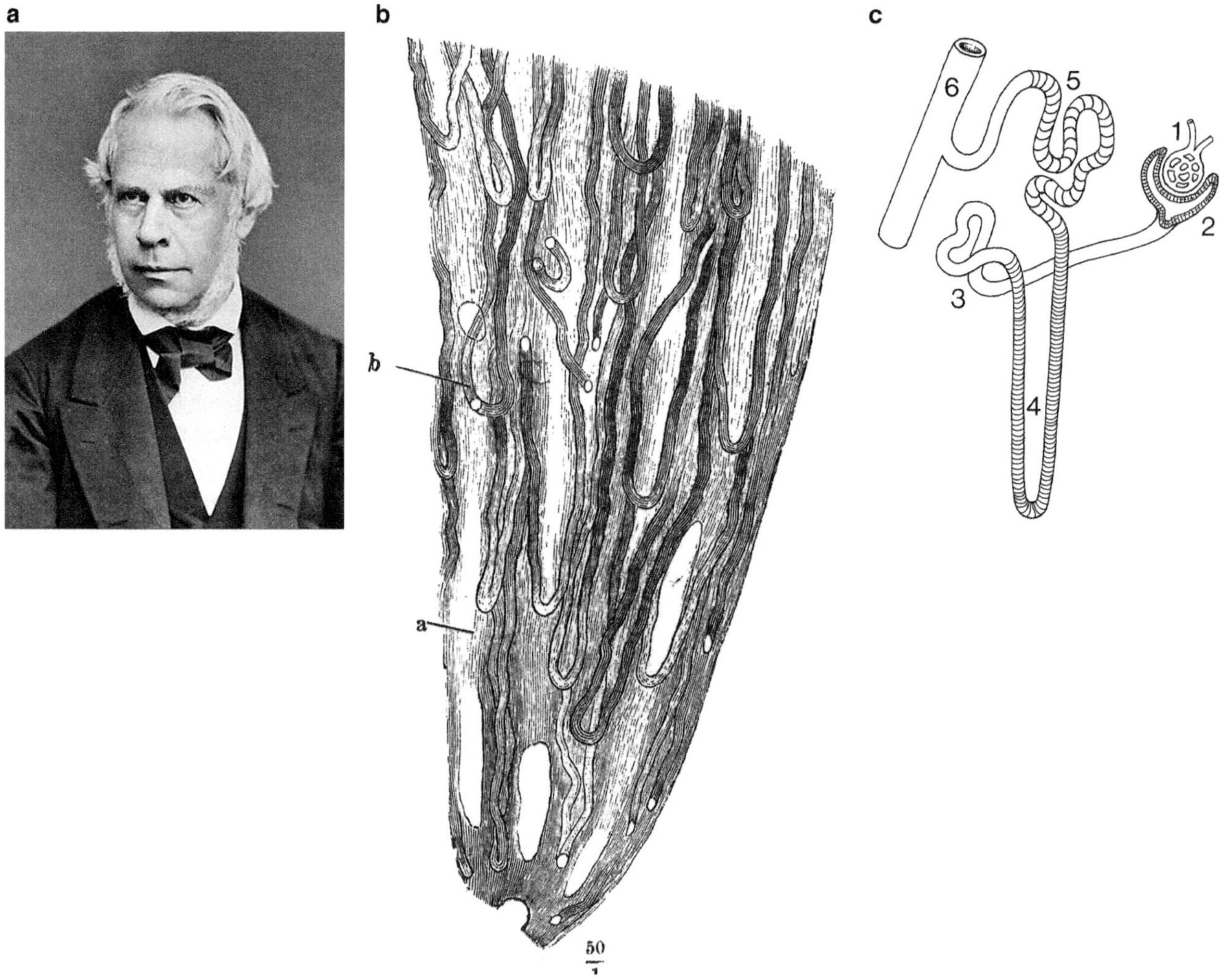

Fig. H8 (**a**) Photo of Henle (from Pagel 1901); (**b**) loop of Henle (from Henle FGJ *Handbuch der systematischen Anatomie des Menschen.* Vieweg, Braunschweig, 1866); (**c**) a nephron: *1* glomerulus, *2* capsule of Bowman, *3* proximal convoluted tubulus of Schachowa, *4* loop of Henle, *5* distal convoluted tubulus, *6* collecting tubulus (after ten Donkelaar HJ, Oostra R-J *Klinische Anatomie en Embryologie,* 4th ed. Springer Media, Houten, 2014)

Lymph node of Gillette-Henle: *Eponym* for the retropharyngeal lymph node.

Membrane of Henle: *Old term* for the basal lamina of the choroid, usually known as membrane of Bruch.

Muscles of Henle: *Eponym* for (a) the anterior auricular muscle and (b) the pubococcygeus muscle (Henle's musculus levator ani).

Nerve of Henle: *Eponym* for an inconstant branch of the ulnar nerve in the forearm, coursing distally with the ulnar artery to supply the ulnar vessels and sometimes also the skin of the dorsal forearm.

Retinacula of Henle: *Eponym* for the inferior and superior fibular (or peroneal) retinacula.

Sheath of Henle: *Eponym* for the endoneurium, the connective tissue of peripheral nerves, also described by Key and Retzius.

Spine of Henle: *Eponym* for the suprameatal spine of the squamous part of the temporal bone.

Triangle of Henle: *Synonym* for the lumbocostal triangle of the diaphragm of Bochdalek.

Trunk of Henle, Gastropancreaticocolic: *Eponym* for a very variable collective trunk of veins coming from the stomach, the right colic flexure and the head of the pancreas.

Tubules of Henle: *Eponym* for the straight portions of the uriniferous tubules that form the nephronic loop of Henle (Fig. H8b, c).

Vein of Henle: *Eponym* for the middle cardiac vein.

Henry, Arnold Kirkpatrick (1886–1962): Irish surgeon and anatomist born in Dublin and working mainly in Dublin.

Master knot or **Chiasm of Henry:** *Eponym* for the plantar chiasm, the plantar crossing of the tendon of the flexor digitorum muscle obliquely over the tendon of the flexor hallucis longus muscle in the midfoot at the level of the navicular bone. Reference: *Exposure of the Long Bones.* 1927.

Hensen, Viktor (1835–1924): German physiologist and embryologist born in Schleswig and working in Kiel (Fig. H9a). He is best known for his studies on embryology (the primitive node) and the histology of sense organs. Reference: *Das Verhalten des Resonanzapparats im menschlichen Ohr.* Sitz Akad Wiss 1902; 2:104–144.

Bodies of Hensen: *Eponym* for whorl-like aggregates of smooth endoplasmic cisternae in outer hair cells of the spiral organ of the cochlea.

Cells of Hensen: *Eponym* for the outer or external border cells of the spiral organ of the cochlea (*see* **Fig. C21**).

Duct or **Canal of Hensen:** *Eponym* for the ductus reuniens of the membranous labyrinth of the inner ear.

Node of Hensen: *Eponym* for the primitive node of the early embryo, also described by Hubrecht (Fig. H9b). Reference: *Ueber die Ableitung der Umkehr der Kernblätter.* 1882.

Stripe of Hensen: *Eponym* for a band on the undersurface of the tectorial membrane.

Hensing, Friedrich Wilhelm (1719–1745): German anatomist born and working in Giessen known for his work on the peritoneum. Reference: *De peritonaeo.* Thesis, Giessen, 1742.

Ligament of Hensing: *Eponym* for the phrenicocolic ligament (Hensing's sustentaculum lienale).

Hering, Heinrich Ewald (1866–1948): Austrian physiologist (Fig. H10). Reference: *Der Karotisdruckversuch.* Münch Med Wiss 1923; 42:1267–1290.

Canals of Hering: *Eponym* for biliary canals with epithelial lining to a hepatic lobule.
Nerve of Hering: *Eponym* for the carotid sinus nerve.

Herophilus (335–280 BC): Physician of Alexandria, born in Chalcedonia. He was one of the most renowned doctors of classical antiquity, founder of scientific anatomy and dedicated to the study of the brain and nerves.

Torcular of Herophilus: *Eponym* for the confluence of the sinuses in the skull.

Herring, Percy Theodore (1872–1967): English physician and physiologist working in Edinburgh, known for his studies on the hypothalamic-posterior pituitary projections.

Bodies of Herring: *Eponym* for spherical thickenings along neurosecretory axons of the hypothalamohypophyseal tract, particularly abundant in the infundibulum and the posterior lobe of the pituitary gland.

Hertwig, Oscar (1849–1922): German anatomist born in Freiburg and working in Jena and later in Berlin.

Epithelial sheath of Hertwig: *Eponym* for the epithelial root sheath, composed of cells of the enamel organ of the developing tooth, persisting postnatally around the roots of teeth. Reference: *Lehrbuch der Entwicklungsgeschichte des Menschen und der Wirbelthiere.* Jena, 1886.

Heschl, Richard Ladislaus (1824–1881): Austrian anatomist born in Wellsdorf, Styria, working in Olomouc, Krakau and finally in Graz (Fig. H11a). Reference: *Über die vordere quere Schläfenwindung des menschlichen Grosshirns.* Braumüller, Vienna, 1878.

Gyri or **Convolutions of Heschl:** *Eponym* for the transverse temporal gyri of the cerebral cortex (Fig. H11b).

Hesselbach, Adam Kaspar (1788–1856): German anatomist and surgeon, son of Franz Kaspar Hesselbach, born and working in Würzburg.

Corona mortis of Hesselbach: *Eponym* for the inconstant aberrant obturator artery, originating from the inferior epigastric artery and coursing over the pecten pubis into the lesser pelvis. Reference: *Ueber den Ursprung und Verlauf der unteren*

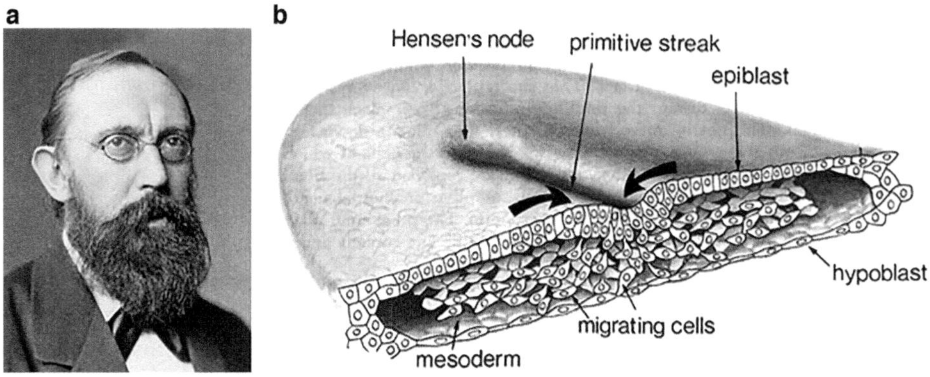

Fig. H9 (**a**) Photo of Hensen (from Pagel 1901); and (**b**) Ingression of mesoderm and entoderm (*arrows*) during gastrulation in the chick embryo. Gastrulation begins with the formation of the primitive streak, a region of proliferating and migrating cells. Future mesodermal and endodermal cells migrate through the primitive streak into the interior of the embryo. The endodermal cells replace the hypoblast. At the rostral end of the primitive streak, the node of Hensen forms (after Balinsky BJ *An Introduction to Embryology,* 2nd ed. Saunders, Philadelphia, PA, 1965; from ten Donkelaar HJ, Lammens M, Hori A *Clinical Neuroembryology,* 3rd ed. Springer, Cham, 2023)

Fig. H10 Photo of Hering (from the National Library of Medicine, Digital Collections; in public domain)

Bauchdeckenschlagader und der Hueftbeinlochschlagader. 1819.

Hesselbach, Franz Kaspar (1759–1816): German surgeon and anatomist born in Hammelburg, Bavaria, and working in Würzburg, father of Adam Kaspar Hesselbach. He is known for his studies on the inguinal canal and its herniations. Reference: *Anatomisch-chirurgische Abhandlung ueber den Ursprung der Leistenbrüche.* Baumgärtner, Würzburg, 1806. For biography, *see* Tubbs RS et al. *Franz Kaspar Hesselbach (1759–1816): Anatomist and surgeon.* World J Surg 2018; 32:2527–2529.

Fascia of Hesselbach: *Eponym* for the cribriform fascia, part of the fascia lata.

Ligament of Hesselbach: *Eponym* for the interfoveolar ligament, a slight thickening in the reduplicated transversal fascia at the medial margin of the deep inguinal ring (Fig. H12).

Triangle of Hesselbach: *Epomym* for a triangular area, which boundaries are the falx inguinalis cranially, the inferior epigastric vessels laterally, the lateral border of the rectus abdominis muscle medially and the inguinal ligament caudally. Here, a medial (direct) inguinal hernia protrudes. In laparoscopic surgery, this triangle forms a safe area for the introduction of cannulae (Baskett 2019).

Heubner, Otto (1843–1926): German pediatrician (Fig. H13) who described the long striate artery (the recurrent artery of Heubner). Reference: *Die luetische Erkrankung der Hirnarterien.* Vogel, Leipzig, 1874.

Recurrent artery of Heubner: *Eponym* for the long striate artery (Fig. H14).

Hey, William (1736–1819): English surgeon born in Pudsey, near Leeds, and working in Leeds.

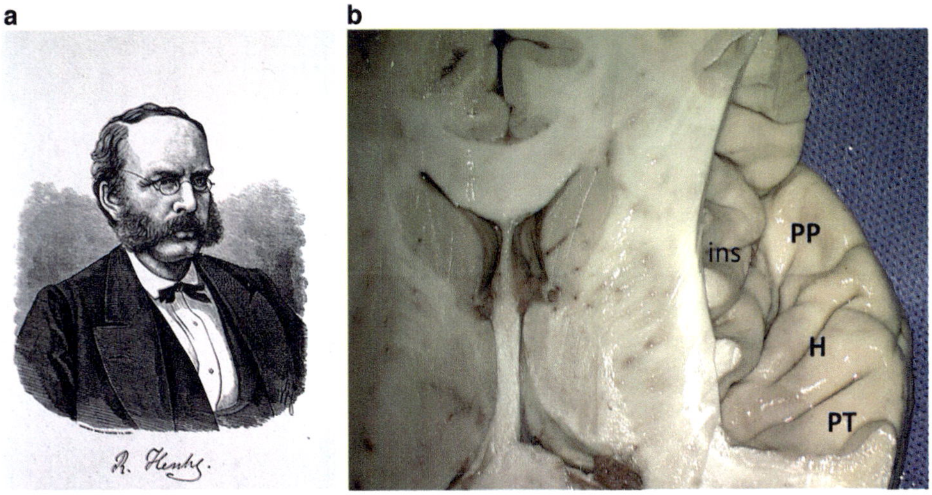

Fig. H11 (**a**) Painting of Heschl (from Wikimedia Commons; in public domain); and (**b**) Superior view of the right superior temporal sulcus (courtesy Robert Bartoš, Ústí nad Labem; from ten Donkelaar HJ, Kachlík D, Tubbs RS *An Illustrated Terminologia Neuroanatomica*. Springer, Cham, 2018). *ins* insula, *H* gyri of Heschl, *PP* planum polare, *PT* planum temporale

Ligament of Hey: *Eponym* for the superior horn of the falciform margin of the saphenous opening, also described in 1806 by Burns. Reference: *Practical Observations in Surgery*. London, 1803.

Protrusion of Cooper-Hey: *Eponym* for a femoral hernia with two sacs, also known as bilocular hernia, also described by Cooper.

Highmore, Nathaniel (1613–1685): English physician born in Fordingbridge, Hampshire, and working in Sherbourne. Although he did not discover the maxillary sinus, which was well known to all older anatomists, his name is still epitomized. Reference: *Corporis humani disquisitio anatomica*. Braun, The Hague, 1651.

Antrum of Highmore: *Eponym* for the maxillary sinus.

Body or Corpus of Highmore: *Eponym* for the mediastinum testis.

Hilgenreiner, Heinrich (1870–1954): Czech orthopedic surgeon born and working in Prague.

Line of Hilgenreiner: *Eponym* for a line drawn on pelvic radiographs to evaluate the acetabular angle.

Hilton, John (1805–1878): English surgeon born in Sible Hedingham, Essex, and working in London known for his studies on the laryngeal nerves and the anorectal canal (Fig. H15). Reference: *On the distribution and probable function of the superior and recurrent laryngeal nerves as demonstrated by dissection, in the human subject*. Guy's Hosp Rep 1837, ii, pp. 514–518.

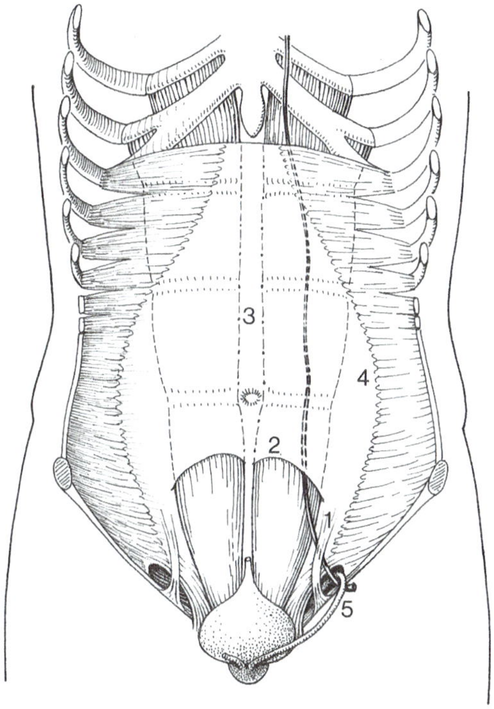

Fig. H12 The Hesselbach interfoveolar ligament (based on a figure by the late Chris van Huijzen). *1* interfoveolar ligament of Hesselbach, *2* arcuate line of Douglas, *3* linea alba, *4* semilunar line of van den Spieghel, *5* ductus deferens crossing the inferior epigastric artery

Muscle of Hilton: *Eponym* for the (inferior) aryepiglotticus muscle or the aryepiglottic part of the oblique arytenoid muscle of the larynx.

Sac of Hilton: *Eponym* for the saccule of the larynx, a small diverticle of the ventricle of the larynx.

Waterbed of Hilton: *Eponym* for the quadrigeminal cistern of the subarachnoid space.

Fig. H13 Photo of Heubner (from Pagel 1901)

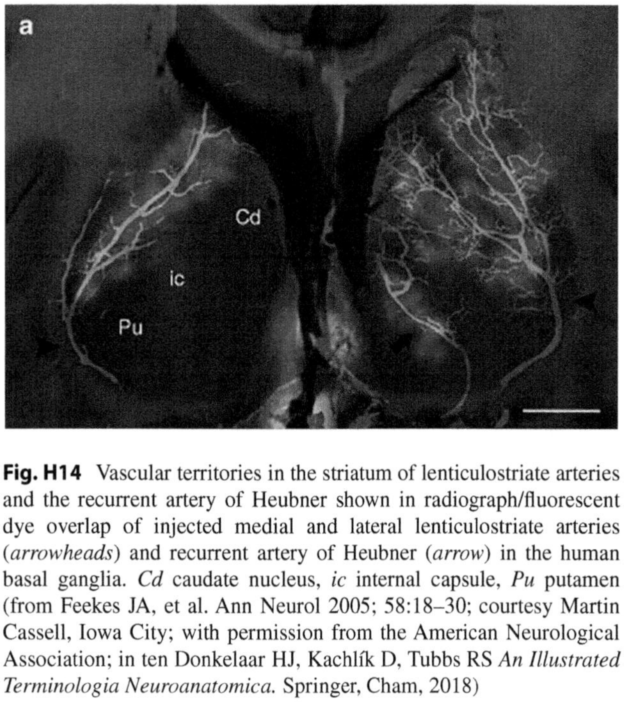

Fig. H14 Vascular territories in the striatum of lenticulostriate arteries and the recurrent artery of Heubner shown in radiograph/fluorescent dye overlap of injected medial and lateral lenticulostriate arteries (*arrowheads*) and recurrent artery of Heubner (*arrow*) in the human basal ganglia. *Cd* caudate nucleus, *ic* internal capsule, *Pu* putamen (from Feekes JA, et al. Ann Neurol 2005; 58:18–30; courtesy Martin Cassell, Iowa City; with permission from the American Neurological Association; in ten Donkelaar HJ, Kachlík D, Tubbs RS *An Illustrated Terminologia Neuroanatomica.* Springer, Cham, 2018)

White line of Hilton: *Eponym* for the anocutaneous line (the white line of the anus) between the anal canal and the anus itself (Fig. H16).

Hippocampus: The first description of the hippocampus is credited to Aranzio. In the *Anatomicarum Observationum Liber*, concerning the lateral ventricle, the choroid plexus and the hippocampus, he described the hippocampus or silkworm and its intraventricular location. In 1729, Duvernoy was the first to illustrate the hippocampus (*see* **Fig. D26**). He compared the protrusion on the floor of the temporal horn to a hippocampus or sea horse, although he hesitated between sea horse and silkworm. In 1732, Winslow suggested the term 'ram's horn' (Winslow 1752), for which de Garengeot (1742) introduced the term cornu Ammonis, after the Egyptian god Ammon (*see* also Engelhardt 2016; Dogan et al. 2022). Earlier illustrations of the hippocampus, mentioning only the horns of the fornix are those of Eustachi (Tabula XVII, Fig. V; Fig. H17a) and Vieussens (Fig. H17b). References: de Garengeot *Splanchnologie ou l'anatomie des viscères.* Paris, 1742; Lewis FT *The significance of the term hippocampus.* J Comp Neurol 1923; 35:213–230; Engelhardt E *Hippocampus discovery. First steps.* Dement Neuropsychol 2016; 10:58–62; Dogan E, Gungor A, Dogulu F, Türe U *The historical evolution of the fornix and its terminology: A review.* Neurosurg Rev. 2022; 45:979–988.

Fig. H15 Photo of Hilton (from the Wellcome Collection CCBY; in public domain)

Hirsch, I Seth (1880–1942): American physician and radiologist born and working in New York, where he was Director of the X-ray Department of the Beth Israel Hospital from 1929 until his death.

Sphincter of Hirsch: *Eponym* for a focal narrowing adjacent and superior to the ileocaecal valve. It may also involve the long segments of the ascending colon producing a coned caecum. Reference: *The cecocolic sphincteric tract.* Med J and Record 1924; 6:541–549.

Hirschfeld, Ludwig Moritz (1816–1876): Polish anatomist born in Rawa and working in Warsaw. Reference: *Névralogie et esthésiologie et iconographie du système*

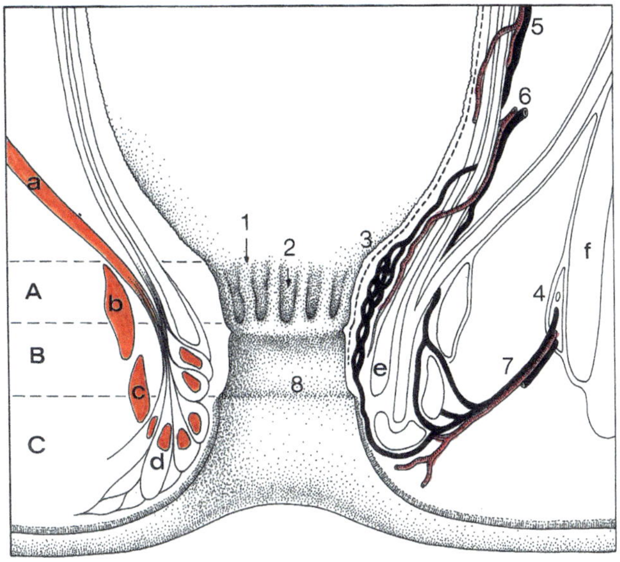

Fig. H16 Schematic frontal section through the anorectal canal (after ten Donkelaar HJ, Oostra R-J *Klinische Anatomie en Embryologie,* 4th ed. Springer Media, Houten, 2014). *A-C* zones of anal canal: *A* haemorrhoidal zone, *B* intermediate zone (pecten analis), *C* cutaneous zone, *a* levator ani muscle, *b-d*: parts of external sphincter ani muscle: *b* deep part, *c* superficial part, *d* subcutaneous part, *e* internal sphincter ani muscle, *1* anal column, *2* anal sinus, *3* submucosal venous plexus (in *black*), *4* pudendal canal with internal pudendal artery and vein, *5* superior rectal artery and vein, *6* middle rectal artery and vein, *7* inferior rectal artery and vein, *8* anocutaneous line (white line of Hilton)

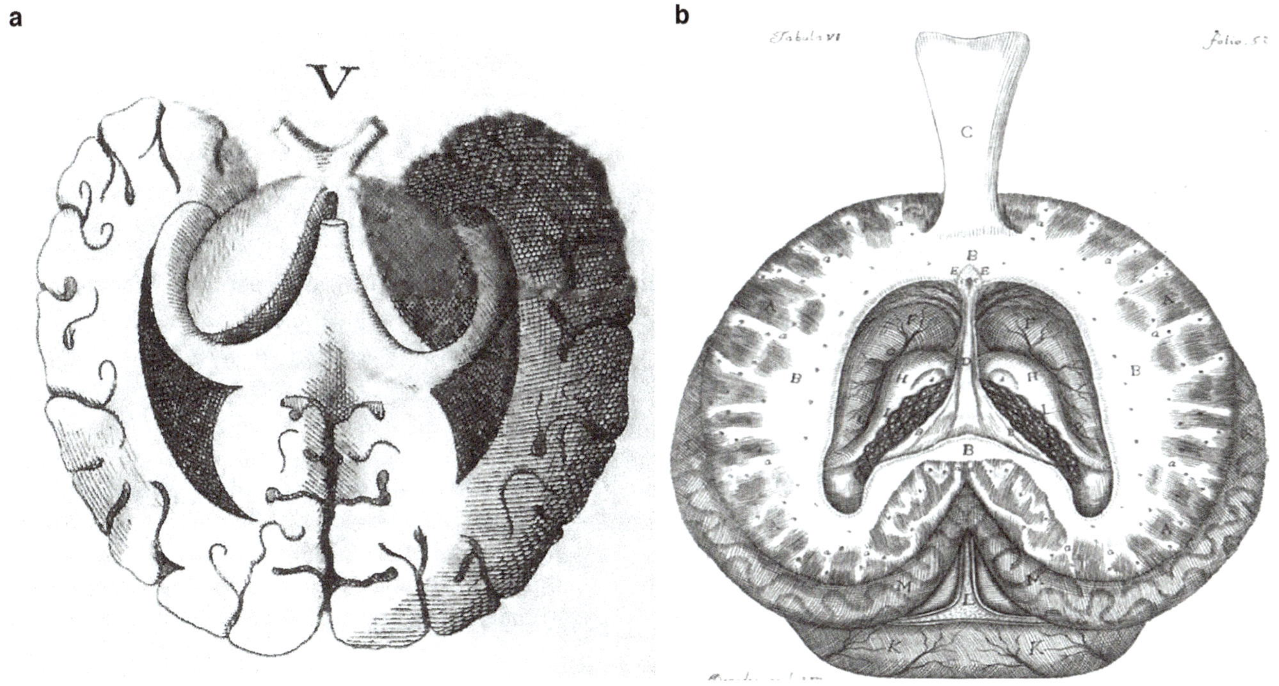

Fig. H17 Hippocampus: Illustrations by (**a**) Eustachi, showing a bilateral dissection seen from above with the median fornix bifurcating into what Lancisi (1714) named the cornua of the fornix (from Tabula XVIII, Fig. V), and (**b**) Vieussens: axial section of the brain. The hippocampus he called the posterior crus of the fornix (*D*; from *Neurographia Universalis*, 1684)

nerveux et des organes des sens de l'homme, avec leur mode de préparation. Paris, 1866.

Canal of Hirschfeld: *Eponym* for the tympanic canaliculus, also described by Jacobson and Sappey.

Ganglion of Hirschfeld: *Eponym* for the superior cervical ganglion of the sympathetic trunk.

Nerve of Hirschfeld: *Eponym* for the lingual branch of the facial nerve

Hirschsprung, Harald (1830–1916): Danish pediatrician born and working in Copenhagen (Fig. H18). He is known for describing congenital pyloric stenosis and congenital megacolon or intestinal aganglionosis, known as Hirschsprung disease.

His, Wilhelm Jr (1863–1934): Swiss anatomist born in Basel and working in Leipzig, Basel, Göttingen and Berlin (Fig. H19a). Reference: *Die Tätigkeit des embryonalen Herzens.* Leipzig, 1893.

Angle of His: *Eponym* for the cardiac notch of the stomach.

Bundle of His: *Eponym* for the atrioventricular bundle of the cardiac conducting system, also described by Kent and Paladino (Fig. H19b).

His, Wilhelm Sr (1831–1904): Swiss anatomist from Basel (Fig. H20a) working in Basel and Leipzig, noted for his studies on the embryological development of the nervous system (Fig. H20b, c). References: *Die Entwickelung des menschlichen Rautenhirns vom Ende des ersten bis zum Beginn des dritten Monats. I. Verlängertes Mark.* Abh Kön Sächs Ges Wiss Math Phys Kl 1890; 29:1–74; *Die Entwickelung des menschlichen Gehirns während der ersten Monate.* Hirzel, Leipzig, 1904. His Sr was the leading authority in preparing the first anatomical terminology (*Basler Nomina Anatomica* or BNA 1895). Reference: *Die anatomische Nomenklatur. Nomina anatomica.* Arch Anat Entwicklgesch 1895, Suppl, pp. 1–180.

Anastomosis of His: *Eponym* for a communicating loop between the right and left hypoglossal nerves, also described by Hyrtl.

Duct of His: *Eponym* for the embryonic thyroglossal duct, also described by Bochdalek, and earlier by Vater.

Isthmus of His: *Eponym* for the rhombencephalic isthmus, a narrowing of the brain stem between the midbrain and the hindbrain.

'Rautenleiste' of His: *Eponym* for the embryonic rhombic lip (**Fig. H21**).

Space of His, Perivascular: *Synonym* for the Virchow-Robin space.

Sulcus limitans of His: *Eponym* for the limiting sulcus (the sulcus limitans), a shallow groove in the wall of the neural tube, indicating the boundary between the alar and basal plates.

Fig. H18 Portrait of Hirschsprung (from Wikimedia Commons; in public domain)

'Zwischenstrang' of His: *Eponym* for the neural crest. Reference: *Untersuchungen über die erste Anlage des Wirbelthierleibes. Die erste Entwicklung des Hühnchen im Ei.* Vogel, Leipzig, 1868.

Hoboken, Nicolaas, van: *See* van Hoboken, Nicolaas.

Hoche, Alfred Erich (1865–1943): German psychiatrist born in Wilderhain and working in Strassburg and Freiburg (Fig. H22). He is known for his studies on variations of the pyramidal tract. References: *Ueber Variationen im Verlaufe der Pyramidenbahn.* Neurol Zbl 1897, 16:993–997; *Beiträge zur Anatomie der Pyramidenbahn und der oberen Schleife, nebst Bemerkungen über die abnorme Bündel in Pons und Medulla oblongata.* Arch Psychiatr Nervenkr 1898; 30:103–136.

Bundle of Hoche: *Eponym* for an aberrant tract of the pyramidal tract leaving it at the level of the facial nerve and crossing the midline to join the contralateral pyramidal tract.

Hochstetter, Ferdinand (1861–1954): Austrian anatomist and embryologist born in Hruschau (now Hrušov, part of Ostrava) and working in Vienna and Innsbrück, known for his extensive embryological studies of the human brain (Fig. H23). References: *Beiträge zur Entwicklungsgeschichte des menschlichen Gehirns, I. Teil.* Deuticke, Vienna, 1919; *II. Teil, Dritte Lieferung: Die Entwicklung des Mittel- und Rautenhirns.* Deuticke, Vienna, 1929.

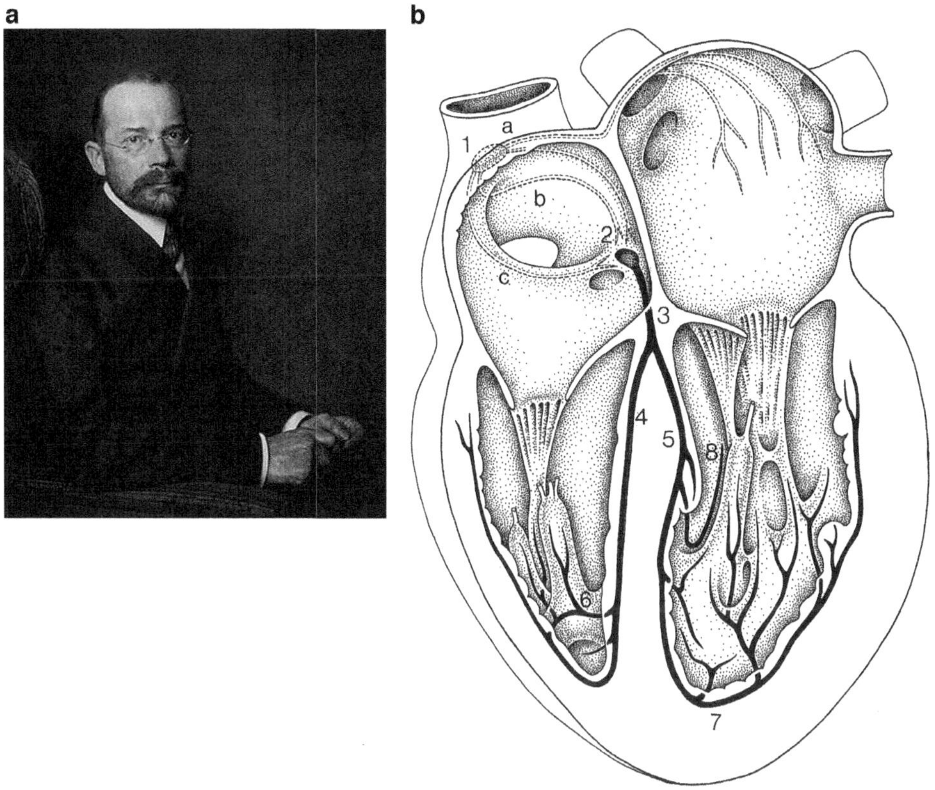

Fig. H19 (**a**) Photo of His Jr. (from Wikimedia Commons; in public domain); (**b**) bundle of His. *1* sinuatrial node, *2* atrioventricular node, *3* bundle of His, *4* crus dextrum, *5* crus sinistrum, *6* moderator band of crus dextrum, *7*, *8* anterior and posterior branches of crus sinistrum, *a-c* internodal bundles (after Agur AM 1991 *Grant's Atlas of Anatomy*, 9th ed. Williams & Wilkins, Baltimore, and ten Donkelaar HJ, Oostra R-J *Klinische Anatomie en Embryologie,* 4th ed. Springer Media, Houten, 2014)

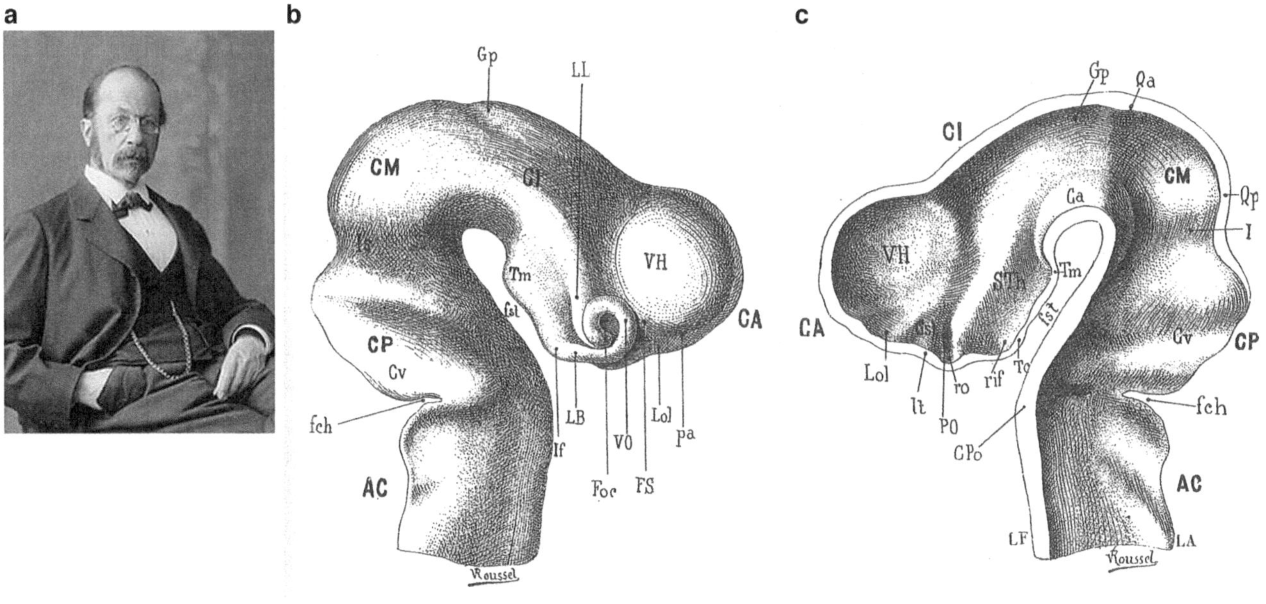

Fig. H20 (**a**) Photo of His Sr. (from Wikimedia Commons; in public domain); (**b**, **c**) drawings of a human embryo of 4 weeks: (**b**) lateral view, **c** medial view (redrawn by Roussel in Dejerine J *Anatomie des centres nerveux, Tome Premier.* Ruef, Paris, 1895). CA, CI, CM, CP anterior (telencephalon), intermediate (diencephalon), middle (mesencephalon) and posterior (hindbrain) parts of the brain

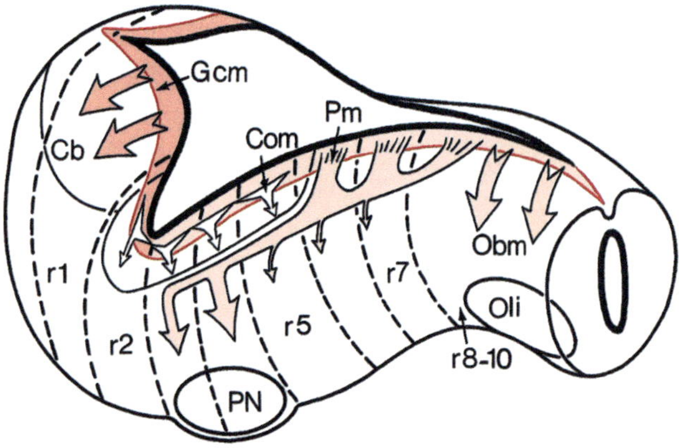

Fig. H21 'Rautenleiste' (rhombic lip) of His: Migratory pathways of the precerebellar nuclei, derivatives of the rhombic lip. The upper rhombic lip (in *medium red*) gives rise to the migration stream of granule precursor cells (*Gcm*) to the cerebellum (*Cb*). The lower rhombic lip (in *light red*) gives rise to the cochlear migration (*Com*) and to the pontine (*Pm*) and olivobulbar (*Obm*) migration streams to the cochlear, pontine (*PN*) and inferior olivary (*Oli*) nuclei. *r1-r10* rhombomeres (after Puelles López L, Martínez Pérez S, Martínez de la Torre M 2008 *Neuroanatomía*. Médica Panamericana, Buenos Aires, Madrid; courtesy Luis Puelles, Murcia; from ten Donkelaar HJ, Lammens M, Hori A *Clinical Neuroembryology,* 3rd ed. Springer, Cham, 2023)

Fig. H22 Photo of Hoche (from Wikimedia Commons; in public domain)

Plate of Hochstetter: *Eponym* for the epithelial layer between the developing medial and lateral nasal processes and the maxillary process.

Tubercle of Hochstetter: *Eponym* for the embryonic cerebellar tubercle (Fig. H24).

Hodge, Hugh Lennox (1796–1873): American obstetrician.

Planes of Hodge: *Eponym* for four planes perpendicular to the pelvic axis used in obstetrics. Reference: *The Principles and Practice of Obstetrics.* Blanchard, Philadelphia, 1864.

Hofbauer, Isfred Isidore (1878–1961): Austrian-American gynaecologist born in Vienna and working in Königsberg, Baltimore and Cincinnati.

Cells of Hofbauer: *Eponym* for macrophages in the placental villi contributing to the suppression of immune responses between mother and foetus. Reference: *Grundzüge einer Biologie der menschlichen Plazenta, mit besonderer Berücksichtigung der Fragen der fötalen Ernährung.* Braumüller, Vienna, 1905.

Hofer, Gustav (1887–1970): Austrian otorhinolaryngologist in Vienna and later in Graz.

Nerve of Hofer: *Eponym* for the depressor nerve, a branch of the superior laryngeal nerve to the cardiac plexus, also described by Cyon.

Hoffa, Albert (1859–1907): German orthopedic surgeon born in Richmond, South Africa, and working in Würzburg (Fig. H25a). Reference: *Lehrbuch der orthopädischen Chirurgie,* 3. Aufl. Stuttgart, 1898.

Fat pad of Hoffa: *Eponym* for the infrapatellar fat pad (Fig. H25b).

Hoffmann, Moritz (1622–1698): German anatomist born in Fürstenwalde and working in Altdorf. Reference: *Synopsis institutionum anatomicarum.* Altdorf, 1661.

Duct of Hoffmann: *Eponym* for the pancreatic duct, demonstrated in fowl, which according to Haller caused Wirsung to look for it in humans (Dobson 1962).

Hole of Tarin: *Eponym* for the interpeduncular fossa of the brain stem; see Tarin, Pierre.

Höhl, Erwin (1861–1917): German physician working in Chemnitz, Saxony.

Cells of Höhl: *Eponym* for bipolar cells localized in the subodontoblastic zone of the peripheral part of the tooth pulp that play a role in the formation of new odontoblast-like cells. Reference: *Beitrag zur Histologie der Pulpa und des Dentins.* Arch Anat Physiol 1896; 32:31–54.

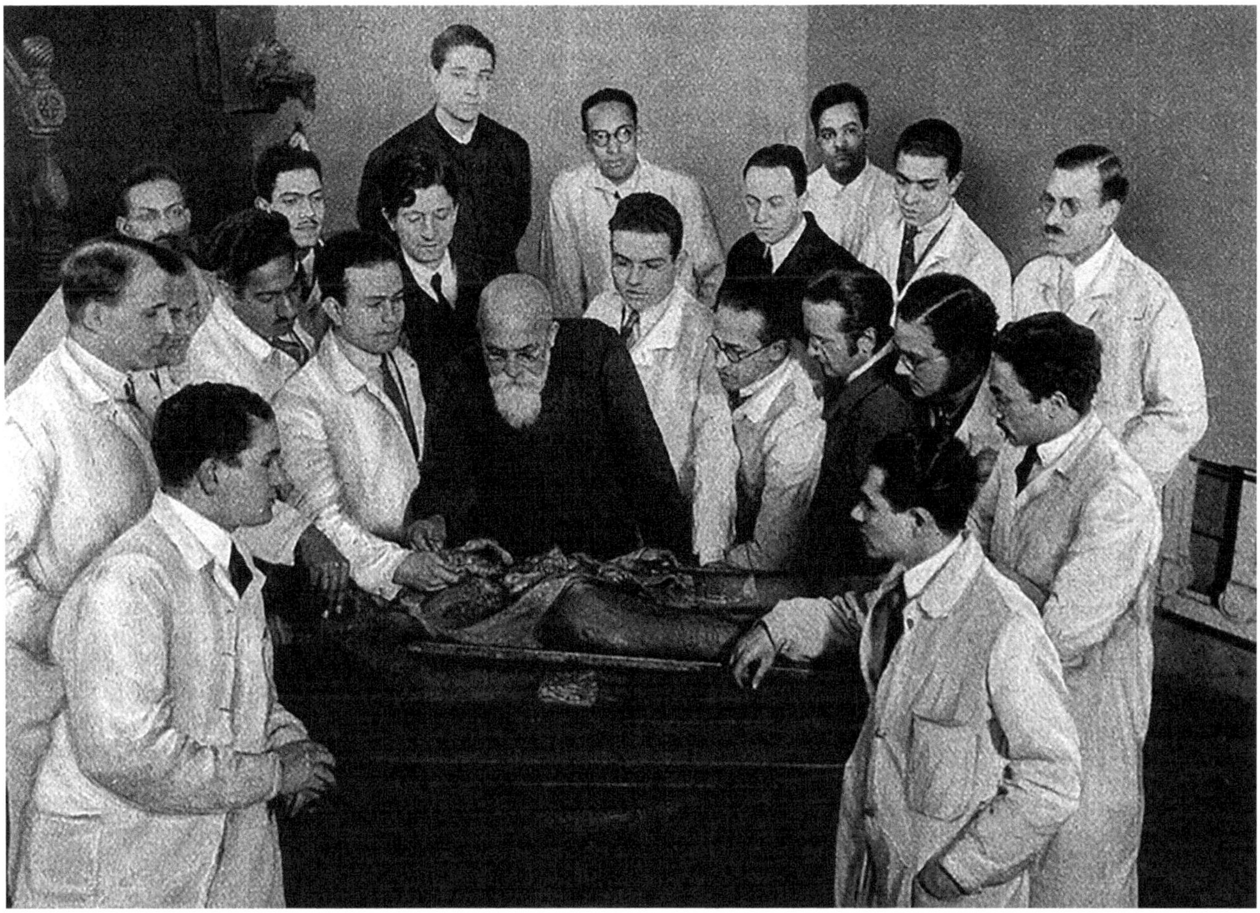

Fig. H23 Photo of Hochstetter's Anatomical Lesson (from the Wellcome Collection CCBY; in public domain)

Holl, Moritz (1852–1920): Austrian surgeon born in Vienna and working in Innsbrück and Vienna.

Ligament of Holl: *Eponym* for the intracrural ligament joining the two cavernous bodies of the clitoris in front of the external urethral opening. Reference: *Zur Homologie und Phylogenie der Muskeln des Beckenausganges des Menschen.* Anat Anz 1896; 12:57–71.

Holtshouse, Carsten (1810–1901): English surgeon.

Hernia of Holtshouse: *Eponym* for an inguinal hernia with extension of the loop of intestine along the inguinal ligament.

Home, Sir Edward (1756–1832): English surgeon born in Hull and working in London.

Lobe of Home: *Eponym* for the median lobe of the prostate. Reference: *On a small lobe of the human prostate gland which has not before been taken notice of by anatomists.* Phil Trans 1806; 96:195–204; *Practical Observations on the Treatment of Diseases of the Prostate Gland.* London, 1811.

Hopewell-Smith, Arthur (1865–1931): English dental histologist born in Boston, Lincolnshire, and working in London.

Hyaline layer of Hopewell-Smith: *Eponym* for the layer of mineralized hyaline material in the developing tooth root, at the dentinocemental junction. References: *Dental Microscopy.* Dental Manufacturing Co, London, 1895; *An Introduction to Dental Anatomy and Physiology, Descriptive and Applied.* Lea & Febiger, Philadelphia, New York.

Horne, Jan van: *see* van Horne, Jan.

Horner, William Edmonds (1793–1853): American anatomist born in Warrington, Virginia, and working in Philadelphia.

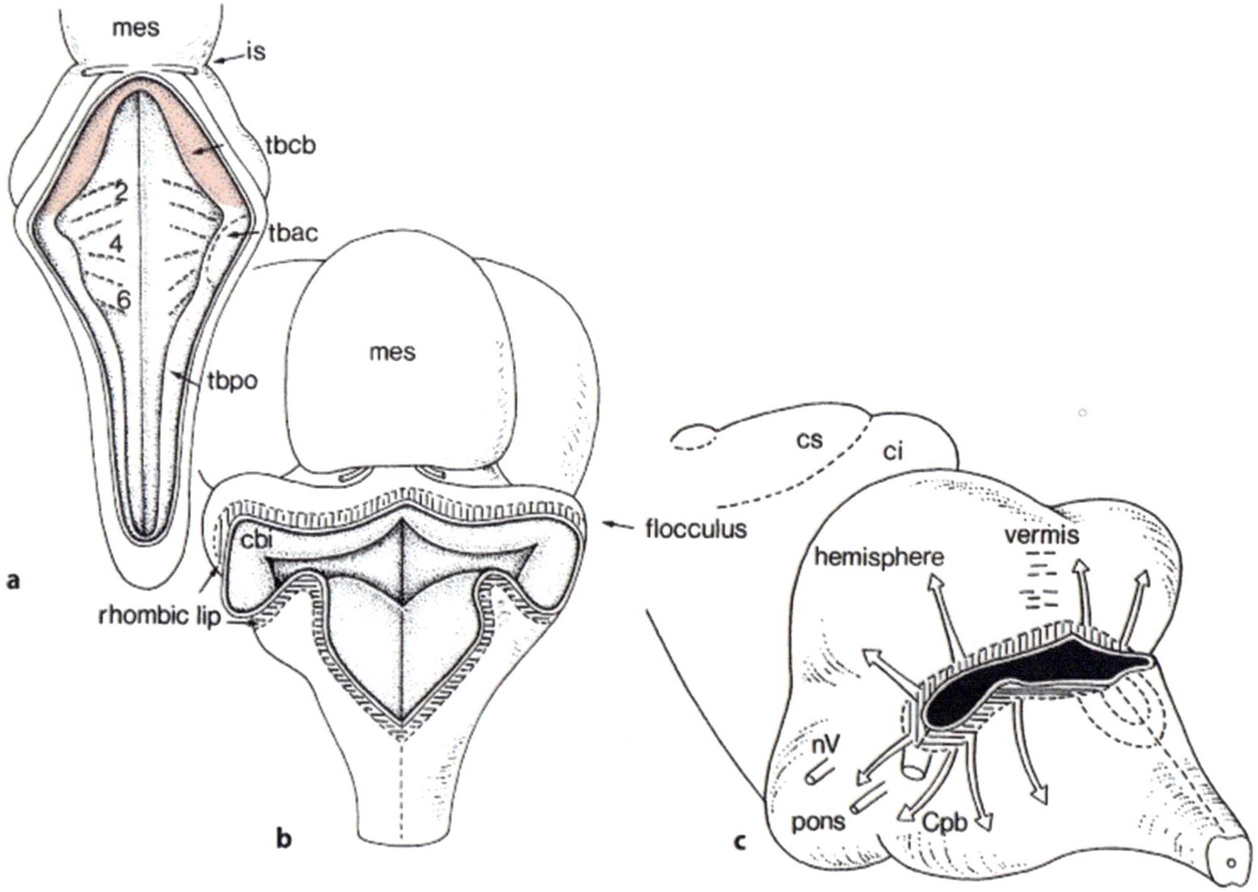

Fig. H24 Embryonic (**a**, **b**) and foetal (**c**) development of the human cerebellum: (**a**) at approximately 4 weeks; (**b**) at the end of the embryonic period (the first 8 weeks); (**c**) at 13 weeks of development. The V-shaped tuberculum cerebelli (*tbcb*) is indicated in *gray*, and the upper and lower rhombic lips in *light red* and *red*, respectively. *cbi* internal cerebellar bulge, *ci* colliculus inferior, *Cpb* corpus pontobulbare, *cs* colliculus superior, *is* isthmus, *IO* inferior olive, *mes* mesencephalon, *nV* trigeminal nerve, *tbac* tuberculum acusticum, *tbpo* tuberculum pontoolivare, *2*, *4*, *6* rhombomeres (from ten Donkelaar HJ, Kachlík D, Tubbs RS *An Illustrated Terminologia Neuroanatomica.* Springer, Cham, 2018)

Fig. H25 (**a**) Photo of Hoffa (from Wikimedia Commons; in public domain); (**b**) fat pad of Hoffa (in *yellow*). In *blue*, the condyles of the femur and the patella are shown, in *green* the patellar ligament (after Čihak R *Anatomie 1.* Grada, Prague, 2011; courtesy Ondřej Naňka, Prague)

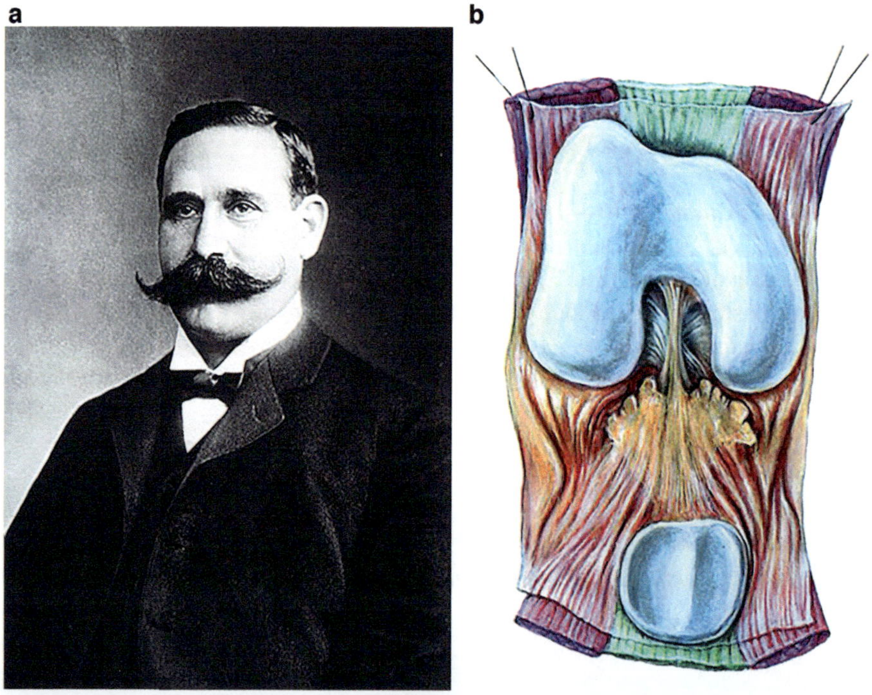

Muscle of Horner: *Eponym* for the lacrimal (deep) part of the orbicularis oculi muscle, previously described by Duverney. Reference: *Description of a small muscle at the internal commissure of the eyelids.* Phil J Med Phys Sci 1824; 8:70–80.

Hortega, Pio del Rio (1882–1944): Spanish neurohistologist born in Portilla, Valladolid, and working in Paris and Oxford. Following the bombing of Britain in World War II, he emigrated to Argentina in 1942.

Cells of Hortega: *Eponym* for microglia.

Houston, John (1802–1845): Irish physician and surgeon born in Northern Ireland and working in Dublin (Fig. H26).

Folds of Houston: *Eponym* for the transverse rectal folds. Reference: *Observations on the mucous membrane of the rectum.* Dublin Hospital Reports 1830, 5:158. These folds were also described by Nélaton and Kohlrausch. Usually, the superior fold is attributed to Houston, the middle fold to Kohlrausch and the inferior fold to Nélaton (*see* **Fig. K11**).

Fold or **Valve of Houston:** *Eponym* for the superior transverse rectal fold.

Muscle of Houston: *Eponym* for a variant of the bulbospongiosus muscle in the penis (compressor venae dorsalis penis muscle). Reference: *An account of the newly discovered muscle for compressing the dorsal vein of the penis.* Dublin Hospital Reports 1830, 5:58.

Hovelacque, André Édouard Emile (1880–1939): French anatomist born and working in Paris (Fig. H27a), known for his extensive work on the peripheral nervous system (Fig. H27b). Reference: *Anatomie des nerfs crâniens et rachidiens et du système grand sympathique chez l'homme.* Doin, Paris, 1927.

Plexus of Hovelacque: *Eponym* for the superior hypogastric plexus.

Howell, William Henry (1860–1945): American physiologist born in Baltimore and working in Michigan, Harvard and Johns Hopkins University (Fig. H28).

Howell-Jolly bodies: *Eponym* for nonextruded nuclei from orthochromatic erythroblasts, also described by Jolly. Reference: *The life-history of the formed elements of the blood, especially the red blood corpuscles.* J Morphol 1890; 4:57–116.

Fig. H26 Portrait of Houston (from the Wellcome Collection CCBY; in public domain)

Howship, John (1781–1841): English surgeon working in London.

Lacunae of Howship: *Eponym* for resorption lacunae in bone and dentin, occupied by osteoclasts and probably formed by their erosive action.

Hoyer, Heinrich (1834–1907): Polish anatomist and histologist born in Inowrazlaw and working in Warsaw.

Anastomoses of Hoyer: *Eponym* for arteriovenous anastomoses, also described by Sucquet.

Hrdlička, Aleš (Alois Ferdinand) (1869–1943): Czech anthropologist born in Humpolec who in 1881 moved with his parents to the USA. In 1903, he became the first Curator of Physical Anthropology at what now is known as the Smithsonian Institution National Museum of Natural History (Fig. H29). In 1916, he founded the American Journal of Physical Anthropology.

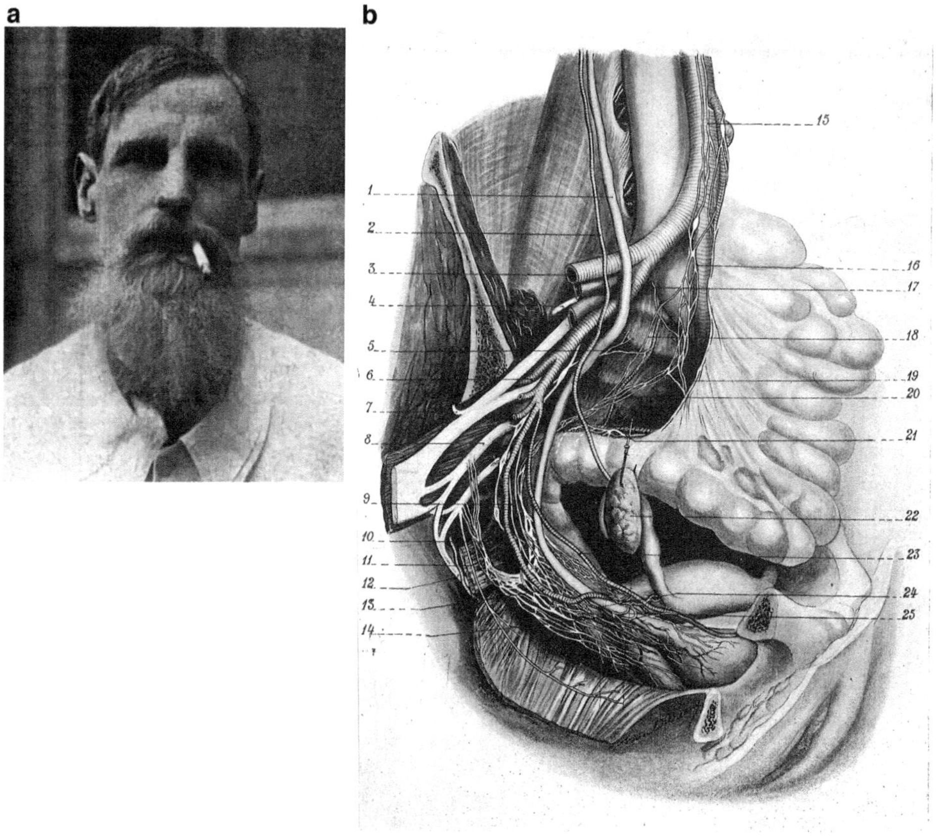

Fig. H27 (**a**) Photo of Hovelacque (from Wikimedia Commons; in public domain); (**b**) the hypogastric plexus in the female (Planche CXX from Hovelacque A *Anatomie des nerfs crâniens et rachidiens et du système grand sympathiue chez l'homme*. Doin, Paris, 1927). Some abbreviations: *1* ureter, *9* sympathetic fibres to the inferior hypogastric plexus to the uterovaginal plexus, *10* secondary fibres of the hypogastric plexus to the uterovaginal plexus, *13* uterine artery, *16* superior hypogastric plexus, *19* left inferior hypogastric plexus, *20* right inferior hypogastric plexus

Fig. H28 Photo of Howell (from Wikimedia Commons; in public domain)

Fig. H29 Photo of Hrdlička. (Courtesy Ondřej Naňka, Prague)

Hubrecht, Ambrosius Arnold Willem (1853–1915): Dutch zoologist and comparative anatomist in Leiden and Utrecht, born in Rotterdam.

Protocordal knot of Hubrecht: *Eponym* for the primitive node of the early embryo, described earlier by Hensen. Reference: *Die Gastrulation der Wirbelthiere.* Anat Anz 1905; 26:353–366.

Hudson, Arthur Cyril (1875–1962): English ophthalmologist born in Bingley, Yorkshire.

Line of Hudson-Stähli: *Eponym* for a brownish line of iron deposition across the lower third of the cornea, also described by Stähli. Reference: *A note on certain peculiar pigmentary markings of the cornea.* Roy Lond Ophthalm Hosp Rep 1911; 18, Part 2:198.

Hueck, Alexander Friedrich (1802–1842): German anatomist and ophthalmologist born in Revel, Estonia, and working in Dorpat (now Tartu).

Ligament of Hueck: *Eponym* for the trabecular meshwork of the sclera, also known as the pectinate ligament, first described by Fontana. Reference: *Die Bewegung der Krystallinse.* Leipzig, 1841.

Hueter, Carl (1838–1882): German surgeon born in Marburg and working in Rostock and Greifswald.

Triangle of Hueter: *Eponym* for the equilateral triangle, formed by the medial and lateral epicondyles of the humerus and the olecranon of the ulna in a 90° flexed elbow.

Huger, WE Jr (twentieth century): American plastic surgeon.

Perfusion zones of Huger, Abdominal: *Eponym* for three abdominal skin perfusion zones of the anterolateral abdominal wall. Reference: *The anatomic rationale for abdominal lipectomy.* Am Surgeon 1979; 45:612–617.

Huguier, Pierre Charles (1804–1874): French surgeon, anatomist and gynaecologist born in Sezanne, Champagne, and working in Paris.

Canal of Huguier: *Eponym* for the anterior canaliculus of the chorda tympani, also described by Civinini; but *see* Mudry (Otol Neurotol 2014; 36:1115–1120) for critique.

Circle of Huguier: *Eponym* for an anastomosis around the isthmus of the uterus between both uterine arteries.

Sinus of Huguier: *Eponym* for the fossula fenestrae vestibuli.

Humphry, Sir George Murray (1820–1896): English surgeon born in Sudbury, Suffolk, and working in Cambridge. He founded the Journal of Anatomy.

Fig. H30 Portrait of John Hunter by John Jackson 1813, after Joshua Reynolds 1786 (from the Wellcome Collection CCBY)

Ligament of Humphry: *Eponym* for the anterior meniscofemoral ligament. Reference: *A Treatise on the Human Skeleton.* MacMillan, Cambridge, 1858.

Hunter, John (1728–1793): Scottish surgeon, anatomist, physiologist and pathologist born in Long Calderwood, near Glasgow, and working in London (Fig. H30), the younger brother of William Hunter.

Canal of Hunter: *Eponym* for the adductor canal of the thigh.

Gubernaculum of Hunter: *Eponym* for the gubernaculum testis. Reference: *Observations on the State of the Testis in the Foetus and on the Hernia Congenita.* Medical Commentaries 1762, Part I:75–88.

Lines of Hunter-Schreger: *Eponym* for alternating dark and clear lines, generally oriented perpendicular to the surface of the enamel, representing sections of neighbouring zones of crossed enamel prisms, also described by Schreger (*see* **Fig. 23.2**).

Vein of Hunter, Perforating: *Eponym* for a perforating vein in the adductor canal, although not described by Hunter himself but named after the canal of Hunter (*see* **Fig. D16**).

Fig. H31 Portrait of William Hunter by Allan Ramsay (from the Wellcome Collection CCBY; in public domain)

Hunter, William (1718–1783): Scottish anatomist and obstetrician born in Long Calderwood, near Glasgow, and working in London (Fig. H31), the older brother of John Hunter.

Ligament of Hunter: *Eponym* for the round ligament of the uterus.

Line of Hunter: *Eponym* for the linea alba.

Membrane of Hunter: *Eponym* for the mucosa of the pregnant uterus (deciduous membrane). Reference: *The Anatomy of the Human Gravid Uterus.* Baskerville, Birmingham, 1774, illustrated by the Dutch artist Jan van Rymsdyk.

Huschke, Emil (1797–1858): German anatomist born in Weimar and working in Jena known for his studies on the skull (Fig. H32). Reference: *Schädel, Hirn und Seele des Menschen und der Thiere nach Alter, Geschlecht und Race.* Mauke, Jena, 1854.

Fig. H32 Photo of Huschke (from Wikimedia Commons; in public domain)

Cartilage of Huschke: *Eponym* for the vomeronasal cartilage, also described by Jacobson.

Foramen of Huschke: *Eponym* for an inconstant opening in the floor of the bony part of the external acoustic meatus near the tympanic membrane.

Teeth of Huschke: *Eponym* for the interdental cells of the spiral organ (*see* **Fig. C21**).

Valve of Huschke: *Eponym* for the prominent lower margin of the opening of the lacrimal canaliculi into the lacrimal sac, opposite the valve of Rosenmüller.

Huxley, Thomas Henry (1825–1895): English anatomist born in Ealing, London, working as surgeon in London (Fig. H33) and Charles Darwin's strongest advocate..

Layer of Huxley: *Eponym* for the middle epithelial layer of cells in the inner sheath of a hair root. Reference: *On a hitherto undescribed structure in the human hair sheath.* London Med Gaz 1845; 36:1340–1342.

Plane of Huxley: *Eponym* for a horizontal plane from the nasion (the midline of the nasofrontal suture) to the basion (the rostral border of the foramen magnum), also known as basicranial axis of Huxley.

Hydatid of Morgagni: *Eponym* for the vesicular appendage of the uterine tube, a remnant of the embryonic mesonephric duct; *see* Morgagni, Giovanni Battista.

Fig. H33 Photo of Huxley (from Pagel 1901)

Hymen: *Eponym* for the vaginal membrane, a thin piece of mucosal tissue that partly covers the vaginal opening. After the Hellenistic God of marriage.

Hyrtl, Joseph (József) (1810–1894): Austrian-Hungarian anatomist born in Eisenstadt and working in Prague and later in Vienna (Fig. H34a). References: *Lehrbuch der Anatomie des Menschen mit Rücksicht auf physiologische Begründung und praktische Anwendung.* Friedrich Ehrlich, Prague, 1846 (Fig. H34b); *Handbuch der topographischen Anatomie.* Braumüller, Vienna, 1847; *Onomatologia anatomica. Geschichte und Kritik der anatomischen Sprache der Gegenwart mit besonderer Berücksichtigung ihres Barbarismen, Widersinnigkeiten, Tropen und anatomische grammatischen Fehler.* Braumüller, Vienna, 1880.

Anastomosis of Hyrtl, Intercostobrachial: *Eponym* for a communicating loop between the lateral cutane-

Fig. H34 (**a**) Photo of Hyrtl (from Pagel 1901); (**b**) Title page *Lehrbuch der Anatomie des Menschen mit Rücksicht auf physiologische Begründung und praktische Anwendung.* Friedrich Ehrlich, 1846. (Courtesy Ondřej Naňka, Prague)

a

b

Lehrbuch

der

Anatomie des Menschen,

mit Rücksicht

auf

physiologische Begründung und praktische Anwendung.

Von

Dr. Jos. Hyrtl,

Professor der Anatomie etc.

Prag, 1846.

Verlag von Friedrich Ehrlich.

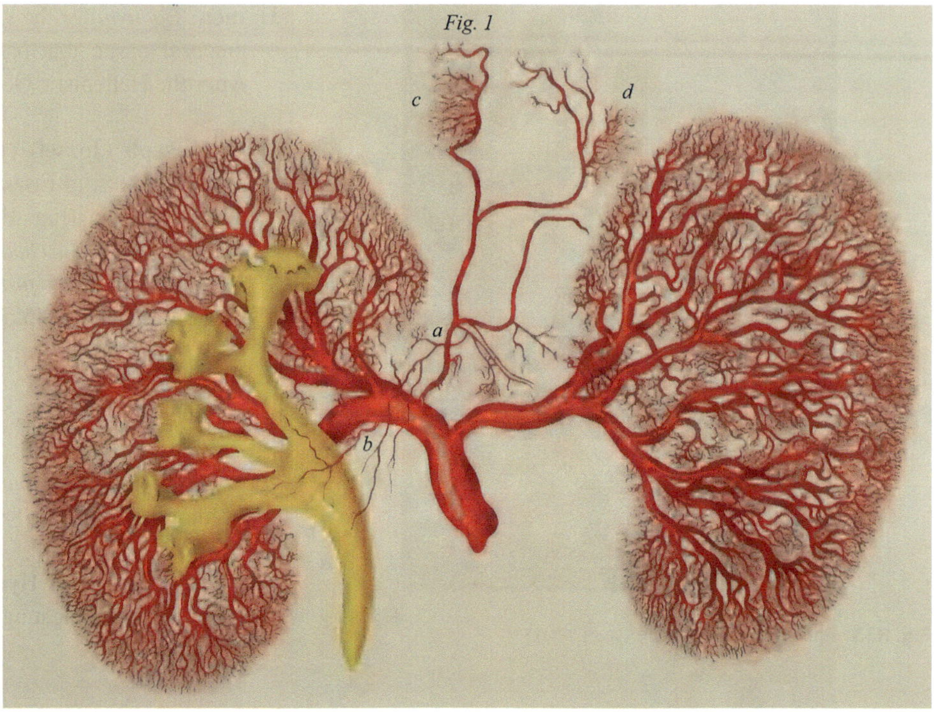

Fig. H35 Bloodless plane of the kidney as described by Hyrtl in a corrosion cast of a left human kidney. Based on the two regularly formed main branches of the renal artery, the kidney can be divided in a ventral and a dorsal shell. The renal pelvis is shown in *yellow*; *a* prominent capsular artery, *b* nutrient branch to the renal pelvis, *c*, *d* two wedge-shaped pieces of renal cortex supplied by the capsular artery (from Hyrtl J *Das Nierenbecken der Säugethiere und des Menschen.* Denkschr Kais Akad Wiss Wien 1872; 31:107–140; courtesy Michael Pretterklieber, Graz)

ous branch of the second or third intercostal nerve and the medial brachial cutaneous nerve.

Anastomosis of Hyrtl, Umbilical: *Eponym* for an anastomosis between the two umbilical arteries. Reference: *Die Blutgefässe der menschlichen Nachgeburt in normalen und abnormed Verhältnissen.* Braumüller, Vienna, 1870. This anastomosis is normally developed in placentas from small for gestational age infants (Ullberg U, Lingman G, Ekman-Ordeberg G, Sandstedt Acta Obstet Gynecol 2003; 82:716–721).

Anastomosis or **Loop of Hyrtl:** *Eponym* for a communicating loop between the right and left hypoglossal nerves, also described by His Sr.

Arch or **Arcade of Hyrtl:** *Eponym* for the gastroomental or gastroepiploic arcade, an anastomosis between the right and left gastroomental or gastroepiploic arteries along the greater curvature of the stomach.

Bloodless plane: *Eponym* for a relatively bloodless plane of the kidney between the segments supplied by the anterior and posterior bundles of the renal artery (Fig. H35), also described by Brödel (*see* **Fig. B39b**).

Canal of Hyrtl: *Eponym* for the musculofibular or musculoperoneal canal between the flexor hallucis longus muscle and the fibula containing the fibular or peroneal vessels. Reference: *Über normale und abnormale Verhältnisse der Schlagadern des Unterschenkels.* Vienna, 1864.

Fascia of Hyrtl: *Eponym* for the fascia transversalis, also described by Velpeau.

Fissure of Hyrtl: *Eponym* for the tympanomeningeal fissure between the round window niche and the meninges of the posterior cranial fossa.

Foramen of Hyrtl: *Eponym* for an occasional foramen in the sphenoidal bone for the motor portion of the trigeminal nerve (his foramen crotaphytico-buccinatorius).

Muscle of Hyrtl: *Eponym* for the iliopsoas muscle.

Ophthalmopetrous sinus of Hyrtl: *Eponym* for an accessory sinus that drains veins of the orbit at the superior orbital fissure to the superior petrosal or transverse sinus.

Plexus of Hyrtl: *Eponym* for the venous plexus beneath the isthmus of the thyroid gland.

Recess of Hyrtl: *Eponym* for the epitympanic recess of the tympanic cavity.

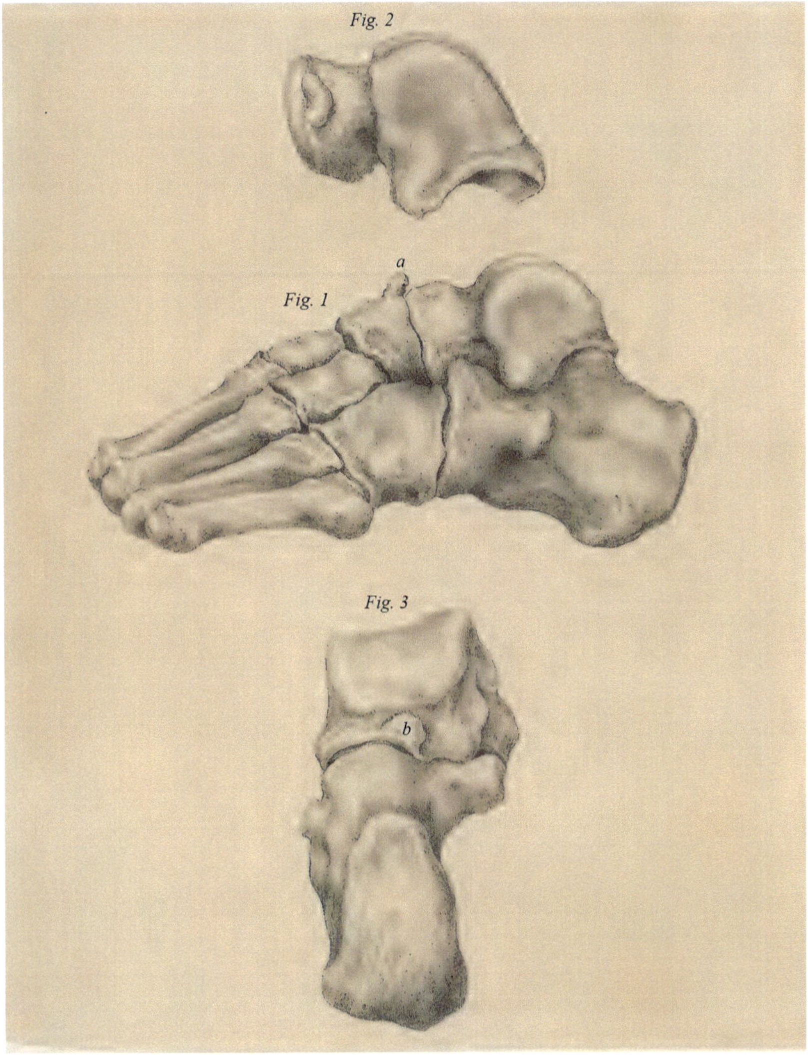

Fig. H36 Trochlear processes of the foot as described by Hyrtl. Fig. 1 shows the lateral aspect with a trochlear process on the navicular bone (**a**), serving as hypomochlion for the tendon of the extensor hallucis longus muscle. In Fig. 2, a lateral trochlear process on an isolated talus is shown, and in Fig. 3, the medial tubercle of the posterior process of the talus is elongated (**b**) and forms a trochlear process for the tendon of the flexor hallucis longus muscle (from Hyrtl J *Über die Trochlearfortsätze der menschlichen Knochen*. Denkschr Kais Akad Wiss Wien 1860; 18:141–155; courtesy Michael Pretterklieber, Graz)

Trochlear processes of Hyrtl: *Eponym* for bony spurs mostly acting as hypomochlia (Fig. H36). Reference: *Über die Trochlearfortsätze der menschlichen Knochen*. Denkschr Kais Akad Wiss Wien 1860; 18:141–155.

Valve of Hyrtl, Spiral: *Eponym* for a fold of the lower part of the nasolacrimal duct, also described by Taillefer.

Vein of Hyrtl: *Eponym* for the femoropopliteal vein.

10 Letters I, J

Ihering, Hermann von (1850–1930): German zoologist and geologist born in Kiel who studied in Berlin and Göttingen. He became Director of the Museu del Estado den Sao Paulo, Brazil (Fig. I1).

Plane of Ihering: Eponym for a preliminary version of the Frankfurt-Virchow plane or 'Frankfurter Horizontale'. Ihering used the centre of the bony external acoustic meatus as a reference point (see Fig. P19). Reference: *Ueber das Wesen der Prognathie und ihr Verhältnis zur Schädelbasis*. Arch Anthropol 1872; 5:359–407.

Imlach, Francis (1819–1891): Scottish anatomist and surgeon in Liverpool (Fig. I2).

Fat pad of Imlach: *Eponym* for adipose tissue surrounding the round ligament of the uterus within the inguinal canal.

Ring of Imlach: *Eponym* for the part of the inguinal canal that contains the round ligament of the uterus. Reference: *On shortening the round ligament of the uterus*. Edinb Med J 1885; 30:913.

Inca bone: *Eponym* for an inconstant interparietal bone, present at the site of the former posterior fontanelle, and named because of its relatively high occurrence among the Inca population in South America.

Incisures:

Incisures of Duverney: *Eponym* for the incisures in the cartilaginous part of the external acoustic meatus, also described by Rivinus and Santorini.

Incisure of Lantermann: *See* Incisure of Schmidt-Lantermann.

Incisure of Rivinus: *Eponym* for the tympanic notch inside the external acoustic meatus; *see* Rivinus, Augustus Quirinus.

Incisure of Santorini: *Eponym* for notches in the cartilaginous part of the external acoustic meatus, also described by Duverney and Rivinus; *see* Santorini, Giovanni Domenico.

Incisure of Schmidt-Lantermann: *Eponym* for intersegmental clefts in the myelin sheaths of peripheral nerves (Fig. I3); *see* Schmidt, Henry D and Lantermann, AJ.

Ingrassia, Giovanni Filippo (1510–1580): Italian anatomist and physician born in Regalbuto, near Palermo, working in Naples and later in Padua (Fig. I4).

Wing or **Apophysis of Ingrassia:** Eponym for the lesser wing of the sphenoidal bone. Reference: In Galeni librum de ossibus commentaria, iconibus insignata, quibus appositus est textus graecus cum nova in latinum versione. Maringhi, Parma, 1603 (published by his grandson Nicola Ingrassia).

Ionesco, Thoma or Jonnesco, Thomas (1860–1926): Rumanian surgeon and anatomist born in Ploesti and working in Bucharest.

Fossa of Ionesco: *Eponym* for the duodenojejunal fossa of the peritoneal cavity, a term sometimes applied also for the retroduodenal fossa (Dobson 1962). Reference: *Anatomie topographique du duodénum et hernies duodénales*. Paris, 1889.

H. J. ten Donkelaar et al., *An Illustrated Guide to Anatomical Eponyms*, https://doi.org/10.1007/978-3-031-91664-9_10

Fig. I1 Photo of Ihering around 1920 (from Wikimedia Commons; in public domain)

Fig. I2 Photo of Imlach (from Wikipedia; in public domain)

Insula of Haller: A doubling of the thoracic duct for part of its thoracic course; *see* Haller, Albrecht von.

Interatrial bundle of Bachmann: *Eponym* for the interatrial bundle of the cardiac conducting system (Fig. I5); *see* Bachmann, Jean George.

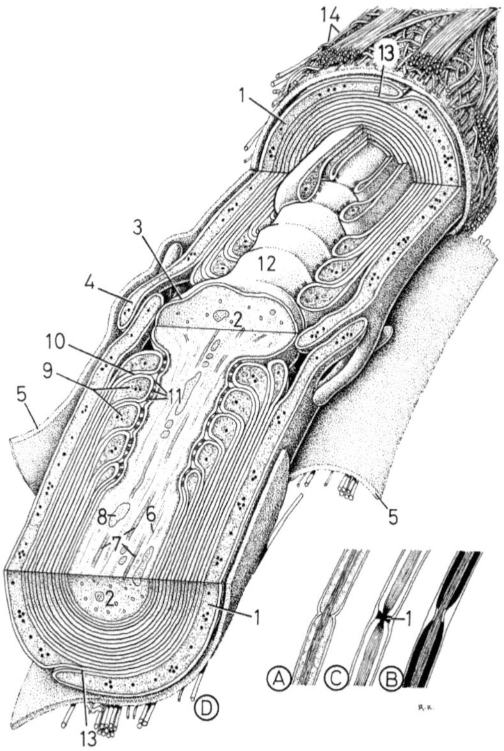

Fig. I3 The myelin sheath and the node of Ranvier (from Krstić R 1988 *Die Gewebe des Menschen und der Säugetiere*, 2nd ed. Springer, Berlin-Heidelberg-New York; with permission; from ten Donkelaar HJ, Kachlík D, Tubbs RS *An Illustrated Terminologia Neuroanatomica*. Springer, Cham, 2018). (**a–c**) light microscopy of the node of Ranvier in Azan staining (**a**), osmium fixation (**b**) and silver impregnation (**c**). (**d**) 3D structure *1* paranodal end parts of two Schwann cells, *2* axon, *3* axolemma, *4* fingerlike protrusions of a Schwann cell, *5* basal lamina, *6* neurofilaments, *7* neurotubuli, *8* endoplasmatic reticulum, *9* cytoplasm of Schwann cell protrusions, *10* internal mesaxon, *11* dense substance connecting the mesaxon to the axolemma, *12* axonal protrusions, *13* external mesaxon, *14* endoneural sheath

Internodal bundles (Fig. I5):

Internodal bundle of James: *Eponym* for the anterior internodal tract of the cardiac conducting system, from which the interatrial bundle of Bachmann arises; *see* James, Thomas Naum.

Internodal bundle of Thorel: *Eponym* for the posterior internodal tract of the cardiac conducting system. Reference: *Über den Aufbau des Sinusknotens und seine Verbindung mit der Cava superior und den Wenckebachschen Bündeln.* Münch Med Wochenschr 1910; 57:183–186; *see* Thorel, C.

Internodal bundle of Wenckebach: *Eponym* for the middle internodal tract of the cardiac conducting system.

Fig. I4 Engraving of Ingrassia (from Wellcome Images, Wellcome Library London; in public domain)

Reference: *Beiträge zur Kenntniss der menschlichen Herztätigkeit.* Arch Anat Physiol 1907; 1:1–2; *see* Wenckebach, Karel Frederik.

Iris: *Eponym* for the anular structure in the eye, responsible for controlling the diameter and size of the pupil, named after the ancient Greek mythological character, the personification of the rainbow.

Islands:

Islands or **Islets of Calleja:** *Eponym* for the olfactory or terminal islands (Fig. I6). Both the ventral striatum and the extended amygdala contain prominent collections of small neurons for which Sanides introduced the term insulae terminales (Sanides F 1957 J Hirnforsch 3:243–273). Calleja described islands of granular neurons in the olfactory tubercle of macrosmatic mammals (Calleja C 1893 *La region olfactoria del cerebro*. Moya, Madrid).

Island of Reil: *Eponym* for the insula, a lobe of the cerebral cortex; *see* Reil, Johann Christian.

Islets of Langerhans: *Eponym* for pancreatic islets; *see* Langerhans, Paul.

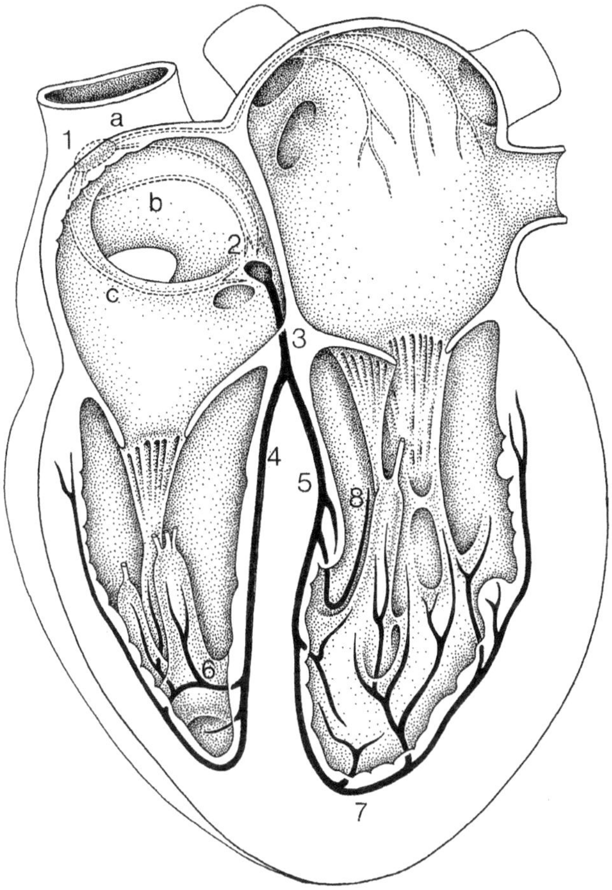

Fig. I5 The three internodal bundles connect the sinuatrial node with the atrioventricular node. *a* anterior internodal bundle of James, giving off the interatrial bundle of Bachmann to the left atrium, *b* middle internodal bundle of Wenckebach, *c* posterior internodal bundle of Thorel, *1* sinuatrial node, *2* atrioventricular node, *3* bundle of His, *4* crus dextrum, *5* crus sinistrum, *6* moderator band of crus dextrum, *7*, *8* anterior and posterior branches of crus sinistrum (after Agur AM *Grant's Atlas of Anatomy*, 9th ed. Williams & Wilkins, Baltimore, 1991; and ten Donkelaar HJ, Oostra R-J *Klinische Anatomie en Embryology,* 4th ed. Springer Media, Houten, 2014)

Isthmus of His: *Eponym* for the rhombencephalic isthmus, a narrowing of the brain stem between the midbrain and hindbrain; *see* His, Wilhelm Sr.

Ito, Toshio (1904–1991): Japanese physician born in Aichi and working in Kejo and later in Tokyo.

Cells of Ito: *Eponym* for hepatic perisinusoidal (fat-storing) cells in the perisinusoidal space of Disse (*see* **Fig. K16**). Reference: *Cytological studies on stellate cells of Kupffer and fat-storing cells in the capillary wall of human liver.* Acta Anat Nipponica 1951; 26:42.

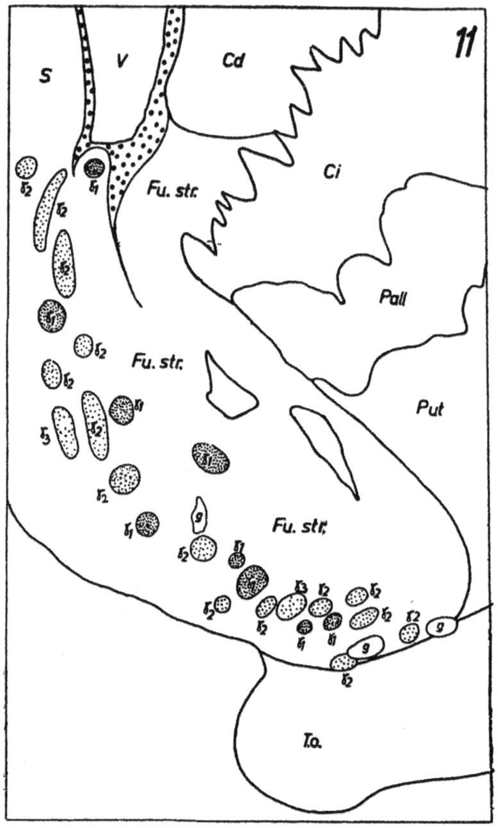

Fig. I6 The islands of Calleja, also known as 'insulae terminales' of Sanides in the fundus striati (*Fu.str.*) or accumbens nucleus (from Sanides F *Die Insulae terminales des erwachsenen Gehirns des Menschen.* J Hirnforsch 1957; 3:243–273; from ten Donkelaar HJ, Kachlík D, Tubbs RS *An Illustrated Terminologia Neuroanatomica.* Springer, Cham, 2018)

Chapter J

Jackson, Jabez North (1868–1935): American surgeon and anatomist born in Labadie, Missouri, and working in Kansas City (Fig. J1a).

Fold or **Membrane of Jackson:** *Eponym* for the pericolic membrane, a peritoneal fold or adhesion between the caecum or ascending colon and the right posterior abdominal wall (Fig. J1b). Reference: *On the developmental topography of the thoracic and abdominal viscera.* Anat Rec 1909; 3:361–396.

Jacob, Arthur (1790–1874): Irish ophthalmic surgeon born in Knockfin, near Maryborough, and working in Dublin.

Membrane of Jacob: *Eponym* for the retinal layer containing the rods and cones. Reference: *An account of a membrane in the Eye now first described.* Phil Trans 1819, pp. 300–307.

Jacobsohn, Louis, later known as Jacobsohn-Lask, Louis (1863–1941): German neurologist and neuroanatomist known for his studies on the human brain stem and spinal cord (Fig. J2). In his 1909 publication, he also first mentioned the pedunculopontine nucleus. References: *Über die Kerne des menschlichen Rückenmarks.* Neurol Zbl 1908; 27:617–626; *Über die Kerne des menschlichen Hirnstamms (Medulla oblongata, Pons und Pedunculus cerebri).* Abh K Preuss Akad Wiss, Phys-Math Kl 1909, Anhang, Abh I:1–70 (with XII Tables, published in Ranson SW, Clark SL *The Anatomy of the Nervous System,* 10th ed. Saunders, Philadelphia, PA, 1959).

Nucleus of Jacobsohn: *Eponym* for the posterior or dorsal paramedian nucleus of the brain stem.

Jacobson, Ludwig Levin (1783–1843): Danish anatomist and physician born and working in Copenhagen (Fig. J3), especially known for his discovery of the vomeronasal organ and its nerve (organ and nerve of Jacobson). References: *Description anatomique d'un organ observé dans les mammifères.* Ann Mus Hist Nat 1809; 18:1412; *Supplementa ad otoiatriam.* Nova Acta Soc Med Hafn 1818; 5:292.

Anastomosis of Jacobson: *Eponym* for a communication between the inferior ganglion of the glossopha-

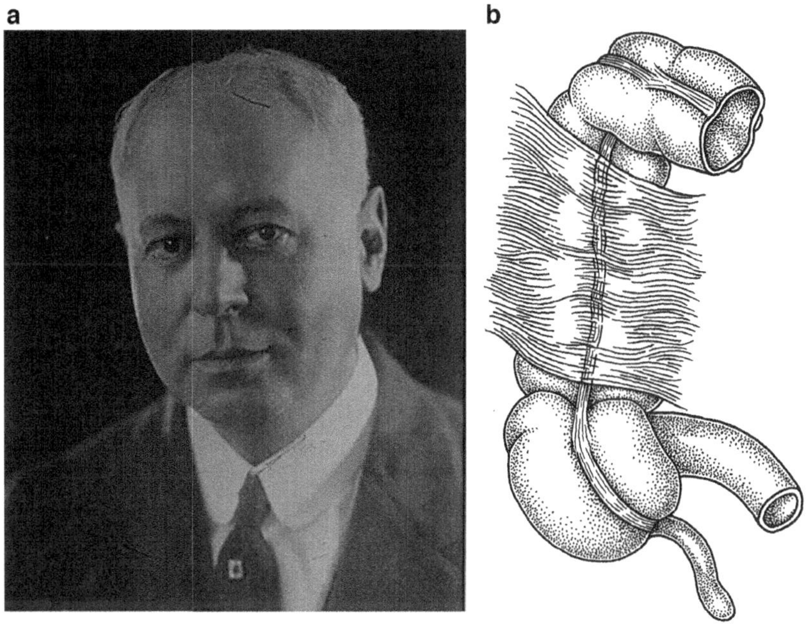

Fig. J1 (**a**) Photo of Jackson (from Wikidata; in public domain); (**b**) membrane of Jackson (after Hollinshead WH *Anatomy for Surgeons, Vol 2: The Thorax, Abdomen, and Pelvis,* 2nd ed. Harper & Row, New York, 1971)

Fig. J2 Photo of Jacobsohn (from Wikimedia Commons; in public domain)

Fig. J3 Portrait of Jacobson (from Wikipedia; in public domain)

ryngeal nerve via the tympanic nerve and plexus with the lesser petrosal nerve and the otic ganglion.

Canal of Jacobson: *Eponym* for the tympanic canaliculus, also described by Sappey and Hirschfeld.

Cartilage of Jacobson: *Eponym* for the vomeronasal cartilage, also described by Huschke.

Foramen of Jacobson: *Eponym* for the opening of the tympanic canaliculus in the temporal bone in the ridge between the jugular foramen and the carotid canal, through which the tympanic nerve from the glossopharyngeal nerve and inferior tympanic vessels enter the tympanic cavity.

Nerve of Jacobson: *Eponym* for (a) the tympanic nerve, earlier described by Andersch; and (b) the vomeronasal nerve (Fig. J4).

Organ of Jacobson: *Eponym* for the vomeronasal organ.

Plexus of Jacobson: *Eponym* for the tympanic plexus on the promontory of the labyrinthine wall of the tympanic cavity, formed by the tympanic nerve, a communicating branch of the facial nerve, and sympathetic branches from the internal carotid plexus via caroticotympanic nerves (Fig. J4).

Jacoby, Josef (1840-?): German ophthalmologist born in Elbing and working in Breslau.

Border tissue of Jacoby: *Eponym* for a layer of glial cells present at the circumference of the scleral lamina, separating the sclera from the axons traversing the optic nerve head. Reference: *Über die Neuroglia des Sehnerven.* Klin Monatsbl Augenheilkd 1905: 43:129–137.

Jacquement, Marcel (1872–1908): French anatomist in Marseille.

Recess of Jacquement: *Eponym* for the hepatocystic recess of the peritoneal cavity between the liver and the gallbladder. Reference: *Le recessus hépatocystique.* Marseille Méd 1897; 35:449.

Jacques, Paul (nineteenth century): French physician in Nancy.

Plexus of Jacques: *Eponym* for a nerve plexus within the muscle layer of the uterine tube. Reference: *Distribution et terminaisons des nerfs dans la trompe utérine.* Bibliogr Anat 1895, pp. 192–195.

James, Thomas Naum (1925–2010): American cardiologist born in Amory, Mississippi, and working in Birmingham, Alabama, and Galveston.

Bundle of James: *Eponym* for the anterior internodal bundle of the cardiac conducting system (see Fig. I5). References: *The connecting pathways between the sinus node and A-V node and between the right and left atrium.* Am Heart J 1963; 66:498–508; *The internodal pathways of the human heart.* Prog Cardiovasc Dis 2001; 43:495–535.

Jarjavay, Jean François (1815–1868): French anatomist and surgeon born in Savignac-les-Église (Dordogne) and working in Paris (Fig. J5).

Ligament of Jarjavay: *Eponym* for the sacrouterine fold, a peritoneal fold, containing the rectouterine muscle. Reference:*Traité d'anatomie chirurgicale, ou de l'anatomie dans ses rapports avec la pathologie externe et la médecine opératoire.* Paris 1852–1854.

Jelgersma, Gerbrandus (1859–1942): Dutch psychiatrist (Fig. J6) born in Doeveren and working in Leiden. He is known for his textbook on psychiatry (in Dutch) and his famous atlas of brain sections (Fig. J7). Reference: *Atlas Anatomicum Cerebri Humani.* Scheltema and Holkema, Amsterdam, 1931. His sections were used in Ranson and Clark's *Anatomy of the Nervous System.* Saunders, Philadelphia, 1963; and formed the basis for sections in Nieuwenhuys/Voogd/van Huijzen's *The Human Central Nervous System.* Springer, Heidelberg, New York, 1978.

Jensen, Julius (1841–1891): German psychiatrist born and working in Kiel and later in Berlin known for his studies of the gyri and sulci of the cerebral cortex. Reference: *Abhandlung über die Furchen und Windungen der menschlichen Grosshirnhemisphären.* Allg Z Psychiatr 1871; 27:473–516.

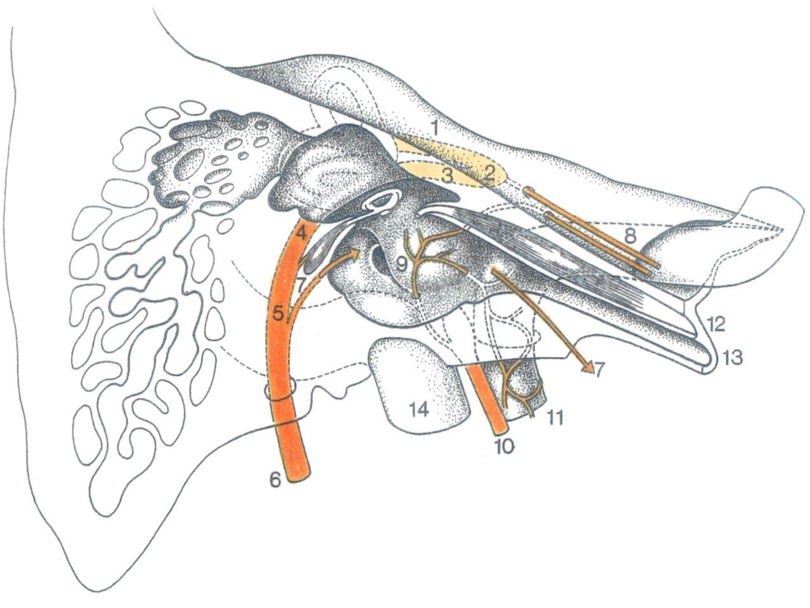

Fig. J4 The nerve and plexus of Jacobson. *1–6* components of facial nerve, *7* chorda tympani, *8* greater and lesser petrosal nerves, *9* tympanic plexus on the promontorium, receiving the nerve of Jacobson from the glossopharyngeal nerve, *10* glossopharyngeal nerve, *11* carotid plexus on the internal carotid artery, *12* semicanal with the tensor tympani muscle, *13* semicanal of the auditory tube (after ten Donkelaar HJ, Oostra R-J *Klinische Anatomie en Embryologie,* 4th ed. Springer Media, Houten, 2014)

Fig. J5 Portrait of Jarjavay (from Wikipedia; in public domain)

Fig. J6 Portrait of Jelgersma (painting by Lissy Ansingh, 1924; University of Leiden; in public domain)

Sulcus of Jensen: *Eponym* for the first intermediate sulcus of the cerebral cortex.

Jesenský, Jan (1870–1947): Bohemian stomatologist born and working in Prague.

Canal of Jesenský, Furcated: *Eponym* for an accessory interradicular root canal in molars.

Jobert de Lamballe, Antoine Joseph (1799–1867): French surgeon born in Matignon and working in Paris.

Fossa of Jobert de Lamballe: *Eponym* for the fossa in the popliteal region bounded proximally by the adductor magnus muscle and distally by the gracilis and sartorius muscles, best visible in a flexed knee and an externally rotated thigh.

Jolly, Justin Marie Jules (1870–1953): French hematologist and histologist born in Melun, Seine-et-Marne, and working at the Collège de France in Paris (Fig. J8).

Bodies of Howell-Jolly: *Eponym* for nonextruded nuclei from orthochromatic erythroblasts, described by both Howell and Jolly. Reference: *Traité technique d'hématologie. Morphologie—Histogenèse—Histophysiologie—Histopathologie.* Tomes I & II. Maloine, Paris, 1924.

Jonáš, Vratislav (1899–1968): Czech physician and cardiologist born and working in Prague.

Point or **Punctum of Jonáš:** *Eponym* for the punctum nervosum of the cervical plexus, located at the midpoint of the posterior border of the sternocleidomastoid muscle, slightly above the point of Erb.

Jonnesco, Thomas; *see* Ionesco, Thoma.

Joints:

Joint of Chopart: *Eponym* for the transverse tarsal joint; *see* Chopart, François.

Joint of Cruveilhier: *Eponym* for the median atlantoaxial joint; *see* Cruveilhier, Jean.

Joint of Lisfranc: *Eponym* for the tarsometatarsal joint; *see* Lisfranc, Jacques.

Jung, Karl Gustav (1793–1864): German anatomist born in Mannheim and working in Basel.

Muscle of Jung: *Eponym* for the pyramidal muscle of the auricle. Reference: *Von dem äussern Ohr und seinen Muskeln beim Menschen.* Verhandl Naturf Ges Basel Ber 1849; 8:54–73.

Juvara, Ernest (1870–1933): Romanian anatomist and physician born in Bârlad and, after studying in Paris with Farabeuf and Poirier, working in Paris, Bucharest and Iasi (Fig. J9). He had a special talent for drawing anatomical dissections (*see* Baran D *Medical significance of Professor*

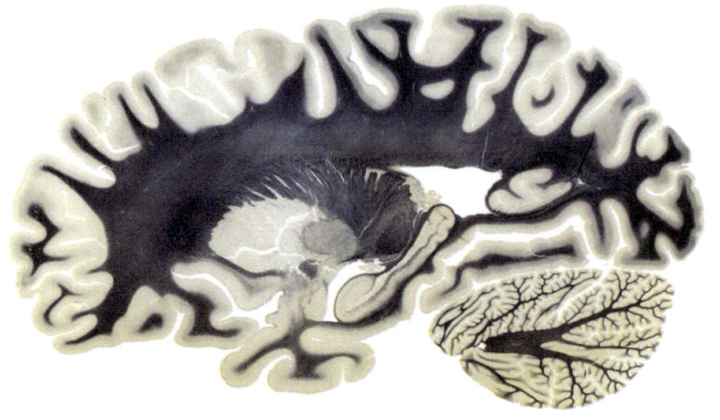

Fig. J7 Sagittal Weigert-stained section of the Jelgersma Collection showing the internal capsule and the hippocampus (from ten Donkelaar HJ *Clinical Neuroanatomy,* 2nd ed. Springer, Cham, 2020)

Fig. J8 Plaquette of Jolly (from Wikimedia Commons; in public domain)

Dr. Ernest Juvara's twelve years in Iasi. J Surg 2022; 18:77–96). Reference: *Leitfaden für die chirurgische Anatomie.* Hirschwald, Berlin, 1899.

Fibres of Rouget-Juvara: *Eponym* for phrenicoo-esophageal muscle fibres, also described by Rouget; usually, the ascending fibres in the thorax are called the fibres of Juvara and the descending fibres in the abdomen the fibres of Rouget.

Fig. J9 Photo of Juvara (from Wikimedia Commons; in public domain)

Septa of Leguen and Juvara: *Eponym* for the vertical septa supporting the palmar fascia on the metacarpal bones, also described by Leguen. Reference: *Des aponevroses de la paume de la main.* Bull Soc Anat Paris 1892; 6:383.

Pterygomaxillary region or **Zone of Juvara:** *Eponym* for a superficial layer of the head through which nerves and vessels pass. Reference: *Anatomie de la région ptérygomaxillaire.* Thesis, No 186, Ecole de Médecine, Paris, 1895. Juvara's work set a standard for many studies to follow (*see* Pessa JE *SMAS fusion zones determine the subfascial and subcutaneous anatomy of the human face: Fascial spaces, fat compartments, and models of facial aging.* Aesthetic Surg J 2016; 6:515–26).

Juxtaoral organ of Chievitz: *Eponym* for an epithelial band on the inner side of the buccinator muscle, originally described by Chievitz as a rudimentary structure. Zenker, however, showed that it persists and contains numerous sensory corpuscles densely innervated by the buccal nerve. Reference: Zenker W. *Juxtaoral Organ (Chievitz Organ).* Urban & Schwarzenberg, Munich-Baltimore, 1982.

11 Letter K

Kadyi, Henryk Karol Klemens (1851–1912): Austro-Hungarian anatomist born in Przemyšl and after training in Cracow and Vienna, working in Lemberg, Galicia (currently Lviv, Ukraine). For biography, see Wincewicz A, Turek-Pietrzykowska M *Scientific and organizational achievements of Professor of Anatomy Henryk Kadyi.* Rom J Morphol Embryol 2017; 58:1623–1629.

Central arteries of Kadyi: Eponym for the sulcomarginal arteries that enter the spinal cord through the anterior median fissure. Reference: *Über die Blutgefässe des menschlichen Rückenmarkes.* Gubrynowice and Schmidt, Lemberg, 1889.

Kaes, Theodor (1852–1913): German neurologist born in Amberg, working in Friedrichsberg (Fig. K1), known for the stripe of Bechterew-Kaes. Reference: *Die Grosshirnrinde des Menschen in ihren Massen und in ihren Fasergehalt.* Jena, 1907.

Stripe of Bechterew-Kaes: *Eponym* for a band of horizontal myelinated fibres in the most superficial part of the third layer of the isocortex, described by Bechterew and Kaes (*see* **Fig. B12**).

Kager, Hans (1910–1941): German surgeon born in Leipzig and working in Meerane, Saxony. He died during World War II.

Fat pad of Kager: *Eponym* for fat within the Kager triangle that in normal conditions appears lucent on radiographs.

Triangle of Kager: *Eponym* for a sharply marginated radiolucent triangle seen posteriorly on lateral radiographs of the ankle. It represents the Kager fat pad. Reference: *Zur Klinik und Diagnostik des Achillessehnenrisses.* Chirurg 1939; 11:691–695; *see also*: Somford MP, Hoogenborg D, Wiegerink JL *Kager's 'Bermuda triangle'.* J Foot Ankle Surg 2014; 53:503–504.

Kaplan Emanuel B (1894–1980): American surgeon born in Kremenchuk (Ukraine) who emigrated to the USA in 1924.

Anastomosis of Kaplan: *Eponym* for an inconstant communicating branch between the superficial and dorsal branches of the ulnar nerve, dorsal to the ulnar canal. Reference: *Variations of the ulnar nerve at the wrist.* Bull Hosp Jt Dis 1963; 24:85–88.

Fibres of Kaplan: *Eponym* for the attachments of the iliotibial tract to the distal femur, recognized as playing an important role in maintaining rotatory stability of the knee, particularly at greater degrees of knee flexion. Reference: *The iliotibial tract, clinical and morphological significance.* J Bone Joint Surg Am 1958; 46A:817–832.

Keith, Sir Arthur (1866–1955): Scottish anatomist born in Old Machar, Aberdeen, and working in London (Fig. K2). Reference: *The form and nature of the muscular connections between the primary divisions of the vertebrate heart.* J Anat (Lond) 1907; 31:172–189 (with Martin W. Flack).

Node of Keith-Flack: *Eponym* for the sinuatrial node of the cardiac conducting system (see **Fig. I5**).

Kent, Albert Frank Stanley (1863–1958): English physiologist born in Stratford Torrey, Willshire, and working in Manchester and Bristol who described the atrioventricular bundle in the same year as Wilhelm His Jr.

Bundle of Kent: *Eponym* for the atrioventricular bundle of the cardiac conducting system, usually known as the bundle of His and in Italy as the bundle

H. J. ten Donkelaar et al., *An Illustrated Guide to Anatomical Eponyms*, https://doi.org/10.1007/978-3-031-91664-9_11

Fig. K1 Photo of Kaes (from Wikimedia Commons; in public domain)

Fig. K2 Photo of Keith (from the Wellcome Collection CCBY; in public domain)

of Paladino. Kent described also an abnormal extra conduction pathway between the atria and the ventricles. Reference: *Researches on the structure and function of the mammalian heart.* J Physiol 1893; 14:233–254.

Kerckring, Dirk or Theodor (1638–1693): Dutch physician and anatomist born in Amsterdam, working in

Fig. K3 Portrait of Kerckring by Jurgen Ovens (1680; Hamburger Kunsthalle, Hamburg; in public domain)

Amsterdam and later in Hamburg (Fig. K3). Reference: *Opera Omnia Anatomica.* Haak & Luchtmans, Leiden, 1729 (Fig. K4).

Bone of Kerckring: *Eponym* for an isolated bone often found in the middle of the posterior margin of the foramen magnum.

Folds or **Valves of Kerckring:** *Eponym* for duodenal folds, circular folds of the small intestine, especially distinct in the duodenum and jejunum; already described by Falloppio.

Nodules of Kerckring: *Eponym* for nodules on the semilunar valve of the aorta.

Kernohan, James (1897–1981): Irish pathologist.

Notch of Kernohan: *Eponym* for an indentation in the cerebral peduncle due to displacement of the midbrain against the incisure of the tentorium cerebelli by a transtentorial herniation.

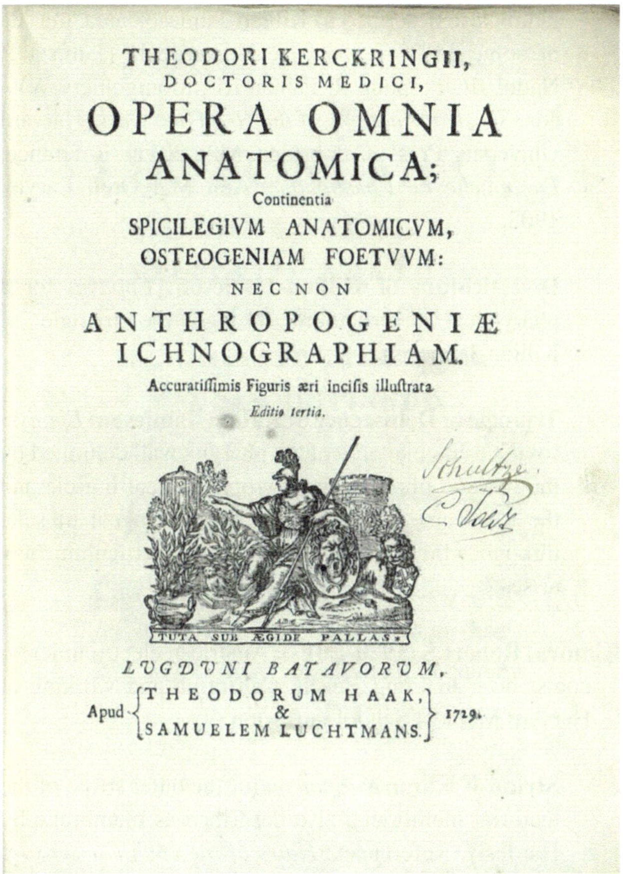
THEODORI KERCKRINGII,
DOCTORIS MEDICI,
OPERA OMNIA
ANATOMICA;
Continentia
SPICILEGIVM ANATOMICVM,
OSTEOGENIAM FOETVVM:
NEC NON
ANTHROPOGENIÆ
ICHNOGRAPHIAM.
Accuratiſſimis Figuris æri inciſis illuſtrata.
Editio tertia.

LVGDVNI BATAVORVM,
Apud THEODORUM HAAK, & SAMUELEM LUCHTMANS. 1729.

Fig. K4 Title page of Kerckring's *Opera Omnia Anatomica*. Haak & Luchtmans, Leiden, 1729

Key, Ernst Axel Henrik (1832–1901): Swedish pathologist born in Smaland and working in Stockholm (Fig. K5) who extensively cooperated with Gustav Magnus Retzius.

Lateral aperture or **Foramen of Key-Retzius:** *Eponym* for the lateral aperture of the fourth ventricle, also described by Luschka and Mierzejewsky. Reference: *Studien in der Anatomie des Nervensystems und des Bindegewebes*, 2 Vols. Samson and Wallin, Stockholm, 1875–1876.

Sheath of Key-Retzius: *Eponym* for the endoneurium, the connective tissue of peripheral nerves, also described by Henle.

Kiernan, Francis (1800–1874): Irish surgeon and physician working in London and known for his studies on the liver. Reference: *The Anatomy and Physiology of the Liver.* Phil Trans 1833, pp. 710–770.

Septa of Kiernan: *Eponym* for fibrous septa of the hepatic lobules (*synonym*: Portobiliary septa of Charcot).

Fig. K5 Photo of Key (from Pagel 1901)

Spaces or **Triangles of Kiernan:** *Eponym* for the interlobular triangles in the liver (Fig. K6; *synonym*: Portobiliary spaces of Charcot).

Veins of Kiernan: *Eponym* for the interlobular veins of the liver.

Kiesselbach, Wilhelm (1839–1902): German laryngologist and rhinologist born in Hanau and working in Erlangen (Fig. K7).

Locus (minoris resistentiae) or **Area of Kiesselbach:** *Eponym* for an area on the nasal septum, site of nasal haemorrhages, also described by Little. Reference: *Ueber spontane Nasenblutungen.* Berl Klin Wschr 1884; 24:375–377.

Kilian, Hermann Friedrich (1800–1863): German gynaecologist and obstetrician born in Leipzig, qualified in Edinburgh and working in St. Petersburg and later in Bonn.

Line of Kilian: *Eponym* for a transverse line marking the level of the promontory of the pelvis. Reference: *Geburtshilflicher Atlas.* Düsseldorf, 1835–1844.

Killian, Gustav (1860–1921): German laryngorhinologist born in Mainz and working in Freiburg and later in Berlin, known for his studies of the pharynx.

Fig. K6 Spaces or triangles of Kiernan (after ten Donkelaar HJ, Oostra R-J *Klinische Anatomie en Embryologie,* 4th ed. Springer Media, Houten, 2014). *1–3* in triangle of Kiernan: *1* interlobular artery, *2* interlobular vein, *3* interlobular bile duct, *4* intralobular artery, *5* liver sinusoids, *6* ventral vein, *7* liver cells, *8* interlobular connective tissue

Fig. K7 Photo of Kiesselbach (from Wikipedia; in public domain)

Bundle or **Muscle of Killian:** *Eponym* for the lowest cricopharyngeal fibres of the inferior constrictor of the pharynx, immediately above the oesophageal musculature, also described as Killian-Jamieson area, the site of some 35% of pharyngeal diverticula (Janfaza P, Nadol JB Jr., Galla R, Fabian RL, Montgomery WW, eds. *Surgical Anatomy of the Head and Neck.* Harvard University Press, Cambridge, MA, 2011). Reference: *La bouche de l'oesophage.* Ann Mal Oreil Larynx 1908.

Diverticulum of Killian-Jamieson: *Eponym* for a pharyngeal diverticulum through the triangle of Killian-Jamieson.

Triangle or **Dehiscence of Killian-Jamieson:** *Eponym* for the triangular area of the pharynx wall delimited by the oblique fibres of the thyropharyngeal muscle and the transverse fibres of the cricopharyngeal muscle, through which a Killian-Jamieson diverticulum may arise.

Kimura, Robert S (1920–2017): American electronmicroscopist born in Long Beach, California and working at Harvard Medical School in Boston.

Stripe of Kimura: *Eponym* for the outer stripe of the tectorial membrane, also described as membrane by Hardesty. References: *Hairs of the cochlear sensory cells and their attachments to the tectorial membrane.* Acta Otolaryngol 1966; 61:55–72; *The ultrastructure of the organ of Corti.* Int Rev. Cytol 1975; 42:173–222.

King, Thomas Wilkinson (1809–1849): English anatomist and surgeon in London born in Dover.

Band of King, Moderator: *Eponym* for the septomarginal trabecula, a muscular band in the right ventricle of the heart extending from the interventricular septum to the base of the anterior papillary muscle, already illustrated by Leonardo da Vinci, and described as ansiform band by Poirier. Reference: *An essay on the safety-valve function in the right ventricle of the human heart and on the gradation of this function in the circulation of warm-blooded animals.* Guy's Hosp Rep 1837; 2:104.

Kirchner, Wilhelm (1849–1936): Austrian otologist.

Diverticulum of Kirchner: *Eponym* for an inconstant diverticulum between the tensor and levator veli palatini muscles.

Klein, Eduard Emmanuel (1844–1925): Hungarian histologist, born in Ersec who after studying and working in Vienna moved to London.

Fig. K8 Photo of Augusta Marie Klumpke (from Wikimedia Commons; in public domain)

Muscle of Klein: *Eponym* for the musculus compressor labii, a marginal bundle of the lips or musculus rectus labii proprius, as also described by Aeby. Reference: *Zur Kenntniss des Baues der Mundlippen.* Vienna, 1869.

Klumpke, Augusta Marie (1859–1927): American neurologist born in San Francisco who moved as a child to Lausanne, and became the first female assistant of Alfred Velpian, head of neurology of the University of Paris (Fig. K8). Here, she published her paper on brachial plexus palsy, named after her Klumpke paralysis. In 1888, she married Joseph Jules Dejerine with whom she co-authored a two-volume book on the anatomy of the nervous system. Reference: *Contributions à l'études des paralysies radiculaires.* Rev. Méd 1885; 5:591–616.

Knoll, Philipp (1841–1900): Bohemian physiologist born in Karlsbad and working in Prague and later in Vienna.

Glands of Knoll: *Eponym* for glands in the vestibular folds of the larynx. Reference: *Ueber Reflexe auf die Athmung bei Zufuhr einiger flüchtiger Substanzen zu den unterhalb des Kehlkopfes gelegenen Luftwegen.* Sitz Akad Wiss Wien 1873; 68:245–273.

Knox, Robert (1791–1862): Scottish surgeon born in Edinburgh and working for a long time as army surgeon.

Foramen of Knox: *Eponym* for an inconstant supracondylar foramen of the humerus. Reference: *On the occasional presence of a supracondyloid process in the human humerus.* Edinb Med Surg J 1841; 56:125.

Kobelt, Georg Ludwig (1804–1857): German anatomist born in Baden and working in Freiburg (Fig. K9a).

Plexus of Kobelt: *Eponym* for the venous plexus of the clitoris (Fig. K9b). Reference: *Die männlichen und weiblichen Wollustorgane des Menschen und einiger Säugethiere in anatomisch-physiologischer Beziehung untersucht und dargestellt.* Freiburg, 1844.

Tubules of Kobelt: *Eponym* for remnants of the mesonephric tubules in females. Reference: *Der Nebeneierstock des Weibes.* Heidelberg, 1847.

Koch, Walter (1880–1931): German physician and pathologist born in Dortmund working in Berlin and later in Freiburg, known for his extensive studies of the heart. In German textbooks, his name is generally used for the atrioventricular node and its position. Reference: *Zur Anatomie und Physiologie der intrakardialen-motorischen Centren des Herzens.* Med Klinik 1912, pp. 108–112.

Node of Koch: *Eponym* for the atrioventricular node of the cardiac conducting system, a few years earlier described by Keith and Flack.

Triangle of Koch: *Eponym* for a triangular area in the wall of the right atrium of the heart, marking the position of the atrioventricular node.

Kocher, Theodor (1841–1917): Swiss surgeon born and working in Bern (Fig. K10), famous for his work on thyroid surgery. He received the Nobel Prize in Physiology and Medicine in 1909. Reference: *Chirurgische Operationslehre.* Fischer, Jena, 1892.

Vein of Kocher: *Eponym* for the middle thyroid vein, also described by Likhačeva..

Koganei, Yoshikiyo (1858–1944): Japanese histologist.

Clump cells of Koganei: *Eponym* for certain cells in the stroma of the iris thought to represent histiocytes. Reference: *Untersuchungen über den Bau der Iris der Menschen und der Wirbelthiere.* Arch Mikr Anat 1885; 25:1–10.

Kohlrausch, Otto Ludwig Bernhard (1811–1854): German physician born in Elberfeld working in Hannover. Reference: *Zur Anatomie und Physiologie der*

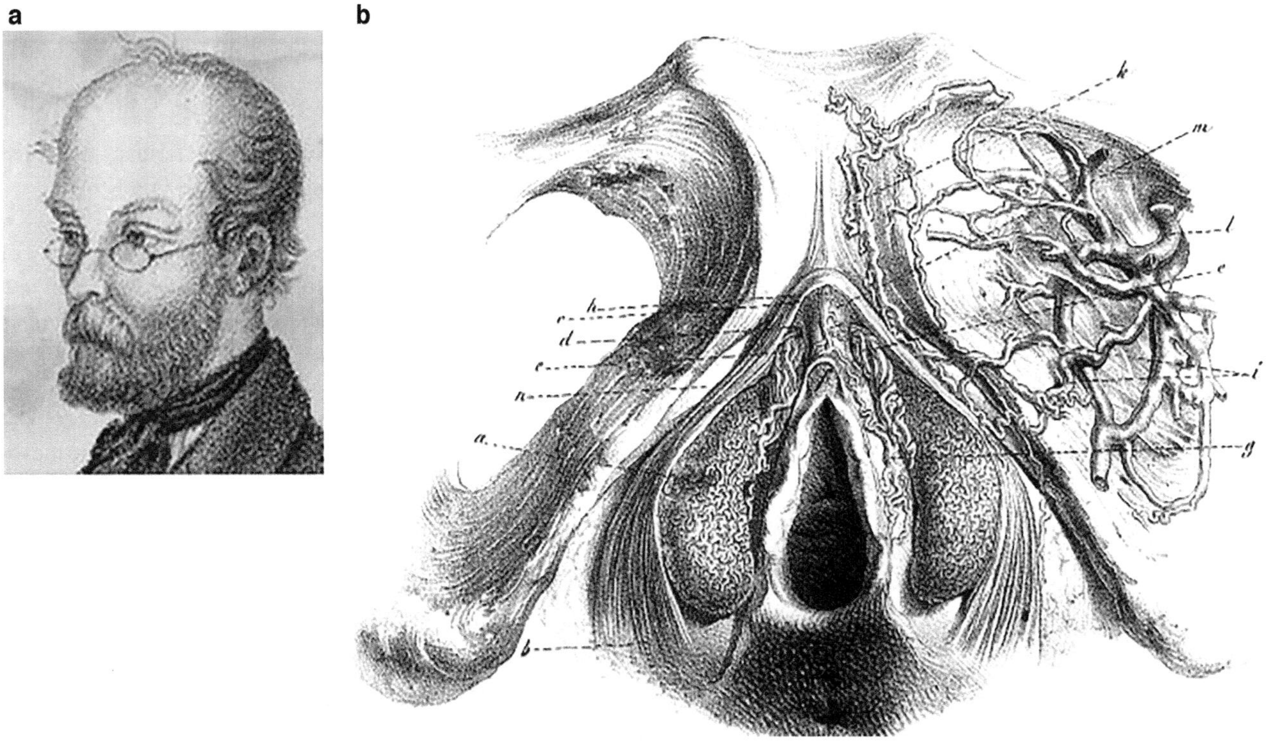

Fig. K9 (**a**) Drawing of Kobelt; (**b**) plexus of of Kobelt: Frontal view of erectile organs in a female: *a* bulbus vestibuli, *b* sphincter vaginae muscles, *e* venous plexus of intermediate part, *f* glans clitoridis, *g* connecting veins, *h* dorsal vein of clitoris, *k* veins passing beneath pubis, *l* obturator vein (both from Wikimedia Commons; in public domain)

Fig. K10 Photo of Kocher (Nobel Prize photo 1909; from Wikimedia Commons; in public domain)

Beckenorgane nebst naturgetreuer Abbildung der Längsdurchschnitte des männlicher und weiblicher Beckens. Leipzig, 1854.

Fold or **Valve of Kohlrausch:** *Eponym* for the middle transverse fold of the rectum, also described by Houston (Fig. K11).

Muscle of Kohlrausch: *Eponym* for longitudinal muscle fibres in the rectum; according to Olry (1995) his sustentaculum mucosae muscle is the corrugator cutis ani muscle of Ellis.

Kohn, Alfred (1867–1959): Bohemian histologist born in Libin and working in Prague (Fig. K12). He is remembered for his discovery of the nature and origin of the parathyroid glands and by his pioneering research on paraganglia.

Bodies of Kohn: *Eponym* for the paraganglia. Reference: *Ueber den Bau und die Entwicklung der*

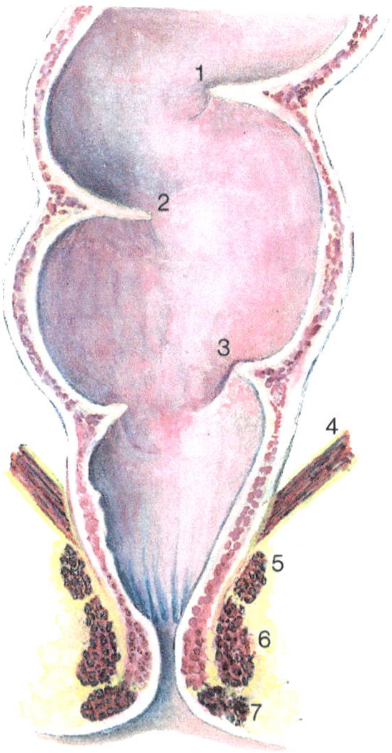

Fig. K11 Fold of Kohlrausch (after Čihak R *Anatomie 2*. Grada, Prague, 2013; courtesy Ondřej Naňka, Prague). *1* superior transverse rectal fold of Houston, *2* middle transverse rectal fold of Kohlrausch, *3* inferior transverse rectal fold of Nélaton, *4* levator ani muscle, *5–7* deep, superficial and subcutaneous parts of external sphincter ani muscle

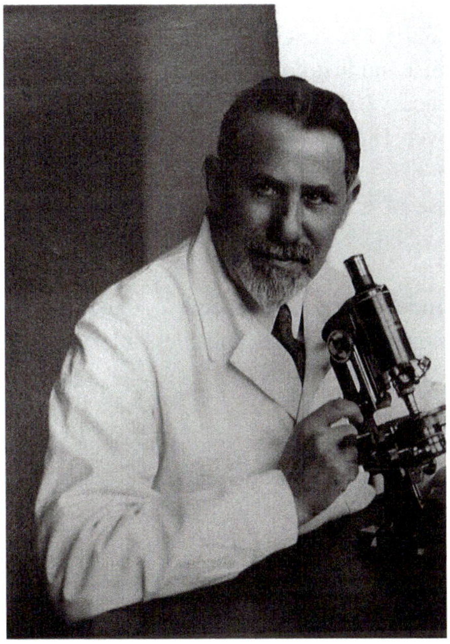

Fig. K12 Photo of Kohn (courtesy Ondřej Naňka, Prague)

Fig. K13 Photo of Kölliker (from Pagel 1901)

sogenannten CaroOtisdrüse. Arch Mikr Anat 1900; 56:81–148.

Kohn, Hans (1866–1935): German pathologist born in Wassertrüdingen, Bavaria, and working in Erlangen and Berlin.

Pores of Kohn: *Eponym* for discrete openings in the wall of adjacent alveoli. Reference: *Zur Histologie der indurierenden fibrösen Pneumonie*. Münch Med Wochenschr 1893; 40:42–45.

Kohnstamm, Oskar (1871–1917): German neurologist and psychiatrist known for his studies of the brain stem, the ascending pathways from the reticular formation and the salivatory nuclei in particular. References: *Der Nucleus salivatorius chordae tympani (nervi intermedii)*. Anat Anz 1902; 21:362–363; *Der Nucleus salivatorius inferior und das cranioviscerale System*. Neurol Zbl 1903; 22:699.

Nuclei of Kohnstamm: *Eponym* for the superior and inferior salivatory nuclei of the brain stem.

Kölliker, Rudolf Albert von (1817–1905): Swiss anatomist and histologist born in Zürich and working in Halle, Leipzig and Würzburg (Fig. K13) known for his extensive histological studies including the brain and the senses. Reference: *Handbuch der Gewebelehre für Ärzte und Studierende*. Engelmann, Leipzig 1852; 6th ed., 1889–1896, 2 Bde.

Cells of Kölliker: *Eponym* for (a) osteoclasts and b) spermatoblasts.

Dental crest of Kölliker: *Eponym* for the incisive bone (the premaxilla) upon which the incisors develop.

Glands of Kölliker: *Eponym* for the olfactory glands of the nasal cavity.

Layer of Kölliker: *Eponym* for the middle, connective tissue layer or stroma of the iris.

Muscle of Kölliker: *Eponym* for the dilator pupillae muscle, later also described by Grynfeltt.

Nucleus of Kölliker: *Eponym* for the intermediate substance of the spinal grey matter.

Nucleus of Kölliker-Fuse: *Eponym* for the subparabrachial nucleus of the brain stem, described together with Fuse.

Retinaculum of Kölliker: *Old term* for neuroglia.

Kolmer, Walter (1859–1931): Austrian histologist who described the epiplexus cell of the choroid plexus as macrophage-like cells on the epithelium of the choroid plexus of lower vertebrates. Reference: *Über eine eigenartige Beziehung von Wanderzellen zu der choroidal Plexus der Gehirne der Wirbeltiere.* Anat Anz 1921; 54:15–19.

Cell of Kolmer: *Eponym* for the epiplexus cell of the choroid plexus.

Konstantinovich, Vikentzi Bonifatiyevich (1845-?): Russian surgeon and anatomist in St. Petersburg. Reference: *Die Anordnung der Gefässe des Mastdarms.* St Petersb Med Zschr 1872–1873; 3:529–546.

Artery of Konstantinovich: *Eponym* for the dorsal rectal artery, a branch of the superior rectal artery.

Vein of Konstantinovich: *Eponym* for the marginal vein of the anus.

Korff, Karl von (1867–1956): German anatomist and histologist born in Hajen near Gröhnde and working first as ship's surgeon and later in Kiel and Tübingen.

Fibres of Korff: *Eponym* for argyrophilic strands of ground substance less mineralized than normal dentin with reticular microfibrils, originating in the tooth pulp. Reference: *Die Entwicklung der Zahnbeingrundsubstanz der Säugetiere.* Arch Mikr Anat 1906; 67:1–17.

Körner, Otto (1858–1935): German otorhinolaryngologist born in Frankfurt am Main and working mainly in Rostock.

Septum of Körner: *Eponym* for the petrosquamosal septum dividing the squamous and petrous portions of the mastoid cavity, already described by Hartmann. Reference: *Die Grenzen zwischen den pneumatischen Zellensystemen des Antrum petrosum und des Antrum squamosum beim Erwachsenen.* Z Hals-Nasen-Ohrenheilkd 1923: 5:425–432.

Kowalewsky, Nikolaus Osipovich (1840–1892): Russian embryologist born and working in Kasan.

Canal of Kowalewsky: *Eponym* for the embryonic neurenteric canal, also described by Braun.

Kraissl, Cornelis J (1902–1999): American plastic and reconstructive surgeon working in North Hackensack.

Lines of Kraissl: *Eponym* for lines of maximal skin tension. Reference: *The selection of appropriate lines for elective surgical incisions.* Plastic Reconstr Surg 1951; 8:1–28.

Krause, Karl Friedrich Theodor (1797–1868): German anatomist and histologist born and working in Hannover. Reference: *Handbuch der Anatomie des Menschen.* Hannover, 1833.

Fossa of Krause: *Eponym* for the ovarian fossa, also described by Claudius and Waldeyer.

Glands of Krause: *Eponym* for the accessory lacrimal glands mostly in the upper eyelid, described also by his son and by Wolfring.

Ligament of Krause: *Eponym* for (a) the lateral costotransversary ligament and (b) the transverse perineal ligament, also described by Henle and Waldeyer.

Muscle of Krause: *Eponym* for the cutaneomucous muscle.

Triangle of Krause: *Eponym* or the superior lumbar triangle, also described by Grynfeltt and Lesshaft.

Valve of Krause: *Eponym* for a fold betwen the lacrimal sac and the nasolacrimal duct, later also described by Béraud.

Fig. K14 Terminal ventricle of spinal cord (courtesy Akira Hori, Hannover; from ten Donkelaar HJ, Lammens M, Hori A *Clinical Neuroembryology,* 3rd ed. Springer, Cham, 2023)

Ventricle of Krause: *Eponym* for the terminal ventricle of the spinal cord (Fig. K14).

Krause, Wilhelm Johann Friedrich (1833–1910): German anatomist born in Hannover and working in Göttingen and later in Berlin, son of Karl Krause. References: *Die terminale Körperchen der einfach sensiblen Nerven.* Hahn, Hannover, 1860; *Handbuch der menschlichen Anatomie.* Hannover, 1879; Leipzig, 1899–1905.

End bulbs of Krause: *Eponym* for bulboid corpuscles, nerve terminals in the skin, mouth, conjunctivae believed to be sensitive to cold.

Glands of Krause: *Eponym* for the accessory lacrimal glands mostly in the upper eyelid, also described by his father and Wolfring.

Membrane of Krause: *Eponym* for the dark line (isotropic band), generally known as Z-line, which runs transversally through bands of myofibrils, and marks the limits of a sarcomere, also described by Amici and Dobie.

Respiratory bundle of Krause: *Eponym* for the solitary tract of the hindbrain.

Kretschmann, Friedrich (1858–1934): German ear surgeon born in Wolmirstedt, Magdeburg, and working in Halle and later in Magdeburg. Reference: *Die operative Eröffnung der Mittelohrräume.* Jena, 1896.

Space of Kretschmann: *Eponym* for a slight depression in the epitympanic recess below the superior recess of the tympanic cavity of Prussak.

Krönlein, Rudolf Ulrich (1847–1910): Swiss surgeon born in Stein am Rhein, Schaffhausen, and working in Berlin.

Hernia of Krönlein: *Eponym* for a complicated inguinal hernia showing a double sac, one in the inguinal canal, the other projecting from the deep inguinal ring into the subperitoneal tissue, also known as preperitoneal inguinal hernia.

Projections of Krönlein: *Eponym* for a system of lines for the projection of the central and lateral sulci onto the lateral side of the skull. The first horizontal line is the Frankfurter horizontal line (FH), the second parallel to the first one through the superior margin of the orbit (the supraorbital horizontal line). Two vertical lines, one through the middle of the zygomatic arch (the zygomatic vertical line), the other along the posterior margin of the mastoid process (the retromastoid vertical line), complete the quadrangle. The stem of the middle meningeal artery crosses the junction of the FH and the zygomatic line, whereas its frontal and parietal branches traverse the crossings of the zygomatic and retromastoid lines with the supraorbital line. At these places, the skull can be opened for ligation of these brances of the middle meningeal artery. Reference: *Ueber Hirnchirurgie.* In: Bergmann E von, Bruns P von, Mikulicz J von, Hrsg *Praktische Chirurgie,* 1900.

Krukenberg, Adolph (1816–1877): German anatomist born in Königslutter and working in Braunschweig and later in Halle.

Vein of Krukenberg: *Eponym* for the central vein of the hepatic lobule. Reference: *Untersuchungem ueber den feineren Bau der menschlichen Leber.* Arch Anat Physiol (Lpz) 1843; 44:318–338.

Kugel, Maurice A (twentieth century): American cardiologist.

Anastomotic artery of Kugel: *Eponym* for a rare variant, an atrial anastomotic branch between branches of the right coronary artery and the circumflex branch of the left coronary artery. Reference: *Anatomical studies on coronary arteries and their branches. I. Arteria anastomotica auricularis magna.* Am Heart J 1927; 3:260–270.

Kühne, Wilhelm (1837–1900): German physiologist born in Hamburg and working in Amsterdam and later in Heidelberg.

Spindle of Kühne: *Eponym* for the neuromuscular endplate. References: *Ueber die peripherischen Endorgane der motorischen Nerven.* Leipzig, 1862; *Die motorischen Nervenendigung.* Verhandl Nat Hist Med Ver 1886; 3:223–231.

Kuhnt, Hermann (1850–1925): German ophthalmologist born in Senftenberg, Brandenburg, and working in Rostock, Heidelberg and finally in Jena. Reference: *Über die Verwertbarkeit der Bindehaut in der praktischen und operativen Augenheilkunde.* 1898.

Intermediary tissue of Kuhnt: *Eponym* for glial spezializations at the optic disc, at the site of fusion of the outer and inner layer of the optic cup. It surrounds the intraocular part of the optic nerve, and acts as a barrier between the optic nerve and the retina.

Meniscus of Kuhnt: *Eponym* for the central supporting fibrous tissue covering the optic disc, forming a fibrous sheath around the hyaloid artery where it leaves the optic disc. In case of incomplete regression, the papilla of Bergmeister persists as a remnant of the fibrous sheath of the hyaloid artery.

Spaces of Kuhnt: *Eponym* for shallow recesses, openings into the posterior chamber of the eye between the ciliary body and the ciliary zonules.

Kulchitsky, Nicholas (1856–1925): Russian anatomist and histologist born in Kronstadt and working in Charkov, Kasan and later in St. Petersburg.

Cells of Kulchitsky: *Eponym* for respiratory neuroendocrine cells, also described by Feyrter.

Kupffer, Karl Wilhelm von (1829–1902): German anatomist and embryologist born in Lesten (Courland, currently Latvia) and subsequently working in Kiel, Königsberg and Munich (Fig. K15).

Fig. K15 Photo of Kuppfer (from Pagel 1901)

Canal of Kupffer: *Eponym* for the ureteric diverticle, an embryonic outgrowth from the mesonephric duct to form the ureter. Reference: *Untersuchungen über die Entwicklung des Harn- und Geschlechtssystems.* Arch Mikr Anat 1865; 2:233–248; 1866; 3:473–488.

Cells of Kupffer: *Eponym* for the stellate cells in the lining of blood channels in the liver (Fig. K16). Reference: *Ueber Sternzellen der Leber.* Arch Mikr Anat 1876; 12:351–358.

Kuster, G (twentieth century): American anatomist.

Vein of Kuster, Perforating: *Eponym* for the medial ankle perforating vein. Reference: *Anatomy of the veins of the foot.* Surg Gynecol Obstet 1968; 127:817–823.

Küttner, Arthur (1862–1945): German surgeon.

Principal lymph node of Küttner: *Eponym* for the jugulodigastric lymph node. Reference: *Über die Lymphgefässe der Zunge, mit Beziehung auf die Verbreitungsweise des Zungencarcinoms.* Verh Deutsch Ges Chir Berlin 1897; 14:120–123.

Kuypers, Henricus (Hans) Gerardus Josephus Maria (1925–1989): Dutch neuroanatomist born in Rotterdam who worked for many years in the USA, and returned in 1966 to Rotterdam (Fig. K17a). From 1984 until 1989, he worked in Cambridge (UK). He was a pioneer of multi-

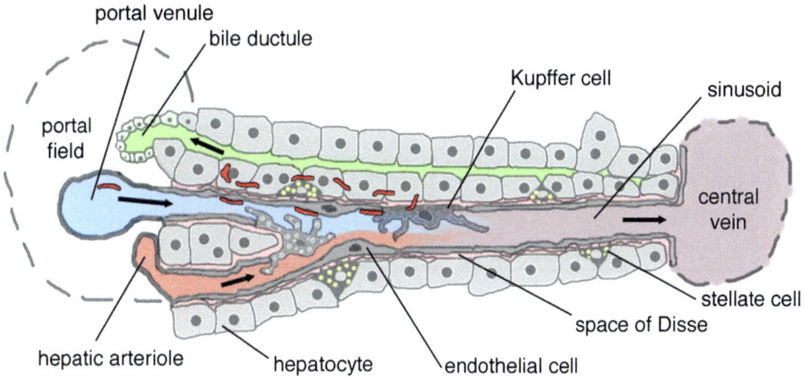

Fig. K16 Schema explaining Kupffer cells, Ito cells and the space of Disse (from Frevert U, Engelmann S, Zougbédé S, Stange J, Ng B et al. *Intravital observation of Plasmodium berghei sporozoite infection of the liver.* PLoS Biology 2005; do 10.1371/journal.pbio.0031192.9011; Open Access)

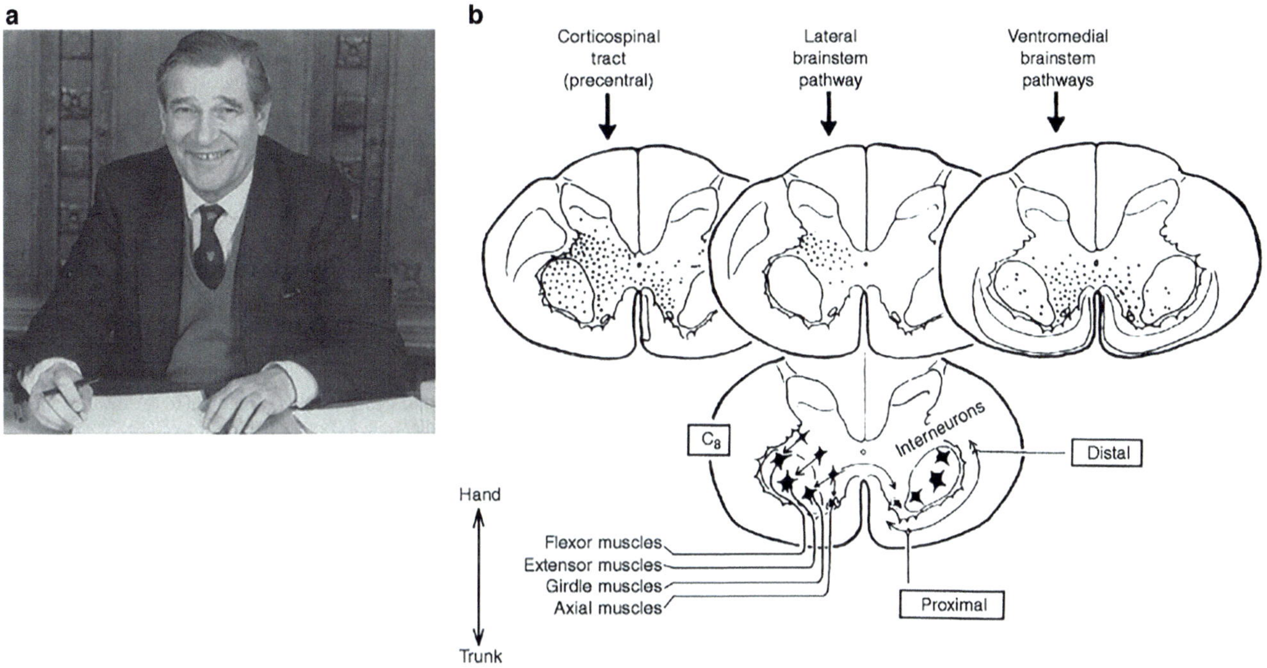

Fig. K17 (**a**) Photo of Kuypers (courtesy Roger Lemon, London); (**b**) Lateral and medial systems: Diagram of the approximate patterns of termination in the monkey lower cervical cord (C8) of the three major categories of descending supraspinal pathways: corticospinal and medial (group A) and lateral (group B) brain stem pathways. The bottom section shows the arrangement of motoneurons and interneurons in the intermediate zone. The lateral brain stem pathway (the rubrospinal tract) terminates in regions of the intermediate zone which connect with motoneurons innervating distal muscles. The medial pathways (reticulospinal and vestibulospinal tracts) terminate in regions which project to proximal and axial motoneurons. The corticospinal tract terminates onto both types of motoneurons as well as on interneurons (after Lawrence DG, Kuypers HGJM *The functional organization of the motor system in the monkey. I. The effects of bilateral pyramidal lesions.* Brain, 1968; 91:1–14; from ten Donkelaar HJ *Clinical Neuroanatomy,* 2nd ed. Springer, Cham, 2020)

disciplinary neuroscience, especially in the field of motor systems, where he introduced the terms lateral and medial motor systems based on experimental work on rhesus monkeys with Donald Lawrence (Lawrence DG, Kuypers HGJM Brain 1968; 91: 1–14, 15–36). Kuypers introduced a number of new tracing techniques, including double-labelling with fluorescent tracers and transneuronal viral tracing. Reference: *Anatomy of the descending pathways.* In: *Handbook of Physiology*, *The Nervous System II.* American Physiological Society, Bethesda, Maryland, 1981; pp. 597–666.

Lateral and medial motor systems: *Eponym* for the motor projection tracts, corticospinal and rubrospinal projections versus vestibulospinal and reticulospinal projections (Fig. K17b).

Letter L

12

Labbé, Charles (1851–1889): French anatomist born in Le Merlerault, Normandy, and working in Paris, nephew of Léon Labbé. Reference: *Note sur la circulation du cerveau et sur le mode de développement des corpuscles de Pacchioni.* Arch Physiol 1879, pp. 135–154. For biography, *see* Bartels RH, van Overbeeke JJ *Charles Labbé (1851–1889).* J Neurosurg 1997; 87:477–480.

Vein of Labbé: *Eponym* for the inferior anastomotic vein of the cerebral cortex (Fig. L1).

Labbé, Léon (1832–1916): French surgeon born in Le Merlerault, Normandy, and working in Paris, uncle of Charles Labbé. Later, he became senator.

Triangle of Labbé: *Eponym* for the area where the stomach is normally in contact with the abdominal wall, comparable to the space of Traube.

Labbock, AJ:

Chain of Labbock: *Eponym* for a sympathetic trunk between the communicating branches of the first and third sacral ganglia. Reference: *Zur Topographie des Truncus sympathicus sacralis.* Anat Anz 1938; 80:42–65.

Lacuna of Morgagni: *Eponym* for a urethral lacuna, one of several recesses in the mucosa of the spongy urethra into which the ducts of the urethral glands empty; *see* Morgagni, Giovanni Battista.

Laignel-Lavastine, M (twentieth century): French pathologist in Paris.

Ansa of Laignel-Lavastine: *Eponym* for a communicating branch between a branch of the right vagus nerve and the left coeliac ganglion. Reference: *Pathologie des sympathiques.* Alcan, Paris, 1924.

Laimer, Edouard (nineteenth century): Prosector in Graz known for his studies of the alimentary tract.

Area of Laimer-Haeckermann: *Eponym* for a triangular area located below the cricopharyngeal fibres in the posterior oesophageal wall. Reference: *Beiträge zur Anatomie des Oesophagus.* Med Jahrb Wien 1883, pp. 333–388.

Diverticulum of Laimer: *Eponym* for a common diverticle of the posterior upper oesophagus, arising from the triangle of Laimer inferior to the transverse fibres of the cricopharyngeal muscle.

Membrane of Bertelli-Laimer: *Eponym* for the phrenicooesophageal membrane or ligament, connecting the oesophagus to the diaphragm, extensions of the diaphragmatic fascia to the oesophagus and stomach, also described by Bertelli.

Lallouette, Pierre (1711–1792): French surgeon born and working in Paris.

Pyramid of Lallouette: *Eponym* for the pyramidal lobe of the thyroid gland, also described by Morgagni (Fig. L2). Reference: *De glandula thyroidea.* Paris, 1743.

Lambert, Margaret Waugh (twentieth century): English pathologist working in Durham.

Canals of Lambert: *Eponym* for microscopic collateral airways between the distal bronchiolar tree and adjacent pulmonary alveoli, developing at the age of 3 to 4 years. Reference: *Accessory bronchiole-alveolar communications.* J Pathol 1955; 70:311–314.

Lancisi, Giovanni Maria (1654–1720): Italian anatomist born and working in Rome and physician to the Popes Innocent XI, Clement XI and Innocent XII (Fig. L3a).

H. J. ten Donkelaar et al., *An Illustrated Guide to Anatomical Eponyms*, https://doi.org/10.1007/978-3-031-91664-9_12

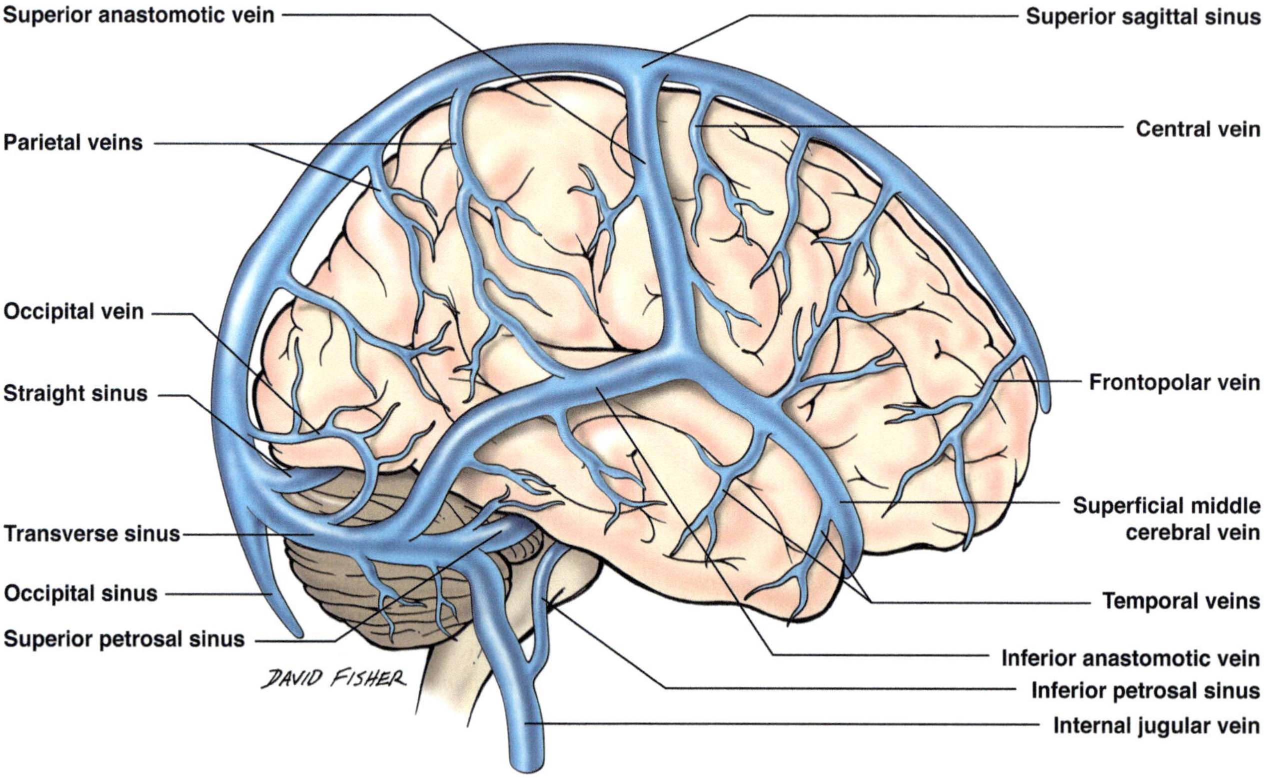

Fig. L1 The vein of Labbé and other superficial veins of the brain. (Ffrom ten Donkelaar HJ, Kachlík D, Tubbs RS *An Illustrated Terminologia Neuroanatomica.* Springer, Cham, 2018)

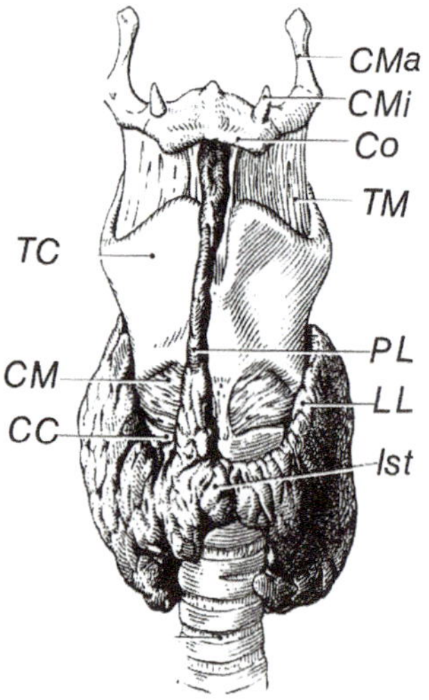

Fig. L2 Lallouette pyramidal lobe of the thyroid gland. (After ten Donkelaar HJ, Oostra R-J *Klinische Anatomie en Embryologie,* 4th ed. Springer Media, Houten, 2014, based on a drawing in Drenckhahn D, Zenker W, Hrsg, *Benninghoff Anatomie, Band 2.* Urban & Schwarzenberg, München, 1994). *CC* corniculate cartilage, *CM* cricothyroid muscle, *CMa, CMi, Co* cornu majus, cornu minus and corpus of hyoid bone, *Ist* isthmus, *LL* left lobe of thyroid gland, *PL* pyramidal lobe of thyroid gland, *TC* thyroid cartilage, *TM* thyrohyoid membrane

Pope Clement XI gave him Eustachi's anatomical plates that were found at the Papal Library, and published them. Reference: *Anatomia corporis humani.* Turin, 1711.

Nerves or **Stripes of Lancisi:** *Eponym* for the longitudinal striae (medial and lateral) overlying the corpus callosum (Fig. L3b).

Landsmeer, Johan Mathijs Frederik (Hans) (1919–1999): Dutch anatomist born in Rotterdam and working in Leiden especially on the anatomy and function of the hand (Fig. L4a). In 1949, he described the oblique retinacular ligament, also known as the 'bandjes' of Landsmeer. Later it became known that Weitbrecht had described the same structure in his *Syndesmologia* (1742) as the retinaculum tendoni longi.

Oblique retinacular ligament of Landsmeer: *Eponym* for the ligament at the level of the proximal interphalangeal joint associated with the extensor mechanism (Fig. L4b). Originally, it was described in 1742 by Weitbrecht as the retinaculum tendoni longi

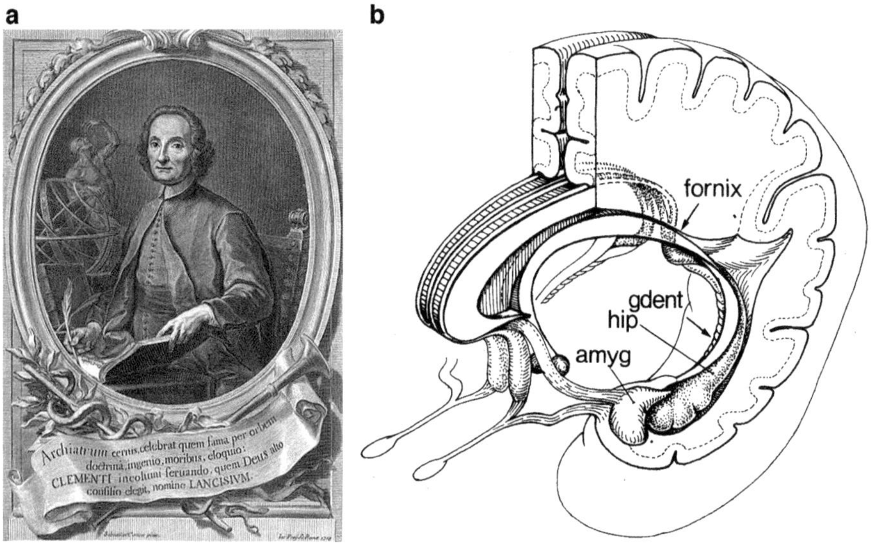

Fig. L3 (**a**) Engraving of Lancisi. (From Wikimedia Commons; in public domain); (**b**) The longitudinal striae or stripes of Lancisi. (From ten Donkelaar HJ, Kachlík D, Tubbs RS *An Illustrated Terminologia Neuroanatomica*. Springer, Cham, 2018). *amygd* amygdala, *gdent* dentate gyrus, *hip* hippocampus

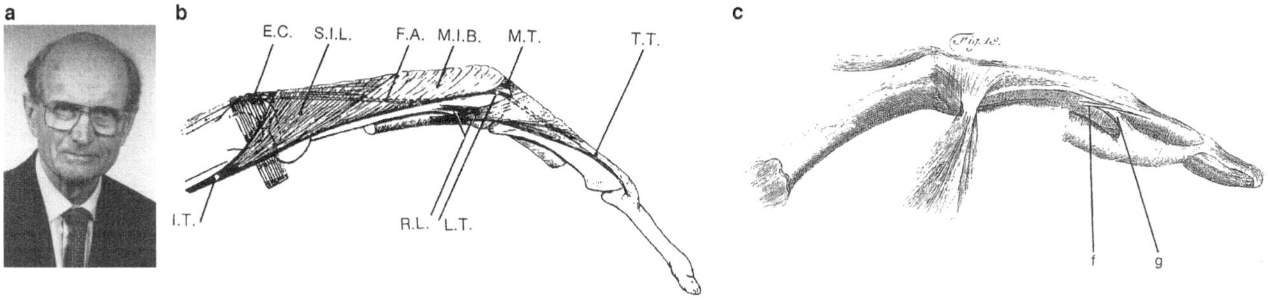

Fig. L4 (**a**) Photograph of Landsmeer. (From Kauer JMG, Huson A *Obituary Professor J.M.F. Landsmeer (1919–1999)*. J Anat (Lond) 2000; 196:301–302; courtesy John Kauer, Nijmegen); (**b**) The oblique retinacular ligament of Landsmeer. (From Landsmeer JMF *The anatomy of the dorsal aponeurosis of the human finger and its functional significance*. Anat Rec 1949; 104:31–44); (**c**) The retinaculum tendini longi of Weitbrecht. (From Weitbrecht J *Syndesmologia sive historia ligamentum corporis humani*. St. Petersburg, 1742)

(Fig. L4c). References: *The anatomy of the dorsal aponeurosis of the human finger and its functional significance*. Anat Rec 1949; 104:31–44; *Landmeer's Atlas of the Anatomy of the Hand*. Churchill Livingstone, Edinburgh, 1976.

Landzert, Theodor (1833–1889): Russian anatomist in St. Petersburg.

Canal of Landzert: *Eponym* for the craniopharyngeal canal, an inconstant canal traversing the body of the sphenoidal bone, also described by Calori. It may contain ectopic hypophysial tissue.

Fossa of Landzert: *Eponym* for the inferior duodenal fossa of the peritoneal cavity, also described by Gruber, so usually defined as fossa of Gruber-Landzert (*see* Fig. G25). Reference: *Ueber die Hernia retroperitonealis (Treitz) und ihre Beziehung zur Fossa duodenojejunalis*. St Petersb Med Ztsch 1871; 2:306, 350–383.

Langenbeck, Bernhard Rudolf Konrad von (1810–1887): German surgeon born in Horneburg and working in Kiel and later in Göttingen, nephew of Conrad von Langenbeck.

Triangle of Langenbeck: *Eponym* for a triangular area formed by lines drawn from the anterior superior iliac spine to the greater trochanter of the femur and to the surgical neck of the femur. Reference: *Ueber Schussverletzungen des Hüftgelenkes*. Berlin, 1874.

Langenbeck, Conrad Johann Martin (1776–1851): German surgeon and anatomist born in Horneburg and working in Würzburg and later in Göttingen, uncle of Bernhard von Langenbeck. References: *Anatomisches Handbuch*. Göttingen, 1806; *Handbuch der Anatomie*. Göttingen, 1831–1847.

Nerve of Langenbeck: *Eponym* for a lateral supraclavicular nerve (his nervus superficialis scapulae).

Langendorff, Oskar (1853–1908): German physiologist born in Breslau and working in Rostock.

Cells of Langendorff: *Eponym* for the follicular cells, the principal cells of the thyroid gland. Reference: *Beiträge zur Kenntniss der Schilddrüsen.* Arch Anat Physiol, Phys Abth 1889, pp. 219–240.

Langer, Karl (Ritter von Edenberg) (1819–1887): Austrian zoologist and anatomist born in Vienna and working in Budapest and later in Vienna (Fig. L5).

Axillary arch of Langer: *Eponym* for a muscle variation that stretches like an arch in the axillary cavity, between the latissimus dorsi and pectoralis major muscles.

Lines of Langer: *Eponym* for cleavage lines of the skin due to the disposition of the subcutaneous connective tissue (Fig. L6a and b). Reference: *Zur Anatomie und Physiologie der Haut.* Sitz Akad Wiss Wien 1862; 44:19–46.

Langerhans, Paul (1847–1888): German physician, anatomist and pathologist born in Berlin and working in Freiburg (Fig. L7) who described cellular masses that form the endocrine portion of the pancreas and are the source of insulin and glucagon.

Bodies or **Granules of Langerhans:** *Eponym* for platelike structures found in the cells of Langerhans.

Fig. L5 Photo of Langer. (From Wikimedia Commons; in public domain)

Cells of Langerhans: *Eponym* for the clear or dendritic cells in the epidermis.

Islets of Langerhans: *Eponym* for pancreatic islets. Reference: *Beiträge zur mikroskopischen Anatomie der Bauchspeicheldrüse.* Berlin, 1869.

Langhans, Theodor (1839–1915): German pathologist and anatomist born in Usingen working in Bern (Fig. L8).

Fibrinoid substance or **Stria of Langhans:** *Eponym* for the subchorionic layer of fibrinoid in the intervillous spaces of the placenta. Rohr described an upper stria, and Nitabuch, one of Langhans' students, a lower stria. Reference: *Untersuchungen über die menschliche Placenta.* Arch Anat Physiol, Anat Abt, pp. 188–267.

Layer and **Cells of Langhans:** *Eponym* for the cytotrophoblast and its cells. Reference: *Zur Kenntniss der menschlichen Placenta.* Arch Gynaekol 1870; 1:317–334.

Langley, John Newport (1835–1925): English physiologist born in Newbury, Berkshire, and working in Cambridge. He is noted for his studies of the autonomic nervous system (Fig. L9). Reference: *The Autonomic Nervous System.* Heffer, Cambridge, 1921. He introduced the terms preganglionic and postganglionic fibres.

Ganglion of Langley: *Eponym* for the submandibular ganglion (Fig. L10).

Nerves of Langley: *Eponym* for the pilomotor nerves.

Lannelongue, Odilon Marc (1840–1911): French surgeon and pathologist born in Catera-Verduzan and working in Paris (Fig. L11).

Foramina of Lannelongue: *Eponym* for venous openings of the smallest cardiac veins into the right atrium, already described by Vieussens and Thebesius. References: *Circulation veineuse des parois auriculaires du coeur.* Thesis, Paris, 1867; *Recherches sur la circulation des parois du coeur.* Arch Physiol Norm Path Paris 1868; 1:24–34.

Ligaments of Lannelongue: *Eponym* for the sternopericardial ligaments, also described by Luschka. Reference: *Note sur un ligament non-décrit du péricarde.* Arch Physiol Norm Path Paris 1868; 1:448–449.

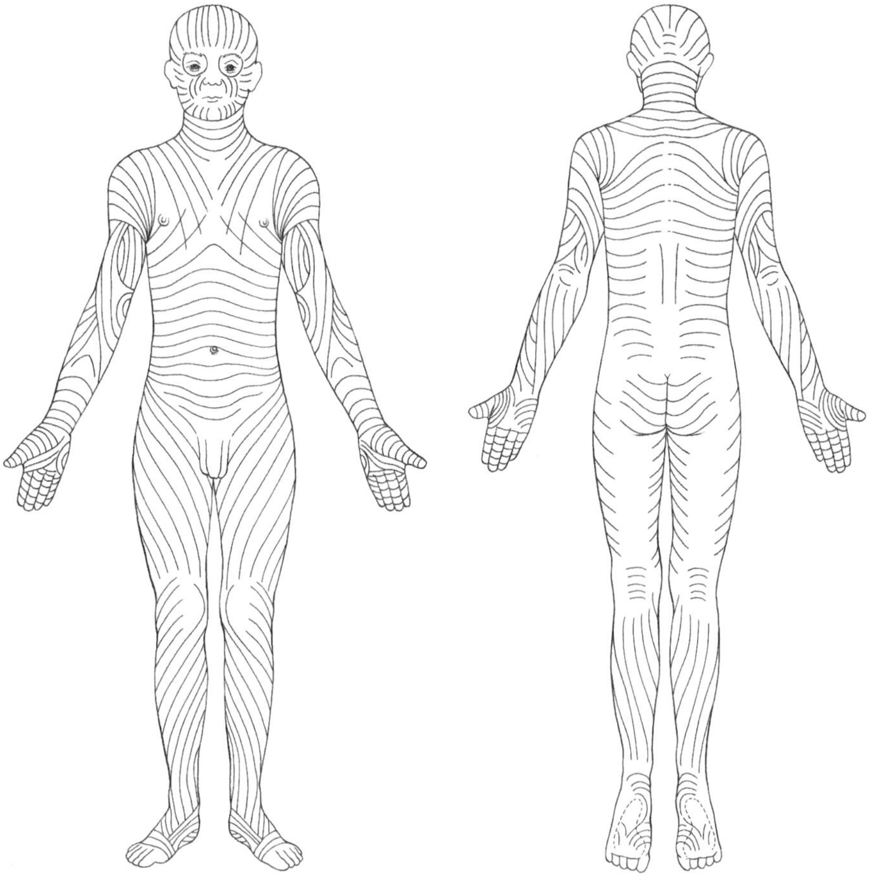

Fig. L6 (**a** and **b**) Lines of Langer. (From ten Donkelaar HJ, Oostra R-J *Klinische Anatomie en Embryologie,* 4th ed. Springer Media, Houten, 2014)

Fig. L7 Photo of Langerhans. (From Wikimedia Commons; in public domain)

Fig. L8 Photo of Langhans. (From the Wellcome Collection CCBY; in public domain)

Fig. L9 Photo of Langley. (From Wikimedia Commons; in public domain)

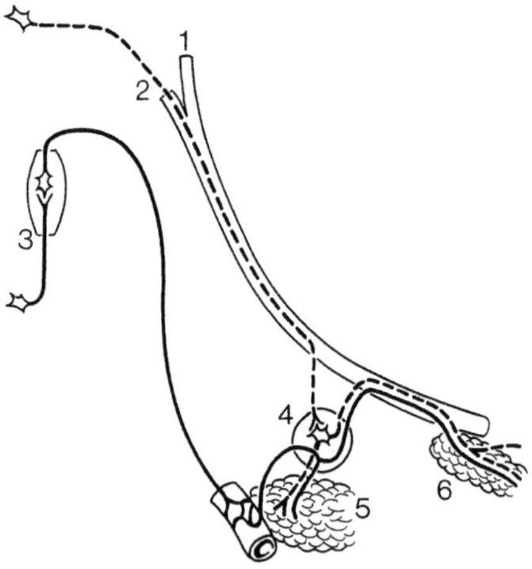

Fig. L10 The submandibular ganglion of Langley. (After O'Rahilly R *Gardner-Gray-O'Rahilly Anatomy*, 5th ed. Saunders, Philadelphia, PA, 1986; from ten Donkelaar HJ, Kachlík D, Tubbs RS *An Illustrated Terminologia Neuroanatomica.* Springer, Cham, 2018). *1* lingual nerve, *2* chorda tympani with preganglionic parasympathetic neuron, *3* superior cervical ganglion with sympathetic innervation, *4* submandibular ganglion, *5* submandibular gland, *6* sublingual gland

Lantermann, A. J. (1855–1910): American anatomist who worked in Strassburg with Waldeyer; *see* Cleft of Schmidt-Lantermann; Reference: *Bemerkungen über den feineren Bau der markhaltigen Nervenfasern.* Zbl Med Wiss 1874; 12:706–709.

Lanz, Otto (1865–1935): Swiss surgeon born in Steffisburg working in Bern and later in Amsterdam.

Fig. L11 Photo of Lannelongue. (From Pagel 1901)

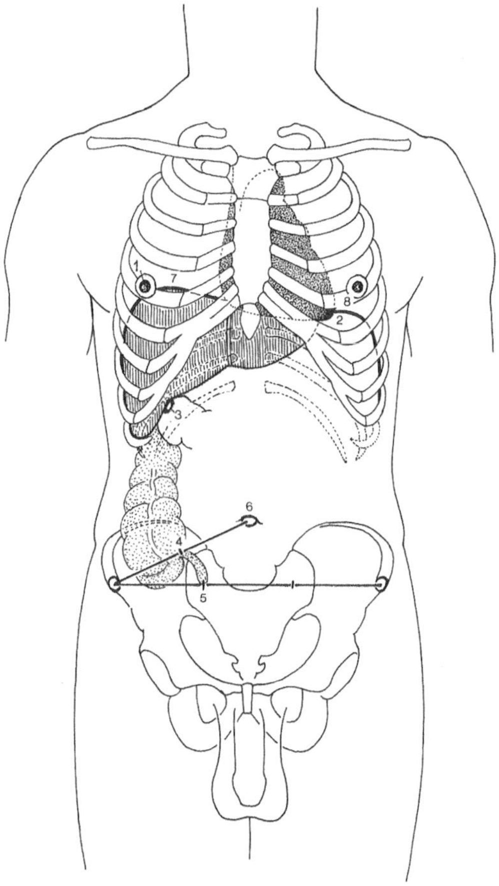

Fig. L12 Point of Lanz. (From an old figure of the late Chris van Huijzen). *4* point of McBurney, *5* point of Lanz, *6* umbilicus

Line of Lanz: *Eponym* for the interspinous line connecting the right and left anterior superior iliac spines.

Point of Lanz: *Eponym* for where the apex of the vermiform appendix of the caecum is projected upon the anterior abdominal wall (Fig. L12), at 1/3 of a line between the right and left anterior superior iliac spines (line of Lanz). Reference: *Der McBurney'sche Punkt.* Zbl Chir 1908; 35:185–190

Larrey, Dominique Jean (1766–1842): French surgeon born in Beaudéan, Pyrenées, and surgeon to Napoléon Bonaparte. He is often considered the first modern military surgeon given his inventions in battlefield medicine and triage (Fig. L13).

Fissure or **Trigone of Larrey:** *Eponym* for the sternocostal triangle of the diaphragm (Fig. L14). This is the place for a parasternal hernia, also noted by Morgagni and usually called hernia of Morgagni.

Fig. L13 Portrait of Larrey by Anne-Louis Girodet. (1805; From Chemin de Mémoire, Ministère des Armées; in public domain)

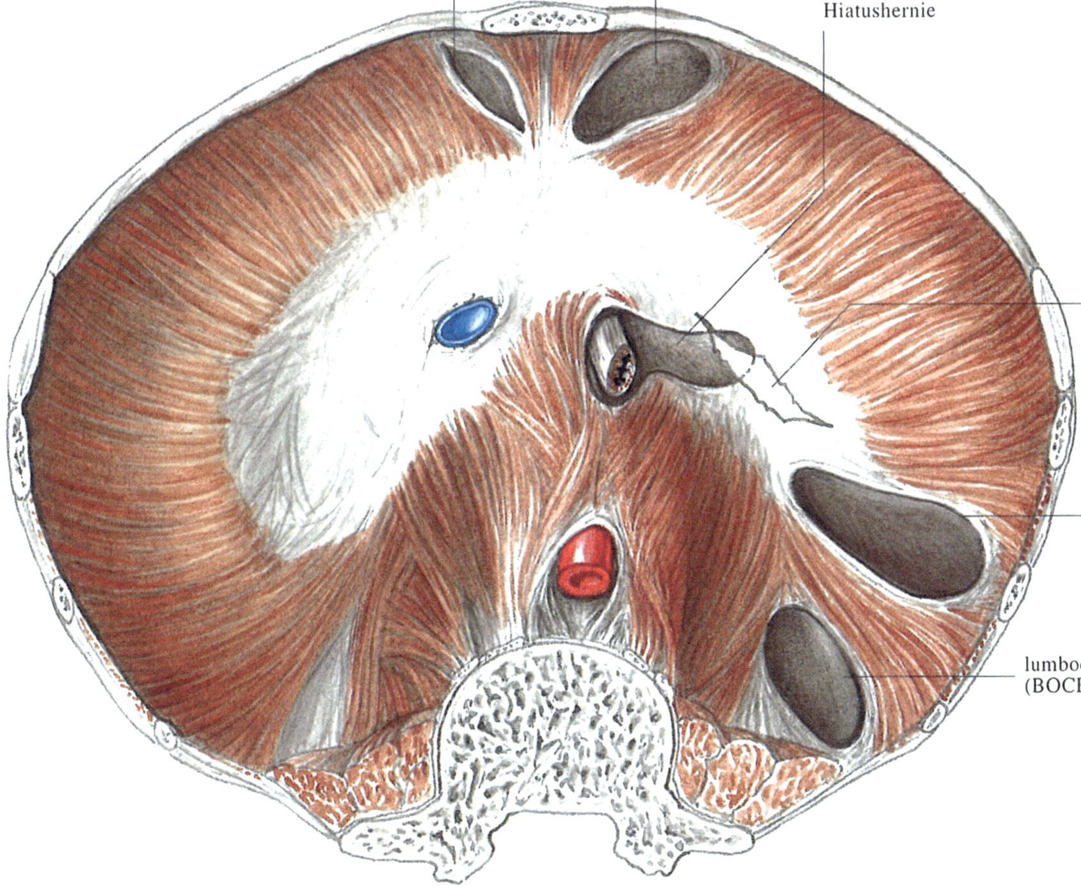

Fig. L14 Fissure of Larrey. (From von Lanz T, Wachsmuth W *Praktische Anatomie.* Loeweneck H, Feifel G *Bauch.* Springer, Berlin-Heidelberg-New York, 1993; with permission)

Point of Larrey: *Eponym* for the femoral compression point, where the femoral artery passes beneath the inguinal ligament, halfway between the anterior superior iliac spine and the pubic symphysis; here, it can be compressed.

Larsell, Olof (1886–1964): Swedish-American neuroanatomist born in Rättvik, Sweden, who moved with his parents to the USA in 1891. He worked in Chicago, Wisconsin and Portland, and focussed on comparative neuroanatomy of the cerebellum (Fig. L15).

Lobules of Larsell: *Eponym* for the lobules numbered I-X of the cerebellum. References: *The Comparative Anatomy of the Cerebellum from Monotremes to Apes.* University of Minnesota Press, Minneapolis, MN, 1970; *The Human Cerebellum, Cerebellar Connections and the Cerebellar Cortex.* University of Minnesota Press, Minneapolis, MN, 1972, with J Jansen).

Latarjet, André Raphael (1877–1947): French anatomist in Lyon (Fig. L16a) known, among others, for his studies on the innervation of the stomach. Reference: *Précis atlas des traveaux, pratiques d'anatomie,* 3 Volumes. Doin, Paris, 1924.

Fig. L15 Photo of Larsell. (Courtesy Oregon Health & Sciences University Historical Collection and Archives, Digital Collection; free to use)

Nerves of Latarjet: *Eponym* for the terminal branches of the anterior and posterior nerves of the lesser curvature of the stomach, branches of the anterior and posterior vagal trunks, respectively (Fig. L16b). Reference: *Les nerfs de l'estomac.* Bull Soc Chir Lyon, 1921.

Principal ureteric nerves of Latarjet: *Eponym* for the inferior and superior principal ureteric nerves. The inferior nerve is a branch of the hypogastric nerve supplying the pelvic part of the ureter. The superior nerve is the greatest ureteric branch of the renal sympathetic plexus.

Vein of Latarjet: *Eponym* for the prepyloric vein, described later also by Mayo.

Latham, Paul Mere (1789–1875): English physician working in London and from 1837 as extraordinary physician to Queen Victoria.

Circle of Latham: *Eponym* for the area on the chest wall corresponding to the area of pericardial dullness.

Laumonier, Jean Baptiste Phillippe Nicolas Réné (1749–1818): French anatomist and surgeon.

Ganglion of Laumonier: *Eponym* for a carotid ganglion, previously described by Petit (1727) and Schmiedel (1747). Reference: *Observation anatomico-physiologique, extraite d'un ouvrage sur les sympathies, nerveuses, considérées dans l'état de santé et de maladies.* Paris, 1793.

Laura, Giovanni Battista (1843–1901): Italian anatomist in Torino who made several contributions to the morphology of the spinal and cranial nerves and of the spinal cord. References: *Sull'origine reale dei nervi spinali e di qualche nervo cerebrale.* Mem Accad Sci Torino 1879; *Sur la structure de la moelle épinière.* Arch Ital Biol 1882; 1:147–175.

Nucleus of Laura: *Eponym* for the lateral vestibular nucleus of the brain stem, earlier described by Deiters.

Lauth, Ernest Alexandre (1803–1837): French anatomist and physiologist born and working in Strassburg (Fig. L17), son of Thomas Lauth.

Canal or **Sinus of Lauth:** *Eponym* for the scleral venous sinus, mostly known as canal of Schlemm, but earlier described by Fontana.

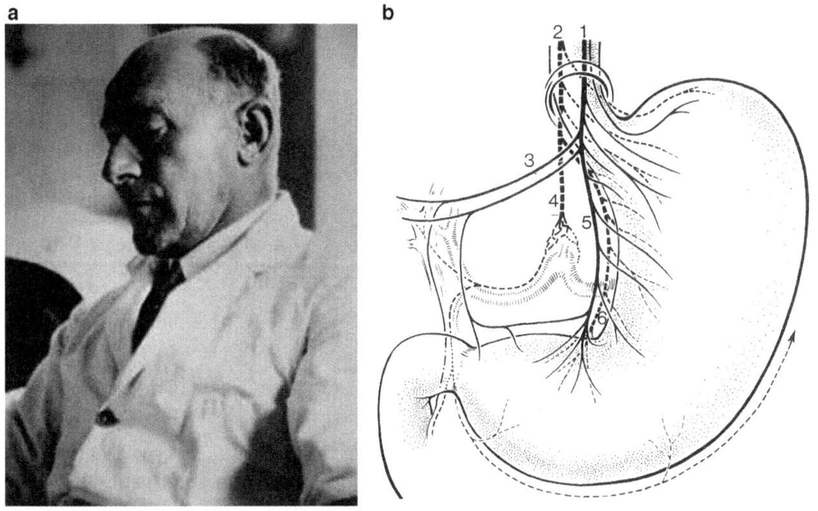

Fig. L16 (**a**) Photo of Latarjet. (From Reveron RR *André Latarjet (1877–1947). Anatomist and surgeon specialized in sports medicine.* Ital J Anat Embryol 2014; 119:250–254); (**b**) The nerves of Latarjet. *1* anterior vagal trunk, *2* posterior vagal trunk, *3* hepatic branches giving off a pyloric branch, *4* celiac branch, *5* anterior nerve of Latarjet giving off anterior gastric nerves, *6* posterior nerve of Latarjet giving off posterior gastric branches. (From ten Donkelaar HJ, Kachlík D, Tubbs RS *An Illustrated Terminologia Neuroanatomica.* Springer, Cham, 2018)

Fig. L17 Portrait of Ernest Lauth. (From Wikimedia Commons; in public domain)

Tunica of Lauth: *Eponym* for the thin connective tissue layer covering the bundles of fibres of the external oblique abdominal muscle and its vast aponeurosis.

Lauth, Thomas (1758–1826): French surgeon and anatomist born and working in Strassburg (Fig. L18), father of Ernest Lauth, after whom many ligaments are named.

Ligament of Lauth: *Eponym* for (a) the inferior pubic ligament; (b) the transverse ligament of the atlas; and (c) the costoclavicular ligament (his rhomboid ligament). Reference: *Elemente der Myologie und Syndesmologie.* Basel, 1798 and 1805.

Fig. L18 Portrait of Thomas Lauth. (From Wikimedia Commons; in public domain)

Layers:

Layer of Cajal: *Eponym* for the molecular layer of the olfactory bulb*; see* Rámon y Cajal, Santiago.

Layer of Chievitz: *Eponym* for a transient anuclear layer of the developing retina; *see* Chievitz, Johan Hendrik.

Layer of Dua: *Eponym* for the preposterior limiting layer, anterior to the posterior limiting layer of Descemet in the cornea, composed of collagen with abundant elastin and almost acellular; see Dua, Harminder.

Layer of Haller: *Eponym* for the vascular layer of the choroid of the eye; *see* Haller, Albrecht von.

Layer of Henle, Fibre: *Eponym* for the layer of inner cone fibres in the central area of the retina; *see* Henle, Friedrich Gustav Jacob.

Layer of Huxley: *Eponym* for the inner layer of the internal root sheath of the hair, forming the middle epithelial layer of cells in the inner sheath of a hair root; *see* Huxley, Thomas Henry.

Layer of Kölliker: *Eponym* for the middle connective tissue layer or stroma of the iris; *see* Kölliker, Rudolf Albert von.

Layer of Langhans: *Eponym* for the cytotrophoblast; *see* Langhans, Theodor.

Layer of Malpighi: *Eponym* for the stratum germinativum, the deep layer of the epidermis comprising the stratum basale and the stratum spinosum; *see* Malpighi, Marcello.

Layer of Meynert: *Old term* for the third layer of the isocortex, the external pyramidal cell layer; *see* Meynert, Theodor.

Layer of Oehl: *Eponym* for the stratum lucidum of the epidermis; *see* Oehl, Eusebio.

Layer of Ollier: *Eponym* for the inner or osteogenic layer of the periosteum; *see* Ollier, Louis Xavier Edouard Leopold.

Layer of Purkinje: *Eponym* for the layer of Purkinje cells in the cerebellar cortex; *see* Purkinje, Jan Evangelista.

Layer of Rauber: *Eponym* for the outer cell mass of the blastodermic vesicle; *see* Rauber, August Antonius.

Layer of Rexed: *Eponym* for the layers of the spinal grey matter; *see* Rexed, Bror.

Layer of Sattler: *Eponym* for the middle layer of the choroid; *see* Sattler, C. Hubert.

Layer of Tomes: *Eponym* for the granular layer of dentin; *see* Tomes, John.

Layer of Tyson: *Eponym* for the epidermal granular layer, later also described by Unna; *see* Tyson, Edward.

Layer of Unna: *Eponym* for the epidermal granular layer, earlier described by Tyson; *see* Unna, Paul Gerson.

Lazorthes, Guy (1910–2014): French neurosurgeon born and working in Toulouse known for his extensive studies of the vascularization of the brain and spinal cord. References: Lazorthes G, Gouazé A, Salomon G *Vascularisation et circulation de l'encéphale.* Masson, Paris, 1976; Lazorthes G, Gouazé A *Vascularisation et circulation de la moelle epinière.* Masson, Paris, 1973.

Arteries of Lazorthes: *Eponym* for arteries supplying the cervical and lumbosacral enlargements of the spinal cord, respectively.

Plexus of Lazorthes: *Eponym* for the anterior and posterior tibial plexuses that supply the distal segment of the anterior and posterior tibial arteries, respectively. Reference: *Le sympathique du membre inférieur. Les nerfs vasculaires. L'innervation des vaisseaux du bassin et du membre inférieur.* Thesis, Toulouse. Maloine, Paris, 1938.

Leber, Theodor Karl Gustav von (1840–1917): German pathologist born in Karlsruhe and working in Göttingen and later in Heidelberg.

Plexus of Leber: *Eponym* for a small venous plexus between the canal of Schlemm and the spaces of Fontana, formed by plexiform veins of the iris associated with the ciliary processes of the ciliary body. Reference: *Anatomische Untersuchungen ueber die Blutgefässe des menschlichen Auges.* Denkschr Akad Wiss Wien 1865; 24:297–330.

Lecat (Le Cat), Claude-Nicholas (1700–1768): French surgeon born in Blérancourt and working in Rouen (Fig. L19).

Gulf of Lecat: *Eponym* for a dilatation of the spongy part of the urethra. Reference: *Pièces concernant l'opération de la taille: Premier recueil qui traite principalement de cette opération pratiquée sur les femmes.* Dumesmil, Rouen, 1749–1753.

Lee, Robert (1793–1877): Scottish gynaecologist and obstetrician born in Melrose and working in London.

Ganglion of Lee: *Eponym* for a parasympathetic ganglion of the cervix uteri, also described by

Fig. L19 Portrait of Lecat. (From the Musée d'histoire de la Médecine, Université Paris V, in Wikipedia Commons, in public domain)

Fig. L20 Portrait of Lenhossék. (From Wikidata; in public domain)

Frankenhäuser (1866). Reference: *On the Ganglia and Other Nervous Structures of the Uterus.* Richard and John E Taylor, London, 1842.

Plexus of Lee-Frankenhäuser: *Eponym* for the neural uterovaginal plexus. Frankenhäuser later gave a more detailed description of the nervous plexus and ganglia (*see* Fig. F21).

Leeuwenhoek, Antonie van: *See* van Leeuwenhoek, Antonie.

Le Fort, Léon Clément (1829–1893): French surgeon in Paris.

Tubercle of Le Fort: *Eponym* for the anterior fibular tubercle, a small prominence on the lateral malleolus of the fibula to which the anterior fibular ligament inserts, also described by Wagstaffe. Reference: *Note sur une variété non descrite de fracture verticale de la malléole externe par attachement.* Bull Gen Thér 1886; 110:193–199.

Veins of Le Fort: *Eponym* for the pleuropulmonary veins. Reference: *Recherches sur l'anatomie des poumons chez l'homme.* Thesis, Paris, 1858.

Lejards, Félix-Marie Louis (1863–1932): French surgeon born in Eure-et-Loire and working in Paris.

Arch of Lejards: *Eponym* for a reno-azygos arch, a variant of the origin of the hemiazygos vein, in which the left lumbar azygos vein is formed by an anastomosis between veins anterior to the lumbar vertebrae and the left renal vein, which may anastomose with the inferior vena cava and the second and third lumbar veins. Reference: *Les voices de sûreté de la veine rénale.* Bull Soc Anat Paris 1888; 43:504–511.

Plexus of Lejards: *Eponym* for the superficial plantar venous plexus, composed of fine venules within the dermis and subcutaneous tissue of the sole, earlier described by Bourceret. Reference: *Les veines de la plante du pied chez l'homme et les grands animaux.* Arch Physiol, 5e série 1890; II:89.

Lenhossék, József (Joseph) von (1818–1888): Hungarian anatomist born in Budapest (Fig. L20) and working subsequently in Basel, Würzburg, Tübingen and Budapest, father of Mihály von Lenhossék. He is known for his studies on the reticular formation and the spinal cord. For biography, *see* Triarhou LC *József von Lenhossék (1818–1888.* J Neurol 2015; doi 10.1007/s00415-015-7810-8.

Reticular process of Lenhossék: *Eponym* for the reticular process of the posterior horn of the spinal cord. Reference: *Ueber den feineren Bau der sogenannten Medulla spinalis.* Vienna, 1858.

Lenhossék, Mihály (Michael) von (1863–1937): Hungarian anatomist and histologist born in Budapest and working subsequently in Basel, Würzburg, Tübingen and in Budapest, son of József von Lenhossék.

Processes of Lenhossék: *Eponym* for short processes of ganglion cells, later also described by Nageotte. Reference: *Der feinere Bau des Nervensystems im Lichte neuester Forschungen.* Fischer, Berlin, 1895.

Lenoir, Camille Alexandre Henri (1867–1928): French surgeon in Bordeaux.

Facet of Lenoir: *Eponym* for an accessory facet (the 'odd facet') on the medial surface of the patella. Reference: *Contribution à l'étude de la fracture de la rotule chez les enfants.* Lanefranque, Bordeaux, 1891.

Leonardo da Vinci (1452–1519): Italian painter and scientist born in Vinci, near Florence, and working in Florence, Milan and Rome. He was the first to illustrate the frontal and maxillary sinuses and to draw accurate figures of the vertebral column and many other structures.

Moderator band of Leonardo: *Eponym* for a structure first so named by King (1837) but figured by Leonardo (his *catena*, beam or band). It is illustrated in his anatomical drawings, currently in Windsor Castle. For a description and the relevant illustrations, see *Leonardo on the Human Body.* Schuman, New York, 1952 (reprint: Dover Publications, New York, 1983). The moderator band is found in the right ventricle and connects the base of the anterior papillary muscles of the tricuspid valve to the ventricular septum. It carries part of the right bundle branch of the atrioventricular bundle (*see* Fig. I5), and has also been described as arcuate tract by Testut and ansiform band by Poirier.

Vein of Leonardo: *Eponym* for the posterior accessory great saphenous vein of the leg, previously known as the posterior arcuate vein of the leg.

Lesser, Ladislaus Leon, Freiherr von (1846–1925): German surgeon.

Triangle of Lesser: *Eponym* for the lingual triangle bounded by the anterior and posterior bellies of the digastric muscle and the hypoglossal nerve. Its floor is formed by the hyoglossus muscle, beneath of which the lingual artery is found.

Lesshaft, Pjotr Frantsowitsch (1836–1909): Russian anatomist in St. Petersburg.

Muscle of Lesshaft: *Eponym* for the levator ani muscle. Reference: *On the endings of the longitudinal muscular fibres of the rectum and on the supplementary strata of the region of the anus and prostate gland in man and some animals.* St. Petersburg, 1884.

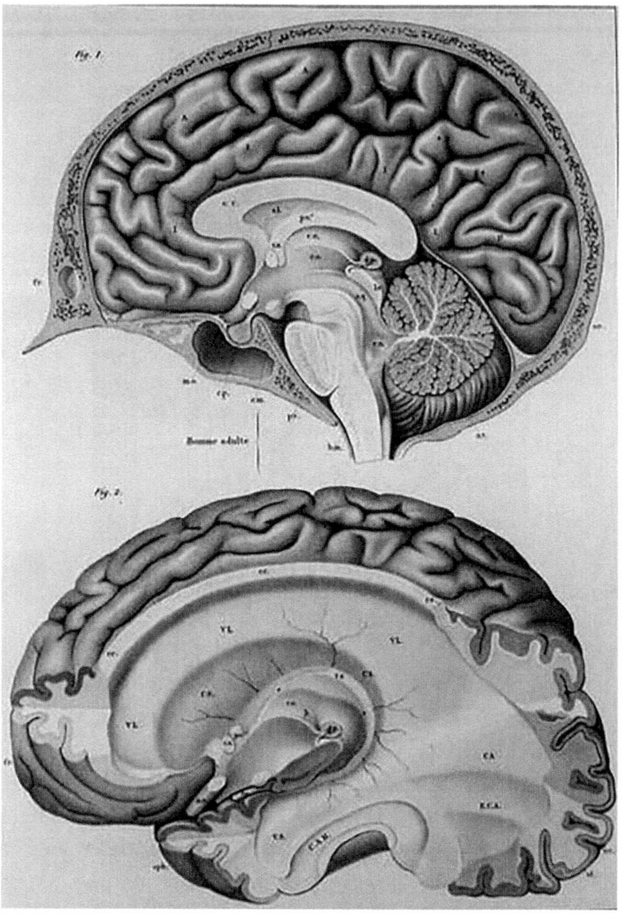

Fig. L21 The brain after Leuret. (From Leuret F, Gratiolet P *Anatomie comparée du système nerveux considéré dans ses rapports avec l'intelligence.* Volume 1. Baillière, Paris, 1839)

Triangle of Lesshaft: *Eponym* for the superior lumbar triangle, described also by Grynfeltt and Krause.

Leuret, François (1797–1851): French anatomist and psychiatrist born in Nancy and working in Paris, teacher of Gratiolet and Broca. He published the first large work on cerebral convolutions (Fig. L21). Leuret coined the name fissure of Rolando after Luigi Rolando (1773–1831). Reference: *Anatomie comparée du système nerveux considéré dans ses rapports avec l'intelligence. Atlas des 32 planches dessinées d'apres nature engravées.* Vol 1. Baillière, Paris, 1839 (with Pierre Gratiolet). Volume 2 was written by Pierre Gratiolet.

Lexer, Erich (1867–1937): German surgeon born in Freiburg and working in Jena, a pioneer of plastic surgery.

Vessels of Lexer: *Eponym* for juxta-epiphyseal vessels of long bones. Reference: *Untersuchungen über Knochenarterien mittelst Röntgenaufnahmen injizierten Knochen.* Berlin, 1904.

Leydig, Franz von (1821–1908): German histologist from Rothenburg working in Würzburg and later in Tübingen and Bonn (Fig. L22). Reference: *Lehrbuch der Histologie des Menschen.* Frankfurt am Main, 1857.

Cells of Leydig: *Eponym* for the interstitial cells of the testis. Reference: *Zur Anatomie des männlichen Geschlechtsorgans und Analdrüsen der Säugethiere.* Z Wiss Zool 1850; 2:1–57.

Lichačeva, Natalia Borisovna (1915–2000): Russian anatomist born in Penza and working in Leningrad (now Saint Petersburg).

Vein of Lichačeva: *Eponym* for the middle thyroid vein (her lateral thyroid vein), also described by Kocher.

Lieberkühn, Joann Nathaniel (1711–1756): German physician and anatomist born and working in Berlin.

Glands of Lieberkühn: *Eponym* for the intestinal glands (Fig. L23), also described by Wepfer, his father-in-law, and Galeazzi. Reference: *Dissertatio de fabrica et actione villorum intestinorum tenuium hominis.* Wishof, Leiden, 1745.

Liebermeister, Carl von (1833–1901): German physician born in Ronsdorf and working in Basel and Tübingen.

Grooves of Liebermeister: *Eponym* for small developmental grooves on the diaphragmatic surface of the liver in the foetal period, which may persist after birth, also described by Zahn.

Lieutaud, Joseph (1703–1780): French physician, anatomist and pathologist born in Aix-en-Provence and working in Versailles.

Trigone of Lieutaud: *Eponym* for the vesical trigone. Reference: *Observations anatomiques sur la Structure de la Vessie.* Mém Acad Sci Paris 1753, pp. 1–26.

Uvula of Lieutaud: *Eponym* for the uvula of the urinary bladder.

Ligaments: *See* **The Classical Anatomical Eponyms.**

Liliequist, Bengt (1923–2008): Swedish neuroradiologist working in Stockholm.

Fig. L22 Photo of Leydig. (From Wikimedia Commons; in public domain)

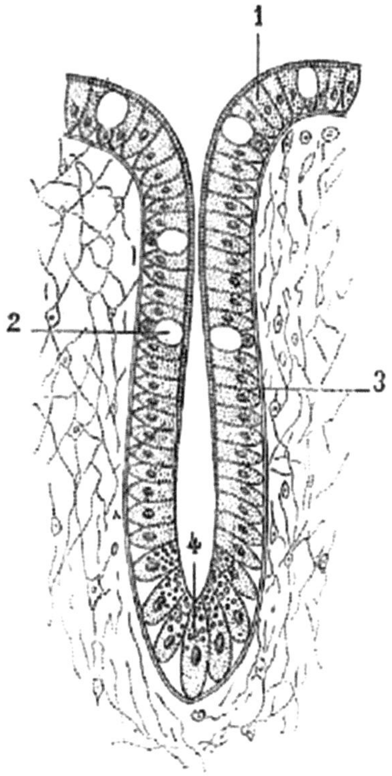

Fig. L23 Gland of Lieberkühn. (Courtesy Veronika Němcová, Prague). *1* superficial cell, *2* calciform cell, *3* proper membrane of gland, *4* granular or Paneth cell

Membrane of Liliequist: *Eponym* for an arachnoid membrane separating the chiasmatic, interpeduncular and prepontine cisterns, originating anteriorly from the diaphragma sellae and extending posteriorly as two sheets. Reference: *The anatomy of the subarachnoid cisterns.* Acta Radiol 1956; 46:61–71.

Lines:

The term **line** is used for many planes and lines, but also to indicate certain structures in teeth, the liver and the internal genital organs. More than 80 cephalic reference lines based on cranial and anthropological landmarks have been described, mainly by anatomists and anthropologists at the end of the nineteenth century. References: Saban R *Les planes d'orientation de la tête.* Cahiers Anthropol 1980; 3–4:285–313; Tamraz JC, Comair YG. *Atlas of Regional Anatomy of the Brain using MRI.* Springer, Berlin-Heidelberg-New York, 2000.

Line of Addison: *Eponym* for the transpyloric plane; *see* Addison, Christopher.

Line of Arlt: *Eponym* for a ventral line corresponding to the projection of the nasolacrimal duct and extending from the medial palpebral ligament to the beginning of the nasolacrimal fold; *see* Arlt, Carl Ferdinand.

Lines of Blaschko: *Eponym* for a system of lines, representing a developmental growth pattern; *see* Blaschko, Alfred.

Lines of Borges: *Eponym* for the resting skin tension lines, rather similar to the lines of Kraissl; *see* Borges, Alfred F.

Line of Cantlie: *Eponym* for a line, which divides the two halves of the liver, extending from the fossa of the gallbladder anteriorly to the inferior vena cava posteriorly, also described as line of Rex-Cantlie; *see* Cantlie, James.

Line of Dobie: *Eponym* for the dark line (isotropic band), generally known as Z-line, which runs transversally through bands of myofibrils and marks the limits of a sarcomere; *see* Dobie, William Murray.

Line of Donaldson: *Eponym* for an imaginary line bisecting the lateral semicircular canal and transecting the posterior semicircular canal to locate the endolymphatic sac, situated inferior to the line; *see* Donaldson, James A.

Line of Douglas, Arcuate: *Eponym* for the semilunar line representing the caudal boundary of the posterior layer of the sheath of the rectus abdominis muscle; *see* Douglas, James.

Lines of Ebner: *Eponym* for poorly defined incremental lines corresponding to growth of dentin; *see* Ebner, Victor von.

Line of Egger: *Eponym* for a circular line, the outer border of the hyalocapsular ligament of Wieger; *see* Egger, Fritz.

Line of Farabeuf: *Eponym* or the iliotrochanteric line; *see* Farabeuf, Louis Hubert.

Line of Farre, White: *Eponym* for the line of attachment of the peritoneum of the mesovarium to the substance of the ovary, also described by Romiti and Waldeyer; *see* Farre, Arthur.

Line of Hilgenreiner: *Eponym* for a line drawn on pelvic radiographs to evaluate the acetabular angle; *see* Hilgenreiner, Heinrich.

Line of Hilton, White: *Eponym* for the anocutaneous line, the white line of the anus; *see* Hilton, John.

Line of Hudson-Stähli: *Eponym* for a brownish line of iron deposition across the lower third of the cornea; *see* Hudson, Arthur Cyril, and Stähli, Jean.

Line of Hunter: *Eponym* for the linea alba of the abdominal wall; *see* Hunter, John.

Lines of Hunter-Schreger: *Eponym* for alternating dark and clear lines, generally oriented perpendicular to the surface of the enamel representing sections of neighbouring zones of crossed enamel prisms; *see* Hunter, John, and Schreger, Christian Heinrich Theodor.

Line of Kilian: *Eponym* for a transverse line marking the level of the promontory of the pelvis; *see* Kilian, Hermann Friedrich.

Lines of Kraissl: *Eponym* for lines of maximal skin tension; *see* Kraissl, Cornelis J.

Line of Langer: *Eponym* for cleavage lines of the skin due to disposition of the subcutaneous fibrous tissue; *see* Langer, Karl.

Line of Lanz: *Eponym* for the interspinous line connecting the right and left anterior superior iliac spines; *see* Lanz, Otto.

Line of Linton: *Eponym* for a line perpendicular to the sole plate and running about 1.5–2 cm behind the medial malleolus, parallel to the great saphenous vein; *see* Linton, Robert R.

Line of Mikulicz-Radecki: *Eponym* for a line from the head of the femur (in Anglosaxon literature from the middle of the inguinal ligament) to the middle of the ankle joint; *see* Mikulicz-Radecki, Johann von.

Line of Monro-Richter: *Eponym* for the right spinoumbilical line between the right anterior superior iliac spine and the umbilicus, later also described by Richter. It serves to determine the beginning of the vermiform appendix, located 6–7 cm medial from the spine, at the point of McBurney; *see* Monro, Alexander, and Richter, August Gottlieb.

Line of Nélaton: *Eponym* for the line between the anterior superior iliac spine and the ischial tuberosity, also known as the line of Roser-Nélaton; *see* Nélaton, Auguste.

Line of Ogston: *Eponym* for an imaginary line from the greater trochanter of the femur to the intercondylar notch of the femur; *see* Ogston, Alexander.

Lines of Owen, Contour: *Eponym* for concentric layers of nonhomogenously calcified dentin occurring as a result of appositional growth; *see* Owen, Richard.

Line of Perkins: *Eponym* for a perpendicular line drawn on an anteroposterior radiograph of the pelvis, intersecting the lateral edge of the acetabular roof; *see* Perkins, George.

Lines of Pickerill, Imbrication: *Eponym* for the external manifestation of the lines of Retzius, parallel impacted lines (perikymata) extending often along the whole crown of the tooth, better visible in young individuals; *see* Pickerill, Henry Percy.

Line of Poupart: *Eponym* for a perpendicular line passing through the midpoint of the inguinal ligament; *see* Poupart, François.

Lines of Retzius, Growth: *Eponym* for growth lines or incremental lines or stripes running obliquely from the surface of the enamel towards the root of the tooth, occurring as a result of appositional growth; *see* Retzius, Anders.

Line of Rex-Cantlie: *Eponym* for the anatomical midline of the liver connecting the fundus of the gallbladder with the inferior vena cava, described by both Rex and Cantlie; *see* Rex, Hugo, and Cantlie, James.

Line of Romiti: *Eponym* for the boundary between the ovarian epithelium and the peritoneal serosa, also described by Farre and Waldeyer; *see* Romiti, Guglielmo.

Line of Roser-Nélaton: *Eponym* for a line from the anterior superior iliac spine to the ischial tuberosity; *see* Nélaton, Auguste, and Roser, Wilhelm.

Lines of Salter, Incremental: *Eponym* for incremental lines sometimes seen in dentin due to hypercalcification; *see* Salter, James Augustus.

Lines of Schreger: *Eponym* for alternating dark and clear lines, generally oriented perpendicular to the surface of the enamel representing sections of neighbouring zones of crossed enamel prisms, also described by Hunter, and known as lines of Hunter-Schreger; *see* Schreger, Christian Heinrich Theodor.

Line of Shenton-Ménard: *Eponym* for a curved line formed by the top of the obturator foramen and the inner side of the neck of the femur, seen on an anteroposterior frontal X-ray of a normal hip joint; *see* Shenton, Edward Warren Hine, and Ménard, Maxime.

Line of van den Spieghel: *Eponym* for the border between the muscle fibres of the transversus abdominis muscle and its aponeurosis; *see* van den Spieghel, Adriaan.

Line of Vesling: *Eponym* for the raphe or middle layer of the scrotum; *see* Vesling, Johann.

Lines of Voight: *Eponym* for the hair tracts; *see* Voight, Christian August.

Line of Waldeyer, White: *Eponym* for a line on the ovary at the junction of the germinal epithelium and the peritoneum, also described by Farre and Romiti; *see* Waldeyer, Heinrich Wilhelm Gottfried.

Lines (Anthropological; *see* Figs. L24 and L25**):**

Line of von Baer: *Eponym* for the Göttingen horizontal line through the superior border of the zygomatic arch; *see* Baer, Karl Ernst von.

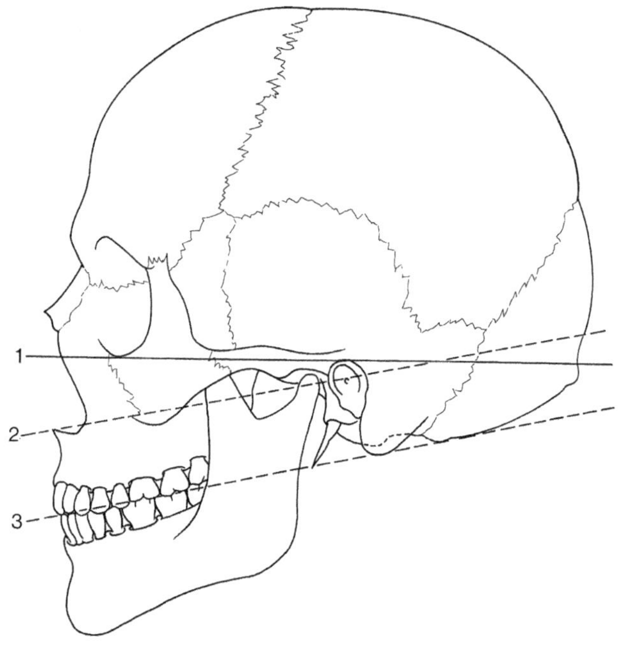

Fig. L24 Anthropological lines I: *1* Frankfurter Horizontale or Line of Virchow-Hoelder, *2* Line of Camper, *3* Line of Blumenbach

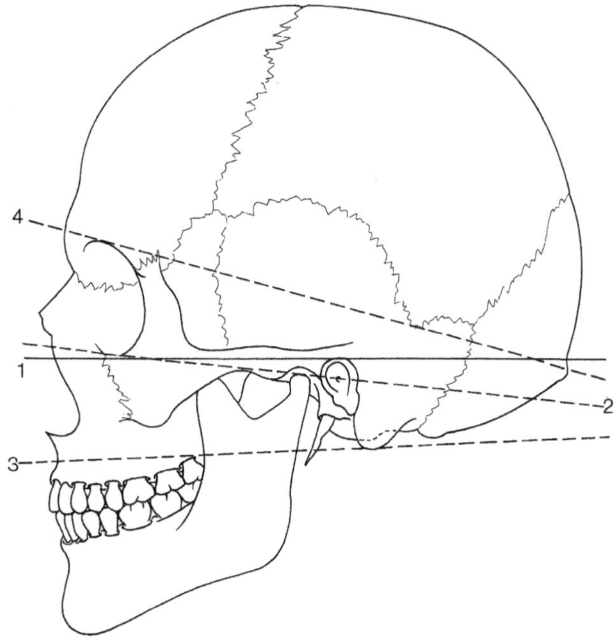

Fig. L25 Anthropological lines II: *1* Frankfurter Horizontale or Line of Virchow-Voelder, *2* Line of von Ihering, *3* Line of Broca, *4* Line of Schwalbe or glabella-inion line

Line of Blumenbach: *Eponym* for the cranium in norma verticalis, lying on its base over a horizontal plane. This plane passes through the points of contact at the maxillary teeth anteriorly and the mastoid processes posteriorly; *see* Blumenbach, Johann Friedrich.

Lines of Broca: Broca distinguished several reference lines and planes.

Horizontal plane of the head (1862): *Eponym* for the alveolar-condylar plane from the alveolar point at the inferior border of the alveolar ridge of the maxilla to the inferior aspect of both occipital condyles.

Plane of horizontal vision (1862) or Visual plane (1873): *Eponym* for the natural attitude of the head that permits the eyes to reach the horizon without muscular contraction. This plane was accepted by British anthropologists as well. German anthropologists preferred the plane of Virchow-Hoelder, which came to be known as the Frankfurt-Virchow plane or 'Frankfurter Horizontale'. In 1877, Broca renamed this plane as **bi-orbital plane**, corresponding to the **neuro-ocular plane** of Cabanis et al. (J Comput Assist Tomogr 1978; 2:150–155).

Line of Camper: *Eponym* for a horizontal plane from the anterior nasal spine to the centre of the external acoustic meatus on both sides; *see* Camper, Petrus.

Line of Cuvier: *Modification* of the horizontal line of Camper; *see* Cuvier, Georges.

Line of Daubenton and Daele: *Eponym* for the plane of the foramen magnum, the first search for a horizontal plane; *see* Daubenton, Louis Jean Marie.

Line of Doornik: *Eponym* for a horizontal line from the upper incisors to the most prominent point of the occiput; *see* Doornik, Jacob Elias.

Line of Hamy: *Eponym* for the glabella-lambda line, roughly parallel to the alveolar-condylar plane of Broca; *see* Hamy, D.

Line of Ihering: *Eponym* for a horizontal line from the inferior border of the orbit to the middle of the external acoustic meatus, a preliminary version of the Frankfurter Horizontale; *see* Ihering, Hermann von.

Line of Lucae: *Eponym* for a horizontal line from the anterior nasal spine to the basion; *see* Lucae, Johann Christian Gustav.

Line of Merkel (1882): *Eponym* for the horizontal orbital-auditory line from the centre of the external acoustic meatus to the inferior border of the orbit, a modification of the Virchow-Hoelder plane; *see* Merkel, Friedrich.

Line of Poirier: *Eponym* for a line from the nasion to the lambda; *see* Poirier, Paul Julien.

Line of Reid, Base: *Eponym* for the auriculoorbital line of the skull passing between the inferior margin of the orbit and the centre of the external acoustic meatus, comparable to the line of Ihering; *see* Reid, Robert William.

Line of Virchow-Hoelder: *Eponym* for the horizontal line of Munich, the supra-auriculo-suborbital plane, from the superior border of the external acoustic meatus to the inferior border of the orbit. This plane was adopted in Frankfurt am Main as the **infraorbital-meatal plane** ('Frankfurter Horizontale').

Linton, Robert R (twentieth century): American surgeon working in Boston.

Line of Linton: *Eponym* for a line perpendicular to the sole plate and running about 1.5–2 cm behind the medial malleolus, parallel to the great saphenous vein. The line serves for identification of the middle paratibial perforating vein of Sherman and the three posterior tibial perforating veins of Cockett. Reference: *The communicating veins of the lower leg and the operative technique for their ligation.* Ann Surg 1938; 107:582–593.

Lisfranc de St. Martin, Jacques (1790–1847): French surgeon from St. Paul, Loire, working in Paris (Fig. L26).

Joint of Lisfranc: *Eponym* for the tarsometatarsal joint (*see* Fig. C11b).

Ligament of Lisfranc: *Eponym* for the interosseous tarsometatarsal ligament between the second metatarsal bone and the medial cuneiform bone. Reference: *Mémoire sur l'amputation du pied dans son articulation tarso-métatarsienne.* Gabon, Paris, 1815.

Tubercle of Lisfranc: *Eponym* for the scalene tubercle on the first rib for the attachment of the scalenus anterior muscle. Reference: *Mémoire sur l'amputation du bras dans l'articulation de l'épaule.* Paris, 1815.

Lissauer, Heinrich (1861–1891): German neurologist born in Neidenburg and working in Breslau (Fig. L27).

Tract or Fasciculus of Lissauer: *Eponym* for the posterolateral or dorsolateral fasciculus of the spinal cord. Reference: *Beitrag zum Faserverlauf im Hinterhorn des menschlichen Rückenmarks und zum Verhalten desselben bei Tabes dorsalis.* Arch Psychiatr Nervenkr 1886; 17:377–438.

Fig. L26 Portrait of Lisfranc. (From Wikidata; in public domain)

Fig. L27 Photo of Lissauer. (From Wikidata; in public domain)

Lister, Lord Joseph (1827–1912): English surgeon born in Stroke Newington, London, working in Glasgow, Edinburgh and finally in London. He is particularly remembered for his application of antiseptic methods in surgery.

Tubercle of Lister: *Eponym* for the dorsal tubercle of the radius, a prominence on the posterior surface of the distal end of the radius between the groove for the tendon of the extensor pollicis longus muscle medially, and the extensor carpi radialis brevis muscle laterally, serving as a pulley of the former. Reference: *On the excision of the wrist.* Lancet 1865.

Little, James Lawrence (1836–1885): American surgeon.

Little area of the nasal septum: *Eponym* for an arterial plexus in the anterior part of the nasal septum, the most common site of epistaxis, also described by Kiesselbach. Reference: *A hitherto undescribed lesion as a cause of epistaxis, with four cases.* The Hospital Gazette, New York 1879; 6:5–6.

Littré, Alexis (1658–1725): French surgeon and anatomist born in Cordes and working in Paris (Fig. L28).

Glands of Littré: *Eponym* for the urethral glands in males, also described by Morgagni. Reference: *Description de l'urèthre de l'homme.* Hist Acad Roy Sci, Paris 1719, pp. 311–316.

Hernia of Littré: *Eponym* for a parietal hernia, in which only a portion of the wall of the intestine is engaged, also described by Richter.

Sinus of Littré: *Eponym* for the venous basilar plexus, an anastomosis between the cavernous sinus and the inferior petrosal and marginal sinuses.

Lobes:

Lobe of Broca, Great limbic: *Eponym* for the cingulate and parahippocampal gyri of the cerebral cortex; *see* Broca, Pierre Paul.

Lobe of Home: *Eponym* for the median lobe of the prostate; *see* Home, Everard.

Lobe of Riedel: *Eponym* for a tonguelike process from the right lobe of the liver lateral to the gallbladder; *see* Riedel, Bernhard Moritz Carl Ludwig.

Lobe of Rokitansky: *Eponym* for an accessory lobe of the lung; *see* Rokitansky, Carl von.

Lobe of van den Spieghel: *Eponym* for the caudate lobe of the liver (Spigelian lobe); *see* van den Spieghel, Adriaan.

Fig. L28 Portrait of Littré. (From Pagel 1901)

Lobe of Wrisberg: *Eponym* for the lobe of the azygos vein, a supernumerary lobe of the right lung related to the azygos vein; *see* Wrisberg, Heinrich August.

Lobstein, Jean Frédéric (1777–1835): French anatomist and pathologist of German origin (born in Giessen), educated in France and working in Strassburg (Fig. L29). References: *Handbuch der Hebammemkunst.* Strassburg, 1827; *Traité d'anatomie pathologique.* Levrault, Paris, 1829.

Ganglion of Lobstein: *Eponym* for an accessory ganglion of the great thoracic splanchnic nerve in connection with the coeliac plexus. Reference: *De nervi sympathetici humani fabrica.* Levrault, Paris, 1823.

Lockwood, Charles Barrett (1856–1914): English surgeon and anatomist born in Stockton-on-Tees and working in London, known for his studies of the eye. He founded the Anatomical Society of Great Britain and Ireland (1887). Reference: *The anatomy of the muscles, ligaments and fasciae of the orbit, including an account of the capsule of Tenon, the check ligaments of the recti and the suspensory ligaments of the eye.* J Anat (Lond) 1886; 20:1–25.

Suspensory ligament of Lockwood: *Eponym* for the suspensory ligament of the eyeball.

Loewenthal, Wilhelm (1850–1894): German physician born in Berlin and working in Lausanne.

Tract or **Bundle of Loewenthal:** *Eponym* for the tectospinal tract, also described by Edinger.

Loop of Hyrtl: *Eponym* for a communicating loop between the right and left hypoglossal nerves; *see* Hyrtl, Joseph.

Fig. L29 Portrait of Lobstein. (From Wikimedia Commons; in public domain)

Fig. L30 Photo of Lorente de Nó. (From the Real Academia de la Historia, Madrid; in public domain)

Lorente de No, Rafael (1902–1990): Spanish histologist who introduced the subdivision of the hippocampus into the CA1-CA4 fields (Fig. L30). Reference: *Studies on the structure of the cerebral cortex.* J Psychol Neurol (Lpz) 1933; 45:381–438; ibid. 1934; 46:113–177.

Louis, Pierre Charles Alexandre (1787–1872): French physician born in Ay, Champagne and working in Paris and a leading authority on pulmonary tuberculosis (Fig. L31). Reference: *Recherches anatomico-pathologiques sur la phtisie.* Gabon, Paris, 1835.

Angle of Louis (Angulus Ludovici): *Eponym* for the sternal angle, the angle between the manubrium and the body of the sternum.

Löwenberg, Benjamin Benno (1836–1905): Austrian otolaryngologist working in Vienna and in Paris.

Canal of Löwenberg: *Eponym* for the cochlear duct, already described by Reissner. Reference: *La lame spirale du limaçon de l'oreille et des mammifères.* Paris, 1864 (also published in J Anat (Paris) 1868, pp. 626–652).

Lower, Richard (1631–1691): English physician and anatomist born in Tremere, Cornwall, and working in London.

Tendinous circles of Lower: *Eponym* for the fibrous anuli of the heart, circular rings for the attachment of the aortic, pulmonary, mitral and tricuspid valves. Reference: *Tractatus de corde.* Redmayne, London, 1669.

Fig. L31 Portrait of Louis. (From Wikimedia Commons; in public domain)

Tubercle of Lower: *Eponym* for the intervenous tubercle, a prominence on the right atrial wall between the two caval openings. Reference: *Tractatus de corde.* Redmayne, London, 1669.

Löwy, R. (19th–twentieth century): Austrian neurologist known for his studies on the flocculus of the cerebellum. Reference: *Ueber die Faseranatomie und Physiologie der Formatio vermicularis cerebelli.* Arb Neurol Inst Wiener Univ 1916; 21:359–382.

Bundle or Tract of Löwy: *Eponym* for the angular bundle of the flocculus.

Lucae, Johann Christian Gustav (nineteenth century): German anthropologist.

Line of Lucae: *Eponym* for a horizontal line from the anterior nasal spine to the basion, the rostral border of the foramen magnum. Reference: *Zur Morphologie der Rassen-Schädel.* Brönner, Frankfurt am Main, 1861.

Lucas, Richard Clement (1846–1915): English surgeon, born in Midhurst, Sussex, and working in London.

Groove or **Sulcus of Lucas:** *Eponym* for the spinous sulcus, a shallow groove on the spine of the sphenoidal bone caused by the chorda tympani.

Ludwig, Karl Friedrich Wilhelm (1816–1895): German physiologist and anatomist (Fig. L32), born in Witzenhausen, subsequently working in Marburg, Zürich, Vienna and Leipzig. Reference: *Lehrbuch der Physiologie des Menschen.* Leipzig, 1862–1869.

Depressor nerve of Ludwig: *Eponym* for the aortic nerve, a branch of the vagus nerve, which ends in the aortic arch and the base of the heart; described together with Elie de Cyon (*Die Reflexe eines der sensiblen Nerven des Herzens auf die Motorischen der Blutgefässe.* Arch Physiol Anst Lpz 1866, pp. 307–328; *Action reflexe d'un der nerfs sensibles du coeur sur les nerfs vasomoteurs.* J Anat Physiol 1867, pp. 472–485); also described by Hofer.

Ganglion of Ludwig: *Eponym* for a small collection of parasympathetic nerve cells in the interatrial septum of the heart.

Labyrinth of Ludwig: *Eponym* for the convoluted part of a nephron.

Lugaro, Ernesto (1870–1890): Italian anatomist from Sicily (Fig. L33).

Cells of Lugaro: *Eponym* for the fusiform horizontal cells of the cerebellar cortex. Reference: *Sulle connessioni tra gli elementi nervosi della corteccia cerebellare con considerazioni generali sul significato fisiologico dei rapporti tra gli elementi nervosi.* Rev. Sper Freniat 1894; 20:297–331.

Fig. L32 Portrait of Ludwig. (From Pagel 1901)

Fig. L33 Photo of Lugaro. (From Wikimedia Commons; in public domain)

Luna, Emerico (1882–1963): Italian anatomist born and working in Palermo.

Olivary vascular capsule of Luna: *Eponym* for a close arteriolar vascular network around the inferior olivary nucleus.

Lund, Fred Bates (1865–1950): American surgeon born in Concord, Merrimack, New Hampshire, and working in Boston.

Node of Lund: *Eponym* for the cystic lymph node, the sentinel lymph node for the gallbladder in the cystohepatic triangle of Calot, in 1787 already described by Mascagni, and later by Calori.

Luschka, Hubert von (1820–1875): German anatomist born in Konstanz and working in Tübingen (Fig. L34) after whom various structures are named, although many were already described by others.

Arterial cone of Luschka: *Eponym* for the conus arteriosus, the outflow part of the right ventricle of the heart, also known as infundibulum of Wolff.

Cartilage of Luschka: *Eponym* for interarytenoid and other accessory laryngeal cartilages connected with the arytenoid cartilages.

Ducts of Luschka: *Eponym* for tubular structures, which sometimes branch from the bile ducts in the wall of the gallbladder, probably aberrant bile ducts.

Fascia of Luschka: *Eponym* for the endothoracic or parietal thoracic fascia. Reference: *Die Brustorgane des Menschen in ihrer Lage.* Laupp, Thüringen, 1857.

Fig. L34 Portrait of Luschka. (From Pagel 1901)

Fold of Luschka: *Eponym* for the ventral vascular ileocaecal fold.

Foramen of Luschka: *Eponym* for the lateral aperture of the fourth ventricle (Fig. L35), also described by Mierzejewsky, Key and Retzius. Reference: *Die Adergeflechte des menschlichen Gehirnes. Eine Monographie.* Reimer, Berlin, 1855.

Fossa of Luschka: *Eponym* for the superior ileocaecal fossa of the peritoneal cavity.

Gland of Luschka: *Eponym* for the coccygeal body (glomus coccygeus).

Joints of Luschka: *Eponym* for the uncovertebral joints between the uncinate processes of the superior surfaces of the bodies of cervical vertebrae and the

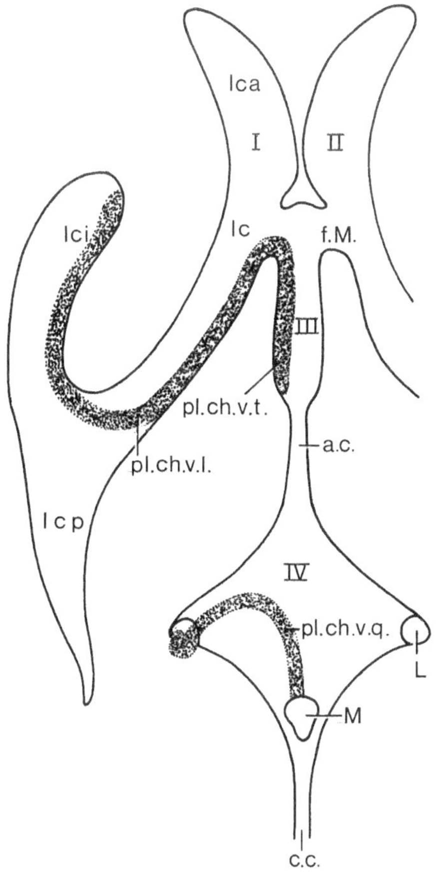

Fig. L35 Overview of the ventricular system in dorsal view. *a.c.* cerebral (mesencephalic) aqueduct, *c.c.* central canal, *f.M.* foramen of Monro, *L*, *M* apertures of Luschka and Magendie, *pl.ch.v.l.*, *pl.ch.v.q.*, *pl.ch.v.t.* choroid plexuses of lateral, fourth and third ventricles, *I-IV* ventricles, *Ica*, *Ic*, *Ici*, *Icp* anterior (frontal) horn, central part, inferior (temporal) horn and posterior (occipital) horn of lateral ventricle. (From ten Donkelaar HJ, Kachlík D, Tubbs RS *An Illustrated Terminologia Neuroanatomica.* Springer, Cham, 2018)

inferior surface of the vertebral bodies superior to them, usually in the upper part of the cervical vetebral column.

Ligaments of Luschka: *Eponym* for (a) the sterno-pericardial ligament (also described by Lannelongue) and (b) the xiphopericardial ligament, the inferior sternopericardial ligament.

Muscle of Luschka: *Eponym* for the rectouterine muscle, smooth muscle fibres in the lesser pelvis.

Nerve of Luschka: *Eponym* for (a) the posterior ethmoidal nerve and (b) the meningeal branch of each spinal nerve, which supplies structures within the vertebral canal, also known as the sinuvertebral nerve.

Recess of Luschka: *Eponym* for (a) the ileocaecal recess of the peritoneal cavity, also described by Hartmann. Reference: *Über peritoneale Umhüllung des Blinddarmes und über die Fossa ileocaecalis.* Arch Path Anat 1861; 21:285–288; and (b) the ileocolic recess of the peritoneal cavity, also described by Lockwood.

Sinus of Luschka: *Eponym* for the petrosquamous sinus, a venous sinus in the petrosquamous suture of the temporal bone, also described by Krause.

Tonsil of Luschka: *Eponym* for the pharyngeal tonsil.

Tubercle of Luschka: *Eponym* for (a) the urethral carina of the vagina, and (b) an inconstant tubercle on the superior angle of the scapula.

Lütkens, Ulrich (1894-?): German surgeon working in Freiburg and Berlin.

Sphincter of Lütkens, Spiral: *Eponym* for a circular bundle of smooth muscle fibres in the neck of the gallbladder, close to the spiral valve of Heister. Reference: *Aufbau und Funktion der extrahepatischen Gallenwege.* Vogel, Berlin, 1926.

Luys, Jules Bernard (1828–1898): French neurologist born and working in Paris (Fig. L36). In 1881, he founded L'Encéphale and was appointed Director of the Asylum at Ivry until 1893. Reference: *Recherches sur le système nerveux cérébrospinal.* Baillière, Paris, 1865. For biography, *see* Parent A, Parent M and Leroux-Hugon V *Jules Bernard Luys: A singular figure of nineteenth century neurology.* Can J Neurol Sci 2002; 29:282–288.

Body or **Nucleus of Luys:** *Eponym* for the subthalamic nucleus (Fig. L37).

Centre médian of Luys: *Eponym* for the centromedian nucleus of the thalamus.

Lyra of David: *Eponym* for the hippocampal commissure or psalterium, named after the biblical figure of David.

Lyra uterina: *Eponym* for the palmate folds of the cervical canal of the uterus.

Fig. L36 Portrait of Luys. (From Wikimedia Commons; in public domain)

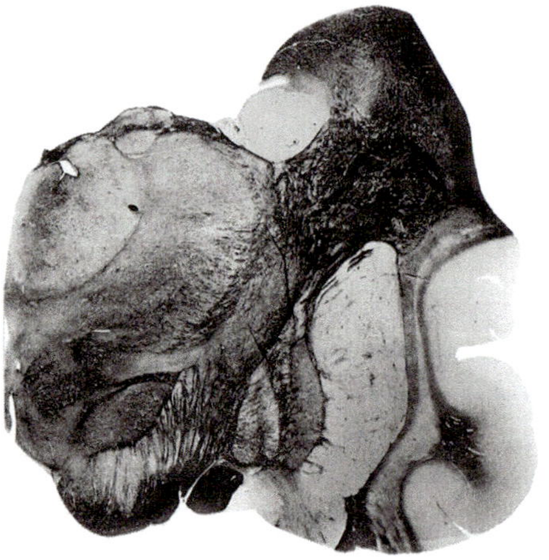

Fig. L37 Frontal Weigert-stained section of the human basal ganglia. The subthalamic nucleus, surrounded by its capsule, is fully developed above the pes pedunculi. The globus pallidus begins to diminish. (Adapted from Jelgersma G *Atlas Anatomicum Cerebri Humani.* Scheltema and Holkema, Amsterdam, 1931; from ten Donkelaar HJ *Clinical Neuroanatomy,* 2nd ed. Springer, Cham, 2020)

13 Letter M

Macalister, Alexander (1844–1919): Irish zoologist, anatomist and surgeon in Dublin and later in Cambridge.

Muscle of Macalister: *Eponym* for an inconstant cricotracheal muscle. Reference: *Additional Observations on Muscular Anomalies.* Trans Roy Irish Acad 1871; 25:1–134. He also described several other accessory muscles.

Macewen, Sir William (1848–1924): Scottish surgeon born in Bute and working in Glasgow.

Triangle of Macewen: *Eponym* for the suprameatal triangle or fossa on the squamous part of the temporal bone.

Mackenrodt, Alwin (1859–1925): German gynaecologist and pathologist born in Nordhausen and working in Berlin (Fig. M1). For biography, *see* Ludwig H *Alwin Mackenrodt (1859–1925). Die quere Beckenfaszie.* Der Gynäkologe 2003; 36:165–166.

Ligament of Mackenrodt: *Eponym* for the cardinal or transverse cervical ligament of the uterus (Fig. M2). Reference: *Ueber die Ursachen der normalen und pathologischen Lagen der Uterus.* Arch Gynaekol 1895; 48:393–421.

MacLean, Paul Donald (1913–2007): American physician and neuroanatomist born in Phelphs, New York, and working in Seattle, Harvard and later at Yale School of Medicine (Fig. M3a). He suggested the term **limbic system** to combine the circuit of Papez with the amygdala and the hypothalamus, and thought responsible for emotional behaviour (the 'visceral' or 'emotional' brain; Fig. M3b). References: *Some psychiatric implications of physiological studies on frontotemporal portions of limbic system (visceral brain).* Electroencephalogr Clin Neurophysiol 1952; 4:407–418; *Contrasting functions of limbic and neocortical systems of the brain and their relevance to psychophysiological aspects of medicine.* Am J Med 1958:25:611–626.

Magendie, François (1783–1855): French physiologist born in Bordeaux and working in Paris (Fig. M4a) who confirmed Bell's earlier findings that the anterior or ventral roots of the spinal nerves are motor and determined that the posterior or dorsal roots are sensory (Law of Bell-Magendie). References: *Mémoire physiologique sur le cerveau.* J Phys Exp Path, Paris 1828; 8:22; *Leçons sur les fonctions et les maladies du système nerveux.* Ebrard, Paris, 1839.

Foramen of Magendie: *Eponym* for the median aperture of the fourth ventricle (Fig. M4b)

Spaces of Magendie: *Eponym* for subarachnoid spaces between the pia mater and the arachnoid at the level of the sulci of the cerebral cortex.

Maier, Rudolf (1824–1888): German pathologist in Freiburg.

Sinus of Maier: *Eponym* for the fornix of the lacrimal sac. Reference: *Ueber den Bau der Thränenorgane.* Freiburg, 1859.

Maissiat, Jacques Henri (1805–1878): French anatomist born in Nantua and working in Paris.

Band or **Bandelette of Maissiat:** *Eponym* for the iliotibial tract of the fascia lata, also described as ligament or tract. Reference: *Etudes de physique animale.* Paris, 1843.

H. J. ten Donkelaar et al., *An Illustrated Guide to Anatomical Eponyms*, https://doi.org/10.1007/978-3-031-91664-9_13

Fig. M1 Photo of Mackenrodt. (From Ludwig H *Alwin Mackenrodt (1859–1925). Die quere Beckenfaszie.* Der Gynäkologe, 2003; 36:165–166)

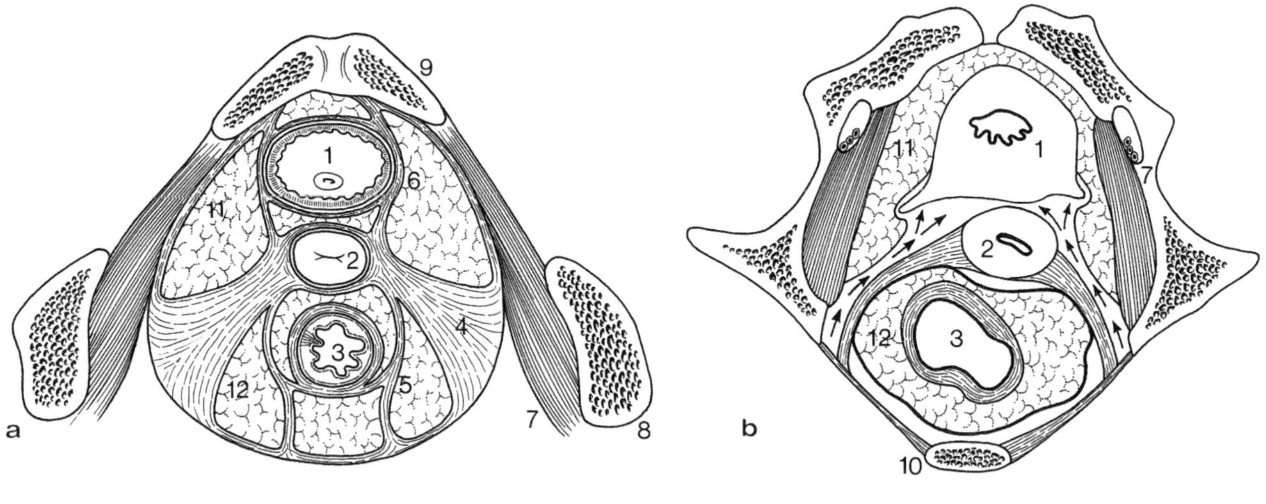

Fig. M2 Ligament of Mackenrodt: (**a**) The classic view; (**b**) modern insights, where this ligament is merely seen as a passage of vessels and nerves. (After Fritsch H, Lienemann A, Brenner E, Ludwikowski B *Clinical Anatomy of the Pelvic Floor.* Adv Anat Embryol Cell Biol 2004; 175:1–66; and ten Donkelaar HJ, Oostra R-J *Klinische Anatomie en Embryologie,* 4th ed. Springer Media, Houten, 2014). *1* bladder, *2* cervix uteri, *3* rectum, *4* cardinal ligament of Mackenrodt, *5* sacrouterine ligament, *6* pubovesical ligament, *7* obturator internus muscle, *8* os ischii, *9* os pubis, *10* os coccygis, *11* paravesical space, *12* pararectal space. The *arrows* in (**b**) indicate the passage of vessels and nerves

Malacarne, Michele Vincenzo Giacinto (1744–1816): Italian anatomist, surgeon and obstetrician born in Saluzzo, and working at Acqui and later in Pavia and Parma known for his studies on the cerebellar lobules.

Pyramid of Malacarne: *Eponym* for a prominence on the inferior surface of the vermis of the cerebellum. Reference: *Nuova esposizione della vera struttura del cervelletto umano.* Briolo, Turin, 1776.

Space of Malacarne: *Eponym* for the posterior perforated substance of the brain stem.

Malassez, Louis Charles (1842–1909): French anatomist and histologist.

Residues of Malassez: *Eponym* for the accumulation of epithelial cells in the alveolodental ligament derived from the primitive sheath of Hertwig. Reference: *Sur*

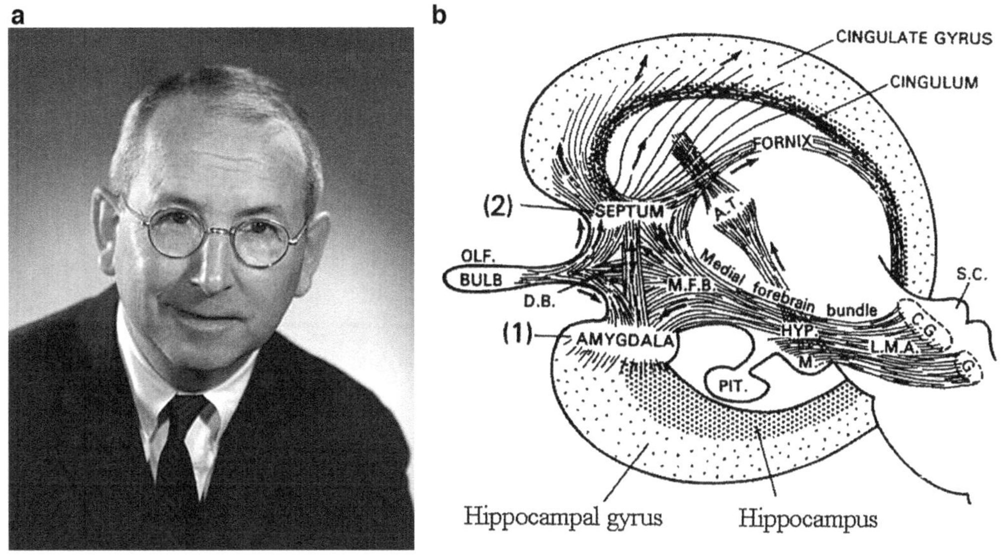

Fig. M3 (**a**) Portrait of MacLean. (From Wikimedia Commons; in public domain); (**b**) MacLean's limbic system. (After MacLean PD *Contrasting functions of limbic and neocortical systems of the brain and their relevance to psychophysiological aspects of medicine.* Am J Med 1958; 23:611–626; from ten Donkelaar HJ, Kachlík D, Tubbs RS *An Illustrated Terminologia Neuroanatomica.* Springer, Cham, 2018). *A.T.* mammillothalamic tract, *C.G.* central gray of rostral mesencephalon, *D.B.* diagonal band nuclei, *G* gray matter of caudal mesencephalon, *HYP* hypothalamus, *L.M.A.* limbic midbrain area of Nauta, *M.* mammillary body, *M.F.B.* medial forebrain bundle, *PIT* pituitary gland, *S.C.* superior colliculus

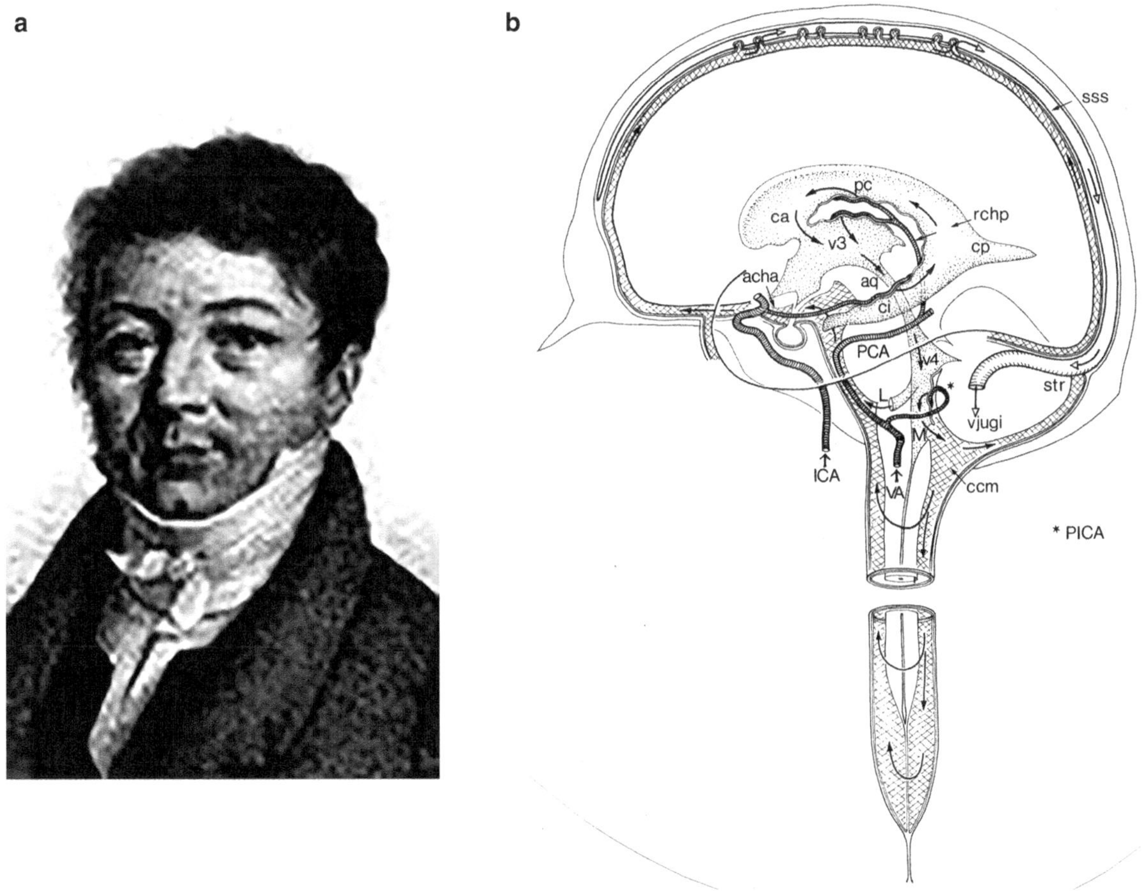

Fig. M4 (**a**) Portrait of Magendie. (From Pagel 1901); (**b**) Diagram of the liquor circulation. *Acha* anterior choroidal artery, *aq* aqueduct, *ca*, *ci*, *cp* anterior, inferior and posterior horn (cornu) of the lateral ventricle, *ccm* cisterna cerebellomedullaris, *ICA* internal carotid artery, *L*, *M* foramina of Luschka and Magendie, *pc* central part of lateral ventricle, *PICA* posterior inferior cerebellar artery, *rchp* ramus choroideus posterior, *sss* superior sagittal sinus, *str* straight sinus, *VA* vertebral artery, *vjugi* internal jugular vein, *v3*, *v4* third and fourth ventricles. (From ten Donkelaar HJ, Kachlík D, Tubbs RS *An Illustrated Terminologia Neuroanatomica.* Springer, Cham, 2018)

Fig. M5 Portrait of Malgaigne. (From Pagel 1901)

l'existence d'amas épithéliaux au tour de la racine des dents chez l'homme adulte et a l'état normal (débris épithéliaux paradentaires). Paris, 1895.

Malgaigne, Joseph François (1806–1865): French anatomist and surgeon born in Charnes-sur-Moselle and working in Paris (Fig. M5). Reference: *Traité d'anatomie chirurgicale.* Paris, 1837.

Accessory ligament of Malgaigne: *Eponym* for a supernumerary ligament of the shoulder joint between the acromion and the greater tubercle of the humerus.

Fossa or **Triangle of Malgaigne:** *Eponym* for the carotid triangle.

Hernia of Malgaigne: *Eponym* for an infantile inguinal hernia prior to the descent of the testis.

Mall, Franklin Paine (1862–1917): American anatomist born in Belle Plaine, Maine, and working in Chicago and as Director of Embryological Studies of the Carnegie Institute (Fig. M6). Reference: *Manual of Human Embryology,* 2 Vols (Keibel F and Mall FP, eds). Lippincott, Philadelphia, 1910–1912.

Space of Mall: *Eponym* for the periportal space between the hepatocytes and the connective tissue of the portal canal of the liver. Reference: *On the origin of lymphatics in the liver.* Bull Hopkins Hosp 1901; 12:146–148.

Fig. M6 Photo of Mall. (From Wikimedia Commons; in public domain)

Malpighi, Marcello (1628–1694): Italian physician born in Crevacuore, near Bologna, and working in Bologna, Pisa, Messina and again in Bologna known for his various anatomical studies (Fig. M7). In 1691, he was appointed physician to the Papal Court in Rome. He is regarded as the founder of microscopical anatomy. Reference: *De Pulmonis et de Viscerum Structura Exercitatio Anatomica.* Montius, Bologna, 1666; Martyn & Allestry, London, 1669. For a biography, *see* Ghosh SK, Kumar A *Marcello Malpighi (1628–1694): Pioneer of microscopic anatomy and exponent of the scientific revolution of the seventeenth century.* Eur J Anat 2018; 22:433–438.

Canal of Malpighi: *Eponym* for the longitudinal duct of the epoophoron, later also described by Gärtner.

Fig. M7 Portrait of Malpighi. (From Pagel 1901)

Capsule of Malpighi: *Eponym* for the hilar part of the splenic capsule.

Corpuscles of Malpighi: *Eponym* for (a) the renal corpuscles (Fig. M8), and (b) splenic corpuscles (lymphatic nodules of the white pulp of the spleen).

Layer of Malpighi: *Eponym* for the stratum germinativum of the epidermis, composed of the stratum basale and the stratum spinosum.

Pyramids of Malpighi: *Eponym* for the medullary pyramids of the kidney.

Mangold, Hilde (1898–1924): German embryologist born in Gotha as Hilde Pröscholdt (Fig. M9) working with Hans Spemann in Freiburg on the **Spemann-Mangold organizer** (*see* Spemann, Hans). Reference: *Ueber die Induktion von Embryonalanlagen durch Implantation artfremder Organisatoren.* Thesis University of Freiburg, 1923. For biography, *see* Fässler PE, Sander K *Hilde Mangold (1898–1924) and Spemann's organizer: Achievement and tragedy.* Roux Arch Dev Biol 1996; 205:323–332.

Marburg, Otto (1874–1948): Austrian neurologist and neuropathologist born in Römerstadt, Moravia (currently Rýmařov, Czech Republic). Reference: *Mikroskopisch-topographischer Atlas des menschlichen Zentralnervensystems mit begleitenden Texte,* 3. Aufl. Deuticke, Leipzig, Vienna, 1927.

Nucleus of Marburg: *Eponym* for the prepositus nucleus of the brain stem.

Marcacci, Arturo (1855–1915): Italian physiologist born in Pisa and working in Pavia.

Muscle of Marcacci: *Eponym* for a sheet of smooth muscle fibres underlying the areola and nipple of the mammary gland.

Marchand, Felix Jacob (1846–1928): German pathologist born in Halle and working in Breslau, Giessen, Marburg and Leipzig.

Glandules of Marchand: *Eponym* for suprarenal glands in abnormal position. Reference: *Beiträge zur Kenntniss der normalen und pathologischen Anatomie der Glandula carotica und der Nebennieren.* Rudolf Virchow's Festschrift gewidmet zur Vollendung seines 70. Lebensjahres 1891; 1:535–581.

Marchant, Gérard (1850–1903) French surgeon in Paris, born in Toulouse.

Zone of Marchant: *Eponym* for an area on the basisphenoid and basioccipital of the base of the skull, from which the dura mater can be easily detached ('zone decollable'). Reference: *Des épanchements sanguins intracraniens consécutifs au traumatisme.* Paris, 1881.

Marchi, Vittorio (1851–1908): Italian physician and anatomist born in Novellara, Reggio Emilia, and working in Modena who developed a technique for staining degenerating nerve fibres (Fig. M10).

Tract of Marchi: *Eponym* for the anterolateral descending tract of the cerebellum. Reference: *Sulle degenerazioni consecutive alla estirpazione totale e parziale del cervelletto.* Riv sper Fren Med Leg 1887; 13:446–452.

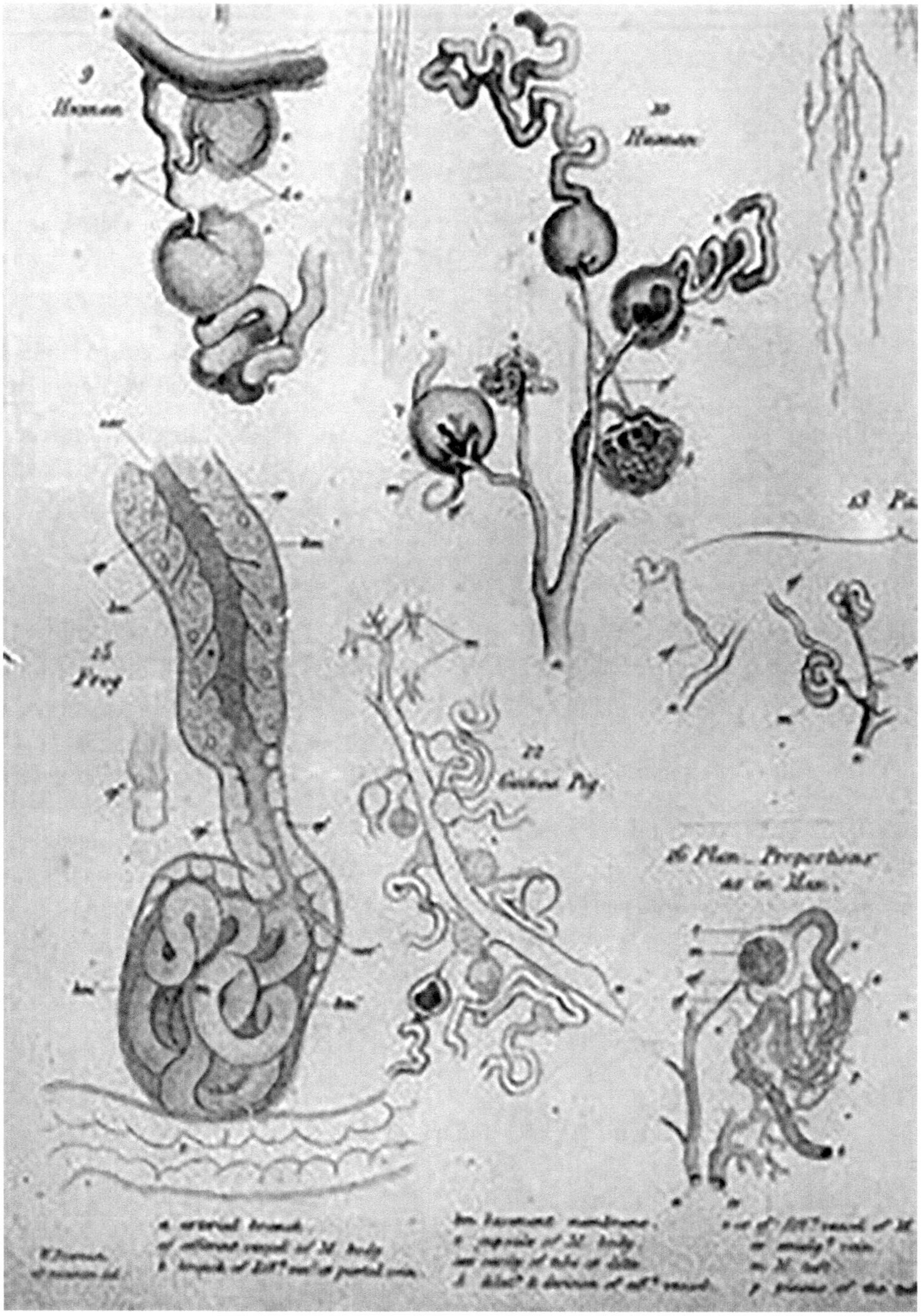

Fig. M8 Corpuscle of Malpighi. (From *De viscerum structura exercitatio anatomica*, 1666, as illustrated by Bowman; available in online version, in public domain)

Marcille, Maurice (1871–1941): French physician who worked with Poirier and Farabeuf in Paris on the iliopelvic area.

Fossa or **Triangle of Marcille:** *Eponym* for the lumbosacral triangle, bounded by the medial border of the psoas major muscle, the lateral margin of the vertebral column and the iliolumbar ligament; in it the obturator nerve and lymph nodes are found. It was also described by Cuéno, and is known as the fossa of Cuéno-Marcille. Reference: *Lymphatiques et ganglions ilio-pelviens.* Paris, 1902.

Marie, Pierre (1853–1940): French neurologist.

Bundle of Marie: *Eponym* for the sulcomarginal fasciculus of the spinal cord.

Marinacci, Albert A. (twentieth century): American neurologist working in Los Angeles.

Anastomosis of Marinacci: *Eponym* for a communicating branch of the ulnar nerve to the median nerve of the forearm. Reference: *The problem of unusual anomalous innervation of hand muscles. The value of electrodiagnosis in its evaluation.* Bull Los Angeles Neurol Soc 1964; 29:133–142.

Fig. M9 Photo of Mangold. (From Fässberg PE, Sander K *Hilde Mangold (1898–1924) and Spemann's organizer: Achievement and tragedy.* Roux Arch Dev Biol 1996; 205:323–332; courtesy Klaus Sander, Freiburg)

Fig. M10 Photo of Marchi. (From Wikimedia Commons; in public domain)

Mariotte, Edme (1620–1684): French priest born in Dijon and Prior of the Cloister of Saint Martin, celebrated for his experiments on vision and his discovery of the blind spot.

Blind spot of Mariotte: *Eponym* for the blind spot of the optic disc of the retina. Reference: *Nouvelles découvertes touchant la vue.* Paris, 1668.

Markowski, Józef (1847–1947): Polish anatomist born and working in Lemberg, Galicia, later known as Lvov, Poland, and currently as Lviv, Ukraine.

Vein of Markowski, Median prosencephalic: *Eponym* for the embryonic median prosencephalic vein. Its persistence may lead to the vein of Galen malformation. References: *Entwicklung der Sinus durae matris und der Hirnvenen des Menschen.* Bull Int Acad Sci Lett, Classe des sciences mathematiques et naturelles. Serie B 1921 (Suppl 1); 1–269; *Über die Entwicklung der Falx cerebri und des Tentorium cerebelli des Menschen mit Berücksichtigung ihres venösen Sinus.* Z Anat EntwGesch 1931; 94:395–439.

Marshall, John (1818–1891): English anatomist and surgeon born in Ely and working in London (Fig. M11a).

Fold of Marshall, Vestigial: *Eponym* for a pericardial fold containing the obliterated remains of the left superior vena cava.

Vein of Marshall: *Eponym* for the left oblique atrial vein, a small vein running on the posterior surface of the heart (Fig. M11b). Reference: *On the development of the great anterior veins in Man and Mammalia.* Phil Trans 1850; 140:133–170.

Martegiani, Francesco (1780–1862): Italian physician and surgeon in Naples.

Area or **Space of Martegiani:** *Eponym* for a funnel-shaped dilatation of the posterior end of the hyaloid canal anterior to the optic disc. Reference: *Novae observationes de oculo humano.* Cajetani Eboli, Naples, 1814.

Martin, Roland (1726–1788): Swedish anatomist who noted the communication between the median and ulnar nerves in the forearm, later also described by Gruber. Reference: *Tal om nervers allmänna egenskaperi människans kropp.* Las Salvius, Stockholm, 1763.

Anastomosis of Martin-Gruber: *Eponym* for an inconstant communicating branch of the median nerve to the ulnar nerve in the forearm, also described by Gruber.

Martinotti, Carlo (1859–1918): Italian pathologist and anatomist. References: *Contributo allo studio della corteccia cerebrale ed all'origine centrale dei nervi.* Ann Freniat Torino, 1888–1889, pp. 314–332; *Beitrag zum Studien der Hirnrinde und der Centralursprung der Nerven.* Int Monatssch Anat Physiol J Int Anat Physiol 1890; 7:69–90.

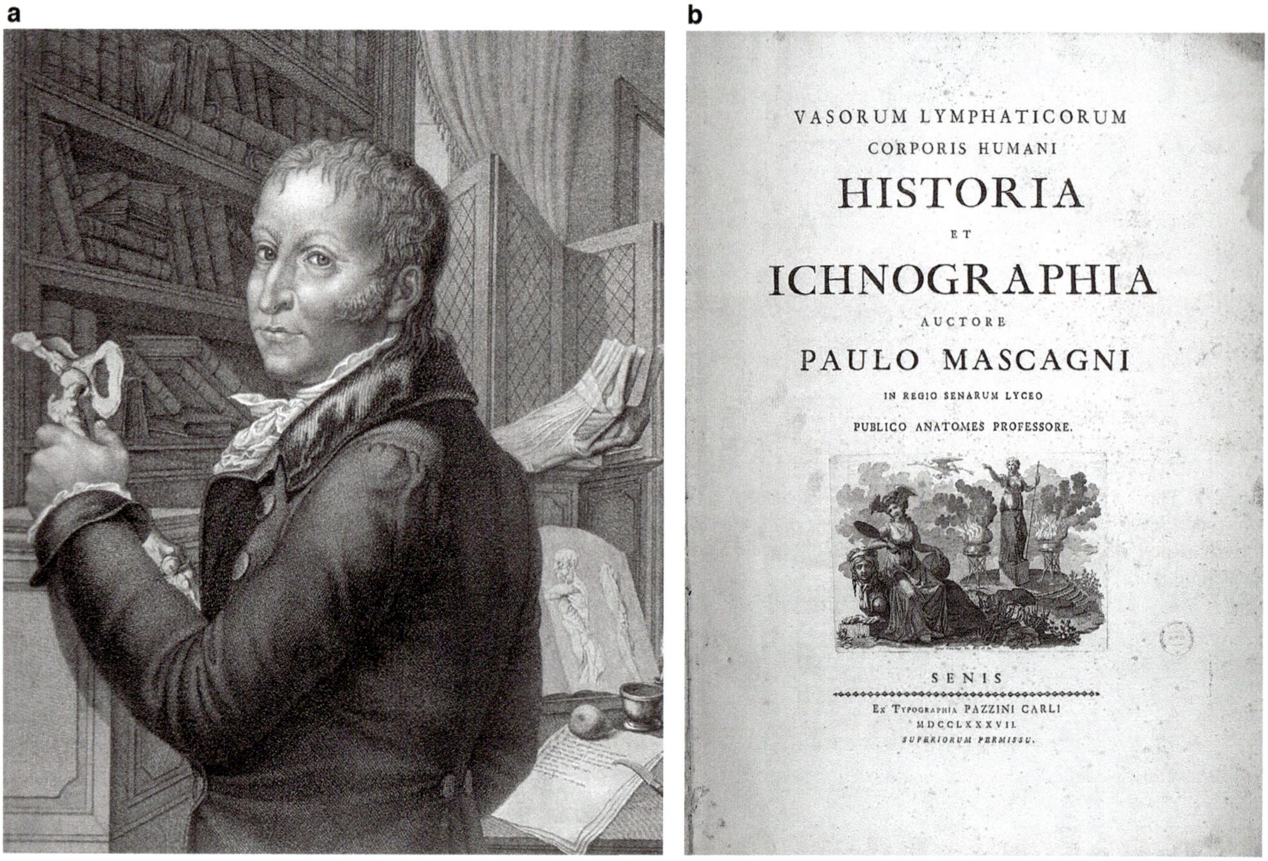

VASORUM LYMPHATICORUM
CORPORIS HUMANI
HISTORIA
ET
ICHNOGRAPHIA
AUCTORE
PAULO MASCAGNI
IN REGIO SENARUM LYCEO
PUBLICO ANATOMES PROFESSORE.

SENIS
EX TYPOGRAPHIA PAZZINI CARLI
MDCCLXXXVII.
SUPERIORUM PERMISSU.

Fig. M11 (**a**) Portrait of Mascagni. (From Wikipedia Commons; in public domain); (**b**) Title page *Vasorum lymphaticorum corporis humani.* Pazzini Carli, Siena, 1787

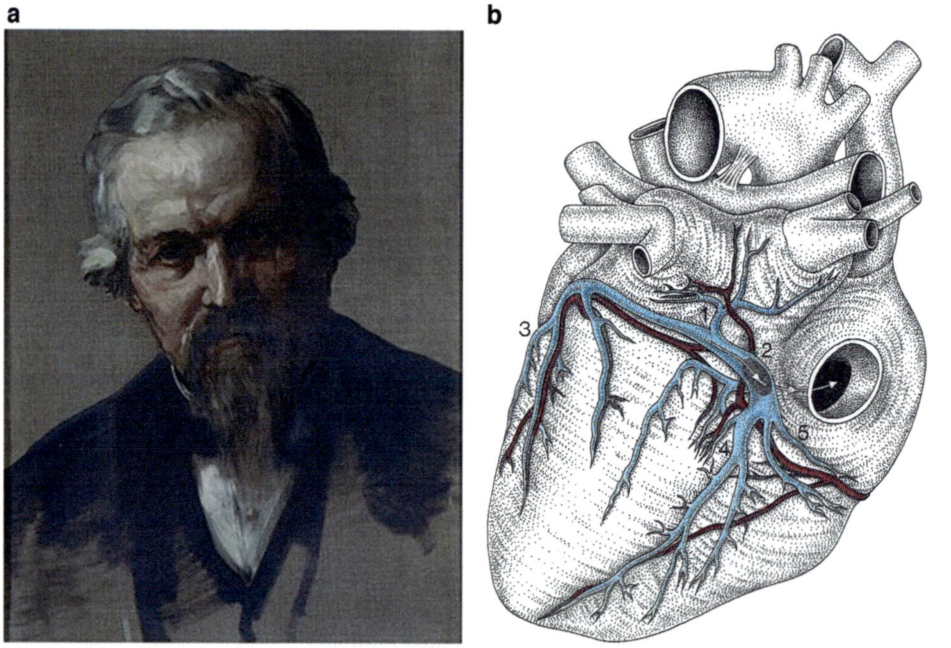

Fig. M12 (**a**) Portrait of Marshall by Alphonse Lapos (Victoria and Albert Museum, London; in public domain); (**b**) Vein of Marshall. (After ten Donkelaar HJ, Oostra R-J *Klinische Anatomie en Embryologie,* 4th ed. Springer Media, Houten, 2014). *1* oblique vein of Marshall, *2* coronary sinus, *3* posterior vein of left ventricle, *4* middle cardiac vein, *5* small cardiac vein

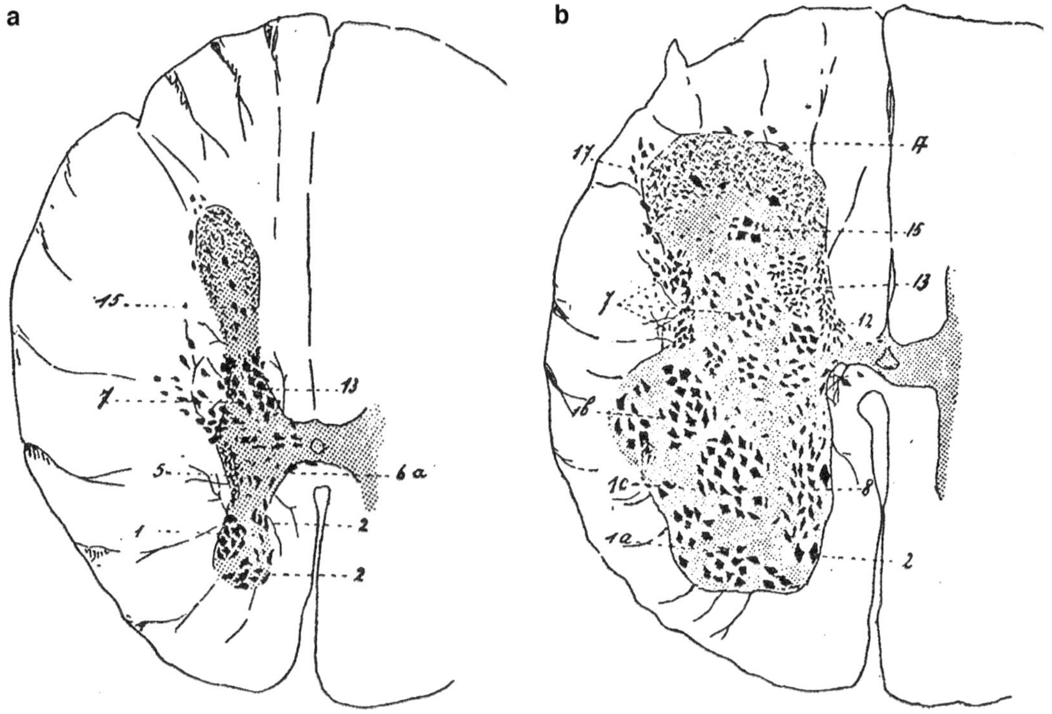

Fig. M13 Massazza's illustrations of the cytoarchitecture of the human spinal cord: (**a**) the fifth thoracic segment: *1* lateral group of motoneurons or intercostal motoneurons, *2* medial group of motoneurons or axial motoneurons; *5* intermediolateral funicular neurons found in the lateral funiculus, *6a* commissural paracentral cell group, *7* intermediolateral nucleus, *13* dorsal nucleus of Clarke, *14* spinothalamic cell group; (**b**) the fourth lumbar segment: *1* lateral (lower limb) motoneurons, *2* medial (axial) motoneurons, *7* intermediolateral nucleus with some funicular neurons, *12* intermediomedial nucleus (between *7* and *12* the intercalated nucleus), *13* dorsal nucleus of Clarke, *15* spinothalamic cell group, *17* marginal cells of Waldeyer in layer I of Rexed. (From Massazza A *La citoarchitettonica del midollo spinale umano I-III.* Arch Anat Histol Embryol 1922–1924; 1:323–410, 2:1–56, 3:115–189)

Cell of Martinotti: *Eponym* for the inverted pyramidal neuron, a distinctive type of cell in the cerebral cortex.

Mascagni, Paolo (1755–1815): Italian anatomist born in Pomerance, Pisa, and working in Siena and Pisa (Fig. M12a). He is known for the first complete description of the lymphatic system. Reference: *Vasorum lymphaticorum corporis humani historia et ichnographia.* Pazzini Carli, Siena, 1787 (Fig. M12b).

Ampulla of Mascagni: *Eponym* for an accessory enlargement of the upper part of the thoracic duct.

Lymph node of Mascagni: *Eponym* for an inconstant gallbladder lymph node in the cystohepatic triangle of Calot, also described by Calori, and later by Lund.

Plexus of Mascagni: *Eponym* for the internal (or hypogastric) and external iliac lymphatic plexus.

Massazza, Adolfo (1898-?): Italian neuropathologist and psychiatrist born in Reggio Calabria and working in Genova. He is known for his studies on the cytoarchitecture of the human spinal cord, the intermediolateral nucleus of the spinal cord and related nuclei in particular (Fig. M13). Apart from the intermediolateral nucleus, present from C8-Co2, he recognized an intermediomedial nucleus and intermediate disseminated cells, both present from C1-Co2. In line with his findings, Petras and Cummings identified four major groups: (a) the principal part of the intermediolateral nucleus; (b) the funicular part of the intermediolateral nucleus in the lateral funiculus; (c) the intermediomedial nucleus surrounding the central canal; and (d) the spinal intercalated nucleus (Penfield JM, Cummings JF *Autonomic neurons in the spinal cord of the rhesus monkey: A correlation of the findings of cytoarchitectonics and sympathectomy with fiber degeneration following dorsal rhizotomy.* J Comp Neurol 1972; 146:189–218).

Nucleus of Massazza: *Eponym* for the intermediolateral nucleus of the spinal cord. Reference: *La citoarchitettonica del midollo spinale umano I-III.* Arch Anat Histol Embryol 1922–1924; 1:323–410, 2:1–56, 3:115–189.

Mauchart, Burkhard David (1696–1751): German anatomist and surgeon born in Marbach and working in Tübingen, best known for his studies on the eye.

Ligaments of Mauchart: *Eponym* for the alar or check ligaments of the extraocular muscles. Reference: *Dissertationes medicae selectae Tubingenses; oculi humani affectus med.-chir. Considerantes sistens.* Tübingen, 1783–1786.

Mauthner, Ludwig (1840–1894): Bohemian ophthalmologist born in Prague and working in Innsbruck and Vienna.

Layer or **Sheath of Mauthner:** *Eponym* for the membrane surrounding the neuronal axis cylinder within the sheath of Schwann. Reference: *Beiträge zur näheren Kenntniss der morphologischen Elemente des Nervensystems.* SB Akad Wiss Wien 1860; 39:583–589.

May, Robert (1863–1902): Austrian surgeon.

Vein of May, Perforating: *Eponym* for the midcalf perforating vein. Reference: *Perforating Veins.* Urban & Schwarzenberg, Munich 1980 (with H Partsch and J Staubesand).

Mayer, Sigmund (1842–1913): German histologist, known for his studies of the sympathetic nervous system. Reference: *Das sympathische Nervensystem.* In: Hermann L, Hrsg, *Handbuch der Physiologie: Spezielle Neurophysiologie*. Leipzig, 1879.

Ganglion of Mayer: *Eponym* for the inconstant sublingual ganglion.

Mayo, Charles Horace (1865–1939): American surgeon born in Rochester and working at the Mayo Clinic in Rochester like his brother William James (1861–1939).

Vein of Mayo: *Eponym* for the prepyloric vein, also ascribed to his brother. Dobson (1962) found this an incorrect eponym: the prepyloric vein should be known as vein of Latarjet.

Mazzoni, Vittorio (1880–1940): Italian anatomist; *see* Golgi-Mazzoni corpuscle. Reference: *Osservazioni microscopische sopra i cosidetti corpuscoli terminali dei tendini dell'uomo e sopra alcune particolari piastre nervose superficiali che si trovano nei medesimi tendini.* Bologna Accad Sci Med 1890; 1:401–431.

Fig. M14 Photo of McBurney. (From the Wellcome Collection CCBY; in public domain)

McBurney, Charles (1845–1913): American surgeon born in Roxbury, Massachusetts, and working in New York (Fig. M14).

Point of McBurney: *Eponym* for the point of maximum tenderness in cases of appendicitis, lying on the line of Monro, the right spinoumbilical line, some 6–7 cm medial from the anterior superior iliac spine (*see* Fig. L12). Reference: *Experience with early operative interference in cases of disease of the vermiform appendix.* New York Med J 1889, pp. 676–684.

Meckel, Johann Friedrich, the Elder (1724–1774): German anatomist, botanist and gynaecologist born in Wetzlar and working in Berlin (Fig. M15) who described among others the pterygopalatine ganglion and the trigeminal cavity. References: *Tractatus anatomico-physiologicus de quinto pare nervorum cerebri.* Vandenhoeck, Göttingen, 1748; *Observation anatomique sur un noeud ou ganglion du second rameau de la cinquième paire des nerfs du cerveau.* Mém Acad Sci Bell Lett, Berlin, 1749, Vol 5; *Description anatomique des nerfs de la face.* Ibid., 1751, Vol 7.

Cave or **Space of Meckel:** *Eponym* for the dural trigeminal cavity in the middle cranial fossa containing the trigeminal ganglion.

Ganglion of Meckel: *Eponym* for (a) the pterygopalatine ganglion, and (b) the submandibular ganglion (Fig. M16).

Fig. M15 Portrait of Meckel JF, the Elder. (From Wikimedia Commons; in public domain)

Groove of Meckel: *Eponym* for the trigeminal impression on the anterior surface of the petrous part of the temporal bone, where the trigeminal ganglion is situated.

Lingula of Meckel: *Eponym* for a bony plate of the greater wings of the sphenoidal bone, located lateral to the carotid sulcus of the sphenoidal bone between the round foramen and the oval foramen.

Meckel, Johann Friedrich, the Younger (1781–1833): German anatomist and surgeon born and working in Halle (Fig. M17), grandson of Johann Friedrich Meckel, the Elder.

Cartilage of Meckel: *Eponym* for the cartilage of the first pharyngeal arch. Reference: *Abhandlungen aus der vergleichenden und menschlichen Anatomie.* Halle, 1805.

Diverticle of Meckel: *Eponym* for the inconstant ileal diverticle. Reference: *Bildungsgeschichte des Darmkanals der Saugtiere und des Menschen.* Deutsch Arch Physiol 1817; 3:1–84.

Ligament of Meckel: *Eponym* for the interosseous tibiofibular ligament.

Meckel, Philipp Friedrich (1756–1803): German anatomist.

Ligament of Meckel: *Eponym* for the anterior mallear ligament. Reference: *Dissertatio anatomico-physiologica de labyrinthi auris contentis.* Thesis 1777.

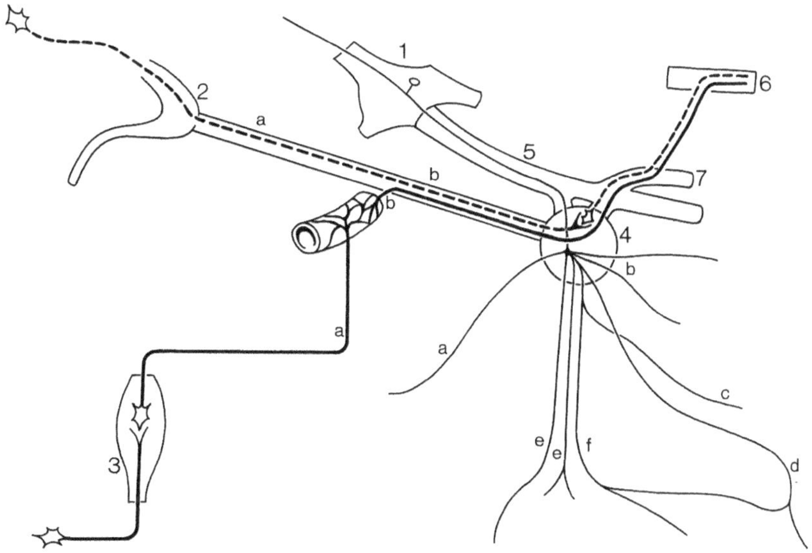

Fig. M16 The pterygopalatine ganglion of Meckel: *1* trigeminal ganglion, *2* geniculate ganglion with *a* greater petrosal nerve, *b* nerve of pterygoid ganglion, *3* superior cervical ganglion with *a* postganglionic sympathetic fibers through *b* the deep petrosal nerve, *4* pterygopalatine ganglion giving off: *a* pharyngeal branch, *b* posterior superior nasal branches, *c* posterior inferior nasal branch, *d* nasopalatine nerve, *e* lesser palatine nerve, *f* greater palatine nerve, *5* maxillary nerve, *6* lacrimal nerve, *7* zygomatic nerve. (After O'Rahilly R *Gardner-Gray-O'Rahilly Anatomy*, 5th ed. Saunders, Philadelphia, PA, 1986; from ten Donkelaar HJ, Kachlík D, Tubbs RS *An Illustrated Terminologia Neuroanatomica.* Springer, Cham, 2018)

Fig. M17 Portrait of Meckel JF, the Younger. (From Wikimedia Commons; in public domain)

***Medusa*:** In Greek mythology one of the three Gorgons, daughter of Phorcys and Ceto, with living snakes in place of hair; her appearance was so hideous that anyone who looked upon her was turned into stone.

***Caput Medusae*:** *Eponym* for the appearance of distended and engorged superficial abdominal veins radiating from the umbilicus and fed by the paraumbilical veins in portal hypertension; the term is derived from its similarity with the head of Medusa.

Meessen, Hubert (1909–1992): German neuroanatomist and neuropathologist in Düsseldorf known for the pars alpha of the gigantocellular nucleus in collaboration with Olszewski. Reference: *A Cytoarchitectonic Atlas of the Rhombencephalon of the Rabbit.* Karger, Basel, 1949 (with Jerzy Olszewski).

Nucleus of Meessen-Olszewski: *Eponym* for the pars alpha of the gigantocellular nucleus of the reticular formation.

Méglin, Jean (1756–1824): French physician born in Sultz, educated in Paris and working in Colmar.

Point of Méglin: *Eponym* for the point of emergence of the greater palatine nerve from the greater palatine foramen. Reference: *Recherches et observations sur la neuralgie facial.* Paris, 1816.

Meibom, Heinrich (1638–1700): German physician born in Lübeck and working in Helmstädt (Fig. M18).

Fig. M18 Portrait of Meibom, Kupferstich by Jacob von Sandrart. (From Wikimedia Commons; in public domain)

Glands of Meibom: *Eponym* for the tarsal glands of the eyelids. Reference: *De vasis palpebrarum novis epistola.* Müller, Helmstädt, 1666.

Meigs, Arthur Vincent (1850–1912): American physiologist born and working in Philadelphia.

Capillaries of Meigs: *Eponym* for the capillary blood vessels found between the muscular fibres of the heart. Reference: *The penetration of the muscular fibres of the human heart by capillaries.* J Anat (Lond) 1899; 33:241.

Meissner, Georg (1829–1905): German anatomist and physiologist born in Hannover and working in Basel, Freiburg and Göttingen (Fig. M19) who described tactile corpuscles in the skin and the submucosal plexus.

Fig. M19 Portrait of Meissner. (From Wikimedia Commons; in public domain)

References: *Ueber Vorhandsein bisher unbekannter eigenthümlicher Tastkörperchen (Corpusculus tactus) in dem Gefühlswärzchen der menschlichen Haut und ueber die Endausbreitung der sensitiver Nerven* (with Rudolf Wagner). Nachr Ges Wiss Göttingen, 1852; *Beiträge zur Anatomie und Physiologie der Haut.* Voss, Leipzig, 1853; *Bemerkungen die Tastkörperchen betreffend.* Z Wiss Zool 1855; 6:296–297.

Corpuscles of Meissner: *Eponym* for tactile corpuscles in the skin (Fig. M20).

Plexus of Meissner: *Eponym* for the submucosal nervous plexus of the digestive tract.

Membranes: *See* **The Classical Anatomical Eponyms.**

Ménard, Maxime (1872–1926): French radiologist born in Saint Michel and working in Paris.

Line of Shenton-Ménard: *Eponym* for a curved line formed by the top of the obturator foramen and the inner side of the neck of the femur, seen on an antero-posterior frontal X-ray of a normal hip joint, also described by Shenton. Reference: *Le bassin normal en radiographie.* Thesis, Paris, 1896.

Mercier, Louis Auguste (1811–1882): French urologist born in Plessis-Saint-Jean and working in Paris well known for his studies on the anatomy of the urinary bladder and urethra.

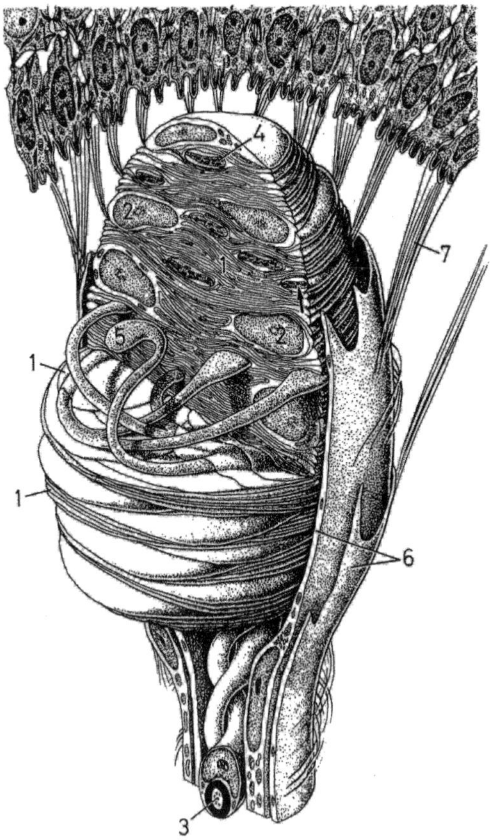

Fig. M20 The structure of a Meissner corpuscle: *1* cytoplasmatic lamellae of terminal Schwann cells, *2* cell bodies of Schwann cells, *3* myelinated axon, *4*, *5* thickenings of terminal axon branches, *6* perineural cells, *7* collagen microfibrils connecting the corpuscle with basal epithelial cells. (From Krstić R *Die Gewebe des Menschen und der Säugetiere,* 2nd ed. Springer, Berlin-Heidelberg-New York, 1988; with permission; from ten Donkelaar HJ, Kachlík D, Tubbs RS *An Illustrated Terminologia Neuroanatomica.* Springer, Cham, 2018)

Bar of Mercier: *Eponym* for the interureteric fold, a mucous fold between the openings of the ureters into the urinary bladder. Reference: *Recherches anatomiques, pathologiques et thérapeutiques sur les valvules du col de la vessie, cause fréquente et peu connue de rétention d'urine, et sur leurs rapports avec les inflammations et les rétrécissements de l'urèthre.* Bechet and Labbé, Paris, 1848.

Valve of Mercier: *Eponym* for an occasional mucous fold of the urinary bladder partially occluding the internal urethral opening.

Merkel, Friedrich Sigmund (1845–1919): German anatomist born in Nürnberg and working, successively in Rostock, Königsberg and Göttingen (Fig. M21). References: *Tastzellen und Tastkörperchen bei den Hausthieren und beim Menschen.* Arch Mikr Anat Entwmech 1875; 11:636–652; *Ueber die Endigungen der sensiblen Nerven in der Haut der Wirbelthiere.* Rostock, 1880.

Fig. M21 Photo of Merkel. (From Pagel 1901)

Cells of Merkel: *Eponym* for oval or rounded cells located among the basal layer cells of the epithelial glandular ridges.

Line of Merkel: *Eponym* for the horizontal orbital-auditory line from the centre of the external acoustic meatus to the inferior margin of the orbit, a modification of the Virchow-Hoelder plane. Reference: *Zur Kenntniss der postembryonalen Entwicklung der menschlichen Schädel.* Festschrift Henle, Bonn, 1882.

Tactile disc of Merkel: *Eponym* for the associated disc-like nerve terminals to the cells of Merkel.

Touch spot ('Haarscheibe') of Merkel: *Eponym* for an aggregate of Merkel discs in the hairy skin (Fig. M22).

Merkel, Karl Ludwig (1812–1876): German anatomist and laryngologist born and working in Leipzig.

Fossa of Merkel: *Eponym* for the central fossa between the two ventricles of the larynx, corresponding to the glottis. Reference: *Anatomie und Physiologie des menschlichen Stimm- und Sprachorgans (Anthropophonik). Nach eigenen Beobachtungen und Versuchen wissenschaftlich begründet und für studirende and ausübende Aertze, Physiologen, Akustiker, Sänger, Gesanglehrer, Tonsetzer, öffentliche Redner, Pädagogen und Sprachforscher dargestellt.* Leipzig, 1857.

Muscle of Merkel: *Eponym* for the inconstant ceratocricoid muscle.

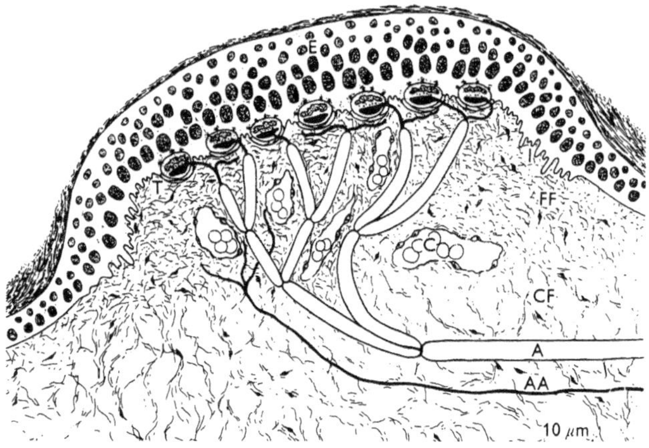

Fig. M22 'Haarscheibe' or Merkel touch spot: *A* single myelinated fibre; *AA* nonmyelinated nerve fibre, *E* thickened epidermis, *T* Merkel cells and associated Merkel discs, *CF*, *FF* collagen fibers. (After Iggo A, Muir AR J Physiol (Lond) 1969; 200:763–796; from ten Donkelaar HJ, Kachlík D, Tubbs RS *An Illustrated Terminologia Neuroanatomica.* Springer, Cham, 2018)

Méry, Jean (1645–1722): French surgeon born in Vatau and working in Paris.

Gland of Méry: *Eponym* for the bulbourethral gland, later also described by Cowper (1699–1700). Reference: *Observations anatomiques.* Mém Acad Sci Paris 1666–1699, 10:461. The term bulbourethral gland was proposed by Gübler (*De glandes de Méry et de leurs maladies chez l'homme.* Thesis, Paris, 1849)

Meyer, Adolf (1866–1950): Swiss-American psychiatrist born in Niederwengen and working at the Johns Hopkins Hospital, Baltimore.

Loop of Meyer: *Eponym* for the temporal loop of the optic radiation, described also by Archambault, Cushing and Flechsig. Reference: *The connections of the occipital lobes and the present status of the cerebral visual affections.* Trans Ass Am Phys 1907; 22:7–16.

Meyer, Edmund Victor (1864–1931): German laryngologist born and working in Berlin.

Cartilages of Meyer: *Eponym* for anterior sesamoid cartilages found in the anterior parts of his inferior thyroarytenoid ligaments. Reference: *Ueber die Luftsäcke der Affen und die Kehlkopfdivertikel beim Menschen.* Arch Laryng Rhinol 1901; 12:1–26.

Meyer, Georg Hermann von (1815–1892): German anatomist, physiologist and histologist born in Frankfurt am

Fig. M23 Portrait of Meynert. (From Pagel 1901)

Main and working in Zürich. Reference: *Stimm- und Sprachbildung*. Berlin, 1871.

Glands of Meyer: *Eponym* for aggregations of glands beneath the tongue within the hyoglossus muscle.

Sinus of Meyer: *Eponym* for a depression in the floor of the external acoustic meatus.

Meynert, Theodor (1833–1892): Austrian psychiatrist and neurologist born in Dresden and working in Vienna (Fig. M23) who made many contributions to the study of the CNS. References: *Vom Gehirne der Säugethiere.* In: Stricker S, Hrsgb, *Handbuch der Lehre von den Geweben des Menschen und der Thiere, Vol 2.* Engelmann, Leipzig, 1872, pp. 694–808 (English translation by Putnam JJ 1872 Stricker S, ed., *Manual of Histology*. William Wood, New York, pp. 650–766); *Psychiatrie. Klinik der Erkrankungen des Vorderhirns begründet auf dessen Bau, Leistungen und Ernährung: 1. Hälfte.* Braumüller, Vienna, 1884 (English translation by B Sachs: *Psychiatry: A clinical treatise on diseases of the fore-brain based upon a study of its structure, functions, and nutrition.* Putnam, New York, 1885).

Basal nucleus or **Ganglion of Meynert:** *Eponym* for the basal nucleus of the forebrain composed of large (cholinergic) neurons, also described by Ganser and Reichert.

Bundle, Fasciculus or **Tract of Meynert:** *Eponym* for the habenulointerpeduncular tract or fasciculus retroflexus.

Cells of Meynert: *Eponym* for the solitary giant cells in the primary visual cortex (BA17).

Commissure of Meynert: *Eponym* for the dorsal supraoptic decussation. References: *Beiträge zur Kenntniss der centralen Projection der Sinnesoberflächen,* 1869; ibid. Sitz Akad Wiss Wien 1870, pp. 547–566.

Decussation of Meynert: *Eponym* for the decussation of the tectobulbospinal tract, the dorsal tegmental decussation.

Layer of Meynert: *Old term* for the third layer of the isocortex, the pyramidal cell layer.

Preoccipital notch of Meynert: *Eponym* for a small notch on the lateral side of the occipital lobe.

Sulcus of Meynert: *Eponym* for the anterior occipital sulcus of the cerebral cortex. Reference: *Die Windungen der convexen Oberfläche des Vorderhirns bei Menschen, Affen und Raubthieren.* Arch Psychiatr Nervenkr 1877; 7:257–296.

U-fibres of Meynert: *Eponym* for short association fibres in the cerebral cortex (Fig. M24).

Michaelis, Gustav Adolf (1798–1848): German physician born in Harburg and working in Kiel.

Rhomboid or **Quadrangular area of Michaelis:** *Eponym* for a rhomboid shaped depression, which is bounded by lines joining the spinal process of the fifth lumbar vertebra to dimples over the posterior superior iliac spines and by lines joining the latter to the point where the medial borders of the gluteus maximus muscles approximate. Reference: *Das enge Becken nach eigenen Beobachtungen und Untersuchungen.* Wigand, Leipzig 1851 (published posthumously by his associate Carl Litemann).

Michels, Nicholas Aloysius (1891–1969): American anatomist in Philadelphia (Fig. M25) known for his studies on the blood supply of the intestines and the liver. References: *Blood Supply and Anatomy of the Upper Abdominal Organs with a Descriptive Atlas.* Lippincott, Philadelphia, Montreal, 1955; *Newer anatomy of the liver and its vari-*

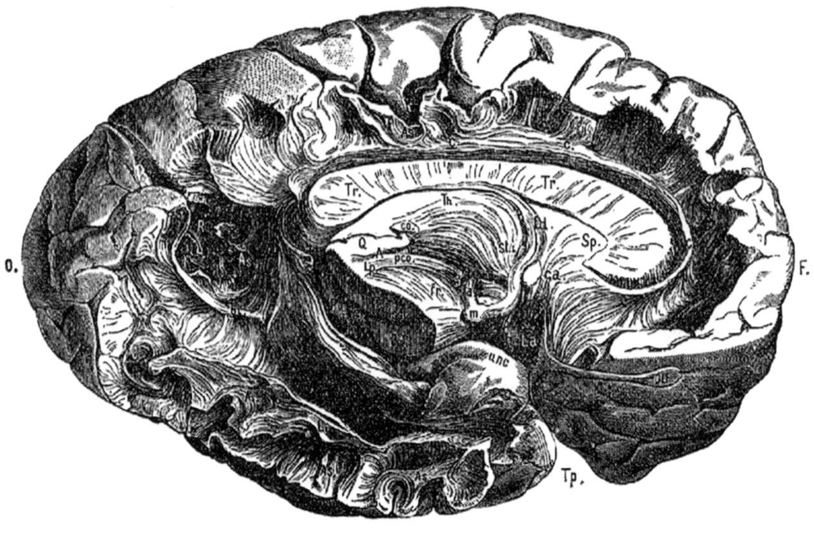

Fig. M24 U-fibres of Meynert in a medial view of the brain. (From Meynert T *Psychiatrie. Klinik der Erkrankungen des Vorderhirns begründet auf dessen Bau, Leistungen und Ernährung*. Braumüller, Vienna, 1884)

Fig. M25 Photo of Michels. (From Wikipedia; in public domain)

ant blood supply and collateral circulation. Am J Surg 1966; 112:337–347.

Arteries of Michels: *Eponym* for the retroduodenal arteries.

Middeldorpf, Albrecht Theodor (1824–1868): German physician and surgeon in Breslau.

Tunica of Middeldorpf: *Eponym* for the muscularis mucosae in the wall of the gastrointestinal canal.

Mierzejewsky, Jan Lucian (1839–1908): Polish neurologist and psychiatrist born in Jendrzejew and working in St. Petersburg.

Foramen of Mierzejewsky: *Eponym* for the lateral aperture of the fourth ventricle, also described by Luschka, Key and Retzius. Reference: *Contributions à l'étude du cerveau des* microcéphales. Bull Soc Anthrop 1875; 10:164–169.

Miescher, Johann Friedrich (1811–1887): German physiologist and pathologist born in Walkringen and working in Berlin, Bern and finally in Basel.

Tubes of Miescher: *Eponym* for tubules in calcifying tissues. Reference: *Die histochemischen und physiologischen Arbeiten.* Leipzig, 1897 (published posthumously by friends).

***Mike's dot*:** In ear-nose-throat surgery and neurosurgery, the macula cribrosa superior is known as 'Mike's dot' (after Michael E. Glasscock). It is an important landmark for translabyrinthine resection of an acoustic neurinoma.

Mikulicz-Radecki, Johann Freiherr von (1850–1905): German-Polish surgeon born in Czerniowce, Austrian-Hungarian Empire (currently Chernivtsi, Ukraine) and working in Krakau, Königsberg and Breslau.

Line of Mikulicz: *Eponym* for a line from the femoral head (in the Anglo-Saxon literature from the middle of the inguinal ligament) to the middle of the ankle joint.

Miller, William Snow (1858–1939): American physician and anatomist born in Stirling, Massachusetts, and working in Madison, Wisconsin.

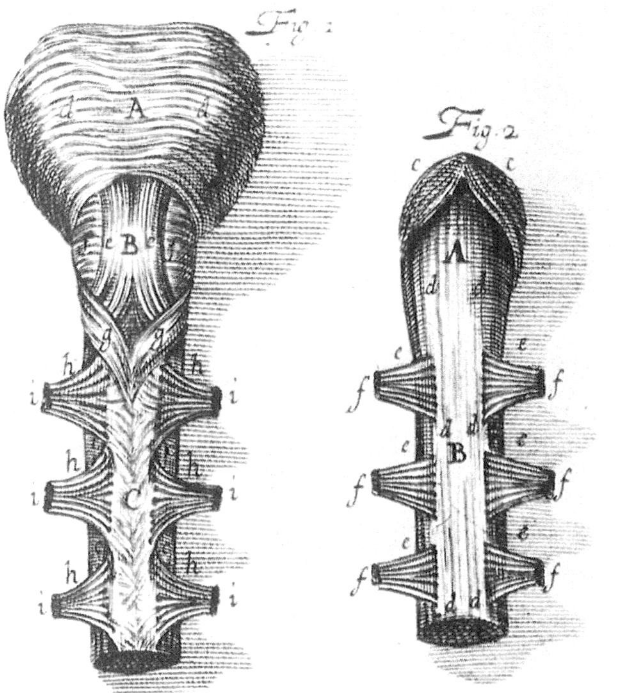

Fig. M26 The decussation of the pyramidal tract as shown by Mistichelli. (From *Trattato dell'apoplessia*. Rossi, Rome, 1709; in Clarke and O'Malley 1996; figure in public domain). Note the crossing fibres at *g* and *c*

Secondary pulmonary lobule of Miller: *Eponym* for a cluster of up to 30 pulmonary acini supplied by a common branch from the pulmonary artery and a bronchiole, bounded by interlobular septa outlining an irregular polyhedron of varying size (1 to 2.5 cm).

Mirizzi, Pablo Luis (1893–1964): Argentinian surgeon working in Cordoba.

Sphincter of Mirizzi: *Eponym* for a physiological sphincter at the confluence of the right and left hepatic ducts. Reference: *Physiologic sphincter of hepatic bile duct.* Arch Surg 1940; 41:1325–1333.

Mistichelli, Domenico (1675–1715): Italian anatomist who described the decussation of the pyramids in 1709 (Fig. M26), 1 year before Pourfour du Petit's report. Reference: *Trattato dell'apoplessia.* Rossi, Rome, 1709.

Mittendorf, Willam Frederick (1844–1919): American ophthalmologist of German origin, born in Königslutter, Brunswick, and working in New York.

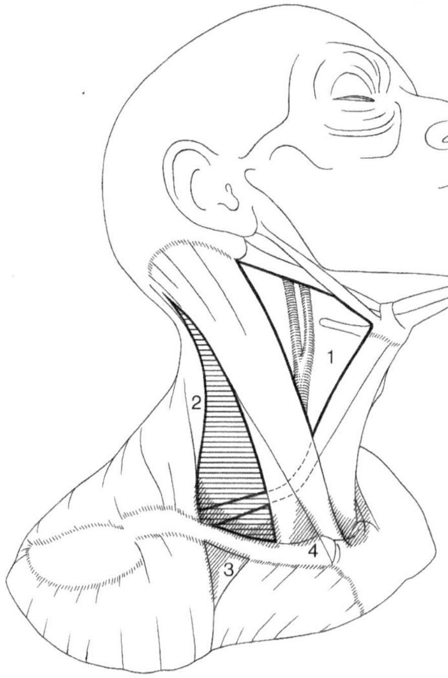

Fig. M27 The fossa or trigone of Mohrenheim. (Based on an old figure of the late Chris van Huijzen). *1* carotid triangle, *2* lateral cervical fossa, *3* infraclavicular fossa of Mohrenheim, *4* supraclavicular fossa

Dot of Mittendorf: *Eponym* for an anterior remnant of the embryonic hyaloid artery at the posterior surface of the lens.

Moderator band: *See* Leonardo da Vinci.

Mohrenheim, Joseph Jacob von (1759–1799): Austrian surgeon in Vienna, working later in St. Petersburg. Reference: *Wienerische Beiträge zur Artzneikunde und Wundarzneikunst und Geburtshilfe.* Vienna, 1781.

Fossa of Mohrenheim: *Eponym* for the infraclavicular fossa, also described by Gerdy (Fig. M27).

Space of Mohrenheim: *Eponym* for the clavipectoral or deltopectoral triangle.

Moll, Jacob Antonius (1832–1914): Dutch ophthalmologist born in Rotterdam and working in The Hague who described the ciliary glands in 1857. Reference: *Dissertatio continens quaedam de anatomia et physiologia palpebrarum.* Utrecht, 1857.

Glands of Moll: *Eponym* for the ciliary or palpebral sweat glands of the eyelids, later also described by Sattler (1877).

Fig. M28 Photo of Monakow. (From the Wellcome Collection CCBY; in public domain)

Monakow, Constantin von (1853–1930): Russian neurologist born in Gute Bobrezowa and working in Zürich (Fig. M28). References: *Neue experimentelle Beiträge zur Anatomie der Schleife.* Neurol Zbl 1885; 4:266–268; *Gehirnpathologie*. Hölder, Vienna, 1897; *Der rote Kern, die Haube und die Regio hypothalamica bei einigen Säugetieren und beim Menschen.* Arb Hirnanat Inst Zürich 1909, Heft III; 1910, Heft IV.

Nucleus of Monakow: *Eponym* for the lateral cuneate nucleus of the brain stem.

Tract of Monakow: *Eponym* for the rubrospinal tract.

Mondino de Luzzi (Liuzzi) or Mundinus (about 1270–1326): Italian physician, anatomist and surgeon in Bologna. He is often viewed as the restorer of anatomy by introducing the practice of public dissection of human cadavers. He also wrote the first modern anatomical text, which is considered the first modern dissection manual. Reference: *Anathomia corporis humani.* 1316, printed in Padua between 1475 and 1478 (Fig. M29a). Examples of his view on the heart and brain are shown (Fig. M29b and c).

Mondor, Henri Jean Justin (1885–1962): French surgeon born in Saint-Cernin, Cantal, and working in Paris known for his studies on rectal cancer.

Arcade of Mondor: *Eponym* for an arterial segment connecting the descending branch of the left colic artery and the ascending branch of the first sigmoid artery.

Monro, Alexander (1733–1817): The second (Secundus) of three Scottish anatomists of the same name born and working in Edinburgh (Fig. M30). Reference: *Of the communication of the ventricles of the brain with each other in Man and Quadrupeds.* Edinburgh, 1797.

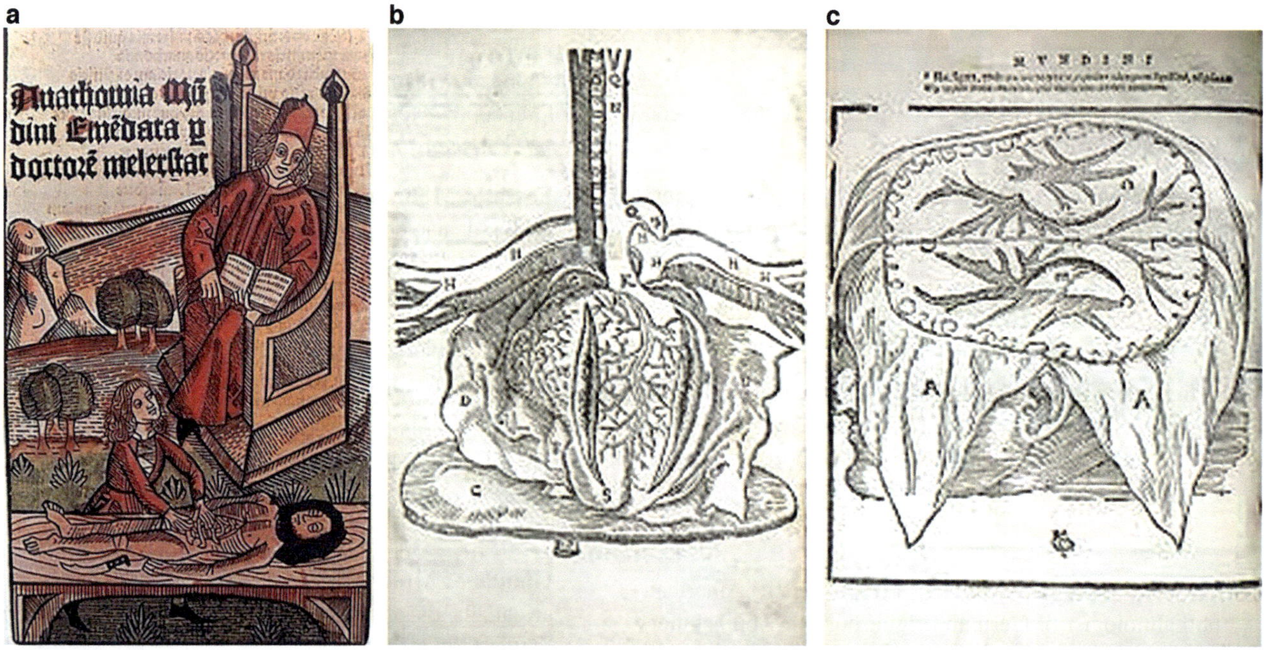

Fig. M29 Mondino de Luzzi; (**a**) Title page *Anathomia corporis humani*, 1493; (**b**) dissection of the heart; (**c**) dissection of the brain. (**b** and **c** From Mondino de Luzzi *Anatomia Mundini, Ad Vetustis.* 1541)

Fig. M30 Portrait of Monro. (From the Wellcome Collection CCBY; in public domain)

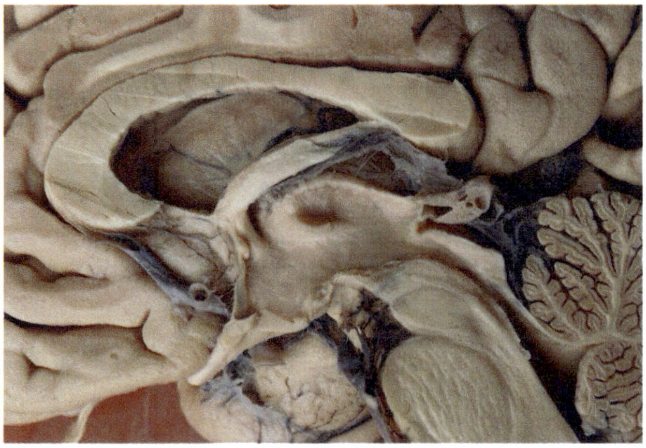

Fig. M31 The right interventricular foramen of Monro connecting the third ventricle and the right lateral ventricle. (Courtesy Veronika Němcová, Prague; from ten Donkelaar HJ, Kachlík D, Tubbs RS *An Illustrated Terminologia Neuroanatomica*. Springer, Cham, 2018)

Foramen of Monro: *Eponym* for the interventricular foramen of the brain (Fig. M31).

Line of Monro-Richter: *Eponym* for the right spinoumbilical line extending between the anterior superior iliac spine and the umbilicus, also described by Richter. It serves to determine the beginning of the vermiform appendix, located 6–7 cm medial from the spine, known as the point of McBurney.

Sulcus of Monro: *Eponym* for the hypothalamic sulcus.

Mons Venus or ***Mons Veneris*:** *Eponym* for the prominence caused by a fat pad over the symphysis pubis in females, named after Venus, the Roman Goddess of love.

Montgomery, William Fetherston Haugh (1797–1859): Irish obstetrician born in Fetenton and working in Dublin.

Glands or **Tubercles of Montgomery:** *Eponym* for areolar glands, small sebaceous glands surrounding the nipple, which become more prominent during pregnancy. Reference: *An Exposition of the Signs and Symptoms of Pregnancy, the Period of Human Gestation and the Signs of Delivery.* Sherwood, Gilbert & Piper, London, 1837.

Morand, Sauveur François (1697–1773): French surgeon born and working in Paris.

'Ergot' of Morand: *Eponym* for the calcar avis, a protrusion in the occipital horn of the lateral ventricle caused by the infolding of the calcarine sulcus (Fig. M32), also known as unguis of Haller and as hippocampus minor. Reference: *Observations anatomiques sur quelques parties du cerveau.* Mém Acad Sci Paris 1744, pp. 312–322.

Foramen of Morand: *Eponym* for the foramen caecum of the skull, which is situated in the anterior cranial fossa between the crista galli and the frontal crest. It usually ends blind but may be patent for a small emissary vein to the nasal cavity.

Morgagni, Giovanni Battista (1682–1771): Italian anatomist born in Forlì and working in Padua (Fig. M33). He is regarded as the founder of morbid anatomy. References: *Adversaria anatomica omnia.* Leiden, 1723; *De sedibus et causis morborum per anatomen indagatis, libri quinque.* Venice, 1761.

Appendages of Morgagni: *Eponym* for (a) the appendix testis, a small sessile body on the superior pole of the testis, a remnant of the paramesonephric duct; and (b) the vesicular appendage of the epoophoron, a small peduncular vesicle attached to the fimbria of the uterine tube.

Appendix of Morgagni: *Eponym* for the inconstant pyramidal lobe of the thyroid gland, later also described by Lalluoette.

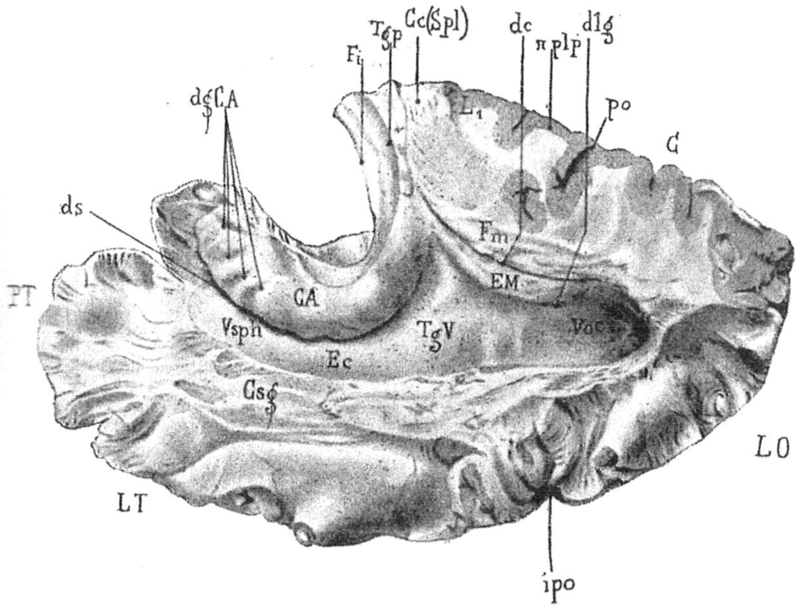

Fig. M32 Morand's 'ergot' in which the lateral ventricle is opened. *CA* cornu ammonis, *EM* ergot of Morand. (From Dejerine J *Anatomie des centres nerveux,* Tome Premier. Rueff, Paris, 1895)

Fig. M33 Portrait of Morgagni. (From Pagel 1901)

Cartilage or **Tubercle of Morgagni:** *Eponym* for the cuneiform cartilage of the larynx, also described by Wrisberg.

Caruncle of Morgagni: *Eponym* for the middle lobe of the prostate.

Columns of Morgagni: *Eponym* for the anal columns, a number of vertical ridges in the mucous membrane of the upper half of the anal canal (Fig. M34).

Concha of Morgagni: *Eponym* for the superior nasal concha.

Crypts or **Sinuses of Morgagni:** *Eponym* for anal sinuses, the grooves between the anal columns (Fig. M34).

Cyst or **Sessile hydatid of Morgagni:** *Eponym* for a vesicular appendage of the uterine tube, a remnant of the embryonic mesonephric duct.

Fissure of Morgagni: *Eponym* for the shallow retrosternal fissure between the sternal parts of the diaphragm.

Foramina of Morgagni: *Eponym* for the foramina nervosa of the cochlea.

Fossa of Morgagni: *Eponym* for the navicular fossa of the male urethra.

Frenulum of Morgagni: *Eponym* for the frenulum of the ileocaecal valve.

Glands of Morgagni: *Eponym* for the paraurethal glands.

Hernia of Morgagni: *Eponym* for a parasternal hernia, also described by Larrey.

Lacunae of Morgagni: *Eponym* for the urethral lacunae of the male urethra.

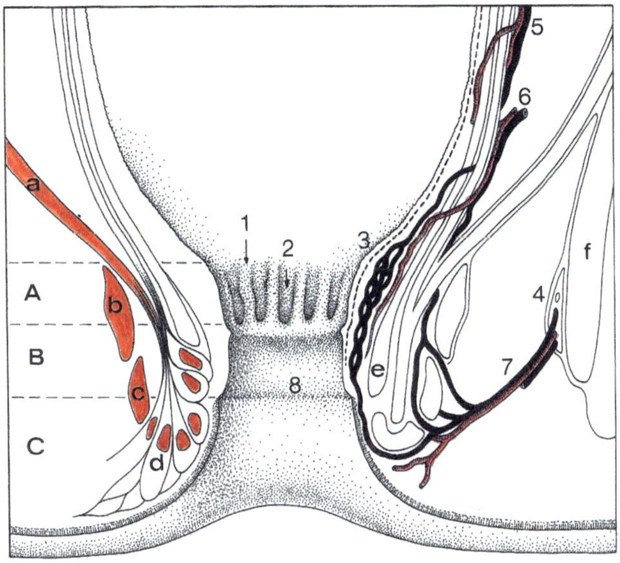

Fig. M34 Columns and valves (sinuses) of Morgagni shown in a schematic frontal section through the anorectal canal. *A-C* zones of anal canal: *A* haemorrhoidal zone, *B* intermediate zone (pecten analis), *C* cutaneous zone, *a* levator ani muscle, *b-d*: parts of external sphincter ani muscle: *b* deep part, *c* superficial part, *d* subcutaneous part, *e* internal sphincter ani muscle, *1* anal column, *2* anal sinus, *3* submucosal venous plexus (in *black*), *4* pudendal canal with internal pudendal artery and vein, *5* superior rectal artery and vein, *6* middle rectal artery and vein, *7* inferior rectal artery and vein, *8* anocutaneous line. (After ten Donkelaar HJ, Oostra R-J *Klinische Anatomie en Embryologie,* 4th ed. Springer Media, Houten, 2014)

Nodules of Morgagni: *Eponym* for the nodes on the semilunar valves of the pulmonary artery.

Pedunculated hydatid of Morgagni: *Eponym* for a structure associated with the fimbriae, usually considered to be a remnant of the paramesonephric duct of Müller (*see* Fig. G4).

Sinus of Morgagni: *Eponym* for the anal sinus. Sometimes also used for the gap between the superior constrictor muscle of the pharynx and the base of the skull.

Triangle of Morgagni-Larrey: *Eponym* for the sternocostal triangle, also described by Larrey, through which a parasternal hernia may occur.

Valves of Morgagni: *Eponym* for the anal valves, mucosal folds that pass between the aboral ends of neighbouring anal columns, also described by Haller and Ball.

Ventricle or **Pouch of Morgagni:** *Eponym* for the laryngeal ventricle.

Morison, James Rutherford (1853–1939): English surgeon born in Hutton-Henry, Durham and working in Hartlepool, Newcastle and finally Durham.

Pouch of Morison: *Eponym* for the right subhepatic space or hepatorenal pouch. Reference: *The anatomy of the right hypochondrium relating especially to operations for gallstones.* Brit Med J 1894; 11:968–971.

Morris, Sir Henry (1844–1926): English surgeon born in Petworth and working in London known for his studies of joints. Reference: *Anatomy of the Joints in Man.* London, 1879. He described alar ligaments and three glenohumeral ligaments, inferior, middle and superior (*see* Fig. F6).

Longitudinal and **Short internal lateral ligaments:** *Eponyms* for the sphenomandibular and medial ligament of the temporomandibular joint, respectively.

Mosher, Harris Peyton (1867–1954): American laryngologist in Boston.

Cell of Mosher: *Eponym* for the bulla ethmoidalis. Reference: *The surgical anatomy of the ethmoid labyrinths.* Ann Otol Rhinol Larynol 1929; 38:870–901.

Cells of Mosher: *Eponym* for extensions of ethmoidal air cells beneath the ethmoidal bulla. Reference: *The Anatomy of the Operation of Reaching the Ethmoidal Cells beneath the Bulla Ethmoidalis.* Philadelphia, 1902.

Moskowitz, Myron (twentieth century): American radiologist working in Cincinnati.

Meandering mesenteric artery of Moskowitz: *Eponym* for an inconstant accessory intermesenteric artery or arcade, connecting the superior and inferior mesenteric arteries directly or via their branches, described earlier in 1920 by Villemin and Huard and later in 1966 by Gonzales. Reference: *The meandering mesenteric artery of the colon* (with H Zimmerman and B Felson). Am J Roentgenol Rad Ther Nucl Med 1964; 92:1088.

Mugnaini, Enrico (1937-today): Italian neuromorphologist. *See* De Zeeuw CI, Jaarsma D, Voogd J, Llinas R,

Fig. M35 Photo of Cécile Mugnier-Vogt and her husband Oskar Vogt. (From the 1903 Archives der Universität Düsseldorf; in public domain)

Martina M. *Editorial on the honorary Cerebellum Issue for the retirement of Enrico Mugnaini.* Cerebellum 2015; 14:487–490.

Unipolar brush cell of Mugnaini: *Eponym* for a particular cell in the cerebellar cortex.

Mugnier, Cécile (1875–1962): French neurologist born in Annecy, Haute-Savoie (Fig. M35), who met Oskar Vogt at the Dejerine's in Paris (*see* Vogt, Oskar and Cécile).

Muir, Sir Robert (1864–1959): Scottish pathologist born in Balfrod, Stirling, and working in Glasgow.

Tract of Muir: *Eponym* for the septomarginal tract of the spinal cord described with Alexander Bruce, also described by Flechsig and as Triangle of Philippe-Gombault. Reference: *On a descending degeneration in the posterior columns in the lumbo-sacral region of the spinal cord.* Brain 1896; 19:333–343.

Müller, Heinrich (1820–1864): German anatomist and pathologist born in Castell and working in Würzburg.

Cells of Müller: *Eponym* for the radial glial cells of the retina. Reference: *Untersuchungen über den Bau der Retina des Menschen.* Leipzig, 1856.

Muscle of Müller: *Eponym* for two smooth muscles: (a) the smooth orbitalis muscle, and (b) the superior tarsal muscle in the upper eyelid, sometimes also used for the inferior tarsal muscle. Reference: *Ueber einen ringformigen Muskel am Ciliar-Körper des Menschen und über den Mechanismus der Accommodation.* Von Graefes Arch Ophthalm 1857; 3:1–24.

Müller, Johannes Peter (1801–1858): German anatomist and physiologist born in Koblenz and working in Bonn and later in Berlin (Fig. M36).

Canal or **Duct of Müller:** *Eponym* for the embryonic paramesonephric duct (Fig. M37). Reference: *Bildungsgeschichte der Genitalien aus anatomischen Untersuchungen an Embryonen des Menschen und der Thiere.* Düsseldorf, 1830.

Tubercle of Müller: *Eponym* for the sinus tubercle, the first evidence of the embryonic uterus and vagina. It remains as the hymen.

Fig. M36 Portrait of Müller, Johannes. (From Pagel 1901)

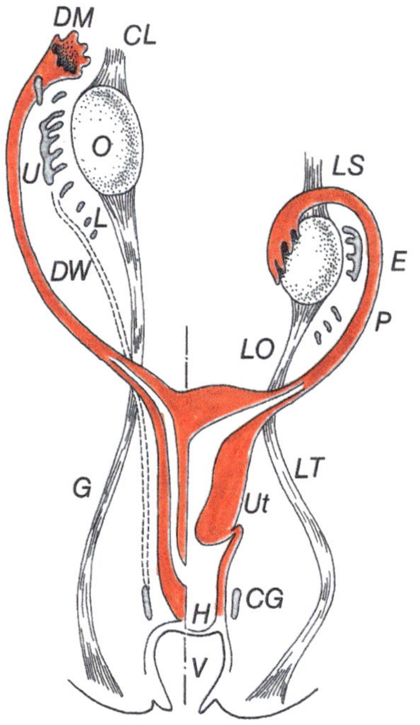

Fig. M37 Development of the duct of Müller, its derivatives (in *red*) and remnants (in *grey*) in females. *CG* cyst of Gärtner, *CL* cranial ligament of ovarium, *DM* duct of Müller, *DW* duct of Wolff, *E* tuba uterina, *G* gubernaculum, *H* hymen, *L* lower caudal mesonephros ducts, *LO* ligamentum ovarium proprium, *LS* ligamentum suspensorium ovarii, *LT* ligamentum teres uteri, *O* ovarium, *P* paroophoron, *left U* cranial part of duct of Wolff and upper mesonephros ducts; *right U* uterus, *V* vagina. (After ten Donkelaar HJ, Oostra R-J *Klinische Anatomie en Embryologie,* 4th ed. Springer Media, Houten, 2014)

Munro, John C (1858–1910): American surgeon.

Point of Munro: *Eponym* for a point at the lateral margin of the right rectus abdominis muscle on the spinoumbilical line to evoke pain and appreciate tenderness on palpation.

Munzer, Egmont (1865–1924): Bohemian physician born in Praschno-Ujezd, near Pilsen, and working in Prague.

Bundle or **Tract of Munzer:** *Eponym* for the tectopontine tract. References: *Beiträge zur Anatomie und Physiologie des Centralnervensystems.* Arch Exp Pathol Pharmak 1895, 35:113–128; *Das Zwischen- und Mittelhirn des Kaninchens und die Beziehungen dieser Teile zum übrigen Centralnervensystems.* Mschr Psychiat Neurol 1902; 12:265.

Muratoff, Wladimir Aleksandrovich (1865–1916): Russian neurologist. Reference: *Secundäre Degenerationen nach Durchschneidung des Balkens.* Neurol Cbl 1893; 12:714–729; *see also* Lichterman B (2010) *A history of Russian and Soviet neuro(patho)logy.* Hb Clin Neurol 95 (3rd Series):737–754.

Bundle of Muratoff: *Eponym* for the subcallosal bundle of the cerebral cortex.

Murphy, John Benjamin (1857–1916): American physician and abdominal surgeon born in Appleton, Wisconsin, and working in Chicago. He early advocated surgical intervention via the removal of the vermiform appendix in cases of appendicitis.

Point of Murphy: *Eponym* for a point at the crossing of the right medioclavicular line with the right axilloumbilical line, marking the projection of the gallbladder on the anterior abdominal wall.

Muscles: *See* **The Classical Anatomical Eponyms.**

Muskens, Louis Jacob Joseph (1872–1937): Dutch neurologist known especially for his studies on epilepsy. Reference: *The central connections of the vestibular nuclei with the corpus striatum and their significance for ocular movements and for locomotion.* Brain 1922; 44:454–478.

Tract of Muskens: *Eponym* for the interstitiospinal tract.

14 N, O

Naboth, Martin (1675–1721): German physician and anatomist born in Kalau, Saxony, and working in Leipzig.

Cysts or **Ovules of Naboth:** *Eponym* for retention cysts of the cervical glands of the uterus. Reference: *De sterilitate mulierum.* Zeidler, Leipzig, 1707.

Naegele, Franz Carl (1778–1851): German surgeon and obstetrician born in Düsseldorf and working in Heidelberg.

Obliquety of Naegele: *Eponym* for the tendency in normal labour for the anterior part of the parietal bone to present as the head engages. Reference: *Ueber den Mechanismus der Geburt.* Dtsch Archiv Physiol 1819; 5:483–531.

Nageotte, Jean (1866–1948): French neuroanatomist born in Dijon and working in Paris.

Cells of Nageotte: *Eponym* for mononuclear cells in the cerebrospinal fluid.

Nerves of Nageotte: *Eponym* for the bundles of spinal nerve rootlets contained in the dural sac before they merge to form the spinal nerve.

Nansen, Fridtjof (1861–1930): Norwegian explorer, oceanographist, and Nobel Prize Winner (1922). As Curator of the Bergen Museum, he made valuable contributions to the anatomy of the spinal cord.

Fibres of Nansen: *Eponym* for the ascending and descending branches of the posterior or dorsal spinal nerve roots. Reference: *The structure and combination of the histological elements of the central nervous system.* Bergens Mus Aarsber 1885.

Nasmyth, Alexander (?-1848): English dentist, surgeon and anatomist in London.

Membrane or **Cuticle of Nasmyth:** *Eponym* for the dental cuticle of the freshly erupted tooth. References: *On the structure, physiology and pathology of the persistent capsular investments and pulp of the teeth.* Med Chir Trans 1839; 22:310–328; *Researches on the development, structure and diseases of the teeth.* London, 1839.

Nauta, Walle Jetze Harinx (1916–1994): Dutch neuroanatomist born in Medan (then Dutch East Indies, now Indonesia) and after working in Leiden and Zürich, he went to the USA (Fig. N1). He is especially known for the introduction of anterograde degeneration techniques and for his studies on the connectivity of the limbic system.

Limbic midbrain area of Nauta: *Eponym* for an extension of the limbic system to the midbrain (Fig. N2). Reference: *Hippocampal projections and related neural pathways to the midbrain in the cat.* Brain 1958; 81:319–340.

Nélaton, Auguste (1807–1873): French surgeon born and working in Paris (Fig. N3). Reference: *Eléments de pathologie chirurgicale,* 5 Vols. Baillière, Paris, 1844–1860.

Line of Nélaton: *Eponym* for the line between the anterior superior iliac spine and the ischial tuberosity, also known as the line of Roser-Nélaton (Fig. N4).

Fold of Nélaton: *Eponym* for the inferior transverse rectal fold (*see* Fig. K11).

Sphincter of Nélaton: *Eponym* for a third sphincter ani muscle, not generally recognized, located at the level of the inferior transverse rectal fold of Nélaton.

Nerves: *See* **The Classic Anatomical Eponyms.**

Neubauer, Johann Ernst (1742–1777): German anatomist and surgeon born in Giessen and working in Jena.

Artery of Neubauer: *Eponym* for the thyroid ima artery. Reference: *Descriptio arteriae innominatae et thyroideae imae.* Jena, 1772 (with AC Erdmann).

H. J. ten Donkelaar et al., *An Illustrated Guide to Anatomical Eponyms*, https://doi.org/10.1007/978-3-031-91664-9_14

Fig. N1 Photo of Nauta. (From the National Academic Press, Washington, DC, Biographical Memoirs, Vol 88, 2006; courtesy the late Ton Lohman, Monnickendam)

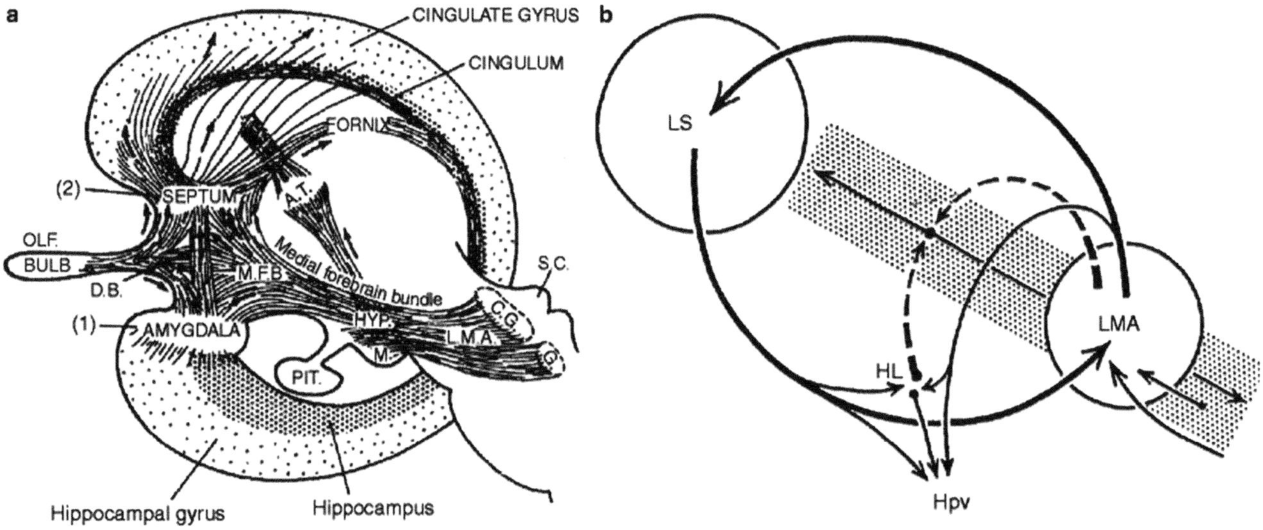

Fig. N2 The limbic midbrain area of Nauta as part of MacLean's limbic system. (After MacLean PD 1958 Am J Med 23:611–626; from ten Donkelaar HJ, Kachlík D, Tubbs RS *An Illustrated Terminologia Neuroanatomica*. Springer, Cham, 2018). *A.T.* mammillothalamic tract, *C.G.* central gray of rostral mesencephalon, *D.B.* diagonal band nuclei, *G* gray matter of caudal mesencephalon, *HYP* hypothalamus, *L.M.A.* limbic midbrain area of Nauta, *M.* mammillary body, *M.F.B.* medial forebrain bundle, *PIT* pituitary gland, *S.C.* superior colliculus

Neumann, Ernst Francis Christian (1834–1918): German anatomist, histologist and pathologist born and working in Königsberg (now Kaliningrad, Russia).

Cells of Neumann: *Eponym* for nucleated red blood cells.

Sheath or **Layer of Neumann:** *Eponym* for the semicalcified layer of matrix surrounding the tubules of dentin. Reference: *Ein Beitrag zur Kenntniss des normalen Zahnbein- und Knochengewebes.* Leipzig, 1843.

Fig. N3 Portrait of Nélaton. (From Pagel 1901)

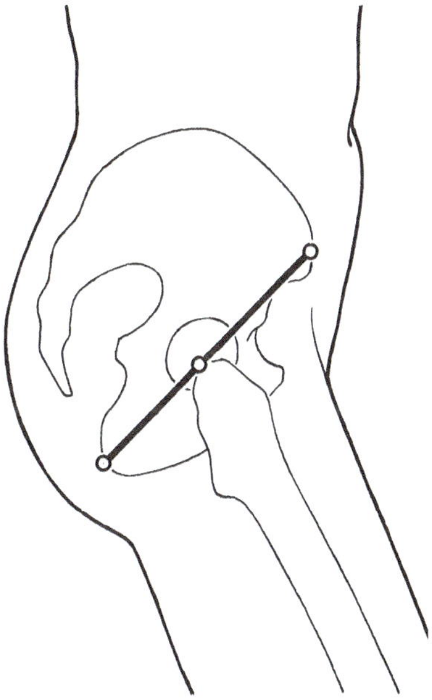

Fig. N4 Line of Roser- Nélaton. (Based on an old figure by the late Chris van Huijzen)

Nieuwenhuys, Rudolf (1927–2024): Dutch neuroanatomist born in Amsterdam. As a medical student, he started working in comparative neuroanatomy at the 'Centraal Instituut voor Hersenonderzoek' in Amsterdam, and continued working there until 1967. From 1967 until 1992, he worked in Nijmegen not only in comparative neuroanatomy, culminating in *The Central Nervous System of Vertebrates* (Springer, Berlin-Heidelberg-New York, 1998; with Hans J. ten Donkelaar and Charles Nicholson), but also on the anatomy of the human brain, resulting in

Fig. N5 Photo of Nieuwenhuys. (Courtesy Loreta Medina, Lleda)

The Human Central Nervous System (Springer, Berlin-Heidelberg-New York, 1978; with Jan Voogd and Chris van Huijzen; 4th ed., 2008). In 1992, he returned to the Dutch Brain Institute in Amsterdam (Fig. N5). He expanded the limbic system by adding extensive medial and lateral tegmental areas throughout the brain stem to his **greater limbic system** (Fig. N6), including monoaminergic structures and a number of brain stem regions from which behavioural and emotional responses can be elicited, such as the vagus-solitary complex, the parabrachial nucleus and other regions of the prepontine tegmentum. More recently, with Cees Broere, he revitalized the myeloarchitectonic work of the Vogts. Reference: *A detailed comparison of the cytoarchitectonic and myeloarchitectonic maps of the human neocortex produced by the Vogt-Vogt School.* Brain Struct Funct 2020; 225:1–7.

Core and Paracore concept of Nieuwenhuys: *Eponym* for a new subdivision of the reticular formation and related cell groups: the medial reticular zone as its core, the lateral reticular formation as the lateral paracore of the brain stem and the raphe nuclei as the median paracore.

Greater limbic system of Nieuwenhuys: *Eponym* for an extensive expansion of Nauta's ideas on the caudal extension of the limbic system (Fig. N6). Reference: *The greater limbic system, the emotional motor system and the brain.* Prog Brain Res 1996; 107:551–580.

Nissl, Franz (1860–1919): German neurologist born in Frankenthal and working in Frankfurt am Main, Heidelberg and later Munich (Fig. N7).

Bodies, Granules or **Substance of Nissl:** *Eponym* for basophilic granules in the cytoplasm of neurons, also known as tigroid substance. Reference: *Ueber die sogenannten Granula der Nervenzellen.* Neurol Zbl 1894; 13:781–789.

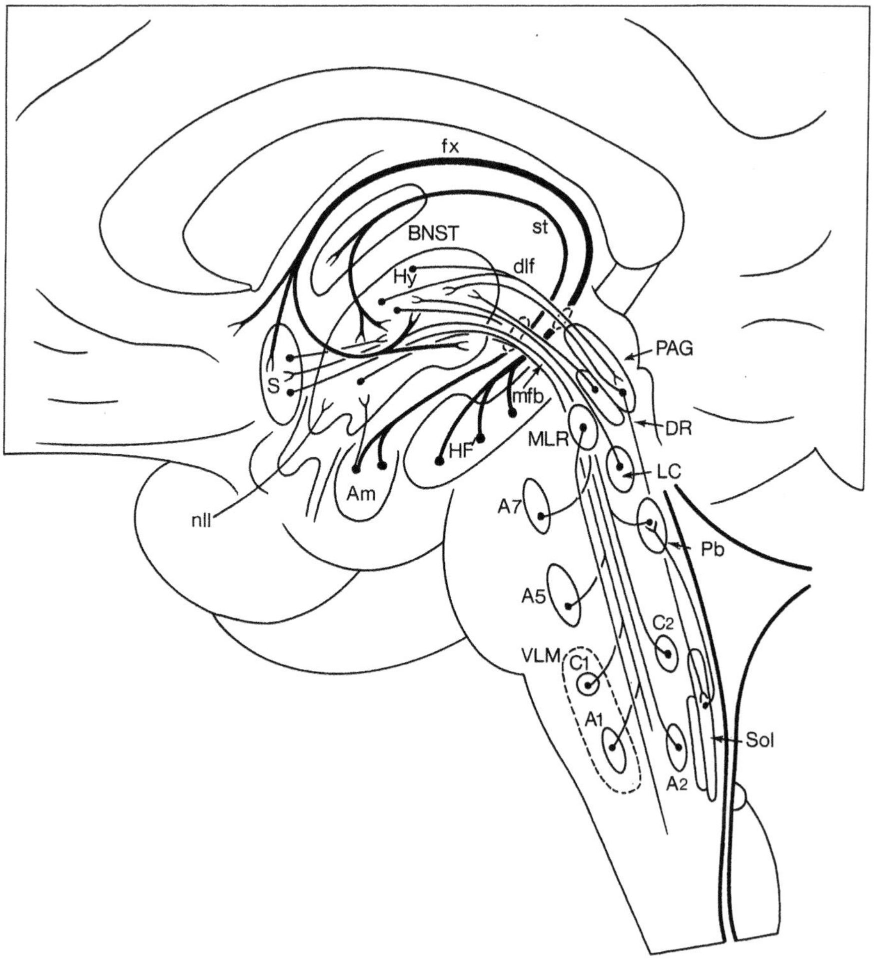

Fig. N6 The greater limbic system of Nieuwenhuys. *Am* amygdala, *A2–A7* noradrenergic cell groups, *BNST* bed nucleus of the stria terminalis, *C1*, *C2* adrenergic cell groups, *dlf* dorsal lateral fasciculus of Schütz, *DR* dorsal raphe nucleus, *fx* fornix, *HF* hippocampal formation, *Hy* hypothalamus, *LC* locus coeruleus, *mfb* medial forebrain bundle, *MLR* mesencephalic locomotor region, *nII* optic nerve, *PAG* periaqueductal gray, *S* septum, *Sol* nucleus of the solitary tract, *st* stria terminalis, *VLM* ventrolateral medulla. (After Nieuwenhuys R *The greater limbic system, the emotional motor system and the brain.* Prog Brain Res 1996; 107:551–580; from ten Donkelaar HJ, Kachlík D, Tubbs RS *An Illustrated Terminologia Neuroanatomica.* Springer, Cham, 2018)

Fig. N7 Portrait of Nissl. (From Wikimedia Commons; in public domain)

Nitabuch, Raissa (1859-?): Russian pathologist born in Vladikavkaz and working with Langhans in Bern as part of a group of Russian women who studied Medicine at the University of Zürich. She is known for her studies on the human placenta (Khasanov AA, Orlov YY, Kuptsova AI *The contribution of R Nitabuch to the World Obstetric science.* Open Access J Reprod System Sex Disorders 2919 1 ISSN: 2641–1644). She also identified the spiral arteries of the uterus.

Stria of Nitabuch: *Eponym* for a layer of fibrin (the lower stria) in the intervillous spaces of the placenta densely placed in the basal decidua (Fig. N8). Reference: *Beiträge zur Kenntniss der menschlichen Placenta.* Thesis, University of Bern, Stämpfli, Bern, 1887.

Nodules:

Nodules of Albini: *Eponym* for minute fibrous nodules on the margins of the mitral and tricuspid valves, representing fetal tissue remnants; *see* Albini, Giuseppe.

Nodules of Aranzio: *Eponym* for nodules on the semilunar valves of the aorta, also described by Bianchi, Kerckring and Valsalva; *see* Aranzio, Julio Caesar.

Fig. N8 Nitabuch, cross section of the central area at the implantation site. (From Thesis, Bern, reproduced in Scott and Cardogan 2021 LITFL, September 22, 2021; Open Access)

Nodules of Bianchi: *Synonym* for the nodules of Aranzio; *see* Bianchi, Giovanni.

Nodule of Kerckring: *Eponym* for nodules on the semilunar valve of the aorta; *see* Kerckring, Dirk.

Nodules of Malpighi: *Eponym* for splenic lymphatic nodules; *see* Malpighi, Marcello.

Nodule of Morgagni: *Eponym* for the nodules on the semilunar valves of the pulmonary artery; *see* Morgagni, Giovanni Battista.

Nodules of Peyer: *Eponym* for solitary lymph nodes of the intestines; *see* Peyer, Johann Conrad.

Nodule of Valsalva: *Synonym* for the nodules of Aranzio; *see* Valsalva, Antonio Maria.

Notches:

Notch of Kernohan: *Eponym* for a notch in the cerebral peduncle due to displacement of the brain stem against the tentorial incisure by a transtentorial herniation; *see* Kernohan, James W.

Notch of Rivinus: *Eponym* for the tympanic notch of the tympanic anulus; *see* Rivinus, Augustus Quirinus.

Notches of Santorini: *Eponym* for notches in the cartilage of the external acoustic meatus; *see* Santorini, Giovanni Domenico.

Nuck, Anton (1650–1692): Dutch anatomist and physician in Leiden, born in Harderwijk (Fig. N9).

Fig. N9 Engraving of Nuck. (From Wikidata; in public domain)

Canal of Nuck: *Eponym* for a patent vaginal process of the peritoneal cavity in females. Reference: *Adenographia curiosa et uteri foeminei anatoma nova.* Luchtmans, Leiden, 1691.

Nuclei: *See* **The Classical Anatomical Eponyms.**

Nuel, Jean-Pierre (1847–1920): Belgian physiologist and otologist born in Tetange, Luxemburg, and working in Louvain (Leuven), Ghent and later in Liège (Fig. N10). Reference: *Recherches microscopiques sur l'anatomie du limaçon des mammifères.* Acad Roy Belg, Brussels, 1873.

Space of Nuel: *Eponym* for the intermediate tunnel of the spiral organ of Corti, located between the outer pillar and outer hair cells (*see* Fig. C21).

Nuhn, Anton Johann (1814–1889): German anatomist born and working in Heidelberg (Fig. N11).

Fig. N10 Portrait of Nuel. (From Wikimedia Commons; in public domain)

Fig. N11 Photo of Nuhn. (From the University Library Heidelberg; in public domain)

Glands of Nuhn: *Eponym* for anterior lingual glands, also described by Bauhin (1590), Blandin (1823) and others. Reference: *Ueber eine bis jetzt noch nicht näher beschriebene Drüse im Innern der Zungenspitze.* Mannheim, 1845.

Nussbaum, Moritz (1850–1915): German biologist and histologist born in Horde, Westphalia, and working in Bonn.

Cells of Nussbaum: *Eponym* for cells in the pyloric glands of the stomach. Reference: *Ueber den Bau und die Thätigkeit der Drüsen.* Arch Mikr Anat 1877; 13:721–753.

Chapter O

O'Beirne, James (1786–1862): Irish surgeon, working in Edinburgh and Dublin.

Sphincter of O'Beirne: *Eponym* for the rectosigmoid sphincter, a thickening of the circular muscle of the most oral part of the rectum. Reference: *New Views of the Process of Defaecation and Their Application to the Pathology and Treatment of Diseases of the Stomach, Bowels and Other Organs, together with an analytical correction of Sir Charles Bell's views respecting the nerves of the face.* Dublin, 1833.

Obersteiner, Heinrich (1847–1922): Austrian neurologist (Fig. O1); *see* Obersteiner-Redlich zone. Reference (with E. Redlich): *Über Wesen und Pathogenese der tabischen Hinterstrangsdegeneration.* Arb Neurol Inst (Wien) 1894; 2:158.

Obersteiner-Redlich zone: *Eponym* for the transitional zone of a nerve root, a segment of a nerve root in the subarachnoid space between the brain or spinal cord and the place of transition from glia to neurolemma.

Ochsner, Albert John (1858–1925): American surgeon born in Baraboo, Wisconsin, and working in Chicago (Fig. O2).

Sphincter of Ochsner: *Eponym* for a sphincter in the descending part of the duodenum below the major duodenal papilla (*see* Fig. O4). Reference: *Further observations on the anatomy of the duodenum.* Amer J Med Sci 1906, 132:1–8.

Fig. O1 Portrait of Obersteiner. (From Wikimedia Commons; in public domain)

Fig. O2 Photo of Ochsner. (From the American College of Surgeons; in public domain)

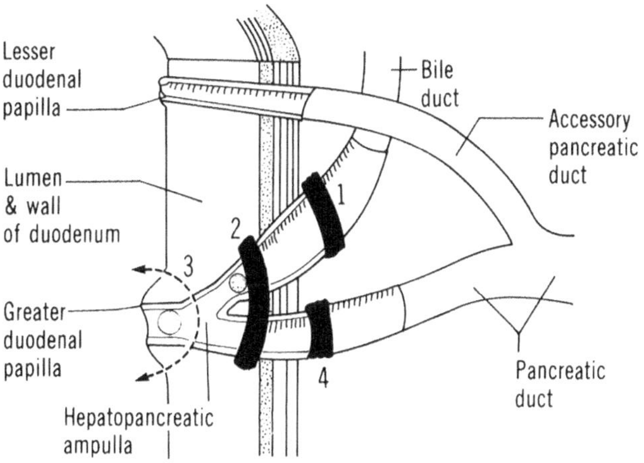

Fig. O4 The sphincter of Oddi. (From ten Donkelaar HJ, Oostra R-J *Klinische Anatomie en Embryologie,* 4th ed. Springer Media, Houten, 2014). *1* proper sphincter of ductus choledochus, *2* circular muscular layer of duodenum, *3* the ampullar sphincter of Oddi, *4* the proper sphincter of the pancreatic duct

Fig. O3 Portrait of Oddi. (From Wikimedia Commons; in public domain)

Oddi, Ruggero (1864–1913): Italian physiologist in Perugia (Fig. O3).

Sphincter of Oddi: *Eponym* for the sphincter of the hepatopancreatic ampulla (Fig. O4), already described by Glisson in 1654. Reference: *D'un disposition à sphincter spéciale de l'ouverture du canal cholédoque.* Arch Ital Biol 1887; 8:317–322.

Oehl, Eusebio (1827–1903): Italian histologist born in Lodi and working in Pavia.

Layer of Oehl: *Eponym* for the stratum lucidum of the epidermis. Reference: *Indagini di anatomia microscopica per servire allo studio dell'epidermide e della cute palmare della mano.* Ann Univ Med Milan 1857; 160:54–130, 280–325, 540–588.

Muscles of Oehl: *Eponym* for strands of muscle fibres in the chordae tendineae of the left atrioventricular valve. Reference: *Sulla presenza di elementi contrattili nelle maggiori corde tendine delle valvole mitrali umane.* Mem Accad Sci Torino 1863; 20:343–354.

Oelsner, L: He described the superior and inferior axillary lymph nodes in the second and third intercostal spaces and in the fourth and fifth intercostal spaces, respectively. Reference: *Anatomische Untersuchungen über die Lymphwege der Brust mit Bezug auf die Ausbreitung des Mammacarcinoms.* Thesis, Breslau, 1901, and Arch Klin Chir 1901; 64:134–158.

Anterior thoracic lymph nodes of Oelsner: *Eponym* for the superior axillary lymph nodes in the second and third intercostal spaces.

Inferior thoracic lymph nodes of Oelsner: *Eponym* for the inferior axillary lymph nodes in the fourth and fifth intercostal spaces.

Ogston, Sir Alexander (1844–1929): Scottish physician and surgeon born and, after studying in Aberdeen, Vienna and Berlin, working in Aberdeen.

Line of Ogston: *Eponym* for an imaginary line from the tubercle of the femur to the intercondylar fossa of the femur. Reference: *The operative treatment in Ogston's operation.* Edinb Med J 1876–1877; 22:782–784.

Oken, Lorenz (1779–1851): German physician born in Bohlsbeck, Würtemberg, and working in Basel, Munich and Zürich. Reference: *Allgemeine Naturgeschichte für alle Stände.* Stuttgart, 1833.

Body of Oken: *Synonym* for the mesonephros, described earlier by Wolff.

Duct of Oken: *Synonym* for the mesonephric duct, described earlier by Wolff.

Ollier, Louis Xavier Edouard Leopold (1825–1901): French surgeon born in Lyon and working in Paris and Lyon.

Layer of Ollier: *Eponym* for the inner or osteogenic layer of the periosteum.

Onodi, Adolf (1857–1920): Hungarian laryngologist.

Cell of Onodi: *Eponym* for a sphenoethmoidal cell, a variant of a posterior ethmoidal air cell in intimate relationship with the optic nerve just distal to the optic chiasm.

Olszewski, Jerzy (1913–1964): Polish-Canadian anatomist and neuropathologist born in Wilno and working from 1948 in Saskatchewan, Toronto and at the Montreal Neurological Institute. He is known especially for his atlas of the human brain stem (with Donald Baxter). Reference: *Cytoarchitecture of the Human Brainstem.* Karger, Basel, 1954; 2nd ed., 1981; third updated version, edited by Jean Büttner-Ennever and Anja Horn, Karger, Basel, 2015.

Nucleus of Meessen-Olszewski: *Eponym* for the pars alpha of the gigantocellular nucleus of the reticular formation.

Onufrowitz (Onuf), Bronislaw Kazimir Arthur (1863–1928): Russian neurologist of Polish origin born in Jeniseysk, Russia, studying in Switzerland and working in the United States (Fig. O5).

Nucleus of Onuf: *Eponym* for the nucleus of the pudendal nerve in the spinal cord (Fig. O6). Reference: *Notes on the arrangement and function of the cell groups in the sacral region of the spinal cord.* J Nerv Ment Dis 1899; 26:498–504.

Onufrowitz, Wladislaw Napoleon Sigismund (1854 or 1856–1899): Russian physician of Polish origin born in Jeniseysk, Russia, studying and working in Switzerland, the older brother of Bronislaw Onufrowitz. He described an acallosal brain for which several sources recommend to use the eponym **Onufrowitz** for the longitudinal callosal fascicles, usually known as the bundle of Probst (*see* Forkel SJ, et al. Cortex 2014; 56:73–84). Reference: *Das balkenlose Mikrocephalengehirn Hofmann. Ein Beitrag zur pathologischen und normalen Anatomie des menschlichen Gehirnes.* Arch Psychiatr Nervenkr 1857; 18:305–328.

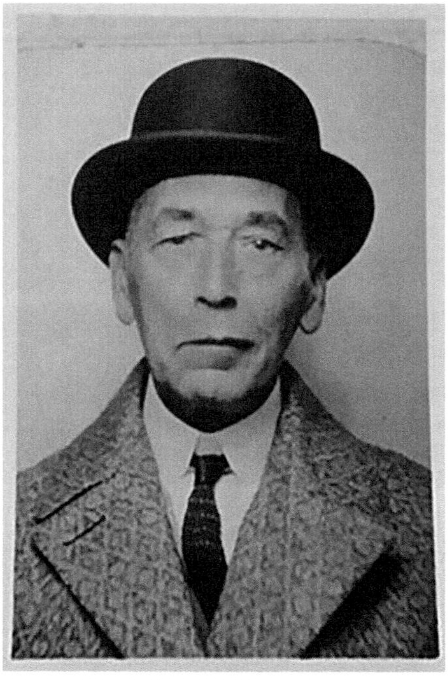

Fig. O5 Photo of Onufrowitz. (From Wikimedia Commons; in public domain)

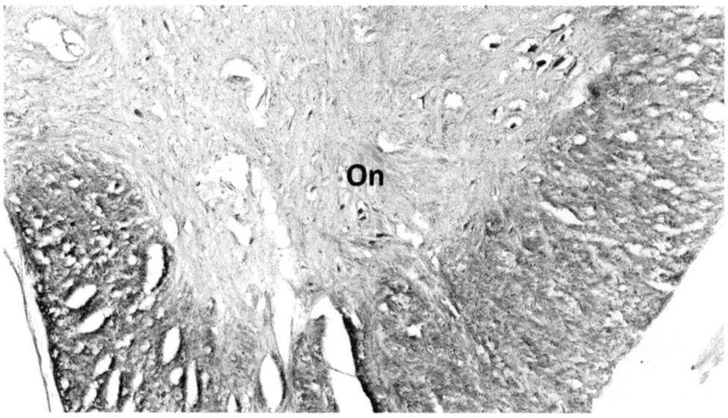

Fig. O6 The nucleus of Onuf (*On*; Courtesy Gulgun Sengul, Izmir; from ten Donkelaar HJ, Kachlík D, Tubbs RS *An Illustrated Terminologia Neuroanatomica*. Springer, Cham, 2018)

Oort, Henricus (1896–1989): Dutch otologist. Reference: *Über die Verästelung des Nervus octavus bei Säugetiere.* Anat Anz 1918; 51:272–280.

Nerve of Oort: *Eponym* for the cochlear communicating branch (vestibulocochlear anastomosis), efferent fibres of the cochlear nerve, which leave the brain stem with the vestibular nerve and join the cochlear nerve to end in the spiral organ of Corti.

Organs:

Organ of Chievitz, Juxtaoral: *Eponym* for an epithelial structure found at the angle of the mandibula with branches of the buccal nerve; *see* Chievitz, Johann H.

Organ of Corti: *Eponym* for the spiral organ of the cochlea; *see* Corti, Alfonso.

Organ of Giraldès: *Eponym* for the paradidymis, later also described by Waldeyer; *see* Giraldès, Joachim Albin Cardazo Cazado.

Organ of Golgi, Tendon: *Eponym* for the tendon organ; *see* Golgi, Camillo.

Organ of Haller: *Eponym* for the caudal aberrant ductule; *see* Haller, Albrecht von.

Organ of Jacobson: *Eponym* for the vomeronasal organ; *see* Jacobson, Ludwig Levin.

Organ of Rosenmüller: *Eponym* for the epoophoron; *see* Rosenmüller, Johann Christian.

Organ of Ruffini, Flower-spray: *Eponym* for the flower-spray ending, one of the two types of sensory nerve endings in the neuromuscular spindle; *see* Ruffini, Angelo.

Organ of Waldeyer: *Eponym* for the paradidymis, earlier described by Giraldès; *see* Waldeyer, Heinrich Wilhelm Gottfried.

Organ of Weber: *Eponym* for the prostatic utricle; *see* Weber, Moritz Ignaz.

Organ of Zuckerkandl: *Eponym* for the paraaortic (chromaffin) bodies; *see* Zuckerkandl, Emil.

Osborne, Geoffrey Vaughan (1918–2005): English surgeon born in North Wales.

Ligament of Osborne: *Eponym* for the fascia investing the ulnar nerve in the cubital tunnel. Reference: *The surgical treatment of tardy ulnar neuritis.* J Bone Joint Surg Br 1957; 39B:782; *Ulnar neuritis.* Postgrad Med J 1959; 35:392–396.

Ostmann, Friedrich Wilhelm Paul (1859–1945): German otologist born in Potsdam and working after military service in Königsberg and later in Marburg.

Fat pads of Ostmann: *Eponym* for lateral and medial fat pads around the pharyngeal opening of the auditory tube. Reference: *Ein Würdigung des Fettpolsters der lateralen Tubenwand.* Arch Ohrenheilk 1893; 34:170–189.

Fig. O7 Photo of Owen. (From Pagel 1901)

Otto, Adolph W (1786–1845): German surgeon.

Intermuscular septum of Otto: *Eponym* for the fibrous septum between the fibularis (peroneus) brevis and fibularis (peroneus) tertius muscles. Reference: *Seltene Beobachtungen an der Anatomie, Physiologie und Pathologie.* Heft I, Breslau, 1816:92.

Overlach, Martin (1860–1912): German physician who worked as assistant in the Histological Laboratory at Munich.

Spines of Overlach: *Eponym* for protoplasmic prolongations of epithelial cells of the cervix of the uterus. Reference: *Die pseudomenstruierende Mucosa uteri nach akuter Phosphorvergiftung.* Arch J Mikr Anat 1885; 25:191.

Owen, Sir Richard (1804–1892): English comparative anatomist born in Lancaster who became Conservator of the Royal College of Surgeons Museum in London, and finally Superintendent of the Natural History Department of the British Museum in London (Fig. O7).

Lines of Owen: *Eponym* for concentric layers of non-homogenously calcified dentin occurring as a result of appositional growth (*see* Fig. 23.1). Reference: *Odontography.* London, 1840–1845.

Spaces of Owen, Interglobular: *Eponym* for incompletely mineralized dentinal areas beneath the dentinoenamelar junction.

15 P, Q

Pacchioni, Antonio (1665–1726): Italian anatomist and physician born in Reggio Emilia and working in Rome (Fig. P1a). Reference: *Dissertatio epistolaris ad Lucam Schroekium de glanduli conglobatis durae meningis humanae, indeque ortis lymphaticis ad piam matrem productis.* Buagni, Rome, 1705.

Bodies or **Granules of Pacchioni:** *Eponym* for the arachnoid granulations in the superior sagittal sinus (Fig. P1b).

Foramen of Pacchioni: *Eponym* for (a) the tentorial notch, and (b) the foramen of the diaphragma sellae.

Foveolae of Pacchioni: *Eponym* for the granular foveolae, small depressions of the inner side of the frontal and parietal bones, which give shelter to the arachnoid granulations.

Pacini, Filippo (1812–1883): Italian anatomist born in Pistoia and working in Pisa and later in Florence (Fig. P2a). The lamellar corpuscles were first described in 1717 by Abraham Vater, and in 1830 rediscovered by Pacini. Reference: *Nuovi organi scoperti nel corpo umano.* Tipografia Cino, Pistoia, 1840.

Corpuscle of Pacini: *Eponym* for the lamellar corpuscle, a large fast-adapting receptor sensitive to vibrations, also known as corpuscle of Vater-Pacini (Fig. P2b).

Palade, George Emil (1912–2008): Romanian-American cell biologist born in Iasi, Romania. He discovered the ribosomes of the endoplasmatic reticulum. For his work on electron microscopy and cell fractionation, he received the Nobel Prize in Physiology and Medicine in 1974.

Bodies of Weibel-Palade: *Eponym* for storage granules of endothelial and endocardial cells. Reference: *New cytoplasmic components in arterial endothelia.* J Cell Biol 1964; 23:101–112.

Paladino, Giovanni Michelangelo (1842–1917): Italian physiologist and histologist born in Potenza and working in Naples.

Bundle of Paladino-His: *Eponym* for the atrioventricular bundle of the cardiac conducting system, also known as bundle of His and of Kent.

System of Paladino: *Eponym* for the system of atrioventricular and ventricular-valvular muscle fibres involved in conduction.

Palfijn, Jean or Palfyn, Jan (1649/1650–1730): Flemish surgeon, obstetrician and anatomist born in Kortrijk and working in Ghent and later in Paris (Fig. P3). The Academic Hospital of Ghent bears his name.

Sinus of Palfijn: *Eponym* for an inconstant air space within the crista galli of the ethmoidal bone, said to communicate with the frontal sinus and ethmoidal cells. Reference: *Beschrijving der Beenderen van 's Menschen Lichaem.* Ghent, 1702; German edition: Breslau, 1730; French edition: Paris, 1731.

Pander, Heinrich Christian von (1794–1865): German anatomist, embryologist and paleontologist born in Riga and working mainly in St. Petersburg. He described the embryonic layers and coined the term 'blastoderm', opening the way for the fundamental research of von Baer.

False amnion of Pander: *Eponym* for the chorionic membrane. Reference: *Beiträge zur Entwicklungsgeschichte des Hühnchens im Ei.* Würzburg, 1817.

Paneth, Joseph (1857–1890): Austrian physiologist born in Vienna and working in Breslau and Vienna (Fig. P4).

Cells of Paneth, Granular: *Eponym* for cells located at the base of glands in the small intestine that contain large acidophilic refractile granules, also described by

H. J. ten Donkelaar et al., *An Illustrated Guide to Anatomical Eponyms*, https://doi.org/10.1007/978-3-031-91664-9_15

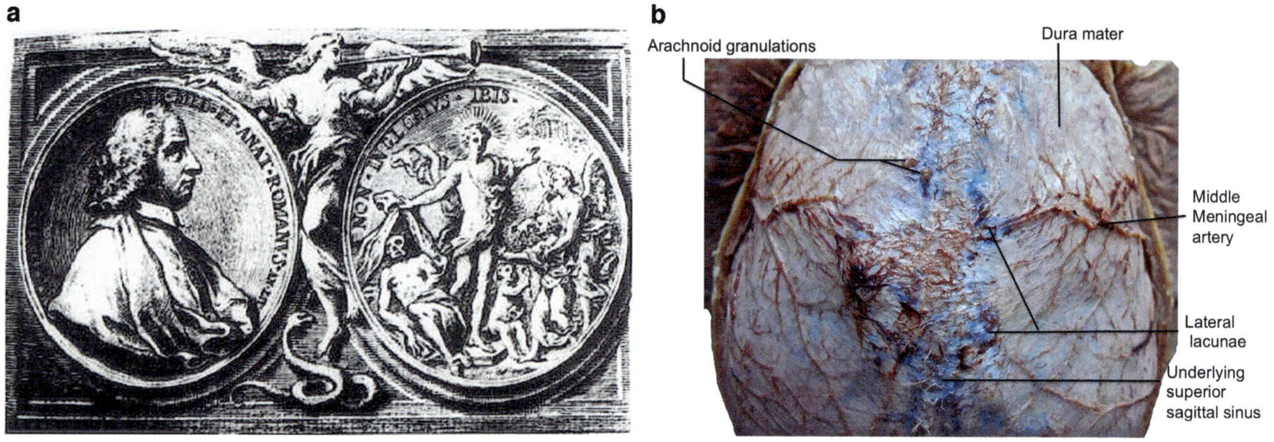

Fig. P1 (**a**) Engraving of Pacchioni. (From Wikipedia; in public domain); (**b**) Arachnoid granulations of Pacchioni around the superior sagittal sinus. (From ten Donkelaar HJ, Kachlík D, Tubbs RS *An Illustrated Terminologia Neuroanatomica*. Springer, Cham, 2018)

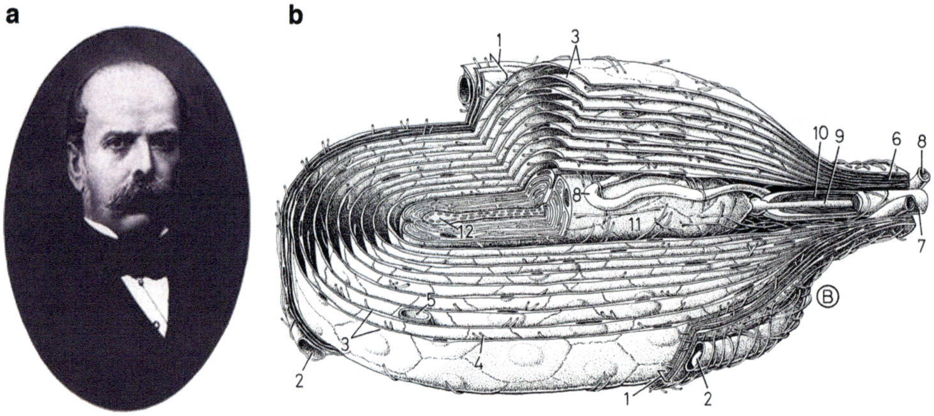

Fig. P2 (**a**) Portrait of Pacini. (From Wikimedia Commons; in public domain); (**b**) The structure of a Pacinian corpuscle. (From Krstić R *Die Gewebe des Menschen und der Säugetiere*, 2nd ed. Springer, Berlin-Heidelberg-New York, 1988; with permission). *1* lamellated capsule, *2*, *5*, *8* capillaries, *3* lamellae, *4* collagen microfibrils, *6* continuation of capsule and lamellae from perineurium, *7* nerve fiber, *9* axon losing its myelin sheath (*10*) but covered by flattened layers of Schwann cells (*11*), *12* thickened end of axon. (From ten Donkelaar HJ, Kachlík D, Tubbs RS *An Illustrated Terminologia Neuroanatomica*. Springer, Cham, 2018)

Davidoff (*see* Fig. L23). Reference: *Ueber die secernierenden Zellen des Dünndarm-Epithels.* Arch Mikr Anat 1887–1888; 31:113–191.

Panizza, Bartolomeo (1785–1867): Italian anatomist born in Vicenza and working in Pavia (Fig. P5) who gave the first description of the visual cortex and of the atrophic degeneration method, subsequently reported by Gudden. Reference: *Osservazioni sul nervo ottico.* Gior I Reale Inst Lombardo 1855; 7:232–252; *see also* Mazzarello P, Della Sala S: *The demonstration of the visual area by means of the atrophic method in the work of Bartolomeo Panizza (1855).* J Hist Neurosci 1993; 2:315–322.

Nerve of Panizza, Gustatory: *Eponym* for the glossopharyngeal nerve.

Plexus of Panizza: *Eponym* for a plexus of lymphatic vessels in the lateral fossae of the prepuce.

Pansch, Adolf Georg (1841–1887): German anatomist and anthropologist born in Eutin, Oldenburg, and working in Kiel who described the intraparietal sulcus in the human brain as Turner did for a chimpanzee brain. Reference: *Ueber die typische Anordnung der Furchen und Windungen auf den Grosshirnhemisphären des Menschen und der Affen.* Arch Anthropol 1868; 3:227–257.

Fissure or **Sulcus of Pansch:** *Eponym* for the intraparietal sulcus of the cerebral cortex.

Papez, James Wenceslaus (1883–1958): American anatomist born in Glencoe, Minnesota, to an American family of Czech descent. He worked in Minnesota and at Cornell University, Ithaca (Fig. P6). In 1937, James Papez proposed his circuit including projections from the hippocampus via the fornix to the mammillary body, then via the mammillothalamic tract of Vicq d'Azyr to the cingu-

Fig. P3 Statue of Palfijn in Ghent. (From Wikipedia; in public domain)

late gyrus, and as last step from the cingulate gyrus to the hippocampus. Papez suggested that his circuit formed the anatomical basis for emotions. MacLean modified it to the limbic circuit.

Circuit of Papez: *Eponym* for the limbic circuit. Reference: *A proposed mechanism of emotion.* Arch Neurol Psychiatr 1937; 38:725–743.

Parchappe de Vinay, Jean Baptiste Maximilien (1800–1866): French physiologist born in Epernay and working in Rouen and Paris.

Valve of Parchappe de Vinay: *Eponym* for the valve of the foramen ovale in the left atrium of the heart.

Paré, Ambroise (1510–1590): French surgeon and obstetrician born in Laval and working in Paris (Fig. P7). Paré introduced internal podalic version and breach retraction. Reference: *De la génération de l'homme.* Paris, 1573.

Fig. P4 Photo of Paneth. (From Wikimedia Commons; in public domain)

Parona, Francesco (1861–1910): Italian surgeon in Milan, specialized in orthopedic surgery.

Space of Parona: *Eponym* for a space between the pronator quadratus muscle and the overlying flexor tendons in the forearm.

Passavant, Philipp Gustav (1815–1893): German surgeon born and working in Frankfurt am Main.

Bar or **Cushion of Passavant:** *Eponym* for a prominence on the posterior wall of the nasopharynx formed by contraction of the superior constrictor of the pharynx during swallowing. Reference: *Ueber die Verschliessung des Schlundes beim Sprechen.* Virchows Arch 1869, 46.

Patches of Peyer: *Eponym* for aggregated lymphatic nodes of the ileum; *see* Peyer, Johann Conrad.

Patey, David Howard (1899–1977): English surgeon born in Monmouth and working in London.

Fig. P5 Portrait of Panizza. (From Wikimedia Commons; in public domain)

Fig. P6 Photo of Papez. (From the National Library of Medicine, Digital Collection; in public domain)

Plexus of Patey: *Eponym* for the neurovenous plexus around the facial nerve within the parotid gland. Reference: *Some points in the surgical anatomy of the parotid gland* (with I Ranger). Br J Sur 1957; 45:250–258.

Pattison, Granville Sharp (1791–1851): Scottish surgeon born in Glasgow and after working there, he emigrated to the USA, where he worked in Philadelphia and Baltimore. He described the septum of Denonvilliers as fascia of the prostatic gland (*see* Schultheiss D *Pattison Faszie; Das vergessene Eponym?* Aktuel Urol 2008; p 39.

Fascia of Pattison, Prostatic: *Eponym* for a fascial struture between the prostate and the rectum, comparable to the septum of Denonvilliers. Reference: *Observations on lithotomy.* Am Med Recorder 1820; 3:1–24.

Fig. P7 Engraving of Paré. (From the Wellcome Collection CCBY; in public domain)

Paulet, Vincent (1828–1906): French anatomist and surgeon.

Fascia of Paulet: *Eponym* for the middle layer of the cervical fascia, also described by Gruber, Richet and Testut. Reference: *Traité d'anatomie topographique, comprenant les principales applications à la pathologie et à la médecine opératoire.* Partie I. Martinet, Paris 1867–1870.

Pavlov, Ivan Petrovich (1849–1936): Russian physiologist in St. Petersburg and 1904 Nobel Laureate (Fig. P8).

Nerve of Pavlov: *Eponym* for the inferior cervical cardiac nerve. Reference: *Zentrobežnije nervij serdna.* St. Petersburg, 1883.

Pavlov, Vladimir Alexejevic (1863-?): Russian histologist, working in Charkov and later in Saratov.

Tract of Pavlov: *Eponym* for the tectoreticular tract, a descending tract from the superior colliculus to the midbrain reticular formation and the nuclei of the oculomotor and trochlear nerves. Reference: *Un faisceau descendant de la substance réticulaire du mesencéphale.* Nevraxe 1900, 1:273–276.

Fig. P8 Portrait of Pavlov. (Official Nobel Prize photo 1904; from Wikimedia Commons; in public domain)

Fig. P9 Photo of Pawlik. (From Wikimedia Commons; in public domain)

Pawlik, Karel (1849–1914): Bohemian gynaecologist born in Klatovy and working in Vienna and later in Prague (Fig. P9).

Triangle or **Vaginal trigone of Pawlik:** *Eponym* for a triangular area of the anterior wall of the vagina in contact with the fundus of the urinary bladder, and distinguished by the absence of vaginal rugae. Reference: *Ueber die Sondirung der Ureteren der weiblichen Blase.* Arch Gynäkol 1881; 18:491–495.

Payr, Erwin (1871–1946): Austrian surgeon born in Innsbrück and working in Graz and later in Germany.

Membrane of Payr: *Eponym* for a peritoneal membrane passing over the left colic flexure of the colon; Reference: *Beiträge zur Pathogenese, pathologischen Anatomie und radicalen operativen Therapie des runden Magengeschwürs.* Verh Dtsch Ges Chir 1910; 10:640–700.

Sphincter of Payr-Strauss: *Eponym* for a colon 'sphincter' at the splenic flexure of the colon.

Pechlin, Johannes Nicolaas (1644–1706): Dutch physician born in Leiden and working later in Kiel and Stockholm, and noted for his studies on the digestive tract.

Glands of Pechlin: *Eponym* for the aggregated lymphatic nodules of the ileum, usually described as glands or patches of Peyer who described them in 1677. Pechlin described these aggregations in: *De purgantium medicamentorum facultatibus exercitatio nova.* Leiden and Amsterdam, 1672.

Pecquet, Jean (1622–1674): French physician born in Dieppe and working in Montpellier and later Paris. In 1647, he described the thoracic duct in animals.

Cystern of Pecquet: *Eponym* for the cisterna chyli, described in animals, but in humans by van Horne. Reference: *Experimentia nova anatomica quibus incognitum hactenus chyli receptaculum, et ab eo per thoracem in ramos usque subclavios vasa lactea deteguntur. Ejusdem dissertatio anatomica, de circulatione sanguinis et chyli motu.* Paris, 1651.

Duct of Pecquet: *Eponym* for the thoracic duct, described in animals, and in humans by van Horne.

Pensa, Antonio (1874–1975): Italian anatomst born in Milan and working in Sassari, Parma and Pavia (Fig. P10).

Lymphatic tissue of Pensa: *Eponym* for the reticular connective tissue of the spleen and of lymph nodes.

Percheron, Gérard (1930–2011): French neurologist born in Pavilly, Seine Maritime, and working in Paris who later became a full-time researcher in the field of the vascularization of the basal ganglia and the thalamus.

Fig. P10 Photo of Pensa. (Courtesy Marina Quartu, Cagliari)

Artery of Percheron: *Eponym* for a variant of the posterior cerebral circulation, characterized by a solitary branch supplying bilaterally the paramedian thalami and the rostral midbrain. Reference: *The anatomy of the arterial supply of the human thalamus and its use for the interpretation of the thalamic vascular pathology.* Z Neurol 1973; 205:1–13.

Perkins, George (1892–1979): English orthopedic surgeon born in Staines, Middlesex, and working in London.

Line of Perkins: *Eponym* for a perpendicular line drawn on an anteroposterior radiograph of the pelvis, intersecting the lateral edge of the acetabular roof.

Perlia, Richard (1860–1931): German ophthalmologist in Krefeld and Frankfurt am Main.

Nucleus of Perlia: *Eponym* for the interoculomotor nucleus. Reference: *Die Anatomie des Oculomotoriuscentrums beim Menschen.* Von Graefe's Arch Ophthalm 1889; 35:287–302.

Perna, Giovanni (1876–1925): Italian anatomist in Bologna.

Ampulla of Perna: *Eponym* for the ampulla of the ductus deferens, the last dilated segment of the ductus deferens which, after having received the excretory duct of the vesicular gland, continues as the ductus ejaculatorius. Reference: *Sullo sviluppo dello vescichetta seminale, della ampollo del canale deferente e del canale ejaculatore nell'uomo.* Ital J Zool 1916; 27:45–49.

Pertick, Otto (1852–1913): Hungarian physician, histologist and pathologist in Budapest.

Diverticulum of Pertick: *Eponym* for an inconstant lateral outgrowing of the posterolateral pharyngeal recess, also described as fossa of Rosenmüller. Reference: *Neues Divertikel des Nasenrachenraums.* Virchow's Arch 1883: 94:1–12.

Petit, Antoine (1718–1794): French anatomist and surgeon in Paris.

Ligament of Petit: *Eponym* for the sacrouterine fold of the peritoneal cavity.

Petit, François Pourfour du (1664–1741): French anatomist born and working in Paris (Fig. P11a) who described the decussation of the pyramids (Fig. P11b), 1 year later than Mistichelli. References: *Lettres d'un médecin des hôpitaux du roy à un autre médecin de ses amis.* Albert, Namur, 1710; *Mémoire sur plusieurs découvertes faites dans les yeux de l'homme, des animaux à quatre pieds, des oiseaux et des poissons.* Mém Acad Sci Paris 1726, pp. 69–96.

Canal of Petit: *Eponym* for a narrow space between the ciliary zonule of Zinn and the vitreous body.

Ganglion of Petit: *Eponym* for the carotid ganglion in the cavernous plexus, later also described by Schmiedel (1747) and Laumonier (1793).

Sinus of Petit: *Eponym* for the aortic sinus, also described by Valsalva.

Petit, Jean Louis (1674–1750): French surgeon and anatomist born and working in Paris (Fig. P12a).

Triangle of Petit, Inferior lumbar: *Eponym* for the inferior lumbar triangle between the iliac crest caudally, the margin of the external oblique abdominal muscle laterally, and the margin of the latissimus dorsi muscle medially (Fig. P12b). The internal oblique abdominal muscle forms its floor.

Petit, Paul (1889-?): French anatomist.

Aponeurosis of Petit: *Eponym* for the posterior layer of the broad ligament of the uterus. Reference: *Aponévrose ombilico-pelvienne ou cache-vaisseaux.* Rev. Gynécol 1900; 4:579–606.

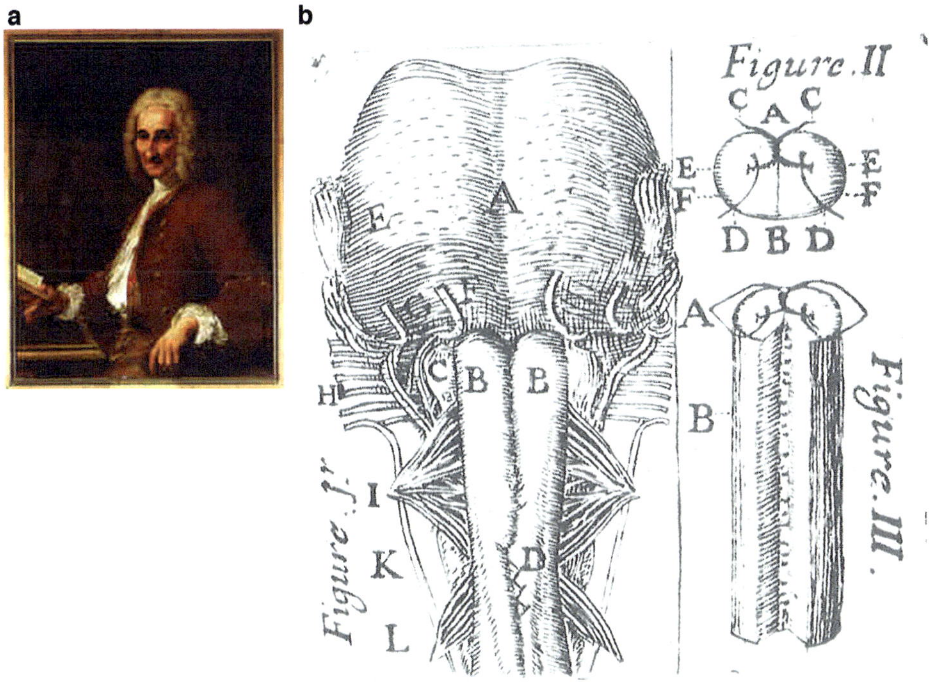

Fig. P11 (**a**) Portrait of Petit, F. Pourfour du. (From the Wellcome Collection CCBY; in public domain); (**b**) Decussation of the pyramidal tract by Pourfour du Petit (*D*; From Clarke E, O'Malley CD *The Human Brain and Spinal Cord. A historical study illustrated by writings from antiquity to the twentieth century.* Norman, San Francisco, 1996; figure in public domain)

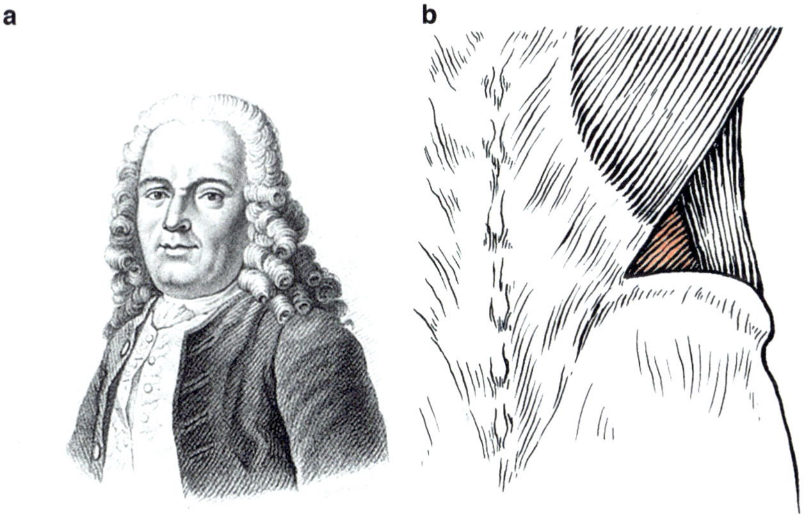

Fig. P12 (**a**) Portrait of Petit, Jean Louis. (From Wikimedia Commons; in public domain). This engraving by I Schuttenberger is found in the Iconographic Collections of the Wellcome Collections; (**b**) The triangle of Petit. (After Čihak R *Anatomie 1.* Grada, Prague, 2011; courtesy Ondřej Naňka, Prague)

Pétrequin, Joseph Pierre Eléonor (1809–1876): French surgeon born in Villeurbanne and working in Lyon.

Ligament of Pétrequin: *Eponym* for a fibrous band connecting the membranous layer of the subcutaneous abdominal tissue to the fascia lata just below the inguinal ligament. Reference: *Traité d'anatomie médico-chirurgicale et topographique.* Baillière, Paris, 1844.

Peyer, Johann Conrad (1653–1712): Swiss physician and anatomist born and working in Schaffhausen. He studied under Duverney in Paris.

Glands or **Patches of Peyer:** *Eponym* for the aggregated lymphatic nodules of the ileum, although described in 1672 by Pechlin. Reference: *Exercitatio anatomico-medica de glandulis intestinorum carumque usu et affectionibus.* Onophrinus and Waldkirch, Schaffhausen, 1677 (Fig. P13).

Pflüger, Edward Friedrich Wilhelm (1829–1910): German physiologist born in Hanau and working in Bonn.

Tubes or **Cords of Pflüger:** *Eponym* for the linear arrangement of sex cells in the developing ovary, also described by Valentin. Reference: *Ueber die Eierstöcke der Säugethiere und des Menschen.* Leipzig, 1863.

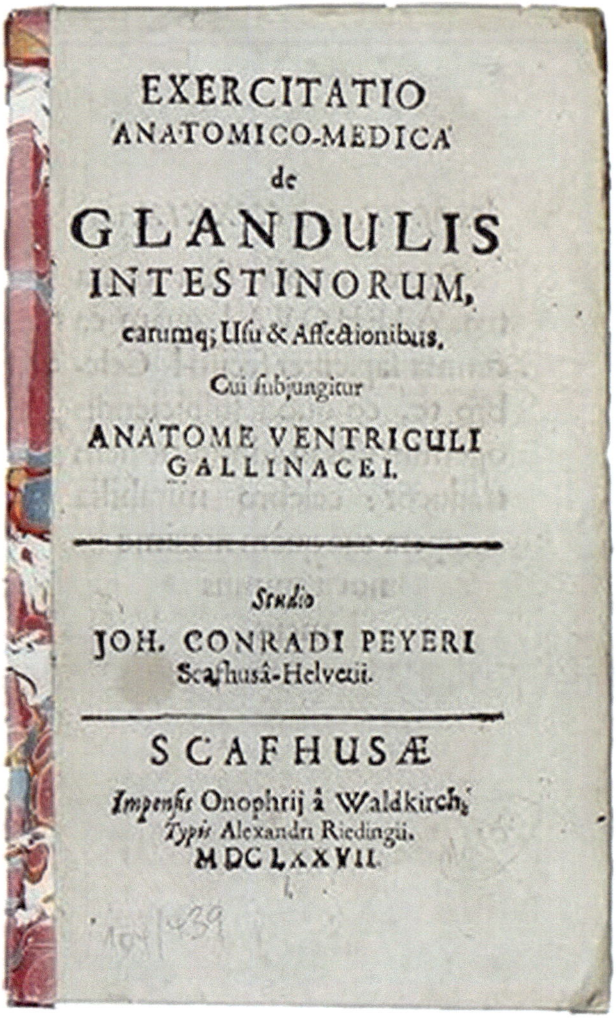

EXERCITATIO
ANATOMICO-MEDICA
de
GLANDULIS
INTESTINORUM,
earumq; Usu & Affectionibus,
Cui subjungitur
ANATOME VENTRICULI
GALLINACEI.

Studio
JOH. CONRADI PEYERI
Scaphusâ-Helvetii.

SCAFHUSÆ
Impensis Onophrij à Waldkirch,
Typis Alexandri Riedingii.
MDCLXXVII.

Fig. P13 Title page *Exercitatio anatomico-medica de glandulis intestinorum carumque usu et affectionibus*. Onophrinus and Waldkirch, Schaffhausen, 1677

Philippe, Claudien (1866–1903): French pathologist in Paris.

Triangle of Philippe-Gombault: *Eponym* for the continuation of the septomarginal tract in the sacral part of the spinal cord, also described by Gombault, his teacher. References: *Contribution à l'étude anatomque et clinque du tabes dorsalis.* Steinheil, Paris, 1897; *Manuel d'histologie pathologique,* Cornil AV, Ranvier LA, eds, 3rd ed., Baillière, Paris, 1901–1912, Vol 2, pp. 695–856.

Piccolomini, Archangelo (1525–1586): Italian anatomist born in Siena and working in Ferrara, Bordeaux and Rome, where he was the personal physician to three successive popes.

Medullary striae of Piccolomini: *Eponym* for the medullary striae on the floor of the fourth ventricle (Fig. P14), also described by Bergmann. Reference: *Anatomicae praelectiones*. Bodfadinus, Rome, 1586.

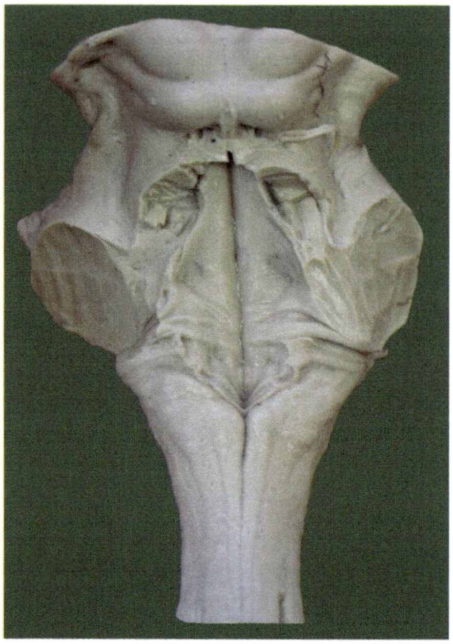

Fig. P14 The striae of Piccolomini of the fourth ventricle shown in a dorsal view. (Courtesy Veronika Němcová, Prague; from ten Donkelaar HJ, Kachlík D, Tubbs RS *An Illustrated Terminologia Neuroanatomica.* Springer, Cham, 2018)

Pick, Arnold (1851–1924): Moravian psychiatrist born in Velké Meziřiče, Moravia, and working in Prague (Fig. P15).

Bundle of Pick: *Eponym* for an anomalous bundle of nerve fibres in the medulla oblongata, accompanying the pyramidal tract but then ascending to the ambiguus nucleus. Reference: *Ueber ein abnormes Bündel der menschlichen Medulla oblongata.* Arch Psychiatr Nervenkr 1890; 21:636–640.

Pickerill, Henry Percy (1879–1956): English dental surgeon born in Hereford and working in Hereford, Birmingham, Dunedin, Sydney and Wellington, New Zealand.

Lines of Pickerill, Imbrication: *Eponym* for the external manifestation of the lines of Retzius, parallel impacted lines (perikymata) extending often along the whole crown of the tooth, better visible in young individuals. Reference: *Stomatology in General Practice: A textbook of diseases of the teeth and mouth.* Frowde, London, 1912.

Pinto, Olympio Faissol (twentieth century): Brasilian dentist working in Rio de Janeiro.

Fig. P15 Portrait of Pick. (From Wikimedia Commons; in public domain)

Ligament of Pinto: *Eponym* for the discomallear ligament, a rudimentary but relatively common ligament, connecting the craniodorsomedial part of the transverse band of the articular disc to the middle ear, passing from the petrotympanic fissure. Reference: *A new structure related to the temporomandibular joint and middle ear.* J Prosth Dent 1962; 12:95–103.

Pirie, George Alexander (1863–1929): Scottish radiologist in Dundee, one of the early pioneers of X-rays.

Bone of Pirie: *Eponym* for the dorsal astragaloscaphoid bone. Reference: *Osseous abnormalities of the foot which cause no discomfort.* Arch Radiol Electrother 1919; 24:12–14.

Pirogoff, Nicolai Ivanovitch (1810–1881): Russian surgeon born and working in Moscow and later St. Petersburg (Fig. P16). He advocated cross-sectional anatomy. Reference: *Anatome topographica sectionibus per corpus humanum congelatum triplici directione ductis illustrata.* Trej, St. Petersburg, 1852–1859.

Angle of Pirogoff: *Eponym* for the venous angle between the internal jugular vein and the subclavian vein, where the main lymphatic ducts terminate.

Aponeurosis of Pirogoff: *Eponym* for the bicipital aponeurosis.

Triangle of Pirogoff: *Eponym* for a triangle bounded by the intermediate tendon of the digastric muscle, the posterior border of the mylohyoid muscle and the hypoglossal nerve, containing the lingual artery.

Fig. P16 Photo of Pirogoff. (From Pagel 1901)

Piskaçek, Ludwig (1854–1932): Austrian-Hungarian obstetrician and gynaecologist born in Karcag, Hungary, and working in Vienna.

Sign of Piskaçek: *Eponym* the phenomenon that the uterus may appear asymmetrical, since the implantation site can enlarge one horn before the other, also described by Dickinson. Reference: *Über Ausladungen umschriebener Gebärmutterabschnitte.* Braumüller, Vienna, 1899.

Pitres, Jean Albert (1848–1927): French physician born in Bordeaux and working in Paris, America and Bordeaux (Fig. P17). He described the areas of the cerebral cortex. Reference: *Note sur la nomenclature des différentes regions du centre ovale des hémisphères cérébraux.* Arch Physiol Norm Pathol 1877; 9:245–258.

Area of Pitres: *Eponym* for the prefrontal area of the frontal lobe of the cerebral cortex.

Planes:

Plane of Addison: *Eponym* for the transpyloric or upper transverse abdominal line; see Addison, Christopher.

Planes of Hodge: *Eponym* for four planes perpendicular to the pelvic arch used in obstetrics; see Hodge, Hugh Lennox.

Planes, Anthropological:

Plane of Ackermann: *Eponym* for a plane passing through the base of the tragus and the infranasal point; *see* Ackermann, Jakob Fidelis.

Plane of Aeby: *Eponym* for a craniometric plane perpendicular to the median plane of the cranium, cutting the nasion and the basion; *see* Aeby, Christopher Theodor.

Fig. P17 Photo of Pitres. (From Wikimedia Commons; in public domain)

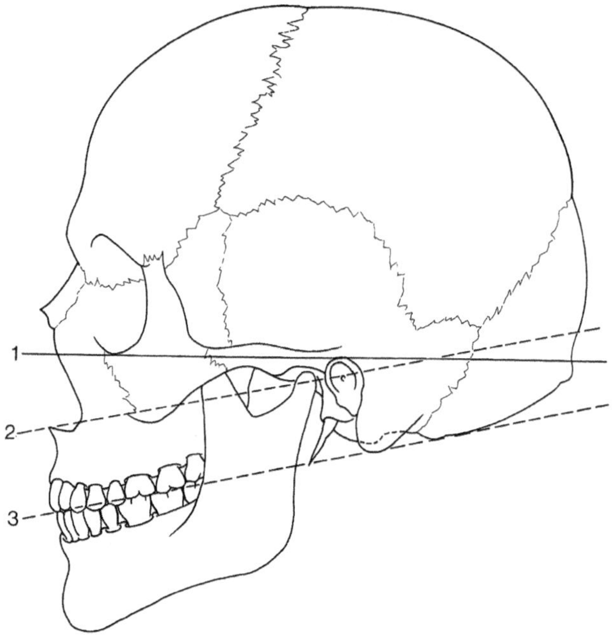

Fig. P18 Anthropological planes I: *1* Frankfurter Horizontale or Plane of Virchow-Voelder, *2* Plane of Camper, *3* Plane of Blumenbach

Plane of von Baer: *Eponym* for a plane through the superior margin of the zygomatic arch; *see* Baer, Karl Ernst von.

Plane of Blumenbach: *Eponym* for a plane passing through the points of contact at the maxillary teeth anteriorly and the mastoid processes posteriorly (Fig. P18); *see* Blumenbach, Johann Friedrich.

Plane of Bolton: *Eponym* for a roentgenographic cephalometric plane connecting a point at the intersection of the occipital condyle and the foramen magnum to the nasion; *see* Bolton, Joseph Shaw.

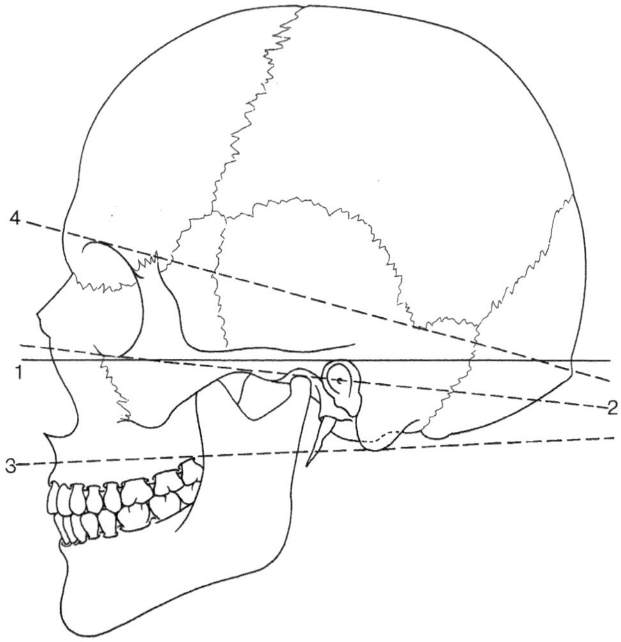

Fig. P19 Anthrolopological planes II: *1* Frankfurter Horizontale or Plane of Virchow-Voelder, *2* Plane of von Ihering, *3* Plane of Broca, *4* Plane of Schwalbe or glabella-inion plane

Planes of Broca: *See* Broca, Pierre Paul:

- **Horizontal plane of the head (1862):** *Eponym* for the alveolar-condylar plane from the alveolar point at the inferior border of the alveolar ridge to the inferior aspect of both occipital condyles (Fig. P19).
- **Orbitofrontal plane:** *Eponym* for the axis passing through the optic canal and the middle of the orbital phase.
- **Plane of horizontal vision (1862)** or **Visual plane (1873):** *Eponym* for the natural attitude of the head that permits the eyes to reach the horizon without muscular contraction. This plane was accepted by British anthropologists as well. German anthropologists preferred the plane of Virchow-Hoelder that came to be known as the Frankfurt-Virchow plane or 'Frankfurter Horizontale'. In 1877, Broca renamed this plane as **bi-orbital plane**, corresponding to the **neuro-ocular plane** of Cabanis et al. (J Comput Assist Tomogr 1978; 2:150–155).

Plane of Camper: *Eponym* for a horizontal plane from the anterior nasal spine to the centre of the external acoustic meatus bilaterally (Fig. P18); *see* Camper, Petrus.

Plane of Cuvier: *Modification* of the plane of Camper; *see* Cuvier, Georges Leopold Chrétien Frédéric Dagobert.

Plane of Daubenton and Daele: *Eponym* for a horizontal plane for the skull; *see* Daubenton, Louis Jean Marie.

Plane of Doornik: *Eponym* for a horizontal line from the incisors to the most prominent part of the occiput; *see* Doornik, Jacob Elisa.

Plane of Huxley, Basicranial: *Eponym* for a horizontal plane from the nasion to the basion; *see* Huxley, Thomas Henry.

Plane of Reid: *Eponym* for a plane through the inferior orbital margin and the centre of the external acoustic meatus, comparable to Ihering's plane; *see* Reid, Robert William.

Plane of Schwalbe: *Eponym* for a horizontal plane from the glabella to the inion (Fig. P19), which rather closely parallels the bicommissural plane; *see* Schwalbe, Gustav Albert.

Plane of Virchow-Hoelder (1875): *Eponym* for the Frankfurt-Virchow plane or 'Frankfurter Horizontale' from the inferior border of the orbit to the superior border of the external acoustic meatus (Figs. P18 and P19). A preliminary version was published by von Ihering (1872) who used the centre of the bony external acoustic meatus as reference point (Fig. P19); *see* Virchow, Rudolf Ludwig Karl.

Plate of Kühne: *Eponym* for the endplate of a motor nerve fibre in a muscle spindle; *see* Kühne, Wilhelm

Plexuses: *See* **The Classical Anatomical Eponyms.**

Points:

Point of Bolton: *Eponym* for a point at the intersection of the occipital condyle and the foramen magnum at the highest notch posterior to the occipital condyle; *see* Bolton, Joseph Shaw.

Point of Demel: *Eponym* for an area of the greater curvature of the stomach where the right and left gastroomental (gastroepiploic) arteries anastomose; *see* Demel, Rudolf.

Point of Desjardins: *Eponym* for the projection point of the major duodenal papilla upon the anterolateral body wall at the crossing of the right medioclavicular line with the right axilloumbilical line; *see* Desjardins, Abel.

Point of Erb: *Eponym* for (a) the punctum nervosum of the upper part of the brachial plexus of the posterior border of the sternocleidomastoid muscle about 3 cm above the clavicula, different from the point of Jonáš; and (b) the auscultatory point located in the third intercostal space at the left sternal border, where sounds of pathologically changed heart valves can be heard; *see* Erb, Wilhelm.

Point of Griffith: *Eponym* for the point at the splenic flexure where the marginal colic artery can be deficient.

Point of Hartmann, Critical: *Eponym* for the point between the last sigmoid artery and the superior rectal artery, also descibed by Sudeck; *see* Hartmann, Henri Albert Charles Antoine.

Point or **Punctum of Jonáš:** *Eponym* for the punctum nervosum of the cervical plexus, located at the midpoint of the posterior border of the sternocleidomastoid muscle; *see* Jonáš, Vratislav.

Point of Lanz: *Eponym* for where the apex of the vermiform appendix is projected upon the anterolateral abdominal wall; *see* Lanz, Otto.

Point of Larrey: *Eponym* for the femoral compression point, where the femoral artery passes beneath the inguinal ligament, halfway between the anterior superior iliac spine and the pubic symphysis; here, it can be compressed; *see* Larrey, Dominique Jean.

Point of Méglin: *Eponym* for the point of emergence of the greater palatine nerve from the greater palatine foramen; *see* Méglin, Jean.

Point of McBurney: *Eponym* for the point of maximum tenderness in cases of appendicitis found on the right spinoumbilical line of Monro, 6–7 cm medial from the anterior superior iliac spine; *see* McBurney, Charles.

Point of Sudeck, Critical: *Eponym* for the point at the rectosigmoid junction where the marginal colic artery may be deficient, also described by Hartmann; *see* Sudeck, Paul Hermann Martin.

Point of Toupet, Critical: *Eponym* for a critical point between the terminal branches of the sigmoid arteries, described by André Toupet, a French surgeon.

Point of Weber: *Eponym* for a point about 1 cm below the sacral promontory that represents the centre of gravity of the body; *see* Weber, Wilhelm Eduard.

Poirier, Paul Julien (1853–1907): French anatomist and surgeon born in Granville and working in Paris (Fig. P20). Reference: *Traité d'anatomie humaine*. Masson, Paris, 1902 (with Adrien Charpy).

Ansiform band of Poirier: *Old term* for the moderator band of the right ventricle of Leonardo and King.

Accessory bursae of Poirier: *Eponym* for (a) the bursa of the conoid lagament, (b) the bursa of the trapezoid ligament, and (c) the internal supracondylar bursa.

Fig. P20 Photo of Poirier. (From Wikimanche; in public domain)

Gland of Poirier: *Eponym* for an inconstant parauterine lymph node at the crossing of the uterine artery with the ureter, described also by Bayer.

Line of Poirier: *Eponym* for a line from the nasion to the lambda.

Lymph nodes of Poirier and Cunéo: *Eponym* for (a) abdominoaortic lymph nodes; (b) juxta-articular lymph nodes, (c) pararectal lymph nodes, also described by Gerota, and (d) popliteal lymph nodes, also described by Bardeleben. Reference: *Système lymphatique.* In: Poirier PJ, Charpy A, eds *Traité d'anatomie humaine*, Tome 2. Masson, Paris, 1902, pp. 1172–1180 (Fig. P21).

Space of Poirier: *Eponym* for the space overlying the palmar aspect of the proximal part of the capitate between the palmar radiocapitate and palmar radiotriquetral ligaments. Here, luxation of the lunate may occur.

Poisson, Francis (1871-?): French surgeon and anatomist born in Thièvres, Pas de Calais, and working in Lille.

Fossa of Poisson: *Eponym* for the infraduodenal recess of the peritoneal cavity. Reference: *Les fossettes péri-duodénales et leur rôle dans la pathogénie des hernies rétroperitonéales* Thesis, Paris 1895.

Politzer, Adam (1835–1920): Austrian otologist in Vienna, born in Alberti.

Cone of light of Politzer: *Eponym* for a triangular area of light reflex on the tympanic membrane, already noted by Wilde in 1853. Reference: *Die anatomische und histologische Zergliederung des menschlichen Gehörorgans im normalen und kranken Zustände.* Stuttgart, 1889.

Popescu, Laurentiu Mircea (1944–2015): Romanian pathologist working in Bucharest.

Cells of Popescu: *Eponym* for telocytes, interstitial, Cajal-like cells first described in the uterus and the uterine tubes. Reference: *Interstitial Cajal-like cells in human uterus and allopian tube* (with SM Ciontea and D Cretou). Ann NY Acad Sci 2007;1101:139–165.

Portal, Baron Antoine (1742–1832): French anatomist born in Gaillac and working in Paris.

Muscle of Portal: *Eponym* for the articularis cubiti muscle, fibres from the shaft of the humerus to the anterior aspect of the capsule of the elbow joint. Reference: *Cours d'anatomie médicale ou élémens de l'anatomie de l'homme avec des remarques physiologiques et pathologiques et les résultats de l'observations sur la siège et la nature des maladies, d'après l'ouverture des corps.* Paris, 1804.

Porter, William Henri (1790–1861): Irish surgeon in Dublin.

Fascia of Porter: *Eponym* for the pretracheal layer of the cervical fascia. Reference: *Observations on the Surgical Pathology of the Larynx and Trachea.* Dublin, 1826.

Pouches:

Pouch of Broca: *Eponym* for the pudendal sac, a pear-shaped encapsulated collection of connective and adipose tissue in the labium majus of the vulva; *see* Broca, Paul Pierre.

Pouch of Browne: *Eponym* for the superficial inguinal pouch, a pocket formed between the membranous layer of the subcutaneous tissue and the superficial abdominal fascia adjacent to the supericial inguinal ring, a common lodging site for an undescended or retractile testis; *see* Browne, Denis John Wolko.

Pouch of Douglas: *Eponym* for the rectouterine pouch of the peritoneal cavity; *see* Douglas, James.

Pouch of Haller: *Eponym* for the oblique sinus of the pericardium; *see* Haller, Albrecht von.

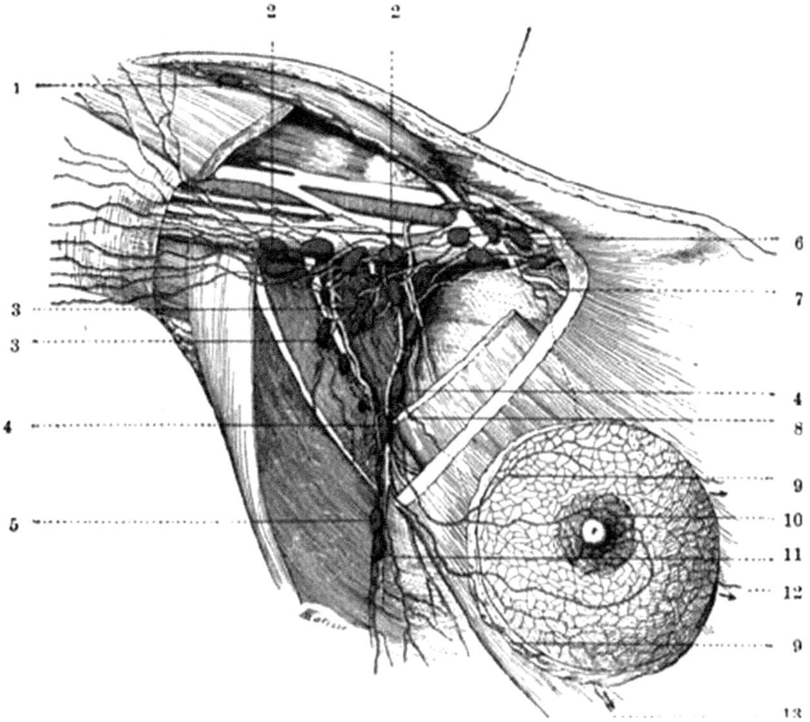

Fig. P21 Lymph nodes of Poirier and Cuéno: Lymphatic drainage of the breast. (From Poirier PJ, Cuéno B *Système lymphatique.* In Poirier PJ, Charpy A, eds *Traité d'anatomie humaine,* Tôme 2. Masson, Paris, 1902)

Pouch of Hartmann: *Eponym* for a dilatation of the neck of the gallbladder; *see* Hartmann, Henri Albert Charles Antoine.

Pouch of Morison: *Eponym* for the right subhepatic space or hepatorenal pouch of the peritoneal cavity; *see* Morison, James Rutherford.

Pouch of Rathke: *Eponym* for the endodermal invagination of the roof of the primitive buccopharyngeal cavity, from which the adenohypophysis arises; *see* Rathke, Martin Heinrich.

Pouch of Seessel: *Eponym* for a diverticle of the embryonic pharynx anterior to the cranial attachment of the buccopharyngeal membrane; *see* Seessel, Albert.

Pouch of Willis: *Old term* for the omental bursa; *see* Willis, Thomas.

Poupart, François (1661–1708): French surgeon born in Le Mans and working in Paris.

Ligament of Poupart: *Eponym* for the inguinal ligament (Fig. P22) that had already been described by Vesalius in 1543 and Falloppio in 1584. Reference: *Suspenseurs de l'abdomen.* Hist Acad Roy Sci Paris, 1705; p 64.

Line of Poupart: *Eponym* for a perpendicular line passing through the midpoint of the inguinal ligament.

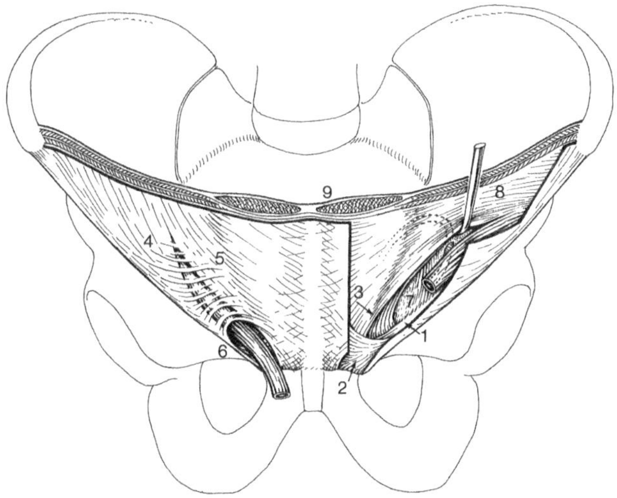

Fig. P22 Ligament of Poupart. *1* inguinal ligament of Poupart, *2* reflected ligament of Colles, *3* inguinal falx, *4* lateral crus, *5* medial crus, *6* superficial inguinal anulus, *7* fascia transversalis, *8* internal oblique abdominal muscle. (After ten Donkelaar HJ, Oostra R-J *Klinische Anatomie en Embryologie,* 4th ed. Springer Media, Houten, 2014)

Pourfour du Petit: *See* Petit, François, Pourfour du.

Pozzi, Samuel Jean (1846–1918): French anatomist and gynaecologist born in Bergerac and working in Paris.

Muscle of Pozzi: *Eponym* for the inconstant extensor digitorum brevis muscle of the hand, previously described in 1734 by Albinus.

Princeteau, Laurent Raphael (1858–1932): French anatomist born in Saint-Loubes, Gironde, and working in Bordeaux.

Tubercle of Princiteau: *Eponym* for the part of the temporal bone at the apex, where the superior petrosal sinus commences.

Pringle, Seton Sidney (1879–1955): Irish surgeon born in Clones, Monaghan, and working in Dublin.

Band of Pringle: *Eponym* for a peritoneal band from the mesocolon to the duodenal flexure. Reference: *Chronic intestinal stasis.* Dublin J Med Sci 1914; 137:81–98.

Pritchard, Urban (1845–1925): English otologist born and working in London.

Membrane of Pritchard, Reticulated: *Eponym* for the intercellular membrane in the ampulla of the semicircular duct. Reference: *The termination of the nerves in the vestibular and semicircular canals of Mammals.* Quart J Micr Sci 1876, 64.

Probst, Moritz (1867–1923): Austrian psychiatrist of German birth known for his studies on agenesis of the corpus callosum (*see* Bundle of Probst, although earlier described by Wladislaw Onufrowitz) and his studies on the brain stem. References: *Ueber den Bau des vollständig balkenloses Grosshirnes sowie über Mikrogyrie und Heterotopie der grauen Substanz.* Arch Psychiatr Nervenkr 1901, 34:709–786; *Experimentelle Untersuchungen über die Anatomie und Physiologie der Leitungsbahnen des Gehirnstammes.* Jahrb Psychiatr 1903, Bd 23.

Bundle or **Tract of Probst:** *Eponym* for a bundle of longitudinal fibres (longitudinal callosal fasciculus) in the absence of the corpus callosum (Fig. P23).

Commissure of Probst: *Eponym* for the commissure of the dorsal nuclei of the lateral lemniscus.

Processes:

Process of Burns, Falciform: *Eponym* for the superior horn of the falciform margin of the saphenous hiatus; *see* Burns, Alan.

Process of Civinini: *Eponym* for the pterygospinous process of the sphenoid bone; *see* Civinini, Filippo.

Process of Folli: *Eponym* for the anterior process of the malleus, also described by Rau; *see* Folli, Cecilio.

Processes of Lenhossék: *Eponym* for short processes of ganglion cells; *see* Lenhossék, Michael von.

Process of Lenhossék, Reticular: *Eponym* for the reticular process of the posterior horn of the spinal cord; *see* Lenhossék, József von.

Process of Rau: *Eponym* for the anterior process of the malleus, also described by Folli; *see* Rau, Johann Jakob.

Process of Stieda: *Eponym* for the posterior process of the talus; *see* Stieda, Alfred.

Proctor, Bruce (1914–1992): American otolaryngologist born at Lake Angelus, Michigan, and working in Detroit.

Ligament of Proctor: *Eponym* for the insertion of the salpingopharyngeal fascia of Troeltsch to the skull base at the sphenopetrosal suture. Reference: *Embryology and anatomy of the Eustachian tube.* Arch Otolaryngol 1967; 86:503–526.

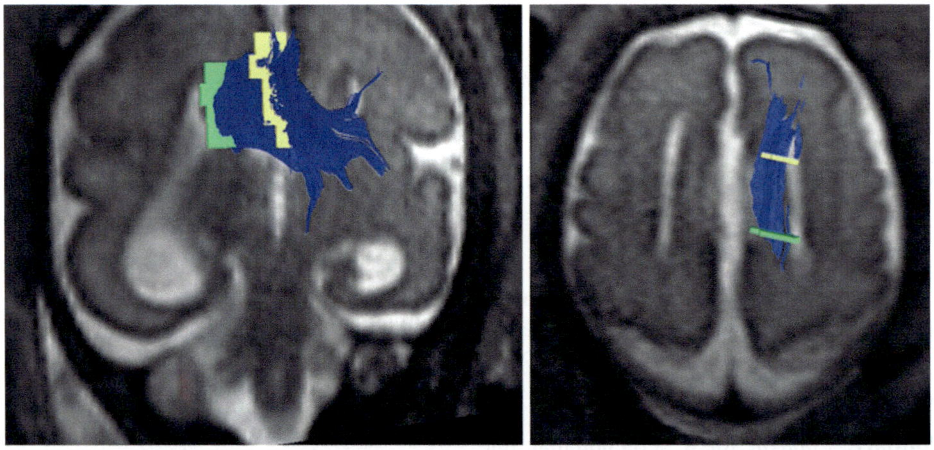

Fig. P23 The bundle of Probst, 3D visualization in the fetal brain. T2-weighted sequences were co-registered in the DTI data set. The fornix is shown in *yellow* and the ventricular atrium in *green*. Tractography showed a fronto-occipitally oriented pathway in the anatomical location of the Probst bundle (in *blue*; Images kindly provided by Gregor Kasprian, Vienna; from ten Donkelaar HJ, Lammens M, Hori A *Clinical Neuroembryology,* 3rd ed. Springer, Cham, 2023)

Fig. P24 Portrait of Prussak (from Wikimedia Commons; in public domain)

Proust, P.T. (nineteenth century): French physician in Champdeniers.

Space of Proust: *Eponym* for the rectovesical pouch of the peritoneal cavity in males. Reference: *Dissertation sur le péritonite.* Thesis, Paris, 1822.

Prussak, Alexander (1839–1894): Russian otolaryngologist born in Moscow and working in St. Petersburg (Fig. P24).

Fibres of Prussak: *Eponym* for connective tissue fibres between the pars flaccida and the pars tensa of the tympanic membrane.

Recess or **Space of Prussak:** *Eponym* for the superior recess of the tympanic membrane (Fig. P25). Reference: *Zur Anatomie des menschlichen Trommelfells.* Arch Ohrenhk 1867; 3:255–280.

Puelles, Luis (1948-today): Spanish neuroembryologist born in Santa Cruz de Tenerife, Canary Islands, and working mainly in Murcia (Fig. P26). He is famous for his genoarchitectonic approach to the developing brain, extensively changing the classical view of the subdivision of the brain into neuromeres to a more modern one (Fig. P27). Especially, his prosomeric approach to the forebrain (prosencephalon), dividing the forebrain into a caudal prosencephalon, giving rise to the diencephalon, and the rostral prosencephalon, giving rise to the telencephalon and the hypothalamus, has to be emphasized (Fig. P28). References: *Expression patterns of homeobox and other putative regularory genes in the embryonic forebrain suggests a neuromeric organization.* Trends Neurosci 1993; 16:472–479 (with John LR Rubenstein); *Survey of the midbrain, diencephalon, and hypothalamus neuroanatomic terms whose prosomeric definition conflicts with columnar tradition.* Front Neuroanat 2019; 13:20.

Purkinje (Purkyně), Jan Evangelista (1787–1869): Bohemian physiologist born in Libochovice and working in Breslau and later in Prague (Fig. P29) who described the Purkinje cells of the cerebellum in 1837. During his time in Breslau, Purkinje wrote his name with 'inje'. Afterwards, in Prague he began to use the Czech name Purkyně.

Cells of Purkinje: *Eponym* for large cells of the cerebellar cortex with many dendritic branches, forming the layer of Purkinje cells (Fig. P30). Reference: *Ueber die gangliösen Körperchen in verschiedenen Theilen des Gehirns.* Ber Naturf Prag 1837, p 179.

Fibres or **Network of Purkinje:** *Eponym* for the subendocardial branches of the atrioventricular conducting system of the heart.

Layer of Purkinje: *Eponym* for the layer of Purkinje cells in the cerebellar cortex.

Vesicle of Purkinje: *Eponym* for the nucleus of the avian ovum. Reference: *Symbolae ad ovi avium historiam ante incubationem.* Leipzig, 1830.

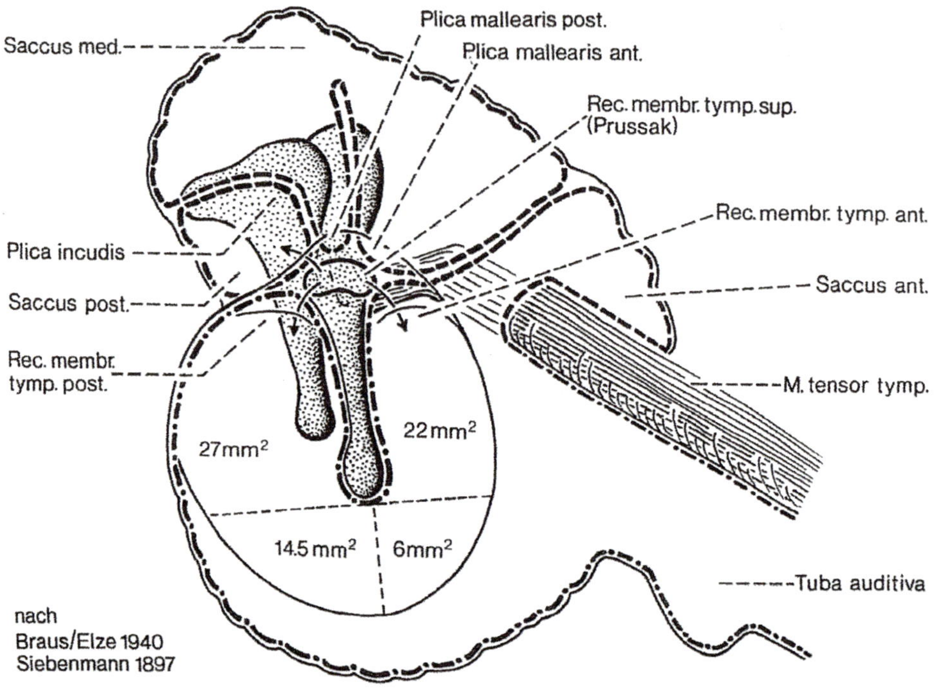

Fig. P25 The superior recess of the tympanic membrane of Prussak. (From Lang J *Klinische Anatomie des Ohres*. Springer, Wien, New York, 1992, after Siebenmann 1897, and Braus and Elze 1940; with permission)

Fig. P26 Photo of Puelles. (Courtesy Loreta Medina, Lleda)

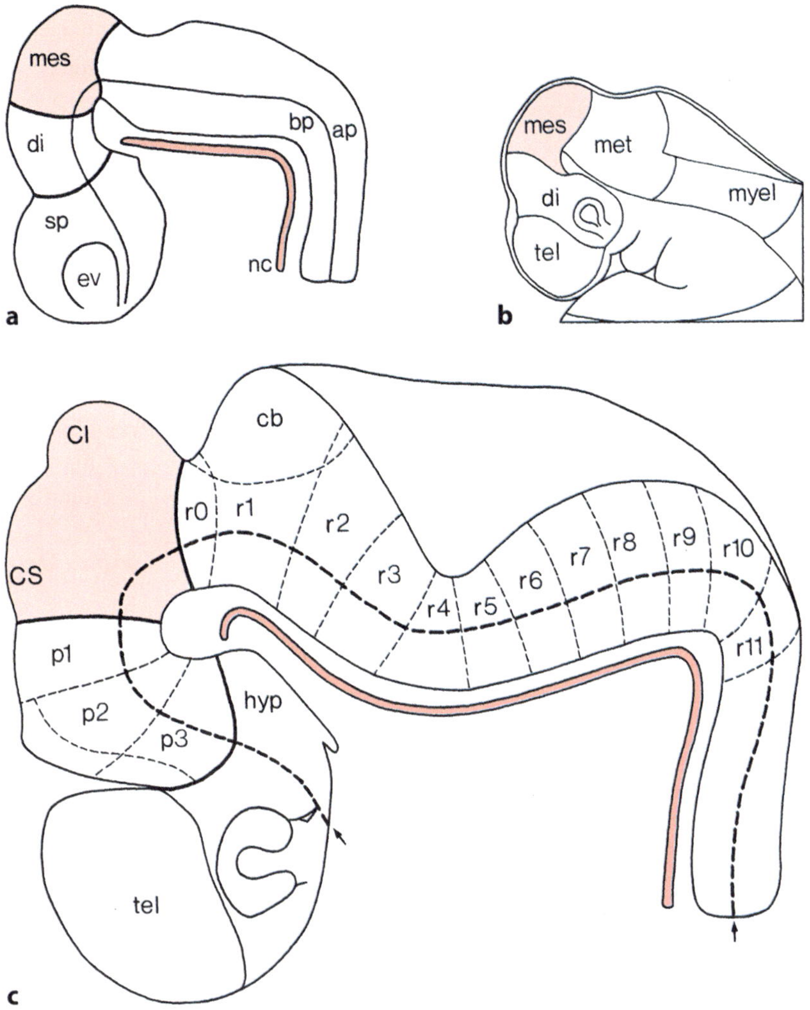

Fig. P27 (**a** and **c**) Segmentation of the human brain; (**b**) shows the classic O'Rahilly and Müller subdivision. *ap* alar plate, *bp* basal plate, *cb* cerebellum, *CI* colliculus inferior, *CS* colliculus superior, *di* diencephalon, *ev* eye vesicle, *hyp* hypothalamus, *mes* mesencephalon (in *light red*), *met* metencephalon, *myel* myelencephalon, *nc* notochord (in *medium red*), *p1-p3* prosomeres, *r0-r11* rhombomeres, *sp* secondary prosencephalon, *tel* telencephalon. (After Puelles L, Martínez S, Martínez de la Torre M *Neuroanatomía.* Panamericana, Buenos Aires, Madrid, 2008; from ten Donkelaar HJ, Lammens M, Hori A *Clinical Neuroembryology,* 3rd ed. Springer, Cham, 2023)

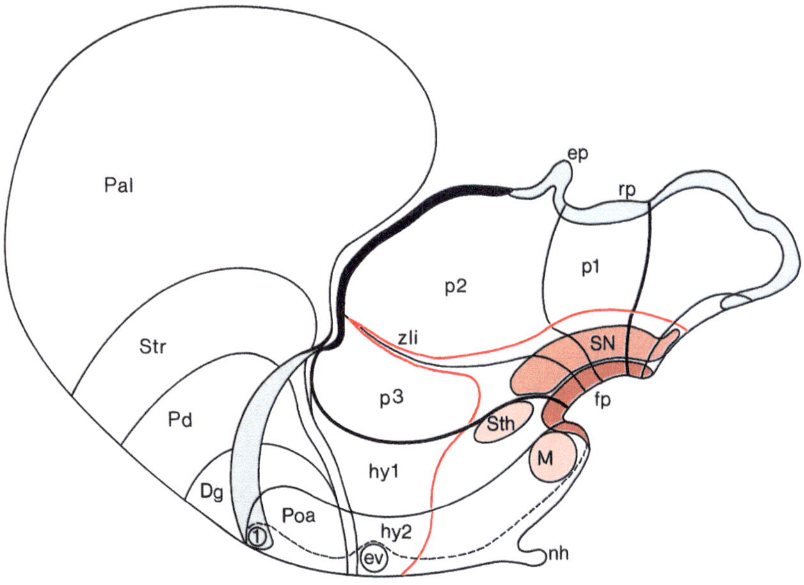

Fig. P28 Prosomeric model of the mouse brain. The roof plate (*rp*) has been indicated in *light grey*, in between the two parts, the choroid plexus of the third ventricle is shown in *black*. The *red line* separates the alar and basal parts. The floor plate (*fp*) is shown in *red*, the substantia nigra (*SN*) in *medium red* and the subthalamic nucleus (*Sth*) and the mamillary body (*M*), both derivatives of the hypothalamic basal plate, in *light red*. *Dg* diagonal band, *ep* epiphysis, *ev* eye vesicle, *hy1*, *hy2* hypothalamic prosomeres, *nh* neurohypophysis, *Pal* pallium, *Pd* pallidum, *Poa* preoptic area, *p1-p3* diencephalic prosomeres, *Str* striatum, *zli* zona limitans intrathalamica. (After Puelles L, Martínez–de-la-Torre M, Bardet S, Rubenstein JLR *Hypothalamus.* In: Watson C, Paxinos G, Puelles L, eds, *The Mouse Nervous System.* Elsevier, Amstredam, 2012, pp. 221–312; from ten Donkelaar HJ, Lammens M, Hori A *Clinical Neuroembryology,* 3rd ed. Springer, Cham, 2023)

Fig. P29 Portrait of Purkinje. (From Pagel 1901)

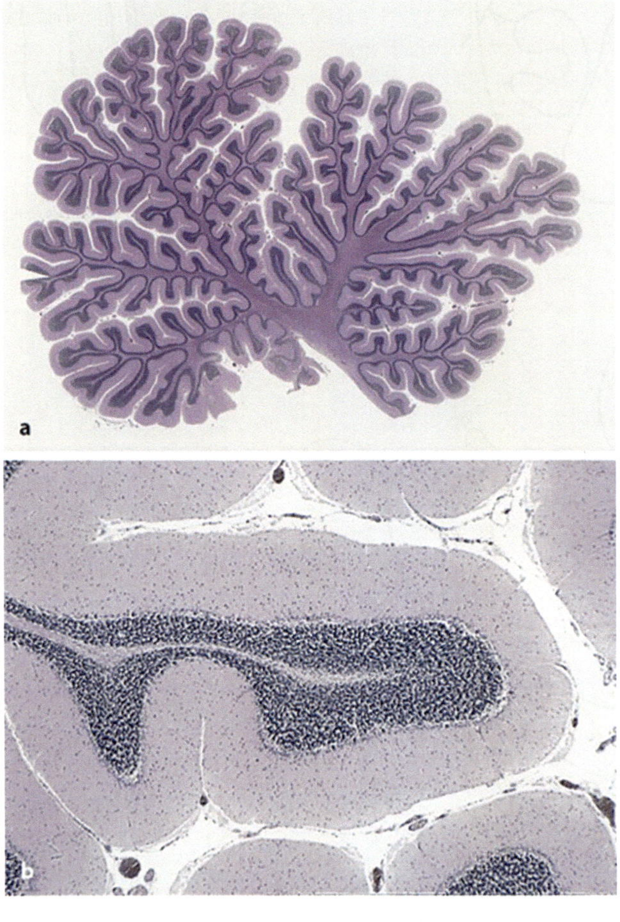

Fig. P30 The Purkinje cell layer shown in a median section (**a**) and in a detail (**b**) of the human cerebellum. (Courtesy Gerard van Noort, Enschede; from ten Donkelaar HJ, Lammens M, Hori A *Clinical Neuroembryology*, 3rd ed. Springer, Cham, 2023)

Pyramids:

Pyramids of Ferrein: *Eponym* for the cortical pyramids of the kidney, corresponding to the medullary pyramids of Malpighi; *see* Ferrein, Antoine.

Pyramid of Lallouette: *Eponym* for the inconstant pyramidal lobe of the thyroid gland; *see* Lalloutte, Pierre.

Pyramid of Malacarne: *Eponym* for a prominence on the inferior surface of the vermis of the cerebellum; *see* Malacarne, Michele Vincenzo Giacinto.

Pyramids of Malpighi: *Eponym* for the medullary pyramids of the kidney, corresponding to the cortical pyramids of Ferrein; *see* Malpighi, Marcello.

Chapter Q

Quain, Jones (1795–1851): Irish anatomist working in London (Fig. Q1).

Iliacus minor of Quain: *Eponym* for an accessory muscle bundle of the iliacus muscle. Reference: *Elements of Descriptive and Practical Anatomy.* 12th ed., Longmans and Green, London, 1908–1929.

Veins of Quain: *Eponym* for the colicospermatic veins, his colicoepigastric veins, collecting blood from the descending colon and draining into the left testicular vein.

Fig. Q1 Portrait of Quain. (From the Wellcome Collection CCBY; in public domain)

Fig. Q2 Portrait of Quatrefages de Bréau. (From Wikipedia; in public domain)

Quatrefages de Bréau, Jean Louis Armand (1810–1892): French anthropologist and zoologist born in Berthezème, Cevennes, and working in Paris (Fig. Q2).

Angle of Quatrefages: *Eponym* for the parietal angle used in craniometry. Reference: *Note sur l'angle pariétal et sur un gonomêtre destiné à le mésurer.* CR Acad Sci Paris 1858; 46:1.

Fig. Q3 Portrait of Quénu. (From Wikimedia Commons; in public domain)

Quénu, Edouard André Victor Alfred (1852–1933): French surgeon, anatomist and pathologist born in Marquise and working in Paris (Fig. Q3).

Lymph nodes of Quénu, Inguinal: Quénu distinguished four groupes of inguinal lymph nodes, 'supéro-interne', 'supéro-externe', 'inféro-interne' and 'inféro-externe'. He also described a 'groupe présaphène', currently known as saphenofemoral lymph node.

Plexus of Quénu, Haemorrhoidal: *Eponym* for a series of lymphatic plexuses in the mucosa and skin of the anus. Reference: *Vaisseaux lymphatiques de l'anus.* Bull Soc Anat Paris 1893, 7:399–409.

16 Letter R

Rainey, George (1801–1884): English anatomist born in Spilsby and working in London.

Corpuscles of Rainey: *Eponym* for calcium nodules occurring in the process of calcification of tissues. Reference: *On the formation of the skeleton of animals and other hard structures formed in connection with living tissues.* Brit For Med Chir Rev. London 1857; 20:451–476.

Tubes of Rainey: *Eponym* for tubules formed in the process of calcification.

Rámon y Cajal, Santiago (1852–1934): Spanish anatomist and histologist born in Petilla, Aragon, and working in Valencia, Barcelona and later in Madrid (Fig. R1), especially known for his splendid use of the Golgi technique on the CNS of various vertebrates (Figs. R2 and R3). In 1906, he received the Nobel Prize in Physiology and Medicine together with Camillo Golgi. His name is remembered in the interstitial nucleus of Cajal, the horizontal cell of the (developing) cerebral cortex (Cajal-Retzius cell) and the double dendritic bouquet neuron. References: *Estructura del cerebelo.* Gac Med Catalana 1888, 11: 449–457; *Sur la structure de l'écorce cérébrale de quelques mammifères.* Cellule 1891; 7:123–176; *Textura del Sistema Nervioso del Hombre y de los Vertebrados,* 3 Vols. Moya, Madrid 1899–1904; French translation by L Azoulay: *Histologie du système nerveux de l'homme et des vertébrés,* 2 Vols. Maloine, Paris, 1909–1911; English translation from the French by N Swanson and LW Swanson: *Histology of the Nervous System of Man and Vertebrates,* 2 Vols. Oxford University Press, New York, 1995.

Cell of Cajal, Horizontal: *Eponym* for the horizontal cell found in the molecular layer of the cerebral cortex.

Cell of Cajal, Interstitial: *Eponym* for the interstitial cell or enteric pacemaker cell, a pacemaker cell of the intestine, found scattered among the smooth muscle cells.

Cell of Cajal-Retzius: *Eponym* for the horizontal cell of the developing cerebral cortex (*see* Fig. R18).

Layer of Cajal: *Eponym* for the molecular layer of the olfactory bulb.

Nucleus of Cajal: *Eponym* for the interstitial nucleus of the medial longitudinal fasciculus in the prerubral tegmentum.

Ramsay Hunt, James (1872–1937): American neurologist born in Philadelphia and working in New York, also known as Hunt, James Ramsay.

Zone of Ramsay Hunt: *Eponym* for a delimited skin area in the external acoustic meatus, supplied by somatosensory fibres of the auricular branch of the vagus nerve, connected to the ganglion geniculi, clinically relevant in otic herpes zoster. Reference: *On herpetic inflammations of the geniculate ganglion. A new syndrome and its complications.* J Nerv Ment Dis 1907; 34:73–96.

Randacio, Francesco (1821–1903): Italian anatomist born in Cagliari, Sardinia, and working in Palermo, Sicily.

Nerves of Randacio: *Eponym* for the pterygopalatine nerves, branches arising from the pterygopalatine ganglion. Reference: *Nuovi rami del ganglio sfeno-palatino.* Palermo, 1863.

Ranke, Johannes (1836–1916): German anthropologist and physiologist born in Thurnau and working in Munich (Fig. R4a).

Angle of Ranke: *Eponym* for a cephalic angle used in craniometry. Reference: *Beiträge zur physischen Anthropologie in Bayern.* Munich, 1883.

Subpial granular layer of Ranke: *Eponym* for a transient layer of the cerebral cortex that develops from the basal periolfactory subventricular zone. It disappears during the last trimester of gestation (Fig. R4b).

H. J. ten Donkelaar et al., *An Illustrated Guide to Anatomical Eponyms*, https://doi.org/10.1007/978-3-031-91664-9_16

Fig. R1 Photo of Rámon y Cajal (Official Nobel Prize photo 1906; from Wikimedia Commons; in public domain)

Reference: *Beiträge zur Kenntnis der normalen und pathologischen Hirnrindenbildung.* Ziegl Beitr Path Anat 1910; 47:51–125.

Ranvier, Louis Antoine (1835–1922): French physician and histologist born in Lyon and working in Paris known for his studies on the structure of nerves (Fig. R5).

Arteries of Ranvier: *Eponym* for medium-calibre arteries with a tunica media rich in smooth muscle cells.

Node or **Constriction of Ranvier:** *Eponym* for gaps of the myelin sheath of a peripheral nerve, where the myelin sheath is discontinuous and at which the axolemma is free from myelin, enabling quicker transmission of the impulse. Reference: *Recherches sur l'histologie et la physiologie des nerfs.* Arch Physiol Norm Pathol 1871; 4:129–149.

Segments of Ranvier: *Eponym* for the segments between the nodes of Ranvier.

Raschkow, Isaac (1809–1872): German physician.

Plexus of Raschkow: *Eponym* for the subodontoblastic plexus, a plexus of nerve fibres located in the subodontoblastic layer (*see* Fig. 23.2). The plexus is responsible for pain transmission from the pulp of the tooth. Reference: Scheinin A, Light EI *Innervation of the dental pulp.* Acta Odont Scand 1969; 27:313–319.

Rasmussen, Grant Lister (1904–1989): American neuroanatomist.

Bundle of Rasmussen: *Eponym* for the olivocochlear bundle of the hindbrain. Reference: *Efferent fibers of the cochlear nerve and cochlear nucleus.* In: Rasmussen, GL, Windle WF, eds, *Neural Mechanisms of the Auditory and Vestibular Systems.* Thomas, Springfield, IL, 1960, pp. 105–115.

Rathke, Martin Heinrich (1793–1860): German anatomist born in Danzig (currently Gdansk, Poland) and working in Dorpat and later in Königsberg (Fig. R6) who described the development of the anterior lobe of the pituitary gland from a depression in the roof of the embryonic stomodeum. Reference: *Ueber die Entstehung der Glandula pituitaria.* Arch Anat Physiol Wiss Med Lpz 1838, pp. 482–485.

Cleft of Rathke: *Eponym* for a fine gap in the intermediate part of the pituitary gland, a remnant of the pouch of Rathke.

Cysts of Rathke: *Eponym* for follicles of the intermediate part of the pituitary gland, which are lined with simple cuboidal epithelium, probably due to fragmentation of the cleft of Rathke.

Folds of Rathke: *Eponym* for two transient lateral folds assumed to form the urogenital septum, dividing the rectum from the urinary bladder. Its role in forming the septum has not been proved. Reference: *Abhandlungen zur Bildungs- und Entwicklungsgeschichte der Thiere.* Leipzig, 1832.

Pouch of Rathke: *Eponym* for the endodermal invagination of the roof of the primitive buccopharyngeal cavity, from which the anterior lobe of the pituitary gland arises (Fig. R7).

Rau, Johann Jakob (1668–1719): German anatomist and surgeon born in Baden who worked in the Dutch Republic (Fig. R8).

Process of Rau: *Eponym* for the anterior process of the malleus, also described by Folli.

Rauber, August Antonius (1841–1917): German anatomist born in Haard and working in Basel, Leipzig and Dorpat.

Artery of Rauber: *Eponym* for the proper hepatic artery.

Ganglion of Rauber: *Eponym* for the sublingual ganglion.

Fig. R2 Neuronal elements of the hippocampal formation. (After Ramón y Cajal S *Histologie du système nerveux de l'homme et des vertébrés*. Maloine, Paris, 1909–1911 and Shepherd GM *The Synaptic Organization of the Brain*, 3rd ed. Oxford University Press, New York, 1990; courtesy the late Gordon Shepherd). The hippocampus proper is divided into the stratum oriens (*1*), the stratum pyramidale (*2*), the stratum radiatum (*3*) and the stratum lacunosum-moleculare (*4*). The giant pyramidal cells in CA3 project to CA1 pyramidal cells via the Schaffer collaterals, to other CA3 pyramidal cells, and through the commissure via fibers in the alveus to the contralateral hippocampus. Granule cells in the dentate gyrus send mossy fibers to CA3 pyramidal cells. The perforant path input is shown to all regions of the hippocampus and dentate gyrus. (From ten Donkelaar HJ, Kachlík D, Tubbs RS *An Illustrated Terminologia Neuroanatomica*. Spriner, Cham, 2018)

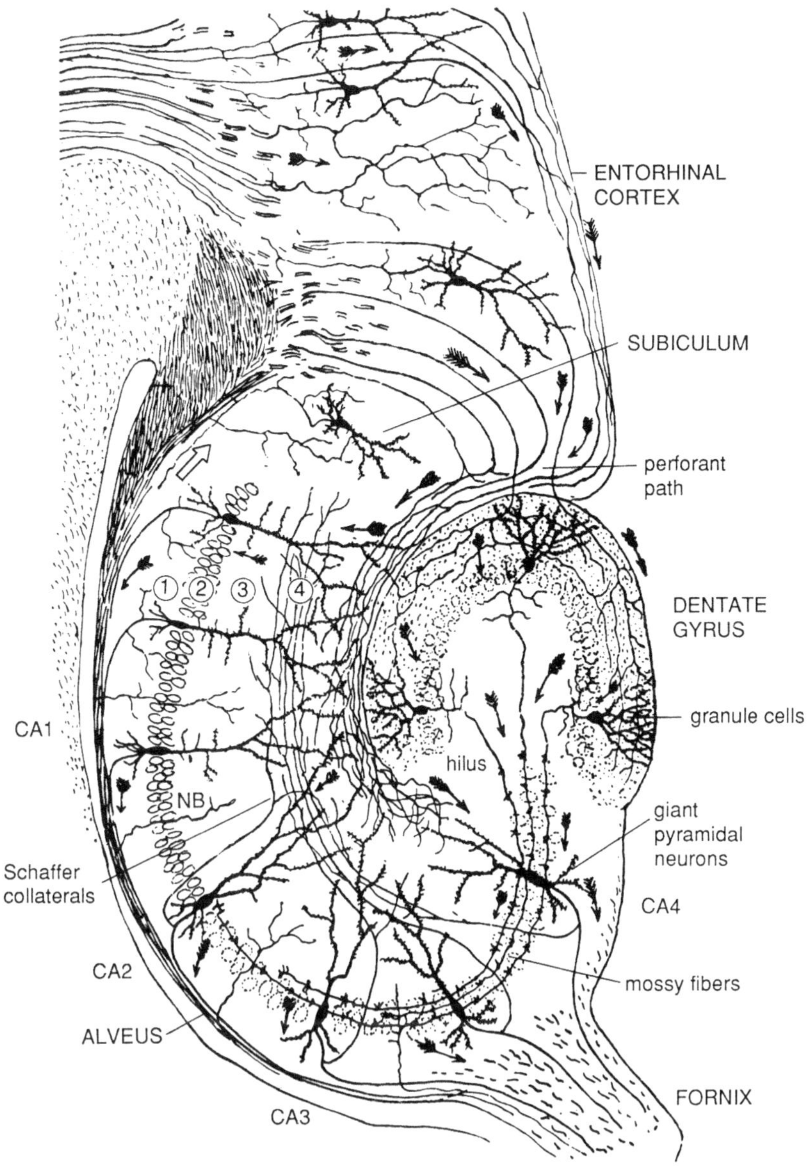

Layer of Rauber: *Eponym* for the outer cell mass of the blastodermic vesicle. Reference: *Formbildung und Formstörung in der Entwicklung von Wirbelthieren*. Leipzig, 1880.

Ligament of Rauber: *Eponym* for the salpingopharyngeal fold.

Recesses:

Recess of Hartmann: *Eponym* for an accessory recess of the peritoneal cavity located anterior to the mesoappendix; *see* Hartmann, Henri Albert Charles Antoine.

Recess of Hyrtl: *Eponym* for the epitympanic recess of the tympanic cavity; *see* Hyrtl, Joseph.

Recess of Jacquement: *Eponym* for the hepatocystic recess, a peritoneal pouch between the liver and the gallbladder; *see* Jacquement, Marcel.

Recess of Prussak: *Eponym* for the superior recess of the tympanic membrane; *see* Prussak, Alexander.

Recess of Rex: *Eponym* for the site of the remnant of the umbilical vein, located in the space between the hepatic segments III and IV; *see* Rex, Hugo.

Recess of Rosenmüller: *Eponym* for the pharyngeal recess; *see* Rosenmüller, Johann Christian.

Recess of Tarin: *Eponym* for the interpeduncular fossa of the midbrain; *see* Tarin, Pierre.

Recesses of Troeltsch: *Eponym* for the anterior and posterior recesses of the tympanic membrane; *see* Troeltsch, Anton Friedrich von.

Recess or **Fossa of Waldeyer:** *Eponym* for an inconstant mesentericoparietal recess of the peritoneal cavity; *see* Waldeyer, Heinrich Wilhelm Gottfried.

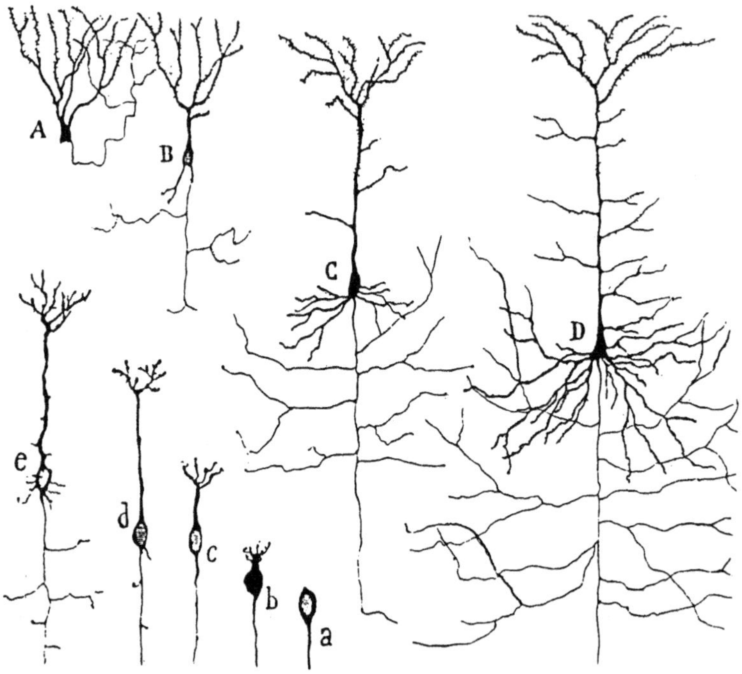

Fig. R3 Pyramidal isocortical neurons. (Redrawn after various sources, from Nieuwenhuys R *The neocortex. An overview of its evolutionary development, structural organization and synaptology.* Anat Embryol (Berl) 1994; 190:307–337; courtesy Rudolf Nieuwenhuys, Abcoude), showing the phylogenetic (*A-D*) and ontogenetic development (*a-c*) of the 'typical pyramidal cell' (Ramón y Cajal S 1893 Arch Anat Physiol Anat Abth (Leipzig) 319–428). *A* Periventricular cell from the pallium of a frog, *B* primitive pyramidal neuron from the cortex of a lizard, *C*, *D* typical pyramidal neurons from the neocortex of the mouse and the human, respectively; *a* neuroblast, *b* beginning of the development of the apical dendrite and its terminal bouquet, *c* lengthening of the apical dendrite, *d* early development of dendritic and axonal processes, *e* further development of dendritic and axonal processes. (From Lorente de Nó R 1922 Trabajos Cajal Madrid 20:41–80; from ten Donkelaar HJ, Kachlík D, Tubbs RS *An Illustrated Terminologia Neuroanatomica.* Springer, Cham, 2018)

Recklinghausen, Friedrich Daniel von (1833–1910): German pathologist born in Gütersloh and working in Königsberg, Würzburg and Strassburg (Fig. R9).

Canals of Recklinghausen: *Eponym* for lymph canaliculi ('Saftkanälchen'), cell spaces in tissues. Reference: *Die Lymphgefässe und ihr Beziehung zum Bindegewebe.* Hirschwald, Berlin, 1862.

Redlich, Emil (1866–1930): Moravian neurologist born in Brno and working in Vienna; *see* Obersteiner-Redlich zone.

Reichel, Friedrich Paul (1858–1934): German surgeon born in Breslau and working in Breslau, Berlin and Würzburg.

Duct of Reichel, Cloacal: *Eponym* for the embryonic communication between the hindgut and the anlage of the urinary bladder. Reference: *Die Entwicklung des Dammes und ihre Bedeutung für die Entstehubg gewisser Missbildungen.* Z Geburtsh Gynaecol 1888; 14:82–94.

Reichert, Karl Bogislaus (1811–1883): German anatomist born in Rastenburg and working in Dorpat and later in Breslau and Berlin (Fig. R10). He described the cartilage of the second pharyngeal arch as well as the innominate substance.

Cartilage of Reichert: *Eponym* for the cartilage of the second pharyngeal arch. References: *De embryonum arcubus sic dictis branchialibus.* Berlin, 1836; *Über die Visceralbogen der Wirbelthiere im Allgemeinen und deren Metamorphosen bei den Vögeln und Säugethieren.* Arch Anat Phys Wiss Med 1837, pp. 120–222.

Membrane of Reichert: *Eponym* for the anterior limiting membrane of the cornea, also described by Bowman.

Substance of Reichert: *Eponym* for the innominate substance of the basal forebrain, also described by Meynert. Reference: *Der Bau des menschlichen Gehirns,* 2 Vols. Engelmann, Leipzig, 1859, 1861.

Reid, Robert William (1851–1939): Scottish anatomist born in Auchindoie, Aberdeen, and working mainly in Aberdeen.

Base line of Reid: *Eponym* for the auriculoorbital horizontal line of the skull, passing between the infe-

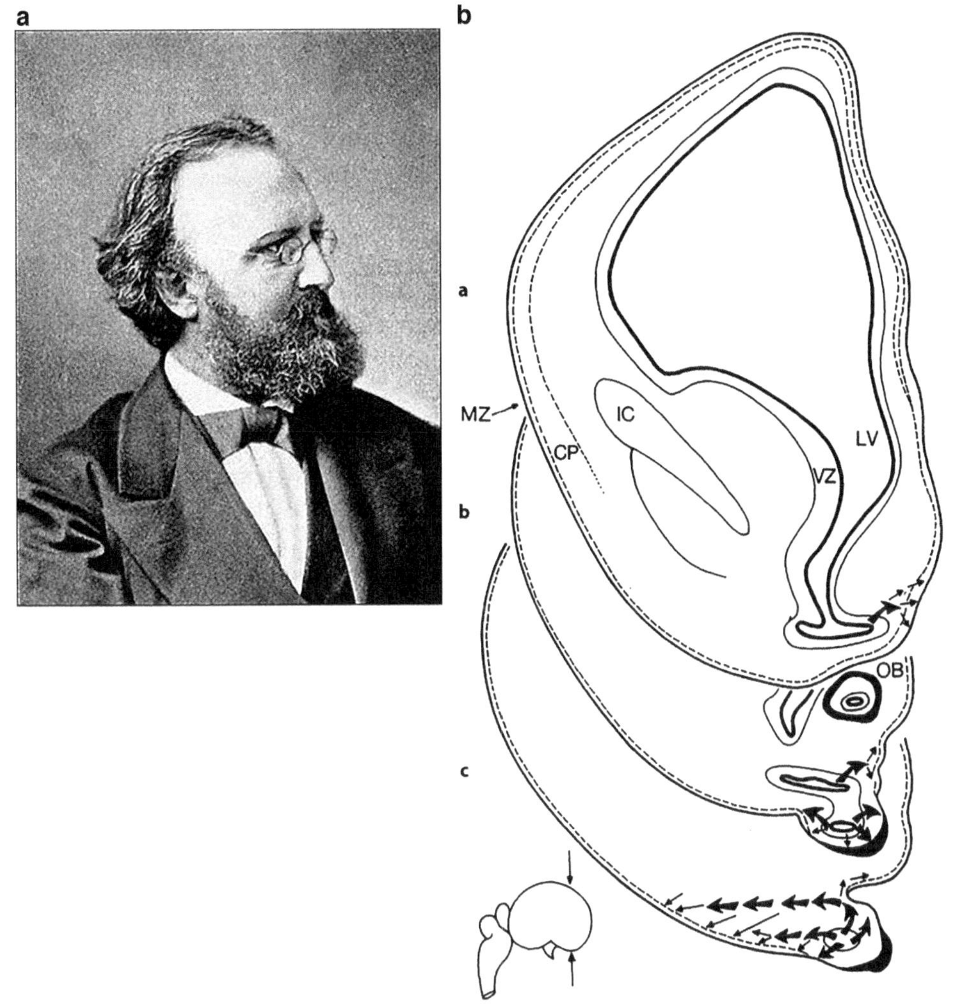

Fig. R4 (**a**) Photo of Ranke. (From Wikimedia Commons; in public domain); (**b**) The subpial granular layer of Ranke in the basal forebrain of a human fetus of 11 gestational weeks. The lateral ventricle (*LV*) is continuous with a ventral dilatation at the level of the paleocortex (**a**). The *arrow* indicates the proposed migration pathway from the medial wall of the paleocortical ventricle towards the pial surface. From here, the cells presumably continue their migration into the marginal zone (*MZ*) of the rostral cortex. The paleocortical ventricle is continuous with the olfactory ventricle inside the olfactory bulb (*OB*; **b**). At this level, cell migrations from the paleocortical ventricle are mainly directed medially. More caudally (**c**), cells migrate laterally in multiple streams towards the marginal zone of the paleocortex. *CP* cortical plate, *IC* internal capsule, *VZ* ventricular zone. (After Meyer G, Wahle P *The paleocortical ventricle is the origin of reelin-expressing neurons in the marginal zone of the foetal human neocortex.* Eur J Neurosci 1999; 11:3937–3944; from ten Donkelaar HJ, Kachlík D, Tubbs RS *An Illustrated Terminologia Neuroanatomica.* Springer, Cham, 2018)

rior margin of the orbit and the centre of the external acoustic meatus, in line with Ihering's proposal. Reference: *Observations on the relation of the principal fissures and convolutions of the cerebrum to the outer surface of the scalp.* Lancet 1884, pp. 539–540.

Reil, Johann Christian (1759–1813): German physician and anatomist born in Rhaude and working in Halle and later in Berlin (Fig. R11). References: *Exercitationum anatomicarum fasciculus primus. De structura nervorum.* Venalis, Halle, 1796; *Untersuchungen über den Bau des grossen Gehirns im Menschen.* Arch Physiol 1809; 9:136–146; *Die Sylvische Grube oder das Thal, das gestreifte große Hirnganglium, dessen Kapsel und die Seitentheile des Gehirns.* Ibid., 9:195–208.

Island of Reil: *Eponym* for the insula or insular lobe of the cerebral cortex (Fig. R12).

Sulcus of Reil, Central: *Eponym* for the sulcus separating the short and long gyri of the insula (Fig. R12).

Sulcus of Reil, Circular or **Limiting:** *Eponym* for the circular or limiting sulcus surrounding the insula (Fig. R12).

Fig. R5 Portrait of Ranvier. (From Wikimedia Commons; in public domain)

Fig. R6 Portrait of Rathke. (From Wikimedia Commons; in public domain)

Tract of Reil: *Eponym* for the medial lemniscus ('Schleife' or 'ruban' of Reil).

Trigone of Reil: *Eponym* for the trigone of the lateral lemniscus, a triangular field located on the lateral surface of the caudal half of the mesencephalon, bordered caudally by the slight prominence of the lateral lemniscus, dorsally by the base of the inferior colliculus and the brachium of the superior colliculus, and ventrally by the cerebral peduncle.

Reinke, Friedrich Berthold (1862–1919): German physician born in Ratzeburg and working in Kiel, Zürich and Wiesbaden.

Crystalloids of Reinke: *Eponym* for rectangular hyaline inclusion bodies in ovoid cells in the hilum of the ovary adjacent to the mesovarium, possibly homologues of the Leydig interstitial cells of the testis. Reference: *Beiträge zur Histologie des Menschen.* Arch Mikrosk Anat 1896; 47:34–44.

Space of Reinke: *Eponym* for loose connective tissue under the epithelium of the vocal cords.

Reisseissen, François Daniel (1773–1828): German anatomist and physician born in Strassburg and working in Strassburg and Berlin.

Muscle of Reisseissen: *Eponym* for smooth muscle fibres of the smallest bronchioli. References: *De pulmonis structura. Specimen inaugurale.* Strassburg, 1803; *Ueber die Struktur, die Verrichtungen und den Gebrauch der Lungen.* Berlin, 1808.

Reissner, Ernst (1824–1878): German anatomist born in Riga and working in Dorpat and later Breslau (Fig. R13). Reference: *De auris internae formatione.* Dorpat (currently: Tartu, Estonia), 1851.

Canal of Reissner: *Eponym* for the cochlear duct of the inner ear.

Membrane of Reissner: *Eponym* for the vestibular membrane of the cochlea (Fig. R14).

Rektorzik, Ernst (1834-?): Austrian anatomist working in Vienna and Lemberg, Galicia (currently Lviv, Ukraine)

Venous plexus of Rektorzik: *Eponym* for the petrooccipital plexus encasing the petrous segment of the internal carotid artery. Reference: *Über das Vorkommen eines Sinus venosus im Canalis caroticus.* Sitz Akad Wiss Math Classe 1858; pp. 466–470.

Remak, Robert (1815–1865): German neurologist born and working in Posen (currently Poznan, Poland; Fig. R15). He described the myelin sheath, identified unmyelinated nerve fibres, and suggested the enteric nervous system.

Band of Remak: *Eponym* for the axon, the axial filament of the nerve fibre.

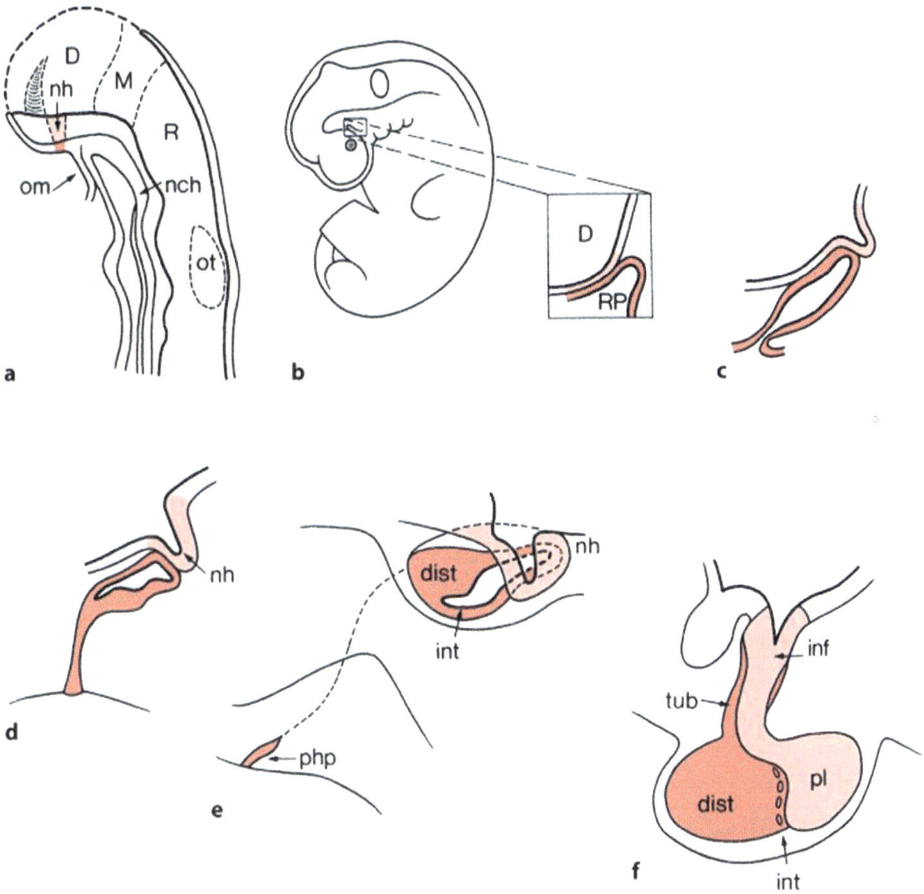

Fig. R7 Development of the pouch of Rathke and the human pituitary gland: (**a**) median section at 4.5 weeks of development (Carnegie stage 11); (**b**) embryo of 4.5 weeks of development (stage 14), showing the pouch of Rathke; (**c**) at 6 weeks (stage 17); (**d**) at stage 19; (**e**) at the end of the embryonic period (stage 23); (**f**) fetal pituitary. The developing neural hypophysis (*nh*) is indicated in light red and the adenohypophysis in red. *D* diencephalon, *dist* distal part, *inf* infundibulum, *int* intermediate part, *M* mesencephalon, *nch* notochord, *om* oral membrane, *ot* otocyst, *php* pharyngeal pituitary, *pl* posterior lobe, *R* rhombencephalon, *RP* Rathke's pouch, *tub* tuberal part. (After O'Rahilly R, Müller F *Human Embryology & Teratology*, 3rd ed. Wiley-Liss, New York, 2001; from ten Donkelaar HJ, Kachlík D, Tubbs RS *An Illustrated Terminologia Neuroanatomica*. Springer, Cham, 2018)

Fibres of Remak: *Eponym* for unmyelinated nerve fibres. Reference: *Observationes anatomicae et microscopicae de systematis nervosi structura.* Reimer, Berlin, 1838.

Ganglion of Remak: *Eponym* for (a) an autonomic ganglion in the opening of the inferior vena cava in the diaphragm, and (b) autonomic ganglia in the nerves of the stomach.

Plexus of Remak: *Eponym* for the submucosal plexus of the digestive tract, also described by Meissner. Reference: *UebereinselbständigesDarmnervensystems.* Berlin, 1847.

Renshaw, Birdsey (1911–1948): American neurophysiologist who described the recurrent inhibitory interneuron in the lamina VII of the spinal cord. Reference: *Influence of discharge of motoneurons upon excitation of neighboring motoneurons.* J Neurophysiol 1941; 4:167–183.

Cells of Renshaw: *Eponym* for recurrent inhibitory interneurons in the spinal cord.

Rete of Haller: *Eponym* for the rete testis; *see* Haller, Albrecht von.

Retzius, Anders (Andreas) Adolf (1796–1860): Swedish anatomist and anthropologist born in Lund and working in Stockholm (Fig. R16). His son Gustav honoured him by suggesting the name gyri of Andreas Retzius.

Band of Retzius: *Eponym* for the fundiform ligament of the penis.

Cave or **Space of Retzius:** *Eponym* for the retropubic or prevesical space; Retzius described a ligament rather than a space (Testut, Vol 4, p 472, 1905)..

Gyri of Andreas Retzius: *Eponym* for the subsplenial gyri of the cerebral cortex, a series of small bumps

Fig. R8 Portrait of Rau. (From Collectiones Icones Leidense; in public domain)

Fig. R9 Portrait of Recklinghausen. (From Pagel 1901)

Fig. R10 Portrait of Reichert. (From Pagel 1901)

Fig. R11 Portrait of Reil. (From the Wellcome Collection CCBY; in public domain)

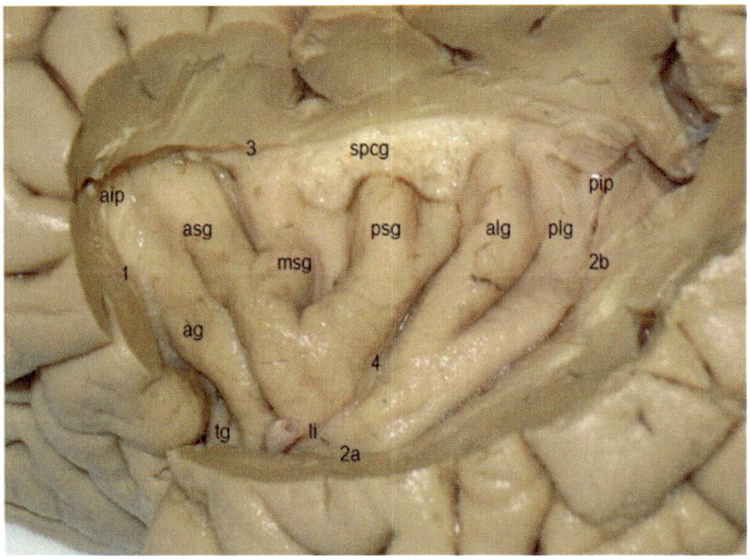

Fig. R12 Lateral view of the insula of Reil, showing three short gyri, an accessory gyrus and two long gyri. (Courtesy Robert Bartoš, Ústí nad Labem). *ag* accessory gyrus, *aip* anterior insular point, *alg* anterior long gyrus, *asg* anterior short gyrus, li limen insulae, *msg* middle short gyrus, *pip* posterior insular point, *plg* posterior long gyrus, *psg* posterior short gyrus, *spcg* gyrus supracentralis, *tg* transverse temporal gyrus, *1–3* circular insular sulcus (*1* anterior periinsular sulcus, *2a*, *2b* inferior insular sulcus, horizontal and posterior parts, *3* superior periinsular sulcus), *4* central insular sulcus. (From ten Donkelaar, Kachlík D, Tubbs RS *An Illustrated Terminologia Neuroanatomica.* Springer, Cham, 2018)

Fig. R13 Portrait of Reissner. (From Wikimedia Commons; in public domain)

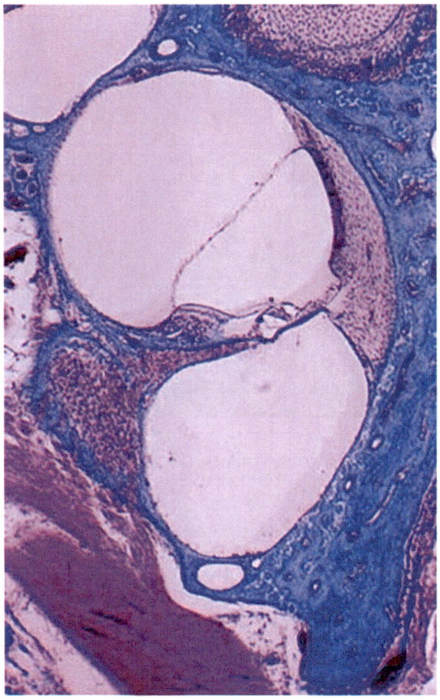

Fig. R14 Photomicrograph of the human cochlea with the membrane of Reissner. (Courtesy Jo Curfs, Nijmegen. (From ten Donkelaar HJ, Lammens M, Hori A *Clinical Neuroembryology,* 3rd ed. Springer, Cham, 2023)

Fig. R15 Portrait of Remak. (From Pagel 1901)

Fig. R16 Portrait of Retzius, Anders. (From Wikimedia Commons; in public domain)

pointing to the end of the CA1 field. This name was given by his son Gustav

Ligament of Retzius: *Eponym* for (a) the deep attachment of the inferior extensor retinaculum in the tarsal sulcus, and (b) the puboprostatic ligament, his pelvoprostatic ligament. Reference: *Ueber das Ligamentum pelvioprostaticum oder den Apparat durch welchen die Harnblase, die Prostata und die Harnröhre an der untern Beckenöffnung befestigt sind.* Arch Anat Physiol (Lpz) 1849, pp. 182–190.

Lines or **Striae of Retzius, Parallel:** *Eponym* for the enamel incremental lines, alternating lighter and darker incremental lines in an enamel section (*see* Fig. 23.1).

Veins of Retzius: *Eponym* for (a) retroperitoneal portocaval anastomoses. Reference: *Bemerkungen über Anastomosen zwischen der Pfortader und der unteren Hohlader ausserhalb der Leber.* Ztsch Physiol 1835; 5:105; and (b) the vorticose veins of the eyeball, earlier described by Stensen and Ruysch.

Retzius, Gustav Magnus (1842–1919): Swedish anatomist born and working in Stockholm (Fig. R17), son of Anders Retzius. He is known for his various studies of the nervous system. Reference: *Das Menschenhirn, Studien in der makroskopischen Anatomie.* Norstedt, Stockholm, 1896. His name is sometimes connected with the ambiens, intralimbic and semilunar gyri as well as with the area postrema. His extensive studies on the inner ear did not earn him an eponym.

Cells of Cajal-Retzius: *Eponym* for what Retzius called the *Cajal'schen Zellen* as described by Cajal (1893) and later became known as cells of Cajal-Retzius (Fig. R18). Reference: *Die Cajal'schen Zellen der Grosshirnrinde beim Menschen und bei Säugetieren.* Biol Unters Neue Folge 1893; 5:1–8.

Foramen of Key-Retzius: *Eponym* or the lateral aperture of the fourth vetricle, also described by Luschka and Mierzejewsky.

Fig. R17 Portrait of Retzius, Gustav. (From Pagel 1901)

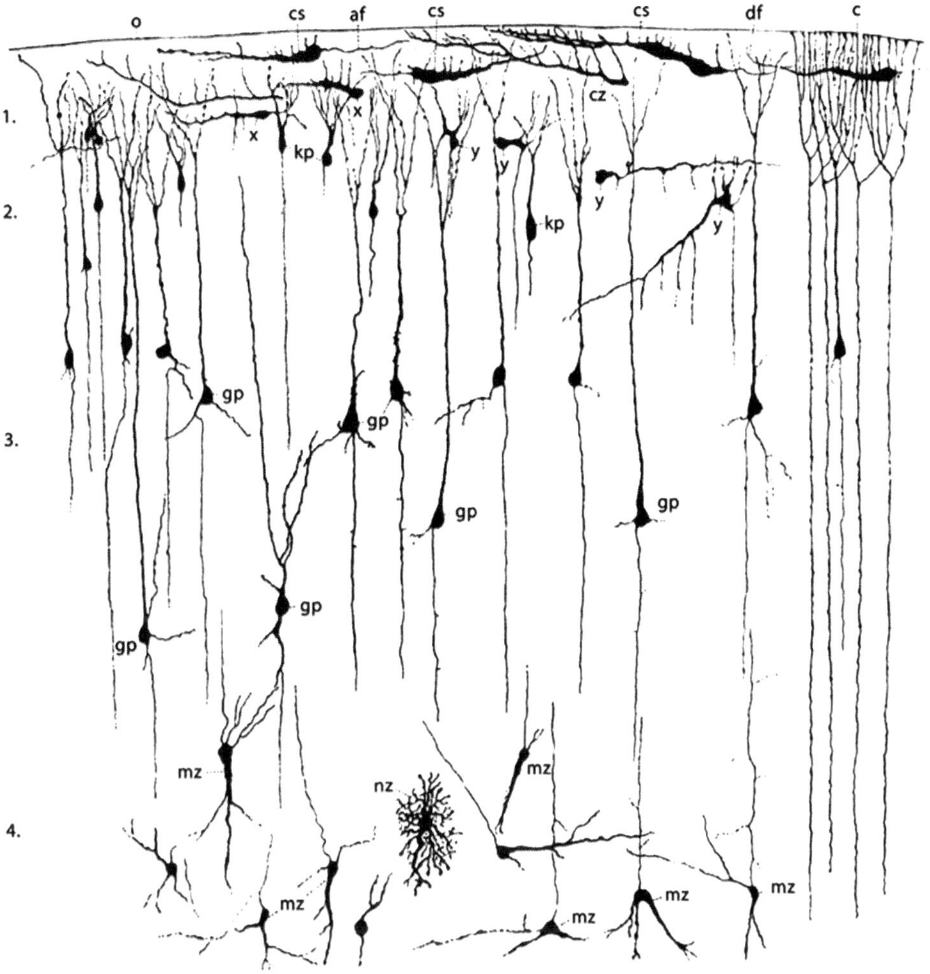

Fig. R18 Retzius data on Cajal-Retzius cells, he described as *Cajal'schen Zellen* (*cz*), that are easily recognized by their horizontal orientation in layer I of the cortex. (From Retzius G *Die Cajal'schen Zellen der Grosshirnrinde beim Menschen und bei Säugetieren.* Biol Unters Neue Folge 1893; 5:1–8; in ten Donkelaar HJ, Lammens M, Hori A *Clinical Neuroembryology,* 3rd ed. Springer, Cham, 2023)

Verrucae hippocampi: Name introduced by Gustav Retzius for hippocampal warts, macroscopically visible on the entorhinal cortex.

Rex, Hugo (1861–1936): German surgeon born and working in Prague. He contributed to the knowledge of the intrahepatic course of secondary branches of the left and right portal veins. Reference: *Beiträge zur Morphologie der Säugerleber.* Morphol Jahrb 1888; 14:517–617.

Line of Rex-Cantlie: *Eponym* for the anatomical midline of the liver connecting the fundus of the gallbladder with the inferior vena cava, also described by Cantlie.

Recess of Rex: *Eponym* for the site of the remnant of the umbilical vein, located in the space between the hepatic segments III and IV.

Rexed, Bror Anders (1914–2002): Swedish neuroanatomist working in Uppsala who described the nine laminae (I-IX) and one area (X) of the spinal gray matter. References: *The cytoarchitectonic organization of the spinal cord in the cat.* J Comp Neurol 1952; 96:415–496; *A cytoarchitectonic atlas of the spinal cord in the cat.* Ibid., 1954; 100:297–379.

Area of Rexed: *Eponym* for the spinal area surrounding the central canal

Laminae (Layers) of Rexed: *Eponym* for the nine layers of the spinal grey matter (*see* Fig. R19).

Ribes, François (1765–1845): French physician and army surgeon born in Baqueres de Bigorre and working in Toulouse, Paris and Rome.

Ganglion of Ribes: *Eponym* for a small sympathetic ganglion on the anterior communicating artery, the so-called uppermost sympathetic ganglion. Reference: *Exposé sommare de quelques recherches anatomiques, physiologiques et pathologiques.* Mém Soc Emul Paris 1817; 8:604–670.

Richard, Félix Dominique Adolphe (1822–1872): French surgeon working in Paris.

Fimbria of Richard: *Eponym* for the ovarian fimbria, i.e. the largest fimbria of the uterine tube that is in

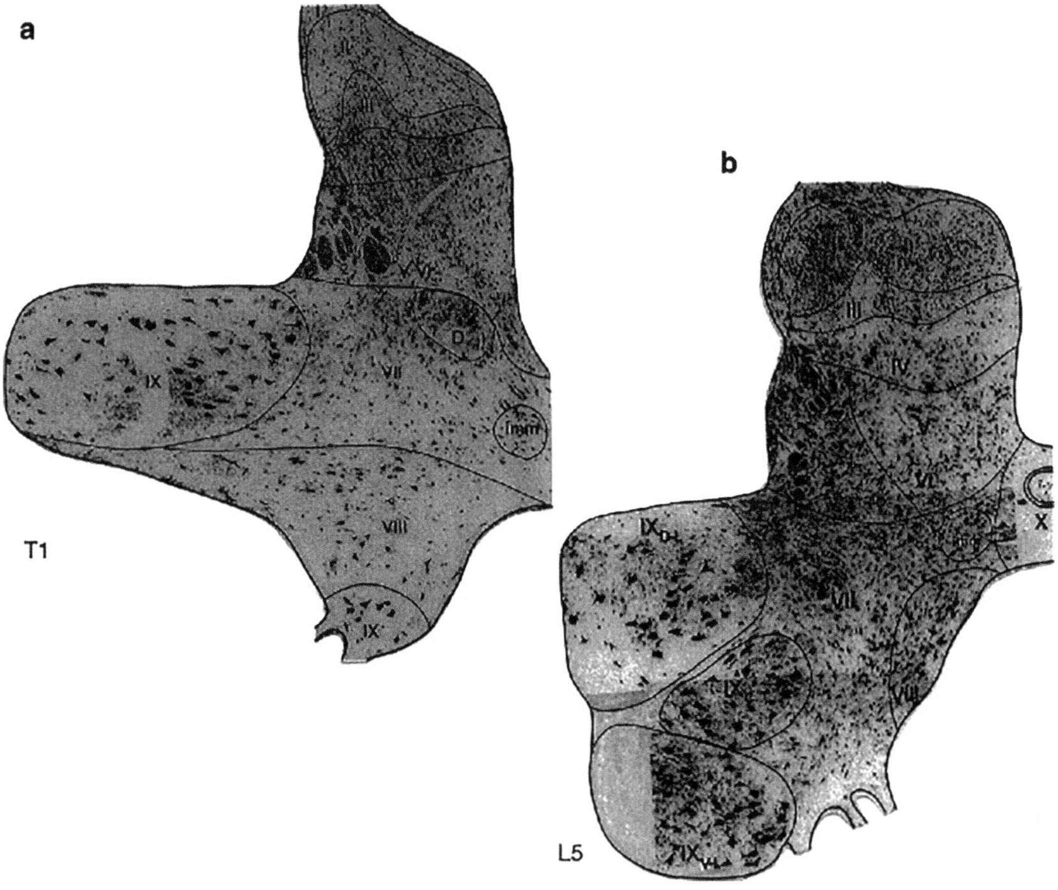

Fig. R19 Nissl-stained sections of the spinal gray matter at (**a**) Th1 and (**b**) L5 showing the layers of Rexed. (From Schoenen J, Faull RLM *Spinal cord.* In: Paxinos G, Mai JK, eds, *The Human Nervous System*, 2nd ed. Elsevier, Amsterdam, 2004; pp. 190–232; courtesy Jean Schoenen, Liège; with permission). The Rexed layers are indicated by Roman numerals. (From ten Donkelaar HJ, Kachlík D, Tubbs RS *An Illustrated Terminologia Neuroanatomica.* Springer, Cham, 2018)

direct contact with the ovary at its superior pole and contracting during ovulation to guide the free egg. Reference: *Anatomies des trompes de l'utérus chez la femme.* Thesis Paris, 1851.

Riché, P (nineteenth century): French physiologist.

Anasomosis of Riché-Cannieu: *Eponym* for a palmar anastomosis between the recurrent branch of the median nerve and the deep branch of the ulnar nerve, described in the same year as Cannieu. Reference: *Le nerf cubital et les muscles de l'eminence thenar.* Bull Mém Soc Anat Paris 1897; 5:251–252.

Richet, Louis Alfred (1816–1891): French surgeon in Paris.

Canal of Richet: *Eponym* for a canal for the umbilical vein on the anterior wall of the abdomen, made by the passage of some fibres of the transversus abdominis muscle deep to the vein. Reference: *Traité pratique d'anatomie médico-chirurgicale.* Paris, 1855–1857.

Fascia of Richet: *Eponym* for (a) a fold of peritoneum enveloping the obliterated umbilical vein, also known as the canal of Richet; and (b) the middle layer of the cervical fascia (his aponévrose omo-claviculaire), also described by Gruber, Paulet and Testut.

Richter, August Gottlieb (1742–1812): German surgeon born in Zörbig, Saxony, and working in Göttingen.

Line of Richter: *Synonym* for the spinoumbilical line of Monro, also known as **line of Monro-Richter**.

Parietal hernia of Richter: *Eponym* for a hernia in which only a portion of the wall of the intestine is engaged, also known as parietal hernia, and also described by Littré.

Ridley, Humphrey (1653–1708): English anatomist and physician born in Mansfeld and working in London.

Sinus of Ridley: *Eponym* for the anterior and posterior intracavernous sinuses. Reference: *The Anatomy of the Brain.* Smith and Walford, London, 1695.

Fig. R20 Photo of Riedel. (From Pagel 1901)

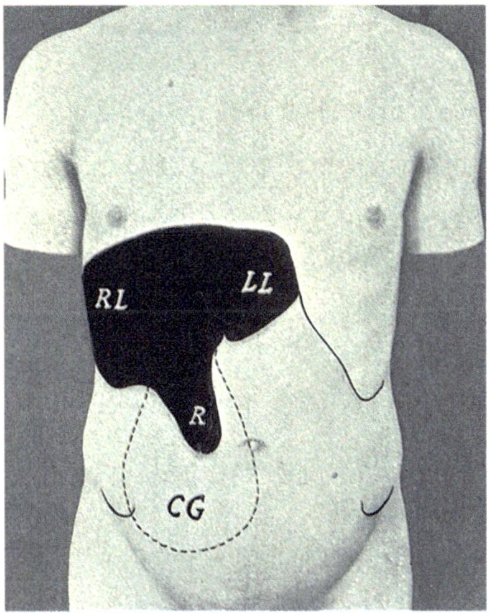

Fig. R21 Lobe of Riedel. (From Eisendrath DN *Surgical Diagnosis*. Saunders, Philadelphia, 1909; in public domain)

Riedel, Bernhard Moritz Carl Ludwig (1846–1916): German surgeon born in Laage, Mecklenburg, and working in Jena (Fig. R20).

Lobe of Riedel: *Eponym* for a tonguelike process from the right lobe of the liver lateral to the gallbladder (Fig. R21). Reference: *Ueber den zungenformigen Fortsatz des rechten Leberlappens und seine pathognostische Bedeutung für die Erkrankung der Gallenblase nebst Bemerkungen ueber Gallensteinoperationen.* Berlin Klin Wschr 1888; 577–581, 602–607.

Rindfleisch, Eduard (1836–1908): German pathologist born in Köthen and working in Breslau, Zürich, Bonn and Würzburg. Reference: *Pathologie der pathologischen Gewebelehre.*Engelmann, Leipzig, 1867–1869.

Folds of Rindfleisch: *Eponym* for the semilunar folds of the serous surface of the pericardium, embracing the aorta.

Riolan, Jean (1577 or 1580–1657): French anatomist and botanist born and working in Paris (Fig. R22). Reference: *Anatomen corporis humani.* Perier, Paris, 1610; *Les oeuvres anatomiques.* Moreau, Paris, 1628–1629.

Arcade or **Anastomosis of Riolan:** *Eponym* for a rare anastomosis between the superior and inferior mesenteric arteries (Fig. R23). Albrecht von Haller suggested the name arcade of Riolan.

Bones of Riolan: *Eponym* for small sutural bones that may be present in the petrooccipital suture.

Fig. R22 Portrait of Riolan. (From Wikimedia Commons; in public domain)

Bouquet or **Septum of Riolan:** *Eponym* for the muscles and ligaments arising from the styloid process of the temporal bone.

Muscle of Riolan: *Eponym* for (a) the cremaster muscle and (b) the palpebral part of the orbicularis oculi muscle.

Rivinus, Augustus Quirinus (1652–1723): German botanist, physiologist and physician born and working in Leipzig (Fig. R24). His father latinized the family name Bachmann to Rivinus.

Ducts of Rivinus: *Eponym* for the minor sublingual ducts.

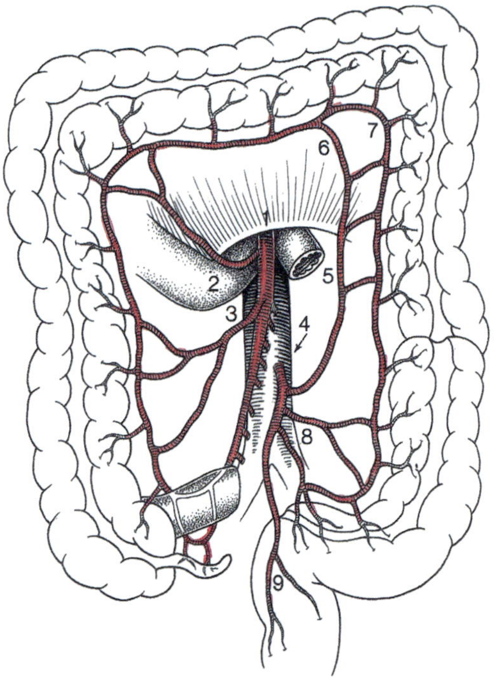

Fig. R23 The arcade of Riolan. *1* superior mesenteric artery, *2* middle colic artery, *3* right colic artery, *4* inferior mesenteric artery, *5* left colic artery, *6* arcade of Riolan, *7* marginal artery of Drummond. (After ten Donkelaar HJ, Oostra R-J *Klinische Anatomie en Embryologie,* 4th ed. Springer Media, Houten, 2014)

Fig. R24 Portrait of Rivinus. (From Wikimedia Commons; in public domain)

Gland of Rivinus: *Eponym* for the sublingual gland.

Incisure or **Notch of Rivinus:** *Eponym* for the tympanic notch of the external acoustic meatus in the superior part of the tympanic ring bridged by the flaccid part of the tympanic membrane..

Mallear folds of Rivinus: *Eponym* for the anterior and posterior mallear folds, usually attributed to Troeltsch.

Membrane of Rivinus: *Eponym* for the flaccid part of the tympanic membrane.

Rizzolatti, Giacomo (1937-today): Italian neuroscientist born in Kiev (former Sovjet Union, now Ukraine) and working in Pisa, Parma and Milan.

Mirror neurons of Rizzolatti: *Eponym* for a particular class of visuomotor neurons in area F5 of the monkey premotor cortex and also reported in humans. Reference: *The mirror neuron system.* Annu Rev. Neurosci 2004: 27:169–192.

Robin, Charles Philippe (1821–1885): French histologist born in Jasseron, Ain, and working in Paris (Fig. R25).

Sheath of Robin: *Eponym* for the cellular sheath surrounding nerves. Reference: *Mémoire sur le périnèvre.* Paris, 1854.

Spaces of Robin: *Eponym* for perivascular spaces around blood vessels entering the brain, mostly known as spaces of Virchow-Robin, also described by His Sr.

Fig. R25 Photo of Robin. (From Pagel 1901)

a b

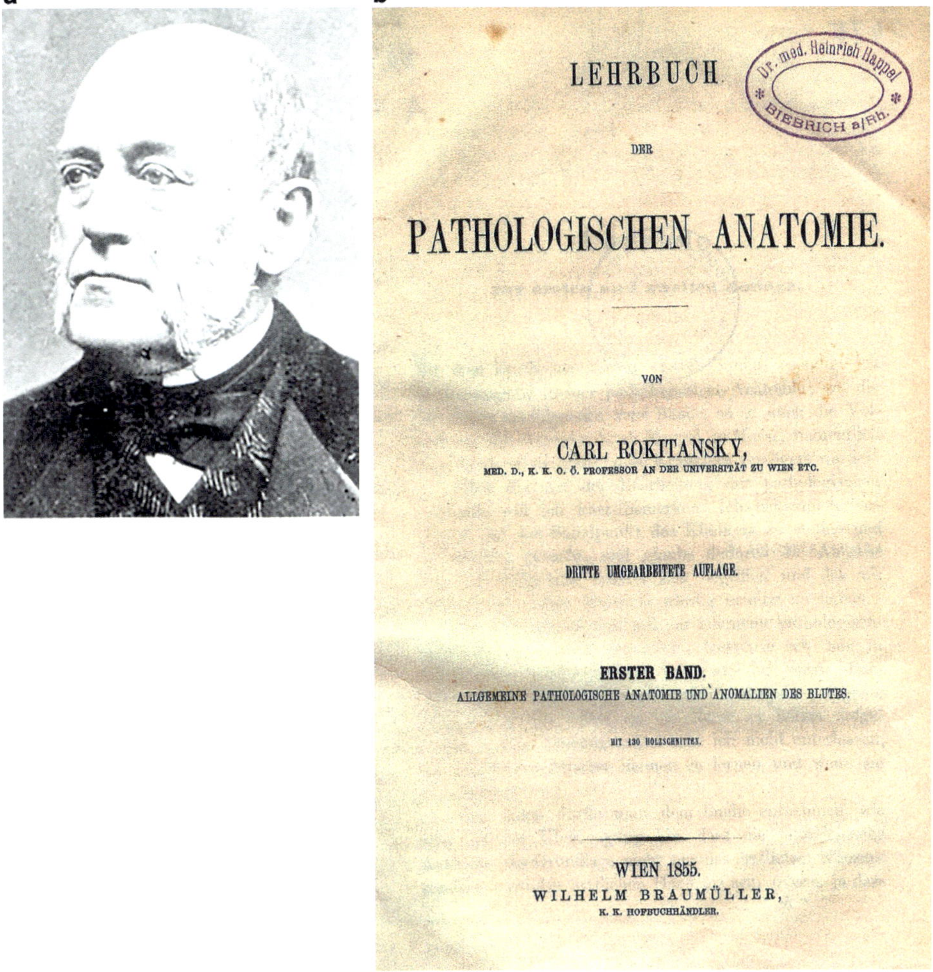
LEHRBUCH

DER

PATHOLOGISCHEN ANATOMIE.

VON

CARL ROKITANSKY,

MED. D., K. K. O. Ö. PROFESSOR AN DER UNIVERSITÄT ZU WIEN ETC.

DRITTE UMGEARBEITETE AUFLAGE.

ERSTER BAND.

ALLGEMEINE PATHOLOGISCHE ANATOMIE UND ANOMALIEN DES BLUTES.

MIT 130 HOLZSCHNITTEN.

WIEN 1855.

WILHELM BRAUMÜLLER,

K. K. HOFBUCHHÄNDLER.

Fig. R26 (**a**) Photo of Rokitansky. (From Pagel 1901); (**b**) Title page *Lehrbuch der pathologischen Anatomie,* 3rd ed. Braumüller, Vienna, 1855

Rohr, Karl (1863–1947): Swiss embryologist and gynaecologist born and working in Bern.

Stria of Rohr: *Eponym* for the external fibrinoid layer of the placenta at the level of the basal plate beneath the stem villi. Reference: *Die Beziehungen der mutterlichen Gefässe zu den intervillösen Raumen der reifen Placenta speciell zur Thrombose derselben ('weisser Infarct').* Virchow's Arch 1889; 115:505–533.

Rokitansky, Carl Joseph Wenzel Prokop von (1804–1878): Austrian pathologist born in Königgrätz (now Hradec Králové, Czech Republic) and working in Vienna (Fig. R26a). Reference: *Lehrbuch der pathologischen Anatomie*, 3rd ed. Braumüller, Vienna, 1855 (Fig. R26b).

Lobe of Rokitansky: *Eponym* for an accessory lobe of the lung.

Sinus of Aschoff-Rokitansky: *Eponym* for mucous crypts in the wall of the gallbladder, probably pathological.

Rolando, Luigi (1773–1831): Italian anatomist born in Torino and working in Sassari (Sardegna) and later in Turin (Fig. R27) who described the central sulcus of the cerebral cortex in his lectures at the University of Turin, already illustrated in Bidloo's 1685 atlas. In 1839, François Leuret (1797–1851) named it the fissure of Rolando in his honour. The gelatinous substance of the posterior horn of the spinal cord is also known as the substance of Rolando. References: *Saggio sopra la vera struttura del cervello dell'uomo e degli' animali e sopra le funzioni del sistema nervoso.* Nella Stamperia da S.S.R.14 Privilegiata, Sassari, 1809; *Della struttura degli emisferi cerebrali.* Memorie della Regia Accademia della Scienze di Torino, 1829; 35:103–145.

Area of Rolando: *Eponym* for the precentral and postcentral gyri of the cerebral cortex adjacent to the central sulcus.

Artery of Rolando: *Eponym* for the middle cerebral artery, already described by Sylvius (François de le Boë).

Fig. R27 Portrait of Rolando. (From Wikimedia Commons; in public domain)

Fig. R28 Lateral view brain by Rolando. (From Wikimedia Commons; in public domain)

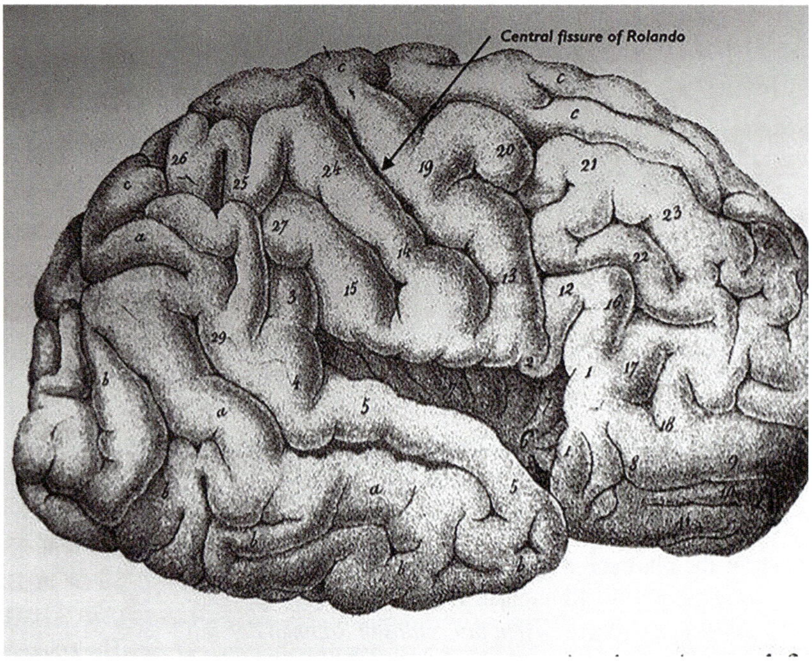

Substance of Rolando: *Eponym* for the gelatinous substance of the posterior horn of the spinal cord, also described by Stilling.

Sulcus of Rolando: *Eponym* for the central sulcus of the cerebral cortex (Fig. R28).

Tubercle of Rolando: *Eponym* for the trigeminal tubercle, a longitudinal prominence on the dorsolateral surface of the medulla oblongata along the lateral border of the cuneate tubercle.

Vein of Rolando: *Eponym* for the central sulcal vein, one of the superior central veins, coursing in the central sulcus of the cerebral cortex.

Roller, Christian Friedrich Wilhelm (1844–1884): German neurologist and psychiatrist born in Baden and working in Strassburg who described the subhypoglossal nucleus. Sometimes, his name is also given to the inferior vestibular nucleus. Reference: *Die centrale Verlauf des Nervus glossopharyngeus - Der Nucleus lateralis medius.* Arch Mikr Anat 1881; 19:347–382.

Nucleus of Roller: *Eponym* for the subhypoglossal nucleus, one of the three nuclei of the perihypoglossal complex of the hindbrain; sometimes also used for the inferior vestibular nucleus.

Rollett, Alexander (1834–1903): Austrian physiologist and histologist born in Baden near Vienna and working in Graz.

Substance or **Stroma of Rollett:** *Eponym* for the colourless stroma of erythrocytes.

Romiti, Guglielmo (1850–1936): Italian anatomist born in Pisa and working in Siena and Pisa.

Line of Romiti: *Eponym* for the boundary between the ovarian epithelium and the peritoneal serosa, also described by Farre and Waldeyer.

Posterior episternal muscle of Romiti: *Eponym* for a variant of the transversus thoracis muscle ('triangular muscle of the sternum').

Rose, Maksymilian (1883–1937): Polish neurologist born in Premyszl, Galicia, then Austria-Hungary. After working in Krakau (currently Cracow, Poland), he went to Germany (Berlin, Tübingen to work with Brodmann), and returned to Krakau (Fig. R29). Later he worked with the Vogt's in Berlin, again Krakau and finally in Vilnius. He introduced the fields **H1-H5** of the hippocampus. Reference: *Cytoarchitektonik und Myeloarchitektonik der Grosshirnrinde.* In: Bumke O, Foerster O (Hrsg) *Handbuch der Neurologie, Allgemeine Neurologie. 1. Anatomie.* Springer, Berlin, 1935, pp. 588–778. For biography, *see* Pekacka-Falkowska K, Owecki MK, Pekacka AM *Maksymilian Rose.* J Neurol 2017; 264:2509–2510.

Fig. R29 Photo of Rose. (From Wikimedia Commons; in public domain)

Fig. R30 Portrait of Rosenmüller. (From the Rijksmuseum Amsterdam; in public domain)

Fields H1-H5 of Rose: *Eponym* for an older subdivision of the hippocampus.

Rosenmüller, Johann Christian (1771–1820): German anatomist, embryologist and surgeon born in Hessberg, Saxony, and working in Leipzig (Fig. R30). Reference: *Chirurgisch-Anatomische Abbildungen für Aerzte und Wundärzte.* Weimar, 1805.

Fossa or **Recess of Rosenmüller:** *Eponym* for the pharyngeal recess.

Gland or **Node of Rosenmüller:** *Eponym* for the proximal deep inguinal lymph node in the lacunar space, adjacent to the inguinal canal, also described by Cloquet.

Organ of Rosenmüller: *Eponym* for the epoophoron. Reference: *Quaedam de ovariis embryonum et foetuum humanorum.* Leipzig, 1802.

Valve of Rosenmüller: *Eponym* for a lacrimal fold central to the ostium of the lacrimal canaliculi to the lacrimal sac.

Rosenthal, Friedrich Christian (1780–1829): German anatomist born in Greifswald and working in Berlin and later in Greifswald (Fig. R31a). Reference: *Versuch einer Pathologie des Gehörs.* Arch Med Erfahr Berlin 1819; 2:1–26.

Canal of Rosenthal: *Eponym* for the spiral canal of the modiolus of the cochlea.

Vein of Rosenthal, Basal: *Eponym* for the basal vein of the brain (Fig. R31b).

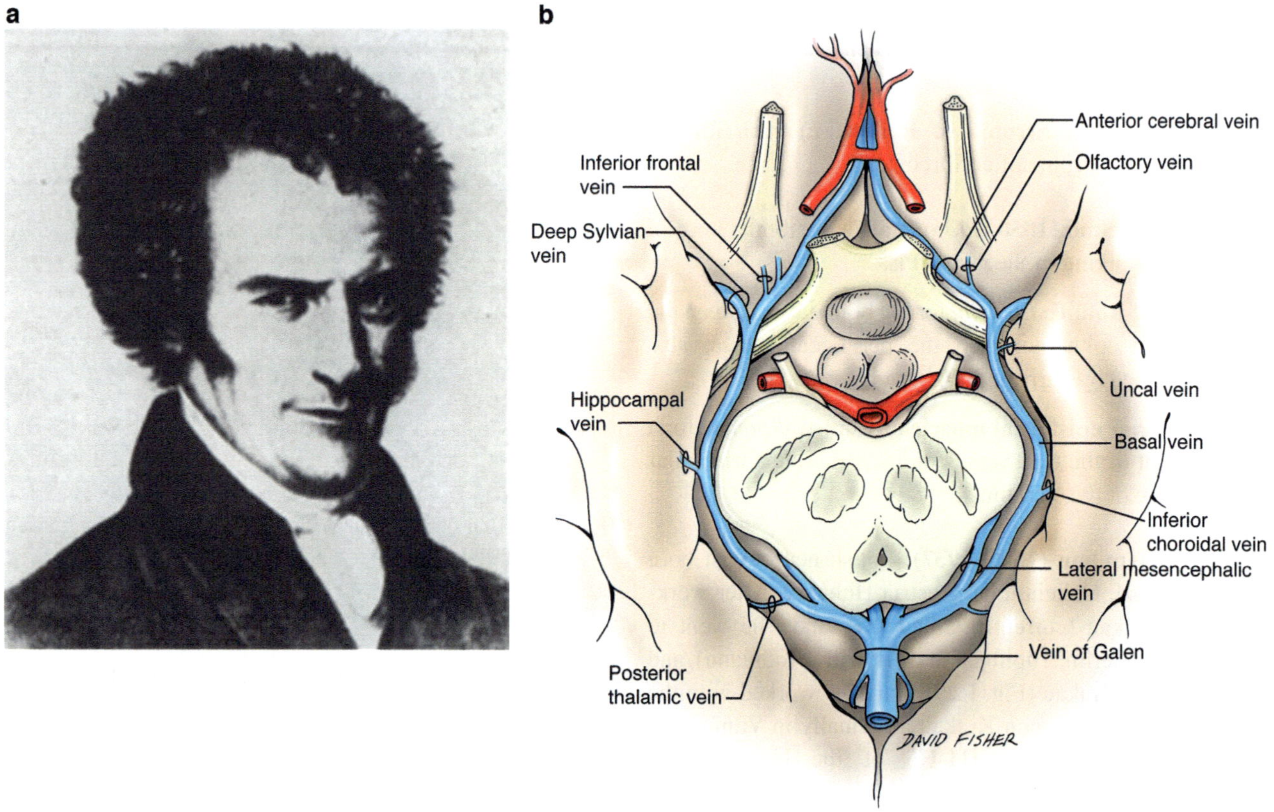

Fig. R31 (**a**) Portrait of Rosenthal. (From Wikimedia Commons; in public domain); (**b**) The basal vein of Rosenthal shown in a ventral view of the brain. (From ten Donkelaar HJ, Kachlík D, Tubbs RS *An Illustrated Terminologia Neuroanatomica.* Springer, Cham, 2018)

Roser, Wilhelm (1817–1888): German surgeon born in Marburg and working in Marburg and later in Zürich and Berlin (Fig. R32).

Line of Roser-Nélaton: *Eponym* for a line between the anterior superior iliac spine and the ischial tuberosity (Fig. R33). Reference: *Handbuch der anatomischen Chirurgie.* Tübingen, 1844.

Roth, Moritz (1839–1915): Swiss pathologist born and working in Basel.

Vasa aberrantia of Roth: *Eponym* for inconstant tubules of the epididymis, connected with the rete testis but not with the ductus deferens. Reference: *Ueber Vasa aberrantia am Rete testis.* Zschr Anat Entwickl 1877; 2:125–126.

Rotter, Josef (1857–1924): German surgeon.

Nodes of Rotter: *Eponym* for the interpectoral axillary lymph nodes between the pectoralis major and minor muscles. Reference: *Günstigere Dauererfolge durch ein verbessertes Operationsverfahren der Mammacarcinome.* Berl Klin Wschr 1896, 33:69–99.

Rouget, Charles Marie Benjamin (1824–1904): French anatomist born in Paris and working in Montpellier and Paris (Fig. R34). Reference: *Note sur la structure de l'oeil et en particulier sur l'appareil iridochoroideum.* C R Soc Biol Paris 1856, pp. 113–132.

Cells of Rouget: *Eponym* for pericytes.

Fibres or **Muscle of Rouget:** *Eponym* for (a) the radial fibres of the ciliary muscle; and (b) the fibres of Rouget-Juvara, the fibres of the phrenooesophageal ligament. Its upper fibres are known as fibres of Juvara, its lower fibres as fibres of Rouget.

Rousseau, Louis François Emanuel (1788–1868): French anatomist and histologist in Paris, born in Belleville.

Bone of Rousseau: *Eponym* for an accessory lacrimal bone anterior to the proper lacrimal bone, also described by Rosenmüller and Weber. Reference: *Description d'un nouvel os de la face chez l'homme.* Ann Sci Nat 1829, 17:86–88.

Rouvière, Henri (1875–1952): French anatomist and embryologist born in Le Bleymard and, after studying in Montpellier, worked in Paris (Fig. R35a). His textbook

Fig. R32 Photo of Roser. (From Pagel 1901)

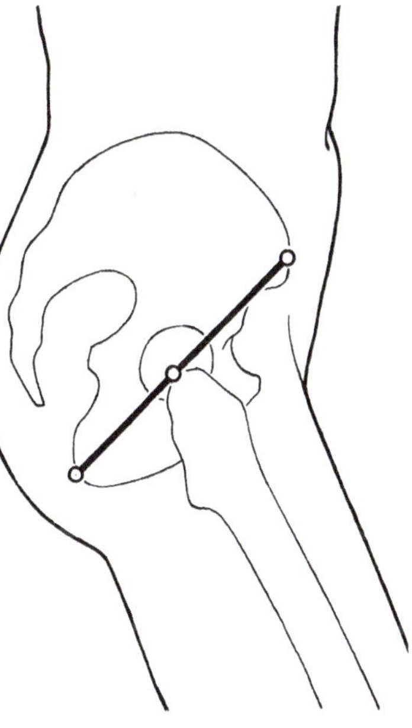

Fig. R33 Line of Roser-Nélaton. (Based on an old figure by the late Chris van Huijzen)

Fig. R34 Photo of Rouget. (From Wikimedia Commons; in public domain)

Anatomie humaine, descriptive et topographique in 3 volumes went through many editions (12th ed., Masson, Paris, 1967).

Collar of Rouvière: *Eponym* for a band of cervical lymph nodes ('cercle ganglionaire péricervical'), including occipital, mastoid, parotid, facial, retropharyngeal, submandibular, submental and sublingual nodes.

Fascia of Rouvière: *Eponym* for the axillary fascia.

Foramen of Rouvière: *Eponym* for a rare dehiscence in the articular capsule of the shoulder joint between the middle and inferior glenohumeral ligaments (Fig. R35b).

Horizontal fissure of Rouvière: *Eponym* for a horizontal fissure of the liver.

Nodes of Rouvière: *Eponym* for various lymph nodes, including (a) the prelaryngeal or cricoid lymph node; (b) the lateral retropharyngeal lymph node; and (c) the xiphoid or prepericardial lymph node, also described by Sappey. Reference: *Anatomie des lymphatiques de l'homme.* Masson, Paris, 1932.

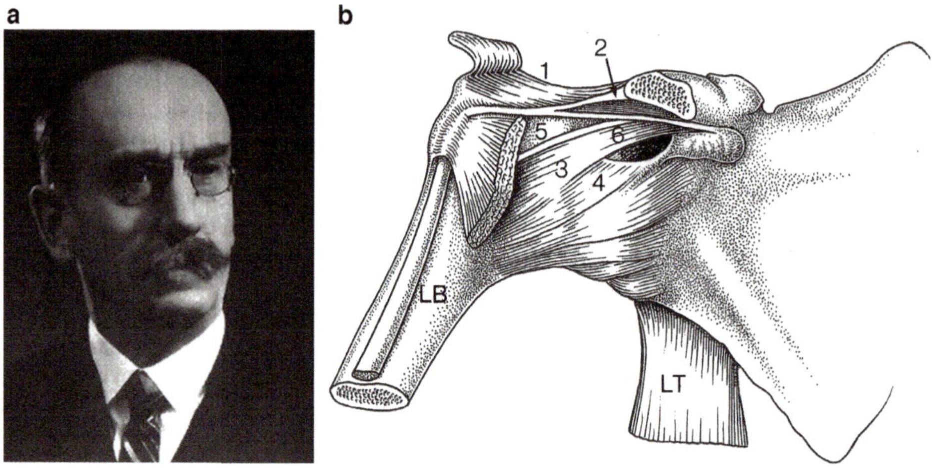

Fig. R35 (**a**) Photo of Rouvière. (From Alchetron; in public domain); (**b**) Foramen of Rouvière. (After Rouvière H *Anatomie humaine, descriptive et topographique*, 12th ed. Masson, Paris, 1967). *1* Coracohumeral ligament, *2–4* superior, middle and inferior glenohumeral ligaments, *5* foramen of Weitbrecht, *6* foramen of Rouvière (in *black*), *LB* long tendon of biceps brachii muscle, *LT* long head of triceps brachii muscle

Fig. R36 Portrait of Rüdinger. (From Wikipedia; in public domain)

Roux, Caesar (1857–1934): Swiss surgeon in Lausanne.

Muscle of Roux: *Eponym* for the rectourethralis muscle. Reference: *Beiträge zur Kenntniss der Aftermusculatur des Menschen.* Arch Mikr Anat 1881; 19:721–732.

Rüdinger, Nicolaus (1832–1896): German anatomist born in Alzey, Rheinhessen, and working in Munich (Fig. R36).

Canal of Rüdinger, Safety: *Eponym* for the upper compartment of the auditory tube between the lateral and the medial laminae of the cartilage, filled with mucus or air. Reference: *Vergleichende Anatomie und Histologie der Ohrtrompete.* JJ Lentner'sche Buchhandlung, Munich, 1870.

Fig. R37 Photo of Ruffini. (From Wikipedia; in public domain)

Muscle of Rüdinger: *Eponym* for a muscle internal to the circular fibres of the rectum (his Musculus dilator internus). Reference: *Ueber die Muskelanordnung im Pfortner des Magens und am Anus.* Allg Wiener Med Zeitung 1879; 24:2–3.

Ruffini, Angelo (1874–1929): Italian anatomist born in Pretare and working in Bologna (Fig. R37).

Corpuscles of Ruffini: *Eponym* for spindle shaped receptors situated parallel to the superficial layer of the dermis. Reference: *Sulla presenza di nuove forme di terminazioni nervose nello strato papillare e subpapillare della cute dell'uomo.* Torino, 1898.

Flower-spray organ of Ruffini: *Eponym* for one of the two types of sensory nerve endings associated with the neuromuscular spindle.

Fig. R38 Portrait of Rusconi. (From Wikimedia Commons; in public domain)

Papillary endings of Ruffini: *Eponym* for papillary endings of nerve endings in the skin.

Rusconi, Mauro (1776–1849): Italian comparative anatomist and embryologist born in Pavia and working in Milan (Fig. R38).

Anus of Rusconi: *Eponym* for the blastopore, the starting point for extensive cell movements inside of the future endoderm and mesoderm during gastrulation. Reference: *Développement de la grenouille commune depuis le moment de sa naissance jusqu'à son ét parfait.* Giusti, Milano, 1826.

Russell, James S. Risien (1863–1939): British physician.

Bundle of Russell, Hook: *Eponym* for the uncinate tract or hook bundle of the cerebellum. Reference: *The origin and destination of certain afferent and efferent tracts in the medulla oblongata.* Brain 1897, 20:409–440.

Ruysch, Frederik (1638–1731): Dutch anatomist, botanist and obstetrician born in The Hague and working mainly in Amsterdam, known especially for his superb preparations of vessels. Some of his preparations are in the Kunstkamera in St. Petersburg. References: *Thesaurus Anatomicis Secundus.* Wolters, Amsterdam, 1702; *Opera omnia anatomico-medico-chirurgica.* Jansonius-Waesberg, Amsterdam, 1721–1737; *Alle de ontleed-, genees- en heelkundige werken.* Amsterdam, 1744. His Anatomical Lesson is shown in Fig. R39; *see* IJpma FFA, Radziun A, van Gulik TM *The anatomy lesson of Dr. Frederik Ruysch.* Eur J Obst Gynec Repr Biol 2013; 170:50–54.

Duct of Ruysch: *Eponym* for a small diverticulum in the foetal nasal septum associated with the vomeronasal organ.

Membrane of Ruysch: *Eponym* for the choroidocapillary layer of the choroid (Fig. R40).

Muscle of Ruysch: *Eponym* for the muscular tissue of the fundus of the uterus.

Veins of Ruysch: *Eponym* for (a) retroperitoneal portocaval anastomoses, later also described by Anders Retzius, and (b) the vorticose veins of the eyeball, also described by Stensen and later by Anders Retzius.

Fig. R39 The Anatomical Lesson by Ruysch, painted by Jan van Neck 1683 (Amsterdam Museum; in public domain)

Fig. R40 The choroid. (From Brüel A, Christensen EI, Tranum-Jensen J, Qvortrup K, Geneser F 2012 *Genesers histologi*. Munksgaard, Copenhagen; with permission). *1* basal complex (membrane of Bruch), *2* choroidocapillary layer (membrane of Ruysch), *3* substantia propria, *4* suprachoroid layer, *5* sclera. (From ten Donkelaar HJ, Kachlík D, Tubbs RS *An Illustrated Terminologia Neuroanatomica*. Springer, Cham, 2018)

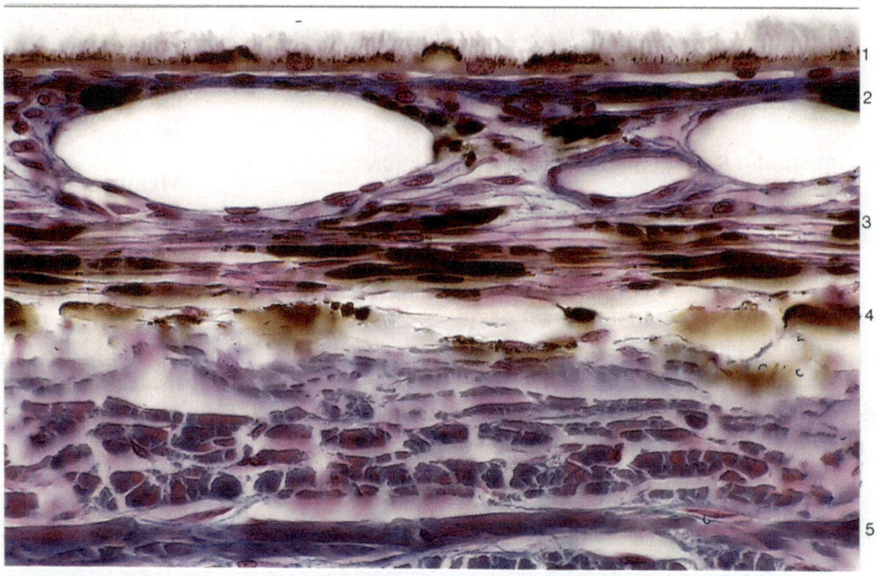

Letter S

17

Sachs, Heinrich (1863–1928): German neurologist born in Halberstadt, Saxony-Anhalt, and working in Breslau. He studied the lobar fibres of the occipital lobe and distinguished six lobar association tracts (Fig. S1). Reference: *Das Hemisphärenmark des menschlichen Grosshirns. I. Der Hinterhauptlappen.* Thieme, Leipzig, 1892; *see* Forkel SJ *Heinrich Sachs (1863–1928).* J Neurol 2015; 262:498–500.

Stratum proprium cunei of Sachs: *Eponym* for one of six association tracts of the occipital lobe of the cerebral cortex, connecting the cuneus with the superior margin of the calcarine sulcus.

Transverse occipital tract of cuneus of Sachs: *Eponym* for a lobar tract connecting the first occipital gyrus with the superior margin of the calcarine sulcus.

Salter, Sir Samuel James Augustus (1825–1897): English dentist born in Poole, Dorset, and working in London.

Lines of Salter: *Eponym* for incremental lines sometimes seen in dentin due to hypercalcification. Reference: *On some points in the anatomy and physiology of the dentinal tubes.* Arch Dent 1865; 1:281–309.

Salvi, Giunio (1869–1943): Italian anatomist working subsequently in Sassari, Parma and Naples.

Artery of Salvi: *Eponym* for the medial sinus tarsi artery, a short branch from the posterior tibial artery, entering the sinus tarsi. Reference: *Arteria dorsalis pedis.* Atti della Società Toscana di Scienze Naturali residenti in Pisa. Memoire 1900; 17:13–52.

Salzmann, Maximilian (1862–1954): Austrian ophthalmologist born in Vienna and working in Vienna and Graz.

Band of Saltzmann: *Eponym* for an anterior orbicular circular band (coronary ligament), part of the ciliary zonule, between the hyaloid membrane and the posterior third of the pars plicata of the ciliary body.

Membrane of Salzmann: *Eponym* for the vitreous membrane, a basal membrane between the vitreous body and the choroid. Reference: *Anatomie und Histologie des menschlichen Augapfels im Normalzustände und sein Altern.* Vienna and Leipzig, 1912.

Sandström, Ivar (1852–1889): Swedish histologist.

Bodies of Sandström: *Eponym* for the parathyroid glands.

Santorini, Giovanni Domenico (1681–1737): Italian anatomist born and working in Venice known for his extensive studies of several parts of the human body (Fig. S2). He described the lateral geniculate body, which he correctly interpreted as the site of central termination of the optic tract. Reference: *Observationes anatomicae.* Recurti, Venice, 1724.

Canal or **Duct of Santorini:** *Eponym* for the accessory pancreatic duct.

Cartilage of Santorini: *Eponym* for the corniculate cartilage of the larynx.

Concha of Santorini: *Eponym* for the supreme nasal concha.

Fissures, Incisures or **Notches of Santorini:** *Eponym* for notches in the cartilage of the the external acustic meatus, also described by Duverney and Huschke.

Labyrinth or **Plexus of Santorini:** *Eponym* for the prostatic (in males) or vesical (in females) venous plexus, a small plexus at the bottom of the retropubic space, draining among others the deep dorsal vein of the penis and the deep dorsal vein of the clitoris, respectively.

Ligament of Santorini: *Eponym* for the cricopharyngeal ligament connecting the corniculate cartilage, the lamina of the cricoid cartilage and the adjacent pharyngeal wall.

H. J. ten Donkelaar et al., *An Illustrated Guide to Anatomical Eponyms*, https://doi.org/10.1007/978-3-031-91664-9_17

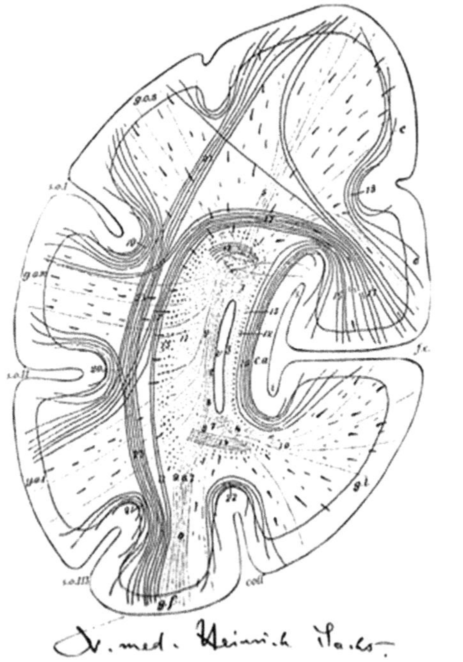

Fig. S1 Figure by Sachs on occipital lobar fibres: frontal section of the right occipital lobe with a schematic representation of its white matter (from Sachs H *Das Haemisphaerenmark des menschlichen Grosshirns. I. Der Hinterhauptlappen.* Thieme, Leipzig, 1892, in Forkel 2015 J Neurol 262:498-500; Open Access). *f.c.* fissura calcarina, *coll.* collateral sulcus, *s.o.I, s.o.II, s.o.III* sulcus occipitalis superior, - medius and inferior, *g.l.* gyrus lingualis, *g.f.* gyrus fusiformis, *g.o.s., g.o.m.* gyrus occipitalis superior, - medius, *1–10* forceps, *11–14* stratum sagittale internum, *15* stratum sagittale externum, *16* stratum calcarinum, *17* stratum cunei transversum, *18* stratum proprium cunei, *19–21* stratum proprium of s.o.I, of s.o.II and s.o.III, respectively, *22* stratum proprium of coll., *23* stratum profundum convexitatis

Fig. S2 Portrait of Santorini (from the Wellcome Collection CCBY; in public domain)

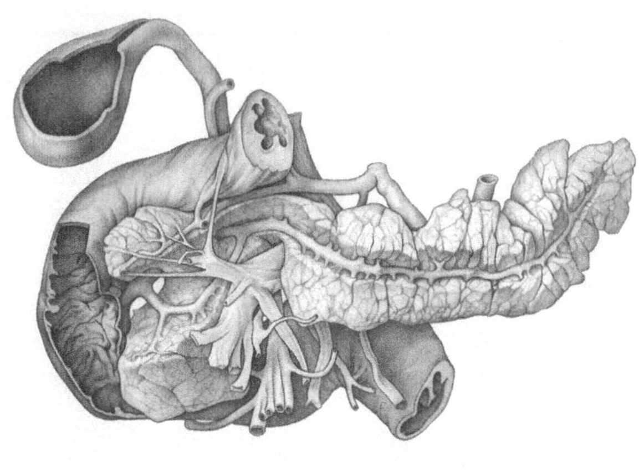

Fig. S3 Duct and papilla of Santorini (from the Anatomisch Museum Nijmegen; courtesy Lucas Boer, Nijmegen)

Major caruncle of Santorini: *Eponym* for the major duodenal papilla, also described by Vater (Fig. S3).

Minor caruncle of Santorini: *Eponym* for the minor duodenal papilla (Fig. S3).

Muscle of Santorini: *Eponym* for (a) the procerus muscle, (b) the risorius muscle, and (c) the incisurae helicis muscle.

Tubercle of Santorini: *Eponym* for the corniculate tubercle in the aryepiglottic fold, elevated by the corniculate cartilage of the larynx.

Vein of Santorini: *Eponym* for the parietal emissary vein.

Sapolini, Giuseppe (1812–1893): Italian neurologist in Milan.

Nerve of Sapolini: *Eponym* for the intermediate nerve of the facial nerve, also described by Wrisberg.

Sappey, Marie Philibert Constant (1810–1896): French anatomist born in Cernon, near Bourg-en-Bresse and working in Paris (Fig. S4).

Anterior accessory cartilage of Sappey: *Synonym* for the vomeronasal cartilage of Jacobson.

Ansa of Sappey: *Eponym* for a communicating branch of the glossopharyngeal nerve with the auricular branch of the vagus nerve, also described by Haller, and known as the ansa of Haller-Sappey.

Canal of Sappey: *Eponym* for the tympanic canaliculus, also described by Jacobson and Hirschfeld.

Deep stylohyoid muscle of Sappey: *Eponym* for an accessory muscle bundle of the stylohyoid muscle.

Fibres of Sappey: *Eponym* for smooth muscle fibres in the check ligaments of the eyeball.

Fig. S4 Photo of Sappey (from the Wellcome Collection CCBY; in public domain)

Foramen of Sappey: *Eponym* for the helicotrema of the cochlea, also described by Scarpa.

Ligament of Sappey: *Eponym* for (a) the anterior triangular ligament of the first sternochondral joint, and (b) the posterior, thickened part of temporomandibular ligament.

Muscle of Sappey: *Eponym* for circular smooth muscle fibres under the areola of the mamma.

Plexus of Sappey, Subareolar: *Eponym* for a lymphatic plexus under the areola of the mamma (Fig. S5). Reference: *Anatomie, physiologie et pathologie des vaisseaux lymphatiques considérés chez l'homme et les vertébrés.* Delahaye and Lecrosnier, Paris, 1874–1885.

Transverse chin muscle of Sappey: *Eponym* for transverse inferior fibres of the depressor anguli oris muscle.

Veins of Sappey: *Eponym* for the paraumbilical veins along the round ligament of the liver, also described by Baumgarten and Burow. Sappey also described **accessory portal veins** entering the hepatic capsule from different locations. Reference: *Mémoire sur les veines portes accessoires.* J Anat Physiol 1883; 19:517–524.

Sattler, Hubert (1844–1928): Austrian ophthalmologist born in Salzburg and working in Giessen, Erlangen and Leipzig (Fig. S6).

Glands of Sattler: *Synonym* for the ciliary glands of the eyelids, described earlier by Moll. Reference: *Beitrag zur Kenntniss der normalen Bindehaut des Menschen.* Von Graefe's Arch Ophthal 1877; 23:1–28.
Layer of Sattler: *Eponym* for the middle layer of the choroid. Reference: *Ueber den feineren Bau der Choroidea des Menschen nebst Beiträge zur pathologischen und vergleichenden Anatomie der Aderhaut.* Von Graefe's Arch Ophthal 1876; 22:1–99.

Savage, Henry (1810–1900): English anatomist and gynaecologist in London.

Perineal body of Savage: *Eponym* for the perineal body, a fibromuscular mass between the anus and the vagina.

Saxer, Franz (1864–1913): German pathologist born in Goslar and working in Marburg and Leipzig.

Cells of Saxer: *Eponym* for 'primary emigrant corpuscles', a primitive form of leucocytes in embryonic mesenchymal tissue. Reference: *Entwicklung und der Bau der normalen Lymphdrüsen und die Entstehung der roten und weissen Blutkörperchen.* Anat Hefte, Wiesbaden 1896; 6:350–352.

Scarpa, Antonio (1747–1832): Italian anatomist and surgeon born in Motta di Livenza and working in Modena and later Pavia (Fig. S7a) who described many structures of the inner ear. References: *De structura fenestrae rotundae aurii et de tympano secundario anatomicae observationes.* Modena, 1772 (Fig. S7b); *Anatomicae disquisitiones de auditu et olfactu.* Galeati, Pavia, 1789; *Anatomicorum annotationum liber primus: De nervorum gangliis et plexibus.* Modena, 1779; *Memoire anatomico-chirurgiche.* Milan, 1809.

Canals or **Foramina of Scarpa:** *Eponym* for accessory canals adjacent to the incisive foramina of Stensen in the maxilla, transmitting the nasopalatine nerves.

Fascia of Scarpa: *Eponym* for the membranous layer of the subcutaneous tissue of the abdomen.

Fluid or **Liquor of Scarpa:** *Eponym* for endolymph.

Fossa major or **Triangle of Scarpa:** *Eponym* or the femoral triangle.

Fossa minor of Scarpa: *Eponym* for the iliopectineal fossa.

Ganglion of Scarpa: *Eponym* for the vestibular ganglion.

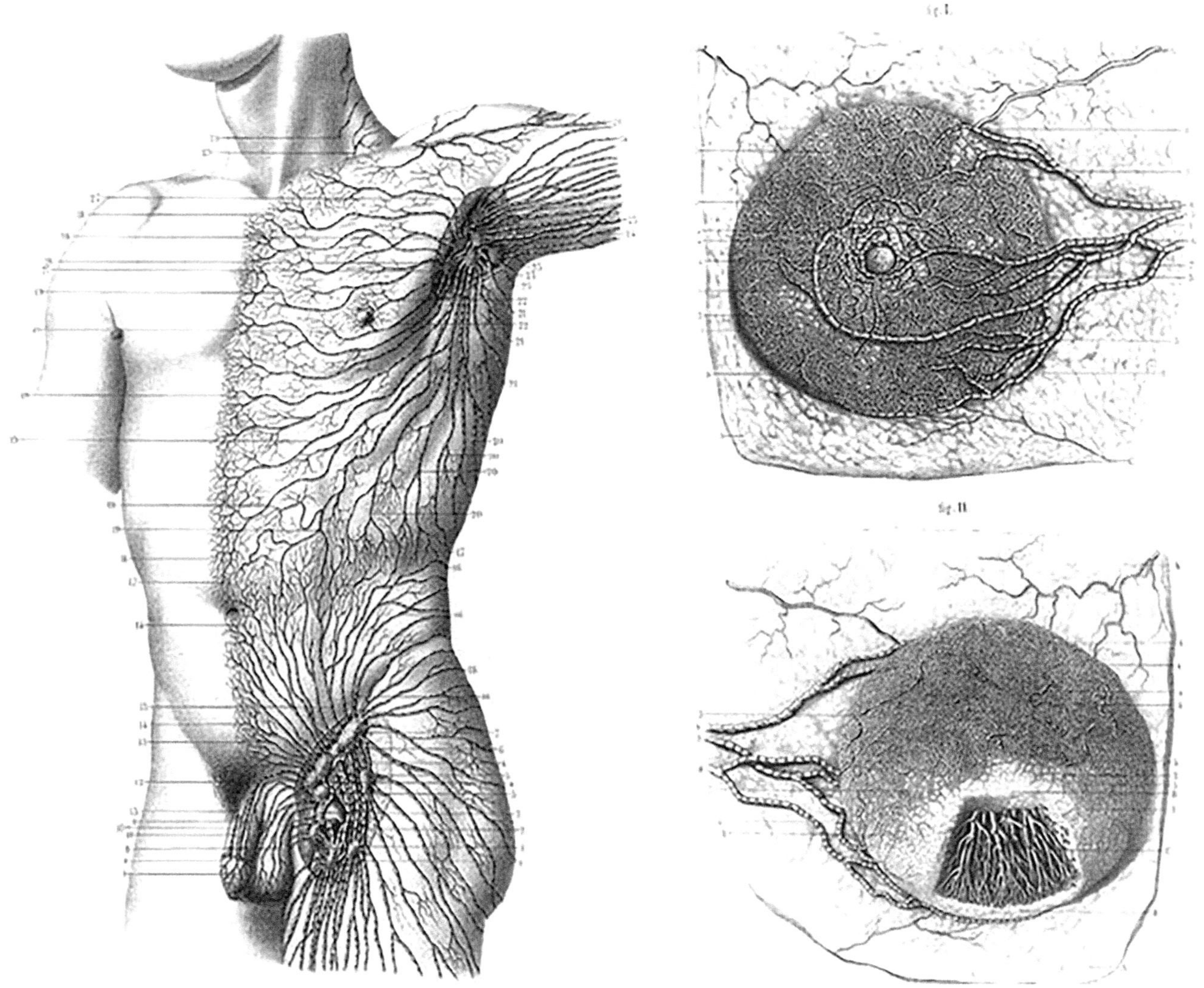

Fig. S5 Sappey's figure on the lymphatic areolar plexus of mamma (from Suami H, Pan WR, Mann GB, Taylor GI *The lymphatic anatomy of the breast and its implications for sentinel lymph node biopsy: A human cadaver study*. Ann Surg Oncol 2007; 15:863–871; from Wikimedia Commons; in public domain) q

Fig. S6 Portrait of Sattler (from Pagel 1901)

Hiatus or **Orifice of Scarpa:** *Eponym* for the helicotrema of the cochlea, also described by Sappey (Fig. S8).

Ligament of Scarpa: *Eponym* for the superior horn of the falciform margin of the saphenous opening.

Membrane of Scarpa: *Eponym* for the secondary tympanic membrane that closes the round window (Fig. S8).

Nerve of Scarpa: *Eponym* for the nasopalatine nerve, also described by Cotugno. Scarpa also described several cardiac nerves: (a) the greatest cardiac nerve (= middle cervical cardiac nerve), and (b) the small cardiac nerve (= inferior cervical cardiac nerve), also described by Pavlov. Reference: *De nervorum gangliis et plexibus*. Modena, 1779.

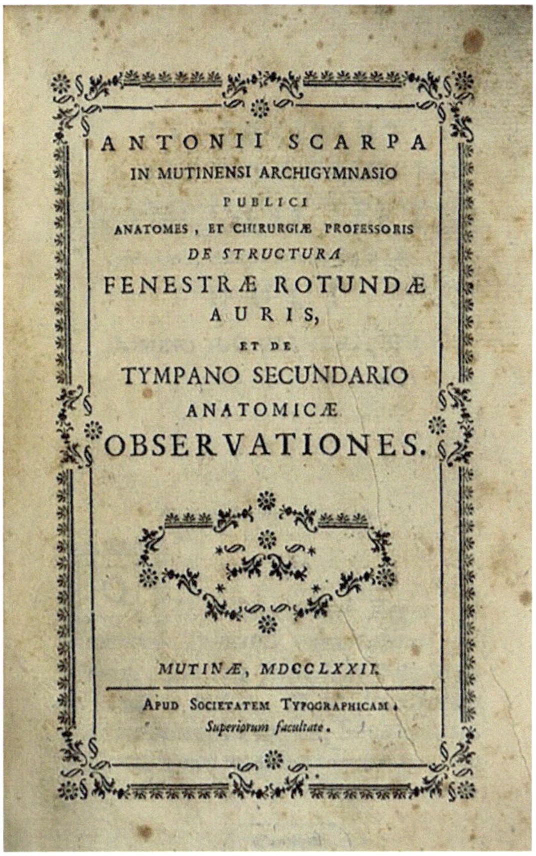

ANTONII SCARPA
IN MUTINENSI ARCHIGYMNASIO
PUBLICI
ANATOMES, ET CHIRURGIÆ PROFESSORIS
DE STRUCTURA
FENESTRÆ ROTUNDÆ
AURIS,
ET DE
TYMPANO SECUNDARIO
ANATOMICÆ
OBSERVATIONES.

MUTINÆ, MDCCLXXII.
APUD SOCIETATEM TYPOGRAPHICAM,
Superiorum facultate.

Fig. S7 (**a**) Portrait of Scarpa (from the Wellcome Collection CCBY; in public domain); (**b**) Title page *De structura fenestrae rotundae aurii et de tympano secundario anatomicae observationes.* Modena, 1772

Sheath of Scarpa: *Eponym* for the cremasteric fascia.

Schabadasch, Arnold Leonowitsch (1898–1978): Russian histologist working in Gorki and Moscow who described the outer submucosal neural plexus. Reference: *Intramurale Nervengeflechte des Darmrohrs.* Z Zellforsch Mikrosk Anat 1930; 10:320–385; *see* also Stach W *In memory of Arnold Leonowitsch Schabadasch.* J Hirnforsch 1979; 20:32–33.

Plexus of Schabadasch: *Eponym* for the external or outer submucosal neural plexus, located in the outer layer of the submucosa (*see* Fig. A18).

Schacher, Polycarp Gottlieb (1674–1737): German physician and anatomist born and working in Leipzig. Reference: *De cataracta.* Leipzig, 1705.

Ganglion of Schacher: *Eponym* for the ciliary ganglion.

Schachowa, Seraphina (1854–?): Russian histologist and physician born in Ekaterinoslav and working in Bern.

Tubules of Schachowa: *Eponym* for the convoluted part of the proximal tubules of the kidney, also described by Ferrein. Reference: *Untersuchungen ueber die Nieren.* Thesis, University of Bern, 1876.

Schaffer, Károly (1864–1939): Hungarian neurologist known for his studies on the hippocampus.

Collaterals of Schaffer: *Eponym* for collaterals of the pyramidal cells in CA3 of the hippocampus (*see* Fig. R2); Reference: *Beitrag zur Histologie der Ammonsformation.* Arch Mikr Anat 1892; 39:611–632.

Schaltenbrand, Georges (1897–1979): German neurologist born in Oberhausen and working in Würzburg. With Paul Bailey he introduced the Intercommissural plane. Reference: *Introduction to Stereotaxis with an Atlas of the Human Brain,* Vol 2. Thieme, Stuttgart, 1959. In 1940, he carried out the notorious 'Schaltenbrand experiment' on the potential viral origin of multiple sclerosis, using patients from a psychiatric institution as experimental subjects (*see* Hopf HC *Georges Schaltenbrand (1897–1979).* J Neurol 1980; 223:153–158).

Schlemm, Friedrich (1795–1858): German anatomist born in Salzgitter and working in Berlin (Fig. S9). He described the scleral venous sinus in 1830, although it had already

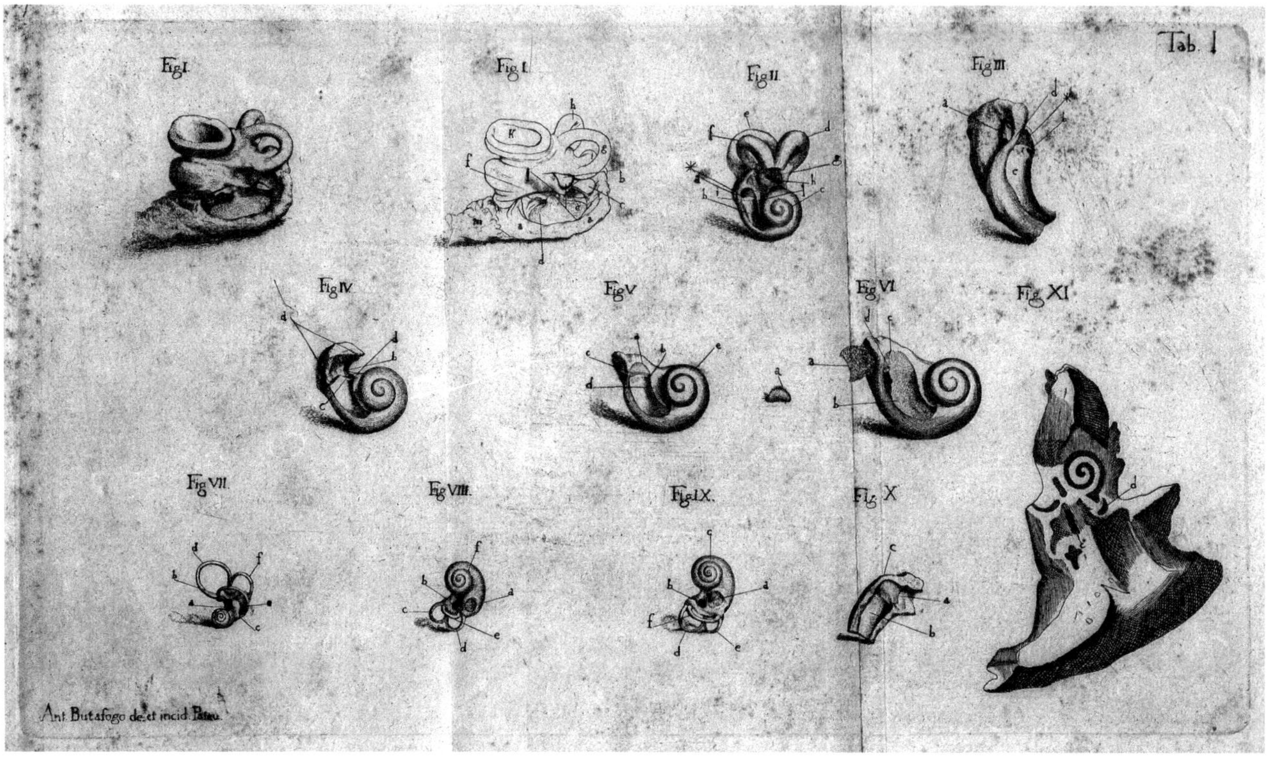

Fig. S8 Scarpa, Table II from *De structura fenestrae rotundae aurii et de tympano secundario anatomicae observationes.* Modena, 1772

Fig. S9 Portrait of Schlemm (from Wikimedia Commons; in public domain)

been described by Lauth in 1829 and noted by Fontana in 1778.

Canal of Schlemm: *Eponym* for the scleral venous sinus (Fig. S10). Reference: *Arteriarum capitis superficialium icon nova.* Berlin, 1830.

Schmidt, Henry D. (1823–1888): American pathologist born in New Orleans; *see* Schmidt-Lantermann cleft. Reference: *On the construction of the dark or double-bordered nerve fibre.* Monthly Micr J London 1874; 11:200–221.

Schmiedel, Kasimir Christoph (1716–1792): German physician, anatomist and botanist born in Bayreuth and working in Bayreuth and later Erlangen.

Ganglion of Schmiedel: *Eponym* for the inferior carotid ganglion within the cavernous plexus. Reference: *Epistola anatomica qua de controversa nervi intercostalis origine quaedam disseruntur; ad D. Wernerum.* Erlangen, 1744. Note from Dobson (1962): The intercostal nerve was then the synonym for the sympathetic nervous system. Schmiedel described sympathetic branches to the carotid and trigeminal ganglia.

Schneider, Conrad Viktor (1614–1680): German physician born in Bitterfeld and working in Wittenberg.

Membrane of Schneider: *Eponym* for the nonolfactory part of the nasal mucosa. Reference: *Liber de ossa cribriformi, et sensu ad organo odoratus, et morbis ad retrumque spectantibus.* Fincel, Wittenberg, 1655.

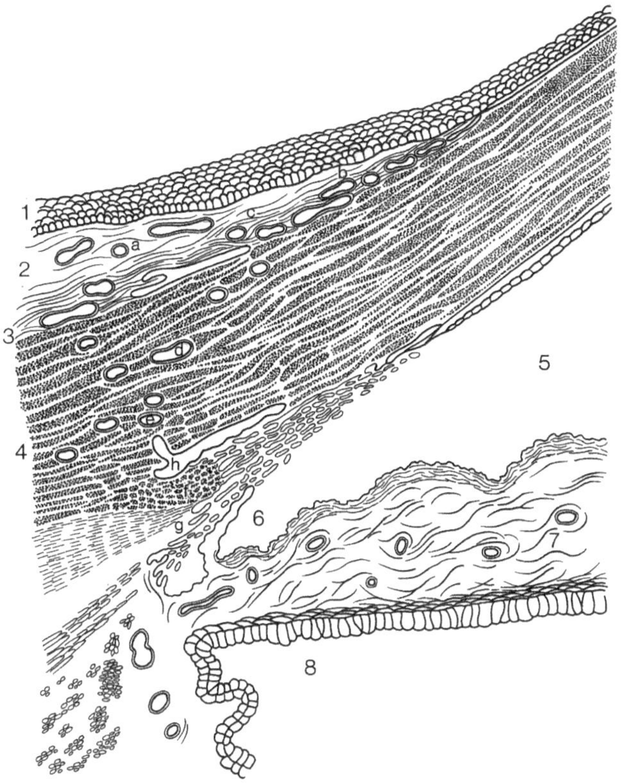

Fig. S10 Canal of Schlemm shown in a schematic drawing of part of a meridional section through the anterior part of the eyeball (after Bloom W, Fawcett DW *A Textbook of Histology*, 10th ed. Saunders, Philadelphia, PA, 1975). *1* anterior corneal epithelium, *2* cornea, *3* anterior chamber, *4* sphincter pupillae muscle, *5* dilator pupillae muscle, *6* iris, *7* sinus venosus sclerae, *8* ocular conjunctiva, *9* ciliary muscle, *10* sclera, *11* ora serrata, *12* posterior chamber, *13* ciliary zonules, *14* ciliary process, *15* lens, *16* vitreous body (from ten Donkelaar HJ, Kachlík D, Tubbs RS *An Illustrated Terminologia Neuroanatomica*. Springer, Cham, 2018)

Schreger, Christian Heinrich Theodor (1768–1833): German anatomist and chemist in Wittenberg and later in Halle.

Line of Hunter-Schreger: *Eponym* for alternating dark and clear lines, generally oriented perpendicular to the surface of the enamel, representing sections of neighbouring zones of crossed enamel prisms, also described by Hunter, and usually referred to as line of Hunter-Schreger (*see* Fig. 23.1). Reference: *Beiträge zur Geschichte der Zähne: Ueber den Zahnschmelz im Knochensubstanz der Zähne.* Beitr Zergleiderungskunst 1800; 1:3–7.

Schroeder van der Kolk, Jacob Ludlow Conrad (1797–1862): Dutch physiologist and pioneer for studies on the nervous system in primates, born in Leeuwarden and working in Utrecht. He is also known for his improvements of the situation of patients in the asylums for the so-called insane.

Fibres of Schroeder van der Kolk: *Eponym* for the reticular formation of the spinal cord and hindbrain. Reference: *Bau und Functionen der Medulla spinalis und oblongata und nächste Ursache und rationelle Behandlung der Epilepsie; aus dem Holländischen übertragen von Friedrich Wilhelm Theile.* Braunschweig, 1859.

Schueller, Karl Heinrich Maximilian (1843–1907): German surgeon born in Molsdorf and working in Berlin.

Ducts of Schueller: *Eponym* for the ducts of the paraurethal glands of Skene. Reference: *Ueber die Localbehandlung des chronischen Blasenkatarrhs, ein Beitrag zur Chirurgie der Harnorgane.* Berlin, 1877.

Schultze, Bernhard Sigismund (1827–1919): German obstetrician and gynaecologist born in Freiberg and working in Berlin, brother of Maximilian Schultze.

Folds of Schultze: *Eponym* for amniotic folds enclosing the yolk stalk at the origin of the umbilical cord from the placenta, the remains of which are normally incorporated in the placenta. Reference: *Das Nabelblaschen, ein constantes Gebildes in der Nachgeburt des ausgetragenen Kindes.* Leipzig, 1861.

Schultze, Maximilian Johann Sigismund (1825–1874): German anatomist born in Halle and working in Halle and later Bonn, brother of Bernhard Schultze. In 1865, he founded the Archiv für mikroskopische Anatomie. Reference: *Allgemeine Untersuchungen ueber die Strukturelemente des Nervensystems.* In: Stricker's Handbuch der Lehre von den Gewebe des Menschen und der Thiere. Leipzig, 1871, Vol 1, Chap 3.

Cells of Schultze: *Eponym* for the olfactory cells of the nasal mucosa.

Membrane of Schultze: *Eponym* for the olfactory mucosa of the nasal cavity.

Schultze, Friedrich (1848–1934): German anatomist born in Rathenow near Berlin and working in Dorpat and later in Bonn (Fig. S11).

Bundle, Fasciculus or **Tract of Schultze:** *Eponym* for the interfascicular fasciculus of the spinal cord (the comma tract of Schultze). Reference: *Beiträge zur Lehre von der secundären Degeneration im Rückenmark des Menschen nebst Bemerkung über die Anatomie der Tabes.* Arch Psychiatr 1883; 14:359–390.

Schütz, Hugo (? - 1928): German psychiatrist in Leipzig.

Fig. S11 Portrait of Schultze (Friedrich; from Wikimedia Commons; in public domain)

Bundle of Schütz: *Eponym* for the posterior or dorsal longitudinal fasciculus in the brain stem. Reference: *Anatomische Untersuchungen über den Faserverlauf im centralen Höhlengrau und den Nervenfaserschwund in demselben bei der progressiven Paralyse der Irren.* Arch Psychiatr Nervenkr 1891; 22:527–587.

Schwalbe, Gustav Albert (1844–1916): German anatomist born in Quedlinburg and working in Amsterdam, Freiberg, Leipzig, Jena and Königsberg, and finally in Strassburg (Fig. S12a). He described various structures in the nervous system. Reference: *Neurologie, in: C.E.E. Hoffmann's Anatomie des Menschen,* Band 2. Besold, Erlangen, 1881.

'Affenspalte' of Schwalbe: *Eponym* for a more pronounced parietooccipital sulcus in the cerebral cortex of monkeys.

Canal or **Hiatus of Schwalbe:** *Eponym* for a tendinous gap, his hiatus pelvicus lateralis, where the levator ani muscle attaches to the obturator fascia, which may serve as a site for a lateral pelvic herniation of pelvic peritoneum into the ischioanal or ischiorectal fossa (perineal hernia).

Fissure of Schwalbe: *Eponym* for the choroid or optic fissure of the developing eye.

Line or **Ring of Schwalbe**: *Eponym* for a vertical line from the end of the anterior limiting membrane of the cornea (of Bowman) to the end of the posterior limiting membrane of the cornea (of Descemet), indicating the corneal border of the corneal limbus.

Nucleus of Schwalbe: *Eponym* for the medial vestibular nucleus.

Plane of Schwalbe: *Eponym* for a horizontal plane from the glabella to the inion, which rather closely parallels the bicommissural plane of the brain (Fig. S12b).

Ponticulus of Schwalbe: *Eponym* for a small depression of the tympanic cavity between the pyramidal eminence and the oval or vestibular window.

Sinus of Schwalbe: *Eponym* for the tympanic sinus, a depression of the medial (labyrinthine) wall of the tympanic cavity between the oval window and the round window.

Spaces of Schwalbe: *Eponym* for the episcleral space of the optic nerve, the spaces within the internal sheath of the optic nerve, filled with cerebrospinal fluid and continuous with the subarachnoid space.

Schwann, Theodor (1810–1882): German anatomist born in Neuss and working in Louvain (Leuven) and later in Liège (Fig. S13). The Schwannocytus, the Schwann cell, is named in his honour. Reference: *Mikroskopische Untersuchungen über die Übereinstimmung in der Struktur und dem Wachstum der Thiere und Pflanzen.* Reimer, Berlin, 1839.

Cells of Schwann: *Eponym* for peripheral glial cells, previously known as neurolemmocytes, but currently as schwannocytes, of which there are three types: myelinating, non-myelinating and terminal (Fig. S14).

Sheath of Schwann: *Eponym* for the neurolemma or neurilemma, the plasmalemma of a neuron.

Substance of Schwann: *Eponym* for the myelin around nerve fibres.

Schweigger-Seidel, Franz (1834–1871): German physiologist born in Halle an der Saale and working in Breslau (now Wróclaw) and Leipzig.

Capillaries of Schweigger-Seidel: *Eponym* for spindle-shaped sleeves that cover the penicillar arteries of the spleen.

'Scissure limbique' of Broca: *Eponym* for the cingulate sulcus separating the limbic lobe from the rest of the cerebral cortex; *see* Broca, Pierre Paul.

Sebileau, Pierre (1860–1953): French anatomist born in Saint-Fort, Charente Inférieur, and working in Paris. His

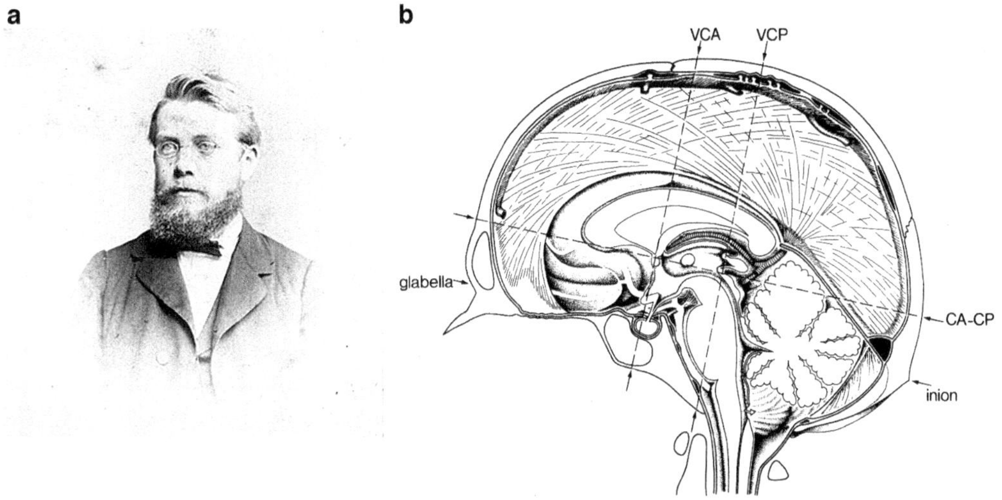

Fig. S12 (**a**) Portrait of Schwalbe (from Pagel 1901); (**b**) Plane of Schwalbe (Fig. A65 from ten Donkelaar et al. 2018 An Illustrated Terminologia Neuroanatomica)

Fig. S13 Portrait Schwann (from Pagel 1901)

early work dealt with the reproductive system, later publications mainly with otolaryngology.

Ganglion of Sebileau: *Eponym* for the cervicothoracic or stellate ganglion.

Hollow of Sebileau: *Eponym* for a depression between the tongue and the sublingual glands. Reference: *Démonstrations d'anatomie.* Paris, 1892.

Ligaments of Sebileau, Suspensory: *Eponym* for the costopleural ligaments. Reference: *L'appareil suspenseur de la plèvre.* Paris, 1891.

Muscle of Sebileau: *Eponym* for the deep layer of the tunica dartos. Reference: *Les envéloppes des testicules.* Paris, 1897.

Sedillot, Charles Emmanuel (1804–1883): French surgeon born in Paris and working in Strassburg (Fig. S15).

Triangle of Sedillot: *Eponym* for the lesser supraclavicular fossa between the two origins of the sternocleidomastoid muscle and the clavicula, also described by Zang.

Seessel, Albert (1850–1910): American embryologist born in Memphis, Tennessee, and working as an assistant to Wilhelm His Sr, where he published his embryological studies. Back in America, he became a neurologist in New York.

Pocket or **Pouch of Seessel:** *Eponym* for a diverticle of the embryonic pharynx anterior to the cranial attachment of the buccopharyngeal membrane. Reference: *Zur Entwickelungsgeschichte des Vorderdarms.* Arch Anat Physiol (Lpz) 1877, pp 449–466.

Seiler, Carl (1849–1905): American laryngologist of Swiss origin, working in Philadelphia.

Cartilage of Seiler: *Eponym* for a small cartilaginous rod attached to the vocal process of the arytenoid cartilage. Reference: *Researches on the anatomy of the vocal cords.* Trans Amer Laryngol Ass 1879, Vol 1.

Sertoli, Enrico (1842–1910): Italian physiologist born in Sondrio and working in Milan (Fig. S16).

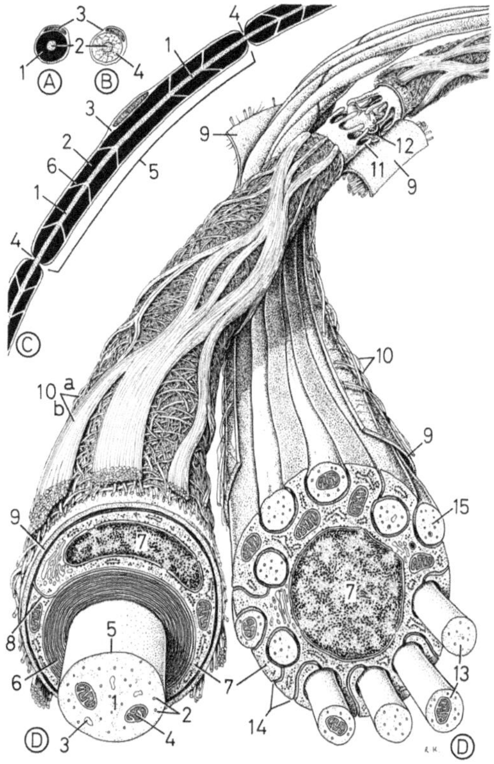

Fig. S14 Schwann cells and nerve fibers (from Krstić R *Die Gewebe des Menschen und der Säugetiere*, 2nd ed. Springer, Berlin-Heidelberg-New York, 1988; with permission). (**a**, **b**) Cross-section of a Schwann cell (*3*) with myelin sheath (*1*) in black in (**a**) and white (*2*, *4*) in (**b**), a routine staining, around an axon (*2*). (**c**) Longitudinal section of a myelinated axon: *1* axon, *2* myelin sheath, *3* Schwann cell, *4* node of Ranvier, *5* internodal segment, *6* Schmidt-Lantermann cleft. (**d**) Drawing after electronmicroscopic pictures: *1* axon with neurofilaments, *2* neurotubuli, *3* endoplasmatic reticulum, *4* mitochondrion, *5* axolemma, *6* myelin sheath, *7* Schwann cell (at the left surrounding a myelinated axon, at the right a large number of unmyelinated axons), *8* external mesaxon, *9*, *10* endoneurium (*9* basal lamina of Schwann cell, *10a*, *b* microfibrils), *11* protrusions of Schwann cell, *12* node of Ranvier, *13* unmyelinated axons, *14* mesaxon, *15* naked axon (from ten Donkelaar HJ, Kachlík D, Tubbs RS *An Illustrated Terminologia Neuroanatomica*. Springer, Cham, 2018)

Cells of Sertoli: *Eponym* for epithelial supporting cells (sustenocytes) in the seminiferous tubules of the testis to which different stages of developing spermatids are attached during spermiogenesis. Reference: *Dell'esistenza di particolari cellule ramificate nei canalicoli seminiferi del testiculo umano.* Il Morgagni 1865; 7:31–40.

Serres, Antoine Etienne Renaud Augustin (1786–1868): French anatomist born in Clayrac and working in Paris.

Glands of Serres: *Eponym* for 'gingival glands', islands of epithelium present in the gums of infants. Reference: *Essai sur l'anatomie et la physiologie des dents, ou nouvelle théorie de la dentition.* Paris, 1817.

Fig. S15 Portrait of Sedillot (from Wikipedia; in public domain)

Fig. S16 Portrait of Sertoli (from Wikimedia Commons; in public domain)

Severino, Marco Aurelio (1580–1656): Italian anatomist born in Tarsia and working in Naples.

Intestinal glands of Severino: *Eponym* for the tubular glands of the small intestine, also described by Galeazzi and Lieberkühn.

Sharpey, William (1802–1880): Scottish anatomist born in Arbroath and working in Edinburgh and later London.

Fibres of Sharpey: *Eponym* for bundles of collagen fibres, which run from the dense collagenous connective tissue of the periosteum to the superficial bone layer (*see* Fig. H3).

Shenton, Edward Warren Hine (1872–1955): English radiologist.

Line of Shenton-Ménard: *Eponym* for a curved line formed by the top of the obturator foramen and the inner side of the neck of the femur, seen on an antero-posterior frontal X-ray of a normal hip joint, already described in 1896 by Ménard.

Sherman, R Shanton (20th century): American surgeon from San Francisco.

Veins of Sherman, Perforating: *Eponym* for inferior and middle paratibial perforating veins of the middle lower leg, located 20 and 24 cm above the sole, respectively, connecting the great saphenous vein and the posterior tibial veins. Reference: *Varicose veins. Anatomic findings and operative procedure based upon them.* Ann Surg 1944; 120:772–784.

Sherren, James (1872–1945): British surgeon born in Weymouth, Dorset, and working in London.

Triangle of Sherren: *Eponym* for an area of skin hyperaesthesia found in acute appendicitis, bordered by lines joining the umbilicus, the pubic tubercle and the anterior superior iliac spine. Reference: *Contributions to Abdominal Surgery.* (Barnard HL, Sherren J, eds). Arnold, London, 1910.

Sherrington, Charles Scott (1857–1952): English neurophysiologist born in London and working in Oxford. He coined the word 'synapse' and received the Nobel Prize in Physiology and Medicine in 1932.

Axial sheath of Sherrington: *Eponym* for the inner sheath, a thin sheath of connective tissue surrounding the intrafusal muscle fibres and fixing them to the inner wall of the capsule of the muscle spindle. Reference: *On the anatomical constitution of the nerves of skeletal muscles; with remarks on recurrent fibres in the ventral spinal nerve root.* J Physiol (Lond) 1894; 17:211–218.

Cells of Cooper-Sherrington: *Eponym* for the spinal border cells in the lumbar spinal cord, the origin of the anterior or ventral spinocerebellar tract. Reference: *Gower's tract and spinal border cells* (with Sybil Cooper). Brain 1940; 63:123–134. In 1932, these cells were already described by Foerster and Gagel (*Die Vorderseitenstrangdurchschneidung beim Menschen. Eine klinisch-patho-physiologisch-anatomische Studie.* Z Ges Neurol 1932; 138:1–92).

Shrapnell, Henry Jones (1761–1834): English military surgeon.

Membrane of Shrapnell: *Eponym* for the flaccid part of the tympanic membrane, already described by Rivinus (1717). Reference: *On the form and structure of the membrane of the tympanum.* Lond Med Gaz 1832, p 120.

Sibson, Francis (1814–1876): English physician born in Maryport, Cumberland, and working in London.

Fascia of Sibson: *Eponym* for the suprapleural membrane.

Muscle of Sibson: *Eponym* for the scalenus minimus muscle. Reference: *On the mechanism of respiration.* Phil Trans 1846, pp 501–550.

Siebenmann, Friedrich (1852–1928): Swiss otolaryngologist in Basel (Fig. S17).

Sinuses:

Sinus of Arlt: *Eponym* for a dilatation of the lacrimal sac; *see* Arlt, Carl Ferdinand.

Sinus of Aschoff-Rokitansky: *See* Sinus of Rokitansky-Aschoff.

Sinus of Breschet: *Eponym* for the sphenoparietal sinus; *see* Breschet, Gilbert.

Fig. S17 Portrait of Siebenmann (from Wikimedia Commons; in public domain)

Sinus of Calori, Renal: *Eponym* for a small enlargement of the inferior vena cava just *above* its junction with the renal veins, *see* Calori, Luigi.

Sinus of English: *Eponym* for the petrooccipital sinus; *see* English, Josef.

Sinus of *Fontana*: *Eponym* for the scleral venous sinus; see Fontana, Felice Gaspar Ferdinand.

Sinus of Grünwald: *Eponym* for the lateral sinus of the ethmoidal cells, locared craniodorsally to the ethmoidal bulla; see Grünwald, Ludwig.

Sinus of Guérin: *Eponym* for the recess of the navicular fossa, a cul-de-sac or diverticle behind the valve of the navicular fossa of the male urethra; *see* Guérin, Alphonse François Marie.

Sinus of Huguier: *Eponym* for the fossula fenestrae vestibuli, a depression on the medial (labyrinthine) wall of the middle ear, which harbours the oval window; *see* Huguier, Pierre Charles.

Sinus of Lauth: *Eponym* for the *scleral* venous sinus, mostly known as canal of Schlemm; *see* Lauth, Ernest Alexandre.

Sinus of Littré: *Eponym* for the venous basilar plexus; *see* Littré, Alexis.

Sinus of Luschka: *Eponym for* the petrosquamous sinus; *see* Luschka, Hubert von.

Sinus of Maier: *Eponym* for the fornix of the lacrimal sac; *see* Maier, Rudolf.

Sinus of Meyer: *Eponym* for a depression in the floor of the external acoustic meatus; *see* Meyer, Georg Hermann von.

Sinus of Morgagni: *Eponym* for anal sinus; *see* Morgagni, Giovanni Battista.

Sinus of Palfijn: *Eponym* for an inconstant space within the crista galli of the ethmoidal bone, said to communicate with the frontal sinus and ethmoidal cells; *see* Palfijn, Jean.

Sinus of Petit: *Eponym* for the sinus aortae; *see* Petit, François Pourfour du.

Sinus of Ridley: *Eponym* for the anterior and posterior intercavernous sinuses; *see* Ridley, Humphrey.

Sinus of Rokitansky-Aschoff: *Eponym* mucous crypts of the wall of the gallbladder; *see* Rokitansky, Carl von.

Sinus of Schwalbe: *Eponym* for the tympanic sinus, a depression of the medial (labyrinthine) wall of the tympanic cavity between the oval window and the round window; *see* Schwalbe, Gustav Albert.

Sinus of Tourtual: *Eponym* for the supratonsillar fossa of the *pharynx*; *see* Tourtual, Kaspar Theobald.

Sinus of Valsalva: *Eponym* for the sinuses of the aorta, also described by Petit; *see* Valsalva, Antonio Maria.

Skene, Alexander Johnston Chalmers (1838–1900): American gynaecologist born in Aberdeenshire, Scotland and working in New York (Fig. S18).

Glands of Skene: *Eponym* for the paraurethral glands of females. Reference: *The anatomy and pathology of two important glands of the female urethra.* Am J Obstet Dis Women Child 1880; 13:265–270.

Slavjanski, Kronid Fedorovich von (1847–1898): Russian obstetrician and gynaecologist in St. Petersburg.

Membrane of Slavjanski: *Eponym* for the basement membrane separating the granulosa from the inner theca of the ovarian follicle. Reference: *On the Normal and Pathological Histology of Human Ovaries* (English translation). Thesis, St. Petersburg, 1871.

Smellie, William (1697–1763): Scottish obstetrician born in Lesmahagow, near Lanark, Scotland, and working in London (Fig. S19). Smellie laid down the principles of clinical pelvimetry, including vaginal examination to estimate the diagonal conjugate. Reference: *A Treatise on the Theory and Practice of Midwifery.* Wilson, London, 1752.

Fig. S18 Portrait of Skene (from the Wellcome Collection CCBY; in public domain)

Fig. S19 Portrait of Smellie (from the Wellcome Collection CCBY; in public domain)

Fig. S20 Portrait of Soemmerring by Wendelin Moosbrugger (1813; from Wikimedia Commons; in public domain)

Soemmerring, Samuel Thomas von (1755–1830): German anatomist born in Thorn an der Weichsel, Prussia, and working subsequently in Kassel, Mainz, Munich and finally in Frankfurt am Main (Fig. S20). He described the substantia nigra (substance of Soemmerring). References: *Vom Baue des menschlichen Körpers.* Varrentrapp & Wenner, Frankfurt am Main, 1791–1796; *De basi encephali et originibus nervorum cranio egredientium libri quinque.* In: *Scriptores neurologici minores selecti sive opera minora ad anatomiam physiologiam et pathologiam nervorum spectantia. Notulis nonnulis illustravit, praefatus est, indicibus auxit Christianus Fridericus Ludwig,* Vol II. Leipzig, 1792, pp 1–112.

Foramen of Soemmerring: *Eponym* for the fovea centralis of the retina.

Ligament of Soemmerring: *Eponym* for the suspensory ligament of the lacrimal gland.

Muscle of Soemmerring: *Eponym* for the levator glandulae thyroideae muscle.

Ring of Soemmerring: *Eponym* for a mass of lenticular fibres enclosed between the anterior and posterior parts of the lenticular capsule, leaving the pupillary area relatively free.

Spot of Soemerring: *Eponym* for the macula lutea of the retina.

Substance or **Ganglion of Soemmerring:** *Eponym* for the substantia nigra (ganglion or locus niger of Soemmerring; Fig. S21).

Sommer, Karl Wilhelm (1852–1900): German psychiatrist known for his studies on Ammon horn sclerosis (Fig. S22). Reference: *Erkrankung des Ammonshorn als aetiologisches Moment der Epilepsie.* Arch Psychiatr Nervenkr 1880; 10:631–675.

Sommer sector: *Eponym* for the CA1 or H1 field and the adjacent subiculum; here, the pyramidal cells are especially sensitive to anoxia and other metabolic disturbances and are often affected in temporal lobe epilepsy.

Sorgius, Wilhelm (19th century): German anatomist.

Gland or **Node of Sorgius:** *Eponym* for the sentinel lymph node of the mammary gland, the caudalmost of the pectoral lymph nodes located on the second or third slip of the anterior serratus muscle. Reference: *Ueber die Lymphgefässe der weiblichen Brustdrüße.* Thesis, University of Strassburg, 1880.

Spaces: *See* **The Classical Anatomical Eponyms.**

Spalteholz, Karl Werner (1861–1940): German anatomist born in Dresden and working in Leipzig. In 1885, he published the first volume of his renowned anatomical atlas.

Arteries of Spalteholz: *Eponym* for arterioles with a wall of endothelial lining and sparse smooth muscle cells.

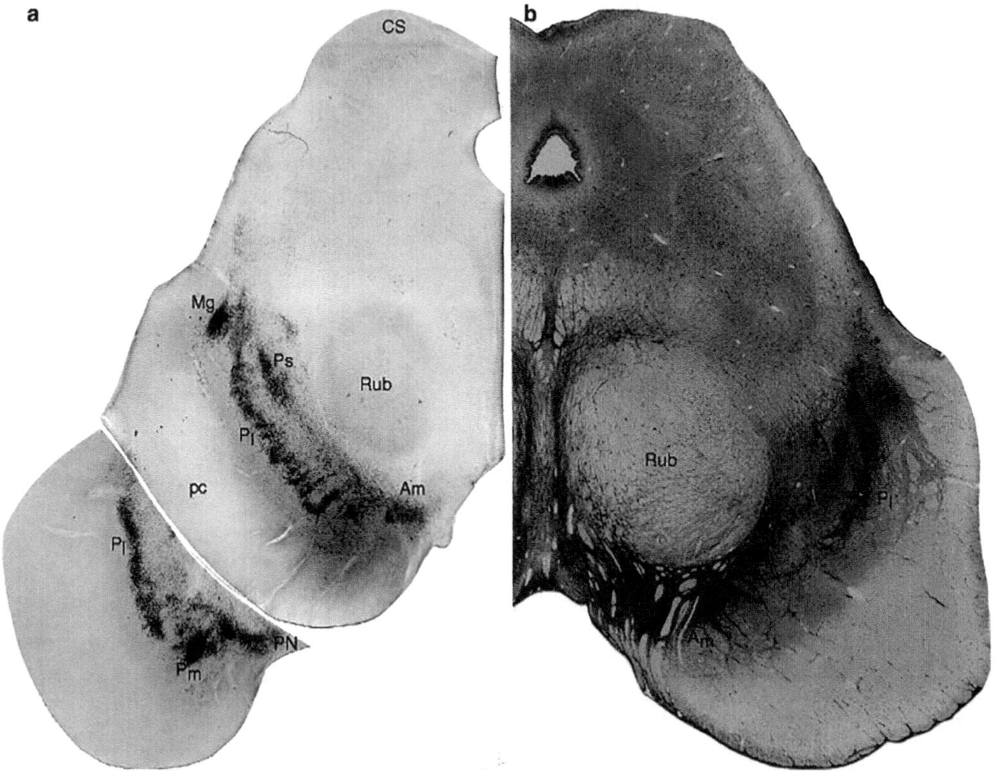

Fig. S21 (**a**) Pigmentoarchitectonics of the human substantia nigra, and (**b**) tyrosine hydroxylase (TH)-stained section of the human mesencephalon. In (**a**), two 800 μ sections, a rostral (above) and a caudal (below) one, show the internal organization of the substantia nugra in transverse section; in (**b**), posterolateral, posterosuperior and anteromedial subnuclei of the substantia nigra are clearly visible as is the midline zone of TH immunoreactivity belonging to the VTA. A_m anteromedial subnucleus; *CS* colliculus superior, *Mg* magnocellular subnucleus, *pc* pedunculus cerebri, P_l, P_m, P_s posterolateral, posteromedial and posterosuperior subnuclei, *PN* paranigral nucleus, *Rub* nucleus ruber (from van Domburg PHMF, ten Donkelaar HJ *The Human Substantia Nigra and Ventral Tegmental Area.* Adv Anat Embryol Cell Biol 1991; 121:1–132; in ten Donkelaar HJ, Kachlík D, Tubbs RS *An Illustrated Terminologia Neuroanatomica.* Springer, Cham, 2018)

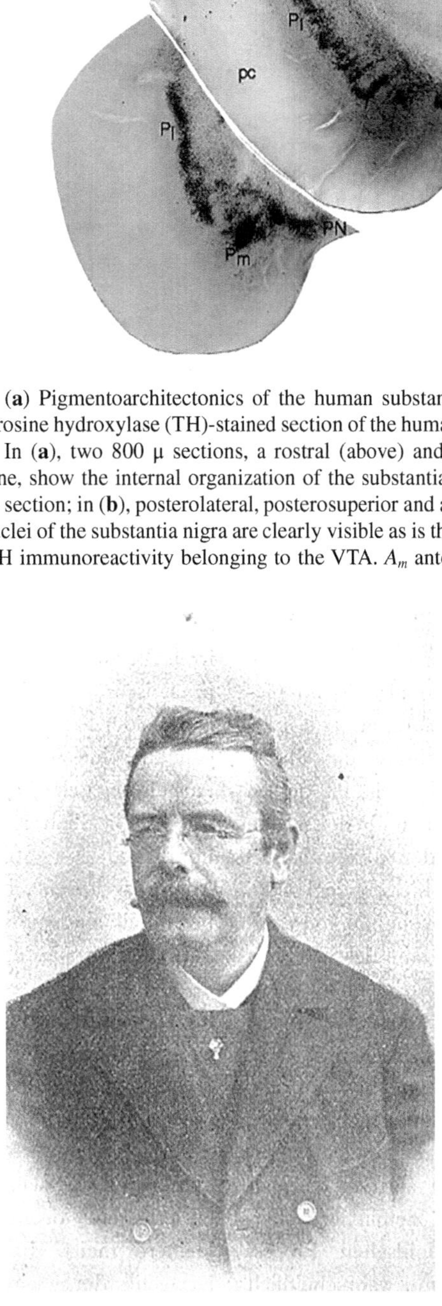

Fig. S22 Portrait of Sommer (from Wikidata; in public domain)

Spee, Ferdinand, Graf von (1855–1937): German anatomist born in Glindfeld, Westphalia, and working in Kiel.

Curve of Spee: *Eponym* for the anatomical curvature of the sagittal occlusal plane. Reference: *Skelettlehre: Kopf.* In: Bardeleben K, Hrsg *Handbuch der Anatomie des Menschen.* Fischer, Jena, 1896.

Spemann, Hans (1869–1941): German embryologist born in Stuttgart who in 1931 received the Nobel Prize for Physiology and Medicine for the discovery of neural induction (Fig. S23), work done in collaboration with Hilde Mangold; *see* Fässler PE *Hans Spemann (1869–1941) and the Freiburg School of Embryology.* Int J Dev Biol 1996; 40:49–57.

Spemann-Mangold organizer: *Eponym* for a region, which is necessary for inducing dorsal ectoderm to form neural tissue (Fig. S24). References: *Die Erzeugung tierischer Chimären durch heteroplastische embryonale Transplantation zwischen Triton cristatus und Triton taeniatus.* Roux Arch Entw Mech Org 1921; 48:533–570; *Über Induktion von*

Fig. S23 Photo of Spemann (from Wikimedia Commons; in public domain)

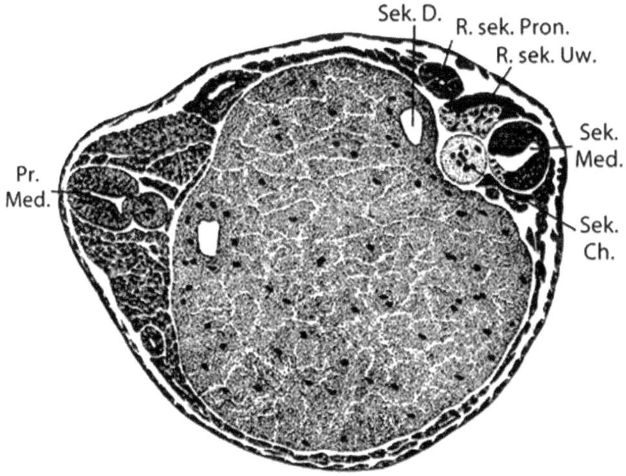

Fig. S24 The 'organizer' experiment carried out in the early 1920s by Hilde Mangold and Hans Spemann. Transplantation of the upper lip of the blastopore from one gastrula of a darkly pigmented newt to another lightly pigmented host often led to the induction of a second nervous system. *Pr.Med.* primary neural tube, *R.sek.Pron* secondary pronephric duct, *R.sek.Uw.* secondary somite, *Sek.Ch.* secondary notochord, *Sek.D.* secondary endoderm, *Sek.Med.* secondary neural tube (from Spemann H, Mangold H *Über Induktion von Embryonalanlagen durch Implantation artfremder Organisatoren.* Roux Arch Entw Mech Org 1924; 100:599–638, in ten Donkelaar HJ, Lammens M, Hori A *Clinical Neuroembryology,* 3rd ed. Springer, Cham, 2023)

Embryonalanlagen durch Implantation artfremder Organisatoren. Roux Arch Entw Mech Org 1924; 100:599–638 (with Hilde Mangold).

Spence, James (1812–1882): Scottish surgeon born and working in Edinburgh.

> **Axillary tail of Spence:** *Eponym* for the axillary process of the mammary gland. Reference: *Lectures of Surgery.* Edinburgh, 1871.

Spincters:

Sphincter of Balli: *Eponym* for a functional sphincter between the descending colon and the sigmoid colon highlighted in radiology; *see* Balli, Ruggero.

Sphincter of Boyden: *Eponym* for the four-part sphincter of the bile duct, the pancreatic duct and the hepatopancreatic ampulla; *see* Boyden, Edward.

Sphincter of Busi: *Eponym* for a true sphincter between the caecum and the ascending colon highlighted in radiology; *see* Busi, Aristide.

Sphincter of Cannon-Böhm: *Eponym* for a 'sphincter' in the transverse colon at the junction of the proximal and middle parts; *see* Cannon, Walter Bradford.

Sphincter of Disse: *Eponym* for circularly thickened muscle layer in the area of the renal papillae; *see* Disse, Joseph Hugo Vincenz.

Sphincter of Hirsch: *Eponym* for a true sphincter in the proximal part of the ascending colon; *see* Hirsch, I Seth.

Sphincter of Lütkens, Spiral: *Eponym* for a circular bundle of smooth muscle fibres in the neck of the gallbladder, close to the spiral valve of Heister; *see* Lütkens, Ulrich.

Sphincter of Mirizzi: *Eponym* for a physiological sphincter at the right and left hepatic ducts; *see* Mirizzi, Pablo Luis.

Sphincter of Moulton: *Eponym* for a true sphincter in the sigmoid colon; *see* Chap. 22.

Sphincter of O'Beirne and Hyrtl: *Eponym* for the rectosigmoid sphincter, a thickening of the circular muscle layer of the most oral or uppermost part of the rectum; *see* O'Beirne, James, and Hyrtl, Joseph.

Spincter of Ochsner: *Eponym* for a sphincter in the duodenum below the opening of the bile duct; *see* Ochsner, Albert John.

Sphincter of Oddi: *Eponym* for the sphincter ampullae; *see* Oddi, Ruggero.

Sphincter of Payr-Strauss: *Eponym* for a 'sphincter' at the splenic flexure of the colon; *see* Chap. 22.

Sphincter of Rossi: *Eponym* for a functional sphincter between the sigmoid colon and the rectum; *see* Chap. 22.

Spieghel (Spigelius): *See* van den Spieghel, Adriaan.

Spinner, Morton (20th century): American orthopaedic surgeon in New York.

Ligament of Spinner: *Eponym* for an inconstant fibrous layer between the flexor digitorum superficialis muscle and the humeral head of the flexor carpi ulnaris muscle, attached to the medial epiondyle of the humerus and the medial surface of the coronoid process of the ulna. Reference: *An anatomic factor significant in transposition of the ulnar nerve* (with S Inserra). J Hand Surg 1986; 11:80–82.

Spitzka, Edward Charles (1852–1914): American neurologist born and working in New York.

Bundle of Spitzka: *Eponym* for fibres of the 'posterior' longitudinal bundle' connecting the oculomotor and abducens nerve nuclei, now believed to be part of the medial longitudinal fasciculus; Reference: *A second origin of the abducens.* Am Psychol J New York 1875–1876; 3:205–209.

Decussation of Spitzka: *Eponym* for the decussation of the medial lemnisci in the medulla oblongata. Reference: *Contributions to the anatomy of the medial lemniscus.* Med Res New York 1884, 26.

Spix, Johannes Baptist (1781–1826): German comparative anatomist born in Hochstädt and working as Conservator of the Museum of Natural History in Munich.

Ossicle or **Spine of Spix:** *Eponym* for the lingula of the mandible.

Staderini, Rutilio (1861–1942): Italian neuroanatomist born in Siena and working in Florence and later in Siena (Fig. S25). Reference: *Sopra un nucleo di cellule nervose intercalato fra i nuclei di origine del vago e dell'ipoglosso.* Monit Zool Ital 1894, Vol 5, p 178.

Nucleus of Staderini: *Eponym* for the intercalated nucleus of the brain stem.

Stähli, Jean (1890–?): Swiss ophthalmologist.

Lines of Stähli: *Eponym* or a brownish line of iron deposition across the lower third of the cornea, occurring in the elderly. Reference: *Über den Fleischerschen Ring beim Keratokonus und eine neue typische Epithelpigmentation von der normalen Kornea.* Klin Monatsbl Augenheilk, Stuttgart 1918; 60:721–741.

Stahr, Hermann (1869–1947): German pathologist born in Lille and working in Kiel and Danzig.

Fig. S25 Photo of Staderini (from Wikimedia Commons; in public domain)

Gland of Stahr: *Eponym* for a facial lymph node located at the facial artery at the inferior margin of the mandibula. Reference: *Ueber den Lymphapparat des ausseren Ohres.* Anat Anz 1899; 15:381–387.

Stanley, Edward (1791–1861): English surgeon born and working in London.

Cervical ligaments of Stanley: *Eponym* for fibres of the capsule of the hip joint reflected onto the neck of the femur. Reference: *A Manual of Practical Anatomy.* London, 1818.

Staubesand, Jochen (1921–2012): German anatomist born in Hammer and working in Freiburg.

Triad of Staubesand: *Eponym* for the triad of a venous perforator, a small artery and a cutaneous nerve coursing together through the muscular fascia in a limb. Reference: *Die Perforanten-Trias: Ein funktionelles System.* Phlebologie 1994; 23:447–458.

Stensen (Steno), Niels (1638–1686): Danish anatomist, geologist and theologican (Fig. S26) known for various structures in the human body. He studied in Copenhagen under Bartholin, in Leiden under Sylvius and in Amsterdam under Blasius to whom he demonstrated the parotid duct. He travelled a lot and became interested in the study of crystals. Finally, he was appointed in Copenhagen but later he became a Catholic bishop. Reference: *Observationes anatomicae quibus varia oris, oculorum, et narium vasa describuntur, novique salivae, lacrymarum et muci fontes deteguntur, et novum nobilissimi Bilsii de lymphae motu et usu commentum examinatur et rejicitur.* Leiden, 1662.

Canals of Stensen: *Eponym* for the paired incisive canals in the maxilla.

Fig. S26 Portrait of Stensen by Joost Sustermans (from the Museum Uffizi, Florence; in public domain)

Duct of Stensen: *Eponym* for the parotid duct, also described by Blasius in the same year.

Foramen of Stensen: *Eponym* for the paired incisive foramen of the maxilla.

Glands of Stensen: *Eponym* for the nasal glands.

Veins of Stensen: *Eponym* for the vorticose veins of the eyeball, also described by Ruysch and Anders Retzius.

Sterzi, Giuseppe Nazzareno (1876–1919): Italian anatomist born in Cittadella, Padua, and working in Pisa and Cagliari on comparative neuroanatomy of vertebrates and on the human nervous system (Fig. S27). He volunteered for the Italian Army as medical officer, and in 1919 as Director of the military hospital in Arezzo died of the postwar Spanish flu. References: *Il sistema nervoso centrale dei vertebrati,* 2 Vols (the second in 2 parts). Draghi, Padua, 1907, 1909, 1912; *Anatomia del sistema nervoso centrale dell'uomo,* 2 Vols. Draghi, Padua, 1914, 1915.

Of special interest is that he rediscovered the contributions of Fabrici and Casseri to the neuroloical sciences. In particular, he found the *Tabulae Anatomicae (Pictae)* of Fabrici in the Marciana Library in Venice, the former library of the Republic. Reference: *Le Tabulae Anatomiche ed i Codici Marciani con note autografe di Hieronymus Fabricius ab Aquapendente.* Anat Anz 1909; 35:338–348.

Apical lobe of Sterzi: *Eponym* for the quadrangular lobe of the cerebellum and the interposed vermis.

Hypothalamic perforant fibres of Sterzi: *Eponym* for a bundle of fibres from the globus pallidus to the subthalamic nucleus, currently known as pallidosubthalamic fibres.

Fig. S27 Photo Sterzi (courtesy Alessandro Riva, Cagliari)

Stieda, Alfred (1869–1945): German surgeon born in Dorpat (now Tartu, Estonia) and working in Königsberg.

Process of Stieda: *Eponym* for the posterior process of the talus. Reference: *Über secundäre Fusswurzelknochen.* Arch Physiol Wissensch Med 1869; pp 108–111.

Stilling, Benedict (1810–1879): German anatomist born in Kirchhain, Hessen, and working in Kassel and later in Vienna (Fig. S28a). He described many features of the spinal cord and cerebellum and greatly improved the illustration of brain sections. References: *Untersuchungen über die Funktionen des Rückenmarks und der Nerven.* Wigand, Leipzig, 1842; *Neue Untersuchungen über den Bau des kleinen Gehirns des Menschen.* Fischer, Kassel, 1878.

Canal or **Duct of Stilling:** *Eponym* for the hyaloid canal of the eyeball, also described by Cloquet.

Column of Stilling: *Synonym* for the column of Clarke of the spinal cord, also described as column of Stilling-Clarke.

Fibres of Stilling: *Eponym* for the reticular formation of the medulla oblongata, also described by Schroeder van der Kolk. Reference: *Ueber die Medulla oblongata.* Erlangen, 1843.

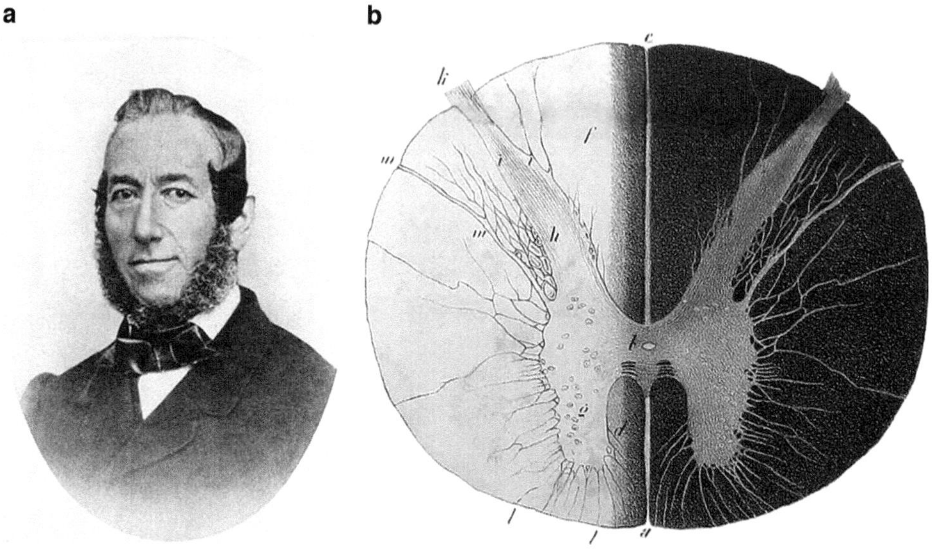

Fig. S28 (**a**) Portrait of Stilling (from Pagel 1901); (**b**) Section of the cervical spinal cord (C3–C4). *a* anterior median fissure, *b* central canal, *c* posterior median sulcus, *d* anterior funiculus, *f* posterior funiculus, *g* anterior horn, *h* posterior horn, *i* gelatinous substance, *k* posterior root, *l* anterior root (from Stilling B *Untersuchungen über den Bau des Nervensystems,* Vol 2. Enke, Stuttgart, 1843)

Fleece of Stilling: *Eponym* for the mesh of myelinated fibres surrounding the dentate nucleus of the cerebellum.

Nucleus of Stilling: *Eponym* for the dorsal or posterior thoracic nucleus of the spinal cord, also described as nucleus of Stilling-Clarke.

Substance of Stilling: *Synonym* for the gelatinous substance of the posterior or dorsal horn of the spinal cord, also described by Rolando (Fig. S28b).

Stohr, Philipp (1849–1911): German anatomist and histologist born and working in Würzburg.

Cells of Stohr: *Eponym* for cells in the pyloric gastric glands, also described by Nussbaum. Reference: *Lehrbuch der Histologie und der microscopischen Anatomie des Menschen mit Einschluss der mikroskopischen Technik.* Jena, 1887.

Stripes:

Stripe of Baillarger, Inner and outer: *Eponym* for the white lines or stripes of nerve fibres in layer V and IV of the isocortex, respectively; *see* Baillarger, Jules Gabriel François.

Stripe of Bechterew-Kaes: *Eponym* for a stripe of nerve fibres in layer II of the isocortex; *see* Bechterew, Vladimir Michailovich von.

Stripe of Gennari: *Eponym* for the occipital stripe, a conspicuous myelinated band in the visual cortex, also described by Vicq d'Azyr; *see* Gennari, Francesco.

Stripe of Kaes: *See* Stripe of Bechterew-Kaes.

Stripe(s) of Lancisi: *Eponym* for the longitudinal striae (medial and lateral) overlying the corpus callosum; *see* Lancisi, Giovanni Maria.

Stripe of Vicq d'Azyr: *Eponym* for the occipital stripe marking the striate cortex, also described by Gennari; *see* Vicq d'Azyr, Felix.

Stroud, Bert Brenette (19th–20th century): American instructor in physiology, zoology and neurology at Cornell University, born in Ithaca, New York.

Pecten of Stroud: *Eponym* for the anal pecten (the squamous anal zone) bordered orally by the pectinate line and aborally by the anocutaneous line of Hilton. Reference: *On the Anatomy of the Anus.* Ann Surg Philad 1896; 24:1–15.

Struthers, Sir John (1823–1899): Scottish anatomist born in Brucefield, Dunfermline, and working in Edinburgh and Aberdeen (Fig. S29).

Arcade of Struthers: *Eponym* for an arcade in the distal third of the arm on its medial aspect. Here, the ulnar nerve passes from the anterior to the posterior compartment of the arm, which may lead to ulnar nerve compression. Reference: *Anatomical and Physiological Observations,* Part I. Sutherland and Knox, Edinburgh.

Ligament of Struthers: *Eponym* for an inconstant fibrous band located on the medial side of the distal part of the humerus; this structure had been noted earlier by Tiedemann, Magendie and Knox (Dobson 1962). Reference: *On a peculiarity of the humerus and humeral artery.* Monthly J Med Sci 1848, New Series XXVIII; 264–267.

Fig. S29 Portrait of Struthers (from Wikimedia Commons; in public domain)

Substances:

Substance of Arnold, Reticular: *Eponym* for the white reticular substance of the entorhinal cortex; *see* Arnold, Friedrich.

Substance of Nissl: *Eponym* for the basophilic substance of neurons; *see* Nissl, Franz.

Substance of Reichert, Innominate: *Eponym* for the innominate substance of the basal forebrain; *see* Reichert, Karl Bogislaus..

Substance of Rolando, Gelatinous: *Eponym* for the gelatinous substance of the posterior or dorsal horn of the spinal cord, also described by Stilling; *see* Rolando, Luigi.

Substance of Schwann: *Eponym* for the myelin around nerve fibres; *see* Schwann, Theodor.

Substance of Soemmerring: *Eponym* for the substantia nigra; *see* Soemmerring, Samuel Thomas von.

Substance of Stilling: *Synonym* for the gelatinous substance of the posterior or dorsal horn of the spinal cord, also described by Rolando; *see* Stilling, Benedict.

Sucquet, JP (1840–1870): French physician and anatomist born and working in Paris.

Canals of Sucquet: *Eponym* for arteriovenous anastomoses, also described by Hoyer. Reference: *De la circulation du sang dans les membres et dans la tête chez l'homme.* Paris, 1860.

Fig. S30 Photo of Sudeck (from Teichmann W, Eggers C, Schröder H-J *100 Jahre Nordwestdeutscher Chirurgen.* Hamburg 2009, p 71; free to use)

Sudeck, Paul Hermann Martin (1866–1945): German surgeon born in Pinneberg, Holstein, and working in Hamburg (Fig. S30).

Point of Sudeck, Critical: *Eponym* for an area in the large intestine between the supply of the sigmoid arteries and that of the superior rectal artery, also described by Hartmann (1909). Reference: *Ueber die Gefässversorgung des Mastdarmes in Hinsicht auf die operative Gangrän.* Münch Med Wochensch 1906; 54:1314–1317.

Suzanne, Jean George (1859-?): French pathologist working in Bordeaux.

Gland of Suzanne: *Eponym* for the mucous gland found in the floor of the mouth close to the midline. Reference: *Recherches anatomiques sur le plancher de la bouche, avec études anatomiques et pathologiques sur la grenouillette commune ou sublinguale.* Arch Physiol Bordeaux 1887; 19:141–197.

Swammerdam, Jan (1637–1680): Dutch physician and theologist born and working in Amsterdam (Fig. S31). During a period in Paris, he discovered valves in lymphatic vessels.

Swan, Joseph English (1791–1875): English anatomist and surgeon born and working in Lincoln.

Foramina of Swan: *Eponym* for the foramina for the superior dental nerves.

Sylvius, François de le Boë (1614–1672): French anatomist born in Hanau who after extensive travels settled in Leiden (Fig. S32). He described the lateral sulcus of the

Fig. S31 Portrait of Swammerdam (from Wikimedia Commons; in public domain)

Fig. S32 Engraving of Sylvius, François de le Boë, by C. van Dalen Jr. (From the Dutch Iconographic Bureau; in public domain)

cerebral cortex and many related structures are also described as 'Sylvian'. Reference: *De spirituum animalium in cerebro cerebelloque confectione, per nervos distributione, atque usu vario.* Leiden, 1660.

Aqueduct of Sylvius: *Eponym* for the mesencephalic or cerebral aqueduct.

Artery of Sylvius: *Eponym* for the middle cerebral artery.

Cistern of Sylvius: *Eponym* for the subarachnoid cistern of the lateral cerebral fossa.

Fissure or **Sulcus of Sylvius:** *Eponym* for the lateral cerebral sulcus o the cerebral cortex. The name 'Sylvian' was given by Thomas Bartholin (1641; Fig. S33a). Collice et al. (2008) showed that the first illustration of the lateral sulcus was given by Fabrici in 1600 (Fig. S33b). *See* Collice M, Collice R, Riva A *Who discovered the Sylvian fissure?* Neurosurgery 2008; 63:623–628.

Fossa of Sylvius: *Eponym* for the lateral cerebral fossa (Sylvian fossa) of the cerebral cortex.

Valve of Sylvius: *Eponym* for the valve of the inferior vena cava, also described by Eustachi.

Vein of Sylvius: *Eponym* for the middle cerebral vein.

Ventricle of Sylvius: *Eponym* for the cavity of the septum pellucidum, also described by Duncan, Verga and Wenzel.

Sylvius, Jacobus (Jacques Dubois, 1478–1555): French classicist born in Louville, Amiens, who later became surgeon in Paris (Fig. S34a) and was a teacher of Vesalius. His *Jacobi Sylvii Medicae Rei* (Aegidium Gorbinum, Paris, 1561) forms part of his commentaries on Galen's work (Fig. S34b).

Bone of Sylvius: *Eponym* for the spatuliform process of the malleus, his lenticular process.

Symington, Johnson (1851–1924): Scottish anatomist in Belfast.

Body of Symington, Anococcygeal: *Eponym* for the anococcygeal ligament. Reference: *The Rectum and Anus.* J Anat (Lond) 1888; 23:106.

System of Havers: *Eponym* for osteon; see Havers, Clopton.

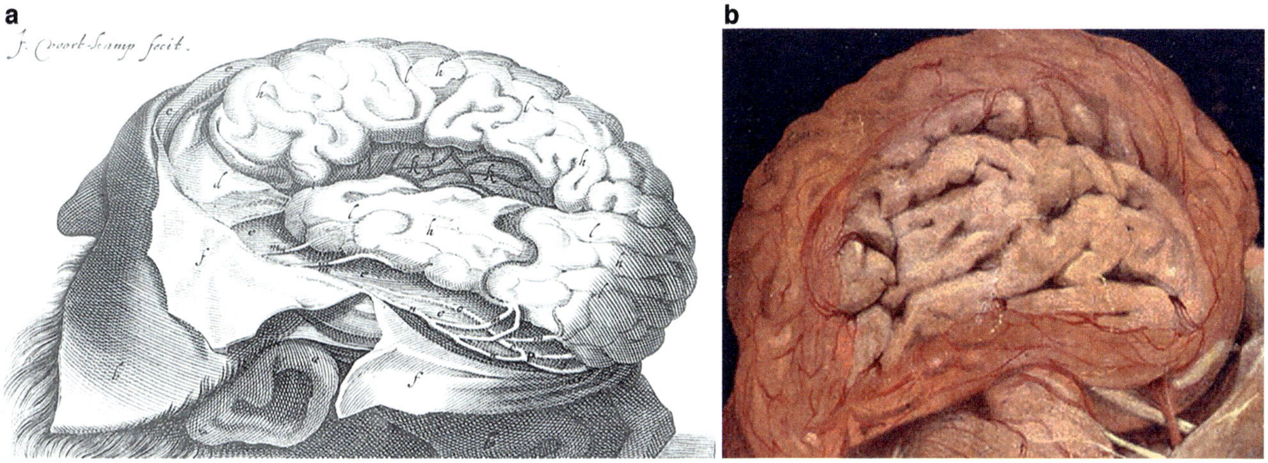

Fig. S33 (**a**) The lateral fissure of Sylvius as illustrated by Bartholin in 1641 (engraving by J. Voort Kamp); (**b**) the first accurate figure of the Sylvian fissure by Fabrici (Tabulae Pictae 112.10; both from Wikimedia Commons; in public domain)

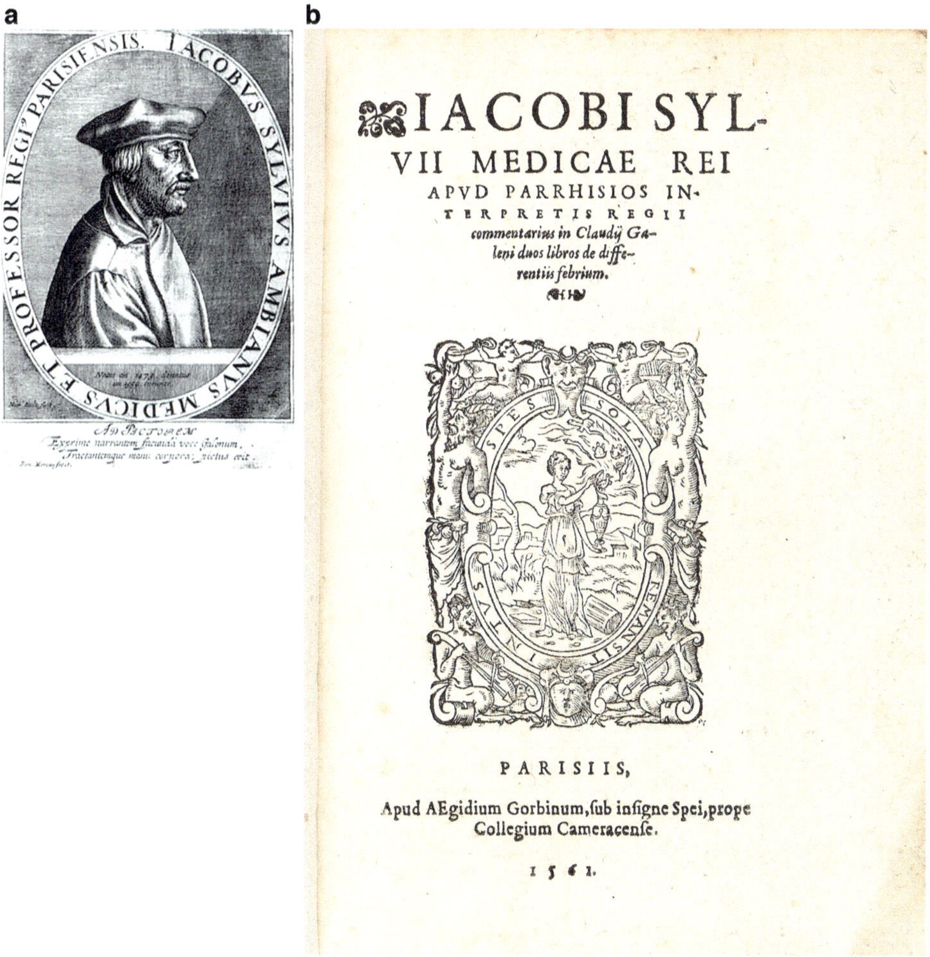

Fig. S34 (**a**) Engraving of Sylvius, Jacques (Wikipedia; free to use); (**b**) Title page *Jacobi Sylvii Medicae Rei.* Aegidium Gorbinum, Paris, 1561

18 T, U

Taillefer, Louis Auguste Horace Sydney Timoléon (1802–1868): French physician.

Valve of Taillefer: *Eponym* for a fold of the lower part of the nasolacrimal duct, also described by Hyrtl. Reference: *Quelques remarques sur la disposition anatomique du canal nasal suivies de la description d'un nouveau procédé pour la cure de la fistule lacrymale.* Arch Gen Med 1826; 11:438–444.

Tandler, Julius (1869–1936): Austrian anatomist born in Jihlava, Moravia, and working in Vienna. He was a pioneer in public health and a Socialdemocratic politician. After the antisemitic terror in Vienna in 1934, he emigrated to China and later to Russia.

Anastomosis of Tandler: *Eponym* for an hypothetic ventral longitudinal anastomosis of the embryonic blood vessels for the intestines. Reference: *Zur Entwicklungsgeschichte der mensclichen Darmarterien.* Anat Hefte 1904; 23:187–208; *Über die Varietäten der Arteria Coeliaca und deren Entwicklung.* Anat Hete 1904; 25:473–500.

Tarin, Pierre (1725–1761): French anatomist born in Courtenay and working in Paris and author of many articles in the *Paris Encyclopaedia* on the anatomy of the brain. Reference: *Dictionnaire anatomique. Suivi d'une bibliothèque anatomique et physiologique.* Briasson, Paris, 1753.

Fascia of Tarin: *Eponym* for the dentate gyrus of the cerebral cortex.

Fossa, Recess or **Space of Tarin:** *Eponym* for the interpeduncular fossa of the brain stem.

Valve of Tarin: *Eponym* for the inferior or posterior medullary velum of the roof of the fourth ventricle.

Tawara, Sunao (1873–1938): Japanese anatomist and pathologist born in Oita and working in Tokyo (Fig. T1). He worked together with Aschoff in Marburg.

Bundles of Tawara: *Eponym* for the right and left bundle branches of the atrioventricular bundle of the cardiac conducting system.

Node of Tawara: *Eponym* for the atrioventricular node, also known as the node of Aschoff-Tawara. Reference: *Ueber die sogenannten abnormen Sehnenfaden des Herzens.* Beitr Path Anat Allg Pathol 1906; 39:563–584.

Teichmann, Ludwig Karl (1823–1895): German anatomist born in Lublin and working in Krakau.

Plexus or **Networks of Teichmann:** *Eponym* for superficial and deep lymphatic plexuses in the wall of the stomach. Reference: *Das Saugadersystem vom anatomischen Standpunkt.* Leipzig, 1861.

Tenon, Jacques René (1724–1816): French surgeon born in Sépeaux, Ivogny, and working in Paris (Fig. T2a). He specialized in ophthalmology. Reference: *Mémoires sur l'anatomie, la pathologie et la chirurgie et principalement sur l'organ de l'oeil.* Paris, 1816.

Capsule of Tenon: *Eponym* for the bulbar fascia of the eyeball.

Space of Tenon: *Eponym* for the episcleral space, the space between the fascial sheath of the eyeball and the sclera (Fig. T2b).

Terni, Tullio (1888–1946): Italian histologist and anatomist born in Livorno and working in Padua. He was a victim of Italian Fascism, despite being a Jewish Member of the Party. After the war, he was removed from the Academia

H. J. ten Donkelaar et al., *An Illustrated Guide to Anatomical Eponyms*, https://doi.org/10.1007/978-3-031-91664-9_18

Fig. T1 Photo of Tawara (from Wikimedia Commons; in public domain)

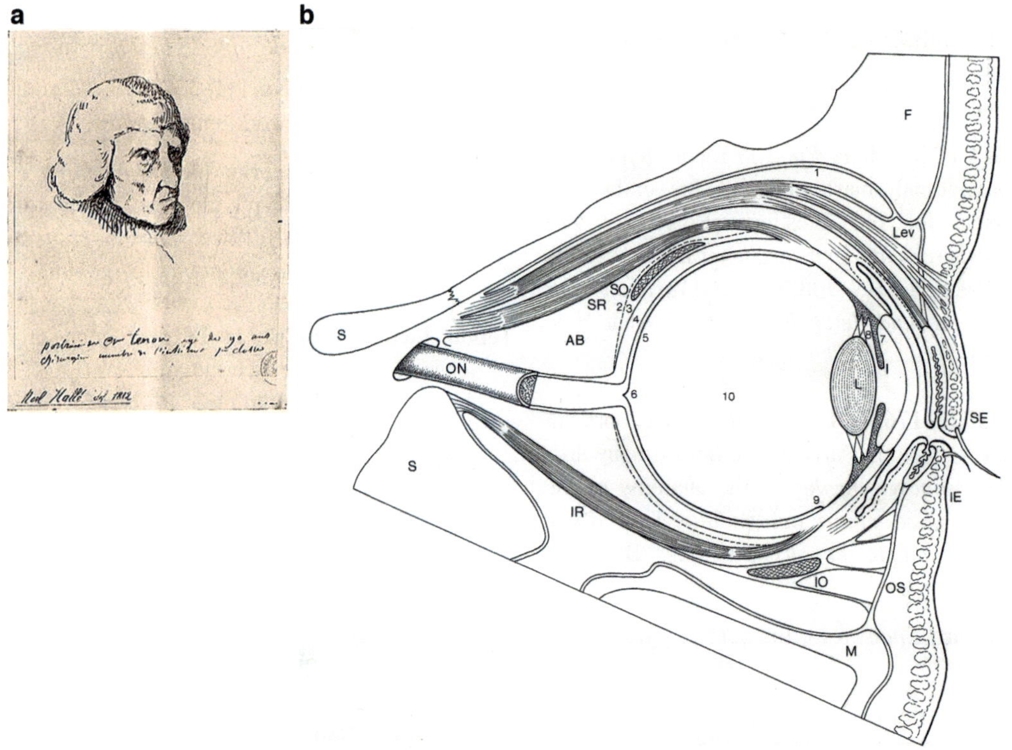

Fig. T2 (**a**) Sketch of Tenon (from Wikimedia Commons; in public domain); **b** Capsule and space of Tenon (from ten Donkelaar HJ, Kachlík D, Tubbs RS *An Illustrated Terminologia Neuroanatomica.* Springer, Cham, 2018), shown in an overview of the eye (after ten Donkelaar HJ, Oostra R-J *Klinische Anatomie en Embryologie,* 4th ed. Springer Media, Houten, 2014). *AB* adipose body, *F* frontal bone, *I* iris, *IE* inferior eyelid, *IO* inferior oblique muscle, *IR* inferior rectus muscle, *L* lens, *Lev* levator palpebrae superioris muscle, *M* maxilla, *ON* optic nerve, *OS* orbital septum, *S* sphenoid bone, *SE* superior eyelid, *SO* superior oblique muscle, *SR* superior rectus muscle, *1* periorbita, *2* vagina bulbi, *3* episcleral space, *4* sclera, *5* retina, *6* optic disc, *7* anterior chamber, *8* posterior chamber, *9* ora serrata, *10* corpus vitreum

Nazionale dei Lincei for his membership and committed suicide. He should be remembered for his studies on the intermediolateral cell column (Fig. T3). For biographical data, see Macchi V, Porzionato A, Stecco C, De Caro R *Tullio Terni (1888–1946): The "column" of spinal cardiovascular regulation.* Clin Anat 2013; 26:544–546; Thiene G, Paradiso B, Zanotta A, Basso C, Zampieri F *Tullio Terni (1888–1946): The life of a neurocardioanatomist with a tragic epilogue.* Int J Cardiol 2019; 289:153–156.

Column of Terni: *Eponym* for the intermediolateral cell column from T1-L2, which he described in chicken embryos. Reference: *Richerche anatomiche sul sistema nervoso autonomo degli ucelli.* Arch Ital Anat Embriol 1923; 20:433–510.

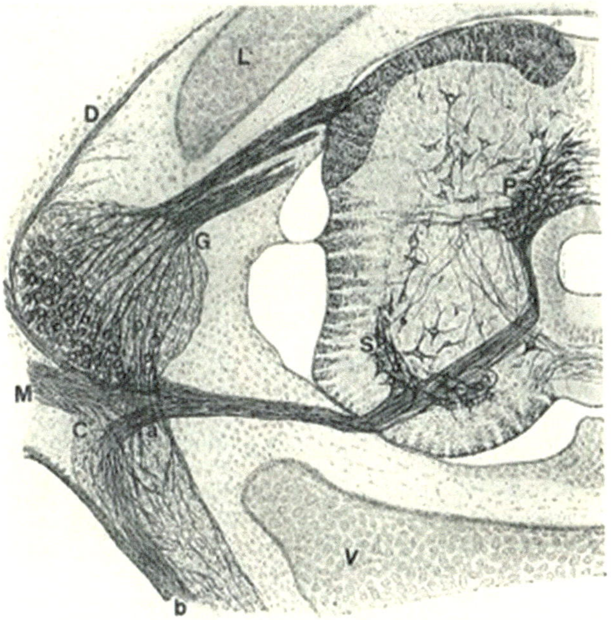

Fig. T3 The intermediolateral cell column at Th5 in the chicken embryo (from Terni T *Richerche anatomiche sul sistema nervoso autonomo degli ucelli.* Arch Ital Anat Embriol 1923; 20:433–510)

Terrier, Louis Felix (1837–1908): French surgeon in Paris.

Valve or **Promontory of Terrier:** *Eponym* for the 'proximal valve' of the gallbladder, a fold between the gallbladder and the cystic duct. Reference: *Cathéterisme des voies biliaires.* Rev. Chir Paris 1891; 11:648–661, 12:136–168 (with M. Dally).

Terson, Albert (1867–1935): French ophthalmologist born and working in Paris.

Glands of Terson: *Eponym* for the accessory lacrimal glands, described before by the Krauses. Reference: *Les glandes lacrymales conjunctivales et orbito-palpébrales; l'ablation des glandes lacrymales palpébrales.* Thesis, Paris, 1892.

Testut, Jean Léon (1849–1925): French anatomist born in Beaumont and working in Bordeaux and Lyon (Fig. T4). Reference: *Traité d'anatomie humaine.* Doin, Paris, 1891, 1894; 5th ed 1905; latest ed. 1948 (with A. Latarjet). To show the excellent descriptions and illustrations from his textbook, examples are given in Figs. T5 and T6.

Arcuate tract of Testut: *Old term* for the septomarginal trabecula or moderator band of the right ventricle of the heart, a band already illustrated by Leonardo, first named by King, and described as ansiform band by Poirier.

Artery of Testut: *Eponym* for the inferior pancreatic artery.

Central vein of Testut: *Eponym* for the principal suprarenal vein.

Fascia of Testut: *Eponym* for the middle layer of the cervical fascia, his 'aponévrose thoraco-hyoïdenne', also described by Gruber, Paulet and Richet.

Fig. T4 Photo of Testut (from Pagel 1901)

Ligament of Testut: *Eponym* for the radioscapholunate ligament of the wrist.

Lumboovarian ligament of Testut: *Synonym* for the suspensory ligament of the ovary, also known as the infundibulopelvic ligament of Henle.

Teutleben, Friedrich Ernst Karl von (1842–?): German anatomist in Leipzig.

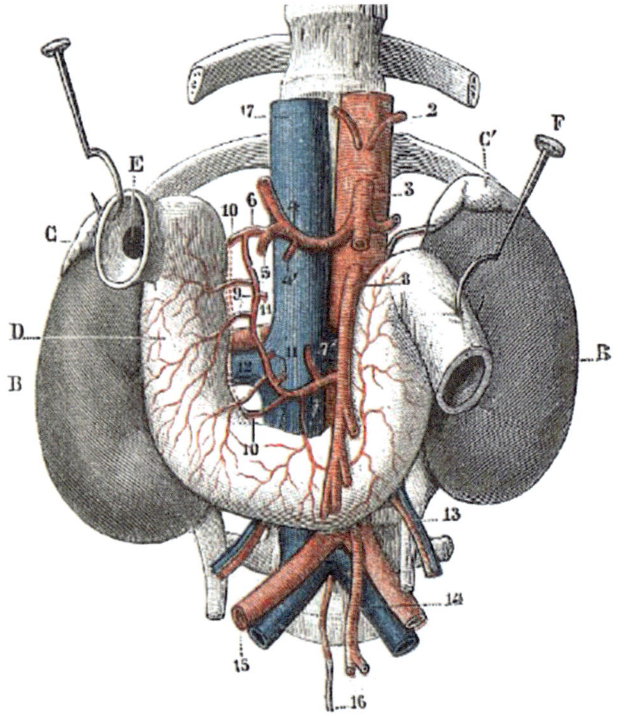

Fig. T5 Arteries of the duodenum (from Testut JL *Traité d'anatomie humaine, Tome III: Organes des sens, appareil de la respiration, appareil uro-génital.* Doin, Paris, 1894). *3* Coeliac trunk, *4* hepatic artery, *5* right gastroomental (gastroepiploic) artery, *6* superior pancreaticoduodenal artery, *7* inferior pancreatic artery, *8* superior mesenteric artery, *9* anterior pancreaticoduodenal arcade, *10, 10′* posterior pancreaticoduodenal arcade, *14* inferior mesenteric artery, *15* common iliac artery, *16* median sacral artery, *17* inferior vena cava

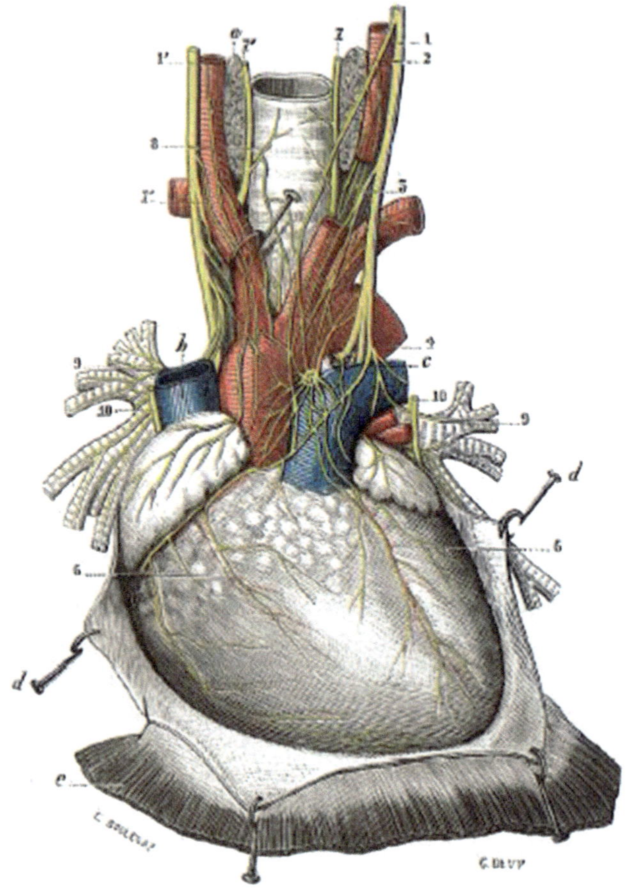

Fig. T6 The cardiac plexus (from Testut JL *Traité d'anatomie humaine, Tome II: Angéiologie, névrologie.* Doin, Paris, 1891). *1, 1′*, Left and right vagus nerve, *2* superior cardiac nerve, *3* middle cardiac nerve, *4* inferior cardiac nerve, *5* ganglion and plexus cardiacus, *6* afferent branches of cardiac plexus, *7, 7′*, left and right recurrent nerve, *8* tracheal branches, *9* anterior pulmonary plexus, *10* phrenic nerve, *a* thyroid gland, *b* superior vena cava, *d* pericard, *e* diaphragm

Ligament of Teutleben: *Eponym* for the pulmonary ligament of the pleura. Reference: *Die Ligamenta suspensoria diaphragmatis des Menschen.* Arch Anat Physiol (Lpz) 1877, pp. 281–287.

Thebesian veins: *Eponym* for the smallest cardiac veins; *see* Thebesius, Adam Christian.

Thebesius, Adam Christian (1686–1732): German anatomist and pathologist born in Sandenwalde, Silesia (currently Sadowel, Poland), and working in Leiden and later in Hirschberg, Silesia (currently Jelenia Góra, Poland), as city physician (Fig. T7). Reference: *De circulo sanguinis in corde.* Elsevier, Leiden, 1708.

Foramina of Thebesius: *Eponym* for the openings of the smallest cardiac veins into the right atrium, also described by Vieussens and Lannelongue (Fig. T8).

Valve of Thebesius: *Eponym* for the valve of the coronary sinus.

Veins of Thebesius: *Eponym* for the smallest cardiac veins (Fig. T8).

Theile, Friedrich Wilhelm (1801–1879): German anatomist born in Buttstadt and working in Jena and Bern, and later as physician in Weimar (Fig. T9).

Angle of Theile: *Eponym* for the acute angle at the confluence of the superior mesenteric and splenic veins, which represent the origin of the portal vein.

Canal of Theile: *Eponym* for the transverse sinus of the pericardial cavity (Fig. T10).

Glands of Theile: *Eponym* for mucous glands in the wall of the cystic duct and the neck of the gallbladder.

Muscle of Theile: *Eponym* for the superficial transverse perineal muscle, also described by Lesshaft.

Pericardiophrenic arteries of Theile: *Eponym* for branches of the superior phrenic artery supplying the pericardium.

Fig. T7 Engraving of Thebesius (from Wikimedia Commons; in public domain)

Thierry, L (20th century): French surgeon.

Vein of Thierry, Perforating: *Eponym* for the popliteal perforating vein, the vein of the popliteal fossa submerging under the fascia. Reference: *Anatomie chirurgicale de la fosse poplitée.* Phlébologie 1986; 39:57–66.

Thoma, Richard (1847–1923): German pathologist born in Bonndorf and working in Heidelberg, Dorpat, and Magdeburg. He invented a microtome.

Ampulla of Thoma: *Eponym* for the terminal expansion of the interlobular artery in the splenic pulp. Reference: *Die Ueberwanderung farbloser Blutkörper von dem Blut in das Lymphgefäss-System.* Heidelberg, 1873.

Thomson, Alexander (1802–1838): English anatomist in London and later in Paris.

Fascia of Thompson: *Eponym* for the iliopubic tract. **Fascia** or **Bandalette of Thomson:** *Eponym* for the iliopectineal fascia.

Tract or **Ligament of Thomson:** *Eponym* for the iliopubic tract, described variously: as a strong band attached to the inguinal ligament and as an aponeurotic band between the iliac fascia and the superior ramus of the pubic bone.

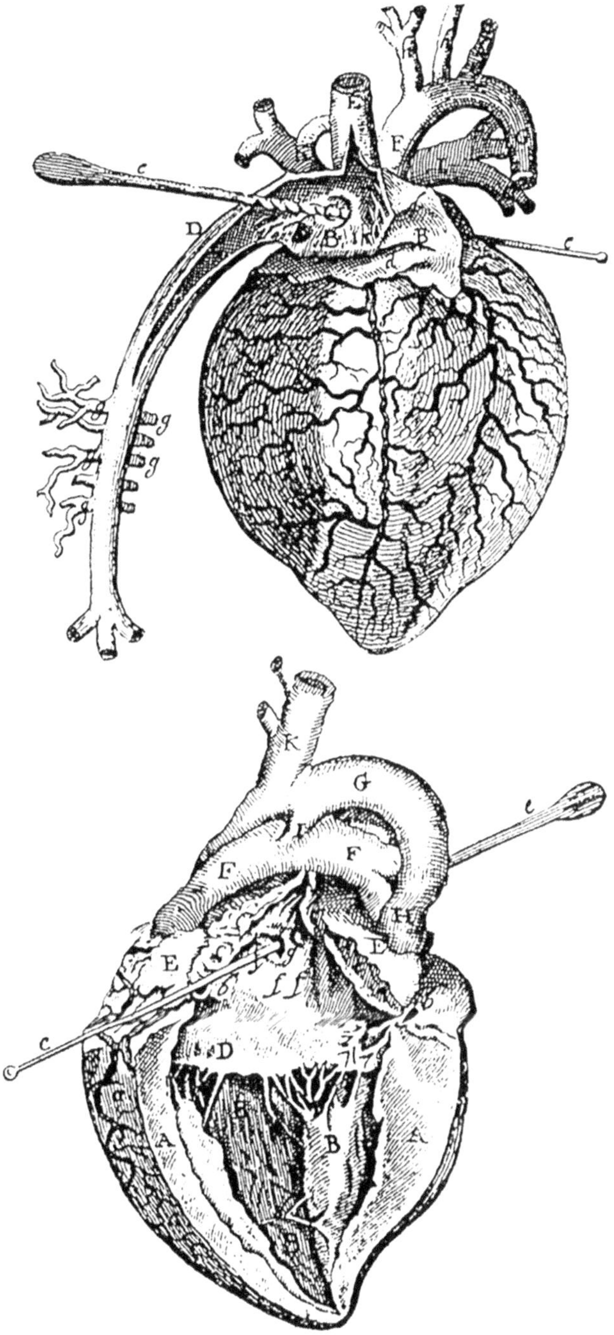

Fig. T8 Thebesian veins (from Rafael Coello, Ecuador, on Research Gate; courtesy Rafael Coello, Guayaquil, Ecuador)

Thorel, Charles (1868–1935): German pathologist.

Bundle of Thorel: *Eponym* for the posterior internodal bundle of the cardiac conducting system between the sinuatrial and atrioventricular nodes (*see* Fig. I5).

Fig. T9 Portrait of Theile (from Wikimedia Commons; in public domain)

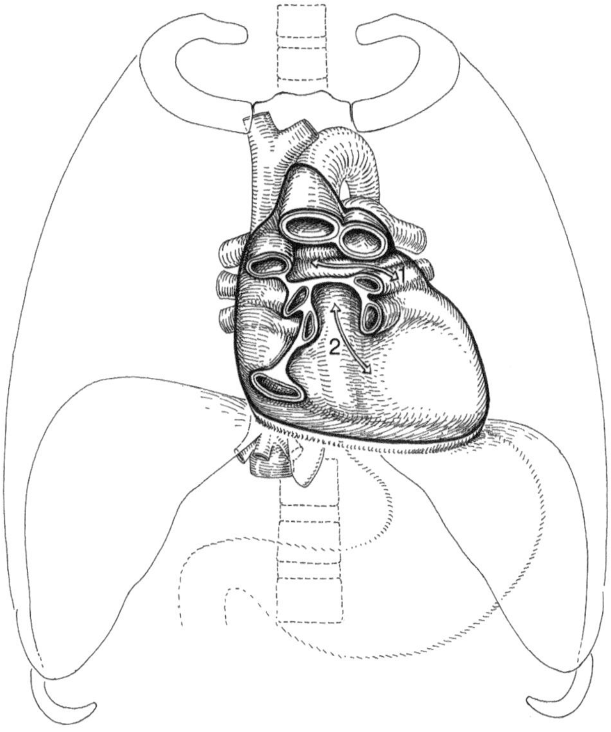

Fig. T10 Canal of Theile (after ten Donkelaar HJ, Oostra R-J *Klinische Anatomie en Embryologie,* fourth ed. Springer Media, Houten, 2014). *1* Transverse sinus or canal of Theile, *2* oblique sinus

Reference: *Über den Aufbau des Sinusknotens und seine Verbindung mit der Cava superior und den Wenckebachschen Bündeln.* Münch Med Wochenschr 1910; 57:183–186.

Fig. T11 Portrait of Tiedemann (from Wikimedia Commons; in public domain)

Tiedemann, Friedrich (1781–1861): German embryologist born in Kassel and working in Landshut and Heidelberg (Fig. T11). Reference: *Anatomie und Bildungsgeschichte des Gehirns im Foetus des Menschen nebst einer vergleichenden Darstellung des Hirnbaues in den Thieren.* Steinisch, Nuremberg, 1816.

Gland of Tiedemann: *Eponym* for the greater vestibular gland, also described by Bartholin and Duverney. Reference: *Von den Duverney'schen, Bartholin'schen und Cowper'schen Drüsen des Weibs und der schiefen Gestaltung und Lage der Gebärmutter.* Heidelberg and Leipzig, 1840.

Nerve of Tiedemann: *Eponym* for the sympathetic nerve plexus surrounding the central retinal artery.

Tigri, Atto (1813–1875): Italian anatomist born in Pistoia and working in Pisa as assistant to Filippo Civinini and later in Siena (Fig. T12).

Reticulum of Tigri: *Eponym* for the reticular framework of the spleen, which acts as a filter for foreign particles (his 'microscopic texture'). Reference: *Nuova disposizione dell'apparecchio vasculare e sanguigno della milza umana.* Annali universali di Medicina 1861; 42, p 544.

Tillaux, Paul Jules (1834–1904): French surgeon born in Aunay-sur-Odon, Calvados, and working in Paris.

Fig. T12 Portrait of Tigri (courtesy Marina Quartu, Cagliari)

Tubercle of Tillaux: *Eponym* for the anterior tibial tubercle, a small prominence on the distal end of the tibia, to which the anterior tibiofibular ligament inserts, also described by Chaput. Reference: *Traité d'anatomie topographique avec applications à la chirurgie*, 2 Vols. Paris, 1875–1877.

Tod, David (1792–1856): English surgeon.

Muscle of Tod: *Eponym* for the oblique auricular muscle. Reference: *The Anatomy and Physiology of the Organ of Hearing*. London, 1832.

Todaro, Francesco (1839–1918): Anatomist born in Turi, Sicily, and working in Messina and finally in Rome (Fig. T13a).

Tendon of Todaro: *Eponym* for an inconstant tendinous structure extending from the right fibrous trigone of the heart towards the valve of the inferior vena cava (Fig. T13b). Reference: *Novelle ricerche sopra la struttura muscolare delle orecchiette del cuore umano et sopra la valvola di Eustachio*. Florence, 1885.

Todd, Robert Bentley (1809–1860): Irish physiologist and anatomist born in Dublin and working in London, and editor of the *Cyclopedia of Anatomy* (1836–1839).

Process of Todd, Ascending: *Synonym* for the fascia of Scarpa, the membranous layer of the subcutaneous tissue (superficial fascia) of the anterolateral abdominal wall.

Toldt, Karl (1840–1920): Austrian anatomist born in Bruneck, South Tyrol, and working in Prague and later Vienna (Fig. T14). Reference: *Lehrbuch der Gewebelehre mit vorzugsweise Berücksichtigung des Körpers*. Stuttgart, 1877.

Fascia of Toldt: *Eponym* for the retropancreatic fascia behind the body of the pancreas.

Membrane of Toldt: *Eponym* for the prerenal layer or fascia, the anterior layer of the renal fascia (Fig. T15).

White line of Toldt: *Eponym* for an avascular plane for incision of the visceral peritoneum of the colon to the parietal peritoneum of the lateral abdominal wall.

Tomes, Sir Charles Sissmore (1846–1928): English anatomist and dentist in London, son of John Tomes.

Processes of Tomes: *Eponym* for inconstant apical processes of ameloblasts in the developing enamel. Reference: *On the development of marsupial and other tubular enamels with notes upon the development of enamel in general*. Phil Trans 1897; 189:107–122.

Tomes, Sir John (1815–1895): English anatomist born and working in London, father of Charles Siismore Tomes. He wrote many papers on comparable odontology.

Fibres of Tomes: *Eponym* for dentinal fibres, projections of odontoblasts, which run inside dentinal tubules perpendicular to the surface of a tooth.

Layer of Tomes: *Eponym* for the granular layer of dentin (*see* Fig. 23.1).

Tonsils:

Tonsil of Gerlach: *Eponym* for the tubal tonsil; *see* Gerlach, Joseph.

Tonsil of Luschka: *Eponym* for the pharyngeal tonsil; *see* Luschka, Hubert von.

Topinard, Paul (1830–1912): French anthropologist born on Isle Adam and working as Curator of the Museum of the Anthropological Society of Paris, founded by Broca.

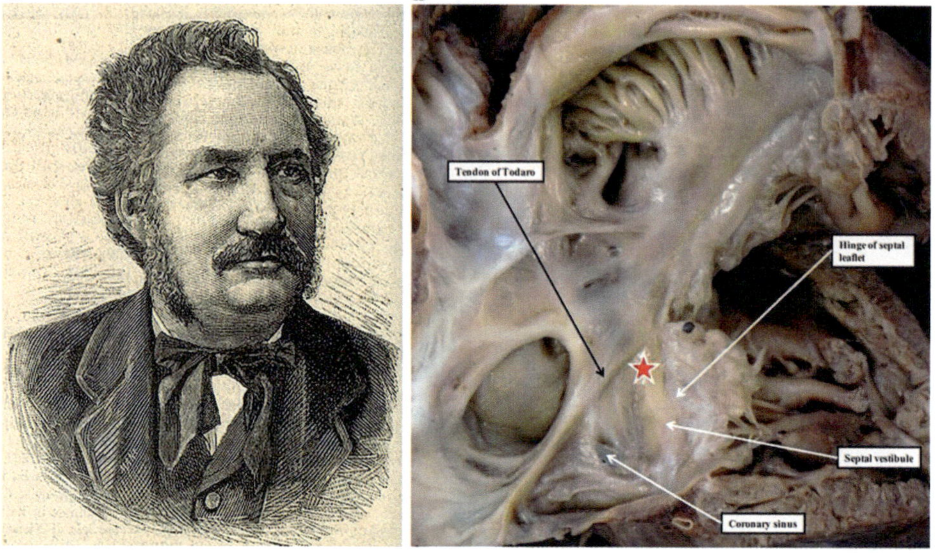

Fig. T13 (**a**) Portrait of Todaro (from Wikimedia Commons; in public domain). (**b**) Tendon of Todaro (from Anderson RH, Mori S, Spicer DE, Sanchez-Quintana D, Jensen B *The anatomy, development, and evolution of the atrioventricular conduction axis.* J Cardiovasc Dev Dis 2018; 5:44; Open Access, CCBY; courtesy Bjarke Jensen, Amsterdam). The *red asterisk* marks the position of the atrioventricular node

Fig. T14 Portrait of Toldt (courtesy Ondřej Naňka, Prague)

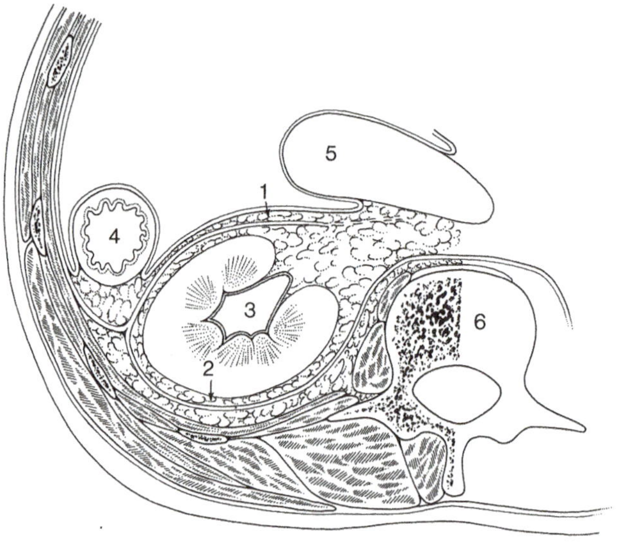

Fig. T15 The prerenal fascia of Toldt (after ten Donkelaar HJ, Oostra R-J *Klinische Anatomie en Embryologie,* 4th ed. Springer Media, Houten, 2014). *1* prerenal fascia, *2* retrorenal fascia, *3* renal pelvis, *4* ascending colon, *5* duodenal flexure, *6* vertebra L2

Angle of Topinard, Facial: *Modification* of the facial angle of Camper, formed between a line joining the glabella and the anterior nasal spine, and another joining the anterior nasal spine and the external acoustic meatus.

Torcular Herophili: *Old term* for the confluence of sinuses.

Tourneux, Frédéric (1852–1922): French embryologist born in Courson, Yonne, and working in Toulouse.

Spur of Tourneux: *Eponym* for the urorectal septum in embryos, which is the fused extraembryonic mesoderm of the allantoic diverticulum and the umbilical vesicle (O'Rahilly R, Müller F *Human Embryology & Teratology,* 3rd ed. Wiley-Liss, New York, 2001). Reference: *Sur les premiers développements du cloaque, du tubercle génital et de l'anus chez l'embryon de mouton.* J Anat (Paris) 1888; 24:503–517.

Tourtual, Kaspar Theobald (1802–1865): German anatomist born in Münster and working in Duisburg and later Münster. Reference: *Neue Untersuchungen ueber den Bau des menschlichen Schlund- und Kehlkopfs, mit vergleichend-anatomische Bemerkingen.* Leipzig, 1846.

Canal of Tourtual: *Eponym* for the major palatine canal.

Fold of Tourtual: *Eponym* for the salpingopharyngeal fold.

Ligament of Tourtual: *Eponym* for the anterior mallear ligament.

Membrane of Tourtual: *Eponym* for the quadrangular membrane of the larynx.

Sinus of Tourtual: *Eponym* for the supratonsillar fossa, also described as palatine recess by Killian.

Toynbee, Joseph (1815–1866): English surgeon and otologist born in Heckington and working in London.

Corpuscles of Toynbee: *Eponym* for the corneal corpuscles, connective tissue cells found between the fibrous layers of the cornea, also described by Virchow.

Muscle of Toynbee: *Synonym* for (a) the tensor tympani muscle, already well known and given its name by Albinus. Reference: *On the structure of the membrana tympani in the human ear.* Phil Trans 1851, pp. 159–168; and (b) the rectococcygeus muscle.

Tracts:

Tract of Arnold: *Eponym* for the frontopontine tract; *see* Arnold, Friedrich.

Tract of Bechterew: *Eponym* for (a) the central tegmental tract, and (b) the spinoolivary tract; *see* Bechterew, Vladimir Michailovich.

Tract of Bruce: *Eponym* for the septomarginal tract of the spinal cord, also described by Flechsig, Muir and Testut, and in the sacral cord as triangle of Philippe-Gombault; *see* Bruce, Alexander.

Tract of Burdach: *Eponym* for the cuneate fasciculus of the spinal cord; *see* Burdach, Karl Friedrich.

Tract of Collier: *Eponym* for the medial longitudinal fasciculus of the brain stem; *see* Collier, James S.

Tract of Deiters: *Eponym* for the lateral vestibulospinal tract; *see* Deiters, Otto Friedrich Karl.

Tract of Dejerine: *Eponym* for the anterior or ventral spinothalamic tract; *see* Dejetine, Joseph Jules..

Tract of Edinger: *Eponym* for the lateral spinothalamic tract; *see* Edinger, Ludwig.

Tract of Flechsig: *Eponym* for (a) the posterior or dorsal spinocerebellar tract; and (b) the septomarginal tract; *see* Flechsig, Paul Emile.

Tract of Goll: *Eponym* for the gracile fasciculus of the spinal cord; *see* Goll, Friedrich.

Tract of Gowers: *Eponym* for the anterior or ventral spinocerebellar tract; *see* Gowers, William Richard.

Tract of Gratiolet: *Eponym* for the geniculocalcarine tract (the optic radiation); *see* Gratiolet, Louis Pierre.

Tract of Gudden: *Eponym* for the mammillotegmental tract; *see* Gudden, Bernhard von.

Tract of Helweg, Triangular: *Eponym* for the olivospinal tract. It stands out in Weigert-Pal preparations at the anterolateral border of the lower medulla oblongata and the first five cervical segments (Helweg H 1888 Arch Psychiatr Nervenkr 19:14–183; Smith MC, Deacon P 1981 Brain 104:249–277), but not when stained with the Klüver-Barrera technique. It consists mainly of extremely fine, myelinated fibers and, therefore, stands out as an area of pallor in the anterior columns of the cervical spinal cord and lower medulla oblongata. It may be reticulospinal, but certainly not an olivospinal tract (*see* Smith and Deacon 1981 for further discussion).

Tract of Lissauer: *Eponym* for the posterolateral or dorsolateral fasciculus of the spinal cord, also desribed by Waldeyer; *see* Lissauer, Heinrich.

Tract of Loewenthal: *Eponym* for the tectospinal tract; *see* Loewenthal, Wilhelm.

Tract of Löwy: *Eponym* for the angular bundle or peduncle of the flocculus; *see* Löwy, R.

Tract of Maissiat: *Eponym* for the iliotibial tract; *see* Maissiat, Jacques Henri.

Tract of Marchi: *Eponym* for the anterolateral descending cerebellar tract, *see* Marchi, Vittorio.

Tract of Meynert: *Eponym* for the habenulointerpeduncular tract; *see* Meynert, Theodor.

Tract of Monakow: *Eponym* for the rubrospinal tract; *see* Monakow, Constantin von.

Tract of Muir: *Eponym* for the septomarginal tract of the spinal cord, also described by Bruce, Flechsig and Testut; *see* Muir, Robert.

Tract of Münzer and Wiener: *Eponym* for the tectopontine tract; *see* Münzer, Egmont, and Wiener, H.

Tract of Muskens: *Eponym* for the interstitiospinal tract; *see* Muskens, Louis Jacob Joseph.

Tract of Pavlov: *Eponym* for the tectoreticular tract, a tract from the superior colliculus to the midbrain reticular formation and the nuclei of the oculomotor and trochlear nerves; *see* Pavlov, Vladimir Alexejevic.

Tract of Probst: *Eponym* for the longitudinal callosal bundle in the cerebral cortex present in the absence of the corpus callosum; *see* Probst, Moritz.

Tract of Reil: *Eponym* for the medial lemniscus ('Schleife' of Reil); *see* Reil, Johann Christian.

Tract of Sachs: *Eponym* for the transverse occipital tract, one of the lobar fibre tracts of the occipital lobe; see Sachs, Heinrich.

Tract of Schultze, Comma: *Eponym* for the interfascicular fasciculus of the spinal cord (the comma tract); *see* Schultze, Friedrich.

Tract of Testut, Arcuate: *Old term* for the septomarginal tract of the spinal cord, also described by Bruce, Muir and Flechsig; *see* Testut, Jean Léon.

Tract of Thomson: *Eponym* for the iliopubic tract of the transversalis fascia; *see* Thomson, Alexander.

Tract of Türck: *Eponym* for the temporal corticopontine tract; *see* Türck, Ludwig.

Tract of Türck, Anterior: *Eponym* for the anterior or ventral corticospinal tract; *see* Türck, Ludwig.

Tract of Vialet: *Eponym* for the transverse occipital tract of the lingual lobe of the occipital lobe; *see* Vialet, N.

Tract of Vicq d'Azyr: *Eponym* for the mammillothalamic tract; *see* Vicq d'Azyr, Felix.

Tract of Waldeyer: *Eponym* for the posterolateral or dorsolateral fasciculus of the spinal cord, also described by Lissauer; *see* Waldeyer, Heinrich Wilhelm Gottfried.

Tract of Wallenberg: *Eponym* for the posterior or dorsal trigeminothalamic tract; see Wallenberg, Adolf.

Traube, Ludwig (1818–1876): German physician and pathologist born in Ratisborn, Silesia (now Racibórz, Poland), and working in Berlin (Fig. T16).

Space of Traube: *Eponym* for the semilunar area on the chest wall over which resonance of the stomach is obtained, comparable to the triangle of Labbé. Reference: *Bemerkungen ueber der halbmondformigen Raum.* Berlin Klin Wschr 1868; 5:510.

Trautmann, Moritz Ferdinand (1832–1902): German ear surgeon born in Wittenberg and working in Berlin.

Triangle of Trautmann: *Eponym* for an area of the temporal bone bounded by the sigmoid sinus, the superior petrosal sinus and the vestibulocochlear nerve, facing the cerebellopontine angle, used during posterior transpetrosal approaches to the posterior cerebral fossa. Reference: *Embolische Processe des Mittelohrs.* Berlin, 1886.

Fig. T16 Portrait of Traube (from Pagel 1901)

Fig. T17 Portrait of Treitz (from Wikipedia; in public domain)

Treitz, Václav or **Wenzel (1819–1872):** Bohemian physician and pathologist born in Hostomice, Bohemia, and working in Krakau and later Prague (Fig. T17).

Arcade of Treitz: *Eponym* for a vascular arcade formed by the inferior mesenteric and the left superior colic artery in the left mesocolon.

Fascia or **Membrane of Treitz:** *Eponym* for the retropancreatic fascia behind the head of the pancreas.

Fossa or **Recess of Treitz:** *Eponym* for (a) the superior and inferior duodenal recesses of the peritoneal cavity (Fig. T18a); (b) the subcaecal fossa; and (c) the inter-

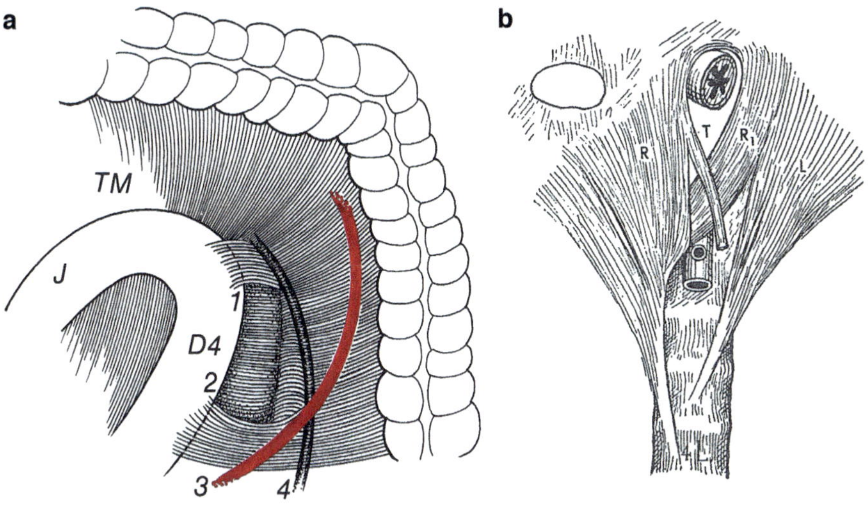

Fig. T18 (**a**) Recessus of Treitz (after Hollinshead WH *Anatomy for Surgeons, Vol 2: The Thorax, Abdomen, and Pelvis,* 2nd ed. Harper & Row, New York, 1971); *D4* fourth part of duodenum, *J* jejunum, *TM* transverse mesocolon, *1* superior recess of Treitz, *2* inferior recess of Treitz, *3* left colic artery, *4* inferior mesenteric vein; (**b**) Suspensory muscle (ligament) of the duodenum, attached to the diaphragm (from Low A *A note on the crura of the diaphragm and the muscle of Treitz.* J Anat & Physiol 1907; 42:93–96; in public domain); *L* left crus, *R*, R_1 parts of right crus, *T* ligament of Treitz

sigmoid recess. Reference: *Hernia retroperitonealis. Ein Beitrag zur Geschichte innerer Hernien.* Credmar, Prague, 1857.

Ligament of Treitz: *Eponym* for the suspensory ligament of the duodenum. Reference: *Ueber einen neuen Muskel am Duodenum des Menschen, ueber elastische Sehnen, und einige andere anatomische Verhältnisse.* Prager Vjschr Prakt Heilk 1853; 37:113–144.

Muscle of Treitz: *Eponym* for (a) the suspensory ligament of the duodenum that may contain muscle fibres (Fig. T18b); (b) the rectococcygeal muscle; and (c) the subcutaneous anal muscle.

Treves, Sir Frederick (1853–1923): English surgeon born in Dorchester and working in London, a pioneer in abdominal surgery.

Bloodless fold or **Fossa of Treves:** *Eponym* for the bloodless inferior ileocaecal fold of the vermiform appendix of the caecum, which forms the inferior ileocaecal fossa or recess. Reference: *The anatomy of the intestinal canal and peritoneum in Man.* Hunterian Lectures, London, 1885.

Triangles: *See* **The Classical Anatomical Eponyms.**

Tripus of Haller: *Eponym* for the coeliac trunk; *see* Haller, Albrecht von.

Troeltsch, Anton Friedrich Freiherr von (1829–1890): German otologist born in Schwabach and working in Nürnberg (Fig. T19a). He is known for his studies on the middle ear and its diseases. References: *Die Krankheiten des Ohres.* Würzburg, 1862; *Lehrbuch der Ohrenheilkunde.* Würzburg, 1868.

Corpuscles of Troeltsch: *Eponym* for minute spaces, resembling corpuscles, between the radial fibres of the tympanic membrane.

Fascia of Troeltsch: *Eponym* for the salpingopharyngeal fascia between the lateral aspect of the tubal cartilage and the salpingopharyngeus muscle.

Folds of Troeltsch: *Eponym* for the anterior and posterior folds of the tympanic membrane (Fig. T19b), described also by Rivinus.

Recesses of Troeltsch: *Eponym* for the anterior and posterior recesses of the tympanic membrane (Fig. T19b).

Trolard, Paulin (1842–1910): French anatomist born in Sedan and working in Algiers (Fig. T20). He is known for his studies on the veins of the cerebral cortex. Reference: *Recherches sur l'anatomie du système veineux de l'encéphale et du crâne.* Thesis, Paris, 1868.

Hypoglossal venous plexus of Trolard: *Eponym* for a dense venous plexus at the extracranial aspect of the hypoglossal canal connecting veins around the occipital condyle with the jugular bulb.

Inferior preoccipital vein of Trolard: *Eponym* for a vein connecting the venous plexus of Rektorzik and the cavernous sinus with the jugular bulb.

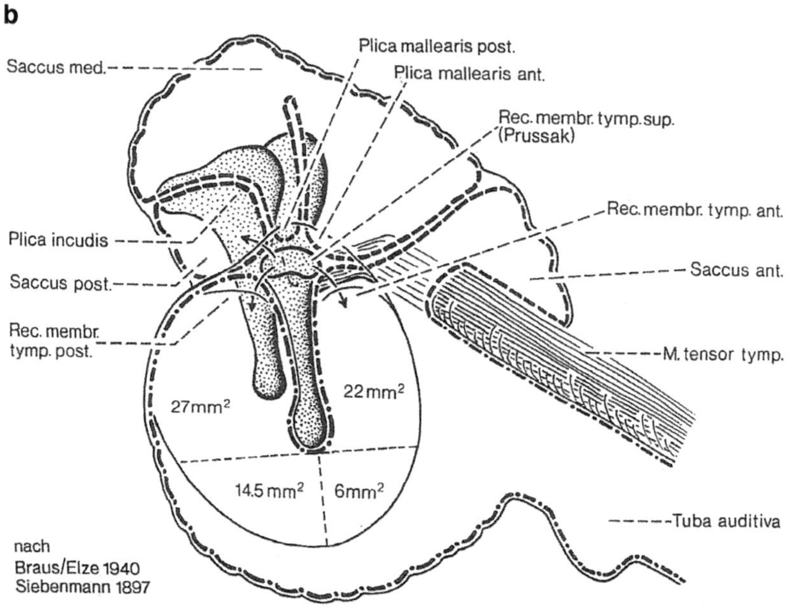

Fig. T19 (**a**) Portrait of Troeltsch (from Pagel 1901); (**b**) Folds and recessus of Troeltsch (after Siebenmann 1897, and Braus/Elze 1940; from Lang J *Klinische Anatomie das Ohres*. Springer, Wien, New York, 1992)

Fig. T20 Portrait of Trolard (from Wikimedia Commons; in public domain)

Vein of Trolard: *Eponym* for the superior anastomotic vein of the cerebral cortex (Fig. T21).

Troisier, Charles-Emile (1844–1919): French physician.

Gland of Troisier: *Eponym* for a lymph node located just above the left clavicle and belonging to the supraclavicular lymph nodes, palpably enlarged as the result of a metastasis from a malignant neoplasm, also described by Virchow, and known as the gland of Virchow-Troisier.

Tsai, Chiao (1897–1990): Chinese-American neuroanatomist born in Xianmei Country, Xinteng Town, Guangdong, who studied from 1921–1925 in California, the University of Columbia and Chicago, where he described the area of Tsai (Fig. T22). Back in China, he became the founder of physiology and avaiation, aerospace and navigation medicine in China.

Area of Tsai: *Eponym* for the ventral tegmental area of the midbrain. Reference: *The optic tracts and centres of the opossum Didelphis virginiana.* J Comp Neurol 1925; 39:173–215.

Tsuchida, Usabro (1867–1932): Japanese anatomist known for his studies of the oculomotor nuclei.

Nucleus of Tsuchida: *Eponym* for the central caudal nucleus of the oculomotor nucleus, already described

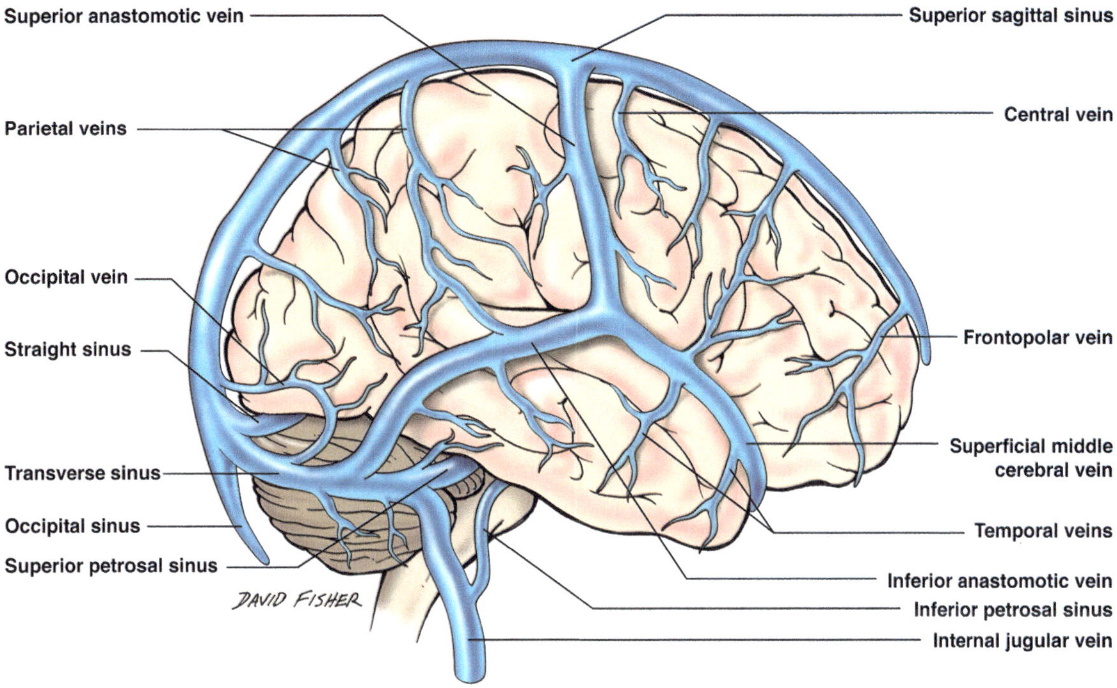

Fig. T21 The superficial cerebral veins and their connections to the venous sinuses (from ten Donkelaar et al. 2018; courtesy David Fisher, Birmingham, Alabama)

Fig. T22 Photo of Tsai (from Wang Y, Li Z, Qian Y, Guo B *Tsai Chiao: The founder of physiology and aviation, aerospace and navigation medicine in China.* Protein & Cell 2022; 13:779–784; Open Access)

in 1898 by Panegrossi as nucleus posterior dorsocentralis (Ric Lab Anat Norm Univ Roma 1898; 6:103–155). Reference: *Ueber Ursprungskerne der Augenbewegungsnerven im Mittelhirn und Zwischenhirn.* Arb Hirnanat Inst Zürich 1906; 1:2; 1:205.

Tubercles: *See* **The Classical Anatomical Eponyms.**

Tubules:

Tubules of Bellini: *See* ducts of Bellini.

Tubules of Henle: *Eponym* for the straight portions of the uriniferous tubules that form the loop of Henle; *see* Henle, Friedrich Gustav Jacob.

Tubules of Kobelt: *Eponym* for remnants of the mesonephric tubules in females, also known as tubules of Wolff; *see* Kobelt, Georg Ludwig.

Tubules of Schachowa: *Eponym* for the convoluted parts of the proximal tubules of the kidney; *see* Schachowa, Seraphina.

Tubules of Skene: *Eponym* for the paraurethral glands of females; *see* Skene, Alexander Johnston Chalmers.

Tubules of Wolff: *Eponym* for remnants of the mesonephric tubules in females, also described by Kobelt; *see* Wolff Kaspar Friedrich.

Tuffier, Theodore (1857–1929): French surgeon born in Bellême, Orne, and working in Paris, a pioneer of renal and pulmonary surgery. Reference: *Petite chirurgie pratique.* Paris, 1903 (with P. Desfosses).

Ligament of Tuffier, Inferior: *Eponym* for the mesentericoparietal fold, the part of the mesentery inserted into the iliac fossa.

Line of Tuffier: *Eponym* for the **intercristal line,** a horizontal line drawn across the highest points of both the iliac crests. In an anteroposterior lumbar radiograph it most often intersects the body of L4 or its inferior endplate in males and the body of L5 or its superior endplate in females and thus it is the most accurate method for numbering the lumbar vertebrae. Reference: *Anesthesie medullaire chirurgicale par injection sousarachnoidienne lombaire de cocaine; technique et results*. Sem Med 1900;20: 167–169.

Tulp, Nicolaes (originally Claes Pieterszoon; 1593–1674): Dutch anatomist born in Amsterdam and working in Amsterdam at the Collegium Chirurgicum Amstelaedamensis, later as Mayor of the city of Amsterdam. He is a central figure in Rembrandt's Lesson in Anatomy (1632; Mauritshuis, The Hague; Fig. T23). Reference: *Observationes medicae.* Amsterdam, 1641. Tulp introduced the term spina bifida. Reference: *Observationum medicorum libri tres.* Elsevier, Amsterdam, 1641.

Valve of Tulp: *Eponym* for the ileocaecal fold, previously considered as a valve, described previously by Falloppio, and later also by Bauhin and Varoli.

Türck, Ludwig (1810–1868): Austrian neurologist and laryngologist born and working in Vienna (Fig. T24). He studied the degeneration of nerve fibres in the spinal cord and described the anterior corticospinal tract. The temporal corticopontine tract is connected with his name, although sometimes questioned (*see* Schmahmann et al. Brain 1992; 115:1911–1924). A recent study by Eliasz Engelhardt (J Hist Neurosci 2013; 22:353–365) justifies the name bundle or tract of Türck. References: *Mikroskopischer Befund des Rückenmarkes eines paraplegischen Weibes.* Z Kais Kön Ges Ärzte Wien 1849; 5:173–176; *Über secundäre Erkrankung einzelnen Rückenmarksstränge und ihren Fortsetzungen zum Gehirne.* Z Kais Kön Ges Ärzte Wien 1852; 8:511–534; *Beobachtungen über das Leitungsvermögen des menschlichen Rückenmarks.* Ztschr Gesellsch Aerzte Wien 1856, p 246.

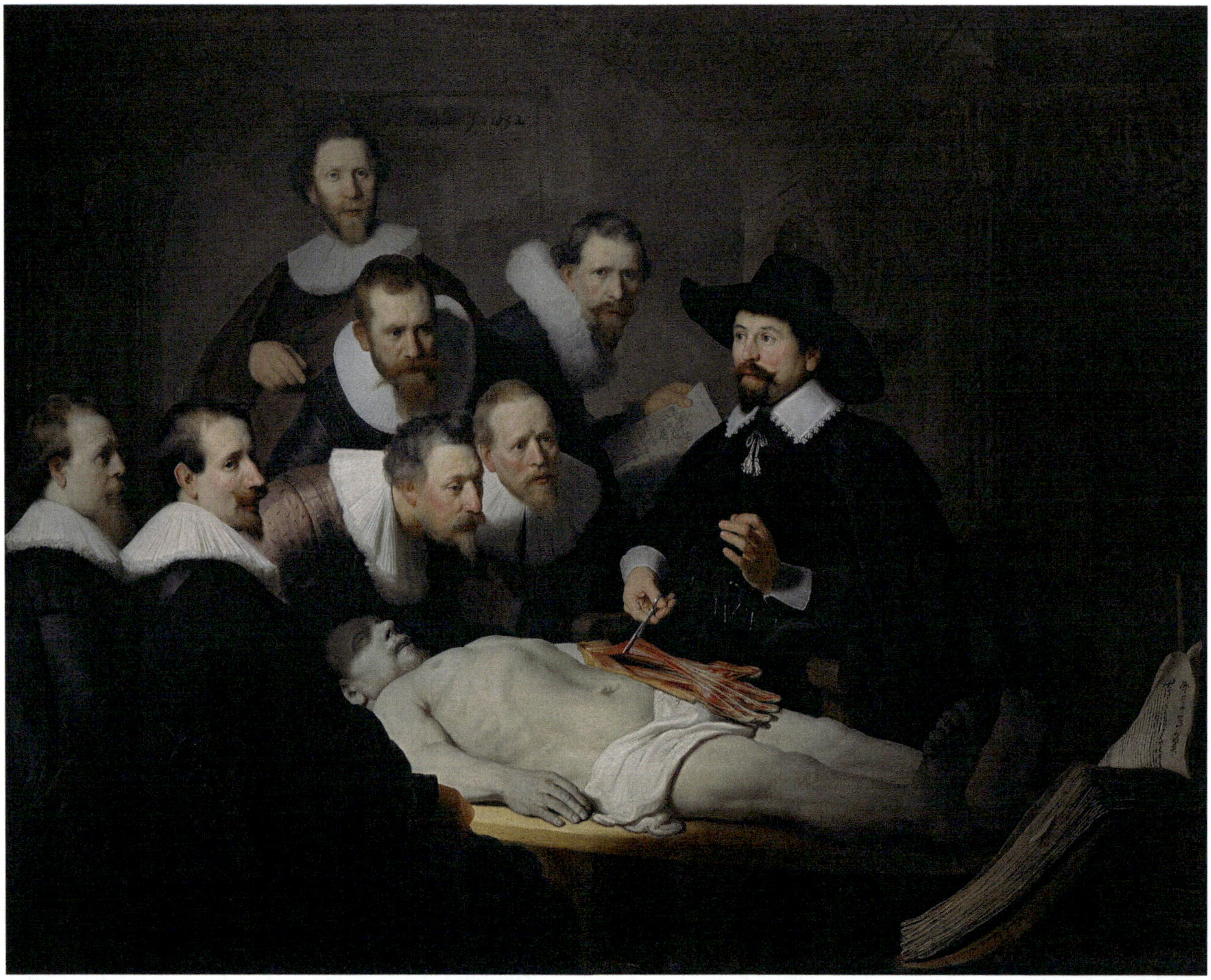

Fig. T23 The Anatomical Lesson of Nicolaes Tulp by Rembrandt (Mauritshuis, The Hague; in public domain)

Fig. T24 Portrait of Türck (from Wikimedia Commons; in public domain)

Fig. T25 Portrait of Turner (from Pagel 1901)

Tract of Türck: *Eponym* for the temporal corticopontine tract.

Tract of Türck, Anterior: *Eponym* for the anterior or ventral corticospinal tract.

Turner, Sir William (1832–1916): English anatomist born in Lancaster and working in Edinburgh (Fig. T25).

Sulcus of Turner: *Eponym* for the intraparietal sulcus of the cerebral cortex, which he described in the chimpanzee; in 1868, Adolph Pansch described this sulcus in the human brain. References: *Notes more especially upon the bridging convolutions in the brain of the chimpanzee.* Proc Roy Soc Edin, February 19, 1866; *The convolutions of the human cerebrum topographically considered.* Edin Med J 1866; 11:1105–1122.

Fig. T26 Portrait of Tyson (from the Royal College of Physicians, LondonArtUK/Wikipedia, in public domain)

Tyrrell, Frederick (1793–1843): English anatomist and surgeon in London.

Fascia of Tyrrell: *Eponym* for the rectovesical fascia or septum between the rectum and the prostate, also described by Denonvilliers.

Tyson, Edward (1650–1708): English physician and anatomist born in Clevedon, near Bristol, and working in London (Fig. T26).

Glands of Tyson, Preputial: *Eponym* for peculiar sebaceous glands in the dermis of the corona of the glans penis and undersurface of the prepuce.

Layer of Tyson: *Eponym* for the granular layer of the epidermis, later also described by Unna.

Chapter U

Unguis of Haller: *Eponym* for the calcar avis, an elevation of the medial wall of the posterior horn of the lateral ventricle; *see* Haller, Albrecht von.

Uchimura, Yushi (1897–1980): Japanese anatomist born in Tokyo and, after studying with Spielmeyer in Munich, working in Tokyo (Fig. U1). He described the arteries supplying the vulnerable sector of the hippocampus. He also was a famous Japanese baseball player. Reference: *Űber die Gefäßversorgung des Ammonshornes.* Z Ges Neurol Psychiatr 1928; 112:1–19. For a more recent study of the arterial vascularization of the hippocampus, *see* Huther G, Dörfl J, Van der Loos H, Jeanmonod D *Microanatomic and vascular aspects of the temporomesial region.* J Neurosurg 1998; 43:1118–1136.

Arteries of Uchimura, Long: *Eponym* for branches of the posterior cerebral artery supplying fields H1-H4 of Rose and the dentate gyrus.

Fig. U1 Portrait of Uchimura (from Wikidata; in public domain)

Fig. U2 Portrait of Unna (from Wikipedia; in public domain)

Arteries of Uchimura, Short: *Eponym* for branches of the posterior cerebral artery supplying the subiculum, the dentate gyrus, and fields H1–H3 of Rose.

Unna, Paul Gerson (1850–1929): German dermatologist born in Hamburg (Fig. U2).

Layer of Unna, Granular: *Eponym* for the granular layer of the epidermis, earlier described by Tyson. Reference: *Zur feineren Anatomie der Haut.* Berl Klin Wochenschr 1921; 58:447.

Uskow, N (19th century): Russian anatomist born in Kronstadt and working for some time in Strassburg.

Pillars of Uskow: *Eponym* for the two embryonic folds or ridges that grow from the dorsolateral part of the body wall and unite with the septum transversum to form the diaphragm. Reference: *Ueber die Entwicklung des Zwerchfells, des Pericardiums und des Coeloms.* Arch Mikr Anat 1883; 22:143–219.

Letter V

19

Valenti, Giulio (1860–1933): Italian anatomist born in Siena and working in Camerino, Perugia, Catania and Bologna.

Lateral dimples of Valenti: *Eponym* for two small blind pits, one on each side of the frenulum of the prepuce.

Valentin, Gabriel Gustav (1810–1883): German physiologist born in Breslau (currently Wróclaw, Poland) and working in Breslau as an assistant of Purkinje and later in Bern (Fig. V1). Reference: *Encyclopédie anatomique.* Baillière, Paris, 1843.

Corpuscles of Valentin: *Eponym* for small amyloid bodies described in nervous tissue

Ganglion of Valentin: *Eponym* for a ganglion on the superior alveolar nerve.

Nerve of Valentin: *Eponym* for a communicating branch between the abducens nerve and the pterygopalatine ganglion. Reference: *De functionibus nervorum cerebralium et nervi sumpathici libri quattuor.* Sumptibus Huber & Socii, Bern, 1839.

Tubes or **Cords of Valentin:** *Eponym* for the linear arrangement of the female sex cells in the developing ovary, also described by Pflüger (1863). Reference: *Ueber die Entwickelung der Follikel in den Eierstöcke der Säugethiere.* Arch Anat Physiol (Lpz) 1838, pp. 526–536.

Valsalva, Antonio Maria (1666–1723): Italian anatomist born in Imola, Emilia Romagna, who succeeded Malpighi in Bologna (Fig. V2). Morgagni was one of his students. He described among others aspects of the aorta, the ear, the taeniae of the colon and the branches of the vagus nerve. Reference: *De aure humana tractatus, in quo integra auris fabrica describitur.* Bologna, 1704.

Antrum of Valsalva: *Eponym* for the mastoid antrum of the tympanic cavity.

Cavity of Valsalva: *Eponym* for the mastoid cavity.
Humour of Valsalva: *Eponym* for the perilymph.

Ligament of Valsalva: *Eponym* for the anterior ligament of the auricle.

Locus of Valsalva: *Eponym* for the mucosa of the anteroinferior area of the nasal septum, where a network of dilated veins is the site of epistaxis. Usually, this is remembered as the Locus of Kiesselbach.

Muscle of Valsalva: *Eponym* for the tragicus muscle of the auricle.

Sinuses of Valsalva: *Eponym* for the aortic sinuses, also described by Petit. Reference: *Tres dissertationes.* Venice, 1740 (1715, 1716 and 1719).

Taeniae of Valsalva: *Eponym* for the taeniae coli.

Valverde de Amusco, Jean (about 1525–1560): Spanish anatomist, born in Castille but working in Italy (Fig. V3a). Reference: *Historia de la composicion del cuerpo humano.* Rome, 1556. All about four of its 42 engraved copperplate illustrations were taken from Vesalius's *De humani corporis fabrica*, but occasionally corrected by Valverde. Famous is the illustration of a muscle man holding his own skin in one hand and a knife in the other (Fig. V3b).

van Deventer, Hendrik (1651–1724): Dutch obstetrician born and working in The Hague (Fig. V4). He was the first to seriously study the pelvis and spinal deformities. Reference: *Operationes chirurgicae novum lumen exhibentes obstetricantibus.* Dyckhuisen, Leiden 1701; English translation: *The Art of Midwifery Improved.* Bettesworth, London, 1716.

van Diemerbroeck, Ijsbrand (1609–1674): Dutch physician and anatomist in Utrecht (Fig. V5). Reference: *L'anatomie du corps humain.* Amisson and Posuel, Lyon, 1695.

H. J. ten Donkelaar et al., *An Illustrated Guide to Anatomical Eponyms*, https://doi.org/10.1007/978-3-031-91664-9_19

Fig. V1 Photo of Valentin (from Pagel 1901)

Fig. V2 Portrait of Valsalva (from Wikimedia Commons; in public domain)

Muscle of van Diemerbroeck: *Eponym* for the adductor quartus muscle, a supernumerary muscle of the medial group of thigh muscles formed by superior fibres of the adductor magnus muscle.

van Helmont, Jean Baptiste (1577–1644): Flemish surgeon from Brussels working in Liège and later in Brussels (Fig. V6).

Mirror or **Speculum of van Helmont:** *Eponym* for the centrum tendineum of the diaphragm. Reference: *Opuscula medica inaudita.* Cologne, 1644.

van Hoboken, Nicolaas (1632–1678): Dutch anatomist and physician born in Utrecht working in Harderwijk and later in Steinfurt, Germany.

Valves of Hoboken: *Eponym* for secondary windings of the vessels of the umbilical cord that form grooves externally, and valve-like projections internally. Reference: *Anatomia secundinae humanae.* Ribbius, Utrecht, 1669.

van Horne, Jan (1621–1670): Dutch surgeon and anatomist born and working in Amsterdam (Fig. V7). He first described the thoracic duct in humans. Stensen and Ruysch were among his students. Reference: *Observationes anatomico-medicae.* Amstredam, 1664.

Duct or **Canal of van Horne:** *Eponym* for the thoracic duct, described in animals by Pecquet (Fig. V8).

Saccus of van Horne: *Eponym* for the cisterna chyli, described in animals by Pecquet (Fig. V8).

van Leeuwenhoek, Antonie (1632–1723): Dutch microscopist born and living in Delft (Fig. V9) who in 1678 discovered the spermatozoa as 'animalcules' in semen (Fig. V10).

Canals of van Leeuwenhoek: *Eponym* for vascular osteonic canals in bone, now known as Haversian canals. They were observed by van Leeuwenhoek in 1675 and described by Havers in 1691. Reference: *Microscopical observations of the structure of teeth and other bones.* Phil Trans 1675; 12:1002–1003.

van den Spieghel, Adriaan (Spigelius; 1578–1625): Flemish anatomist born in Brussels and working in Venice and later in Padua where he succeeded Fabrici and Casseri (Fig. V11a). Reference: *De humani corporis fabrica libri decem. Opus posthumus Daniel Bucretius jussu authoris in lucem profert.* Deuchinus, Venice, 1627. Casseri prepared nine embryological tables, illustrating *De formatio foetu*, published posthumously in 1626 by van den Spieghel's son-in-law Liberale Crema (Fig. V12).

Line of van den Spieghel: *Eponym* for the semilunar line of the muscles of the abdominal wall (Fig. V11b).

Lobe of van den Spieghel (Spigelian lobe): *Eponym* for the caudate lobe of the liver.

Varolio, Costanzo (Varolius, Constantius; 1543–1575): Italian anatomist born in Bologna and working in Bologna

a b

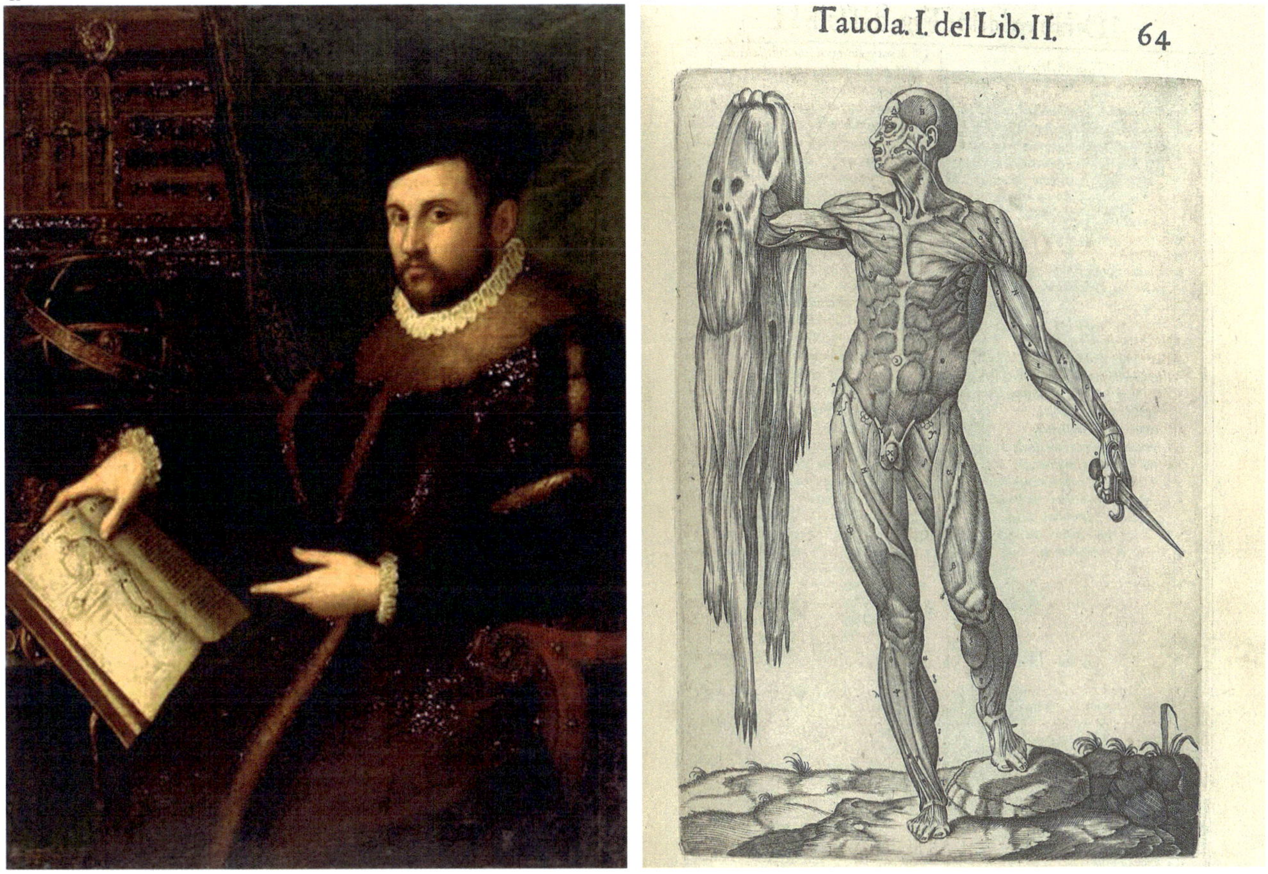

Fig. V3 Valverde de Amusco, (**a**) Portrait; (**b**) Skinned man (both Wellcome Collection CCBY; in public domain)

Fig. V4 Engraving of van Deventer (Wellcome Collection CCBY; in public domain)

and later in Rome and physician to the Pope (Fig. V13). He described the pons. Reference: *De Nervis Opticis nonnullesque aliis praeter communem opinionem in Humano capite observatis.* Meitti, Parma, 1573; *Anatomiae sive de resolutione corporis humani.* Wechel and Fischer, Frankfurt am Main, 1591. For biography, *see* Zago S, Meraviglia MV *Costanzo Varolio (1543–1575).* J Neurol 2009; 256:1195–1196.

Pons of Varolio: *Eponym* for the pons (Fig. V14).

Valve of Varolio: *Eponym* for the ileocaecal fold, previously described as a valve, described for the first time but later also by Falloppio, Bauhin and Tulp.

Vastarini-Cresi, Giovanni (1870–1925): Italian anatomist born in Taranto and working in Naples. He is especially known for his studies of the human tongue.

Intracaecal vallate papilla of Vastarini-Cresi: *Eponym* for the median circumvallate papilla protruding from the slope of the foramen caecum of the tongue.

Fig. V5 Engraving of van Diemerbroeck by Johannes Edelinck, Atlas of Stolk, Rotterdam (from Wikimedia Commons; in public domain)

Fig. V6 Portrait of van Helmont (from the Wellcome Collection CCBY; in public domain)

Fig. V7 Engraving of van Hoboken (from the Wellcome Collection CCBY; in public domain)

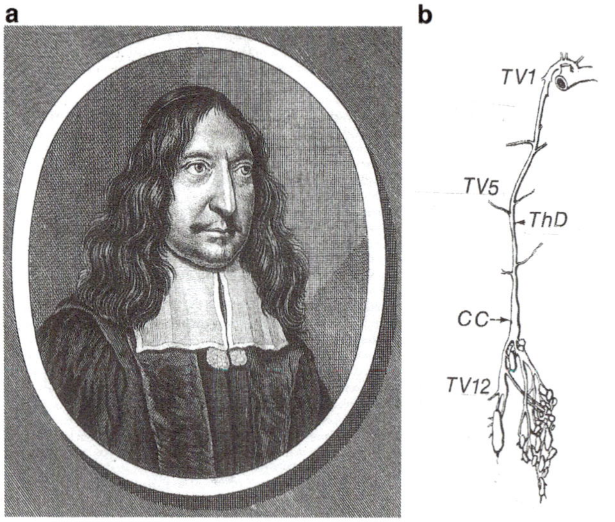

Fig. V8 (a) Engraving of van Horne (from the Dutch Iconographic Bureau; in public domain); (b) Thoracic duct and cysterna chyli (after ten Donkelaar HJ, Oostra R-J *Klinische Anatomie en embryologie,* fourth ed. Springer Media, Houten, 2014). *CC* Cysterna chyli, *TD* thoracic duct, *TV1*, *TV5*, *TV12* level of first, fifth and 12th thoracic vertebrae

Vater, Abraham (1684–1751): German anatomist and pathologist born and working in Wittenberg (Fig. V15) who first noted the large encapsulated nerve endings, later rediscovered and described by Pacini. Reference: *Dissertatio qua oeconomiam sensuum ex speciali organorum sensoriorum et sigillatum ex papillarum nervearum*

Fig. V9 Portrait of van Leeuwenhoek by J. Verkolje, Rijksmuseum, Amsterdam (lent to Museum Boerhaave, Leiden; from the Wellcome Collection CCBY; in public domain)

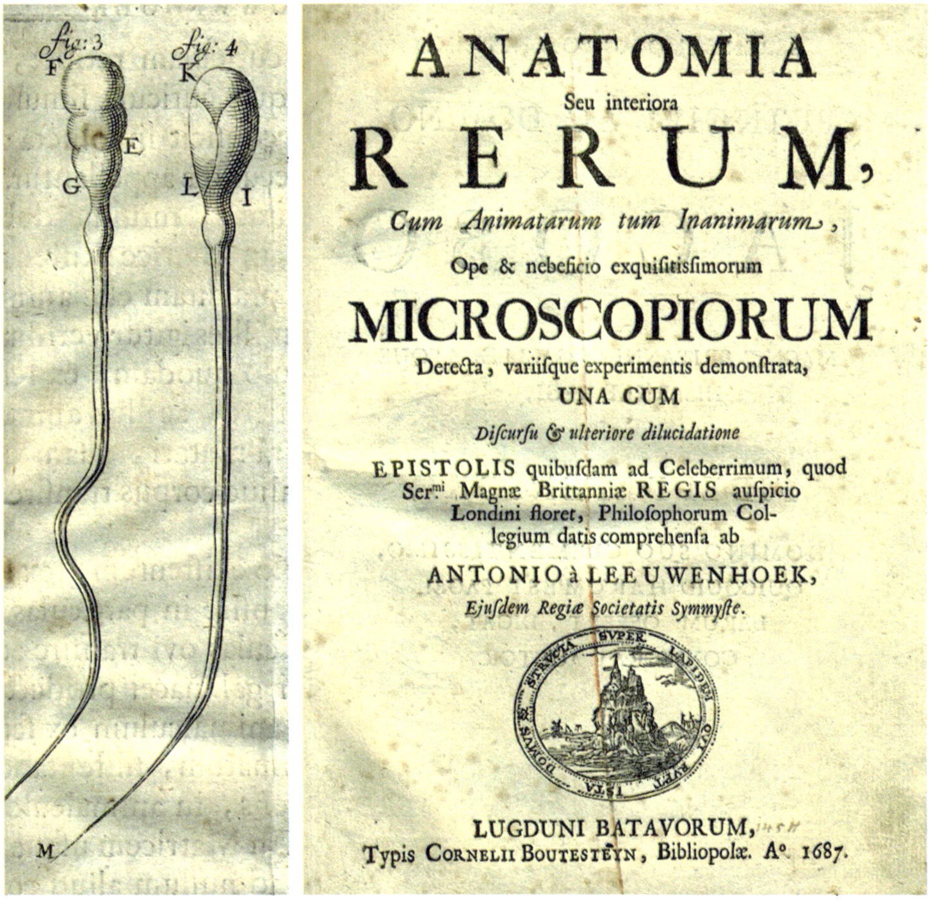

ANATOMIA
Seu interiora
RERUM,
Cum Animatarum tum Inanimarum,
Ope & nebeficio exquiſitiſſimorum
MICROSCOPIORUM
Detecta, variiſque experimentis demonſtrata,
UNA CUM
Diſcurſu & ulteriore dilucidatione
EPISTOLIS quibuſdam ad Celeberrimum, quod Ser.mi Magnæ Brittanniæ REGIS auſpicio Londini floret, Philoſophorum Collegium datis comprehenſa ab
ANTONIO à LEEUWENHOEK,
Ejuſdem Regiæ Societatis Symmyſte.
LUGDUNI BATAVORUM,
Typis CORNELII BOUTESTEYN, Bibliopolæ. A° 1687.

Fig. V10 (**a**) Title page *Anatomia rerum microscopiorum*. Leiden, 1687, and (**b**) Spermatozoa (courtesy Ondřej Naňka, Prague)

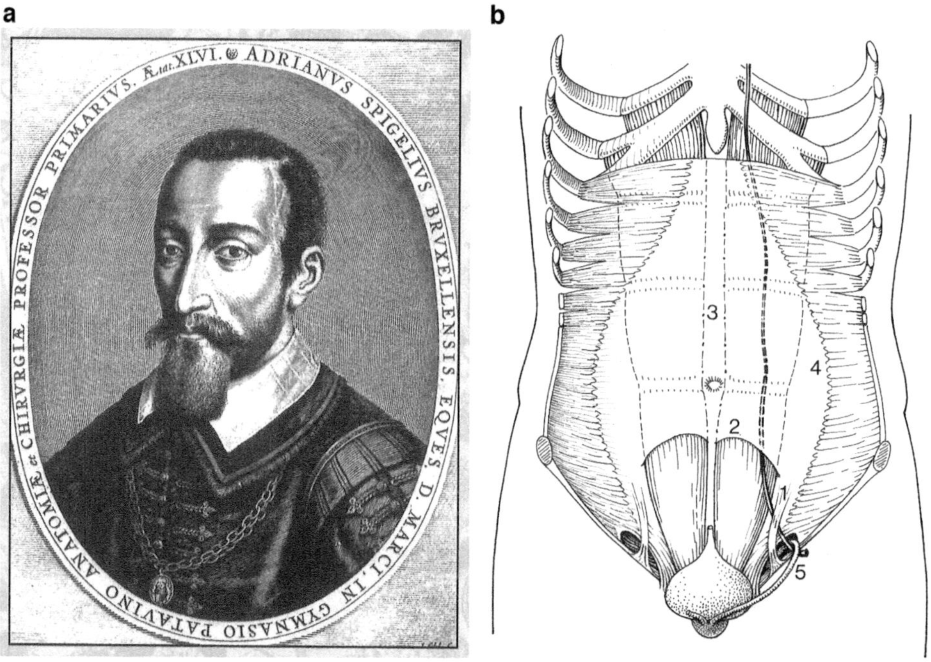

Fig. V11 (**a**) Engraving van den Spieghel (from Wikimedia Commons; in public domain); (**b**) Semilunar line of van den Spieghel (after an old figure by the late Chris van Huijzen). *1* Interfoveolar ligament of Hesselbach, *2* arcuate line of rectus sheath, *3* linea alba, *4* semilunar line of van den Spieghel, *5* ductus deferens crossing the inferior epigastric artery

Fig. V12 (**a**) Pregnant woman with opened uterus; (**b**) neonate with placenta (both by Casseri in van den Spieghel *De formatio foetu.* 1626; from Wikimedia Commons, in public domain)

Fig. V13 Statue of Varolio (from Wikimedia Commons; in public domain)

Fig. V14 Varolio's original figure of the pons (courtesy Marina Quartu, Cagliari). The pons is indicated by *h*

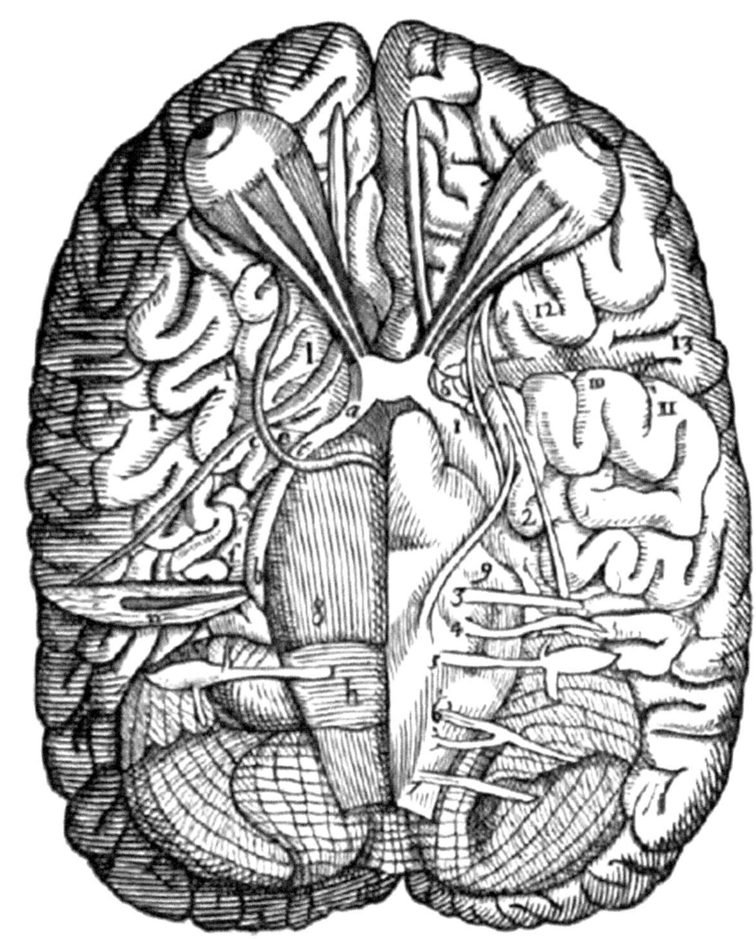

Fig. V15 Portrait of Vater (from the Portraitsammlungen des Herzogs August Bibliothek, Wolfenbüttel, Germany in Wikimedia Commons; in public domain)

textura mechanica demonstrata et placidae eruditorum disquisitioni offert. Wittenberg, 1717.

Ampulla of Vater: *Eponym* for the hepatopancreatic ampulla, the common termination of the bile duct and the pancreatic duct.

Corpuscles of Vater: *Eponym* for the lamellar corpuscles, large encapsulated nerve endings, also known as corpuscles of Vater-Pacini (Fig. V16).

Duct of Vater: *Eponym* for the embryonic thyroglossal duct also described by Bochdalek and His Sr.

Fold of Vater: *Eponym* for a mucous fold in the duodenum just above the major duodenal papilla.

Tubercle of Vater: *Eponym* for the major duodenal papilla.

Veins: *See* **The Classical Anatomical Eponyms.**

Velpeau, Alfred Armand Louis Marie (1795–1867): French anatomist and surgeon born in Brèches, Indre-et-Loire, and working in Paris. Reference: *Traité complet d'anatomie chirurgicale, générale et topographique, du corps humain, ou anatomie considerée dans ses rapports avec la pathologie chirurgicale et la médecine opératoire.* Crevot, Paris, 1825–1826. There is a painting by Auguste Feyen-Perrin of his Anatomical Lesson *La leçon d'anatomie de Velpeau à la Charité* (Fig. V17).

Canal of Velpeau: *Eponym* for the inguinal canal.

Fascia of Velpeau: *Eponym* for the fascia transversalis, also described by Hyrtl.

Foramen or **Space of Velpeau, Quadrangular:** *Eponym* for the humerotricipital foramen, a space determined by the teres major muscle caudally, the teres minor muscle cranially, the humerus laterally and the long head of the triceps brachii muscle medially (Fig. V18).

Fossa of Velpeau: *Eponym* for the ischioanal or ischiorectal fossa.

Hernia of Velpeau: *Eponym* for a femoral hernia, in which the intestine is found in front of the blood vessels.

Ventricles: For the cavity of the septum pellucidum eponyms were ascribed to Duncan, Sylvius, Verga, Vieussens and Wenzel, but only the eponym Ventricle of Verga remains in use.

Ventricle of Aranzio (Arantius): *Eponym* for the posterior part of the fourth ventricle; *see* Aranzio, Giulio Caesare.

Ventricle of Duncan: *Eponym* for the cavity of the septum pellucidum; *see* Duncan, Daniel.

Ventricle of Krause: *Eponym* for the terminal ventricle of the spinal cord; *see* Krause, Karl Friedrich Theodor.

Ventricle of Morgagni: *Eponym* for the laryngeal ventricle; *see* Morgagni, Giovanni Battista.

Ventricle of Sylvius: *Eponym* for the cavity of the septum pellucidum; *see* Sylvius, François de le Boë.

Ventricle of Verga: *Eponym* for a space in the caudal part of the septum pellucidum, the cavum Vergae; *see* Verga, Andrea.

Ventricle of Vieussens: *Eponym* for the cavity of the septum pellucidum; *see* Vieussens, Raymond de.

Ventricle of Wenzel: *Eponym* for the cavity of the septum pellucidum; *see* Wenzel, Joseph.

Veraguth, Otto (1870–1944): Swiss neurologist born in Thusis.

Fold of Veraguth: *Eponym* for a skin fold that runs obliquely from the lower outside to the upper inside of the upper eyelid, assumed to be characteristic for patients suffering from depression.

Verga, Andrea (1811–1895): Italian anatomist and neurologist born in Milan (Fig. V19).

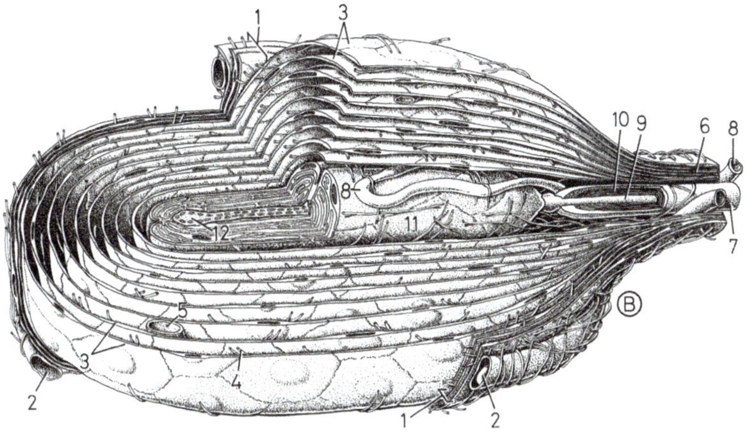

Fig. V16 The structure of a corpuscle of Vater-Pacini (from Krstić R 1988 *Die Gewebe des Menschen und der Säugetiere*, 2nd ed. Springer, Berlin-Heidelberg-New York; with permission). *1* lamellated capsule, *2*, *5*, *8* capillaries, *3* lamellae, *4* collagen microfibrils, *6* continuation of capsule and lamellae from perineurium, *7* nerve fiber, *9* axon losing its myelin sheath (*10*) but covered by flattened layers of Schwann cells (*11*), *12* thickened end of axon (from ten Donkelaar HJ, Kachlík D, Tubbs RS *An Illustrated Terminologia Neuroanatomica*. Springer, Cham, 2018)

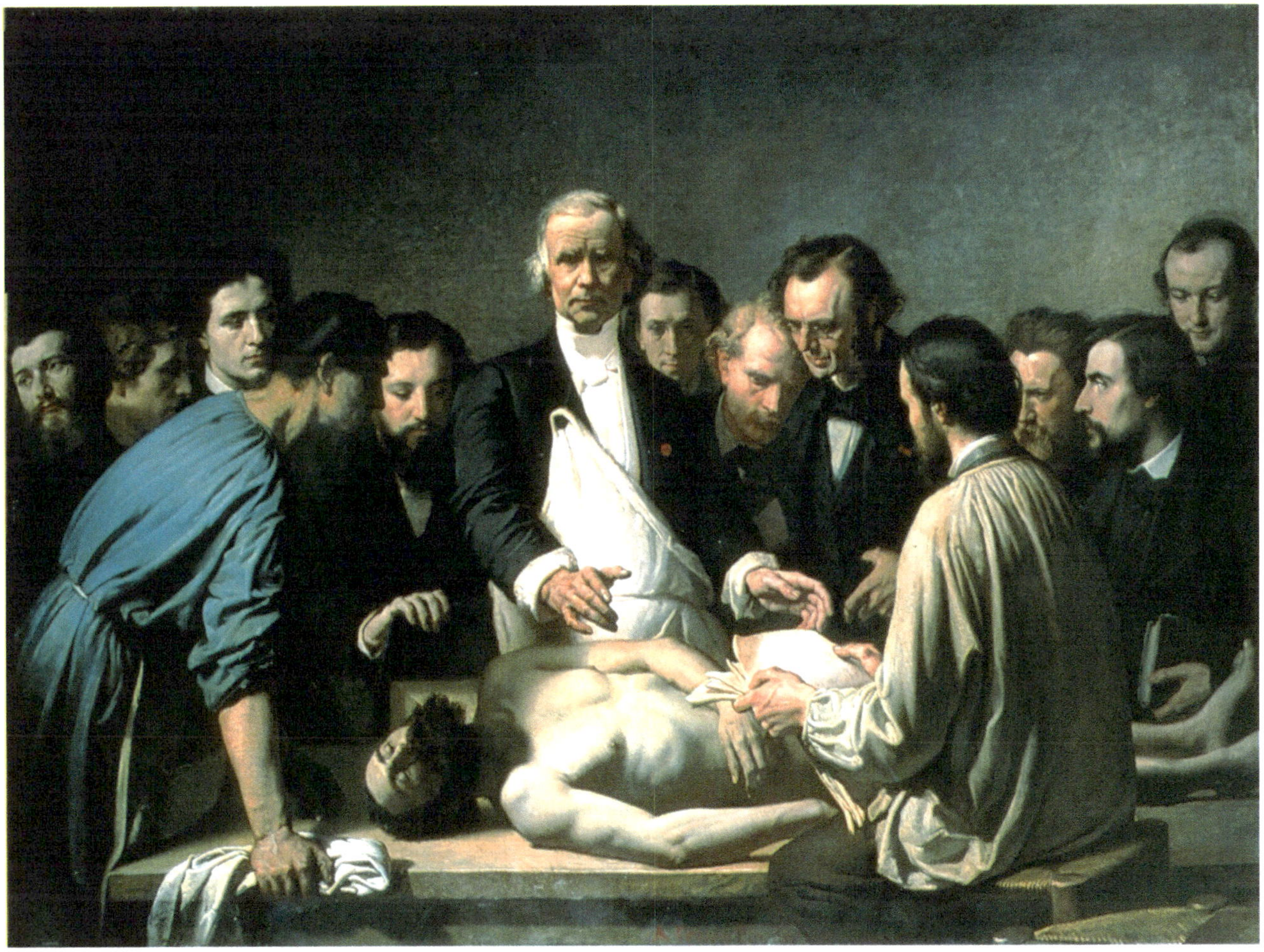

Fig. V17 The Anatomical Lesson by Velpeau, painting by François Nicolas Augustin (known as Auguste) Feyen-Perrin (Hôpital Musée de l'Assistance publique, Hôpitaux de Paris; in public domain)

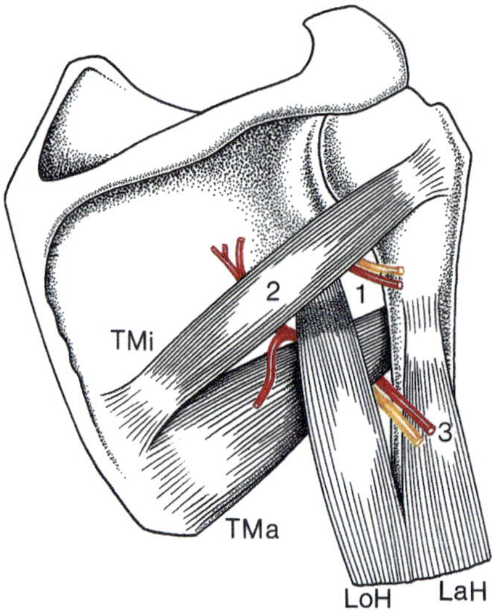

Fig. V18 Trigone or quadrangular space of Velpeau for the passage of the posterior circumflex humeral artery (in *red*) and the axillary or circumflex nerve (after O'Rahilly R *Gardner-Gray-O'Rahilly Anatomy. A regional study of human structure,* 5th ed. Saunders, Philadelphia, PA, 1986; *1* quadrangular space with axillary nerve and posterior circumflex humeral artery, *2* triangular space through which the circumflex scapular artery passes, *3* radial nerve and profunda brachii artery, *TMa* teres major muscle, *TMi* teres minor muscle, *LaH*, *LoH* lateral and long heads of triceps brachii muscle

Fig. V19 Buste of Verga in Milan (from Wikimedia Commons; in public domain)

Accessory sinus of Verga: *Eponym* for a variable communication between the ophthalmic vein or cavernous sinus anteriorly with the transverse sinus posteriorly.

Ventricle of Verga (cavum Vergae): *Eponym* for the caudal part of the cavity of the septum pellucidum (Fig. V20). Reference: *Dell'apparato ventricolare del setto lucido e della volta a tre pilastri.* G Ist Lomb 1856; 8:89–104.

Verheyen, Philip (1648–1710): Flemish anatomist born in Verbroeck and working in Leuven (Louvain), after being a farmer and theologian in early life (Fig. V21). Reference: *Corporis humani anatomia in qua omnia tam veterum quam recentiorum anatomicorum inventa methodo nove et intellecta facillima describuntur, ac tabulis aeniis repraesentantur.* Louvain, 1699; Brussels, 1710.

Veins or **Stars of Verheyen:** *Eponym* for the star-shaped arrangement of the stellate veins on the surface of the kidney.

Verhoeff, Frederick H (1874–1968): American ophthalmologist.

Membrane of Verhoeff: *Eponym* for the outer hyperreflective band or interdigitation zone of the retina (zone 13 in optical coherence tomography or OCT). Reference: *A hitherto undescribed membrane of the eye and its significance.* Boston Med Surg J 1903; 149:456–458; for OCT, *see* Staurenghi G et al. *Proposed lexicon for anatomic landmarks in normal posterior segment spectral-domain optical coherence tomography.* Ophthalmology 2014; 121:1572–1578.

Versari, Riccardo (1878–1945): Italian anatomist and Senator born in Milan and working in Rome, Palermo and Naples.

Loop muscle bundle or **Valve of Versari:** *Eponym* for a muscular band surrounding the point where the ureter opens into the bladder.

Vesalius, Andreas (Andries van Wezel, 1514–1564): Flemish anatomist born in Brussels and working in Padua (Fig. V22) whose *De humani corporis fabrica* (Oporinus, Basel, 1543), illustrated by Jan Stefan van Calcar, revolutionized Anatomy. Later, he became imperial physician of Charles V.

Bone of Vesalius: *Eponym* for (a) an accessory bone of the foot (**Os vesalianum pedis**), a result of failed fusion of the secondary ossification centre of the base of the fifth metatarsal, located within the insertional tendon of the fibularis brevis muscle; (b) an accessory bone of the hand (**Os vesalianum manus** or **Os vesalianum carpi**), a result of failed fusion of the secondary ossification centre of the base of the fifth metacarpal,

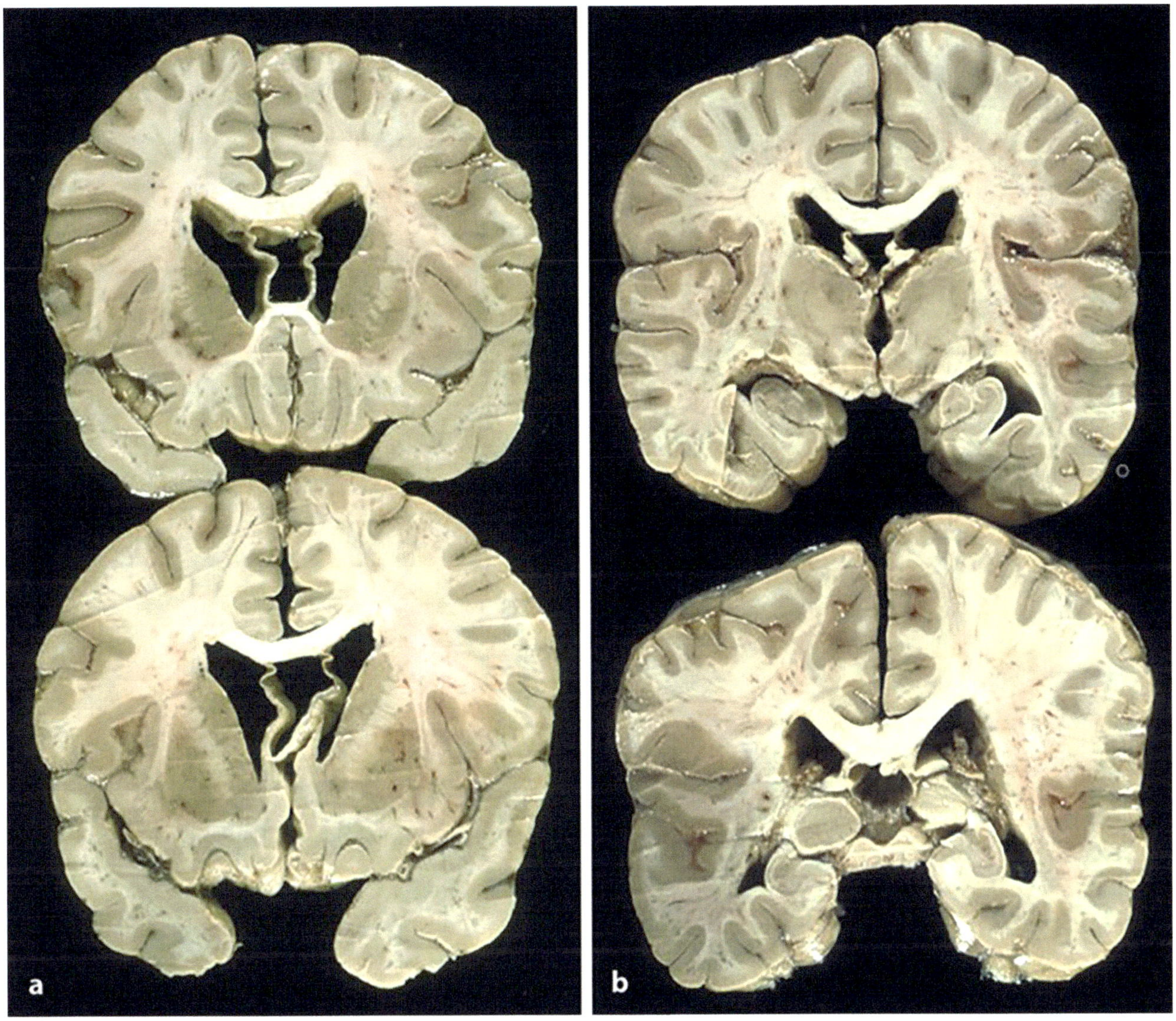

Fig. V20 Cavum septi pellucidi (**a**) and cavum Vergae (**b**) found as incidental findings at autopsy in a 54-year-old patient (from the Department of Neuropathology, Medizinische Hochschule Hannover; courtesy Akira Hori, Wedemark; from ten Donkelaar HJ, Lammens M, Hori A *Clinical Neuroembryology,* 3rd ed. Springer, Cham, 2023)

Fig. V21 Portrait of Verheyen by Jan Baptist Bertenham (from Wikimedia Commons; in public domain)

located between the metacarpal and the hamate; and (c) the fabella, an inconstant sesamoid bone in the tendinous origin of the lateral head of the gastrocnemius muscle.

Foramen of Vesalius: *Eponym* for a small inconstant foramen in the sphenoidal bone anteromedial to the oval foramen through which an emissary vein from the cavernous sinus may pass to the pterygoid plexus.

Glands of Vesalius: *Eponym* for bronchial lymph nodes.

Ligament of Vesalius: *Eponym* for the inguinal ligament, also described by Falloppio and Poupart.

Vein of Vesalius: *Eponym* for the sphenoidal vein, an emissary vein from the cavernous sinus to the pterygoid plexus.

Fig. V22 Wood cut of Vesalius (from Pagel 1901)

Vesling, Johann (1598–1649): German anatomist and botanist born in Minden, Westphalia, and working in Padua.

Line of Vesling: *Eponym* for the raphe or septum of the scrotum. Reference: *Syntagma anatomicum.* Frambotti, Padua, 1641.

Vialet, N (19th century-early 20th century): French neurologist.

Tract of Vialet: *Eponym* for the transverse occipital tract of the lingual lobe, one of the lobar association tracts of the occipital lobe, connecting the second and third occipital gyri with the inferior margin of the calcarine sulcus. Reference: *Les centres cérébraux de la vision et l'appareil visuel intra-cérébral.* Thesis, Paris, 1893.

Vicq d'Azyr, Felix (1748–1794): French physician and anatomist born in Valognes (Fig. V23) who described the mammillothalamic tract (tract of Vicq d'Azyr) and beautifully illustrated the brain (Fig. V24). Reference: *Traité d'anatomie et de physiologie.* Vol 1: *Anatomie et physiologie du cerveau.* Didot l'Aine, Paris, 1786.

Fig. V23 Portrait of Vicq d'Azyr (from the Wellcome Collection CCBY; in publi domain)

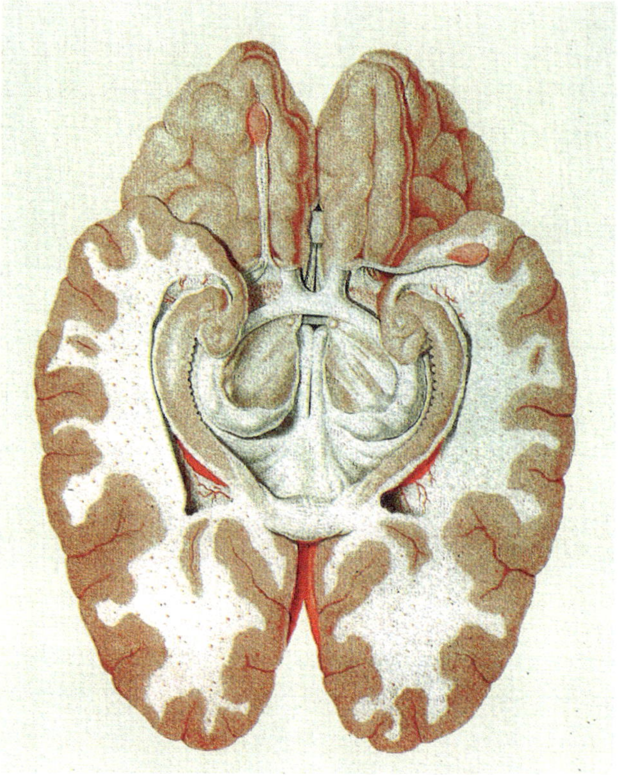

Fig. V24 Vicq d'Azyr, horizontal section with hippocampus (from *Traité d'anatomie et de physiologie.* 1786)

Band or **Stripe of Vicq d'Azyr:** *Eponym* for the occipital stripe, a conspicuous band of myelinated fibres, marking the striate cortex of the occipital lobe, also described by Gennari.

Bundle or **Tract of Vicq d'Azyr:** *Eponym* for the mammillothalamic tract of the brain.

Foramen of Vicq d'Azyr: *Eponym* for the foramen caecum of the medulla oblongata.

Vidian artery: *Eponym* for the artery of the pterygoid canal; *see* Vidius, Vidus (Guidi, Guido).

Vidian nerve: *Eponym* for the nerve of the pterygoid canal composed of the great and deep petrosal nerves; *see* Vidius, Vidus (Guidi, Guido).

Vidian vein: *Eponym* for the vein of the pterygoid canal; *see* Vidius, Vidus (Guidi, Guido).

Vidius, Vidus, *see* Guidi, Guido.

Vieussens, Raymond de (1641–1716): French anatomist and physician born in Vieussens and working in Montpellier and later in Paris (Fig. V25). Reference: *Neurographia universalis.* Certe, Lyons, 1685.

Ansa of Vieussens: *Eponym* for the ansa subclavia, a loop of nerve fibres around the subclavian artery connecting the middle cervical and the cervicothoracic (stellate) sympathetic ganglia (Fig. V26).

Anulus or **Ring of Vieussens:** *Eponym* for the limbus of the fossa ovalis in the right atrium of the heart.

Fig. V25 Portrait of Vieussens (from the Wellcome Collection CCBY; in publi domain)

Anus of Vieussens: *Old term* for the opening of the mesencephalic or cerebral aqueduct into the third ventricle, also described by Bartholin.

Centre of Vieussens: *Eponym* for the semioval centre as found in horizontal sections of the brain (Fig. V27).

Foramina of Vieussens: *Eponym* for the openings of the smallest cardiac veins into the right atrium, also described by Thebesius and Lannelongue.

Ganglion of Vieussens: *Eponym* for the coeliac ganglion.

Valve of Vieussens: *Eponym* for (a) the superior medullary velum of the cerebellum, and (b) an inconstant valve-like structure at the transition of the great cardiac vein into the coronary sinus.

Veins of Vieussens: *Eponym* for the smallest cardiac veins.

Ventricle of Vieussens: *Eponym* for the cavity of the septum pellucidum, also described by Duncan, Verga and Wenzel.

Vignal, Guillaume or William (1852–1893/94): French physiologist and histologist born in La Felche and, after studying in Bonn and Edinburgh, working in Paris.

Cells of Vignal: *Eponym* for embryonic mesenchymal cells lying upon the axis cylinders of foetal nerve fibres, forming a complete sheath. Reference: *Développement des éléments du système nerveux cérebro-spinal; nerfs péripheriques; moelle; couches corticales du cerveau et du cervelet.* Paris, 1889.

Villemin, Fernand (late 19th century-20th century): French anatomist born in Reims and working in Lyon and Bordeaux.

Arcade or **Intermesenteric artery of Villemin:** *Eponym* for an inconstant connection between both mesenteric arteries. Reference: *La constitution de l'arc de Treitz* (with P Huard). CR Ass Anat Strasbourg, pp. 263–267.

Sphincter of Villemin: *Eponym* for sphincteric fibres at the termination of the duodenum. References: *The lower limit of the duodenum.* Lancet 1913; ii:543; *Recherches d'anatomie comparée sur le duodénum de l'homme et des mammifères; sa signification morphologique et fonctionelle.* Paris, 1922.

Virchow, Rudolf Ludwig Karl (1821–1902): German pathologist born in Schievelbein, Pommerania, and working in Würzburg and Berlin (Fig. V28). He is regarded as the founder of cellular pathology. He noted the spaces

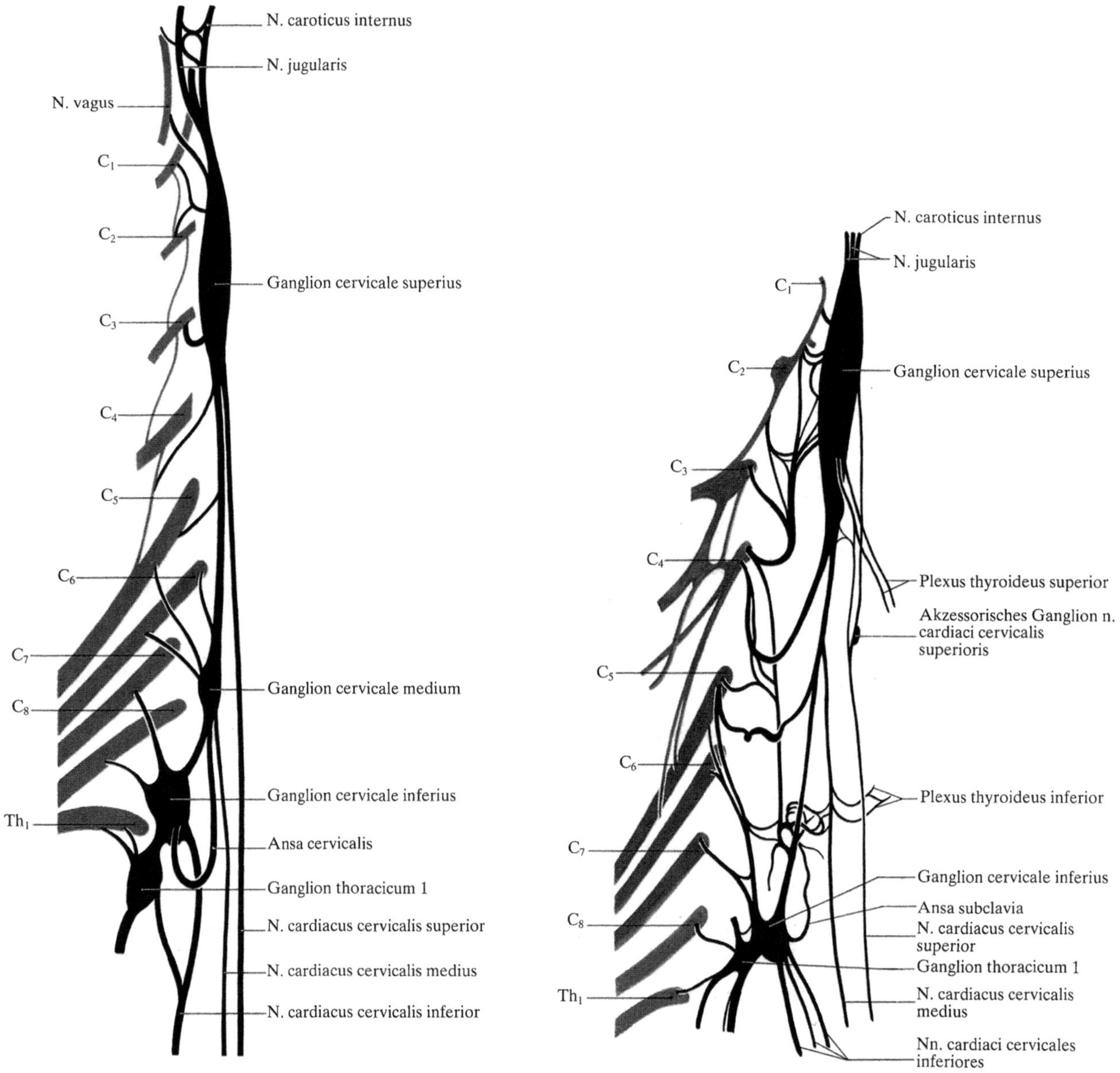

a Ganglion cervicale inferius und thoracicum 1 nicht verschmolzen

b Ganglion cervicale medium fehlt

Fig. V26 Ansa subclavia in cases of variations of the cervical ganglia: (**a**) separate inferior cervical and first thoracic ganglia; (**b**) no middle cervical ganglion present (from Lang J 1985 *Kopf A: Übergeordnete Systeme.* In: von Lanz T, Wachsmuth W, eds, *Praktische Anatomie.*Springer, Berlin-Heidelberg-New York; with permission; from ten Donkelaar HJ, Kachlík D, Tubbs RS *An Illustrated Terminologia Neuroanatomica.* Spriner, Cham, 2018)

around blood vessels in sections of the brain, later described in great detail by Robin (Virchow-Robin spaces). Reference: *Die Cellularpathologie in ihrer Begründung auf physiologische und pathologische Gewebelehre.* Hirschwald, Berlin, 1856.

Corpuscles of Toynbee-Virchow: *Eponym* for the corneal corpuscles, connective tissue cells between the fibrous layers of the stroma of the cornea, also described by Toynbee.

Corpuscles of Virchow-Hassall: *Eponym* for the thymic corpuscles, also described by Hassall.

Line of Virchow-Hoelder: *Eponym* for the horizontal line of Munich, the supraauriculo-suborbital plane, from the superior border of the external acoustic

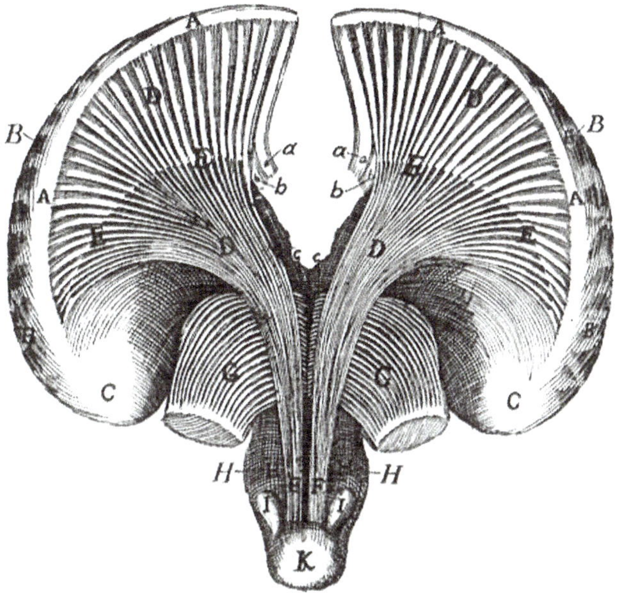

Fig. V27 Preparation of the brain by Vieussens depicting the base of the brain after removing the cerebellum, showing white matter tracts projecting from the centrum ovale (*A*) through the corpora striata (*C*, *D*). (Tabula XV from *Neurographia*. 1684)

Fig. V28 Portrait of Virchow (from Pagel 1901)

meatus to the inferior border of the orbit. This plane was adopted in Frankfurt am Main as the infraorbital-meatal plane ('Frankfurter Horizontale').

Node of Virchow-Troisier *Eponym* for the left supraclavicular lymph node, also described by Troisier.

Plane of Virchow-Hoelder: *Eponym* for the Frankfurt-Virchow plane or Frankfurter Horizontale from the inferior border of the orbit to the superior border of the external acoustic meatus.

Space of Virchow-Robin: *Eponym* for the perivascular space, a tunnel-like extension of the subarachnoid space surrounding blood vessels entering the brain and spinal cord (Fig. V29).

Vitali, Giovanni (1876–1963): Italian anatomist born in Pozzo, Tuscany, and working in Cagliari and later in Pisa (Fig. V30).

Ganglia of Vitali: *Eponym* for small visceromotor ganglia in the inferior orbital fissure, depending on the pterygopalatine ganglion.

Glands of Vitali: *Eponym* for ungual sweat glands present on the anterior fourth of the nail bed.

Vogt, Alfred (1879–1943): Swiss ophthalmologist born in Aargau and working in Aargau, Basel and Zürich.

Arcuate line of Vogt, White: *Eponym* for a circular line of the anterior aspects of the hyaloid canal at the posterior aspect of the lens capsule.

Ring of Vogt: *Eponym* for the anular glial tissue attachment of the posterior aspect of the vitreous body to the optic nerve, visible as a partial or complete ring after complete detachment of the vitreous body, described earlier by Weiss. Reference: *Lehrbuch und Atlas der Spaltlampenmikroskopie,* Vol XI. Springer, Berlin, 1930.

Vogt, Oskar (1870–1959): German physician and neurologist born in Husum. He and his wife Cécile Mugnier (1875–1962) born in Annecy, Haute-Savoie, are known for their extensive cytoarchitectonic and particularly myeloarchitectonic studies of the brain (Fig. V31). In 1897, they met in Paris, where they worked with Jules Dejerine and his wife. They worked in Berlin and in Düsseldorf, and in between, after he was dismissed in

Fig. V29 The space of Virchow-Robin and its relation to the glia limitans. *1* protoplasmic astrocyte, *2* superficial glial limiting membrane, *3* pia mater, *4* arachnoid, *5* trabeculae of arachnoid, *6* blood vessels in subarachnoid space (*7*), *8* dura mater, *9* subdural space, *10* Virchow-Robin space, *11* artery penetrating the cerebral cortex, *12* capillaries, *13* perivascular glial limiting membrane. For magnification of the quadrangles (**a**) and (**b**), see Krstić (1988). (from Krstić R *Die Gewebe des Menschen und der Säugetiere*, 2nd ed. Springer, Berlin-Heidelberg-New York, 1988; with permission; from ten Donkelaar HJ, Kachlík D, Tubbs RS *An Illustrated Terminologia Neuroanatomica.* Springer, Cham, 2018)

1936 from his position as Director of the Kaiser Wilhelm Brain Research Institute, in a small hospital near Neustadt im Schwarzwald. The Vogts distinguished the iso- and allocortex, and promoted a six-layer pattern for the isocortex, instead of the five layers of Meynert and the seven layers of Cajal. Reference: *Allgemeine Ergebnisse unserer Hirnforschung.* J Psychol Neurol (Lpz) 1919; 25:361–376. The studies from the Vogt and Vogt School were used for new maps of the human cerebral cortex; *see* Nieuwenhuys R, Broere CAJ *A detailed comparison of the cytoarchitectonic and myeloarchitectonic maps of the human neocortex produced by the Vogt-Vogt School.* Brain Struct Funct 2020; 225:1–7.

Fig. V30 Photo of Vitali (from Wikimedia Commons; in public domain)

Voight, Christian August (1809–1890): Austrian anatomist in Vienna.

Lines of Voight: *Eponym* for the hair tracts. Reference: *Abhandlung ueber die Richtung der Haare am menschlichen Körper.* Vienna, 1857.

Voit, Max (1876–1920): German anatomist in Freiburg and later Göttingen.

Nerve of Voit: *Eponym* for a branch of the anterior ramus of the cochlear nerve, supplying the macula of the sacculus. Reference: *Zur Frage der Verästelung des Nervus acusticus bei den Säugethieren.* Anat Anz 1907; 31:635–655.

Nucleus of Voit: *Eponym* for an accessory dentate nucleus in the cerebellum.

Volkmann, Alfred Wilhelm (1800–1877): German physiologist in Dorpat and later Halle (Fig. V32), father of Richard von Volkmann.

Canals of Volkmann: *Eponym* for perforating canals in compact bone connecting the osteonic canals of Havers with the periosteum, containing blood vessels. Reference: *Ueber die näheren Bestandtheile der menschlichen Knochen.* Ber Sachs Ges Wiss Math Phys Kl 1873; 25:275.

Fig. V31 Oskar and Cécile Vogt in 1903 with among others Max Lewandowsky at the right and Korbinian Brodmann at the left (Archiv Universität Düsseldorf; in public domain)

Fig. V32 Portrait of Volkmann (from Pagel 1901)

Volkmann, Richard von (1830–1889): German surgeon born in Leipzig and working in Berlin and Halle, son of Alfred Wilhelm Volkmann.

Triangle of Volkmann: *Eponym* for the posterolateral corner of the distal end of the tibia.

Tubercle of Volkmann: *Eponym* for the posterior tibial tubercle, a small prominence on the distal end of the tibia, to which the posterior tibiofibular ligament inserts.

Voogd, Jan (1932-today): Dutch neuroanatomist born in Schiedam working since the 4th year of his medical studies in Leiden (Fig. V33a) and later in Rotterdam as successor of Hans Kuypers. He focussed on the longitudinal and zonular organization of the cerebellar cortex. References: *The Cerebellum of the Cat. Structure and Fibre Connections.* Thesis, University of Leiden; van Gorcum, Assen, 1964; *Comparative aspects of the structure and fibre connections of the mammalian cerebellum.* Prog Brain Res 25:94–134; *Cerebellar zones: A personal history.* Cerebellum 2011; 10:334–350; *The cerebellum*

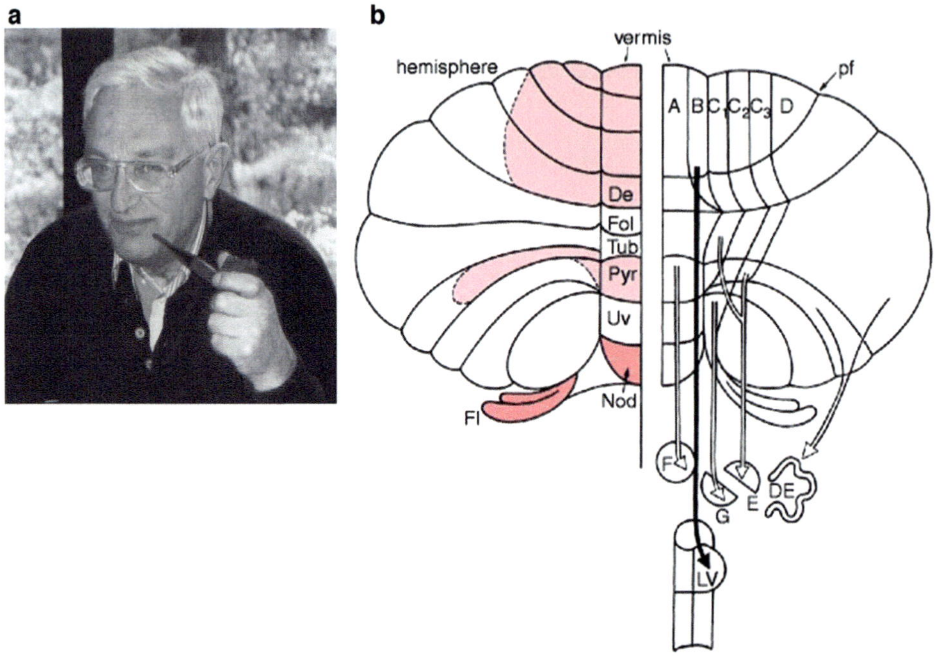

Fig. V33 (**a**) Photo of Jan Voogd (courtesy Tom Ruigrok, Rotterdam); (**b**) The longitudinal subdivision of the human cerebeelum as pioneered by Jan Voogd (from ten Donkelaar HJ, Kachlík D, Tubbs RS *An Illustrated Terminologia Neuroanatomica.* Springer, Cham, 2018)

and precerebellar nuclei. In: Mai JK, Paxinos G (eds) *The Human Nervous System,* 3rd ed. Elsevier, Amsterdam, pp. 471–545 (with Tom JH Ruigrok).

Longitudinal zones of Voogd: *Eponym* for the longitudinal zones of Purkinje cells in the cerebellar cortex, A-, B-, C- and D-zones, each projecting to its own cerebellar nucleus and receiving input from different parts of the inferior olive (Fig. V33b).

20 Letter W

Wachendorff, Eberhard Jacob (1702–1758): Dutch botanist and surgeon born and working in Utrecht (Fig. W1).

Membrane of Wachendorff: *Eponym* for the embryonic pupillary membrane, earlier described by Albinus in 1737. Reference: *Commercii litterarii ad rei medicae et scientiae naturalis incrementum instituti.* Nuremberg, 1740, Hebdomas 18:137–141.

Wagner, Rudolf (1805–1864): German zoologist born in Bayreuth and working in Erlangen and later as zoologist, comparative anatomist and physiologist in Göttingen.

Corpuscles of Wagner: *Eponym* for tactile nerve endings, described together with Meissner. Reference: *Ueber Vorhandsein bisher unbekannter eigenthümlicher Tastkörperchen in dem Gefüllswärzchen der menschlichen Haut, und ueber die Endausbreitung der sensitiver Nerven.* Nachr Ges Wiss Göttingen, 1852.

Spot of Wagner: *Eponym* for the nucleolus of the ovum. Reference: *Prodomus historiae generationes.* Leipzig, 1836.

Wagstaffe, William Warwick (1843–1910): English surgeon.

Tubercle of Wagstaffe: *Eponym* for the anterior fibular tubercle, a small prominence on the lateral malleolus, to which the anterior tibiofibular ligament inserts; also described by Le Fort. Reference: *An unusual form of fracture of the fibula.* St Thomas's Hospital Reports 1875; 6:43–49.

Waldeyer, Heinrich Wilhelm Gottfried (1836–1921): German anatomist, pathologist and embryologist born in Hehlen and working in Strassburg and later in Berlin (Fig. W2). Later, he added his mother's name von Hartz. Waldeyer is well known for his studies on the sex glands. He was the first to apply the term neuron. Reference: *Über einige neuere Forschungen im Gebiete der Anatomie des Centralnervensystems.* Dtsch Med Wochenschr 1891; 17:1213–1218, 1244–1246, 1267–1269, 1287–1289, 1331–1332, 1352–1356.

Anulus or **Ring of Waldeyer:** *Eponym* for the lymphatic pharyngeal ring formed by the lingual, palatine, tubal and pharyngeal tonsils, also described by Bickel.

Body of Waldeyer: *Eponym* for the retrosternal adipose tissue, formed by thymic residues.

Epithelium of Waldeyer: *Eponym* for the surface epithelium of the ovary.

Fascia of Waldeyer: *Eponym* for the floor of the retrorectal space, above the anococcygeal ligament, formed by the distal and posterior usion of the pelvic and rectal fascia; sometimes also referred to as retrorectal or presacral fascia of Waldeyer.

Fossa of Waldeyer: *Eponym* for (a) the ovarian fossa of the peritoneal cavity, a slight depression on the pelvic wall between the external and internal iliac vessels, also described by Claudius and Krause, and (b) the mesentericoparietal fossa (Fig. W3).

Ligament of Waldeyer: *Eponym* for the transverse perineal ligament, also described by Henle and Krause.

Neurons of Waldeyer: *Eponym* for large marginal neurons of the spinal cord.

Organ of Waldeyer: *Eponym* for the paradidymis, earlier described by Giraldès.

Recessus of Waldeyer: *Eponym* for a mesentericoparietal recess of the peritoneal cavity (Fig. W3).

Sheath of Waldeyer: *Eponym* for an external layer of longitudinal smooth muscle surrounding the ureter. It extends downwards and forms the deep layer of the trigone of the urinary bladder.

Space of Waldeyer: *Eponym* for the tubular space between the bladder wall and the intramural part of the ureter.

H. J. ten Donkelaar et al., *An Illustrated Guide to Anatomical Eponyms*, https://doi.org/10.1007/978-3-031-91664-9_20

Fig. W1 Painting of Wachendorff (from Wikimedia Commons; in public domain)

Fig. W2 Photo of Waldeyer (from Pagel 1901)

Tract of Waldeyer: *Synonym* for the tract of Lissauer.

White line of Waldeyer: *Eponym* for a line on the ovary at the junction of the surface epithelium and the peritoneum, also described by Farre and Romiti. Reference: *Eierstock und Ei. Ein Beitrag zur Anatomie und Entwicklungsgeschichte der Sexualorgane.* Engelmann, Leipzig, 1870 (Fig. W4).

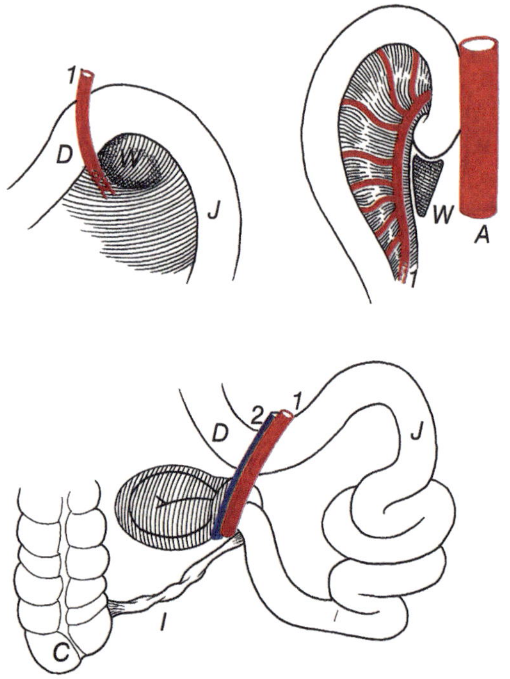

Fig. W3 Mesentericoparietal fossa of Waldeyer (after Gray DW, Skandalakis JE *Embryology for Surgeons.* Saunders, Philadelphia, PA, 1972); (**a**, **b**) Relationships among the mesentericoparietal fossa, **c** mesenteriocoparietal hernia; *A* abdominal aorta, *C* caecum, *D* duodenum, *I* ileum (collapsed), *J* jejunum, *W* fossa of Waldeyer, *1* superior mesenteric artery, *2* superior mesenteric vein

Zone of Waldeyer: *Eponym* for the marginal zone of the spinal cord.

Wallenberg, Adolf (1862–1949): German physician who described the Wallenberg syndrome and the posterior or dorsal trigeminothalamic tract. Reference: Anat Anz 1896; 12:95.

Artery of Wallenberg: *Old term* for the posterior inferior cerebellar artery.

Tract of Wallenberg: *Eponym* for the posterior or dorsal trigeminothalamic tract.

Walter, Johann Gottlieb (1734–1818): German anatomist born in Königsberg and working in Frankfurt am Main and later in Berlin. Reference: *Observationes anatomicae.* Berlin, 1775.

Nerve of Walter: *Eponym* for a branch of the least thoracic splanchnic nerve to the renal plexus. Reference: *Tabulae nervorum thoracis et abdominis.* Decker, Berlin, 1783.

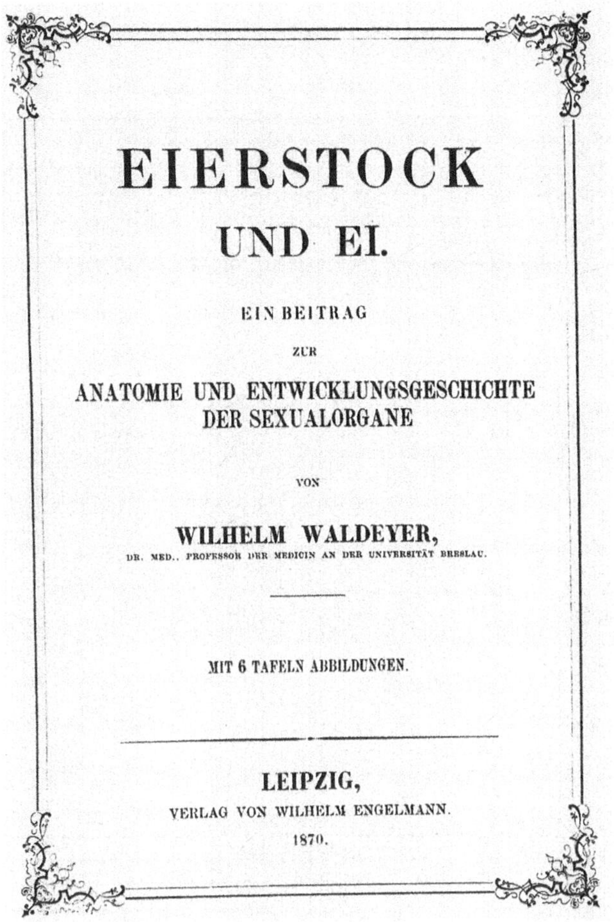
EIERSTOCK

UND EI.

EIN BEITRAG

ZUR

ANATOMIE UND ENTWICKLUNGSGESCHICHTE
DER SEXUALORGANE

VON

WILHELM WALDEYER,

DR. MED., PROFESSOR DER MEDICIN AN DER UNIVERSITÄT BRESLAU.

MIT 6 TAFELN ABBILDUNGEN.

LEIPZIG,

VERLAG VON WILHELM ENGELMANN.

1870.

Fig. W4 Title page of *Eierstock und Ei* (Engelmann, Leipzig, 1870)

Walther, Augustin Friedrich (1688–1746): German anatomist and surgeon born in Wittenberg and working in Leipzig (Fig. W5).

Canals or **Ducts of Walther:** *Eponym* for the minor sublingual ducts, also described by Bartholin and Rivinus.

Ganglion of Walther: *Eponym* for the ganglion impar (Fig. W6). Reference: *Ad anatomiam publicam cadaverii feminae morbo convulsivo affectae invitat.* Langenheim, Leipzig, 1727.

Nerve of Walther: *Eponym* for the greater thoracic splanchnic nerve.

Plexus of Walther: *Eponym* for the intracavernous sympathetic plexus.

Ward, Frederick Oldfield (1818–1877): English physician born in London who as a student published his *Outlines of Human Osteology* (Renshaw, London, 1838).

Triangle of Ward: *Eponym* for an area of diminished density in the trabecular pattern of the neck of the femur. He also described what is currently known as **triangle of Babcock**: a relatively translucent triangular area seen in the subcapital region of the femoral neck.

Fig. W5 Portrait of Walther (from Wikimedia Commons; in public domain)

Wasmann, Adolphus (19th century): German anatomist.

Glands of Wasmann: *Eponym* for the gastric glands.

Weber, Eduard Friedrich (1806–1871): German physiologist and anatomist born in Wittenberg and working in Leipzig (Fig. W7) who wrote extensively on locomotion and body mechanics.

Weber, Ernst Heinrich (1795–1878): German anatomist and physiologist born in Wittenberg, brother of Eduard Friedrich Weber, and working in Leipzig.

Glands of Weber: *Eponym* for mucous glands in the root of the tongue.

Weber, Moritz Ignatz (1795–1875): German anatomist and pathologist born in Landshut and working in Bonn.

Artery of Weber: *Eponym* for a tympanic branch of the external carotid artery.

Organ of Weber: *Eponym* for the prostatic utricle, discovered by Albinus (1704). Reference: *Über das Rudiment eines Uterus beim männlichen Säugethiere.* Leipzig, 1846.

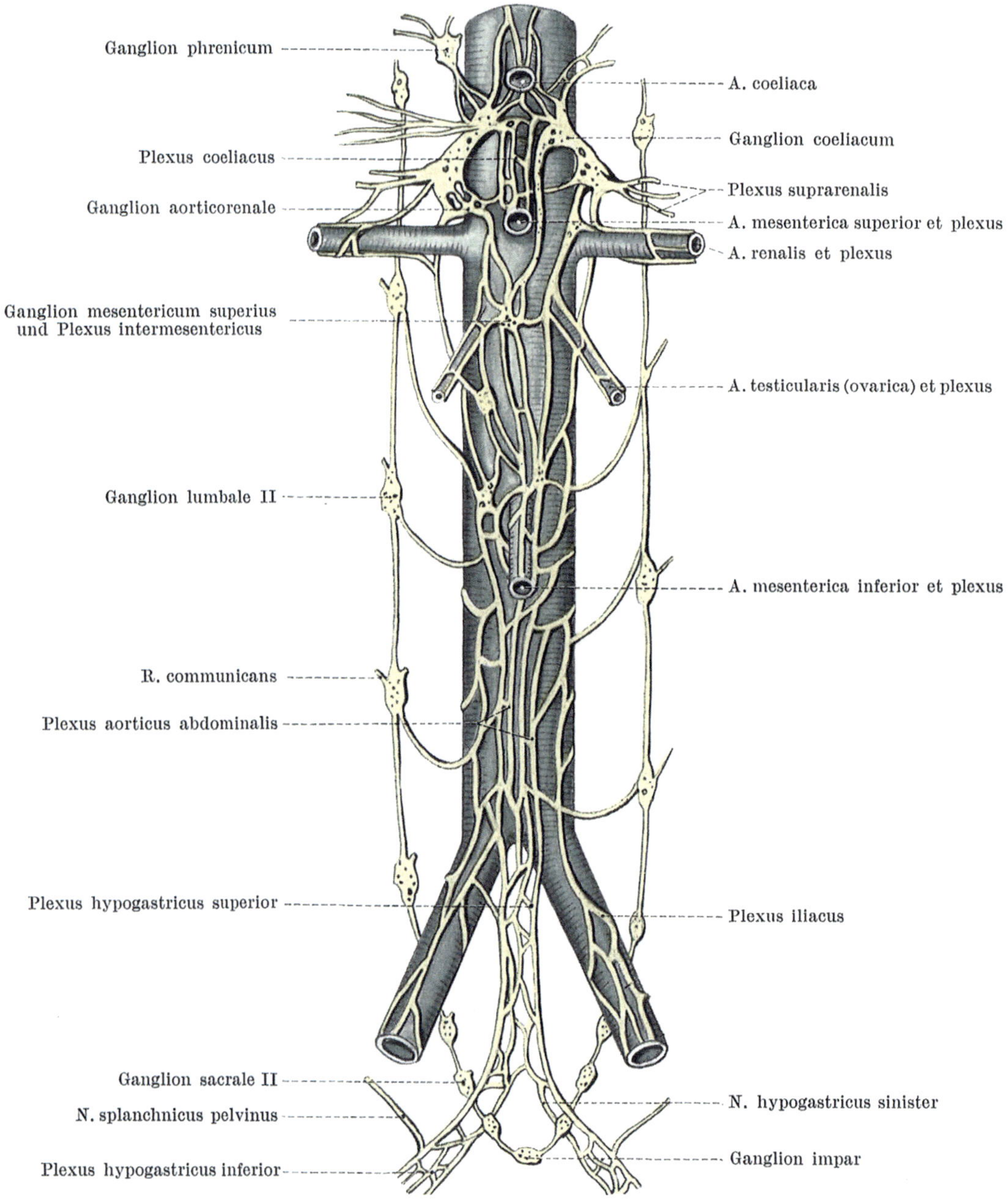

Fig. W6 The ganglion of Walther or ganglion impar (from Hafferl A *Lehrbuch der topographischen Anatomie.* Springer, Berlin-Heidelberg, 1957; with permission)

Weber, Wilhelm Eduard (1804–1891): German physicist in Göttingen (Fig. W8) who collaborated with Eduard Friedrich Weber and published together *Mechanik der menschlichen Gehwerkzeuge; eine anatomisch-physiologische Untersuchung.* Göttingen, 1836.

Area or **Triangle of Weber:** *Eponym* for an area on the sole of the foot formed by the heads of the first and fifth metatarsals and the midpoint of the plantar surface of the heel.

Point of Weber: *Eponym* for a point about 1 cm below the promontory of the sacrum that represents the centre of gravity of the body.

Zone of Weber: *Eponym* for the orbicular zone, circular fibres of the capsule of the hip joint.

Weber-Liel, Friedrich Eugen (1832–1891): German ear surgeon in Berlin and later in Jena.

Fascia of Weber-Liel: *Eponym* for the salpingopalatal fascia or the lateral fascia of the tensor veli palatini,

Fig. W7 Portrait of Weber, EF (from the Wellcome Collection CCBY; in public domain)

Fig. W8 Portrait of Weber, WE (from Wikimedia Commons; in public domain)

separating the medial pterygoid and tensor veli palatini muscles. Reference: *Ueber das Wesen und die Heilbarkeit der häufigsten Form progessiver Schwerhörigkeit.* Hirschwald, Berlin, 1873.

Weibel, Ewald Rudolf (1929–2019): Swiss anatomist who described the endothelial organelles with Palade.

Bodies of Weibel-Palade: *Eponym* for multitubular bodies, storage granules of endothelial and endocardial cells. Reference: *New cytoplasmic components in arterial endothelium.* J Cell Biol 1964; 23:101–112.

Weil, Ludwig Adolf (1849–1895): German dentist working in Munich.

Zone of Weil: *Eponym* for a relatively cell-poor zone of the tooth pulp underlying the layer of odontoblasts, containing cytoplasmic processes of fibroblasts, extracellular matrix, vessels and nerves.

Weiss, Leopold (1849–1901): German ophthalmologist born in Giessen and working in Heidelberg.

Ring of Weiss: *Eponym* for the anular glial tissue attachment of the posterior aspect of the vitreous body to the optic nerve, visible as a partial or complete ring after complete detachment of the vitreous body, later also described by Vogt. Reference: *Über das Vorkommen von scharfbegrenzten Ektasien im Augengrunde und über partielle Farbenblindheit bei hochgradiger Myopie.* Wiesbaden, 1897.

Weissmann, August (19th century): Worked for a time at the Anatomical Institute at Frankfurt-am-Main.

Bundle of Weissmann: *Eponym* for the aggregation of intrafusal muscular fibres of a neuromuscular spindle. Reference: *Ueber das Wachsen der quergestreiften Muskeln, nach Beobachtungen am Frosch.* Ztschr Rat Med 1861; 10:263–284.

Weitbrecht, Josias (1702–1747): German-Russian anatomist born in Schorrdorf and working in St. Petersburg known for his studies on ligaments. Reference: *Syndesmologia sive historia ligamentorum corporis humani quam secundum observationes anatomicas concinnavit.* St. Petersburg, 1742 (Fig. W9). For biography, *see* Bartoníček, Naňka O. *Josias Weitbrecht, the founder of syndesmology and the history of the retinacula of Weitbrecht.* Surg Radiol Anat 2019; 41:1103–1111.

Cartilage of Weitbrecht: *Eponym* for the articular disc of the acromioclavicular joint.

Fibres or **Retinacula of Weitbrecht:** *Eponym* for the retinacula of the capsule of the hip joint, three longitudinal folds of the articular capsule, containing vessels to the head of the femur.

Foramen of Weitbrecht, Oval: *Eponym* for an opening in the articular capsule of the shoulder joint between the superior and middle glenohumeral ligaments (Fig. W10).

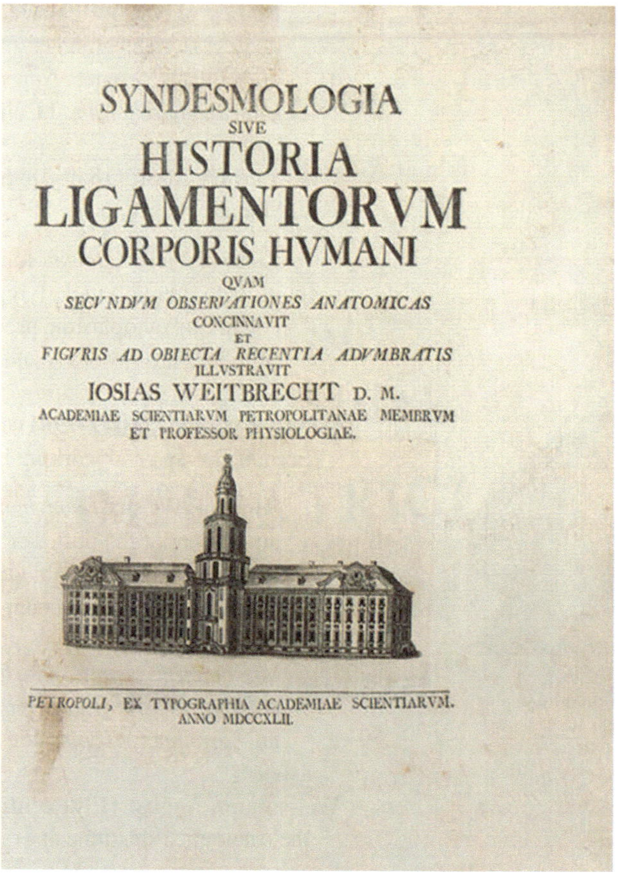
SYNDESMOLOGIA
SIVE
HISTORIA
LIGAMENTORVM
CORPORIS HVMANI
QVAM
SECVNDVM OBSERVATIONES ANATOMICAS
CONCINNAVIT
ET
FIGVRIS AD OBIECTA RECENTIA ADVMBRATIS
ILLVSTRAVIT
IOSIAS WEITBRECHT D. M.
ACADEMIAE SCIENTIARVM PETROPOLITANAE MEMBRVM
ET PROFESSOR PHYSIOLOGIAE.

PETROPOLI, EX TYPOGRAPHIA ACADEMIAE SCIENTIARVM.
ANNO MDCCXLII.

Fig. W9 Weitbrecht: Title page *Syndesmologia*

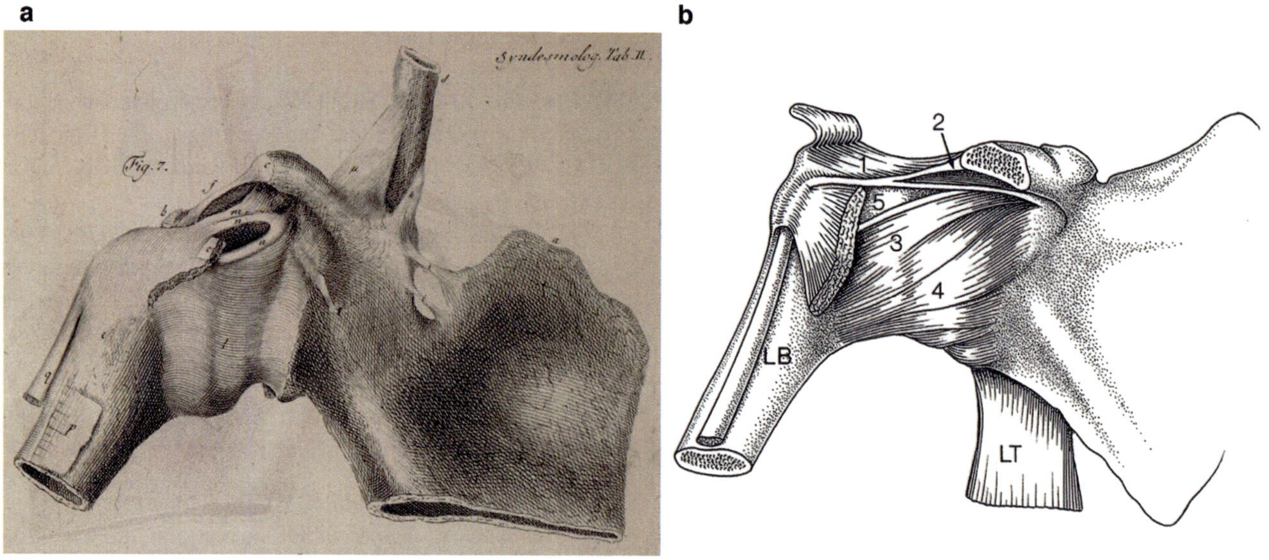

Fig. W10 Foramen of Weitbrecht: (**a**) Original figure from Weitbrecht's *Syndesmologia* (1742); (**b**) Interpretation by Rouvière H *Anatomie humaine, descriptive et topographique*, 12th ed. Masson, Paris, 1967). *1* Coracohumeral ligament, *2–4* superior, middle and inferior glenohumeral ligaments, *5* foramen of Weitbrecht, *LB* long tendon of biceps brachii muscle, *LT* long head of triceps brachii muscle

Ligament of Weitbrecht: *Eponym* for (a) the oblique radioulnar ligament; (b) the anterior ligament of the radiocarpal joint; and (c) the inferior extensor retinaculum of the foot. He also described the skin ligament, which today is called the ligament of Cleland (*see* Fig. C14).

Retinaculum of Weitbrecht: *Eponym* for Weitbrecht's retinaculum tendoni longi (*see* Fig. L4c), which is currently known as the oblique component of the retinacular ligament of Landsmeer.

Vincula tendinum of Weitbrecht: *Eponym* for the vincula or mesotendons of the flexors of the fingers.

Welcker, Hermann (1822–1898): German anthropologist and anatomist born in Giessen and working in Halle, Heidelberg and Giessen.

Angle of Welcker, Sphenoidal: *Eponym* for the angle of the basocranial axis. Reference: *Die Asymmetrien der Nase und des Nasenkehles.* Stuttgart, 1882.

Welsh, David Arthur (1875–1948): Scottish pathologist in Edinburgh and later in Sydney.

Cells of Welsh: *Eponym* for oxyphil cells of the parathyroid gland. Reference: *Concerning the parathyroid glands: A critical anatomical and experimental study.* J Anat (Lond) 1898; 32:292–307.

Wenckebach, Karel Frederik (1864–1940): Dutch physician born in The Hague and working in Groningen and later in Vienna (Fig. W11). He is primarily remembered for his work in cardiology.

Fig. W11 Photo of Wenckebach (Wikipedia; free to use)

Bundle of Wenckebach: *Eponym* for the middle internodal tract of the cardiac conducting system (*see* Fig. I5). Reference: *Beiträge zur Kenntnis der menschlichen Herztätigkeit.* Arch Anat Physiol 1907; 1:1–2.

Wenzel, Joseph (1768–1808): German anatomist and physiologist in Mainz.

Ventricle of Wenzel: *Eponym* for the cavum of the septum pellucidum, also described by Duncan, Sylvius, Verga and Vieussens. Reference: *Prodomus einer Werkes ueber das Hirn der Memshen und Thieren.* Tübingen, 1806.

Wepfer, Johann Jacobus (1620–1695): Swiss physician born in Schaffhausen and working in Basel.

Glands of Wepfer: *Eponym* for duodenal glands, also described by Brunner, his son-in-law. Reference: *Cicutae aquaticae historia & novae.* Basel, 1679.

Wernekinck, Friedrich Christian Gregor (1798–1835): German anatomist in Strassburg who described the decussation of the superior cerebellar peduncles; the term was introduced by his student Franz Joseph Julius Wilbrand in 1840 (*Anatomie und Physiologie der Centralgebilde im Nervensystems.* Rieker, Giessen, 1840; *see* Voogd J, van Baarsen K Cerebellum 2014; 13:113–120).

Decussation of Wernekinck: *Eponym* for the decussation of the superior cerebellar peduncles (Fig. W12).

Wernicke, Carl (1848–1905): German neurologist born in Tarnowitz, Silesia (currently Tarnowskia Góry, Poland) who described several neurological disorders including

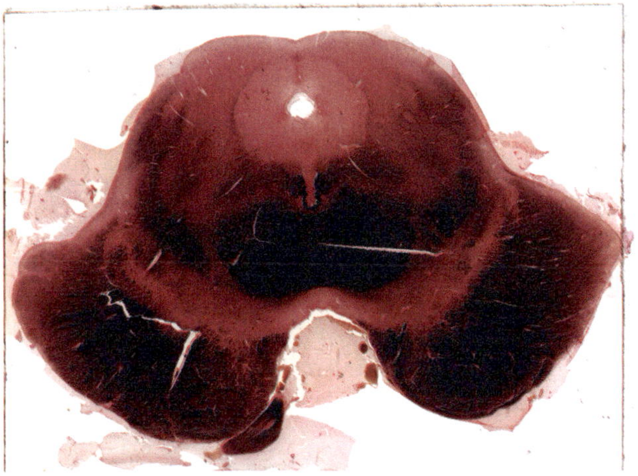

Fig. W12 The Wernekinck decussation of the superior cerebellar peduncles shown in a Weigert-carmine-red myelin-stained transverse section of the mesencephalon (courtesy Veronika Němcová, Prague; from ten Donkelaar HJ, Kachlík D, Tubbs RS *An Ilustrated Terminologia Neuroanatomica.* Springer, Cham, 2018)

Fig. W13 Photo of Wernicke (from the Wellcome Collection CCBY; in public domain)

sensory aphasia (Fig. W13). References: *Der aphasische Symptomencomplex. Ein psychologische Studie auf anatomischer Basis.* Cohn & Weigert, Breslau, 1874; *Das Urwindungssystem des menschlichen Gehirns.* Arch Psychiatr Nervenkr 1876; 6:298–326.

Area of Wernicke: *Eponym* for the sensory speech area on the posterior third of the superior temporal gyrus of the cerebral cortex.

Sulcus of Wernicke: *Eponym* for the frontomarginal sulcus of the cerebral cortex.

Westphal, Carl Friedrich Otto (1833–1890): German neurologist born in Berlin (Fig. W14).

Nucleus of Edinger-Westphal: *Eponym* for the preganglionic part of the accessory oculomotor nucleus, a parasympathetic nucleus of the oculomotor complex. Reference: *Ueber einen Fall von chronischer progressiver Lähmung der Augenmuskeln (Ophthalmoplegia externa) nebst Beschreibung von Ganglienzellengruppen im Bereiche des Oculomotoriuskerns.* Arch Psychiatr Nervenkr 1887; 18:846–871.

Wharton, Thomas (1614–1673): English physician born in Winston on Tees, Durham, and working in London.

Duct or **Canal of Wharton:** *Eponym* for the submandibular duct. Reference: *Adenographia sive glandularum totius corporis descriptio.* Marriot, London, 1656.

Fig. W14 Portrait of Westphal (from Pagel 1901)

Jelly of Wharton: *Eponym* for the connective tissue of the umbilical cord.

Whitnall, Samuel Ernest (1876–1950): English anatomist born in Eccles, Manchester, and working in Montreal and later in Bristol.

Superior transverse ligament of Whitnall: *Eponym* for a condensation of fascial sheaths around the levator muscle of the upper eyelid; it inserts medially over the trochlea and laterally passes between the two parts of the lacrimal gland to insert on the superior part of the lateral orbital wall.

Tubercle of Whitnall: *Eponym* for the orbital tubercle of the zygomatic bone to which several structures attach: the check ligament of the lateral rectus muscle, the suspensory ligament of the eyeball of Lockwood, the lateral palpebral ligament and the lateral horn of the levator aponeurosis. Reference: *On the tubercle of the malar bone and on the lateral attachments of the tarsal plates.* J Anat (Lond) 1911; 45:426–432.

Wieger, Germain or Friedrich (1821–1890): German pathologist born and working in Strasbourg).

Ligament of Wieger: *Eponym* for the ring-shaped hyalocapsular ligament, a thickening of the vitreous body, attached to the posterior surface of the lens. Reference: *Über den Canalis Petiti und ein Ligamentum hyaloideocapsularis.* Thesis, University of Strassbourg, 1883.

Wiesel, Joseph Heinrich (1876–1928): Austrian physician born and working in Vienna.

Paraganglion of Wiesel, Cardiac: *Eponym* for the chromaffin body situated in the cardiac plexus. Reference: *Beiträe zur Anatomie und Entwicklung der menschlichen Nebenniere.* Anat Hefte 1902; 19:481–522.

Wilde, Sir William Robert Wills (1815–1876): Irish ophthalmologist born in Castlereagh and working in Dublin.

Cone of light or **Triangle of Wilde:** *Eponym* for the pyramid of light, the area at the anterior inferior part of the external surface of the tympanic membrane, also known as cone of light of Politzer. Reference: *Practical Observations on Aural Surgery and the Nature and Treatment of Diseases of the Ear.* London, 1853.

Wilkie, David Percival Dalbreck (1882–1938): Scottish surgeon in Edinburgh.

Artery of Wilkie: *Eponym* for the supraduodenal artery. Reference: *The blood supply of the duodenum with special reference to the supraduodenal artery.* Surg Gyn Obst 1911; 13:399–405.

Willis, Thomas (1621–1675): English physician born in Great Bedwyn, Wiltshire, and working in Oxford and later in London (Fig. W15). He is especially known for his contributions to the anatomy of the brain. He reclassified the cranial nerves and described the anastomosing arteries on the base of the brain (the cerebral arterial circle of Willis) and the spinal accessory nerve (the accessory nerve of Willis), and was one of the founders of the Royal Society. Reference: *Cerebri anatome, cui acessit nervorum descriptio et usus.* Martyn & Allestry, London, 1664 (Fig. W16a, b). For biography, *see* Molnár Z *Thomas Willis (1621–1675), the founder of clinical neuroscience.* Nat Rev. Neurosci 2004; 5:329–335.

Antrum of Willis: *Eponym* for the pyloric antrum of the stomach.

Arterial circle of Willis: *Eponym* for the cerebral arterial circle, an anastomotic circle of arteries at the base of the brain (Fig. W16a).

Fig. W15 Engraving of Willis by Thomas Joseph Pettigrew (from the Wellcome Collection CCBY; in public domain)

Centrum nervosum of Willis: *Eponym* for the coeliac ganglia.

Cords of Willis: *Eponym* for dural trabeculae in the superior sagittal sinus.

Fold of Willis: *Eponym* for the inferior medullary velum of the fourth ventricle.

Loop of Willis: *Eponym* for the transition of the circular muscle fibres of the oesophagus to the oblique fibres of the stomach at the cardiac notch, also described by Helvetius.

Nerve of Willis: *Eponym* for (a) the accessory nerve, and (b) the ophthalmic nerve.

Pancreas of Willis: *Eponym* for the uncinate process of the pancreas, also described by Winslow.

Pouch of Willis: *Old term* for the omental bursa.

Wilson, George H (1855–1922): American dentist.

Curve of Wilson: *Eponym* for the curvature of the frontal occlusal plane. Reference: *Proving the articulation, contour and expression.* Dent Regist 1911; 60:382–388.

Wilson, James (1765–1821): English surgeon born in Beith, Ayrshire, and working in London.

Muscle of Wilson: *Eponym* for the external urethral sphincter, also described by Guthrie. Reference: *A description of two muscles surrounding the membranous part of the urethra.* Med-Chir Trans 1812; 1:175–180.

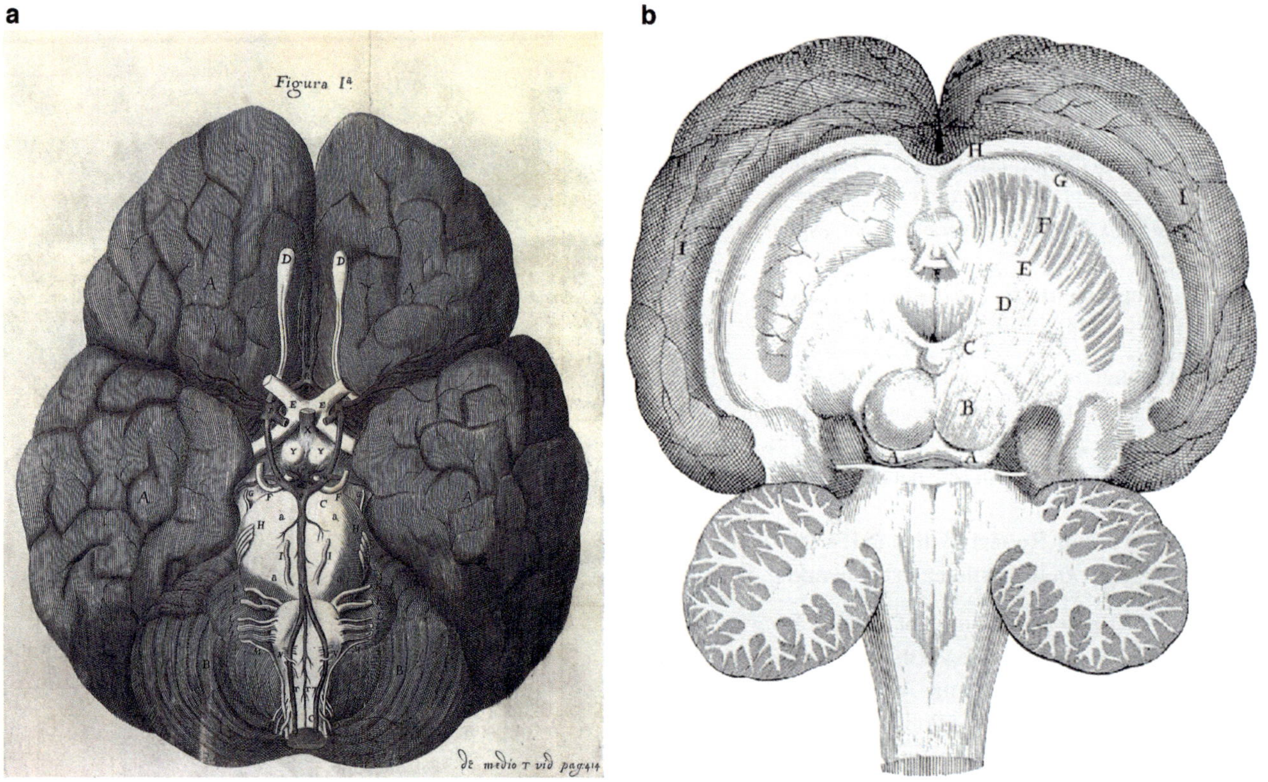

Fig. W16 (**a**) Brain with arterial circle of Willis; (**b**) drawing brain (from Willis)

Wilson, Samuel Alexander Kinnier (1877–1937): American physician known for his anatomical and pathological studies of the basal ganglia.

Pencil bundles of Wilson: *Eponym* for bundles of striatopallidal fibres of the basal ganglia. Reference: *An experimental research into the anatomy and physiology of the corpus striatum.* Brain 1914; 36:427–492.

Winslow, Jacob Benignus (1669–1760): Danish anatomist born in Odense and working mainly in Paris (Fig. W17). Reference: *Exposition anatomique de la structure du corps humain.* Desprez, Paris, 1732.

Foramen of Winslow: *Eponym* for the omental or epiploic foramen, the entrance to the omental or epiploic bursa (Fig. W18).

Ligament of Winslow: *Eponym* for (a) the fibular collateral ligament, (b) the transverse ligament of the knee joint, and (c) the posterior meniscofemoral ligament of the knee joint, also described by Wrisberg.

Pancreas of Winslow: *Eponym* for the uncinate process of the pancreas, described earlier by Willis.

Pathway of Winslow: *Eponym* for an arterial collateral communication between the subclavian and external iliac arteries via the superior and inferior epigastric arteries.

Stars of Winslow: *Eponym* for capillary whorls in the choroidocapillary layer of the choroid, from which the vorticose veins arise.

Winterhalter, Elisabeth Hermine (1856–1952): German gynaecologist, one of the first female doctors in Germany (Fig. W18). Born in Munich, she went to Zürich to study Medicine, since that was not allowed in Germany until 1902. After internships in Paris and Munich, she and her companion Odille Roederstein went to Frankfurt am Main and later to Hofheim am Taunus.

Ganglion of Winterhalter: *Eponym* for an aggregation of sympathetic nerve cells forming the ovarian ganglion. Reference: *Ein sympathisches Ganglion im menschlichen Ovarium.* Arch Gynaekol 1896: 51:49–55.

Wirsung, Johann Georg (1600–1643): German anatomist born in Augsburg and working in Padua (Fig. W19).

Duct of Wirsung: *Eponym* for the pancreatic duct (Fig. W20). Reference: *Figura ductus cuiusdam cum multiplicibus suis ramulis noviter in pancreate inventis in diversis corporibus humanis observati.* Padua, 1642.

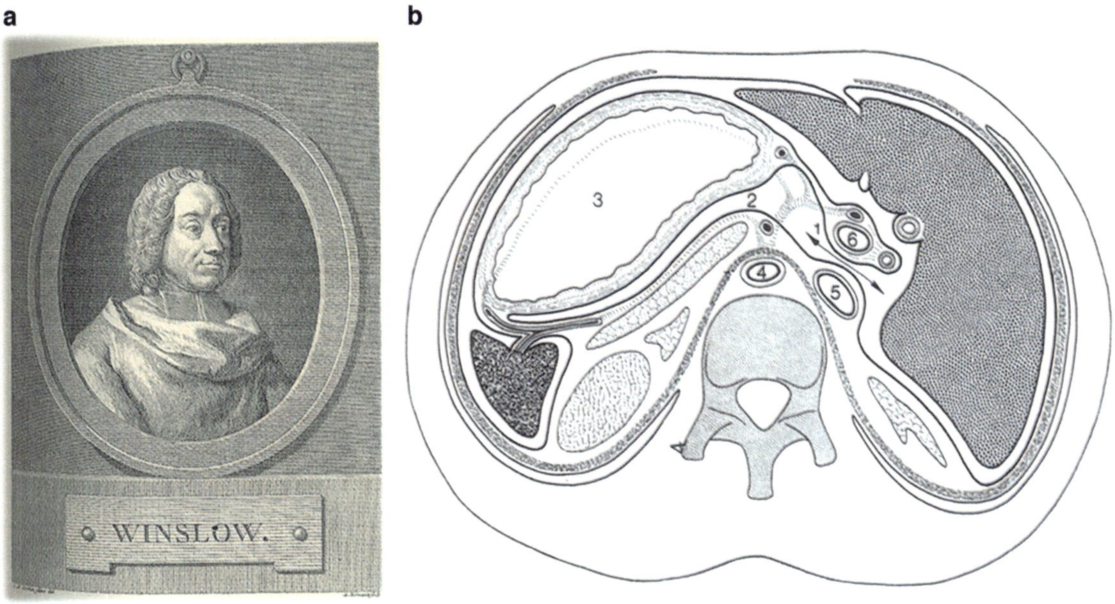

Fig. W17 (**a**) Engraving of Winslow (from Wikimedia Commons; in public domain); (**b**) Foramen of Winslow shown in a transversal section of the abdomen (after ten Donkelaar HJ, Oostra R-J *Klinische Anatomie en Embryologie,* 4th ed. Springer Media, Houten, 2014). *1* Foramen epiploicum of Winslow, *2* bursa omentalis, *3* stomach, *4* abdominal aorta, *5* inferior vena cava, *6* portal vein

Fig. W18 Portrait of Winterhalter by Ottile Roederstein (1918; from Wikipedia; in public domain)

Fig. W19 Fresco at the University of Padua showing Wirsung

Woelfler, Anton (1850–1917): Bohemian surgeon born in Klatovy, Bohemia, and working in Graz and later in Prague.

Gland of Woelfler: *Eponym* for an accessory thyroid gland. Reference: *Ueber die Entwicklung und den Bau der Schilddrüse mit Rücksicht auf die Entwicklung der Kropfe.* Berlin, 1880.

Wolff, Caspar Friedrich (1733–1794): German anatomist born in Berlin and working in St. Petersburg and one of the founders of embryology (Fig. W21). He was the first to describe the embryogenesis of the kidney in the chick embryo. Rathke (1793–1860) named the mesonephros the Wolffian body and its excretory duct the Wolffian duct. Reference: *Theoria generationis.* Halle, 1759.

Body of Wolff: *Eponym* for the mesonephros, later also described by Oken.

Duct of Wolff: *Eponym* for the mesonephric duct, later also described by Oken (Fig. W22).

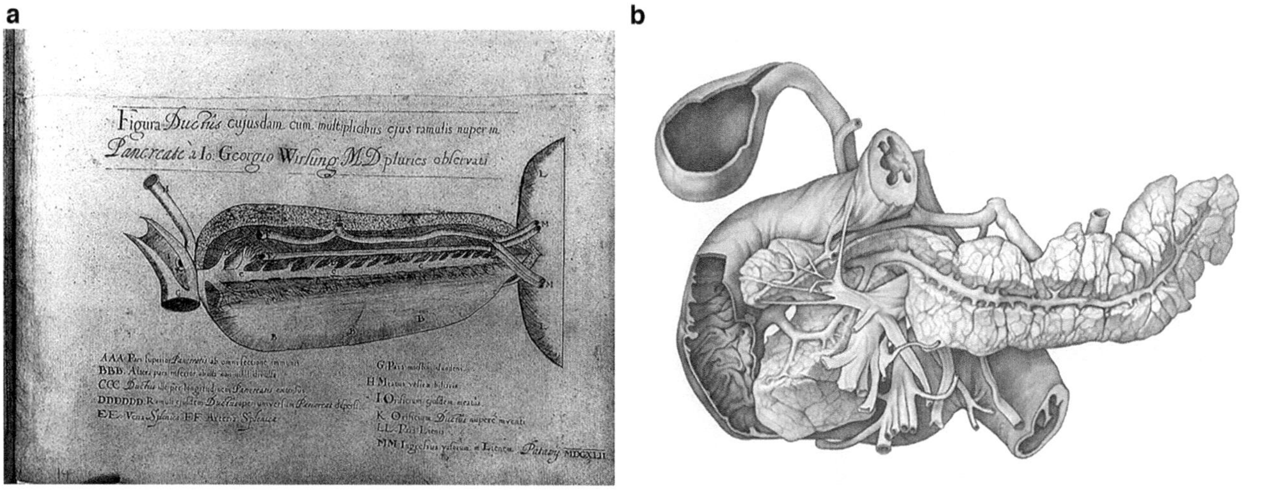

Fig. W20 Duct of Wirsung: (**a**) Original drawing (from *Figura ductus cuiusdam cum multiplicibus suis ramulis noviter in pancreate inventis in diversis corporibus humanis observati.* Padua, 1642); (**b**) Preparation Anatomisch Museum Nijmegen (courtesy Lucas Boer, Nijmegen)

Fig. W21 Portrait of Wolff (from Wikimedia Commons; in public domain)

Rests or **Tubules of Wolff:** *Eponym* for mesonephric nests, remnants of the Wolffian duct in the female genital tract that may give rise to cysts, also described by Kobelt.

Wolfring, Ernst von (1832–1905): Polish ophthalmologist in Warsaw.

Glands of Wolfring: *Eponym* for accessory lacrimal glands mostly in the upper eyelid, also described by Ciaccio, Krause en Henle. Reference: *Untersuchungen über die Drüsen der Bindehaut des Auges.* Zbl Med Wiss 1872; 10:852–854.

Wollschlaeger, Paul B (1920–1981): German-American radiologist born in Berlin and working in Cleveland, New York and Detroit.

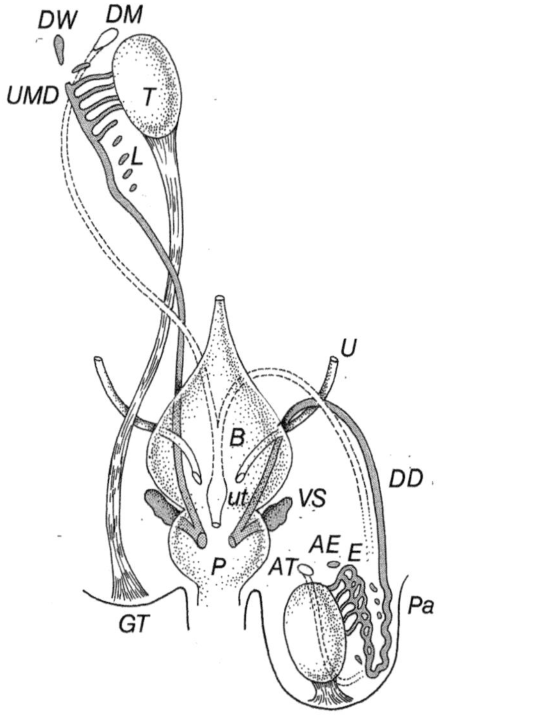

Fig. W22 Duct of Wolff and its remnants (in *gray*). *AE* Appendix of epidydimis, *AT* appendix of testis, *B* bladder, *DD* ductus deferens, *DM* duct of Müller, *DW* duct of Wolff, *E* epidydimis, *GT* gubernaculum, *T* testis, *L* lower caudal mesonephric ducts, *P* prostate, *Pa* paradidymis, *T* testis, *U* ureter, *UMD* upper caudal mesonephric ducts, *ut* utriculus prostaticus, *VS* vesicula seminalis (after ten Donkelaar HJ, Oostra R-J *Klinische Anatomie en Embryologie,* 4th ed. Springer Media, Houten, 2014)

Artery of Wollschlaeger and Wollschlaeger: *Eponym* for an inconstant dural branch of the superior cerebellar artery, supplying parts of the falx cerebri and tentorium cerebelli, described together with his wife Gertraud. Reference: *Eine infratentorielle meningeale*

Arterie [An infratentorial meningeal artery]. Radiologe 1965; 5:451–452.

Wood, Alexander (1817–1884): Scottish physician.

Node of Wood: *Eponym* for the tonsillar lymph node.

Wood, John (1827–1891): English surgeon born in Bradford, Yorkshire, and working in London known for his studies on variations in human myology.

Muscle of Wood: *Eponym* for the abductor ossis metatarsi quinti pedis muscle. Reference: *Variations in human myology.* Proc Roy Soc 1864; 13:303.

Woolner, Thomas (1825–1892): English sculptor and poet who drew Darwin's attention to the tubercle on the helix of the auricle known as tubercle of Darwin.

Worm, Ole (1588–1654): Danish anatomist and theologian born in Aarhus and working in Copenhagen (Fig. W23).

Wormian bones: *Eponym* for the variable sutural bones.

Wrisberg, Heinrich August (1739–1808): German anatomist and gynaecologist born in St.Andreasberg, Harz, and working in Göttingen (Fig. W24). Several structures are linked to his name (*see* Tubbs RS, et al. Clin Anat 2014; 27:10–13). Reference: *Sylloge Commentationum Anatomicarum.* Dieterich, Göttingen, 1786; *Commentationes medici, physiologici, anatomici et obstetrici.* Dieterich, Göttingen, 1800.

Fig. W23 Engraving of Worm (from Wikimedia Commons; in public domain)

Fig. W24 Portrait of Wrisberg (from Wikipedia; in public domain)

Anastomosis of Wrisberg: *Eponym* for an interconnection between the medial brachial cutaneous nerve and the intercostobrachial nerve.

Ansa or **Loop of Wrisberg:** *Eponym* for a loop formed by a branch of the right vagus nerve, the coeliac ganglion and the greater thoracic splanchnic nerve.

Cartilage or **Tubercle of Wrisberg:** *Eponym* for (a) the epiglottis, and (b) the cuneiform cartilage and tubercle of the larynx.

Ganglion of Wrisberg: *Eponym* for a large cardiac ganglion located below the arch of the aorta to the right of the ligamentum arteriosum.

Ligament of Wrisberg: *Eponym* for the posterior meniscofemoral ligament of the knee joint, also described by Winslow.

Lobe of Wrisberg: *Eponym* for the lobe of the azygos vein, a supernumerary lobe of the right lung related to the arch of the azygos vein.

Nerve of Wrisberg: *Eponym* for (a) the intermediate nerve, and (b) the medial brachial cutaneous nerve. Reference: *Observationes anatomicae de quinto pars nervorum encephali.* Dieterich, Göttingen, 1777.

Fig. W25 Portrait of Wutzer (from Universitätsarchiv Bonn; Wikimedia Commons, in public domain)

Wutzer, Karl Wilhelm (1789–1863): German surgeon born in Bonn and working in Münster (Fig. W25).

Ganglion of Wutzer: *Eponym* for the inferior or nodose ganglion of the vagus nerve, also described by Bendtz. Reference: *De corporis humani ganglionorum fabrica atque usu.* Nicolai, Berlin, 1817.

X–Z

21

Young, Robert Bruce (1858–1927): Scottish anatomist in Glasgow.

Ligament of Young: *Eponym* for the trapeziometacarpal ligament between the trapezium and the second and third metacarpals.

Zaglas, John (19th century): Scottish anatomist in Edinburgh.

Ligament of Zaglas: *Eponym* for a short fibrous part of the posterior sacroiliac ligament, extending from the posterior superior iliac spine to the second transverse tubercle of the sacrum.

Zahn, Friedrich W (1845–1904): German pathologist.

Groove of Zahn: *Eponym* for a groove on the diaphragmatic surface of the liver, predominantly on the right lobe of the liver; such grooves were also described during the foetal period by Liebermeister. Reference: *Note sur les plis respiratoires du diaphragme et les sillons diaphragmatiques du foie.* Rev. Méd Suisse Romana 1882; 2:531–535.

Zang, Christoph Bonifacius (1772–1835): German surgeon born in Frickenhausen, Prussia, working in Vienna and later in Würzburg.

Space of Zang: *Eponym* for the lesser supraclavicular fossa between the two origins of the sternocleidomastoid muscle and the clavicula, also described by Sedillot.

Zaufal (Coufal), Emanuel (1833–1910): Bohemian otologist born in Pluschwitz (currently Buškovice), near Cheb, Bohemia, and working in Prague (Fig. Z1). For biography, *see* Piffl O *Emanuel Zaufel (1837–1910).* Arch Ohrenheilk 1910; 82:132–141.

Fold of Zaufal: *Eponym* for the salpingopharyngeal fold. Reference: *Die Plica salpingo-pharyngea (Wulstfalte).* Arch Ohr-Heilkunde 1880; 15:96–144.

Zeis, Eduard (1807–1868): German physician in Dresden, also working as a surgeon in Marburg.

Glands of Zeis: *Eponym* for the ciliary glands. Reference: *Anatomische Untersuchungen der Meibomischen Drüsen des Menschen und der Thiere mit besonderer Berücksichtigung ihres Verhaltens zum Tarsus.* Zschr Ophthal 1835, 3:231–271.

Zeissl, Maximilian von (1853–1925): Austrian dermatologist born and working in Vienna.

Membrane of Zeissl: *Eponym* for a doubtful compact layer of the subglandular layer of the gastric mucosa.

Zenker, Friedrich Albert von (1825–1898): German pathologist born in Dresden and working in Dresden and Erlangen (Fig. Z2).

Diverticulum of Zenker: *Eponym* for a common diverticulum of the oesophagus, arising between the internal pharyngeal constrictor muscle and the cricopharyngeal muscle. Reference: *Krankheiten des Oesophagus* (with von Ziemssen H). In: von Ziemssen H, Hrsg *Handbuch der speciellen Pathologie und Therapie*. Siebenter Band, Erste Hälfte. Vogel, Leipzig, pp. 1–87.

Zenker, Wolfgang (1925–2022): German-Swiss anatomist born in Brno and, after studying in Prague and Vienna, working in Vienna, Bochum and Zürich. For several editions, he was one of the editors of *Benninghoff Anatomie* (Urban & Schwarzenberg, Munich). He is also known for his studies on the juxtaoral apparatus of Chievitz.

Retroarticular cushion or **Pillow of Zenker:** *Eponym* for the loose connective tissue and a venous plexus located between the bifurcated lamellae of the bilaminar zone of the temporomandibular articular disc, already described as plexus articularis in 1867 by Luschka. Reference: *Das retroartikulare Polster des Kiefergelenks und seine mechanische Bedeutung.* Z Anat Entwicklungsgesch 1956;119:375–388.

H. J. ten Donkelaar et al., *An Illustrated Guide to Anatomical Eponyms*, https://doi.org/10.1007/978-3-031-91664-9_21

Fig. Z1 Portrait of Zaufal (from Pagel 1901)

Fig. Z2 Portrait of Zenker (from Wikimedia Commons; in public domain)

Ziehen, Theodor (1862–1850): German neurologist known for his studies of the human brain (Fig. Z3).

Nucleus of Ziehen: *Eponym* for the sagulum nucleus of the midbrain. Reference: *Anatomie des Centralnervensystems.* In: Bardeleben, K von, Hrsg *Handbuch der Anatomie des Menschen,* Vol 4, Part 1. Fischer, Jena, 1899.

Fig. Z3 Photo of Ziehen (from Wikimedia Commons; in public domain)

Zimmermann, Gustav Heinrich Eduard (1817–1866): German physician born in Stettin who died of cholera in the Prusso-Austrian war in 1866.

Granules of Zimmermann: *Eponym* for blood platelets, three years earlier described by Bizzozero. Reference: *Die Methode der Blutanalyse nebst Untersuchungen über die Constitution des gesunden Blutes.* Hamm, 1855.

Zimmermann, Karl V (1861–1935): German histologist.

'Polkissen' of Zimmermann: *Eponym* for mesangial cells filling the triangular surface between the macula densa and the afferent and efferent arterioles of the juxtaglomerular apparatus (*synonym*: extraglomerular mesangium), also described as juxtaglomerular cells by Goormaghtigh. Reference: *Über den Bau des Glomerulus der Säugerniere.* Z Microsc Anat Forsch 1933; 3:176–278.

Zinn, Johann Gottfried (1727–1759): German anatomist and physician born in Ansbach, Bavaria, and working in Göttingen (Fig. Z4). He published a classical treatise on the eye. Reference: *Descriptio anatomica oculi humani.* Vandenhoeck, Göttingen, 1755.

Fig. Z4 Portrait of Zinn (from Pagel 1901)

Anulus, Ring or **Tendon of Zinn:** *Eponym* for the common tendinous ring of the extraocular muscles (Fig. Z5).

Artery of Zinn: *Eponym* for the central retinal artery.

Ligament of Zinn, Suspensory or **Zonule of Zinn:** *Eponym* for the suspensory ligament of the lens.

Membrane of Zinn: *Eponym* for the anterior layer of the iris.

Vascular circle or **Corona of Zinn:** *Eponym* for the vascular circle of the optic nerve, a network of branches of the short ciliary arteries on the sclera around the exit of the optic nerve, also described by Haller.

Zoja, Giovanni (1833–1899): Italian anatomist born in Castelbelforte and working in Pavia.

Glands of Zoja: *Eponym* for small accessory thyroid glands.

Promontory of Zoja: *Eponym* for the ethmoidal bulla protruding into the middle nasal meatus.

Zones:

Zone of Lissauer: *Eponym* for the posterolateral fasciculus of the spinal cord, sometimes also connected with Spitzka; *see* Lissauer, Heinrich.

Zone of Marchant: *Eponym* for an area on the skull base (basisphenoid and basioccipital parts), from which the dura mater can be easily detached; *see* Marchant, Felix.

Zone of Obersteiner-Redlich: *Eponym* for the true boundary between the central and the peripheral nervous system.

Zone of Spitzka: *Synonym* for the zone of Lissauer; *see* Spitzka, Edward C.

Zone of Waldeyer: *Eponym* for the marginal zone of the spinal cord; see Waldeyer, Heinrich Wilhelm Gottfried.

Zone of Weber: *Eponym* for the orbicular zone, circular fibres of the capsule of the hip joint; *see* Weber, Wilhelm Eduard.

Zone of Weil: *Eponym* for a relatively cell-poor zone of the tooth pulp underlying the layer of odontoblasts; *see* Weil, Ludwig.

Zuckerkandl, Emil (1849–1910): Austrian anatomist born in Raab (now Györ, Hungary) and working in Graz and Vienna, older brother of Otto Zuckerkandl. He is well known for his exhaustive work on topographical anatomy (Fig. Z6). Reference: *Atlas der topographischen Anatomie des Menschen.* Vienna-Leipzig, 1904.

Air cells of Zuckerkandl: *Eponym* for anterior ethmoidal air cells related to the frontal sinus.

Bodies of Zuckerkandl: *Eponym* for the chromaffin or paraaortic bodies.

Concha of Zuckerkandl: *Eponym* for an accessory concha of the nasal cavity.

Convolution of Zuckerkandl: *Eponym* sometimes used for the subcallosal gyrus of the cerebral cortex.

Fascia of Zuckerkandl: *Eponym* for the posterior layer of the renal fascia. Reference: *Beiträge zur Anatomie des menschlichen Körpers. Über den Fixations-Apparat der Nieren.* Med Jahrb 1883, pp. 59–67.

Gland of Zuckerkandl: *Eponym* for an ectopic thyroid gland located in front of the hyoid bone. Reference: *Über eine bisher noch nicht beschriebene Drüse in der Regio suprahyoidea.* Stuttgart, 1879.

Recess of Zuckerkandl: *Eponym* for the salpingopharyngeal recess.

Tubercle of Zuckerkandl: *Eponym* for an inconstant tubercle of the thyroid gland, his processus posterior

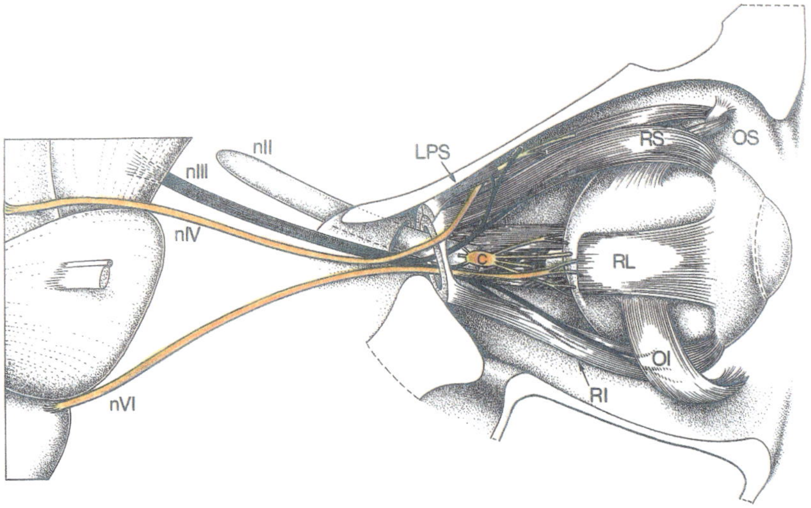

Fig. Z5 Ring of Zinn (after ten Donkelaar HJ, Oostra R-J *Klinische Anatomie en Embryologie,* 4th ed. Springer Media, Houten, 2014). *c* Ciliary ganglion, *nII*, *nIII*, *nIV*, *nVI* cranial nerves, *LPS* levator palpebrae superioris muscle, *OI*, *OS* inferior and superior oblique muscles, *RI*, *RL*, *RS* inferior, lateral and superior rectus muscles

glandulae thyreoideae, projecting from the posterior or posteromedial margin of the thyroid gland and extending posterior into the tracheoesophageal groove. It was first described by Otto Madelung in 1867 as the posterior horn of the thyroid. Reference: *Die Epithelkörperchen von Didelphys azara nebst Bemerkungen über die Epithelkörperchen des Menschen*. Anat Hefte 1902; 19: 59–84.

Zuckerkandl, Otto (1861–1921): Austrian urologist born in Raab (now Györ, Hungary) and working in Vienna, younger brother of Emil Zuckerkandl. Reference: *Handbuch der Urologie*. 1905.

Fascia of Zuckerkandl: *Eponym* for the rectovaginal fascia or septum.

Fig. Z6 Portrait of Zuckerkandl (from Wikimedia Commons; in public domain)

Part III

Anatomical Eponyms Used in Clinical Disciplines

Introduction

In most clinical disciplines, the use of anatomical eponyms is more frequent than in anatomy. It is beyond the scope of this treatise to list all such eponyms, the more since many of the clinical eponyms concern operations, techniques or instruments. A few examples will be given in the following chapters, including:

Chapter 22: Anatomical Eponyms in Abdominal Surgery (David Kachlík and Hans J. ten Donkelaar);
Chapter 23: Anatomical Eponyms in Dentistry (David Kachlík, Jiři Šedý and Hans J. ten Donkelaar);
Chapter 24: Anatomical Eponyms in Neurology (Hans J. ten Donkelaar, Veronika Němcová and Marina Quartu);
Chapter 25: Anatomical Eponyms in Obstetrics and Gynaecology (Hans J. ten Donkelaar, Veronika Němcová, Marina Quartu);
Chapter 26: Anatomical Eponyms in Oncology (David Kachlík and Hans J. ten Donkelaar);
Chapter 27: Anatomical Eponyms in Ophthalmology (Hans J. ten Donkelaar, David Kachlík and Johannes R.M. Cruysberg);
Chapter 28: Anatomical Eponyms in Orthopaedics (Hans J. ten Donkelaar, Marina Quartu, Ondřej Naňka and Michael Pretterklieber);
Chapter 29: Anatomical Eponyms in Otology (Hans J. ten Donkelaar, David Kachlík and Vedat Topsakal);
Chapter 30: Anatomical Eponyms in Phlebology (David Kachlík, Alberto Caggiati and Hans J. ten Donkelaar);
Chapter 31: Anatomical Eponyms in Radiology of the Digestive System (Marina Quartu and Hans J. ten Donkelaar).

Anatomical Eponyms in Abdominal Surgery

22

David Kachlík and Hans J. ten Donkelaar

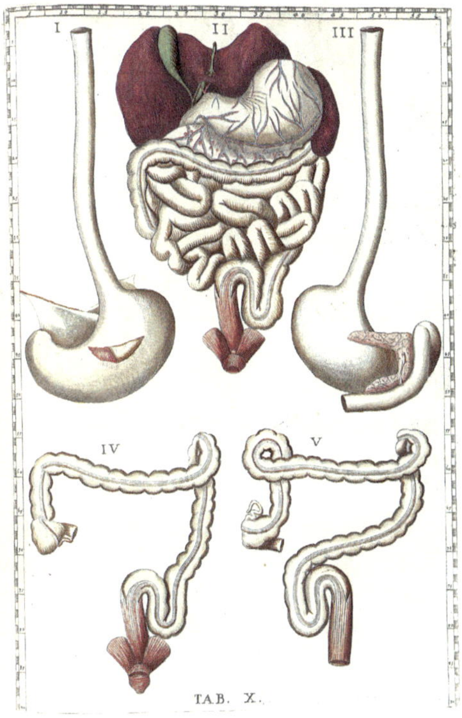

Vignette The abdominal organs according to Eustachi (1552), published 1714 by Lancisi.

22.1 Introduction

In this section, an overview will be presented of anatomical eponyms used in abdominal surgery, including projections of abdominal organs upon the anterior body wall and certain lines in use for that, the fasciae of the abdomen and perineum, the peritoneal cavity with its pouches and recesses, the inguinal canal and its herniae, the abdominal organs, the anorectal canal, and the vascularization and lymphatic drainage of the intestines. For further data on the various authors quoted, reference is made to Part II (*The Anatomical Eponyms from A to Z*) and figures in this part. The **Vignette** shows the abdominal organs according to Eustachi (1552, published 1714 by Lancisi; *see* Eustachi, Bartolomeo).

For the cleavage lines of the skin due to the disposition of the subcutaneous tissue, the *eponym* **lines of Langer** (*see* Langer, Karl, and Fig. L5) was recommended as a guideline for surgical incisions. More recently, these lines have been replaced by other lines: (a) the **lines of Kraissl**, an *eponym* for lines of maximal skin tension (Kraissl 1951; *see* Kraissl, Cornelius J); (b) the **lines of Borges**, an *eponym* for the resting skin tension lines, rather similar to the lines of Kraissl (Borges 1984; *see* Borges, Albert F); and (c) the **lines of Blaschko**, an *eponym* for a system of lines on the surface of the human body representing a developmental growth pattern (Blaschko 1901; *see* Blaschko, Alfred). For several incisions of the abdominal wall, *eponyms* are in use: **subcostal of Kocher** (*see* Kocher, Theodor) for hepatobiliary surgery**,** **McBurney** (*see* McBurney, Charles) for exposure of the caecum and its vermiform appendix, and **Pfannenstiel**, a transverse lower abdominal incision used mainly in gynaecology (after Hermann Johann Pfannenstiel, a German gynaecologist, 1862–1909).

22.2 Eponyms in Use for Projection of Organs

The level of the first lumbar vertebra marks the **transpyloric plane** or **line of Addison**, found halfway between the xiphoid process of the sternum and the umbilicus (Addison 1899; *see* Addison, Christopher). On this plane, the gallbladder, the pylorus, the neck of the pancreas, the kidneys and the aortic hiatus of the diaphragm are located. The right spinoumbilical line or **line of Monro-Richter** (Fig. 22.1; *see* Monro, Alexander; and Richter, August Gottlieb) extends from the right anterior superior iliac spine to the umbilicus. On this line, some 6–7 cm medial to the spine, the **point of McBurney** is found, an *eponym* for the point of maximum tenderness in appendicitis (McBurney 1889; *see* McBurney,

H. J. ten Donkelaar et al., *An Illustrated Guide to Anatomical Eponyms*, https://doi.org/10.1007/978-3-031-91664-9_22

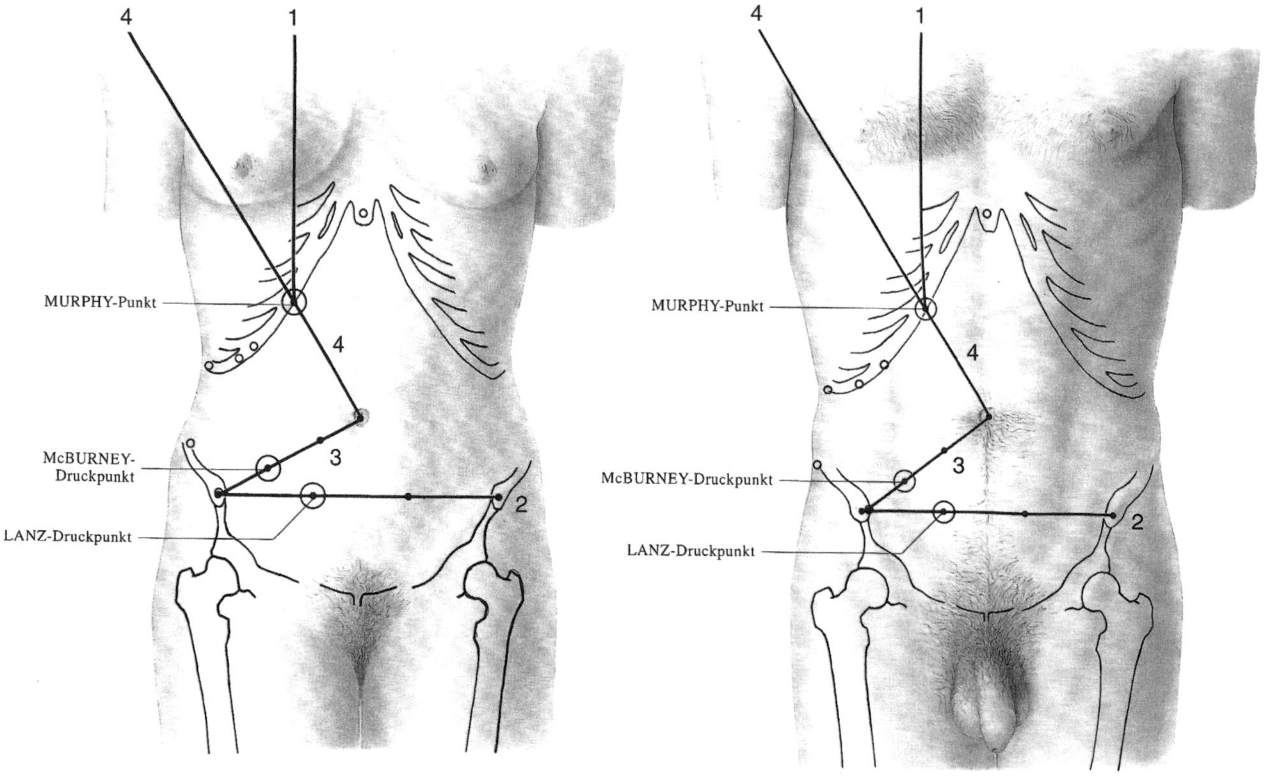

Fig. 22.1 Lines and points for the projection of abdominal organs (from von Lanz und Wachsmuth, *Praktische Anatomie*: *Bauch*, von H Loeweneck und G Feifel. Springer, Berlin, Heidelberg, New York, 1993; with permission)

Charles). On the interspinous line, also known as the *eponym* **line of Lanz**, at 1/3 of a line between the right and left anterior superior iliac spines, the **point of Lanz** marks the projection of the apex of the vermiform appendix of the caecum upon the anterior abdominal wall, another point of tenderness in appendicitis (Lanz 1908; *see* Lanz, Otto). The **triangle of Sherren** (*see* Sherren, James) is an area of hyperaesthesia present in acute appendicitis, bordered by lines connecting the umbilicus, the pubic tubercle and the right anterior superior iliac spine. Several other points or signs, such as the **point of Clado** (*see* Clado, Spiro) and the **point of Munro** (*see* Munro, John), both at the lateral margin of the right rectus abdominis muscle, have been described indicative for peritoneal irritation and appendicitis. The point of Clado is found on the interspinous line, that of Munro on the spinoumbilical line. At the crossing of the right medioclavicular line with the right axilloumbilical line (Fig. 22.1), the **point of Murphy** marks the projection of the gallbladder (*see* Murphy, John Benjamin). On the right axilloumbilical line, the **point of Desjardins** has been described as the projection site of the major duodenal papilla of Vater (*see* Desjardins, Abel). The anterior wall of the stomach projects onto the anterior abdominal wall, for which the *eponyms* **triangle of Labbé** (*see* Labbé, Léon) and **space of Traube** (Traube 1868; *see* Traube, Ludwig) are used.

22.3 The Fasciae of the Abdomen and the Perineum

The fasciae of the abdomen are a complex and intricate structure. In the Anglosaxon literature, usually the **superficial abdominal fascia** of the lower part of the anterolateral abdominal wall means its **subcutaneous tissue**, and it is divided into two parts or layers, a superficial and a deep one but the accuracy of this subdivision has been questioned (Hollinshead 1971), and this subdivision is hardly used in other languages (Wendell-Smith 1997):

(1) The more fatty and superficial part or layer is known as the **fascia of Camper** (*see* Camper, Petrus) or as the superficial adipose layer of the subcutaneous layer of the anterolateral abdominal wall, although Camper may not have described it himself (Tobin and Benjamin 1949).
(2) The deeper layer of the superficial fascia is described as more fibrous and usually called the subcutaneous abdominal fascia or the **fascia of Scarpa** (*see* Scarpa, Antonio) or as the membranous layer of the subcutaneous tissue of the anterolateral abdominal wall. The apparent fibrous layer of Scarpa may not necessarily form the deep surface of the superficial fascia but may be enclosed in fat as well.

Below the subcutaneous tissue, there remains a certain amount of connective tissue bound to the external oblique abdominis and rectus abdominis muscles, and termed the **innominate fascia of Gallaudet** (Gallaudet 1931; *see* Gallaudet, Bern).

The fascia of Scarpa is continuous across the inguinal ligament with the superficial fascia of the thigh (emphasized as the saphenous fascia), and in the pubic region with the superficial fascia of the perineum and the tunica dartos of the scrotum. The **superficial perineal fascia** (**fascia of Cruveilhier**; Cruveilhier 1871; *see* Cruveilhier, Jean Baptiste) undergoes a change in texture at the root of the penis and the root of the scrotum; here, it loses its fat. It is usually also described as having two layers, an outer, more fatty layer without an eponym, and a deeper, membranous part that in the urogenital part of the perineum is known as **fascia of Colles** (Colles 1811; *see* Colles, Abraham). The **deep perineal fascia** bears the *eponyms* **fascia of Buck** (Buck 1848; *see* Buck, Gordon) and **fascia of Gallaudet.** It is closely associated with the membranous layer of the superficial perineal fascia. Along the penis, the superficial perineal fascia continues as a loose envelope, known as the superficial fascia of the penis, and at the glans penis fuses with the deep fascia of the penis, also known as **fascia of Buck**. At the base of the scrotum, the superficial perineal fascia is thickened by the acquisition of smooth muscle fibres and is highly attached to the skin, forming the tunica dartos, also described as the **superficial scrotal fascia of Colles**.

22.4 The Abdominal Walls

The semilunar line of the muscles of the anterior abdominal wall bears the *eponym* **line of van den Spieghel** (*see* van den Spieghel, Adriaan). Behind the abdominal muscles, the **retromuscular fossa of Charpy** is found (*see* Charpy, Adrien), a fat-filled submuscular space between the posterior aspect of the rectus abdominis muscle and the transversalis fascia, bordered medially by the fusion of the transversalis fascia and the linea alba.

The diaphragm contains two weakened triangles (Fig. 22.2), the **sternocostal triangle of Larrey** (*see* Larrey, Dominique Jean), also described by **Morgagni** (*see* Morgagni, Giovanni Battista), and the **lumbocostal triangle of Bochdalek** (Bochdalek 1848; *see* Bochdalek, Vincenz Alexander), also described by **Henle** (*see* Henle, Friedrich Gustav Jacob). Through both triangles diaphragmatic herniae may occur, a parasternal hernia through the sternocostal triangle, known as the *eponym* **hernia of Morgagni**, and a posterior one through the triangle of Bochdalek. The lower part of the anterior abdominal wall is discussed in Sect. 22.7. The posterior abdominal wall contains two weak spots, in which very rarely, especially after a trauma, a lumbar hernia may occur: the **superior lumbar triangle of Grynfeltt-Lesshaft** (Grynfeltt 1866; *see* Grynfeltt, Joseph Casimir, and Lesshaft, Pjotr Frantsowitsch, and Fig. G26) and the **inferior lumbar triangle of Petit** (Petit and Lesne 1774; *see* Petit, Jean Louis; and Fig. P12b).

22.5 The Abdominopelvic Cavity

The abdominopelvic cavity is covered by **parietal fasciae**, called the parietal abdominal or endoabdominal fascia and the parietal pelvic or endopelvic fascia, which can be further subdivided as the **transversalis fascia of Velpeau** and **of Hyrtl** (*see* Velpeau, Alfred Armand Louis Marie, and Hyrtl, Joseph), the umbilical **fascia of Richet** (*see* Richet, Louis Alfred), a fold of peritoneum enveloping the obliterated umbilical vein, the iliopsoas fascia, the obturator fascia, the superior fascia of the pelvic diaphragm, the **presacral** or **retrorectal fascia**, i.e. the caudal part of the parietal pelvic fascia, sometimes attributed to **Waldeyer** (Waldeyer 1899; *see* Waldeyer, Heinrich Wilhelm Gottfried), and the anterior layer of the thoracolumbar fascia, covering the quadratus lumborum muscle. Inside the cavity, there are other fasciae of different origin and function such as the renal **fascia of Gerota** (*see* Gerota, Dumitru, and Fig. G11) with its **anterior layer of Toldt** (*see* Toldt, Karl) and its **posterior layer of Zuckerkandl** (*see* Zuckerkandl, Emil), and fasciae behind the ascending and descending colon and the pancreas: the **retrocolic fascia of Toldt** behind the ascending and descending colon, the **retropancreatic fascia of Treitz** (*see* Treitz, Václav) behind the head of the pancreas, and the preduodenopancreatic **fascia of Fredet** in front of the duodenum and the head of the pancreas (*see* Fredet, Pierre, and Brachini et al. 2023 for discussion). Between the fascia transversalis and the parietal peritoneum between the umbilicus and the urinary bladder, a thin triangular layer of connective tissue is found, known as the *eponym* vesicoumbilical **fascia of Delbet** (Delbet 1907; *see* Delbet, Paul; and Elliot Smith 1908 for discussion). It is bordered laterally by the lateral umbilical ligament and continuous with the connective tissue covering the urinary bladder. The **white line of Toldt** represents an avascular plane for incision during surgery of the ascending and descending colon. It forms a vertical line at the transition of the visceral peritoneum of the colon to the parietal peritoneum of the lateral abdominal wall, and is the entry to the retrocolic fascia of Toldt.

22.6 The Peritoneal Fossae

The peritoneal cavity, due to the rotation of the gut, becomes a space with many smaller outpouchings and pockets, some of them constant and some of them variable, which may

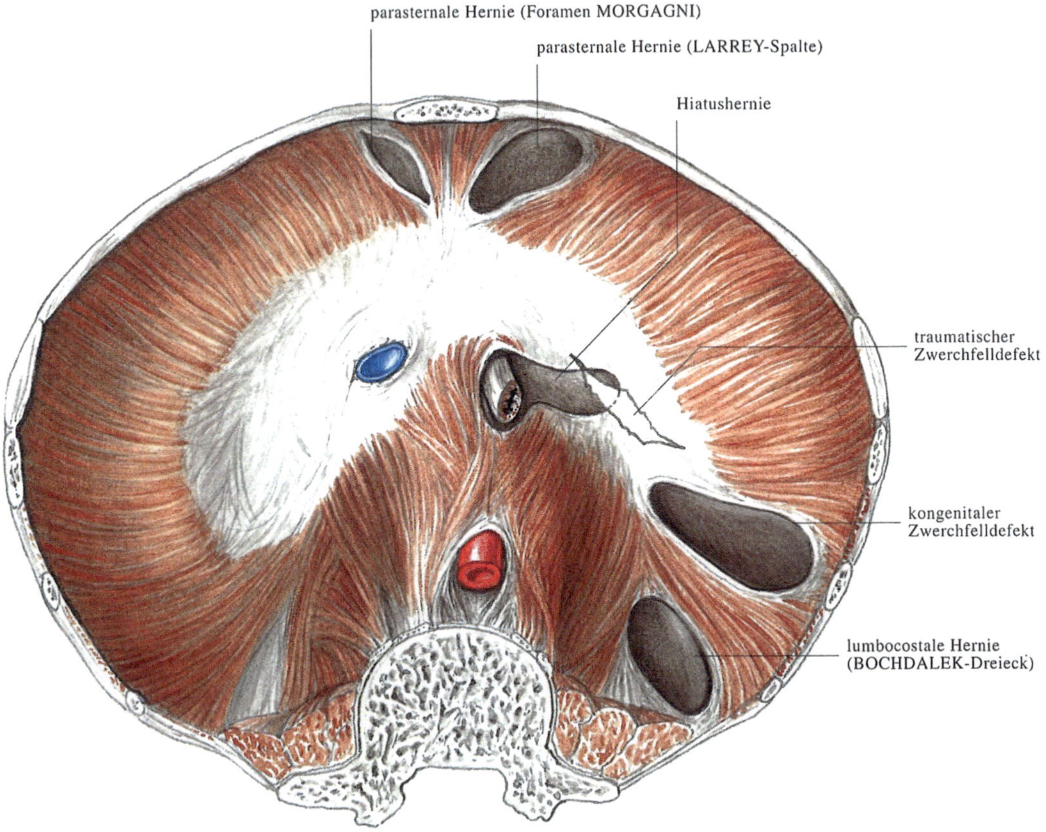

Fig. 22.2 The abdominal side of the diaphragm with the sites of diaphragmatic herniae (from von Lanz und Wachsmuth, *Praktische Anatomie: Bauch*, von H Loeweneck und G Feifel. Springer, Berlin, Heidelberg, New York, 1993; with permission)

serve as places of fluid collection during pathological processes and positions of the patient after surgery, or as places of rare internal herniae. This may result in intestinal obstruction (Hollinshead 1971; Gray and Skandalakis 1972). To be mentioned are (a) the hepatorenal or right subhepatic space, known as the *eponym* **pouch of Morison** (*see* Morison, James Rutherford); (b) the splenorenal recess, known as the *eponym* **pouch of Koller** (after the German surgeon H-J Koller); (c) the omental bursa, communicating with the peritoneal cavity through the **foramen of Winslow** (*see* Winslow, Jacob Benignus); and various duodenal, paraduodenal, ileocaecal, subcaecal and intersigmoid recesses. This large group includes: (a) the intermesocolic **fossa of Brösike** (Fig. 22.3a; *see* **Brösike, Gustav**); (b) the inferior duodenal **recess of Gruber-Landzert** (Fig. 22.3b; *see* **Gruber, Wenzel**, and **Landzert, Theodor**); (c) the duodenojejunal **fossa of Ionesco** (*see* Ionesco, Thoma); (d) the superior ileocaecal **fossa of Luschka** (*see* Luschka, Hubert von); (e) the infraduodenal **recess of Poisson** (*see* Poisson, Francis); (f) the superior and inferior duodenal **fossae of Treitz** (*see* Fig. T18a *see* Treitz, Václav); (g) the subcaecal **fossa of Treitz**; (h) the inferior ileocaecal **fossa of Treves** (*see* Treves, Frederick); (i) the mesentericoparietal **fossa of Waldeyer** (*see* Fig. W3); (j) the subcaecal **fossa of Treitz**; (k) the intersigmoid **recess of Treitz**; and (l) the hepatocystic **recess of Jacquement** (Jacquement 1897; *see* Jacquement, Marcel). The clinically more important are the fossae of Brösike, Gruber-Landzert, Treitz and Waldeyer (Gray and Skandalakis 1972). The fossae of Treitz, Waldeyer and the rare fossa of Brösike may contain a right paraduodenal hernia, and the fossa of Gruber-Landzert a left paraduodenal hernia (Kelahan et al. 2021). In a Littré-Richter hernia, only part of the intestinal wall is herniated (*see* Littré, Alexis, and Richter, August Gottlieb).

The larger excavations or pouches are the **pouch** or **cul-de-sac of Douglas**, an *eponym* for the rectouterine pouch in females (Douglas 1730; *see* Douglas, James), and the **space of Proust**, an *eponym* for the rectovesical pouch in males (*see* Proust, PT). The pelvic organs are surrounded by visceral pelvic fasciae, encompassing the corresponding blood vessels and nerves. Below the level of the rectovesical pouch of Proust, the rectum is separated from the urinary bladder by the **rectovesical fascia** or **septum of Denonvilliers** (Denonvilliers 1836; *see* Denonvilliers, Charles Pierre) also known as **fascia of Tyrrell** (*see* Tyrrell, Frederick), but actually this fascia was first described as **prostatic fascia of Pattison** (Pattison 1820; *see* Pattison, Granville Sharp). Below the rectouterine pouch of Douglas, the rectum is separated from the vagina by the **retrovaginal fascia** or **septum of Denonvilliers**. More ventrally, the **vesicovaginal fascia**

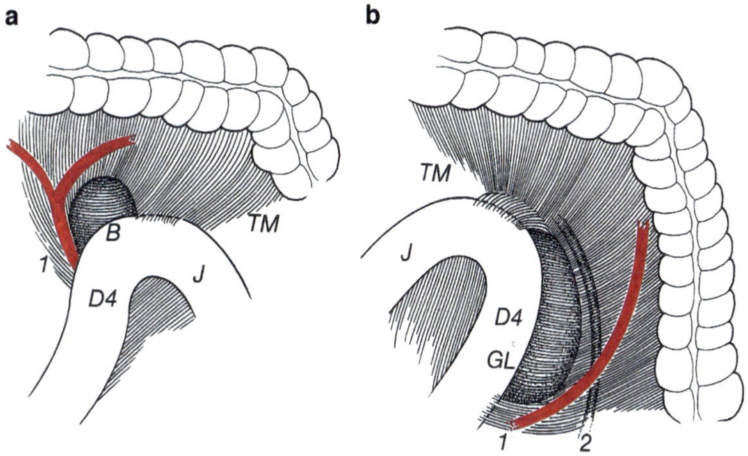

Fig. 22.3 (**a**) The fossa of Brösike, *B* fossa of Brösike, *D4* fourth part of duodenum, *J* jejunum, *TM* transverse mesocolon, *1* middle colic artery; and (**b**) the fossa of Gruber-Landzert, *D4* fourth part of duodenum, *GL* fossa of Gruber-Landzert, *J* jejunum, *TM* transverse mesocolon, *1* left colic artery, *2* inferior mesenteric vein (after Gray and Skandalakis 1972)

or **septum of Halban** (*see* Halban, Josef) is found, an *eponym* for fascial strips between the trigone of the urinary bladder and the anterior part of the vaginal wall. The fossae for the ovary (Claudius, Krause and Waldeyer) are discussed and illustrated in Chap. 25.

The peritoneum forms several folds, sometimes called ligaments: (a) the **band of Pringle** (*see* Pringle, Seton Sidney), a peritoneal fold between the transverse mesocolon and the antimesenterial border of the beginning of the jejunum; (b) the **fold** or **membrane of Jackson** (Jackson 1909; *see* Jackson, Jabez North, and Fig. J1b), a pericolic membrane, fold or adhesion between the caecum or ascending colon and the right posterior abdominal wall; (c) the **inferior ligament of Tuffier** (*see* Tuffier, Theodore), a mesenterico-parietal fold, a part of the mesentery inserted into the iliac fossa; (d) the **ligament of Charpy** (*see* Charpy, Adrien), the right phrenicocolic ligament (his abdominocolic legament), a peritoneal fold between the right lateral abdominal wall and the right colic flexure; and (e) the **bloodless fold of Treves**, an *eponym* for the bloodless inferior ileocaecal fold (Treves 1885; *see* Treves, Frederick). The preperitoneal space bordered ventrally by the fascia transversalis and dorsally by the anterior parietal peritoneum is known as the *eponym* retroinguinal **space of Bogros** (Bogros 1823; *see* Bogros, Annet Jean) and contains the external iliac vessels and adipose tissue. Caudally, it is continuous with the **retropubic space of Retzius** (*see* Retzius, Anders).

22.7 The Inguinal Canal, Its Ligaments and Herniae

The lower part of the anterior abdominal wall presents the inguinal canal with several weak spots. The superficial inguinal ring overlays the **triangle of Hesselbach**, an *eponym* for a triangular area, which boundaries are cranially the falx inguinalis, laterally the inferior epigastric vessels, medially the lateral border of the rectus abdominis muscle, and caudally the inguinal ligament (Hesselbach 1806; *see* Hesselbach, Franz Kaspar). It may become the internal opening for a direct (medial) inguinal hernia. The proposed **triangle of Henke** forms a triangular area between the descending and lateral parts of the inguinal fold and the lateral border of the rectus abdominis muscle (*see* Henke, Wilhelm). A larger, oval potential opening in the weak inguinal area, where all inguinal and femoral herniae originate, received the *eponym* **myopectineal orifice of Fruchaud**, and is bounded cranially by the internal oblique and transverse abdominis muscles, caudally by the superior pubic ramus and medially by the sheath of the rectus abdominis muscle (Fruchaud 1956; *see* Fruchaud, Henri René).

In the inguinal region (Fig. 22.4), the **inguinal ligament**, as described by Vesalius, Falloppio and Poupart, predominates. Related to it are: (a) the **reflected inguinal ligament of Colles** (Colles 1811; *see* Colles, Abraham), an *eponym* for a small band detached from the pubic attachment of the lateral crus of the superficial inguinal ring, extending upwards and medially, behind the medial crus and blending with the opposite aponeurosis of the external oblique abdominis muscle; (b) the **lacunar ligament of Gimbernat**, an *eponym* for a triangular ligament formed by the lowermost fibres of the aponeurosis of the external oblique abdominis muscle or inguinal ligament (Gimbernat 1793; *see* Gimbernat, Manuel Louise Antonio), turning and continuing as the **pectineal ligament of Cooper**, inserting on the pecten pubis (Cooper 1804–1807; *see* Cooper, Astley Paston); and (c) the **interfoveolar ligament of Hesselbach**, an *eponym* for a slight thickening in the reduplicated transversalis fascia at the medial margin of the deep inguinal ring (*see* Figs. C18b, G14 and H2). Other structures in this area are the iliopubic **tract of Thomson**, the thickened caudal margin of the transversalis fascia (*see* Thomson, Alexander) and the so-called **ligament of Henle**, a lateral extension of the sheath of the rectus abdominis muscle. Erroneously, the term ligament of Henle is also given to the conjoint tendon or falx inguinalis (*see* Anson, McVay 1938 for discussion). Superficially, a fibrous band connecting the membranous layer of the subcutaneous abdominal tissue to the fascia lata has received the *eponym*

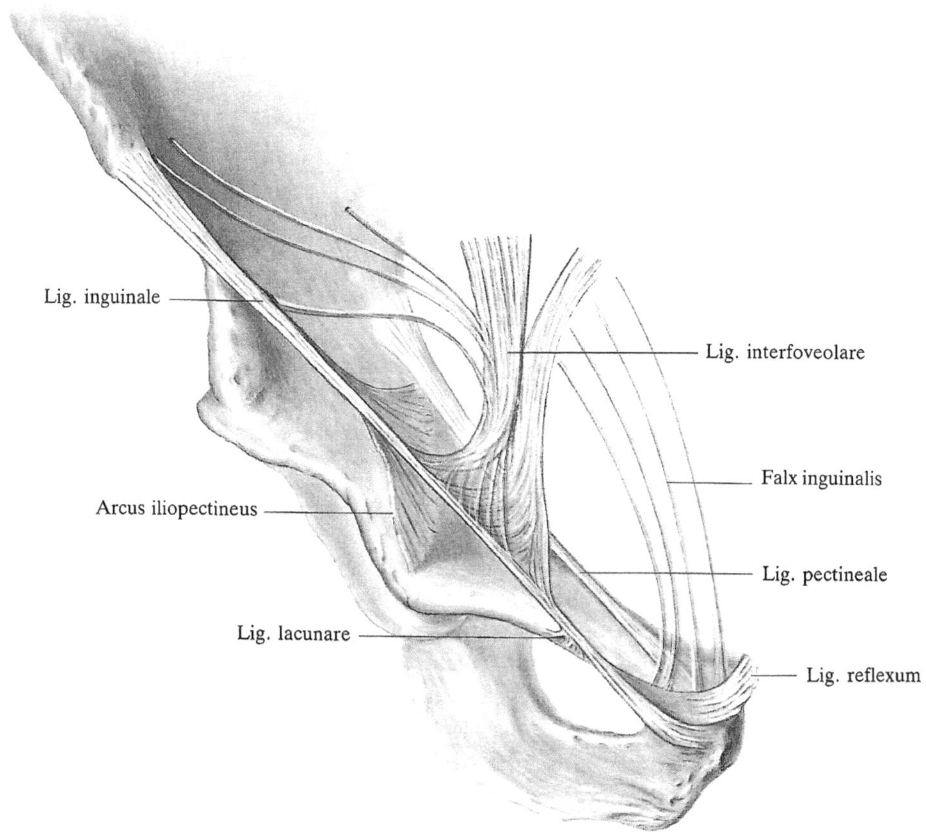

Fig. 22.4 The inguinal ligament with related structures (from von Lanz und Wachsmuth, *Praktische Anatomie*: *Bauch*, von H Loeweneck und G Feifel. Springer, Berlin, Heidelberg, New York, 1993; with permission)

ligament of Pétrequin (*see* Pétrequin, Joseph Pierre Eléonor). The part of the female inguinal canal that contains the round ligament of the uterus is termed the **ring of Imlach** and the fat surrounding this ligament within the inguinal canal bears the *eponym* **fat pad of Imlach** (Imlach 1885; *see* Imlach, Francis).

Inguinal herniae are usually subdivided into congenital, lateral or indirect and medial or direct inguinal herniae. A few inguinal herniae are known as an *eponym*: (a) the **hernia of Amyand** (Amyand 1735; *see* Amyand, Claudius), in which the vermiform appendix of the caecum is involved; (b) the **hernia of Krönlein**, a complicated hernia with a double sac, one part in the inguinal canal, the other projecting from the deep inguinal ring into the subperitoneal tissue, also known as preperitoneal inguinal hernia (*see* Krönlein, Rudolf Ulrich); (c) the **hernia of Holthouse**, an inguinal hernia with extension of the loop of the intestine along the inguinal ligament (*see* Holthouse, Carsten); and (d) the **hernia of Malgaigne**, an infantile inguinal hernia before the descent of the testis (*see* Malgaigne, Joseph François). Related to the inguinal region are the **femoral herniae**, also a few of which received an *eponym*: (a) the **Cooper-Hey protrusion**, a femoral hernia with two sacs, also known as bilocular femoral hernia (Cooper 1804–1807, *see* Cooper, Astley Paston; Hey 1803, *see* Hey, William); (b) the **femoral hernia of Velpeau**, in which the intestine is in front of the blood vessels (*see* Velpeau, Alfred Armand Louis); and (c) the **femoral hernia of Cloquet**, in which the intestine is behind the blood vessels (Cloquet 1817; *see* Cloquet, Jules Germain).

22.8 Anatomical Eponyms on Abdominal Organs

The contents of the abdominal cavity are arranged in a complex way and there are many clinically important structures highlighted with an eponym. The **Budde-Rocko triangle**, an *eponym* for the hepatocystic triangle, is bounded by the inferior border of the liver, laterally by the gallbladder and the cystic duct, and medially by the common hepatic duct (Rocko et al. 1981; *see* Fig. 22.5). Its lower part is known as the **triangle of Calot**, the cystohepatic triangle, bounded by the cystic artery, the cystic duct and the common hepatic duct (Calot 1890; *see* Calot, Jean-François, and Fig. C2). Nowadays, the *eponym* triangle of Calot is often imprecisely used as *synonym* for the triangle of Budde-Rocko. On the liver, the **line of Rex-Cantlie** marks the anatomical midline of the liver connecting the fundus of the gallbladder with the inferior vena cava (Fig. 22.6). Rex (1888; *see* Rex, Hugo) advocated the concept of the hemiliver based on the branching of the portal vein according to this line. Independently, Cantlie (1897; *see* Cantlie, James) found the same anatomi-

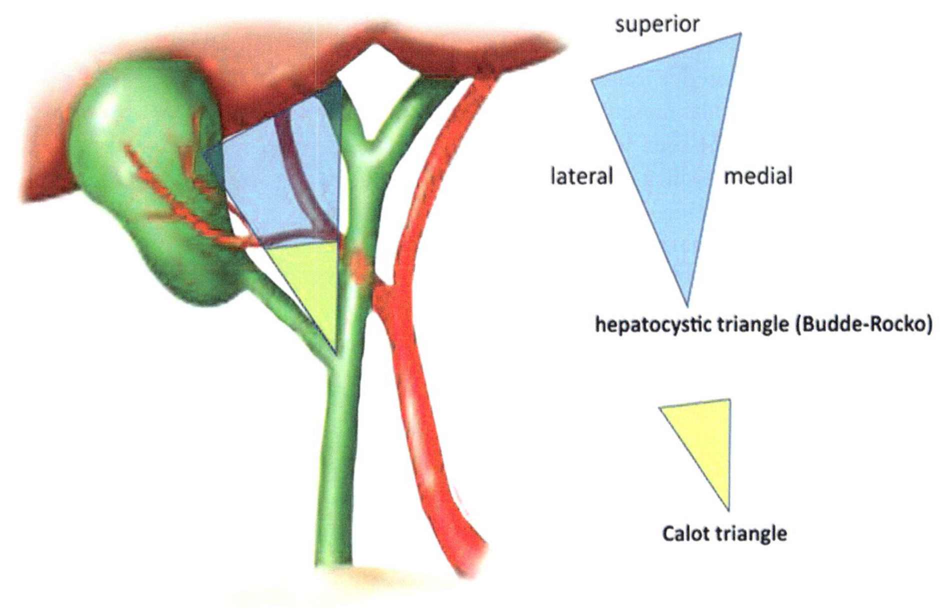

Fig. 22.5 The triangle of Budde-Rocko (from Mischinger H-J, Wagner D, Korprat P, Becker H, Werkgartner G. Eur Surg *The "critical view of safety (CVS)" cannot be applied – What to do? Strategies to avoid bile duct injuries.* 2020; doi.org/10.1007/s10353-020-00660-1; CC-BY 4.0, courtesy Hans-Jörg Mischinger, Graz)

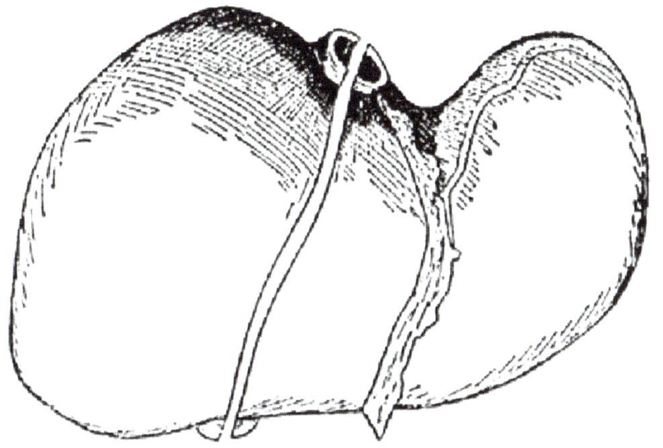

Fig. 22.6 The line of Cantlie (from Cantlie J *On a new arrangement of the right and left lobes of the liver.* Proc Anat Soc Great Britain & Ireland. J Anat Physiol 1897; 32:4–9)

cal characteristics and predicted clinical application for bleeding control during hepatectomy or portal vein ligation.

As for the gut, the duodenojejunal flexure is fixed by the **suspensory ligament of Treitz** (*see* Treitz, Václav), and the ileum can rarely (in 2%) show an ileal **diverticulum of Meckel** (Meckel 1817; *see* Meckel, Johann Friedrich The Younger), a remnant of the embryonic omphalomesenteric or vitelline duct, which should be obliterated at the end of the embryonic period.

On the kidney, there exists a relatively **avascular zone** between the segments supplied by the anterior and posterior branches of the renal artery near the midline, which bears the *eponyms* **bloodless line of Brödel** and **of Hyrtl** (Brödel 1901; *see* Brödel, Max; Hyrtl 1872; *see* Hyrtl, Joseph).

22.9 The Anorectal Canal

The anorectal canal consists of three parts: it begins orally at the rectosigmoid junction, continues as the long and wide rectal ampulla and ends aborally as a short and narrow anal canal, passing into the anus. The transition from the sigmoid colon to the rectum is not equally defined among medical disciplines. The anatomical border is found at the linea terminalis, the transition between the greater and lesser pelvis, whereas the surgical border is placed either at the level of the intervertebral disc S2/S3, more broadly at the level of the S3 vertebra or it is defined by the inferior border of the peritoneum in the deepest point of the peritoneal cavity, i.e. in the **rectovesical pouch of Proust** (Proust 1822; *see* Proust, PT) in males and in the **rectouterine pouch of Douglas** (Douglas 1730; *see* Douglas, James) in females. The rectosigmoid transition is smooth and without a defined or visible border, but functionally by a zone of higher pressure controlling the filling of the rectum and preventing backflow. This place is located about 2–3 cm above the superior transverse rectal fold, and contains a slightly strenghtened incomplete circular muscle, known as the **rectosigmoid sphincter muscle of O'Beirne** and **of Hyrtl** (O'Beirne 1833; *see* O'Beirne, James; Hyrtl 1846; *see* Hyrtl, Joseph; *see* also Chap. 31).

The rectal ampulla follows the curvature of the sacrum and coccyx and shows another bending, the anorectal flexure. Moreover, it curves three times laterally and these lateral flexures are matched by three inner **transverse rectal folds of Houston**, formed by mucosa, submucosa and a circular muscle layer (Houston 1830; *see* Houston, John). Usually, the superior fold is described as **fold of Houston**, the middle fold as **fold of Kohlrausch** (Kohlrausch 1854; *see*

Kohlrausch, Otto Ludwig Bernhard) and the inferior fold as **fold of Nélaton** (Nélaton 1844–1860; *see* Nélaton, Auguste). At the level of the inferior transverse rectal fold, there may exist a thickened circular muscular ring, called the **sphincter of Nélaton**. It has not been identified later. The rectum is covered by adventitia, which in the newborn is remodeled by small fat lobules occurring between connective tissue lamellae. The outer lamella covers the perirectal subcompartment and is called the **rectal fascia** (Fritsch 1990), and may be attributed to Waldeyer (Waldeyer 1899; *see* Waldeyer, Heinrich Wilhelm Gottfried). Behind the rectum, the retrorectal space is found bordering on the presacral or retrorectal fascia, probably erroneously also attributed to Waldeyer. In front of the rectum, the urinary bladder is separated by the rectovesical pouch of the peritoneal cavity and below in males by the **septum of Denonvilliers** (*see* Denonvilliers, Charles-Pierre) or **fascia of Tyrrell** (see Tyrrell, Frederick) but it was earlier described as **prostatic fascia of Pattison** (Pattison 1820; *see* Pattison, Granville Sharp). In females, the rectovaginal pouch separates the rectum from the uterus, and below from the vagina by the **retrovaginal septum** or **septum of Denonvilliers**.

The anatomical border of the rectal ampulla and the anal canal is the sharp anorectal transition line at the level of the pectinate or dentate line. The anorectal canal can be divided into an upper haemorrhoidal zone, an intermediate zone (the **pecten of Stroud**; Stroud 1896; *see* Stroud, Bert Brenette), and a cutaneous zone (Fig. 22.7). In clinical disciplines, the aboral, narrowing part of the ampulla is often described as the anal canal. The anatomical anal canal, however, extends from the pectinate line to the anocutaneous or **white line of Hilton** (*see* Hilton, John). In the haemorrhoidal zone, the **anal columns of Morgagni** (*see* Morgagni, Giovanni Battista) are found, connected by the **valves of Ball** (*see* Ball, Charles Bent) and **of Morgagni**, and between the columns there are depressions called the anal **sinuses of Morgagni**. At the transition from the anal canal to the skin, the **circumanal glands of Gay** are found (Gay 1871; *see* Gay, Alexander Heinrich).

The anorectal canal is surrounded by a system of muscle loops and sphincters (Fig. 22.7). The system includes the smooth internal anal sphincter, the skeletal external anal sphincter, consisting of deep, superficial and subcutaneous parts, and pelvic floor muscles (together the levator ani muscle), consisting of puborectalis, puboanalis and pubococcygeus (the **muscle of Henle**) muscles. Some peculiar muscles are: (a) the **subcutaneous anal muscle of Treitz** (*see* Treitz, Václav), an *eponym* for fibrous bands radiating through the external anal sphincter into the submucosa of the anal canal; (b) the **corrugator cutis ani muscle of Ellis** (*see* Ellis, George Viner), an *eponym* for smooth muscle fibres reaching the dermis of the skin, also described as **sustentaculum mucosae muscle of Kohlrausch** (Kohlrausch 1854; *see* Kohlrausch, Otto Ludwig Bernhard); and (c) the **internal dilatator muscle of Rüdinger**, an *eponym* for a muscle internal to the circular fibres of the rectum (Rüdinger 1879; *see* Rüdinger, Nicolaus). The *eponym* **subsphincteric space of Courtney** is used for the deep postanal space located posterior to the deep part of the external anal sphincter muscle between the levator ani muscle and the superficial part of the external anal sphincter muscle (Courtney 1949; *see* Courtney, Harold).

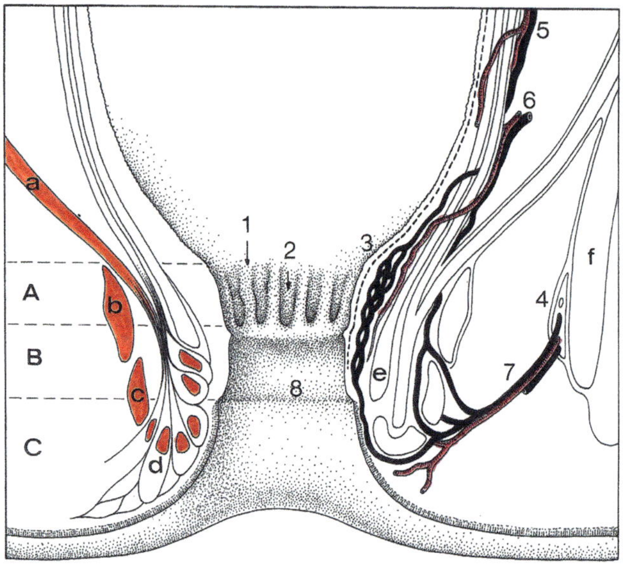

Fig. 22.7 The anorectal canal: Schematic frontal section through the anorectal canal (after ten Donkelaar HJ, Oostra R-J *Klinische Anatomie en Embryologie*, 4th ed. Springer Media, Houten, 2014). *A-C* zones of anal canal: *A* haemorrhoidal zone, *B* intermediate zone (pecten analis), *C* cutaneous zone, *a* levator ani muscle, *b-d*: parts of external sphincter ani muscle: *b* deep part, *c* superficial part, *d* subcutaneous part, *e* internal sphincter ani muscle, *f* obturator internus muscle, *1* anal column, *2* anal sinus, *3* submucosal venous plexus (in *black*), *4* pudendal canal with internal pudendal artery and vein, *5* superior rectal artery and vein, *6* middle rectal artery and vein, *7* inferior rectal artery and vein, *8* anocutaneous line (white line of Hilton)

22.10 The Vascularization of the Intestines

The intestines are supplied by an intricate network of blood vessels, fed by three principal arterial trunks: the coeliac trunk, known as the *eponym* **tripus Halleri** (*see* Haller, Albrecht von), the superior mesenteric artery and the inferior mesenteric artery (Fig. 22.8). These three trunks supplying the derivatives of the foregut, the midgut and the hindgut, respectively, are joined together via arterial trunk anastomoses (Douard et al. 2006). The longitudinal arterial anastomosis theory suggested by Mackay (1889) and Tandler (1904a, b) may account for the development of arterial trunk anastomoses and the many anatomical variations, but *see* Feigl et al. (1975) for abandoning the Mackay-

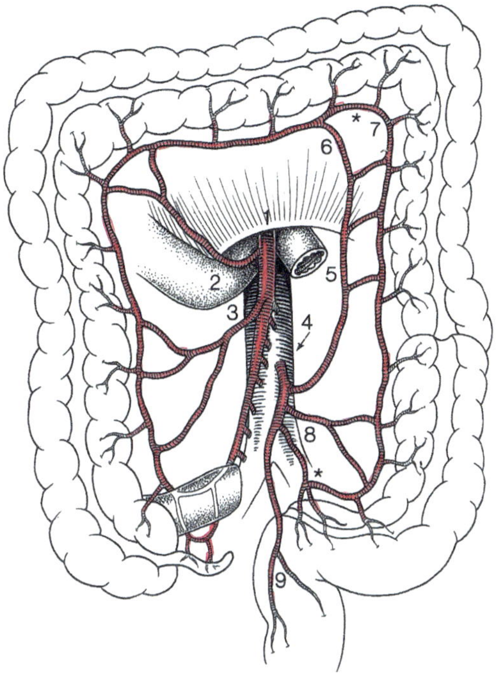

Fig. 22.8 The blood supply of the colon and the critical ponts of Griffiths and Sudeck-Hartmann (marked by *asterisks*). *1* superior mesenteric artery, *2* middle colic artery, *3* right colic artery, *4* inferior mesenteric artery, *5* left colic artery, *6* arc of Riolan, *7* marginal artery of Drummond (after ten Donkelaar HJ, Oostra R-J *Klinische Anatomie en Embryologie,* 4th ed. Springer Media, Houten, 2014)

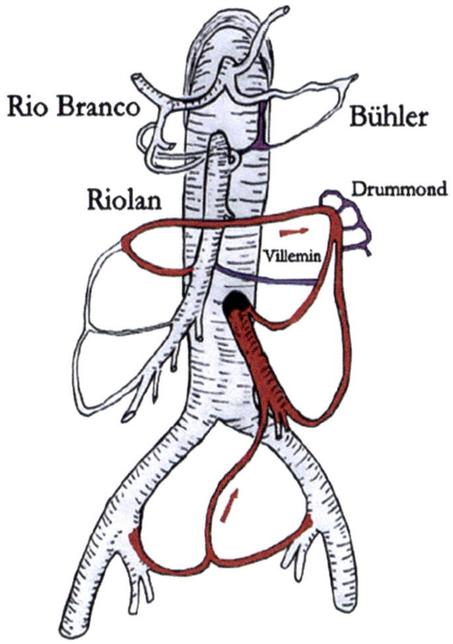

Fig. 22.9 Collateral arcades between both mesenteric arteries (from Douard R, Chevallier J-M, Delmas V, Cugnenc P-H *Clinical interest of digestive arterial trunk anastomoses.* Surg Radiol Anat 2006; 28:219–227; with permission; courtesy Richard Douard, Paris)

Tandler theory. The coeliac trunk and the superior mesenteric artery are rarely directly interconnected by the coeliacomesenteric **arcade of Bühler** (Bühler 1904; Fig. 22.9). Between some of the branches of the coeliac trunk and the superior mesenteric artery, anastomoses exist in the form of an arch or arcade: (a) between the right and left gastroomental or gastroepiploic arteries as the gastroomental **arch of Hyrtl** at the greater curvature of the stomach (see Hyrtl, Joseph); (b) between the right, middle and left omental branches of the gastroomental arteries at the free margin of the greater omentum, called the transomental or transepiploic **arcade of Haller-Barkow** (Haller 1778; *see* Haller, Albrecht von; and Barkow 1868; *see* Barkow, Hans Karl Leopold); (c) between the superior and inferior pancreaticoduodenal arteries as the **arch of Rio Branco** (Rio Branco da Silva 1912); (d) between the dorsal pancreatic artery and the superior pancreaticoduodenal artery by the **arch of Kirk** (Kirk 1931); (e) between the superior and inferior mesenteric arteries as the **anastomosis maxima** or **magna of Haller** (*see* Haller, Albrecht von); (f) as a chain of anastomoses along the mesocolic margin, known as the **marginal colic artery of Drummond** (Drummond 1914; *see* Drummond, David); (g) as a comparable system to that of Drummond along the mesenteric margin of the jejunum and ileum bearing the *eponym* **parallel artery of Dwight**; (h) as an inconstant **intermesenteric artery of Villemin** between the stems of both mesenteric arteries (Villemin and Huard 1924; *see* Villemin, Fernand), corresponding to the **meandering mesenteric artery of Moskowitz** (Moskowitz et al. 1964) and the **artery of Gonzales** (Gonzales, Jaffe 1966), if pathologically enlarged; and (i) as the **arcade of Mondor**, an *eponym* for a connection between the descending branch of the left colic artery and the ascending branch of the first sigmoid artery (*see* Mondor, Henri). Considering the fact that Riolan never specifically wrote about the blood supply of the colon (Lange et al. 2007), it has been assumed that the *eponym* **arcade of Riolan**, used unspecifically for different levels of communication between the superior and inferior mesenteric arteries, was probably introduced by Albrecht von Haller out of respect for this great anatomist. For the arch of Riolan, various opinions are found in the literature (van Gulik, Schoots 2005; Lange et al. 2007): (1) as a *synonym* for the marginal colic artery of Drummond; and (2) as a rare anatomical entity connecting the superior and inferior mesenteric arteries. Among the *synonyms* mentioned are the arteries described by Villemin, Moskowitz and Gonzales. Lange et al. (2007) suggested to abandon the term arch of Riolan. The arterial supply of the rectum is twofold: (a) the oral part is supplied by the superior rectal artery from the inferior mesenteric artery; and (b) the aboral parts by middle and inferior rectal arteries from the internal iliac artery. A thicker branch from the superior rectal artery received the *eponym* **dorsal rectal artery of Konstantinovich** (Konstantinovich 1872–1873; *see* Konstantinovich, Vikentzi Bonifatiyevich).

The marginal colic artery may be absent in some areas of the large intestine and these 'watershed' areas are clinically viewed as critical points (Fig. 22.8). The *eponym* **critical point of Griffiths** (Griffiths 1956; *see* Griffiths, John Daniel) was given by Michels et al. (1962). It is located at the left colic flexure. Another problematic area is the **critical point of Sudeck** and **of Hartmann** (Sudeck 1907; *see* Sudeck, Paul Hermann Martin; Hartmann 1909; *see* Hartmann, Henri Albert Charles Antoine) between the last sigmoid artery and the superior rectal artery. According to Griffiths (1956) and Michels et al. (1965), it has no practical significance. A critical point between the terminal branches of the sigmoid arteries bears the *eponym* **critical point of Toupet** (after André Toupet, a French surgeon). At the terminal ileum, the **avascular zone of Treves** may be found for the last aboral segment of the terminal ileum (Treves 1885; *see* Treves, Fredrick). The *eponym* **point of Demel** is used for the area at the greater curvature of the stomach where the right and left gastroomental (or gastroepiploic) arteries anastomose (*see* Demel, Rudolf).

Several smaller arteries also bear *eponyms*: (a) the **arteries of Michels**, an *eponym* for the retroduodenal arteries (Michels 1955; *see* Michels, Nicholas Aloysius); (b) the **artery of Wilkie**, an *eponym* for the supraduodenal artery (Wilkie 1911; *see* Wilkie, David Percival Dalbreck); (c) the **arcade of Treitz**, an eponym for a vascular arcade formed by the inferior mesenteric artery and the left superior colic artery in the left mesocolon (*see* Treitz,Václav); and (d) the **corona mortis of Hesselbach**, an inconstant aberrant obturator artery, arising from the inferior epigastric artery and coursing over the pecten pubis into the lesser pelvis (Hesselbach 1819; *see* Hesselbach, Adam Kaspar). There appear to be three skin perfusion or vascular zones of the anterolateral abdominal wall, bearing the *eponym* **zones of Huger** (Huger 1979; *see* Huger, WE Jr). An arterial collateral communication between the subclavian and external iliac arteries via the superior and inferior epigastric arteries received the *eponym* **pathway of Winslow** (*see* Winslow, Jacob Benignus).

For several **abdominal veins**, *eponyms* are in use: (a) the **prepyloric vein of Mayo**, draining the anterior surface of the pylorus (*see* Mayo, Charles Horace); (b) the **paraumbilical veins of Sappey**, usually two small veins within the falciform ligament of the liver (Sappey 1883; *see* Sappey, Marie Philibert Constant), which enlarge during portal hypertension; (c) the **gastropancreaticocolic trunk of Henle**, a very variable short collective trunk of veins coming from the stomach, the right colic flexure and the head of the pancreas (*see* Henle, Friedrich Gustav Jacob); (d) the **venous circle of Bendavid**, an *eponym* for an anastomotic network among the deep inferior epigastric, suprapubic and pubic veins and veins for the rectus abdominis muscle, located in the lower preperitoneal space (Bendavid 1992; *see* Bendavid, Robert); and (e) a **marginal vein of anus of Konstantinovich** (Konstantinovich 1872–1873; *see* Konstantinovich, Vikentzi Bonifatiyevich).

Portal hypertension leads to opening of pre-existing small communications between the tributaries of the portal vein and those of the caval veins. Such **portocaval anastomoses** are found in different layers of the trunk, ranging from tributaries between the left gastric vein and the oesophageal veins to anastomoses between the inferior mesenteric vein and the inferior rectal veins (Fig. 22.10). Some of these portocaval anastomoses received an *eponym*. The **paraumbilical veins** include the veins of Burow and the superior and inferior veins of Sappey, connecting the portal vein with the anterior abdominal wall. In late foetuses, Burow (1838) described a pair of veins, which ascend from the inferior epigastric veins alongside the umbilical vein and, after uniting, they drain into the upper part of the umbilical vein. They are known as the *eponym* **veins of Burow** (Burow 1838; *see* Burow, Karl August von). Sappey (1883) studied postmortem specimens from cases of advanced portal hypertension, in which the veins were extended. These veins became known as **veins of Sappey** (Sappey 1883; see Sappey, Marie Philibert Constant). **Baumgarten** (1891) showed that the umbilical vein and the veins of Burow may become dilated in portal hypertension. The dilated veins of Burow communicate with dilated inferior epigastric veins after piercing the sheath of the rectus abdominis muscle (Baumgarten 1891; *see* Baumgarten, Paul Clemens). The superficial abdominal veins, sometimes may distend and radiate from the umbilicus, and are then termed the **caput Medusae**. In front of the posterior abdominal wall, the **veins of Ruysch** and **of Retzius** (*see* Ruysch, Frederik; Retzius 1835, Retzius, Anders) form a system of many tiny veins within the retroperitoneal space, between the mesenteries and the retroperitoneal tissue itself, including the **veins of Quain** (*see* Quain, Jones), his colicospermatic veins, draining blood from the descending colon to the left testicular vein. The **ductus venosus of Aranzio** (*see* Aranzio, Giulio Cesare), a foetal shunt from the umbilical vein to the inferior vena cava bypassing the liver, can become recanalized.

22.11 Lymphatic Drainage of the Abdominal Organs

In the **abdomen**, lymph nodes are found mainly along the arteries and are discussed in Chap. 26 (Anatomical Eponyms in Oncology). A few of them received an *eponym*: (a) an inconstant **cystic lymph node**, found in the cystohepatic triangle of Calot, was described by Mascagni (1787; *see* Mascagni, Paolo) and Calori (1850–1855; *see* Calori,

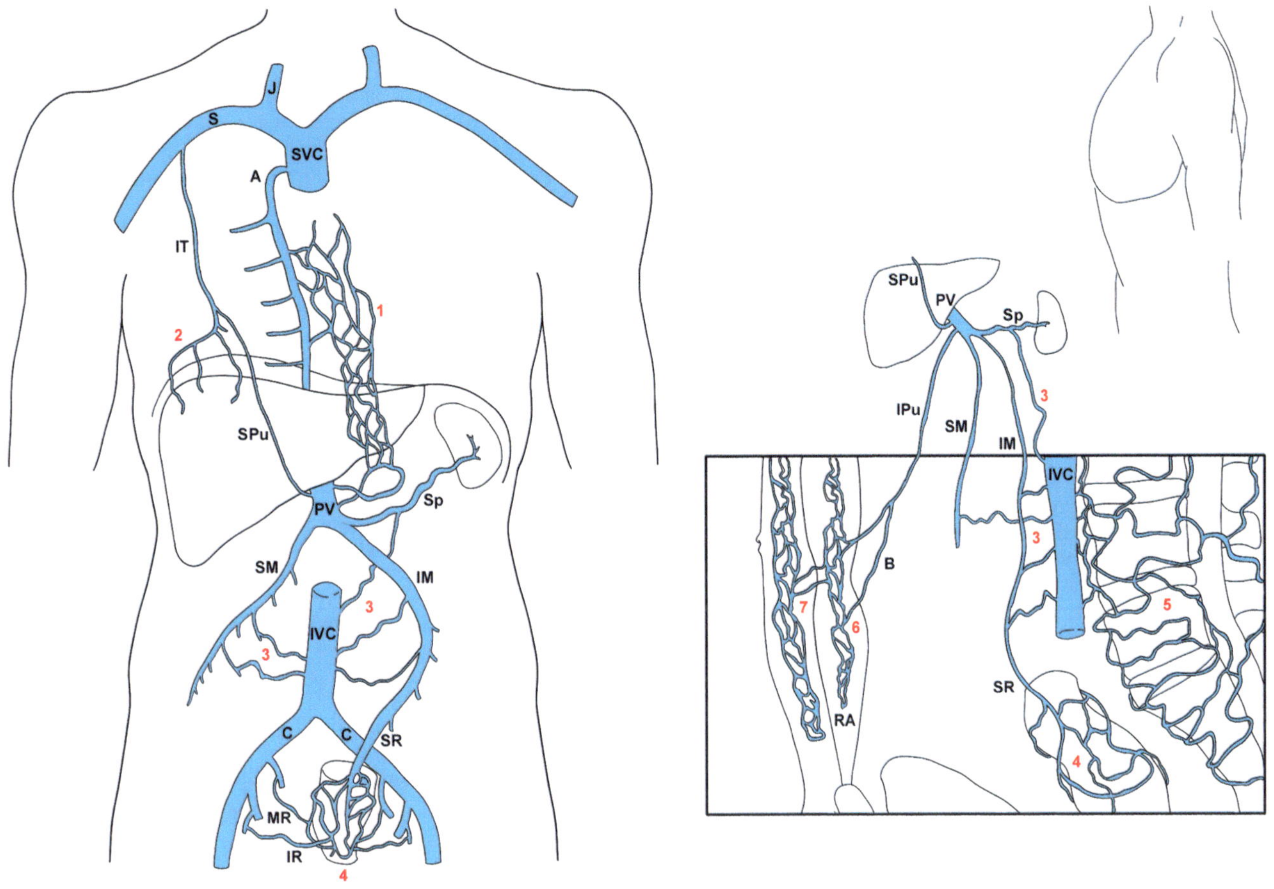

Fig. 22.10 Portocaval anastomoses viewed in a frontal (at the *left*) and a lateral view (at the *right*; courtesy David Kachlik and Anhelina Khadanovich, Prague). *A* Azygos vein, *B* veins of Burow, *C* common iliac vein, *IM*, *SM* inferior and superior mesenteric vein, *IPu*, *SPu* inferior and posterior paraumbilical vein of Sappey, *IR*, *MR*, *SR* inferior, middle and superior rectal vein, *IT* internal thoracic vein, *IVC*, *SVC* inferior and superior vena cava, *J* internal jugular vein, *PV* portal vein, *RA* rectus abdominis muscle, *S* subclavian vein, *Sp* splenic vein, *1* oesophageal veins, *2* veins of liver capsule and diaphragm, *3* retroperitoneal veins of Ruysch and Retzius, *4* rectal venous plexuses, *5* venous vertebral plexus of Batson, *6* superior and inferior epigastric veins, *7* subcutaneous abdominal veins (caput Medusae)

Luigi), and more recently described as sentinel lymph node of the gallbladder by the American surgeon Fred Bates **Lund** (1865–1950); (b) an inconstant parauterine lymph node at the crossing of the uterine artery and the ureter was described by Bayer (*see* Bayer, Heinrich; *eponym*: **parauterine lymph node of Bayer**); (c) for inguinal lymph nodes, various names and eponyms were given. Cloquet (1825) described the deepest or proximal deep inguinal lymph node located in the lacunar space and adjacent to the inguinal canal (*eponym*: **gland** or **node of Cloquet**), earlier descibed by Rosenmüller (1805), so the *eponym* **node of Cloquet-Rosenmüller**. The deep inguinal nodes lie deep to the fascia lata, the superficial lymph nodes with horizontal and vertical tracts subcutaneously. Quénu (1893) distinguished not only four groups of superficial inguinal nodes (*eponym*: **inguinal lymph nodes of Quénu**), 'supérior-interne' (superomedial), 'supérior externe' (superolateral), 'inférior-interne' (inferomedial) and 'inférior-externe' (inferolateral) but also a 'groupe présaphène', currently known as saphenofemoral lymph nodes. His subdivision has been used in modern studies on the lymphatic drainage of the penis and its sentinel node in penile cancer. A series of lymphatic plexuses in the mucosa and skin of the anus is known as the **haemorrhoidal plexus of Quénu**. Several eponyms were given to lymph nodes of the pelvic organs: **anorectal lymph nodes of Gerota** (Gerota 1895; *see* Gerota, Dumitru), comparable to the **pararectal lymph nodes of Poirier and Cunéo** (1902; *see* Poirier, Paul Julien). Poirier and Cunéo (1902) also described abdominoaortic lymph nodes, currently known as lumbar lymph nodes.

More recently, emphasis has been given to the **sentinel lymph node** for the **penis** (Cabanas 1977, 1992). The superomedial superficial inguinal lymph node appeared to be the sentinel lymph node for the penis (*eponym*: **lymph node of Cabanas**). The skin and preputium of the penis as well as the scrotum drain to the superficial inguinal lymph nodes, and the glans penis and the spongy urethra to the deep inguinal and external iliac lymph nodes.

References

Addison C (1899) On the topographical anatomy of abdominal viscera in Man. J Anat (Lond) 33:565

Amyand C (1735) Of an inguinal rupture, with a pin in the appendix caeci, incrusted with stone; and some observations on wounds in the guts. Philos Trans R Soc Lond 39:329–355

Anson BJ, McVay CB (1938) Inguinal hernia. I. The anatomy of the region. Surg Gynec Obst 66:186–191

Barkow HCL (1868) Comparative Morphologie des Menschen und der menschenähnliche Thiere. Breslau

Baumgarten P (1891) Über die Nabelvene des Menschen und ihre Bedeutung für die Circulationsstörung bei Lebercirrhose. Arb Geb Path Anat Bakt Path Anat Inst Tübingen 1:1–46

Bendavid R (1992) The space of Bogros and the deep inguinal venous circulation. Surg Gynecol Obst 174:356–357

Blaschko A (1901) Die Nervenverteilung in der Haut in ihrer Beziehung zu den Erkrankungen der Haut. Beilage Verhandl Dtsch Dermatol Gesellsch, VII Congress, Breslau

Bochdalek VA (1848) Einige Betrachtungen über die Entstehung des angeborenen Zwergfellbruches. Als Beitrag zur Anatomie der Hernien. Vjschr Prakt Heilk 5:89–97

Bogros AJ (1823) Essai sur l'anatomie chirurgical de la région iliaques et description d'un nouveau procédé pour faire la ligature des artères épigastrique et iliaque externe. Didot Le Jeune, Paris

Borges AF (1984) Relaxed skin tension lines (RSTL) versus other skin lines. Plast Reconstr Surg 73:144–150

Brachini G, Cirillo B, Matteuci M, Cirocchi R, Tebala GD, Cavaliere D et al (2023) A systematic review of varying definitions and the clinical significance of Fredet's fascia in the era of complete mesocolic excisions. J Clin Med 12:6233

Branco R, da Silva P (1912) Essai sur l'anatomie et la médecine opératoire du tronc coeliaque et de ses branches de l'artère hépatique en particulier. Thesis Univerity of Paris

Brödel M (1901) The intrinsic blood vessels of the kidney and their significance in nephrotomy. Johns Hopkins Hosp Bull 12:10–13

Buck G (1848) A new feature in the anatomical structure of the genitourinary organs not hitherto described. Trans Amer Med Ass 1:367

Bühler A (1904) Über eine Anastomose zwischen den Stammen der Art. Coeliaca und der Art. Mesenterica superior Gegenbauers Morphol Jahrb 32:185–188

Cabanas RM (1977) An approach for the treatment of penile carcinoma. Cancer 39:456–466

Cabanas RM (1992) Anatomy and biopsy of sentinel lymph nodes. Urol Clin North Am 19:267–276

Calori L (1850–1855) Tavole anatomiche rappresentanti la struttura del corpo umano. Tipografia Sassi nelle Spaderie, Bologna

Calot JF (1890) De la cholécystectomie (Ablation de la vésicule biliaire). University of Paris, Steinheil, Paris, Thesis

Cantlie J (1897) On a new arrangement of the right and left lobes of the liver. Proc Anat Soc Great Britain Ireland 32:4–9

Cloquet JG (1817) Recherches anatomiques sur les hernies de l'abdomen. Méquignon Marvis, Paris

Cloquet JG (1825) Manuel d'anatomie descriptive, Paris

Colles A (1811) A treatise of surgical anatomy, Edinburgh

Cooper AP (1804–1807) The anatomy and surgical treatment of abdominal Hernia. Longman, London

Courtney H (1949) The posterior subsphincteric space; its relation to posterior horseshoe fistula. Surg Gynec Obst 89:222–226

Cruveilhier JB (1871) Traité d'anatomie descriptive. Asselin, Paris

Delbet P (1907) Vessie, Urètre, Prostate, Verge, Périnée. In: Poirier PJ, Charpy A (eds) Traité d'anatomie humaine. Tome V. Masson, Paris

Denonvilliers CP (1836) Anatomie du périnée. Bull Soc Anat Paris, pp 105–107

Douard R, Chevallier J-M, Delmas V, Cugnenc P-H (2006) Clinical interest of digestive arterial trunk anastomoses. Surg Radiol Anat 28:219–227

Douglas J (1730) A description of the peritoneum and of that part of the Membrana cellularis which lies on the outside. Roberts, London

Drummond D (1914) The arterial supply of the rectum and pelvic colon. Brit J Surg 1:677–685

Elliot Smith G (1908) Studies in the anatomy of the pelvis, with special reference to the fasciae and visceral supports. J Anat Physiol 42:252–270

Eustachi B (1714) Tabulae anatomicae (of 1552), edited and published by Lancisi. Gonzao, Rome

Feigl W, Firbas W, Sinzinger H, Wicke L (1975) Various forms of the celiac trunk and its anastomoses with the superior mesenteric artery. Acta Anat 92:272–284

Fritsch H (1990) Development of the rectal fascia. Anat Anz 170:273–280

Fruchaud HR (1956) L'anatomie chirurgicale des hernies de l'aine. Doin, Paris

Gallaudet B (1931) A description of the planes of Fascia of the human body with special reference to the Fascia of the abdomen, pelvis and perineum. Columbia University, New York

Gay AH (1871) Die Circumanaldrüsen des Menschen. Sitz Akad Wiss Wien 63:329–332

Gerota D (1895) Der anorectale Lymphapparat. Sitz Preus Akad Wiss, pp 253–256

Gimbernat MLA (1793) Nuevo metodo de operar en la hernia crural, Madrid

Gonzales LL, Jaffe MS (1966) Mesenteric arterial insufficiency following abdominal aortic resection. Arch Surg 93:10–20

Gray SW, Skandalakis JE (1972) Embryology for Surgeons. Saunders, Philadelphia

Griffiths JD (1956) Surgical anatomy of the blood supply of the distal colon. Ann Roy Coll Surg 19:241–256

Grynfeltt J (1866) Quelques mots sur la hernie lombaire. Montp Med 16:32

Hartmann HACA (1909) Some considerations upon high amputation of the rectum. Ann Surg (Philadelphia) 50:1091

Hesselbach FK (1806) Anatomisch-chirurgische Abhandlung ueber den Ursprung der Leistenbrüche. Baumgärtner, Würzburg

Hesselbach AK (1819) Ueber den Ursprung und Verlauf der unteren Bauchdeckenschlagader und der Hüftbeinlochschlagader

Hey W (1803) Practical observations in surgery, London

Hollinshead WH (1971) Anatomy for surgeons, Vol 2: The thorax, abdomen and pelvis. Harper & Row, New York

Houston J (1830) Observations on the mucous membrane of the rectum. Dublin Hospital Reports 5:158

Huger WE Jr (1979) The anatomic rationale for abdominal lipectomy. Am Surg 45:612–617

Hyrtl J (1846) Lehrbuch der Anatomie des Menschen mit Rücksicht auf physiologische Begründung und praktische Anwendung. Friedrich Ehrlich, Prague

Hyrtl J (1872) Das Nierenbecken der Säugethiere und des Menschen. Denkschr Kais Akad Wiss Wien 31:107–140

Imlach F (1885) On shortening the round ligament of the urerus. Edinb Med J 30:913

Jackson JN (1909) On the developmental topography of the thoracic and abdominal viscra. Anat Rec 3:361–396

Jacquement M (1897) Le recessus hépatocystique. Marseilles Méd 35:449

Kelahan L, Menias CO, Chow L (2021) A review of internal hernias related to congenital peritoneal fossae and apertures. Abdom Radiol 46:1825–1836

Kirk E (1931) Untersuchungen über die gröbere und feinere topographische Verteilung der Arterien, Venen und Ausführungsgänge in der menschlichen Bauchspeicheldrüse. Z Ges Anat 94:822–875

Kohlrausch OLB (1854) Zur Anatomie und Physiologie der Beckenorgane nebst naturgetreuer Abbildung der Längsdurchschnitte des männlichen und weiblichen Beckens. Leipzig

Konstantinovich VB (1872–1873) Die Anordnung der Gefässe des Mastdarms. St Petersb Med Zschr 3:529–546

Kraissl C (1951) The selection of appropriate lines for elective surgical incisions. Plast Reconstr Surg 8:1–28

Lange JF, Komen N, Akkerman G, Nout E, Horstmanshoff H, Schlesinger F, Bonjer J, Kleinrensink G-J (2007) Riolan's arch: Confusing, misnomer, and obsolete. A literature survey of the connection(s) between the superior and inferior mesenteric arteries. Am J Surg 193:742–748

Lanz O (1908) Der McBurney'sche Punkt. Zbl Chir 35:185–190

Mackay JY (1889) On the arterial system of vertebrates homologically considered. In: Cleland J, Mackay JY, Young RB (eds) Memoirs and Memoranda, vol I. Williams and Norgate, Edinburgh

Mascagni P (1787) Vasorum lymphaticorum corporis humani historia et ichnographia. Pazzini Carli, Siena

McBurney C (1889) Experience with early operative interference in cases of disease of the vermiform appendix. New York Med J:676–684

Meckel JF (1817) Bildungsgeschichte des Darmkanals der Saugtiere und namentlich des Menschen. Deutsch Arch Physiol 3:1–84

Michels NA (1955) Blood supply and anatomy of the upper abdominal organs with a descriptive atlas. Lippincott, Philadelphia, Montreal

Michels NA, Siddharth P, Kornblith PL, Parke WW (1962) The variant blood supply to the small and large intestines: Its importance in regional resections. A new anatomic study based on four hundred dissections with a complete review of the literature. J Int Coll Surg 39:127–190

Michels NA, Siddharth P, Kornblith PL, Parke WW (1965) The variant blood supply to the descending colon, rectosigmoid and rectum based on 400 dissections. Its importance in regional resections: a review of medical literature. Dis Colon Rectum 8:251–278

Moskowitz M, Zimmerman H, Felson H (1964) The meandering mesenteric artery of the colon. Am J Roentgenol Radium Therapy, Nucl Med 92:1088–1099

Nélaton A (1844–1860) Eléments de pathologie chirurgicale, 5 Vols. Baillière, Paris

O'Beirne J (1833) New views of the process of defecation, and their application to the pathology and treatment of diseases of the stomach, bowels, and other organs. Together with an analytical correction of Sir Charles Bell's views respecting the nerves of the face. Dublin

Pattison GS (1820) Observations on lithotomy. Am Med Recorder 3:1–24

Petit JL, Lesne F-D (1774) Traité des maladies chirurgicales et les opérations qui leur conviennient, vol 2. Didot, Paris

Poirier P, Cuéno B (1902) Système lymphatique. In: Poirier P, Charpy A (eds) Traité d'anatomie humaine, Tome 2. Masson, Paris

Proust PT (1822) Dissertation sur le péritonite. Thesis, University of Paris

Quénu EAVA (1893) Vaisseaux lymphatiques de l'anus. Bull Soc Anat Paris 7:399–409

Retzius A (1835) Bemerkungen über Anastomosen zwischen der Pfortader und der unteren Hohlader ausserhalb der Leber. Ztsch Physiol 5:105

Rex H (1888) Beiträge zur Morphologie der Säugerleber. Morphol Jahrb 1:517–615

Rocko JM, Swan KG, Di Gioia JM (1981) Calot's triangle revisited. Surg Gynec Obst 153:40–44

Rosenmüller JC (1805) Chirurgisch-Anatomische Abbildungen für Aertze und Wundärzte. Weimar

Rüdinger N (1879) Ueber die Muskelanordnung im Pfortner des Magens und am Anus. Allg Wiener Med Zeitung 24:2–3

Sappey MPC (1883) Mémoire sur les veines portes accessoires. J Anat Physiol 19:517–525

Stroud BB (1896) On the Anatomy of the anus. Ann Surg Philad 24:1–15

Sudeck P (1907) Ueber die Gefässversorgung des Mastdarmes in Hinsicht auf die operative Gangrän. Münch Med Wschr 54:1314–1317

Tandler J (1904a) Zur Entwicklungsgeschichte der menschlichen Darmarterien. Anat Hefte 23:187–208

Tandler J (1904b) Über die Varietäten der Arteria Coeliaca und deren Entwicklung. Anat Hefte 25:473–500

Tobin CE, Benjamin JA (1949) Anatomic and clinical reevaluation of Camper's, Scarpa's and Colles' fasciae. Surg Gynec Obst 88:545

Traube L (1868) Bemerkungen ueber der halbmondformigen Raum. Berlin Klin Wschr 5:510

Treves F (1885) The anatomy of the intestinal canal and peritoneum in Man. Hunterian Lectures, London

van Gulik TM, Schoots I (2005) Anastomosis of Riolan revisited; the meandering mesenteric artery. Arch Surg 140:1225–1229

Villemin F, Huard P (1924) La contribution de l'arc de Treitz. CR Ass Anat Strasbourg, pp 263–267

von Bürow KA (1838) Beitrag zur Gefässlehre des Fötus. Arch Anat Physiol:44–45

von Haller A (1778) Epiploon. In: Encyclopédie ou dictionnaire raisonné des sciences, arts et des métiers. Pellet, Geneva

Waldeyer W (1899) Das Becken. Cohen, Bonn

Wendell-Smith CP (1997) Fascia: An illustrative problem in international terminology. Surg Radiol Anat 19:273–277

Wilkie DPD (1911) The blood supply of the duodenum with special reference to the supraduodenal artery. Surg Gynec Obst 13:399–405

23 Anatomical Eponyms in Dentistry

David Kachlík, Jiři Šedý, and Hans J. ten Donkelaar

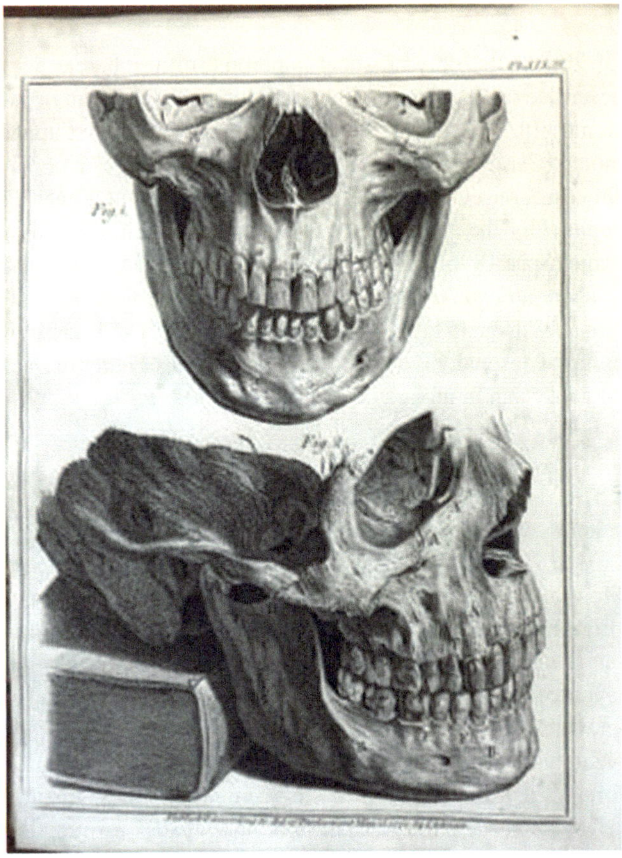

Vignette The teeth as illustrated by John Hunter in his *The Natural History of the Human Teeth*, published 1771 in London. (Courtesy The Royal College of Surgeons of England)

23.1 Introduction

In Dentistry, many anatomical eponyms are in use, most of them on histology. We present those eponyms according to the development, morphology, histology and innervation of the teeth. For further data on the authors quoted, reference is made to Part II (*The Anatomical Eponyms from A to Z*) and figures in this part. The **Vignette** shows the teeth as illustrated by John Hunter in his *The Natural History of Teeth* (1771).

23.2 Eponyms on the Development of the Teeth

During embryonic development, two ectodermal thickenings, arranged as parallel arches, appear in the underlying ectomesenchyme in each of the developing upper and lower jaws. The external thickening, the **vestibular band** or **labiogingival lamina**, forms a sulcus between the lip and the gum. The internal thickening, the **dental lamina**, is the ectodermal component from which the teeth are developing. The epithelium will produce enamel, whereas the ectomesenchyme forms predentin. Thickenings appearing in the dental lamina form the enamel organs, of which five have appeared in each quadrant by 9 weeks of development, forming the deciduous teeth (Kraus and Jordan 1956; Oöe 1981; O'Rahilly and Müller 2001). Each **enamel organ** grows around an ectomesenchymal proliferation known as the **dental papilla**. Together they are described as a **tooth bud**. The cap-shaped enamel organ progresses bell-shaped and shows external and internal epithelial layers. The basal lamina of the internal enamel epithelium separates it from the dental papilla as the amelodentinal junction, the boundary between the future enamel and dentin. The **membrane of Hannover** is an *eponym* for the stratum intermedium of the enamel, the inner layer of cells within the enamel organ of the dental germ in the foetus (Hannover and Jones 1857; *see* Hannover, Adolph).

The ectomesenchymal cells of the dental papilla form the **odontoblasts**. The remaining cells of the dental papilla form the pulp of the tooth. The internal enamel epithelium becomes a layer of **ameloblasts**, which form the **enamel**. An ectomesenchymal condensation, the dental sac or follicle, surrounds the dental papilla and the enamel organ, and forms the periodontal ligament, which anchors the tooth in its socket. The **root bud function** consists in the proliferation

H. J. ten Donkelaar et al., *An Illustrated Guide to Anatomical Eponyms*, https://doi.org/10.1007/978-3-031-91664-9_23

of the epithelium of the enamel organ. The most peripheral cells of the dental papilla along the internal surface of the root sheath differentiate into odontoblasts from which the root dentin is formed. From the innermost layer of the dental sac, the **cement** that covers the root originates through apposition, whereas the epithelial sheath of the root regresses and disappears.

Several structures in the developing teeth received an eponym (Fig. 23.1): (a) the **fibres of Tomes**, an *eponym* for dentinal fibres, projections of odontoblasts, which run inside dentinal tubules perpendicular to the surface of a tooth (*see* Tomes, John); (b) the **processes of Tomes**, an *eponym* for inconstant apical processes of ameloblasts in the developing enamel (Tomes 1897; *see* Tomes, Charles Sissmore); (c) the **epithelial sheath of Hertwig**, an *eponym* for cells of the epithelial root sheath, composed of cells of the enamel organ of the developing tooth, persisting postnatally around the roots of teeth (Hertwig 1886; *see* Hertwig, Oscar); and (d) the **hyaline layer of Hopewell-Smith**, an *eponym* for the thin layer of mineralized hyaline material at the dentinocemental junction, secreted over newly formed dentin, and lying superjacent to the granular layer of Tomes (Hopewell-Smith 1895, 1913; *see* Hopewell-Smith, Arthur); and (e) the **dental crest of Kölliker**: an *eponym* for the incisive bone upon which the incisors develop (Von Kölliker 1852; *see* Kölliker, Rudolf Albert von).

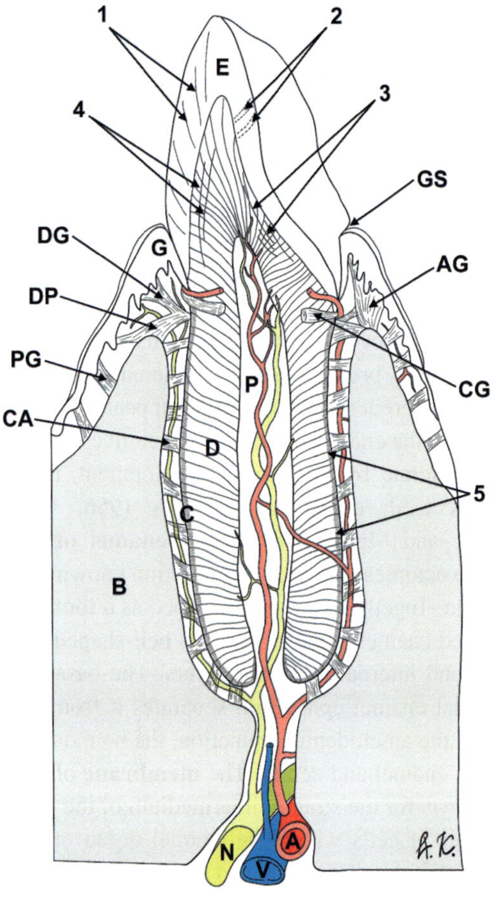

Fig. 23.1 Schematic overview of a tooth. (After Drenckhahn D, Zenker W, eds, *Benninghoff Anatomie*, 16th ed. Urban & Schwarzenberg, Munich, 2004; courtesy David Kachlík and Anhelina Khadanovich, Prague). *A*, *N*, *V* alveolar artery, nerve and vein, *A* alveologingival fibres, *B* alveolar bone, *C* cement, *CA* cementoalveolar fibres, *CG* circular gingival fibres, *D* dentin, *DG* dentogingival fibres, *DP* dentoperiosteal fibres, *E* enamel, *G* gingival papilla (free or marginal gingiva), *GS* gingival sulcus, *PG* periostogingival fibres, *1* lines of Retzius, *2* lines of Hunter-Schreger, *3* lines of Ebner, *4* lines of Owen, *5* granular layer of Tomes (in *grey*)

23.3 Eponyms on the Morphology of the Teeth

Of the teeth, only a supernumerous fifth tubercle was described as **tubercle of Carabelli**, an *eponym* for an occasional fifth tubercle on the lingual surface of the first upper molar (Carabelli 1831; *see* Carabelli, George). For occlusion, several curves are in use: (a) the **curve of Spee**: an *eponym* for the anatomical curvature of the sagittal occlusal plane (Spee 1896; *see* Spee, Ferdinand); and (b) the **curve of Wilson**, an *eponym* for the curvature of the frontal occlusal plane (Wilson 1911; *see* Wilson, George H). The **furcation canal of Jesenský** is an *eponym* for an accessory interradicular root canal in molars (*see* Jesenský, Jan).

23.4 Eponyms on the Histology of the Teeth

Most of the eponyms in use in dentistry concern histology (Fig. 23.1). For **dentin**: (a) the **peritubular** or **intertubular dentin of Neumann**, an *eponym* for the semicalcified layer of matrix surrounding the tubules of dentin, also known as sheath or layer of Neumann (Neumann 1843; *see* Neumann, Ernst Francis Christian); (b) the **incremental lines of Ebner**, an *eponym* for cementing lines, distinct refractile borders of collagen-poor bone matrix forming the outer limit of secondary osteons (von Ebner 1902; *see* Ebner, Victor von); (c) the **contour lines of Owen**, an *eponym* for concentric layers of nonhomogenously calcified dentin occurring as a result of appositional growth (Owen 1840–1845; *see* Owen, Richard); (d) the **interglobular spaces of Owen**, an *eponym* for incompletely mineralized dentinal areas beneath the dentinoenamelar junction (Owen 1840–1845); (e) the **interglobular lacunae** or **spaces of Czermak** (or **Čermák**), an *eponym* for hypomineralized areas of interglobular dentin, which should not be considered as lacunae or spaces (Czermak 1850; *see* Czermak, Johann Nepomuk); (f) the **granular layer of Tomes**, an *eponym* for the granular layer of dentin (Tomes 1897; *see* Tomes, John); (g) the **incremental lines of**

Salter, an *eponym* for incremental lines sometimes seen in dentin due to hypercalcification (Salter 1865; *see* Salter, Samuel James Augustus); (h) the **fibres of Korff**, an *eponym* for argyrophilic strands of ground substance less mineralized than normal dentin with reticular microfibrils, originating in the pulp of the tooth (von Korff 1906; *see* Korff, Karl von); and (i) the **cells of Höhl**, an *eponym* for postmitotic and undifferentiated daughter cells of the pre-odontoblasts that may differentiate into odontoblast-like cells and form secondary dentin and tertiary or reparative dentin after decline or injury of primary odontoblasts (Höhl 1896; *see* Höhl, Erwin).

For **enamel**, the following eponyms are in use: (a) the **lines** or **stripes** of **Hunter-Schreger**, an *eponym* for alternating dark and clear lines, generally oriented perpendicular to the surface of the enamel, representing sections of neighbouring zones of crossed enamel prisms (Hunter 1771, *see* Hunter, John; Schreger 1800, *see* Schreger, Christian Heinrich Theodor); (b) the **incremental lines of Retzius**, an *eponym* for the alternating lighter and darker incremental lines in an enamel section, representing a period of 6–12 years of enamel growth referred to as an individual's periodicity (Retzius 1837; *see* Retzius, Anders); and (c) the **processes of Tomes**, an *eponym* for the inconstant apical processes of ameloblasts in the developing enamel (Tomes 1897; *see* Tomes, John). The **imbrication lines of Pickerill** are an *eponym* for the external manifestation of the course lines of Retzius, parallel impacted lines (perikymata) extending often along the whole crown of the tooth, better visible in young individuals (Pickerill 1912; *see* Pickerill, Henry Percy).

For the **cementum**, the **fibres of Sharpey** (*see* Sharpey, William) should be mentioned, an *eponym* for bundles of collagen fibres, which run from the dense collagenous connective tissue of the periosteum to the superficial bone layer, and for the **pulpa**, the basal **zone of Weil** (*see* Weil, Ludwig Adolf), an *eponym* for a relatively cell-poor zone of the tooth pulp underlying the layer of odontoblasts, containing processes of fibroblasts, extracellular matrix, vessels and nerves. For the periodontium, three eponyms are in use: (a) the **residues** or **epithelial rests of Malassez**, also known as cementicles, an *eponym* for the accumulation of epithelial cells in the dentoalveolar ligament derived from the primitive sheath of Hertwig (Malassez 1895; *see* Malassez, Louis Charles); (b) the **glands of Serres**, an *eponym* for 'gingival glands', islands of epithelium present in the gums of infants (Serres 1817; *see* Serres, Antoine Etienne Renaud Augustin); and (c) the **epithelial attachment of Gottlieb** (*see* Gottlieb, Bernhard), an *eponym* for a collar of epithelial cells attached to the tooth surface and the gingival connective tissue that forms a protective barrier between the periodontal tissue and materials in the oral cavity. Today, it is considered only a functional part, not histologically definable, of the gingivo-dental closure, dynamically changing over time, based also on dental hygiene.

23.5 Eponyms on the Innervation of the Teeth

For pain transmission of the teeth, two plexuses received special attention (Fig. 23.2): (a) the **plexus of Raschkow**, an *eponym* for the subodontoblastic plexus, a plexus of nerve fibres located in the subodontoblastic layer, responsible for pain transmission from the pulp of the tooth (Raschkow 1969; *see* Raschkow, Isaac); and (b) the **predentinal plexus of Bradlaw**, an *eponym* for the marginal nerve plexus of free nerve endings between the odontoblast layer and the dentin in permanent teeth (Bradlaw 1939; *see* Bradlaw, Robert Vivian).

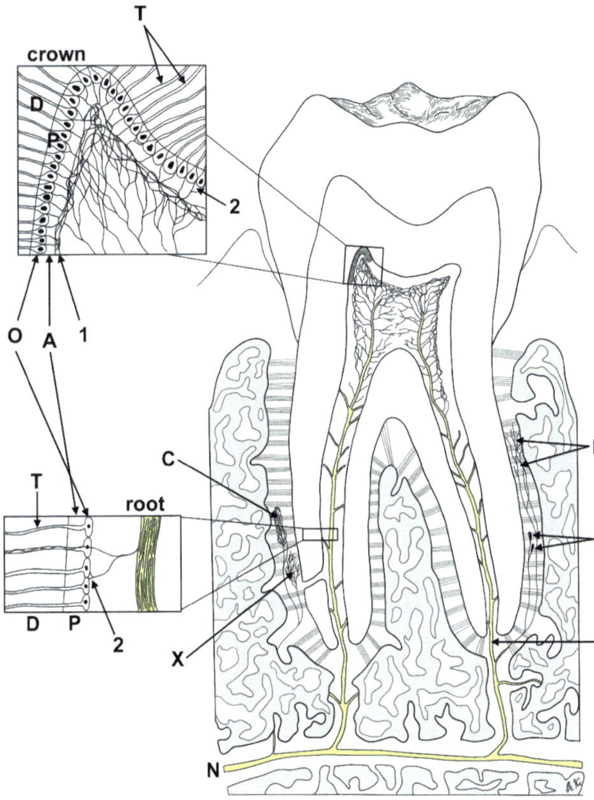

Fig. 23.2 Diagram showing the distribution of nerve endings associated with the teeth. Free nerve endings are found in the pulp cavity and the periodontium. Two nerve plexuses are formed, the plexus of Raschkow and the predentinal plexus of Bradlaw. (After Waite PME, Ashwell KWS *Trigeminal Sensory System*. In: Paxinos G, Mai JK, eds, *The Human Nervous System*, 2nd ed. Elsevier, Amsterdam, 2004; courtesy David Kachlík and Anhelina Khadanovich, Prague). *A* Acellular layer, *C* coiled nerve endings, *D* dentin, *E* expanded spindle-like endings, *F* free nerve endings, *O* odontoblast layer, *N* alveolar nerve, *P* predentin, *T* dentinal tubules, *U* pulpal nerve, *X* dendritic nerve endings, *1* plexus of Raschkow, *2* plexus of Bradlaw

References

Bradlaw R (1939) The histology and histopathology of dental innervation. Proc Roy Soc Med 32:1040–1053

Carabelli G (1831) Systematisches Handbuch der Zahnheilkunde. Vienna

Czermak JN (1850) Beiträge zur mikroskopischen Anatomie der menschlichen Zähne. Z Wiss Zool:61–64

Hannover A, Jones TW (1857) On the development and structure of the teeth of Mammifera. Br Foreign Med Chir Rev 19:229–240

Hertwig O (1886) Lehrbuch der Entwicklungsgeschichte des Menschen und der Wirbelthiere. Jena

Höhl E (1896) Beitrag zur Histologie der Pulpa und des Dentins. Arch Anat Physiol 32:31–54

Hopewell-Smith A (1895) Dental microscopy. Dental Manufactering Co, London

Hopewell-Smith A (1913) An introduction to dental anatomy and physiology, descriptive and applied. Lea & Febiger, Philadelphia

Hunter J (1771) The natural history of the human teeth. Johnson, London

Kraus B, Jordan R (1956) The human dentition before birth. Lea & Febiger, Philadelphia

Malassez LC (1895) Sur l'existence d'amas épithélaux au tour de la racine des dents chez l'homme adulte et à l'état normal (débris épithéliaux paradentaires). Paris

Neumann EC (1843) Ein Beitrag zur Kenntniss des normalen Zahnbein- und Knochengewebes. Leipzig

O'Rahilly R, Müler (2001) Human embryology & teratology, 3rd edn. Wiley, New York

Oöe T (1981) Human tooth and dental arch development. Ishiyaku, Tokyo

Owen R (1840–1845) Odontography. London

Pickerill HP (1912) Stomatology in general practice: a textbook of diseases of the teeth and mouth. Frowde, London

Raschkow I (1969) Innervation of the dental pulp. Acta Odont Scand 27:313–319

Retzius A (1837) Bemerkungen über den inneren Bau der Zähne, mit besonderer Rücksicht auf dem in Zahnknochen vorkommenden Röhrenbau. Arch Anat Physiol 1837:486–566

Schreger CHT (1800) Beiträge zur Geschichte der Zähne: Ueber den Zahnschmelz im Knochensubstanz der Zähne. Beitr Zergliederungskunst 1:3–7

Serres AERA (1817) Essai sur l'anatomie et la physiologie des dents, ou nouvelle théorie de la dentition. Paris

Spee F (1896) Skelettlehre: Kopf. In: Bardeleen K (Hrsg) Handbuch der Anatomie des Menschen. Fischer, Jena

Tomes CS (1897) On the development of marsupial and other tubular enamels with notes upon the development of enamel in general. Phil Trans 189:107–122

von Ebner V (1902) Histologie der Zähne mit Einschluss der Histogenese. In: Scheff J (ed) Handbuch der Zahnheilkunde. Holder, Wien, pp 243–299

von Kölliker A (1852) Handbuch der Gewebelehre des Menschen, Leipzig

von Korff K (1906) Die Entwicklung der Zahnbeingrundsubstanz der Säugetiere. Arch Mikr Anat 67:1–17

Wilson GH (1911) Proving the articulation, contour and expression. Dent Regist 60:382–388

Anatomical Eponyms in Neurology

24

Hans J. ten Donkelaar, Veronika Němcová,
and Marina Quartu

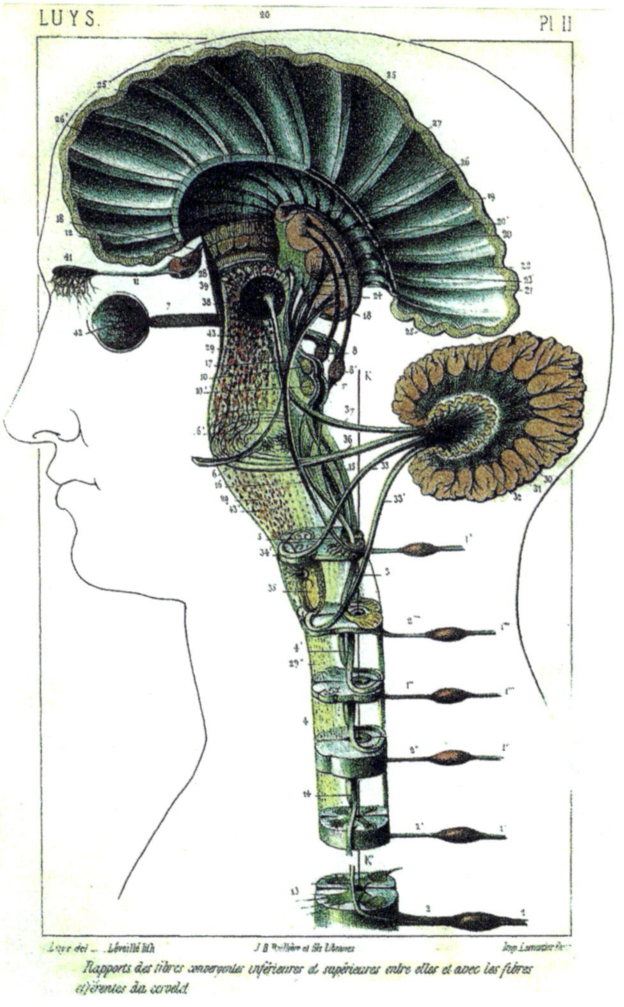

Vignette Illustration by Jules Bernard Luys in his *Recherches sur le système nerveux cérébrospinal*, published 1865 by Baillière in Paris

24.1 Introduction

Many anatomical eponyms are in use in Neurology for Nuclei, Fasciculi or Tracts in the central nervous system (*see* Part I, *The Classical Eponyms*, and ten Donkelaar et al. 2018a). For the functional systems, *eponyms* are in use for the motor system, like the **body of Luys** (the subthalamic nucleus; *see* Luys, Jules Bernard), the **tract of Arnold** (the frontopontine tract; *see* Arnold, Friedrich) and the **tract of Türck** (the temporal corticopontine tract; *see* Türck, Ludwig), and for the optic system, the *eponyms* **optic radiation of Gratiolet** (*see* Gratiolet, Louis Pierre) and **loop of Meyer** for its temporal loop (*see* Meyer, Adolf; also described by Archambault, Cushing and Flechsig) are used as well as the term **band of Gennari**, an *eponym* for the occipital stripe, a conspicuous band of myelinated fibres marking the striate cortex of the occipital lobe (*see* Gennari, Francesco). Here, we will focus on eponyms for blood vessels of the brain and the spinal cord, some macroscopical terms for the brain and its ventricular system, planes of the brain, and especially on the somatosensory system and the spinal cord, and on areas, gyri and sulci of the cerebral cortex. For further data on the various authors quoted, reference is made to Part II (*The Anatomical Eponyms from A to Z*) and figures in this part. The **vignette** shows an illustration by Jules Bernard Luys in his *Recherches sur le système nerveux cérébrospinal*, published in 1865.

24.2 Eponyms for Blood Vessels of the Brain and the Spinal Cord

Although the arteries and veins of the brain and spinal cord were extensively described, only few eponyms are in use. For the **arteries**, there are *eponyms* known as the **artery of Adamkiewicz** (the great radicular artery of the spinal cord; Adamkiewicz 1881, 1882; *see* Adamkiewicz, Albert Wojciech, and Fig. A3), the **artery of Bernasconi and Cassinari** (the marginal tentorial branch; Bernasconi and Cassinari 1956, 1957; *see* Bernasconi, Vittorio, and Cassinari, Valentino), the **artery of Charcot** (any of the lenticulostriate arteries, known as 'artery of haemorrhage'; *see* Charcot, Jean Martin), the **arteries of Foix** (thalamogeniculate, thalamoperforated and thalamotuberal arteries; *see* Foix, Charles), the **arteries of Foix and Hillemand** on the arterial

H. J. ten Donkelaar et al., *An Illustrated Guide to Anatomical Eponyms*, https://doi.org/10.1007/978-3-031-91664-9_24

supply of the brain stem (Foix and Hillemand 1925), the **recurrent artery of Heubner** (the long striatal or long central artery; Heubner 1874; *see* Heubner, Johann Otto, and Fig. H4), the **arteries of Kadyi** (the sulcomarginal arteries of the spinal cord; *see* Kadyi, Henryk Karol Klemens), the **arteries of Lazorthes** (arteries supplying the cervical and lumbosacral enlargements of the spinal cord; *see* Lazorthes, Guy), the **artery of Rolando** and the **artery of Sylvius** (both for the middle cerebral artery; Rolando 1829, Sylvius 1660; *see* Rolando, Luigi, and Sylvius, François de le Boë), the **arteries of Uchimura** (for long and short branches of the posterior cerebral artery, supplying the hippocampal formation; Uchimura 1928; *see* Uchimura, Yushi), the **artery of Wallenberg** (for the posterior inferior cerebellar artery; *see* Wallenberg, Adolf), the **circle of Willis** (*see* Willis, Thomas), and the **artery of Wollschlaeger and Wollschlaeger** (for an inconstant dural branch of the superior cerebellar artery, supplying parts of the falx cerebri and tentorium cerebelli; Wollschlaeger and Wollschlaeger 1965). Although the many variants of the cerebral arteries are beyond the scope of this section, one should be mentioned here: the **artery of Percheron**, a variant of the posterior cerebral circulation, in which one branch supplies the paramedian thalami and the rostral midbrain bilaterally (Percheron 1973; *see* Percheron, Gérard). The anterior, middle and posterior cerebral arteries can be divided into segments, which are usually indicated by *acronyms* (Fig. 24.1; ten Donkelaar et al. 2018a, 2020c): for the anterior cerebral artery, A1-A5, meaning A1 as the precommunicating segment, A2 the infracallosal segment, A3 the precallosal segment, A4 the supracallosal segment, and A5 the postcallosal segment; for the middle cerebral artery: M1-M4, M1 for the sphenoidal segment, M2 for the insular segment, M3 for the opercular segment, and M4 for the cortical segment; and for the posterior cerebral artery: P1-P4, P1 for the precommunicating segment, P2 for the postcommunicating segment, P3 for the quadrigeminal segment, and P4 for the cortical segment. For the internal carotid artery, giving rise to the anterior and middle cerebral arteries, currently the subdivision of Bouthillier et al. (1996; Fig. 24.1) is used: C1 for the cervical segment, C2 for the petrous segment, C3 for the lacerum segment, C4 for the cavernous segment, C5 for the clinoid segment, C6 for the ophthalmic segment, and C7 for the communicating segment. For the vertebral arteries, giving rise to the basilar artery, the origin of the posterior cerebral arteries, four segments are distinguished: V1 for the prevertebral segment, V2 for the cervical segment, V3 for the atlantic segment, and V4 for the intracranial segment.

For the **veins** (Benner et al. 2021; ten Donkelaar et al. 2020c), there are *eponyms* including the **plexus of Batson**, an *eponym* for the venous plexus extending from the cranial dural sinuses along the vertebral column downwards towards the lower sacral vertebrae (Batson 1940 *see* Batson, Oscar Vivian), the **veins of Breschet** (the diploic veins of the skull; Breschet 1829; *see* Breschet, Gilbert), the **vein of Browning** (part of the superior anastomotic vein of Trolard; *see* Browning, William), the **vein of Dandy** (the petrosal vein; *see* Dandy, Walter Edward), the **vein of Galen** (the great cerebral vein; *see* Galen, Claudius, and Fig. 24.2a), the **torcular of Herophilus** (the confluence of sinuses; *see* Herophilus), the **vein of Labbé** (the inferior anastomotic vein of the cerebral cortex; Labbé 1879; *see* Labbé, Charles, and Fig. 24.2b), the **vein of Markowski** (the median prosencephalic vein; Markowski 1921, 1931; *see* Markowski,

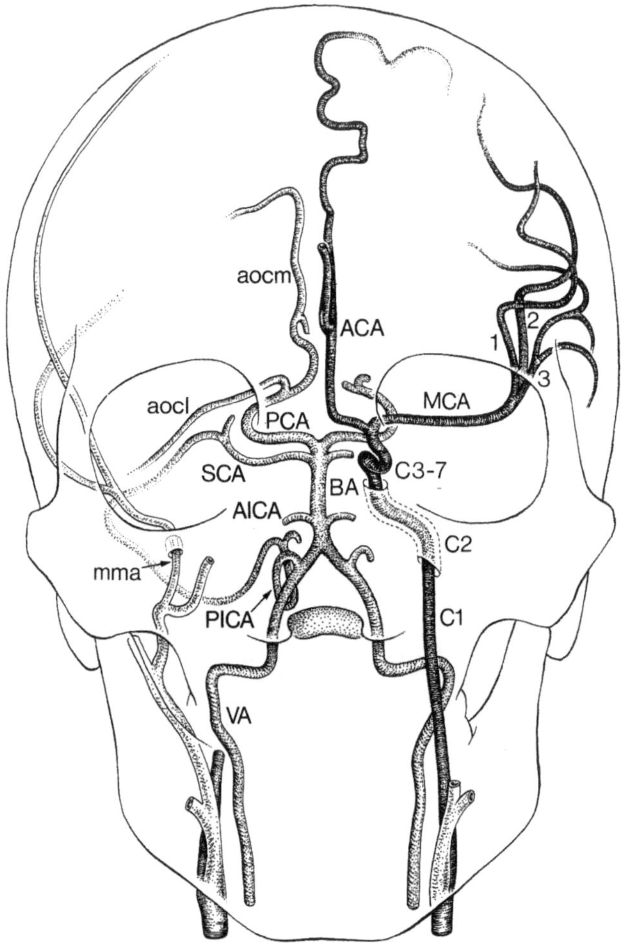

Fig. 24.1 Orthogonal frontal projection of the main cranial arteries (carotid and vertebrobasilar systems). Note trifurcation of the MCA. *ACA* anterior cerebral artery, *acalc* calcarine artery, *acmg* callosomarginal artery, *aocl, aocm* lateral and medial occipital arteries, *AICA* anterior inferior cerebellar artery, *BA* basilar artery, *C1-C7* cervical, petrous, lacerum, cavernous, clinoid, ophthalmic and communicating segments of internal carotid artery, *MCA* middle cerebral artery, *mma* middle meningeal artery, *PCA* posterior cerebral artery, *PICA* posterior inferior cerebellar artery, *SCA* superior cerebellar artery, *VA* vertebral artery, *1–3* refer to the trifurcation of the MCA. (After Nieuwenhuys R, Voogd J, van Huijzen C *The Human Central Nervous System.* Springer, Berlin, Heidelberg, New York 1988; from ten Donkelaar HJ *Clinical Neuroanatomy*, 2nd ed. Springer, Cham)

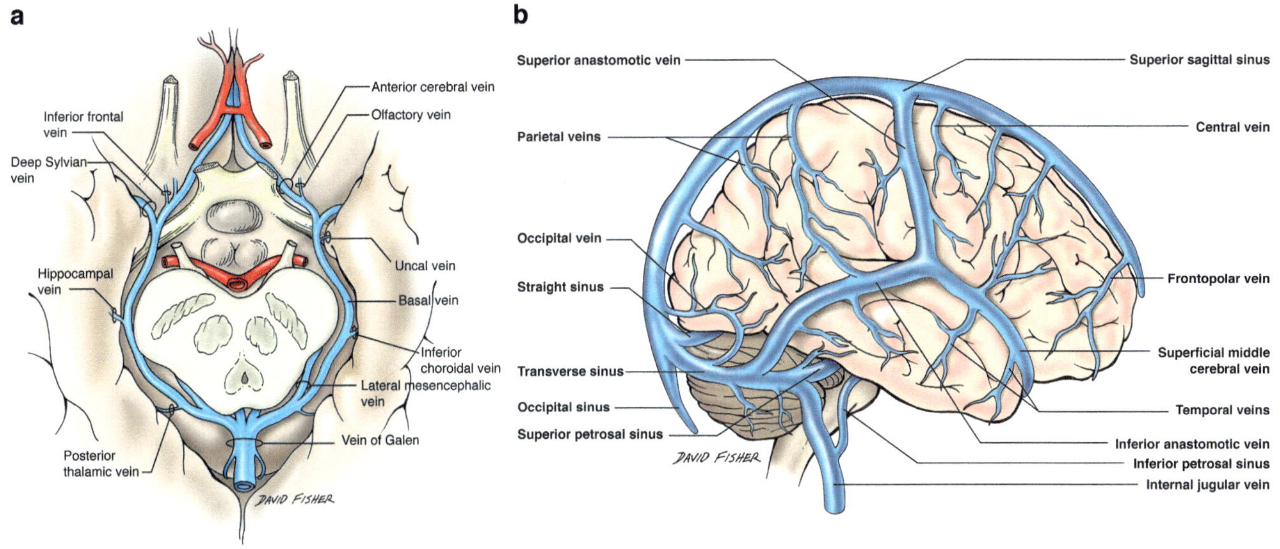

Fig. 24.2 (**a**) The deep cerebral veins shown in a ventral view of the brain, and (**b**) the superficial cerebral veins and their connections to the venous sinuses. (From ten Donkelaar HJ, Kachlik D, Tubbs RS *An Illustrated Terminologia Neuroanatomica*. Springer, Cham, 2018; courtesy R. Shane Tubbs, New Orleans)

Józef), the **plexus of Rektorzik** (the petrooccipital venous plexus encasing the petrous segment of the internal carotid artery; Rektorzik 1858; *see* Rektorzik, Ernst), the **vein of Rolando** (the central vein; *see* Rolando, Luigi), the **vein of Rosenthal** (the basal vein; *see* Rosenthal, Friedrich Christian, and Fig. 24.2a), the **vein of Sylvius** (the middle cerebral vein; *see* Sylvius, François de le Boë), the **vein of Trolard** (the superior anastomotic vein of the cerebral cortex; Trolard 1868; *see* Trolard, Paulin, and Fig. 24.2b), and the **vein of Vesalius** (the sphenoidal emissary vein; *see* Vesalius, Andreas). Trolard is also known for the *eponyms* **hypoglossal plexus of Trolard**, a dense venous plexus at the extracranial aspect of the hypoglossal canal, and the **petrooccipital vein of Trolard**, a vein connecting the venous plexus of Rektorzik and the cavernous sinus with the jugular bulb.

24.3 Eponyms for Some Macroscopical Structures of the Brain and Its Ventricular System

Some *eponyms* frequently used for particular structures of the brain are: (a) the **semioval centre of Vieussens**, an *eponym* for the semioval structure of white matter found in horizontal sections of the brain (Vieussens 1685; *see* Vieussens, Raymond de; Fig. 24.3); (b) **spaces:** apart from epidural, subarachnoid, subdural and subpial spaces, the *eponym* **space of Virchow-Robin** is used for the perivascular space, a tunnel-like extension of the subarachnoid space surrounding blood vessels entering the brain and spinal cord (*see* Virchow, Rudolf Ludwig Karl, and Fig. V29), and the *eponym* **spaces of Magendie** for spaces between the pia mater and the arachnoid at the level of sulci of the cerebral cortex (*see* Magendie, François); (c) **bodies** or **granules of Pacchioni:** an *eponym* for the arachnoid granulations in the superior sagittal sinus (*see* Pacchioni, Antonio, and Fig. P1b); and (d) for the cavum of the septum pellucidum several *eponyms* are in use: **ventricle of Verga** (**cavum Vergae**; Verga 1856; *see* Verga, Andrea, and Fig. V20) for the caudal part of the septum pellucidum, and **ventricle of Duncan** (*see* Duncan, Daniel), **Sylvius** (*see* Sylvius, François de le Boë) and **Wenzel** (*see* Wenzel, Joseph) for a more general cavity in the septum pellucidum.

Eponyms for the ventricular system (Fig. 24.4) in use are: (a) **apertures** or **foramina of the fourth ventricle**, including: **lateral: Luschka** (*see* Luschka, Hubert von), **Key-Retzius** (*see* Key, Ernst Axel Hentik, and Retzius, Gustav Magnus) and **Mierzejewsky** (*see* Mierzejewsky, Lucian); and **median: Magendie** (*see* Magendie, François); (b) the **flower basket** or **bouquet of Bochdalek**, an *eponym* for the fruticulus, that part of the choroid plexus of the fourth ventricle that extends through its lateral aperture, originally described as 'Blumenkörbchen' (Bochdalek 1833; *see* Bochdalek, Vincenz Alexander); (c) the **plexus of Bochdalek**, an *eponym* for the plexus choroideus of the fourth ventricle; (d) the **aqueduct of Sylvius**, an *eponym* for the mesencephalic or cerebral aqueduct (Sylvius 1660; *see* Sylvius, François de le Boë); its opening into the third ventricle received the old *eponyms* **anus of Bartholin** (*see* Bartholin, Thomas, The Elder) and **anus of Vieussens** (*see* Vieussens, Raymond de); and (e) the **foramen of Monro**, an *eponym* for the interventricular foramen between the third and lateral ventricles (*see* Monro, Alexander).

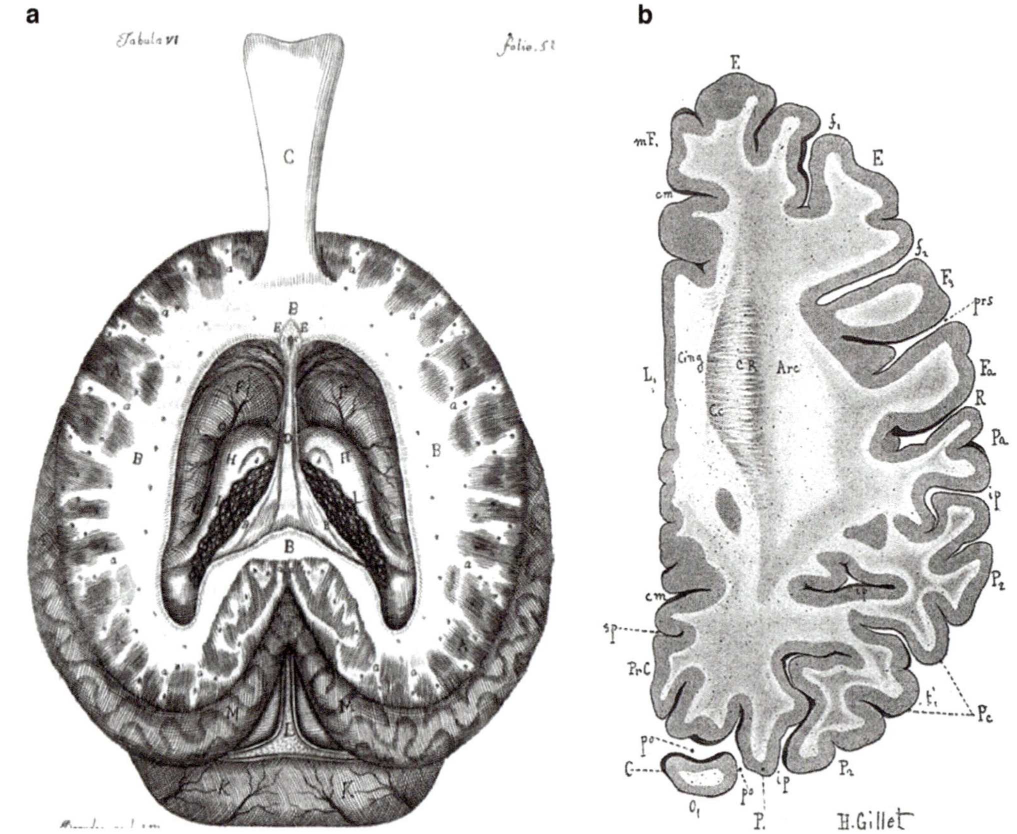

Fig. 24.3 The semioval centre shown in (**a**) the original drawing of Vieussens, labelled as *E*, and (**b**) a horizontal section of the brain. (From Dejerine J *Anatomie des centres nerveux*, Vol 1. Rueff, Paris, 1895)

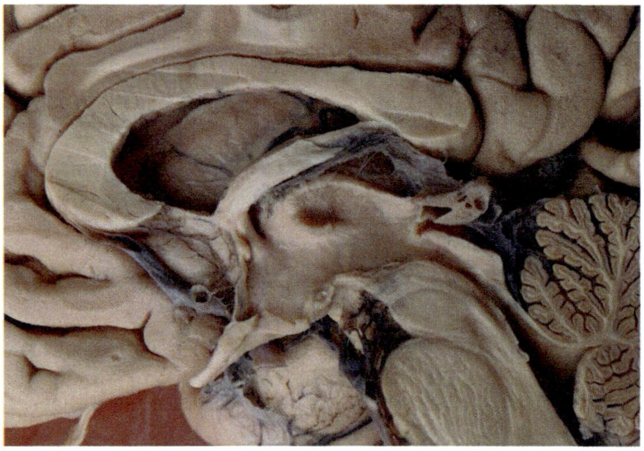

Fig. 24.4 The ventricular system showing the fourth ventricle below the cerebellum, the mesencephalic aqueduct, the third ventricle and the right lateral ventricle, connected by the interventricular foramen of Monro. (From ten Donkelaar HJ, Kachlík D, Tubbs RS *An Illustrated Terminologia Neuroanatomica*. Springer, Cham, 2018; courtesy Veronika Němcová, Prague)

24.4 Planes of the Brain

The most commonly used horizontal reference planes of the brain in neuroimaging are the **bicommissural plane of Talairach** (Talairach et al. 1952, 1967; Talairach, Tournoux 1988) and the **intercommissural plane of Schaltenbrand and Bailey** (Schaltenbrand and Bailey 1959). The **bicommissural plane** is the plane through the line joining the upper border of the anterior commissure and the lower border of the posterior commissure (Figs. 24.5 and 24.6). Two cranial reference lines rather closely parallel the bicommissural plane: the canthomeatal line, connecting the outer canthus or lateral angle of the eye with the centre of the external acoustic meatus (Szikla et al. 1977), and the glabella-inion line of Schwalbe (Tokunaga et al. 1977). The major brain sulci seem to maintain relatively constant relationships with respect to the bicommissural line. The **intercommissural plane** is the plane joining the centres of the anterior and posterior commissures. The terms bicommissural and intercommissural planes are often used as synonyms. For particular reasons, two other planes may be used: (a) the **chiasmatic-commissural plane**, a plane joining the superior border of the optic chiasm and the inferior border of the posterior commissure (Tamraz et al. 1991); this plane closely parallels the lateral sulcus and is particularly suitable for studying the hippocampus; (b) the **commissural-obex plane**, a plane vertical to the chiasmatic-commissural plane, which roughly follows the direction of the corticospinal tract and the medial longi-

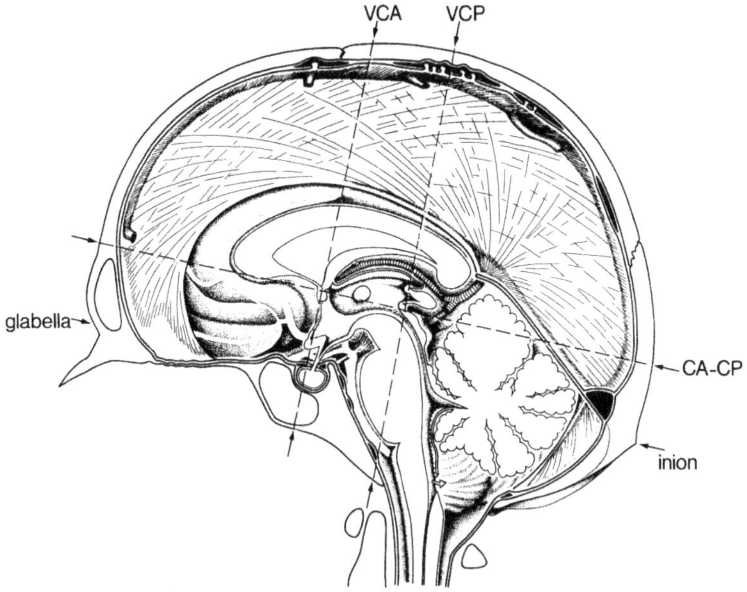

Fig. 24.5 Cephalic reference planes. (Based on Nieuwenhuys R, Voogd J, van Huijzen C *The Human Central Nervous System*. Springer, Berlin, Heidelberg, New York, 1988; from ten Donkelaar HJ, Kachlík D, Tubbs RS *An Illustrated Terminologia Neuroanatomica*. Springer, Cham, 2018). *VCA* and *VCP* are planes through the anterior and posterior commissure, vertical to the bicommissural plane (*CA-CP*)

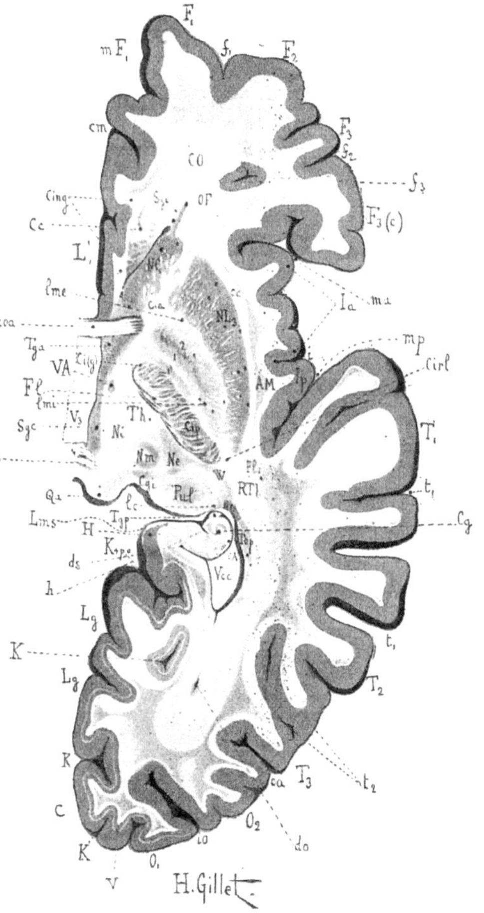

Fig. 24.6 A horizontal section of the right human cerebral hemisphere at the bicommissural plane (from Dejerine J *Anatomie des centres nerveux*, Vol 1. Rueff, Paris, 1895). *coa*, *cop* anterior and posterior commissures

tudinal fasciculus (Tamraz, Comair 2000); and (c) the **neuro-ocular plane**, a plane particularly useful for imaging the visual system. It was defined as the plane passing through the lenses, the optic nerves and the optic canals, and corresponds to Broca's visual plane (*see* Broca, Pierre Paul).

24.5 Eponyms for the Somatosensory System and the Spinal Cord

Henry Head proposed that the cutaneous senses can be divided into two broad categories (Head 1920): the **protopathic system** for the mediation of pain and the extremes of temperature sensation, and the **epicritic** or **gnostic system** for touch, size, and two-point discrimination and detection of small thermal gradients. This terminology has also been applied to the somatosensory pathways. Head's dual system has been superseded by two other dualities: **lemniscal** versus **non-lemniscal** (Willis and Coggeshall 1991) and **large-fibred** versus **small-fibred systems** (Mountcastle 2005). The large-fibred somatosensory afferent system deals with the discriminative sensory aspects of somaesthesis and the small-fibred system with the affective-vegetative aspects. The discriminative sensory system includes the dorsal column–medial lemniscus pathway and the trigeminal lemniscus that deal with mechanoreceptive aspects of somaesthesis, whereas the affective-vegetative system conveys somatosensory information from all spinal columns, especially from the ventral quadrants of the spinal cord and the spinal trigeminothalamic tract.

Cutaneous sensory receptors can be classified as mechanoreceptors, thermoreceptors and nociceptors, many of

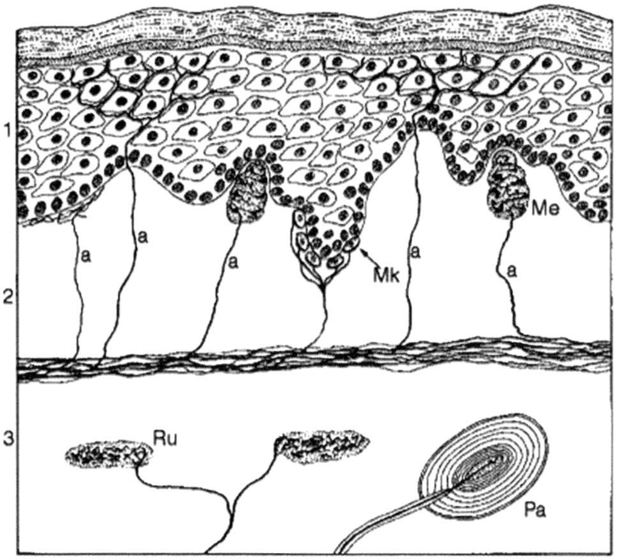

Fig. 24.7 Receptors in the glabrous skin. *a* axon, *Me* Meissner corpuscle, *Mk* Merkel cell, *Pa* Pacinian corpuscle, *Ru* Ruffini corpuscle, *1* epidermis, *2* dermis, *3* subcutis. (From ten Donkelaar HJ, Broman J, van Domburg P *The Somatosensory System.* In: ten Donkelaar HJ *Clinical Neuroanatomy,* 2nd ed. Springer, Cham, pp. 171–255)

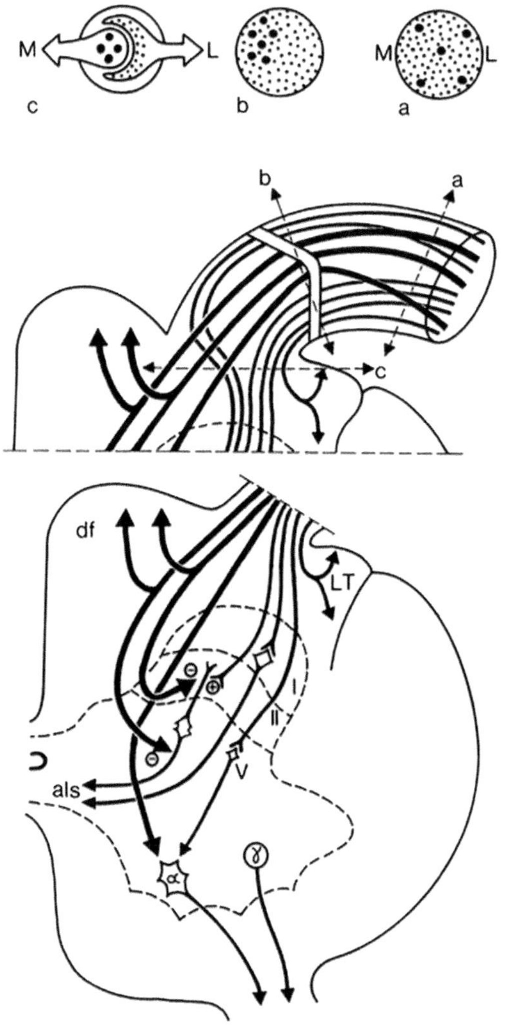

Fig. 24.8 Reorganization of dorsal root fibres from lateral (**a**) to medial (**c**) in the dorsal root entry zone in humans. *als* anterolateral system, *df* dorsal funiculus, *L* lateral, *LT* Lissauer tract, *M* medial, *I-V* Rexed layers of the spinal grey, α alpha-motoneurons, γ gamma-motoneurons. (After Sindou M, Quoex C, Baleydier C *Fiber organization at the posterior spinal cord—rootlet junction in man.* J Comp Neurol 1974; 153:15–26; from ten Donkelaar HJ, Broman J, van Domburg P *The Somatosensory System.* In: ten Donkelaar HJ *Clinical Neuroanatomy,* 2nd ed. Springer, Cham, pp. 171–255)

them bear an *eponym* (Fig. 24.7). The skin receptors include unencapsulated and encapsulated receptor terminals. **Unencapsulated terminals** include the free nerve endings, the **Merkel discs** and associated **Merkel cells** (Merkel 1875; *see* Merkel, Friedrich Sigmund). **Encapsulated terminals** include the **Meissner corpuscles** (Meissner 1852; *see* Meissner, Georg) in the papillary layer of the dermis, and the **Vater-Pacini corpuscles** (Pacini 1840; *see* Pacini, Filippo; but already noted in 1741 by Vater; *see* Vater, Abraham) and the **Ruffini corpuscles** (Ruffini 1898; *see* Ruffini, Angelo). Two types of **cutaneous nociceptors** are found in the human skin: Aδ-mechanical nociceptors and C-polymodal nociceptors. In muscles, stretch receptors (muscle spindles and Golgi tendon organs) and pressure-pain endings are present. The primary and secondary endings of the muscle spindles are innervated by Ia and group II axons, respectively. The **Golgi tendon organs** (Golgi 1878; *see* Golgi, Camillo) are innervated by group Ib axons.

The central branches of the primary afferent fibres enter the spinal cord at the **dorsal root entry zone** (**DREZ**). Peripherally, the dorsal root fibres do not show a particular organization, but close to the DREZ, the small fibres become situated increasingly laterally (Sindou et al. 1974; Fig. 24.8), and enter the posterolateral or dorsolateral fasciculus (*eponym*: **tract of Lissauer**; Lissauer 1886; *see* Lissauer, Heinrich). The large fibres become so arranged that fibres from muscle receptors are located centrally and fibres to enter the dorsal funiculus or column are situated medially. Thus, dorsal root fibres segregate into a **lateral component**, composed of thin myelinated Aδ-fibres from the skin and unmyelinated C-fibres from the skin and viscera, which enter through the lateral part of the tract of Lissauer, and a **medial component**, composed of thicker myelinated Aβ-fibres from the skin and proprioceptive afferent fibres from the muscle spindles (Ia- and II-fibres) and Golgi tendon organs (Ib-fibres). Each fibre category terminates in specific layers of the spinal cord (*eponym*: **layers of Rexed**; Rexed 1952, 1954; *see* Rexed, Bror Anders): (a) Aδ- and C-fibres, which innervate most of the receptors responding to noxious stimuli, terminate on the **marginal cells of Waldeyer** (*see* Waldeyer, Heinrich Wilhelm Gottfried) in layer I of Rexed (also known as **zone of Waldeyer**) and in layer II (the superficial part of the **substantia gelatinosa of Rolando**; *see*

Rolando, Luigi), respectively; (b) collaterals of the thicker myelinated Aδ-fibres, innervating touch receptors in the skin, terminate in the deep layers of the substantia gelatinosa, the nucleus proprius and the adjacent intermediate grey (layers III-V); (c) Ib-fibres from the Golgi tendon organs end in layer VI; and (d) Ia-fibres from the muscle spindles end on motoneurons in layer IX.

The **large-fibred dorsal column-medial lemniscus pathway** consists of two fasciculi, the dorsal column nuclei and the medial lemniscus (Fig. 24.9a). The medial, gracile fasciculus (*eponym*: **tract of Goll**; Goll 1860; *see* Goll, Friedrich) contains primary afferent fibres from the lower extremity and the trunk, whereas the lateral cuneate fasciculus (*eponym*: **tract of Burdach**; Burdach 1819–1826; *see* Burdach, Karl Friedrich) contains afferents from the upper extremity and associated parts of the trunk and neck. Both tracts terminate in dorsal column nuclei, the gracile fasciculus in the gracile nucleus (*eponym*: **nucleus of Goll**) and the cuneate fasciculus in the cuneate nucleus (*eponym*: **nucleus of Burdach**). Axons of the dorsal column nuclei cross the midline in the caudal hindbrain as the internal arcuate fibres, ascend in the medial lemniscus (*eponym*: **tract of Reil**; Reil 1796; *see* Reil, Johann Christian) and terminate in the ventroposterior nucleus of the thalamus. The **ventral quadrant**

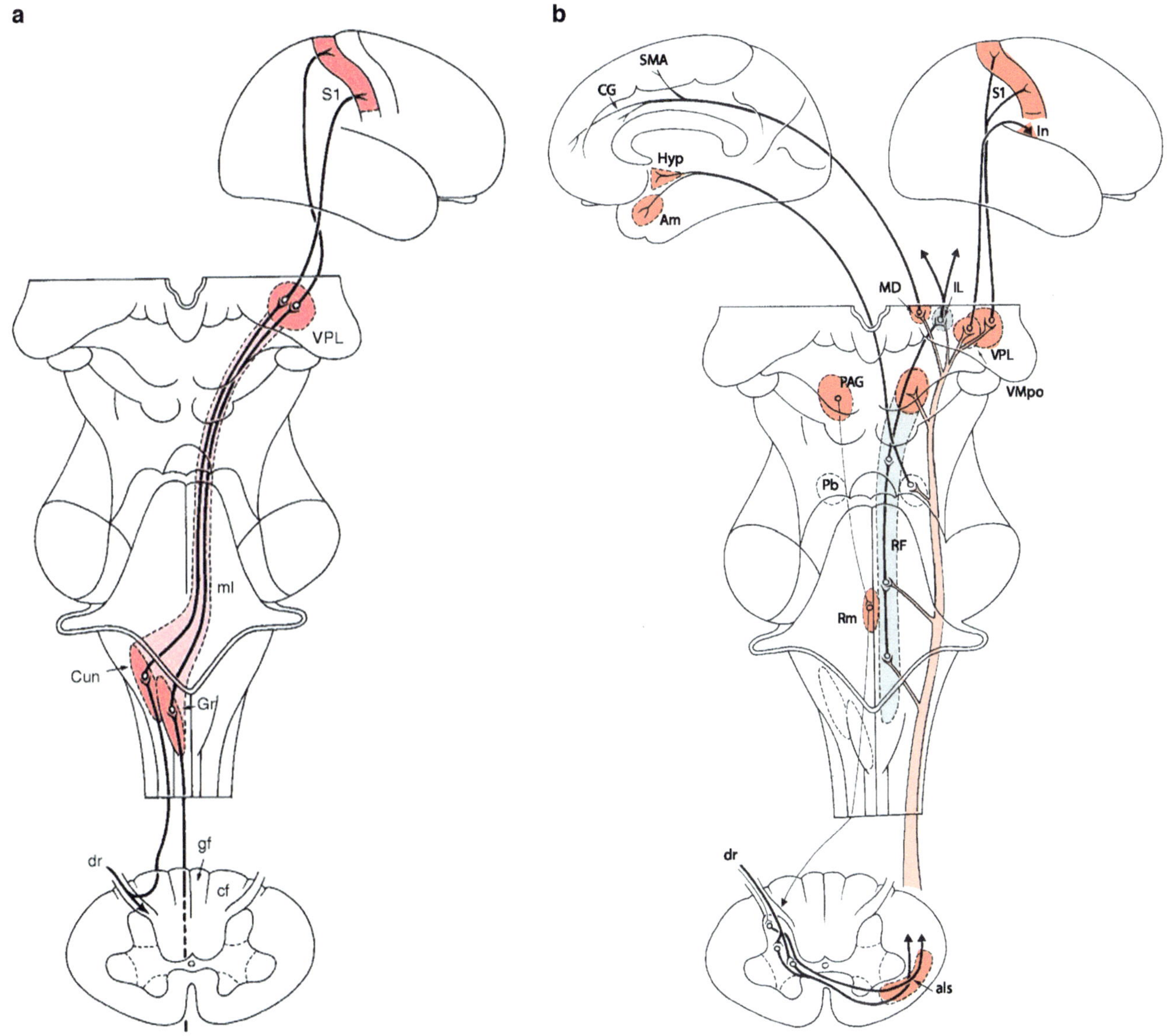

Fig. 24.9 (**a**) Overview of the human ascending spinal system for gnostic sensibility. *cf* cuneate fasciculus, *Cun* cuneate nucleus, *dr* dorsal root, *gf* gracile fasciculus, *Gr* gracile nucleus, *ml* medial lemniscus, *S1* primary somatosensory cortex, *VPL* ventroposterior lateral nucleus; (**b**) Overview of the human ascending spinal systems for vital sensibility. *Als* anterolateral system, *Am* amygdala, *CG* cingulate gyrus, *dr* dorsal root, *Hyp* hypothalamus, *IL* intralaminar nuclei, *In* insula, *MD* mediodorsal nucleus, *PAG* periaqueductal grey, *Pb* parabrachial nucleus, RF reticular formation, *Rm* raphe magnus nucleus, *SMA* supplementary motor area, *S1* primary somatosensory cortex, *VMpo* posterior part of vental medial nucleus, *VPL* ventroposterior lateral nucleus. (From ten Donkelaar HJ *Clinical Neuroanatomy,* 2nd ed. Springer, Cham, 2020)

of the human spinal cord contains pathways that are crucial for pain and temperature sensations (Fig. 24.9b). A number of pathways ascending via the ventral quadrant of the spinal cord to the brain are activated by noxious stimuli. The spinothalamic tract has been regarded as the major pathway responsible for evoking pain sensations in humans. An anterolateral spinothalamic tract (*eponym*: **tract of Dejerine**; Dejerine 1895; *see* Dejerine, Jules) and a lateral spinothalamic tract (*eponym*: **tract of Edinger**; Edinger 1893; *see* Edinger, Ludwig) have been distinguished. Other pathways involved in pain transmission are spinoreticular, spinoparabrachial, spinomesencephalic, spinohypothalamic and spinolimbic tracts. Together with the spinothalamic tracts they are usually described as the **anterolateral system** (for further data, *see* Mountcastle 2005; ten Donkelaar et al. 2020a).

In the spinal cord (Fig. 24.10), two spinocerebellar tracts bear an *eponym*: the posterior or dorsal spinocerebellar tract (*eponym*: **tract of Flechsig**; Flechsig 1876; *see* Flechsig, Paul Emil) arising from the dorsal nucleus of the spinal cord in the thoracic cord (*eponym*: **nucleus of Clarke** and **of Stilling**; Clarke 1851; *see* Clarke, Jacob Augustus Lockhart; and Stilling 1842; *see* Stilling, Benedict), and the anterior or ventral spinocerebellar tract (*eponym*: **tract of Gowers**; Gowers 1880; *see* Gowers, William Richard), arising, amomg others, from the spinal border cells in the lumbar cord, sometimes bearing the *eponym* **cells of Cooper-Sherrington** (Cooper and Sherrington 1940), but already noted by Foerster and Gagel (1932) and later by Kuru (1949). In and between the dorsal funiculi, some peculiar small fibre tracts have been described, composed of descending collaterals of dorsal column fibres: the interfascicular fasciculus in the cervical and upper thoracic dord (*eponym*: **comma tract of Schultze**; Schultze 1883; *see* Schultze, Friedrich), the septomarginal fasciculus in the thoracic and lumbar cord (*eponym*: **oval area of Flechsig**; Flechsig 1876), and its caudal continuation in the sacral cord, bearing the *eponym* **triangle of Philippe-Gombault** (Philippe 1897; *see* Philippe, Claudien; Gombault 1894, 1895; *see* Gombault, François Alexis Albert).

The **descending supraspinal pathways** from the cerebral cortex and the brain stem to the spinal cord have been subdivided into lateral and medial systems (Kuypers 1981; *see* Kuypers, Hans; Fig. 24.11). The **medial system of Kuypers** consists of interstitiospinal, reticulospinal and lateral and medial vestibulospinal tracts, passing via the anterior funiculus and the ventral part of the lateral funiculus. The lateral vestibular tract bears the *eponym* **tract of Deiters** (Deiters 1865; *see* Deiters, Otto Friedrich Karl). The **lateral system of Kuypers** consists of corticospinal, rubrospinal (*eponym*: **tract of Monakow**; Monakow 1909; *see* Monakow, Constantin von), some reticulospinal and raphespinal tracts, all passing via the dorsal part of the lateral funiculus. Various aberrant pyramidal tract bundles have been observed: (a) the

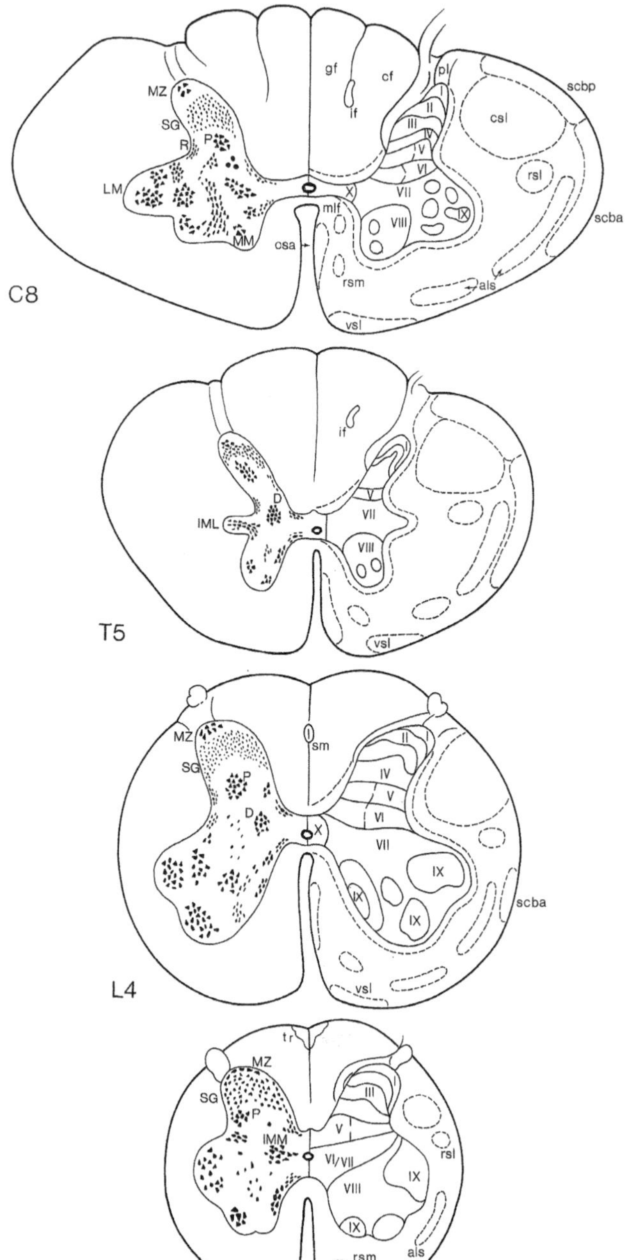

Fig. 24.10 Diagrams of four transverse sections through the spinal cord, at the left the cell structure, at the right the Rexed layers and fiber tracts demonstrated by degeneration studies. *als* anterolateral system, *cf* cuneate fasciculus, *csa*, *csl* anterior and lateral corticospinal tracts, *D* dorsal nucleus of Clarke, *gf* gracile fasciculus, *if* interfascicular fasciculus, *IML* intermediolateral nucleus, *IMM* intermediomedial nucleus, *LM* lateral motoneuron column, *mlf* medial longitudinal fasciculus, *MM* medial motoneuron column, *MZ* marginal zone, *P* nucleus proprius, *pl* posterolateral zone of Lissauer, *R* reticular process of von Lenhossék, *rsl*, *rsm* lateral and medial reticulospinal tracts, *scba*, *scbp* anterior and posterior spinocerebellar tracts, *SG* substantia gelatinosa, *sm* septomarginal fasciculus, *tr* triangle of Philippe-Gombault, *vsl* lateral vestibulospinal tract, *I-IX* Rexed layers I-X, *X* Rexed area X. (After ten Donkelaar HJ, Kachlík D, Tubbs RS *An Illustrated Terminologia Neuroanatomica*. Springer, Cham, 2018)

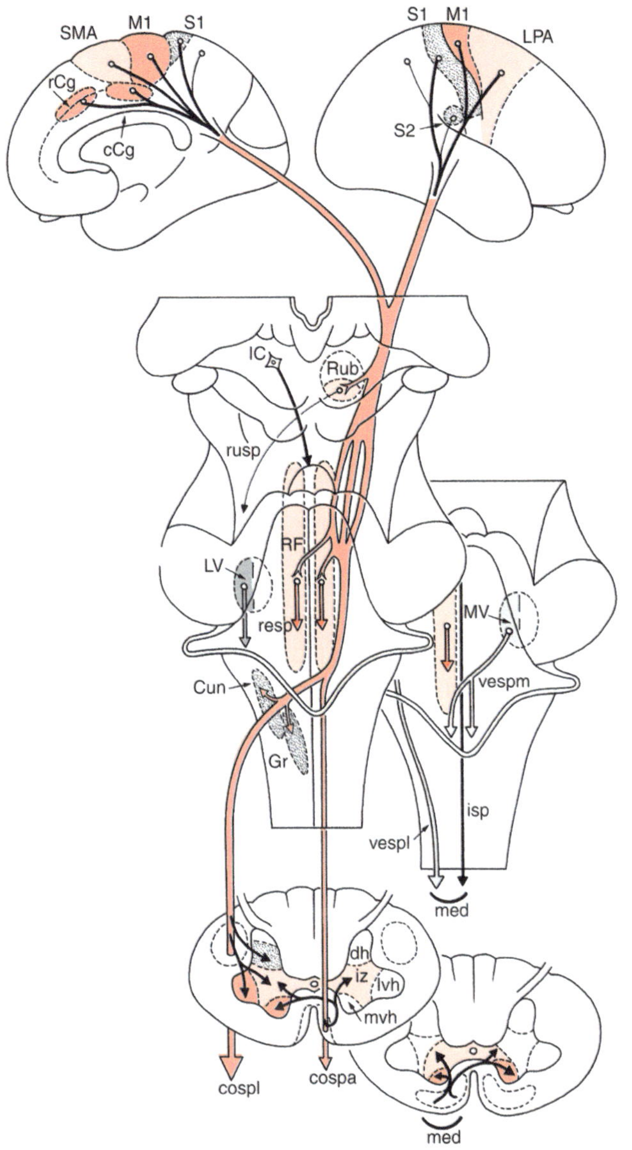

Fig. 24.11 Overview of human descending supraspinal systems *cCg* caudal cingulate motor area, *cospa*, *cospl* anterior and lateral corticospinal tract, *Cun* cuneate nucleus, *dh* dorsal horn, *rCg* rostral cingulate motor area, *Gr* gracile nucleus, *IC* interstitial nucleus of Cajal, *isp* interstitiospinal tract, *iz* intermediate zone, *LPA* lateral premotor area, *lvh*, *mvh* lateral and medial parts of ventral horn, *LV*, *MV* lateral and medial vestibular nuclei, *med* medial system, *M1* primary motor cortex, *resp* reticulospinal tract, *RF* reticular formation, *Rub* red nucleus, *rusp* rubrospinal tract, *SMA* supplementary motor area, *S1*, *S2* primary and secondary somatosensory cortices, *vespl*, *vespm* lateral and medial vestibulospinal tracts. (From ten Donkelaar HJ *Clinical Neuroanatomy*, 2nd ed. Springer, Cham, 2020)

pes pedunculi bundles of Dejerine (Dejerine 1901; *see* Dejerine, Jules), two fibre bundles, which descend partly within the cerebral peduncle, and partly in the region of the medial lemniscus (*see* Fig. D8); (b) the **bundle of Barnes**, in which corticospinal fibres leave the pyramidal tract at the pontine level and descend lateral to the inferior olive (Barnes 1901; *see* Barnes, Stanley); (c) the **bundle of Hoche**, in which a bundle of fibres leave the pyramidal tract at the level of the facial nerve, crossing in the hindbrain, to join the contralateral pyramidal tract (Hoche 1897; *see* Hoche, Alfred Erich); and (d) the **bundle of Pick**, recurrent pyramidal tract fibres, which after decussating at the medullospinal junction, ascend in the bulbar tegmental field (Pick 1890; *see* Pick, Arnold).

24.6 Eponyms for Areas, Gyri and Sulci of the Cerebral Cortex

The gyri and sulci of the cerebral cortex are schematically shown in Fig. 24.12. They were named and defined by Gratiolet (1854), Huschke (1854), Ecker (1869), Pansch (1868, 1879), Jensen (1871), Wernicke (1876), Eberstaller (1890) and Brissaud (1893), and extensified by Dejerine (1895), Retzius (1896), von Economo and Koskinas (1925), Rose (1935) and Testut and Latarjet (1948). A more complete account can be found in ten Donkelaar et al. (2018b). Of all the **areas** of the cerebral cortex distinguished by Brodmann (1909) and von Economo and Koskinas (1925), only a few received an *eponym*: (a) the **area** of **Broca**, an *eponym* for the triangular and opercular parts of the inferior frontal gyrus, also known as motor speech area of Broca (Broca 1878a; *see* Broca, Pierre Paul); (b) the **areas of Brodmann**, an *eponym* for the cortical areas distinguished cytoarchitectonically by Brodmann (Brodmann 1909; *see* Brodmann, Korbinian, and Fig. B38); (c) the **area of Campbell**, an *eponym* for the precentral motor area (Campbell 1905; *see* Campbell, Alfred Walter); (d) the **areas of von Economo, an** *eponym* for the cortical areas distinguished by von Economo and Koskinas (1925; *see* Economo, Constantin von, and Fig. E6); (e) the **area of Pitres**, an *eponym* for the prefrontal area of the frontal lobe (*see* Pitres, Jean Albert); (f) the **areas of Rolando**, an *eponym* for the precentral and postcentral gyri of the cerebral cortex, adjacent to the central sulcus (Rolando 1829; *see* Rolando, Luigi); and (g) the **area of Wernicke**, an *eponym* for the sensory speech area on the posterior third of the superior temporal gyrus of the cerebral cortex (Wernicke 1874; *see* Wernicke, Carl). For certain centres also *eponyms* are in use: (a) the **centre of visual images of words of Dejerine**, an *eponym* for part of the angular gyrus (Dejerine 1892); (b) the **visual word form area**, in the middle part of the left lateral occipitotemporal sulcus, first described by Cohen et al. (2000); (c) the **graphic motor image centre**, also described with the *eponym* **writing centre of Exner** (Exner 1881; *see* Exner, Sigismund); and (d) the **territory of Geschwind** (Geschwind 1965; see Geschwind, Norman), an *eponym* for part of the inferior parietal lobule, involved in the perisylvian language networks (Catani et al. 2005; ten Donkelaar et al. 2020b).

Also only a few of the **gyri** of the cerebral cortex, for which in the past, the term convolutions was used, received

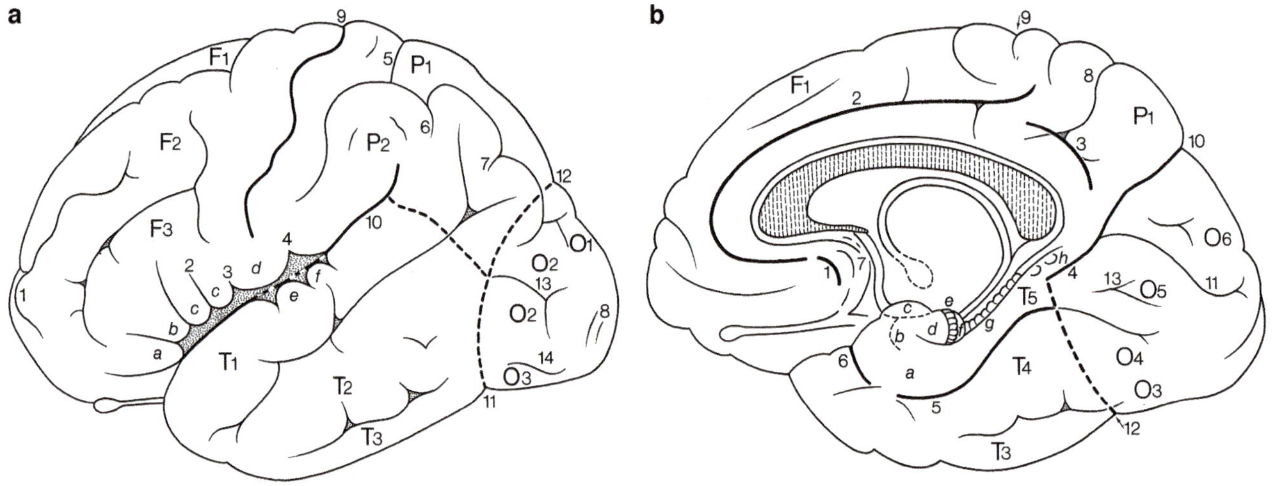

Fig. 24.12 (**a**) Gyri and sulci on the superolateral surface of the cerebral hemisphere. (After Duvernoy H *Le cerveau humain. Surface, coupes sériée tridimensionelles et IRM*. Springer, Berlin, Heidelberg, New York, 1992; and ten Donkelaar HJ, Kachlik D, Tubbs RS *An Illustrated Terminologia Neuroanatomica*. Springer, Cham, 2018). *a, b, c* orbital, triangular and opercular (divided into two parts by *2*) parts of the inferior frontal gyrus, *d* subcentral gyrus, *e, f* transverse temporal gyri of Heschl, *F1, F2, F3* superior, middle and inferior frontal gyri, *O1, O2, O3* superior, middle and inferior occipital gyri, *P1, P2* superior and inferior parietal lobules, *T1, T2, T3* superior, middle and inferior temporal gyri, *1* frontomarginal sulcus of Wernicke, *2* diagonal sulcus of Eberstaller, *3* anterior subcentral sulcus, *4* posterior subcentral sulcus, *5* transverse parietal sulcus of Brissaud, *6* first intermediate sulcus of Jensen, *7* second intermediate sulcus of Eberstaller, *8* descending occipital gyrus of Ecker, *9* central sulcus, *10* lateral sulcus, *11* pre-occipital notch of Meynert, *12* parieto-occipital sulcus, *13* lunate sulcus. *14* inferior occipital sulcus. (**b**) Gyri and sulci on the inferomedial surface of the cerebral hemisphere. *a* entorhinal cortex, *b* ambient gyrus, *c* semilunar gyrus, *d* uncinate gyrus, *e* band of the dentate gyrus of Giacomini, *f* intralimbic gyrus or uncal apex, *g* dentate gyrus, *h* gyri of Andreas Retzius or subsplenial gyri, *F1* superior frontal gyrus, *P1* precuneus, *O3* inferior occipital gyrus, *O4* fusiform gyrus (occipital part), *O5* lingual gyrus, *O6* cuneus, *T3* inferior temporal gyrus, *T4* fusiform gyrus (temporal part), *T5* parahippocampal gyrus, *1–6* parts of the "limbic sulcus" or "scissure limbique": *1* anterior paraolfactory sulcus, *2* cingulate sulcus, *3* subparietal sulcus, *4* anterior part of parieto-occipital sulcus, *5* collateral sulcus, *6* rhinal sulcus, *7* posterior paraolfactory sulcus, *8* transverse parietal sulcus of Brissaud, *9* central sulcus, *10* parieto-occipital sulcus, *11* calcarine sulcus, *12* pre-occipital notch of Meynert, *13* lingual sulcus

an *eponym*: (a) the **gyrus of Broca**, an *eponym* for the inferior frontal gyrus (Broca 1878a; *see* Broca, Pierre Paul); (b) the **gyrus of Eberstaller**, an *eponym* for a short annecting gyrus, joining the three short insular gyri to the orbital part of the inferior frontal gyrus (Eberstaller 1890; *see* Eberstaller, Oscar, and Fig. 24.13); (c) the **gyrus of Ecker**, an *eponym* for the descending occipital gyrus (Ecker 1869; *see* Ecker, Alexander); (d) the **gyri of Heschl**, an *eponym* for the transverse temporal gyri (Heschl 1878; *see* Heschl, Richard Ladislaus); (e) the **gyrus of Meynert, fornicate**, an obsolete *eponym* for the cingulate and parahippocampal gyrus (Meynert 1872; *see* Meynert, Theodor); (f) the **gyri of Retzius**, an *eponym* for the subsplenial gyri, a series of small bumbs pointing to the end of the CA1 field; *see* Retzius, Anders; this name was given by his son Gustav (Retzius 1896); (g) the **gyrus of Rolando**, an *eponym* for the precentral gyrus (Rolando 1829; *see* Rolando, Luigi); (h) the **gyrus of Tarin**, an *eponym* for the dentate gyrus (Tarin 1753; *see* Tarin, Pierre); and (i) the **insula of Reil**, an *eponym* for the insular lobe (Reil 1809; *see* Reil, Johann Christian). The insular lobe is usually subdivided into three short, an accessory gyrus and two long gyri (*see* Fig. 24.13).

More *eponyms* are in use for the **sulci** of the cerebral cortex, for many of which in the past the term fissure was used:

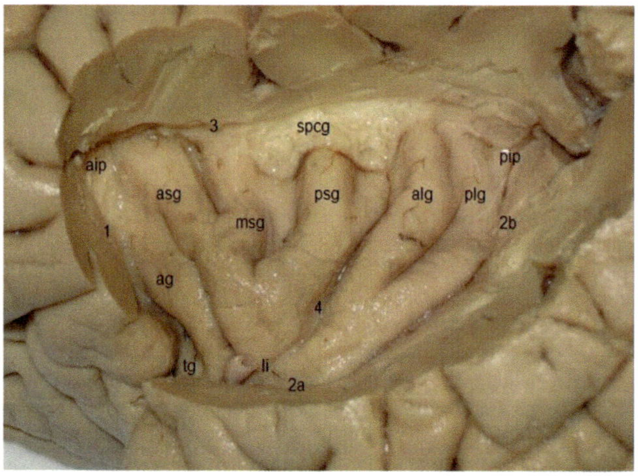

Fig. 24.13 Lateral view of the insula, showing three short gyri, an accessory gyrus and two long gyri. (From ten Donkelaar HJ, Kachlik D, Tubbs RS *An Illustrated Terminologia Neuroanatomica*. Springer, Cham, 2018; courtesy Robert Bartoš, Ústí nad Labem). *ag* accessory gyrus, *aip* anterior insular point, *alg* anterior long gyrus, *asg* anterior short gyrus, li limen insulae, *msg* middle short gyrus, *pip* posterior insular point, *plg* posterior long gyrus, *psg* posterior short gyrus, *spcg* gyrus supracentralis, *tg* transverse temporal gyrus, *1–3* circular insular sulcus (*1* anterior periinsular sulcus, *2a, 2b* inferior insular sulcus, horizontal and posterior parts, *3* superior periinsular sulcus), *4* central insular sulcus

(a) the **sulcus of Brissaud**, an *eponym* for the transverse parietal sulcus (Brissaud 1893; *see* Brissaud, Edouard); (b) the **'scissure limbique' of Broca**, an *eponym* for the sulcus separating the limbic lobe from the rest of the brain, usually composed of the collateral and rhinal sulci (Broca 1878b; *see* Broca, Pierre Paul); (c) the **sulcus of Broca, metopic**, an *eponym* for a sulcus of the inferior part of the inferior frontal gyrus, parallel to the cingulate gyrus (Broca 1878a); (d) the **sulci of Dejerine**, an *eponym* for the anterior and posterior subcentral sulci (Dejerine 1895; *see* Dejerine, Jules); (e) the various **sulci of Eberstaller**, *eponyms* for (1) the diagonal sulcus of the frontal lobe, a variable branch from the lateral sulcus, dividing the opercular part into two triangular parts; (2) the second or posterior intermediate sulcus of the parietal lobe; and (3) a transverse insular sulcus, a *synonym* for the central sulcus of Reil (Eberstaller, 1884, 1890; *see* Eberstaller, Oscar); (f) the **sulcus of Ecker**, an *eponym* for the anterior or transverse occipital sulcus (Ecker 1869; *see* Ecker, Alexander); (g) the **sulcus of Gratiolet**, an *eponym* for the parietooccipital sulcus (Gratiolet 1854; *see* Gratiolet, Louis Pierre); this sulcus is more pronounced in monkeys and received the *eponym* 'Affenspalte' of Schwalbe (*see* Schwalbe, Gustav Albert); (h) the **sulcus of Jensen**, an *eponym* for the first or intermediate sulcus of the parietal lobe (Jensen 1871; *see* Jensen, Julius); (i) the **sulcus of Meynert**, an *eponym* for the anterior occipital sulcus (Meynert 1877; *see* Meynert, Theodor); (j) the **sulcus of Reil, central**, an *eponym* for the sulcus separating the short and long gyri of the insula (Reil 1809; *see* Reil, Johann Christian; Figs. 24.12a and 24.13); (k) the **sulcus of Reil, circular** or **limiting**, an *eponym* for the circular or limiting sulcus surrounding the insula (Reil 1809; *see* Reil, Johann Christian; Fig. 24.13); (l) the **sulcus of Rolando**, an *eponym* for the central sulcus; *see* Rolando, Luigi. The name central fissure of Rolando was given by Leuret (Leuret 1839; *see* Leuret, François). The central sulcus is not a straight line but forms two arches, the genu superius and the genu inferius (Broca 1878a; *see* Broca, Pierre Paul); (m) the **sulcus of Sylvius**, an *eponym* for the lateral sulcus, previously known as Sylvian fissure (Sylvius 1660; *see* Sylvius, François de le Boë); (n) the **sulcus of Turner**, an *eponym* for the intraparietal sulcus, described in the chimpanzee (Turner 1866; *see* Turner, William) and by Pansch for the human brain (Pansch 1868; Pansch, Adolf Georg); and (o) the **sulcus of Wernicke**, an *eponym* for the frontomarginal sulcus (Wernicke 1876; *see* Wernicke, Carl).

24.7 Eponyms for the Limbic Lobe

The cingulate and parahippocampal gyri form a border (limbus) around the corpus callosum and the brain stem (Broca 1878b). Broca subdivided his **'grand lobe limbique'** into

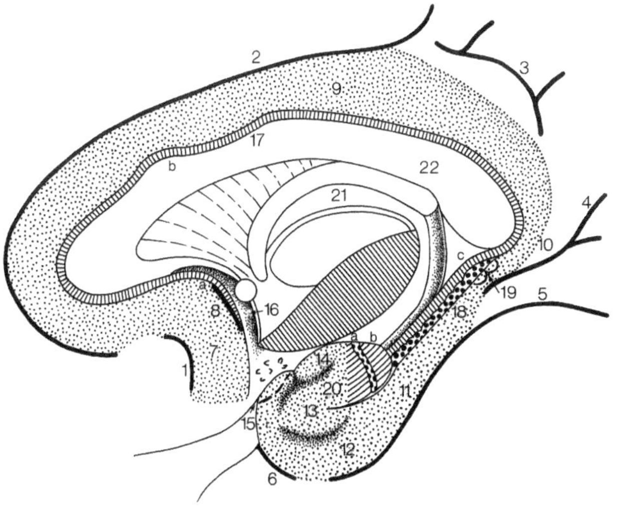

Fig. 24.14 The limbic lobe. (After Duvernoy HM *The Hippocampus*, 2nd ed. Springer, Berlin-Heidelberg-New York, 1998; from ten Donkelaar HJ, Kachlik D, Tubbs RS *An Illustrated Terminologia Neuroanatomica*. Springer, Cham, 2018). *1* anterior paraolfactory sulcus, *2* cingulate sulcus, *3* subparietal sulcus, *4* rostral part of calcarine sulcus, *5* collateral sulcus, *6* rhinal sulcus, *7* subcallosal gyrus, *8* posterior paraolfactory sulcus, *9* cingulate gyrus, *10* isthmus, *11* parahippocampal gyrus, *12* entorhinal cortex, *13* ambient gyrus, *14* semilunar gyrus, *15* piriform cortex, *16* paraterminal gyrus, *17* indusium griseum, *18* dentate gyrus, *19* gyri of Andreas Retzius, *20* uncinate gyrus, *21* fornix, *22* corpus callosum, *a* bandelette of Giacomini, *b* apex of uncus

inner (the hippocampal formation, known in the French literature as intralimbic gyrus; *see* Testut and Latarjet 1948) and outer (the cingulate and parahippocampal gyri) rings for which the general term **limbic lobe** is used (Heimer et al. 2008; Nieuwenhuys et al. 2008). The **'scissure limbique'** separates the limbic lobe from the rest of the cerebral cortex (Fig. 24.14) and can be viewed as an interlobar sulcus (Duvernoy 1992; ten Donkelaar et al. 2018a). The **limbic gyrus** includes the subcallosal area, the cingulate gyrus, the isthmus of the cingulate gyrus, the parahippocampal gyrus, the entorhinal cortex and the uncus. The **entorhinal cortex** has eight subfields (Insausti et al. 1995) and shows two peculiarities: the **white substance of Arnold**, an *eponym* for the white reticular matter (*see* Arnold, Friedrich), and the hippocampal warts or **verrucae hippocampi of Retzius** (Retzius 1896; see Retzius, Gustav). The **uncus** includes a number of bulges: the uncinate gyrus, the **band of Giacomini** (Giacomini 1883; *see* Giacomini, Carlo), also known as the band of the dentate gyrus, and the intralimbic gyrus or apex (see Fig. 24.14).

The **hippocampal formation** consists of three allocortical (the 'other' cortex) areas, the dentate gyrus, the hippocampus proper or **Ammon horn**, and the subiculum. The hippocampal formation develops from the medial pallium, and during the outgrowth of the cerebral hemispheres, first caudalwards and then ventralwards and rostralwards (ten Donkelaar et al. 2023): the retrocommissural part becomes situated in the temporal lobe, the precommissural part disappears and the middle, supracommissural part can be found above the corpus callosum as the indusium griseum, a thin cell layer. This is flanked by the **lateral longitudinal stria of Lancisi** or taenia tecta, and the **medial longitudinal stria of Lancisi** or taenia libera (*see* Lancisi, Giovanni Maria, and Fig. L3b).

References

Adamkiewicz AW (1881) Die Blutgefässe des menschlichen Rückenmarks. I. Teil: Die Gefässe der Rückenmarkssubstanz. Sitz Kais Akad Wiss Wien, 84. Band

Adamkiewicz AW (1882) Ibid, II. Teil: Die Gefässe der Rückenmarksoberfläche. Ibid, 85. Band

Barnes S (1901) Degenerations in hemiplegia; with special reference to a ventrolateral pyramidal tract, the accessory fillet and Pick's bundle. Brain 24:463–501

Batson OV (1940) The function of the vertebral veins and their role in the spread of metastasis. Ann Surg 112:138–149

Benner D, Hendriks BK, Benet A, Lawton MT (2021) Eponyms in vascular neurosurgery: comprehensive review of 18 veins. World Neurosurg 151:190–200

Bernasconi V, Cassinari V (1956) Un segno carotidografico tipico do meningioma del tentorio. Chirurgia (Milan) 11:586–588

Bernasconi V, Cassinari V (1957) Angiographical characteristics of meningiomas of the tentorium. Radiol Med 43:1015–1026

Bochdalek VA (1833) Anleitung zur praktischen Zergliederung des menschlichen Gehirns nebst einer anatomischen Beschreibung deselben mit besonderer Rücksicht auf das kleine Gehirn. Gottlieb Haase Söhne, Prague

Bouthillier A, van Loveren HR, Keller JT (1996) Segments of the internal carotid artery: a new classification. Neurosurgery 38:425–533

Breschet G (1829) Recherches anatomiques, physiologiques et pathologiques sur le système nerveux. Villaret, Paris

Brissaud E (1893) Anatomie du cerveau de l'homme: Morphologie des hémisphères cérébraux, ou cerveau proprement dit. Masson, Paris

Broca P (1878a) Nomenclature cérébrale: Dénomination et subdivision des hémisphères et des anfractuosités de la surface. Rev Anthropol 2:193–236

Broca P (1878b) Anatomie comparée des circonvolutions cérébrales. Le grand lobe limbique et la scissure limbique dans le série des mammifères. Rev Anthropol 2:385–498

Brodmann K (1909) Vergleichende Lokalisationslehre der Grosshirnrinde in ihren Prinzipen dargestellt auf Grund des Zellenbaues. Barth, Leipzig; translated by Lawrence Garey as Brodmann's Localisation in the Cerebral Cortex. Imperial College Press, London, 1999

Burdach KF (1819–1826) Vom Baue und Leben des Gehirns, 3 Vols. Dyk, Leipzig

Campbell AW (1905) Histological studies of the localisation of cerebral function. Cambridge University Press, Cambridge

Catani M, Jones DK, Ffytche DH (2005) Perisylvian language networks of the human brain. Ann Neurol 57:8–16

Clarke JAL (1851) Researches into the structure of the spinal cord. Phil Trans 141:607–621

Cohen L, Dehaene S, Naccache I, Lehéricy S, Dehaene-Lambertz G, Hénaff MA et al (2000) The visual word form area: spatial and temporal characterization of an initial step of reading in normal subjects and posterior split-brain patients. Brain 123:291–307

Cooper S, Sherrington CS (1940) Gower's tract and spinal border cell. Brain 63:123–134

Deiters OFK (1865) Untersuchungen über Gehirn und Rückenmark des Menschen und der Säugetiere. Vieweg, Braunschweig

Dejerine J (1892) Contributions à l'étude anatomicopathologique et clinique des différences variétés de cécité verbale. Mem Soc Biol 44:61–90

Dejerine J (1895) Anatomie des centres nerveux, vol 1. Rueff, Paris

Dejerine J (1901) Anatomie des centres nerveux, vol 2. Rueff, Paris

Duvernoy H (1992) Le cerveau humain. Surface, coupes sériée tridimensionelles et IRM. Springer, Berlin/Heidelberg/New York. (English edition1992, 2nd English ed, 1998)

Eberstaller O (1884) Zur Oberflächenanatomie der Grosshirnhemisphären. Wien Med Bl 7:479–482, 542–582, 644–646

Eberstaller O (1890) Das Stirnhirn. Ein Beitrag zur Anatomie der Oberfläche des Grosshirns. Urban & Schwarzenberg, Wien

Ecker A (1869) Die Hirnwindungen des Menschen. Vieweg, Braunschweig

Edinger L (1893) Vorlesungen über den Bau der nervösen Centralorgane des Menschen und der Thiere. Vogel, Leipzig

Exner S (1881) Untersuchungen über die Localization der Functionen in der Grosshirnrinde des Menschen. Braumüller, Wien

Flechsig PE (1876) Die Leitungsbahnen im Gehirn und Rückenmark des Menschen auf Grund entwicklungsgeschichtlicher Untersuchungen. Engelmann, Leipzig

Foerster O, Gagel O (1932) Die Vorderseitenstrangdurchschneidung beim Menschen. Eine klinisch-pathophysiologisch-anatomische Studie. Z Ges Neurol Psychiatr 138:1–92

Foix C, Hillemand P (1925) Les artères de l'axe encéphalique jusqu'au diencéphale inclusivement. Rev Neurol 2:705–739

Geschwind N (1965) Disconnection syndromes in animals and man. I and II. Brain 88(237–294):585–644

Giacomini C (1883) Fascia dentata del grande ippocampo nel cervello umano. G Accad Med Torino

Golgi C (1878) Della terminazione del nervi tendini et di un nuovo apparato nervoso terminale musculotendineo. Atti Soc Ital Sci Nat:272–275

Goll F (1860) Beiträge zur feineren Anatomie des menschlichen Rückenmarks. Denkschr Med Chir es Zürich, p 130

Gombault FAA (1894) Contribution à l'étude des lésions systématisées des cordons blancs de la moelle épinière. Arch Med Exp Anat Pathol 6:365–424

Gombault FAA (1895) Etat actuel des connaissances sur la systématisation des cordon postérieures de la moelle épinière. Sem Med 15:161–166

Gowers WR (1880) The Diagnosis of Disease of the Spinal Cord. London

Gratiolet LP (1854) Mémoire sur les plis cérébraux de l'homme et des primates. Bertrand, Paris

Head H (1920) Studies in Neurology. H Frowde, London

Heimer L, Van Hoesen GW, Trimble M, Zahm DS (2008) Anatomy of neuropsychiatry. Elsevier, Amsterdam

Heschl RL (1878) Über die vordere quere Schläfenwindung des menschlichen Grosshirns. Braumüller, Wien

Heubner JO (1874) Die luetische Erkrankung der Hirnarterien. Vogel, Leipzig

Hoche AE (1897) Beiträge zur Anatomie der Pyramidenbahn und der oberen Schleife, nebst Bemerkungen über die abnorme Bündel in Pons und Medulla oblongata. Arch Psychiatr Nervenkr 30:103–136

Huschke E (1854) Schädel, Hirn und Seele des Menschen und der Thiere nach Alter, Geschlecht und Race. Mauke, Jena

Insausti R, Tuñón T, Sobrevieda T, Insausti AM (1995) The human entorhinal cortex: a cytoarchitectonic analysis. J Comp Neurol 355:171–198

Jensen J (1871) Abhandlung über die Furchen und Windungen der menschlichen Grosshirnhemisphären. Allg Z Psychiatr 27:473–516

Kuru M (1949) Sensory paths in the spinal cord and brain stem of man. Sogensya, Tokyo

Kuypers HGJM (1981) Anatomy of the descending pathways. In: Handbuch of physiology, the nervous system II. American Physiological Society, Bethesda, pp 597–666

Labbé C (1879) Notes sur la circulation du cerveau et sur le mode de développement des corpuscles de Pacchioni. Arch Physiol: 135–154

Leuret F (1839) Anatomie comparée du système nerveux considéré dans ses rapports avec l'intelligence. Atlas des 32 planches dessinées d'apres nature engravées, Vol 1 (with P Gratiolet). Baillière, Paris

Lissauer H (1886) Beitrag zur Faserverlauf im Hinterhorn des menschlichen Rückenmarks und zum Verhalten desselben bei Tabes dorsalis. Arch Psychiatr Nervenkr 17:377–438

Markowski J (1921) Entwicklung der Sinus durae matris und der Hirnvenen des Menschen. Bull Int Acad Sci Lett, Classe des sciences mathematiques et naturelles. Serie B (Suppl 1):1–269

Markowski J (1931) Über die Entwicklung der Falx cerebri und des Tentorium cerebelli des Menschen mit Berücksichtigung ihres venösen Sinus. Z Anat EntwGesch 94:395–439

Meissner G (1852) Ueber Vorhandsein bisher unbekannter eigenthümlicher Tastkörperchen (Corpusculi tactus) in dem Gefühlswärzchen der menschlichen Haut und ueber die Endausbreitung der sensitiver Nerven. Nachr Ges Wiss Göttingen

Merkel FS (1875) Tastzellen und Tastkörperchen bei den Hausthieren und beim Menschen. Arch Mikr Anat Entmech 11:636–652

Meynert T (1872) Vom Gehirne der Säugethiere. In: Stricker, Hrsg, Handbuch der Lehre von den Geweben des Menschen und der Thiere, Vol 2. Engelmann, Leipzig, pp 694–808 (English translation by JJ Putnam, 1872 Stricker, ed, Manual of Histoloy. William Wood, New York, pp 650–766)

Meynert T (1877) Die Windungen der convexen Oberfläche des Vorderhirns bei Menschen, Affen und Raubthieren. Arch Psychiatr Nervenkr 7:257–296

Mountcastle VB (2005) The sensory hand. Neural mechanisms of somatic sensation. Harvard University Press, Cambridge, MA

Nieuwenhuys R, Voogd J, van Huijzen C (2008) The human central nervous system, 4th edn. Springer, Berlin/Heidelberg/New York

Pacini F (1840) Nuovi organi scoperti nel corpo umano. Tipografia Cino, Pistoia

Pansch AG (1868) Ueber die typische Anordnung der Furchen und Windungen auf den Grosshirnhemisphären des Menschen und des Affen. Arch Anthropol 3:227–257

Pansch AG (1879) Die Furchen und Wülste am Grosshirn des Menschen. Oppenheim, Berlin

Percheron G (1973) The anatomy of the arterial supply of the human thalamus and its use for the interpretation of the thalamic vascular pathology. Z Neurol 205:1–13

Philippe C (1897) Contribution à l'étude anatomique et clinique du tabes dorsalis. Steinheil, Paris

Pick A (1890) Ueber ein abnormes Bündel der menschlichen Medulla oblongata. Arch Psychiatr Nervenkr 21:636–640

Reil JC (1796) Exercitationum anatomicarum fasciculus primus. De struttura nervorum. Venalis, Halle

Reil JC (1809) Die Sylvische Grube oder das Thal, das gestreifte groβe Hirnganglium, dessen Kapsel und die Seitentheile des Gehirns. Arch Physiol 9:195–208

Rektorzik E (1858) Über das Vorkommen eines Sinus venosus im Canalis caroticus. Sitzber Akad Wiss Math Klasse, pp 466–470

Retzius G (1896) Das Menschenhirn: Studien in der mikroskopischen Morphologie. Norstedt, Stockholm

Rexed B (1952) The cytoarchitectonic organization of the spinal cord in the cat. J Comp Neurol 96:415–496

Rexed B (1954) A cytoarchitectonic atlas of the spinal cord of the cat. J Comp Neurol 100:297–379

Rolando L (1829) Della struttura degli emisferi cerebrali. Memorie della Regia Accademia della Scienze di Torino

Rose M (1935) Cytoarchitektonik und myeloarchitektonik der Groβhirnrinde. In: Bumke O, Foerster O, Hrsg, Handbuch der Neurologie, Bd 1. Springer, Berlin, pp 588–778

Ruffini A (1898) Sulla presenza di nuove forme di terminazioni nervose nello strato papillare e subpapillare della cute del'uomo, Torino

Schaltenbrand G, Bailey P (eds) (1959) Introduction to Stereotaxis with an atlas of the human brain, vol 2. Thieme, Stuttgart

Schultze F (1883) Beiträge zur Lehre von der secundären Degeneration im Rückenmark des Menschen nebst Bemerkungen über die Anatomie der Tabes. Arch Psychiatr 14:359–390

Sindou M, Quoex C, Baleydier C (1974) Fiber organization at the posterior spinal cord-rootlet function in man. J Comp Neurol 153:15–26

Stilling B (1842) Untersuchungen über die Funktionen des Rückenmarks und der Nerven. Wigand, Leipzig

Sylvius F (1660) De spirituum animalium in cerebro cerebelloque confectione, per nervos distributione, atque usu vario. Leiden

Szikla G, Bouvier G, Hori T, Petrov V (1977) Angiography of the human brain cortex. Springer, Berlin/Heidelberg/New York

Talairach J, Tournoux P (1988) Co-planar Stereotaxic Atlas of the Human Brain. Three-dimensional proportional system: The approach to cerebral imaging. Thieme, Stuttgart

Talairach J, De Ajuriaguerra J, David M (1952) Etudes stéréotaxiques des structures encéphalique chez l'homme. Presse Med 28:605–609

Talairach J, Szikla G, Tournoux P, Prossalentis A, Bordas-Ferrier M, Covello I et al (1967) Atlas d'anatomie stéréotaxique du télencéphale. Masson, Paris

Tamraz JC, Comair YG (2000) Atlas of regional anatomy of the brain using MRI. Springer, Berlin/Heidelberg/New York

Tamraz JC, Saban R, Répérant J, Cabanis EA (1991) A new cephalic reference plane for use with magnetic resonance imaging: the chiasmatico-commissural plane. Surg Radiol Anat 13:197–201

Tarin P (1753) Dictionnaire anatomique. Suivi d'une bibliothèque anatomique et physiologique. Briasson, Paris

ten Donkelaar HJ, Kachlik D, Tubbs RS (2018a) An illustrated Terminologia Neuroanatomica. Springer, Cham

ten Donkelaar HJ, Tzourio-Mazoyer N, Mai JK (2018b) Toward a common terminology for the gyri and sulci of the human cerebral cortex. Front Neuroanat 12:93

ten Donkelaar HJ, Broman J, van Domburg P (2020a) The somatosensory system. In: ten Donkelaar HJ (ed) Clinical Neuroanatomy, 2nd edn. Springer, Cham, pp 171–255

ten Donkelaar HJ, Catani M, van Domburg P, Eling PATM, Küsters B, Hori A (2020b) The cerebral cortex and complex cerebral functions. In: ten Donkelaar HJ (ed) Clinical Neuroanatomy, 2nd edn. Springer, Cham, pp 831–952

ten Donkelaar HJ, Cömert A, van der Vliet T, van Domburg P, Wesseling P (2020c) Vascularization of the brain and spinal cord. In: ten Donkelaar HJ (ed) Clinical Neuroanatomy, 2nd edn. Springer, Cham, pp 71–126

ten Donkelaar HJ, Lammens M, Hori A (2023) Clinical Neuroembryology, 3rd edn. Springer, Cham

Testut L, Latarjet A (1948) Traité d'anatomie humaine, vol 2. Doin, Paris

Tokunaga A, Takase M, Otani K (1977) The glabella-inion line as a baseline for CT scanning of the brain. Neuroradiology 14:67–71

Trolard P (1868) Recherches sur l'anatomie du système veineux de l'encéphale et du crâne. Thesis, Paris

Turner W (1866) The convolutions of the human cerebrum topographically considered. Edinb Med J 11:1105–1122

Uchimura Y (1928) Über die Gefäβversorgung des Ammonshornes. Z Ges Neurol Psychiatr 112:1–19

Verga A (1856) Dell'apparato ventricolare del setto lucido e della volta a tre pilastri. G Ist Lomb 8:89–104

Vieussens R (1685) Neurographia universalis. Certe, Lyons

von Economo C, Koskinas GN (1925) Die Cytoarchitektonik der Hirnrinde des erwachsenen Menschen. Springer, Berlin; English translation by LC Triarhou Cytoarchitectonics of the Adult Human Cerebral Cortex. Karger, Basel, 2008

von Monakow C (1909) Der rote Kern, die Haube und die Regio hypothalamica bei einigen Säugetieren und beim Menschen. Arb Hirnanat Inst Zürich, Heft III

Wernicke C (1874) Der aphatische Symptomencomplex. Ein psychologische Studie auf anatomische Basis. Cohn & Weigert, Breslau

Wernicke C (1876) Das Urwindingssystem des menschlichen Gehirns. Arch Psychiatr Nervenkr 6:298–326

Willis WD, Coggeshall RE (1991) Sensory mechanisms of the spinal cord, 2nd edn. Plenum, New York

Wollschlaeger P, Wollschlaeger G (1965) Eine infratentorielle meningeale Arterie [An infratentorial meningeal artery]. Rdiologe 5:451–452

Anatomical Eponyms in Obstetrics and Gynaecology

25

Hans J. ten Donkelaar, Veronika Němcová, and Marina Quartu

Vignette Illustration by Casseri, published 1626 by Adriaan van den Spieghel in *De formatio foetu* (posthumously by van den Spieghel's son-in-law Liberale Crema), showing a pregnant woman with opened uterus

25.1 Introduction

Baskett (2019) published an exceptional book on eponyms and names in obstetrics and gynaecology, most of them concerning instruments, signs, tests and techniques, but also many anatomical and embryological eponyms. Among the founders of modern obstetrics, at least four names should be mentioned:

1. Ambroise Paré (1510–1590): French surgeon and obstetrician born in Laval and working in Paris. Reference: *De la génération de l'homme*. Paris, 1573 (Dutch edition, 1592). Paré introduced internal podalic version and breech retraction.
2. Hendrik van Deventer (1651–1724): Dutch obstetrician born and working in The Hague. Reference: *Manuale Operatiën, zijnde een Nieuw Ligt voor Vroed-meesters en Vroedvrouwen*. Dyckhuisen, Leiden, 1701 (also in Latin: *Operationes chirurgicae novum lumen exhibentes obstetricantibus*. Dyckhuisen, Leiden, 1701, and an English version: *The Art of Midwifery Improved*. Bettesworth, London, 1716). Van Deventer was the first to seriously study the pelvis and its deformities.
3. Smellie, William (1697–1763): Scottish obstetrician born in Lesmahagow, near Lanark, Scotland, and working in London. Reference: *A Treatise on the Theory and Practice of Midwifery*. Wilson, London, 1752 (illustrated by Jan van Rymsdyk, later also to be used by John Hunter in his *The Anatomy of the Human Gravid Uterus*. Baskerville, Birmingham, 1774). Smellie laid down the principles of clinical pelvimetry, including vaginal examination to estimate the diagonal conjugate.
4. Jean-Louis Baudelocque (1746–1810): French obstetrician born in Heilly, Picardie, and working in Paris.

H. J. ten Donkelaar et al., *An Illustrated Guide to Anatomical Eponyms*, https://doi.org/10.1007/978-3-031-91664-9_25

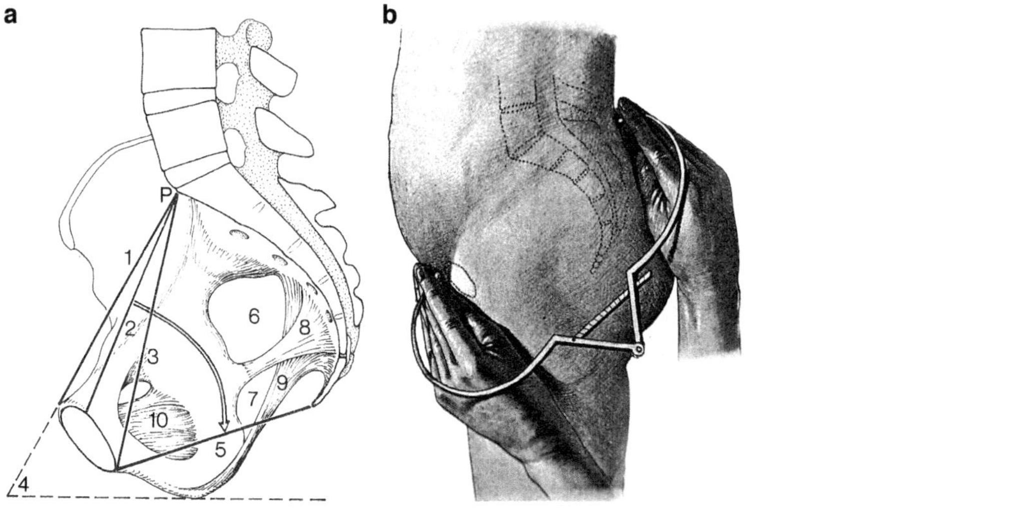

Fig. 25.1 (**a**) The conjugates of the pelvis. (After ten Donkelaar HJ, Oostra R-J 2014 *Klinische Anatomie en Embryologie,* 4th ed. Springer Media, Houten). *P* Promontorium, *1, 2, 3* conjugata anatomica, – vera, and – diagonalis of the superior pelvic aperture, *4* inclinatio of pelvis, *5* conjugata of of inferior pelvic aperture, *6, 7* major and minor ischiadic foramen, *8* sacrospinal ligament, *9* sacrotuberal ligament, *10* obturator membrane, in front of which the obturator canal is seen; (**b**) the external conjugate of the pelvis (from Baudelocque 1781 *L'art des accouchements.* Méquignon, Paris; from the Wellcome Collection CCBY)

Reference: *L'art des accouchements.* Méquignon, Paris, 1781. Baudelocque emphasized the importance of pelvic measurements and introduced the diameter or line of Baudelocque (*see* Fig. 25.1b).

In this part, anatomical eponyms used in various aspects of obstetrics and gynaecology will be discussed, including: eponyms for diameters and signs in obstetrics, development of the follicle and the placenta, development of the female genital organs and their remnants, the female pelvic organs, neurovascular sheaths in the female pelvis, the innervation of the female pelvic organs, and finally their lymphatic drainage. For further data on the various authors quoted, reference is made to Part II (*The Anatomicals Eponyms from A to Z*) and figures in this part. The **vignette** shows an illustration by Casseri, published 1626 by Adiaan van den Spieghel in *De formatio foetu.*

25.2 Eponyms for Diameters and Signs in Obstetrics

The **internal diameters** of the pelvis, the **conjugates**, are shown in Fig. 25.1a. The curved arrow shows the middle of the bony birth canal. Carus published another curve (*eponym*: **curve of Carus**; Carus 1820; *see* Carus, Carl Gustav): a curve having its centre at the pubic symphysis and embracing the pelvic outlet. A transverse line marking the level of the promontory is known as the *eponym* **line of Kilian** (Kilian 1835–1844; *see* Kilian, Hermann Friedrich). Caldwell and Moloy (1933) published a subdivision of the female pelvis into android, anthropoid, gynecoid and platypelloid, using radiological pelvimetry, but by 1960, its clinical uselessness became apparent. The rhomboid or quadrangular area of Michaelis (*eponym*: **rhomboid of Michaelis**; Michaelis 1851; *see* Michaelis, Gustav Adolph) is a rhomboid-shaped depression in females, bounded by lines joining the spinous process of the fifth lumbar vertebra to dimples over the posterior superior iliac spines and by lines joining the latter to the point where the medial borders of the gluteus maximus muscles approximate. In cases of abnormal body configuration, in his time rachitis in particular, he showed that the upper angle became more obtuse, suggesting pelvic malformation.

In early pregnancy, Piskaçek noted that the uterus may appear asymmetrical, since the implantation site can enlarge one horn before the other (*eponym*: **sign of Piskaçek**; Piskaçek 1899; *see* Piskaçek, Ludwig). Dickinson (1892; *see* Dickinson, Robert Latou) noted the same phenomenon. Naegele (1819) noted the tendency in normal labour for the anterior part of the parietal bone to present as the head engages (*eponym*: **obliquity of Naegele**; *see* Naegele, Franz Carl). Bandl (1875) observed a prominent muscular ring at the junction of the upper and lower parts of the uterus, resulting from obstructed labour (*eponym*: **ring of Bandl**; *see* Bandl, Ludwig). The **joint of Budin** is an *eponym* for the posterior intraoccipital synchondrosis, a cartilaginous connection of the squamous and lateral parts of the occipital bone in the newborn (*see* Budin, Pierre Constant).

25.3 Eponyms Used for the Development of the Follicle and the Placenta

The mature or Graafian follicle was described by Reinier de Graaf (*eponym*: **follicle of de Graaf**; de Graaf 1672; *see* de Graaf, Reinier). Small cavities found among the ovarian granulosa cells of growing and mature ovarian follicles are known with the *eponym* **bodies of Call-Exner** (Call and Exner 1875; *see* Call, Emma Louise, and Exner, Siegmund). The implantation of the embryo is shown in Fig. **25.2**. The *eponym* **membrane of Slavjanski** is used for the basement membrane separating the granulosa from the inner theca of the follicle (von Slavjanski 1871; *see* Slavjanski, Kronid Fedorovich von). The chorionic membrane received the *eponyms* **membrane of von Baer** (von Baer 1820–1837; *see* Baer, Karl Ernst von) and **false amnion of Pander** (von Pander 1817; *see* Pander, Heinrich Christian von). Langhans described the cytotrophoblast and its layers (*eponym*: **layer** and **cells of Langhans**; Langhans 1870; *see* Langhans, Theodor). The *eponym* **cells of Hofbauer** (Hofbauer 1905;

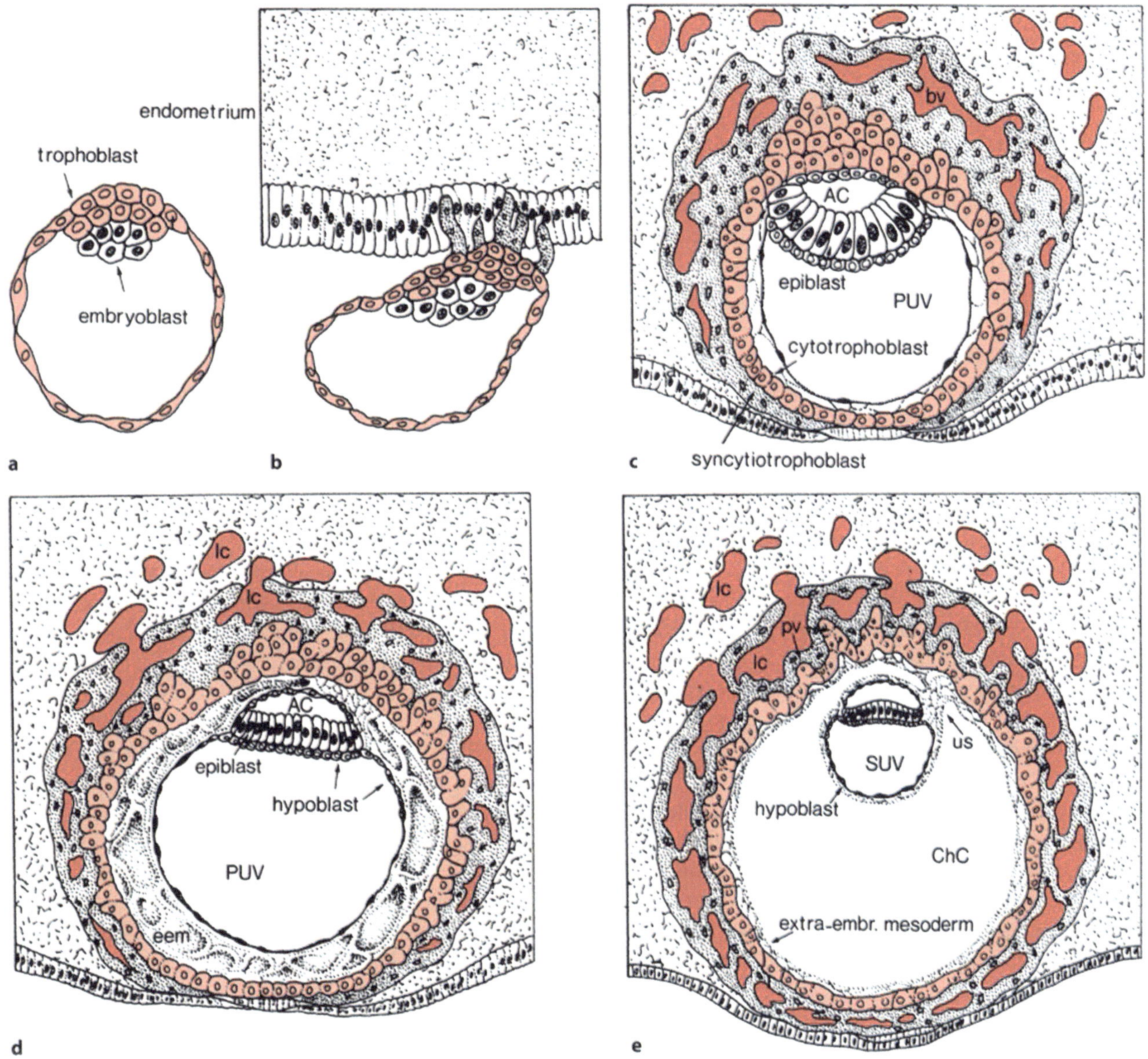

Fig. 25.2 Implantation and the formation of the bilaminar embryo: (**a**) 107-cell blastocyst; (**b–e**) blastocysts of approximately 4.5, 9, 12 and 13 days, respectively. The trophoblast and the cytotrophoblast are indicated in *light red*, the syncytiotrophoblast is *stippled* and maternal blood in lacunae is shown in *red*. *AC* amniotic cavity, *ChC* chorionic cavity, *eem* extra-embryonic mesoderm, *lc* lacuna, *pv* primary villi, *PUV* primary umbilical vesicle, *SUV* secondary umbilical vesicle (yolk sac), *us* umbilical stalk. (After Langman J 1962 *Medical Embryology*. Williams & Wilkins, Baltimore; from ten Donkelaar HJ, Lammens M, Hori A 2023 *Clinical Neuroembryology. Development and developmental disorders of the human central nervous system*, 3rd ed. Springer, Cham)

see Hofbauer, Isfred Isidore) is used for macrophages in the placental villi contributing to the suppression of immune responses between mother and foetus. For the fibrinoid substance of the placenta (*see* Kaufmann et al. 1996), Langhans, Rohr and Nitabuch each described a layer: Langhans the subchorionic layer (*eponym*: **stria of Langhans**; Langhans 1877), Nitabuch the lower layer, deeply placed in the basal decidua (*eponym*: **stria of Nitabuch**; Nitabuch 1887; *see* Nitabuch, Raisa) and Rohr the upper layer at the level of the basal plate beneath the stem villi (*eponym*: **stria of Rohr**; Rohr 1889; *see* Rohr, Karl). John Hunter described the mucosa of the pregnant uterus (*eponym*: **membrane of Hunter**; Hunter 1774). Of note are two eponyms for the umbilical cord. Van Hoboken described valves or folds for secondary windings of the vessels of the umbilical cord that form grooves externally, and valve-like projections internally (*eponym*: **valves of Hoboken**; van Hoboken 1669; *see* van Hoboken, Nicolaas). Wharton described the connective tissue of the umbilical cord (*eponym*: **jelly of Wharton**; *see* Wharton, Thomas). Mucous retention cysts of the cervix are rather common (*eponym*: **cyst of Naboth**; Naboth 1707; *see* Naboth, Martin).

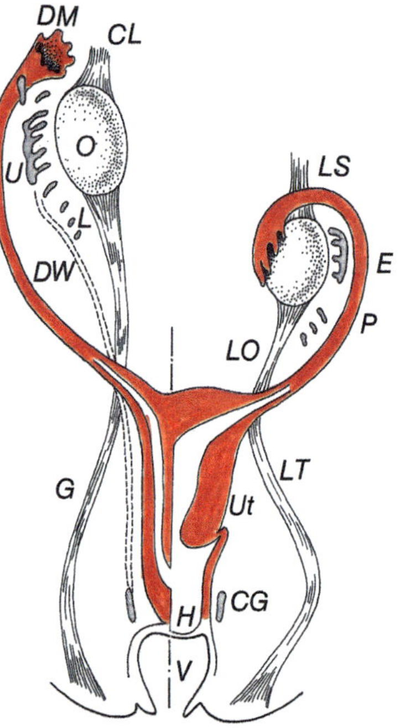

Fig. 25.3 Development of the duct of Müller, its derivatives (in *red*) and remnants (in *grey*) in females. *CG* cyst of Gärtner, *CL* cranial ligament of ovary, *DM* duct of Müller, *DW* duct of Wolff, *E* uterine tube, *G* gubernaculum, *H* hymen, *L* lower caudal mesonephros ducts, *LO* ligamentum ovarium proprium, *LS* ligamentum suspensorium ovarii, *LT* ligamentum teres uteri, *O* ovary, *P* paroophoron, *left U* cranial part of duct of Wolff and upper mesonephros ducts; *right Ut* uterus, *V* vagina. (After ten Donkelaar HJ, Oostra R-J 2014 *Klinische Anatomie en Embryologie*, 4th ed. Springer Media, Houten)

25.4 Eponyms for Embryological Structures and Their Remnants

For certain embryological structures, eponyms are still in use, such as the paramesonephic or Müllerian duct (*eponym*: **duct of Müller**; Müller 1830; *see* Müller, Johannes Peter; Fig. 25.3), the mesonephric or Wolffian duct (*eponym*: **duct of Wolff**; Wolff 1759; *see* Wolf, Caspar Friedrich; later also described by **Oken**) and their derivatives. The mesonephric ducts are important for the development of the male genital organs, whereas the paramesonephric ducts form the uterine tubes, the uterus and the upper part of the vagina. The vagina has a dual origin (Mijsberg 1924; Koff 1933). The lower one-third of the vagina appears to develop from epithelial proliferations of the urogenital sinus, together forming the vaginal plate. It seems likely that these sinuvaginal proliferations develop from the mesonephric ducts. The term **tubercle of Müller** is used for the sinus tubercle, the first evidence of the embryonic uterus and vagina. It remains as the hymen. With the decline of the mesonephros, the mesonephric duct and its tubules in females regress, but not completely. Several remnants occur particularly in the broad ligament of the uterus. Gardner et al. (1948) found 65 cystic structures in routine sections of 598 adult broad ligaments and verified them histologically. Five were remnants of the mesonephric duct and 33 of tubular origin. The rest were believed to be of paramesonephric origin. The course of the obliterated mesonephric duct in adult females begins near the abdominal ostium of the uterine tube and runs within the broad ligament. It enters the uterine wall above the cervix and continues downwards in the walls of the cervix and vagina to end near the vaginal orifice. The following remnants may appear (Gray, Skandalakis 1972; Fig. 25.4):

1. A structure associated with the fimbriae, usually considered to be a remnant of the paramesonephric duct (*eponym*: **pedunculated hydatid of Morgagni** (Morgagni 1764; *see* Morgagni, Giovanni Battista);
2. the vesicular appendage or sessile hydatid, usually considered as the blind end of the mesonephric duct;
3. aberrant ductules, separate from the epoophoron;
4. the epoophoron (*eponym*: **organ of Rosenmüller**; Rosenmüller 1802; *see* Rosenmüller, Johann Christian), which consists of a portion of the mesonephric duct with some 8–13 tubules entering it;
5. the paroophoron (*eponym*: **tubules of Wolff**; Wolff 1759, and **tubules of Kobelt**; Kobelt 1847; *see* Kobelt, Georg Ludwig), a group of mesonephric tubules caudal to the epoophoron;
6. the distal portion of the mesonephric duct in the walls of the cervix and vagina may remain as a longitudinal duct of the epoophoron, and received the *eponym* **duct of**

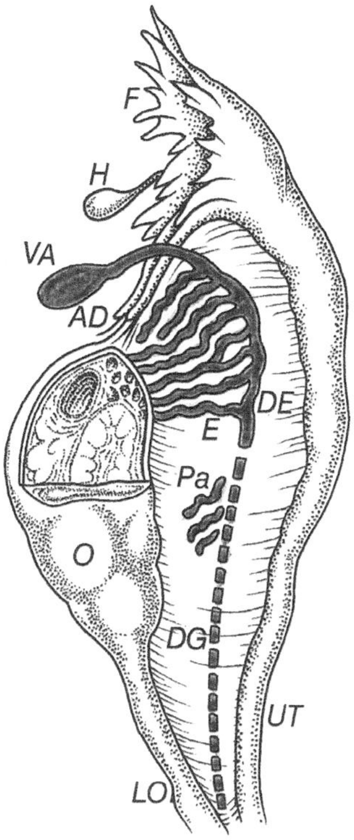

Fig. 25.4 Duct of Gärtner and other remnants of the mesonephric duct of Wolff in the female (in *grey*). *AD* aberrant ductule, *DE* duct of epoophoron, *DG* duct of Gärtner, *E* epoophoron, *F* fimbriae, *H* pedunculated hydatid of Morgagni, *LO* proper ligament of ovary, *O* ovary, *Pa* paroophoron, *UT* uterine tube, *VA* vesicular appendage. (After Arey LB *Developmental Anatomy,* 7th ed. Saunders, Philadelphia, PA, 1965)

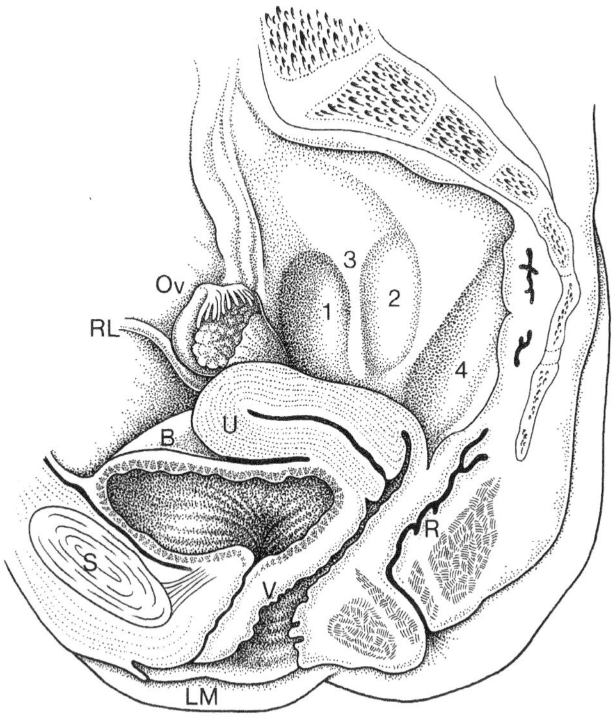

Fig. 25.5 The fossa of Claudius. (After Rouvière H *Anatomie humain: Descriptive et topographique*. Tôme II, Tronc. 8th ed, revised by G Cordier. Masson, Paris, 1959). *B* bladder, *LM* labium majus, *Ov* ovary with uterine tube, *R* rectum, *RL* round ligament of uterus, *S* pubic symphysis, *U* uterus, *1* ovarian fossa, *2* fossa of Claudius, *3* prominence of ureter, *4* pouch of Douglas

Gärtner (Gärtner 1822; *see* Gärtner, Hermann Treschow), although already described by Malpighi in 1681. Rieder (1884) observed a presence of about 20% in his small series of adult females. Sneeden (1958) found cervical remnants in 15 of 216 hysterectomies, but believed that many more were undetected.

25.5 Eponyms for the Female Pelvic Organs

In gynaecology, still many eponyms are in use for the female pelvic organs. The uterine tubes are usually known as Falloppian tubes (*eponym*: **tube of Falloppio**; Falloppio 1561; *see* Falloppio, Gabriele). The ovarian fimbria, the largest fimbria of the uterine tube, is in direct contact with the superior pole of the ovary and contracts during ovulation to guide the free egg (*eponym*: **fimbria of Richard**; Richard 1851; *see* Richard, Félix Dominique Adolphe). As to the relations of the female pelvic organs, *eponyms* such as the **space of Retzius**, an extraperitoneal retropubic (prevesical) space located between the pubic symphysis and the urinary bladder, and the **rectouterine pouch** or **cul-de-sac of Douglas** (Douglas 1730; *see* Douglas, James) are in use. The upper part of the vagina without rugae contacts the urinary bladder (*eponym*: **triangle of Pawlik**; Pawlik 1881; *see* Pawlik, Karel). The sacrouterine folds were described by **Jarjavay** and **Petit**. Special attention has been given to the ovarian fossa (Fig. 25.5). Claudius (1865) described the ovarian fossa as a peritoneal fossa, housing the ovary in multipara, between the internal iliac artery and the sacrum (*eponym*: **fossa of Claudius**; Claudius 1865; *see* Claudius, Friedrich Matthias). In nullipara, the ovary lies in the proper ovarian fossa between the external and internal iliac vessels (*eponyms*: **fossa of Waldeyer** and **fossa of Krause**), but the ovary may lie almost anywhere against the lateral pelvic wall, and in some cases, it lies in the pouch of Douglas (Hollinshead 1971). The line of attachment of the peritoneum of the mesovarium to the substance of the ovary is known as the *eponym* **white line of Farre** (Farre 1859; *see* Farre, Arthur) also described by **Romiti** (*see* Romiti, Guglielmo) and **Waldeyer**. For the arrangement of sex cells in the ovary, the *eponyms* **tubes of Pflüger** (1863; *see* Pflüger, Edward Friedrich) and **tubes of Valentin** (1838; *see* Valentin, Gabriel Gustav Wilhelm) have been used. For the suspensory ligament of the ovary, several *eponyms* are in use: **band of Clado** (Clado 1892; *see* Clado, Spiro), **infundibulopelvic ligament of Henle** (Henle 1855–1872; *see*

Henle, Friedrich Gustav Jacob) and **lumboovarian ligament of Testut** (Testut 1891; *see* Testut, Jean Léon). Clado (1892) also described the appendiculoovarian ligament, a mesenteric fold from the vermiform appendix of the caecum to the right broad ligament of the uterus, through which an anastomosis between the appendicular and ovarian arteries passes. During development, the caudal ligament of the ovary is known as the **gubernaculum of Hunter**, who originally described the gubernaculum testis (Hunter 1762). Haller (1748; *see* Haller, Albrecht von) had named the caudal ligament of the ovary the gubernaculum vaginae. The female caudal ligament forms the proper ovarian ligament and the round ligament of the uterus (*see* Fig. 25.3). The round ligament of the uterus received the name of William Hunter (*eponym*: **ligament of Hunter**; Hunter 1774). It extends into the inguinal canal, surrounded by adipose tissue (*eponym*: **fat pad of Imlach**; Imlach 1885; *see* Imlach, Francis). The part of the inguinal canal that contains the round ligament is also known as the *eponym* **ring of Imlach**. A patent vaginal process of the peritoneal cavity in females is known as the *eponym* **canal of Nuck** (Nuck 1691; *see* Nuck, Anton).

The greater vestibular glands in the perineum are known as the *eponym* **glands of Bartholin** (Bartholin 1677; *see* Bartholin, Caspar, The Younger), but they were also described by Duverney in cows, and by Tiedemann (1840; *see* Tiedemann, Friedrich). The female paraurethral glands received the *eponyms* **glands of Morgagni** (1723; *see* Morgagni, Giovanni), **Guérin** (1864; *see* Guérin, Alphonse François Marie), **Schueller** (1877; *see* Schueller, Karl Heinrich Maximilian), and **Skene** (1880; *see* Skene, Alexander Johnston Chalmers), but usually Skene gets the credits.

For gynaecological surgery, Pfannenstiel's incision provides good exposure (*eponym*: **incision of Pfannenstiel**; Pfannenstiel 1900). The various fasciae of the anterior abdominal wall (Camper, Colles, Cooper, Hesselbach, Poupart and Scarpa) are discussed in Chap. 22 on anatomical eponyms used in abdominal surgery. Here, it should be noted that the triangle of Hesselbach forms in laparoscopic surgery a safe area for the introduction of cannulae (Baskett 2019). It is bounded cranially by the falx inguinalis, the inferior epigastric vessels laterally, the lateral border of the rectus abdominis muscle medially and the inguinal ligament caudally (Hesselbach 1806; *see* Hesselbach, Franz Kaspar).

25.6 Neurovascular Sheaths of the Female Pelvis and the Innervation of the Pelvic Organs

Neurovascular sheaths of the female pelvis are shown in Fig. 25.6, but whether they are ligaments is long debated. Since Cameron (1908), uterosacral and cardinal (transverse cervical) ligaments are regarded as perivascular tissue rather than as ligaments. For the cardinal ligament the *eponym* **ligament of Mackenrodt** (Mackenrodt 1895; *see* Mackenrodt, Alwin; Fig. 25.6a) is still used. The extensive studies by Fritsch et al. (2004) oncemore confirmed Cameron's statement (Fig. 25.6b). On the pelvic floor, the vesical venous plexus (*eponym*: **veins** or **plexus of Santorini**; Santorini 1724; *see* Santorini, Giovanni Domenico) should be mentioned. In females, the vesical venous plexus communicates with the anterior part of the vaginal plexus, and receives directly the deep dorsal vein of the clitoris, draining the venous plexus of the clitoris (*eponym*: **plexus of Kobelt**; Kobelt 1844; see Kobelt, Georg Ludwig). In males, the vesicoprostatic venous plexus receives blood from the prostate and the urinary bladder and is found anterior and lateral to the prostate. The deep dorsal vein of the penis drains to the venous plexus anterior to the prostate, sometimes referred to as the pudendal plexus or as the plexus of Santorini (Hollinshead 1971; Dietrich 1997). More frequently, the entire prostatic venous plexus is described as plexus of Santorini. For the innervation of the female pelvic organs, the uterovaginal plexus (Fig. 25.7) should be mentioned (*eponym*: **plexus of Frankenhäuser**; Frankenhäuser 1867; *see* Frankenhäuser, Ferdinand; also described by Lee; Lee 1842; *see* Lee, Robert). The pudendal nerve and vessels pass via the **canal of Alcock** (Alcock 1836–1839; *see* Alcock, Benjamin), the pudendal canal in the ischioanal or ischiorectal fossa.

25.7 Eponyms in Use for the Lymphatic Drainage of the Female Genital Organs

The **lymphatic drainage** of the female pelvic organs and external genitalia is shown in Fig. 25.8. The lumbar lymph nodes drain the retroperitoneal area and large part of the internal genitalia (the ovary, the uterine tube and the fundus of the uterus). The three groups of iliac lymph nodes and the sacral lymph nodes are the regional lymph nodes for the body and cervix of the uterus, the vagina, the urinary bladder and the lower part of the ureter and urethra. The superficial and deep inguinal lymph nodes are the regional lymph nodes for the lower extremity, the floor of the pelvis, the external genitalia (including the lower part of the vagina), the anterior abdominal wall beneath the level of the umbilicus, part of the uterus and part of the anal canal. In the past, several names and eponyms were given to these lymph nodes. Cloquet's node, the proximal deep inguinal lymph node located in the lacunar space, is a landmark in the operation of radical vulvectomy, first described by Cloquet in his monograph on the surgical anatomy of inguinal and femoral hernia (Cloquet 1817). It was earlier described by Rosenmüller (1805), therefore the *eponym* **gland** or **node of Cloquet-Rosenmüller**.

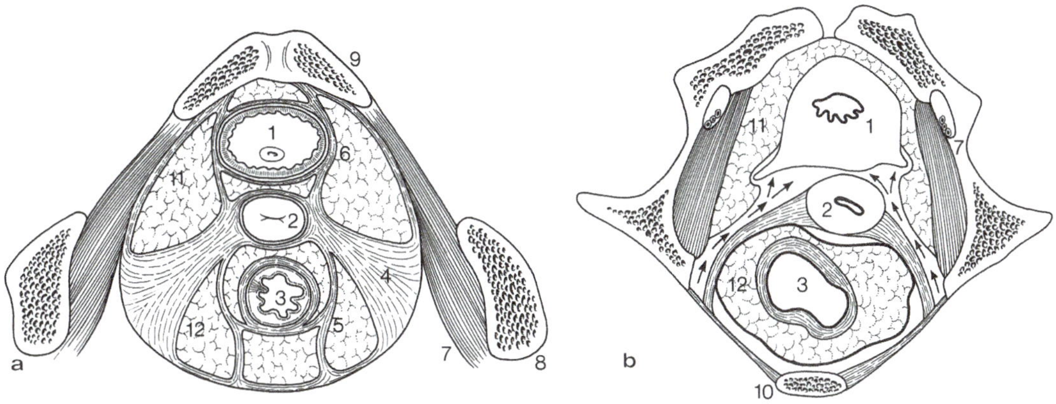

Fig. 25.6 Ligament of Mackenrodt: (**a**) The classical view; (**b**) modern insights, where this ligament is merely seen as a passage of vessels and nerves. (After Fritsch H, Lienemann A, Brenner E, Ludwikowski B *Clinical anatomy of the pelvic floor.* Adv Anat Embryol Cell Biol 2004; 175:1–66; and ten Donkelaar HJ, Oostra R-J 2014 *Klinische Anatomie en Embryologie*, 4th ed. Springer Media, Houten). *1* bladder, *2* cervix uteri, *3* rectum, *4* cardinal ligament of Mackenrodt, *5* sacrouterine ligament, *6* pubovesical ligament, *7* obturator internus muscle, *8* ischium, *9* pubis, *10* coccyx, *11* paravesical space, *12* pararectal space. The *arrows* in (**b**) indicate the passage of vessels and nerves

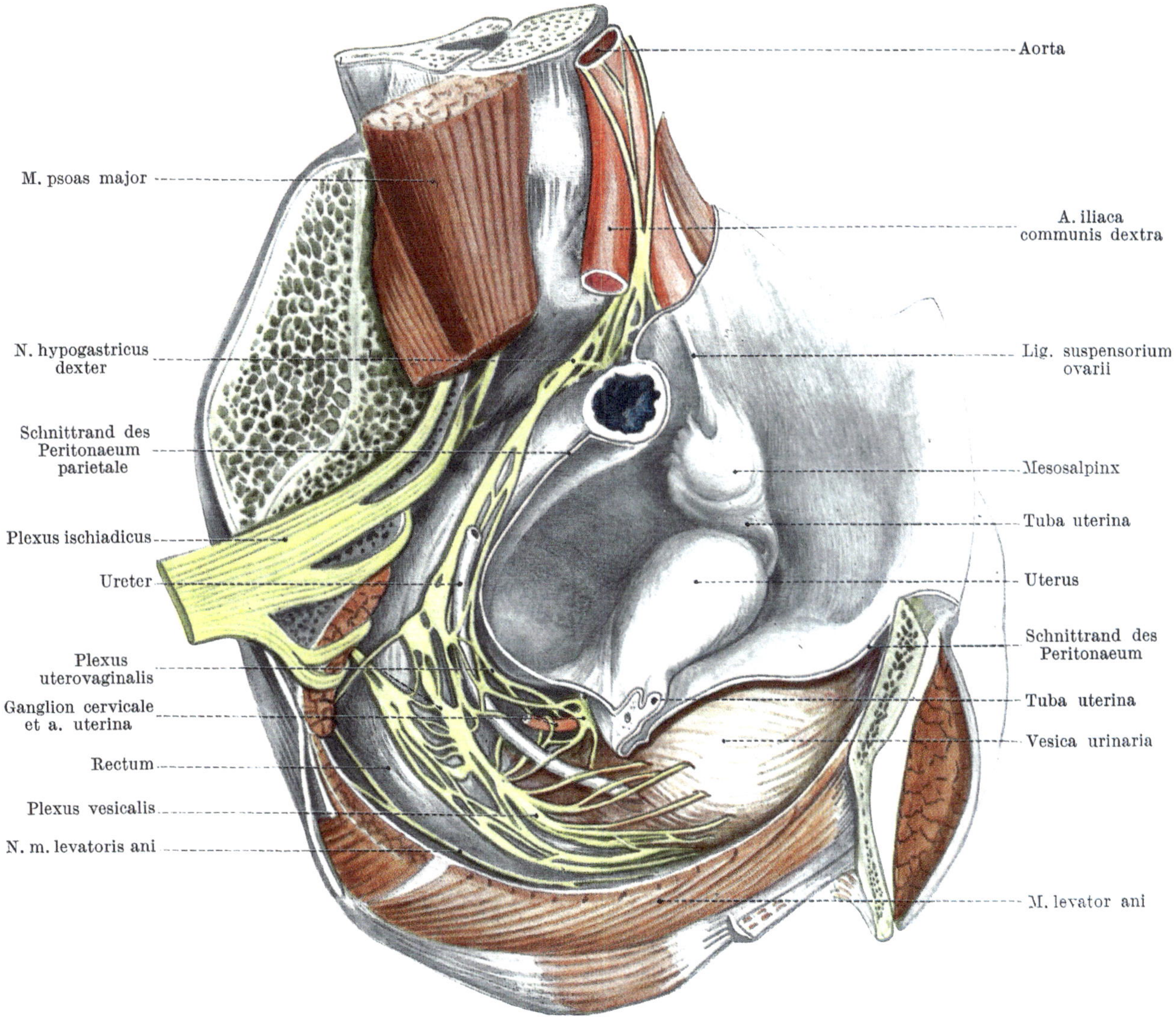

Fig. 25.7 The uterovaginal plexus. (From Hafferl A *Lehrbuch der topographischen Anatomie,* 2nd ed. Springer, Berlin, Göttingen, Heidelberg, 1957; with permission)

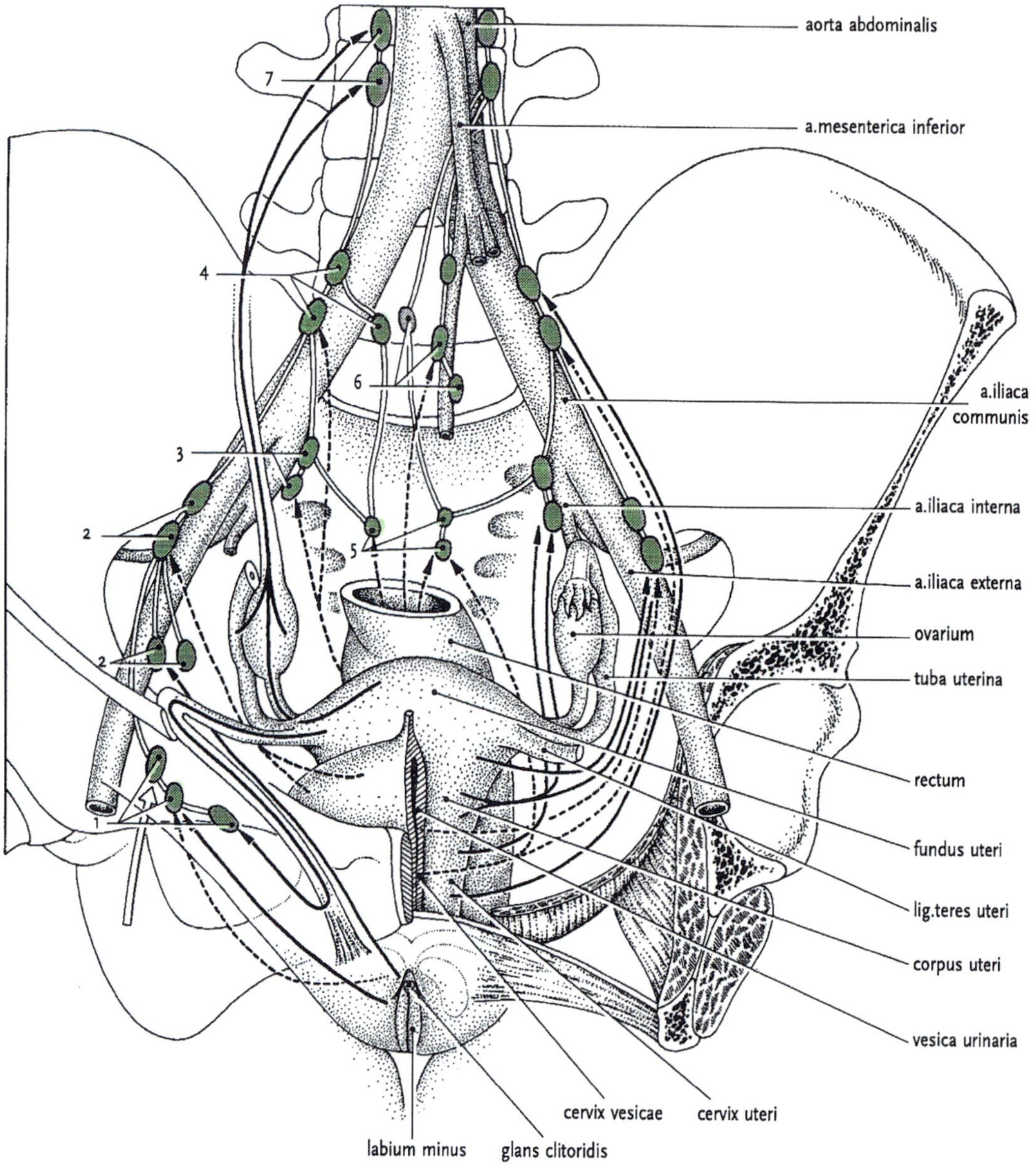

Fig. 25.8 Lymphatic drainage of the pelvic organs in females. (After ten Donkelaar HJ, Oostra R-J 2014 *Klinische Anatomie en Embryologie*, 4th ed. Springer Media, Houten). *1* Superficial and deep inguinal lymph nodes, *2* external iliac lymph nodes, *3* internal iliac lymph nodes, *4* common iliac lymph nodes, *5* sacral lymph nodes, *6* inferior mesenteric lymph nodes, *7* lumbar lymph nodes

An inconstant parauterine node at the crossing of the uterine artery and the ureter received the *eponym* **node of Bayer**.

A peculiar nodule received the *eponym* **nodule of Sister Mary Joseph**, the Head nurse of William Mayo at St Mary's Hospital in Rochester. It means a hard or ulcerated umbilical nodule in cases of widespread metastasic disease from primary intraabdominal tumours, usually from the stomach or ovary (*see* Joseph, Sister Mary). William Mayo honoured her in 1928 (Mayo 1928). The Wertheim operation and its modifications are designed to completely remove all the lymph nodes receiving lymph from the cervix uteri (*eponym*: **radical hysterectomy of Wertheim**; Wertheim 1900).

References

Alcock B (1836–1839) Iliac arteries. In: Todd's cyclopedia of anatomy and physiology, Vol II: 835

Bandl L (1875) Über Ruptur der Gebärmutter und ihre Mechanik. Czermak, Vienna

Bartholin C (1677) De ovariis mulierum et generationis historia. Zieger, Rome

Baskett T (2019) Eponyms and names in obstetrics and Gynaecology, 3rd edn. Cambridge University Press, Cambridge

Caldwell WE, Moloy HC (1933) Anatomical variations in the female pelvis and their effect on labor, with a suggested classification. Am J Obst Gyn 26:479–505

Call EL, Exner S (1875) Zur Kenntniss des Graafschen Follikels und des Corpus luteum beim Kaninchen. SB Akad Wiss Wien Math-Naturw 71:321–328

Cameron J (1908) The fascia of the perineum and pelvis of the female, with special reference to the mechanical supports of the pelvic viscera. J Anat Physiol 42:438–455
Carus CG (1820) Lehrbuch der Gynaecologie, oder systematische Darstellung der Lehren von Erkentniss und Behandlung eigentümlicher, gesunder und krankhafter Zustände, sowohl der Nichtschwangern, Schwangern und neugeborenen Kinder. Fleischer, Leipzig
Clado S (1892) Appendice caecal, anatomie, embryologie. CR Soc Biol Paris, 9th Ser, IV, Mem, pp 133–172
Claudius FM (1865) Über die Lage des Uterus. Ztsch Rat Med 22:248
Cloquet JG (1817) Recherches anatomiques sur les hernies de l'abdomen, Paris
de Graaf R (1672) De mulierum organis generatione. Hackiana, Leiden
Dickinson RL (1892) The diagnosis of pregnancy between the second and seventh weeks by bimanual examination. NY J Gynaecol Obstet 10:544–555
Dietrich H (1997) Giovanni Domenico Santorini (1681-1737), Charles-Pierre Denonvilliers (1808-1872) first description of urosurgically relevant structures in the small pelvis. Eur Urol 32:124–127
Douglas JC (1730) A description of the peritoneum and of that part of the Membrana cellularis which lies on its outside. Roberts, London
Falloppio G (1561) Gabrielis Falloppii medici Mutinensis Observationes Anatomicae. Ulmus, Venice
Farre A (1859) The uterus and its appendages. In: Todd RE (ed) The cyclopaedia of anatomy and physiology, p 548
Frankenhäuser F (1867) Die Nerven der Gebärmutter und ihre Endigung in den glatten Muskelfasern: Ein Beitrag zur Anatomie und Gynäkologie. Mauke, Jena
Fritsch H, Lienemann A, Brenner E, Ludwikowski B (2004) Clinical anatomy of the pelvic floor. Adv Anat Embryol Cell Biol 175:1–66
Gardner GH, Greene RR, Peckham BM (1948) Normal and cystic structures of the broad ligament. Am J Obstet Gynecol 55:917–937
Gärtner HT (1822) An anatomical description of a glandular organ observed in the uterus of some lower animals. Edinb Med Surg 21:460–463
Gray SW, Skandalakis JE (1972) Embryology for surgeons. Saunders, Philadelphia
Guérin AFM (1864) Maladies des organes génitaux externe de la femme. Paris
Henle FGJ (1855–1872) Handbuch der systematischen Anatomie des Menschen, 3 Vols. Vieweg, Braunschweig
Hesselbach FK (1806) Anatomisch-chirurgische Abhandlung ueber den Ursprung der Leistenbrücke. Baumgärtner, Würzburg
Hofbauer II (1905) Grundzüge einer Biologie der menschlichen Plazenta, mit besonderer Berücksichtigung der Fragen der fötalen Ernährung. Braumüller, Vienna
Hollinshead WH (1971) Anatomy for surgeons, Vol 2: the Thorax, Abdomen and Pelvis, 2nd edn. Harper & Row, New York
Hunter J (1762) Observations on the State of the Testis in the Foetus and on the Hernia Congenita. Medical Commentaries, Part 1:75–88
Hunter W (1774) The anatomy of the human gravid uterus. Baskerville, Birmingham
Imlach F (1885) On shortening the round ligament of the uterus. Edinb Med J 30:913
Kaufmann P, Huppertz B, Frank H-G (1996) The fibrinoids of the human placenta, compositions and functional relevance. Ann Anz 178:485–501
Kilian HF (1835–1844) Geburtshilflicher Atlas. Düsseldorf
Kobelt GL (1844) Die männlichen und weiblichen Wollustorgane des Menschen und einiger Säugethiere in anatomisch-physiologischer Beziehung untersucht und dargestellt. Freiburg
Kobelt GL (1847) Der Nebeneierstock des Weibes. Würzburg
Koff AK (1933) Development of the vagina in the human fetus. Contrib Embryol Carnegie Inst 24:59–90
Langhans T (1870) Kenntniss der menschlichen Placenta. Arch Gynaekol 1:317–314
Langhans T (1877) Untersuchungen über die menschliche Placenta. Arch Anat Physiol, Anat Abt, 188–267
Lee R (1842) On the ganglia and other nervous structures of the uterus. Richard and John E Taylor, London
Mackenrodt A (1895) Ueber die Ursachen der normalen und pathologischen Lagen der Uterus. Arch Gynaekol 48:393–421
Mayo W (1928) Metastasis in cancer. Proc Staff Meetings Mayo Clinic 3:227
Michaelis GA (1851) Das enge Becken nach eigenen Beobachtungen und Untersuchungen. Wigand, Leipzig (published posthumously by his associate Carl Litemann)
Mijsberg WA (1924) Über Entwicklung der Vagina, des Hymen und des Sinus urogenitalis beim Menschen. Z Anat Entwicklungsgeschichte 74:684–760
Morgagni GB (1723) Adversaria anatomica omnia. Leiden
Morgagni GB (1764) De sedibus et causis morborum per anatomen indagatis, libris quinque. Venice
Müller JP (1830) Bildungsgeschichte der Genitalien aus anatomischen Untersuchungen an Embryonen des Menschen und der Thiere. Düsseldorf
Naboth M (1707) De sterilitate mulierum. Zeidler, Leipzig
Naegele FC (1819) Ueber den Mechanismus der Geburt. Deutsche Arch Physiol 5:483–531
Nitabuch R (1887) Beiträge zur Kenntniss der menschlichen Placenta. Thesis, University of Bern. Stämpfli, Bern
Nuck A (1691) Adenographia curiosa et uteri foeminei anatoma nova. Luchtmans, Leiden
Pawlik K (1881) Ueber die Sondierung der Ureteren der weiblichen Blase. Arch Gynakol 18:491–495
Pfannenstiel HJ (1900) Über die Vortheile des suprasymphysären Fascienquerschnitts für die gynäkologischen Köliotomien zugleich ein Beitrag zu der Indikationsstellung der Operationswege. Samml Klin Vortschr Leipzig 268:1735–1756
Pflüger EFW (1863) Ueber die Eierstöcke der Säugethiere und des Menschen. W. Engelmann, Leipzig
Piskaçek L (1899) Über Ausladungen umschriebener Gebärmutterabschnitte. Braumüller, Vienna
Richard FDA (1851) Anatomie des trompes de l'uterus chez la femme. Thesis, Paris
Rieder C (1884) Ueber die Gartner'sche (Wolff'schen) Kanäle beim menschlichen Weibe. Virchow's Arch Path Anat 96:100–130
Rohr K (1889) Die Beziehungen der mutterlichen Gefässe zu den intervillösen Raumen der reifen Placenta speciell zur Thrombose derselben ('weisser Infarct'). Virchows Arch 115:505–533
Rosenmüller JC (1802) Quaedam de ovariis embryonum et foetuum humanorum. Leipzig
Rosenmüller JC (1805) Chirurgisch-Anatomische Abbildungen für Aerzte und Wundärzte. Landes Industrie Comptoir, Weimar
Santorini GD (1724) Observationes anatomicae. Recurti, Venice
Schueller KHM (1877) Ueber die Localbehandlung des chronischen Blasenkatarrhs, ein Beitrag zur Chirurgie der Harnorgane. De Gruyter, Berlin
Skene AJC (1880) The anatomy and pathology of two important glands of the female urethra. Am J Obstet Dis Women Child 13:265–270
Sneeden VD (1958) Mesonephric lesions of the cervix. Cancer 11:334–336
Testut JL (1891) Traité d'anatomie humaine. Doin, Paris
Tiedemann F (1840) Von den Duverney'schen, Bartholin'schen und Cowper'schen Drüsen des Weibes und der schiefen Gestaltung und Lage der Gebärmutter. Heidelberg, Leipzig
Valentin GG (1838) Ueber die Entwickelung der Follikel in den Eierstöcke der Säugethiere. Arch Anat Physiol (Lpz), pp 526–536
van Hoboken N (1669) Anatomia secundinae humanae. Ribbius, Utrecht

von Baer KE (1820–1837) Entwickelungsgeschichte der Thiere, 3 Vols. Bornträger, Königsberg

von Haller A (1748) De viis feminis. Göttingen

von Pander HC (1817) Beiträge zur Entwickelungsgeschichte des Hühnchens im Ei. Würzburg

von Slavjanski KF (1871) On the normal and pathological histology of human ovaries (English translation). Thesis, St. Petersburg

Wertheim E (1900) Zur Frage der Radicaloperation beim Uteruskrebs. Arch Gynaekol 61:627–628

Wolff CF (1759) Theoria generationis. Halle

Anatomical Eponyms in Oncology

26

David Kachlík and Hans J. ten Donkelaar

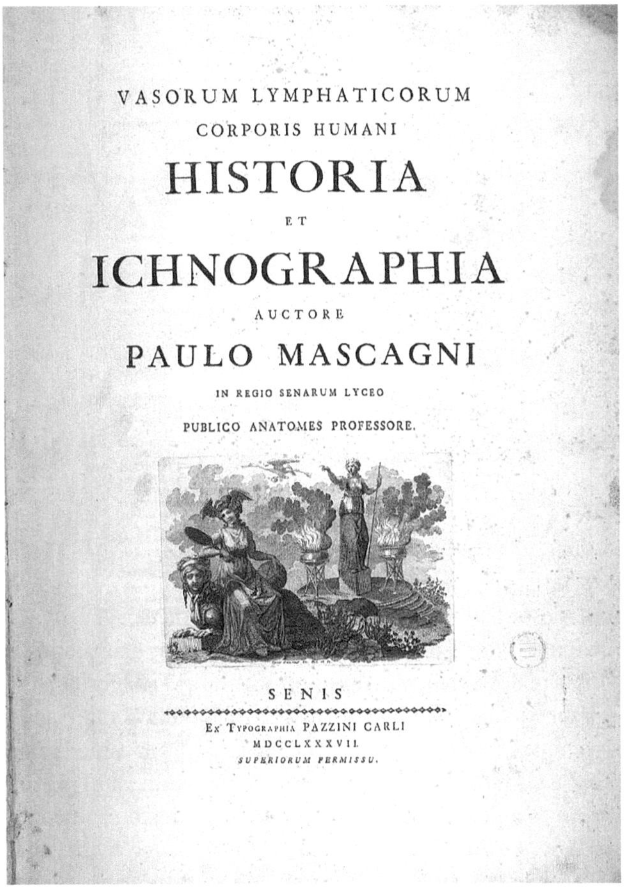

Vignette Title page of *Vasorum lymphaticorum corporis humani historia et ichnographia* by Paolo Mascagni.

26.1 Introduction

Knowledge of the lymphatic drainage of the human organs is of great importance in oncology. The lymphatic system consists of vessels connected with masses of lymphatic tissue known as lymph nodes. The first or regional lymph nodes drain to other, secondary and tertiary, lymph nodes and finally to lymphatic trunks, the thoracic duct and the right lymphatic duct. The thoracic duct enters the venous system at the junction of the left internal jugular and subclavian veins, and the right lymphatic duct at the junction of the right internal jugular and subclavian veins, i.e. at the left and right venous **angles of Pirogoff** (*see* Pirogoff, Nicolai Ivanovich. The right lymphatic duct drains the right side of the head, neck and thorax, the thoracic duct the rest of the body. A dilated sac, the cisterna chyli is found at the beginning of the thoracic duct, present in about 50% of cases (Hollinshead 1971). The intestinal and lumbar lymph nodes drain to the cisterna chyli. The thoracic duct and cisterna chyli were discovered in the seventeenth century by Aselli, Pecquet and van Horne, whereas anatomical studies on the lymph nodes and lymphatic drainage of organ systems date largely from the nineteenth century and early twentieth century, especially the French School of anatomists (Sappey, Quénu, Poirier and Cuéno, Rouvière). The Italian Paolo Mascagni is known for the first complete description of the lymphatic system. The **vignette** shows the title page of his *Vasorum lymphaticorum corporis humani historia et ichnographia* (Mascagni 1787; *see* Mascagni, Paolo). Several lymph nodes received eponyms from various European countries. More recent clinical studies pointed at the sentinel lymph nodes, the first lymph nodes affected by metastases of a malignancy.

Here, we summarize the lymphatic drainage from the head and neck, the breast, the lungs, the abdominal and pelvic organs, and the external genital organs, where most of the anatomical eponyms became popular. For further data on the authors quoted, reference is made to Part II (*The Anatomical Eponyms from A to Z*) and figures in this part.

26.2 Eponyms on the Lymphatic Drainage of the Head and Neck

The lymphatic drainage of the **head** and **neck** goes primarily to the **deep cervical lymph nodes** (Fig. 26.1), which receive lymph from all other lymph nodes in the head and

H. J. ten Donkelaar et al., *An Illustrated Guide to Anatomical Eponyms*, https://doi.org/10.1007/978-3-031-91664-9_26

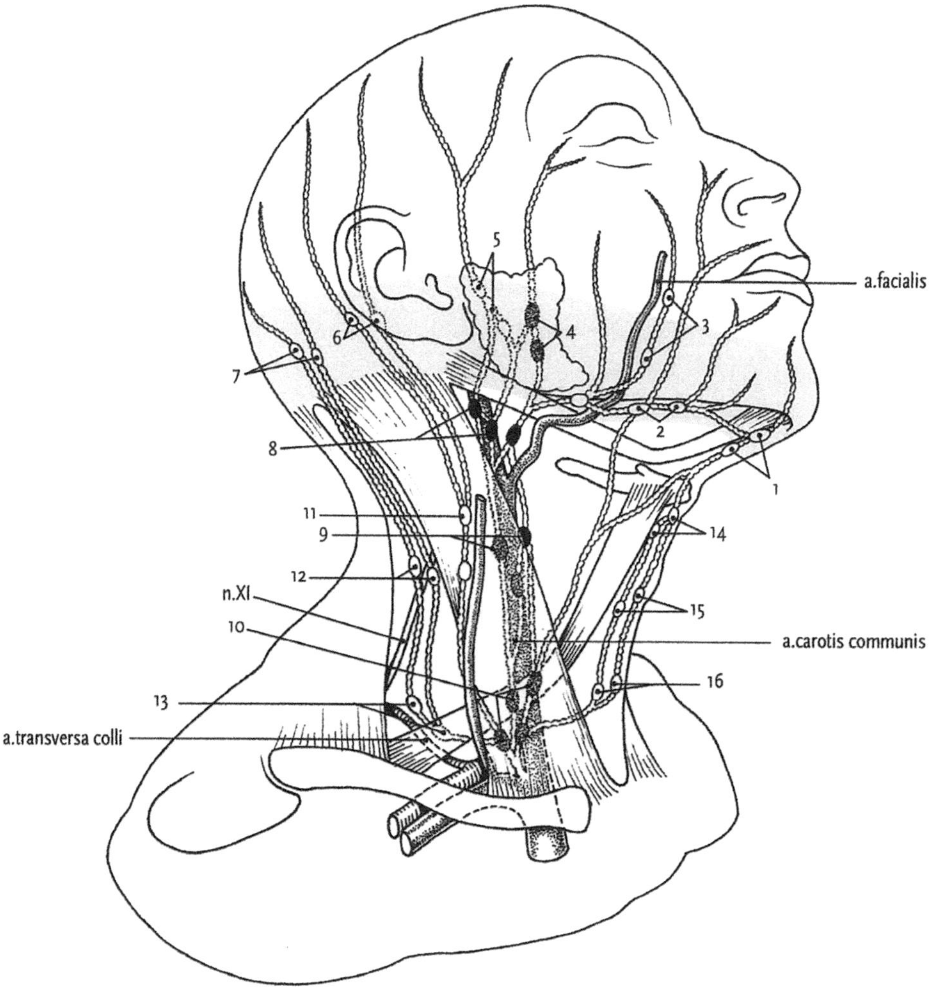

Fig. 26.1 Lateral view of the lymphatic drainage from the head and neck. (After ten Donkelaar HJ, Oostra R-J 2014 *Klinische Anatomie en Embryologie*, 4th ed. Springer Media, Houten). Superficial lymph nodes are in *white*, deep lymph nodes in *grey* or *black*. The pericervical collar is shaded in *grey*. *nXI* accessory nerve, *1* Submental lymph nodes, *2* submandibular lymph nodes, *3* buccal lymph nodes, *4*, *5* deep and superficial parotid lymph nodes, *6* posterior auricular lymph nodes, *7* occipital lymph nodes, *8–10* superior, middle and inferior groups of deep cervical lymph nodes, *11* superficial cervical lymph nodes, *12* posterior cervical lymph nodes, *13* supraclavicular lymph nodes, *14* infrahyoid lymph nodes, *15* prelaryngeal lymph nodes, *16* pretracheal lymph nodes

neck. The deep cervical lymph nodes drain to the jugular trunk, of which the right one ends in the right lymphatic duct, and the left one directly in the thoracic duct. The lymph nodes of the head and neck can be grouped into the pericervical ring, a vertical tract and visceral lymph nodes. The lymph nodes of the **pericervical ring** form a collar at the border of the head and neck (the **'cercle ganglionaire péricervical'** of Rouvière; Rouvière 1932; *see* Rouvière, Henri), and include occipital, mastoid, parotid, facial (*eponym*: **lymph node of Stahr**; Stahr 1899; *see* Stahr, Hermann), submandibular, submental and sublingual nodes (Fig. 26.1). The **vertical tract** consists of superficial and deep cervical lymph nodes. The superficial cervical lymph nodes are found on the sternocleidomastoid muscle and drain to the deep cervical nodes. The deep cervical lymph nodes lie along the internal jugular vein and form three groups, superior, middle and inferior. The superior group is found between the skull base and the hyoid bone and includes the jugulodigastric lymph node (*eponym*: **principal lymph node of Küttner**; Küttner 1897; *see* Küttner, Arthur). The middle group is found between the hyoid bone and the border between the thyroid and cricoid cartilages. The inferior group includes the juguloomohyoid lymph node described by **Poirier and Cuéno** (1902). The supraclavicular lymph nodes belong to the inferior deep cervical lymph nodes (*eponym*: **node of Virchow-Troisier** for the supraclavicular lymph node above the left clavicula, palpably enlarged as the result of a metastasis from a malignant neoplasm; Virchow 1856; *see* Virchow, Rudolf, and Troisier, Charles Emile). Involvement of this node in stomach carcinoma is attributed to the predominant drainage by the thoracic duct. The **visceral group** includes the tonsillar lymph node, which bears the *eponym* **lymph node of Wood** (*see* Wood, Alexander), the **retropharyngeal lymph node of Gilletter-Henle** (*see* Gillette, Eugène Paulin, and Henle, Friedrich Gustav Jacob, and prelaryngeal ('Delphian'), pretracheal and paratracheal lymph nodes, several of these were described by **Rouvière** (1932), and all drain to the deep cervical lymph nodes. Bourgery (1837–1840; *see* Bourgery, Jean-Baptiste Marc) described accessory lymph nodes in front of the thyroid gland (*eponym*: **preisthmic lymph nodes of Bourgery**).

Clinically, for the head and neck a **lymph node level** approach is advocated based on research on patterns of met-

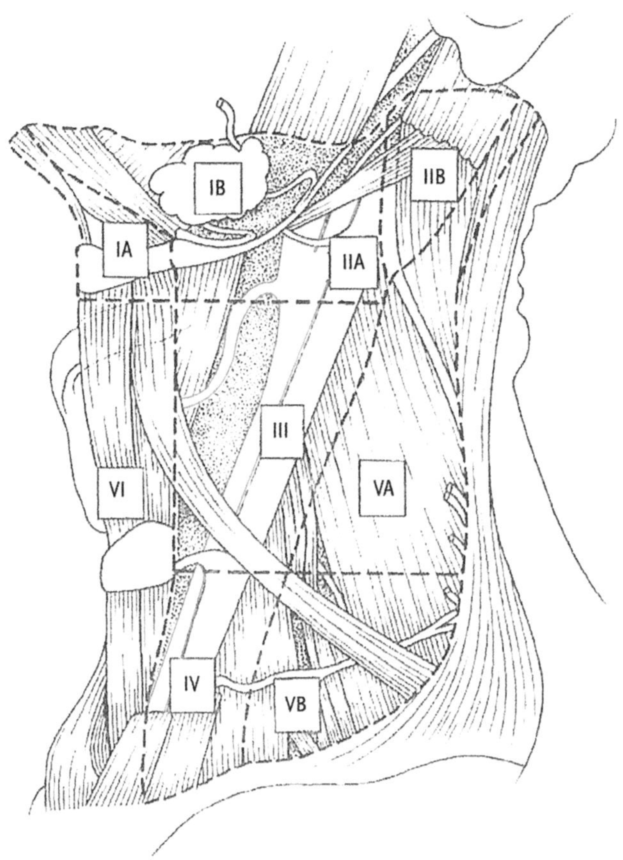

Fig. 26.2 Clinical lymph node levels of the head and neck. (After Harisinghani MG, ed. 2013 *Atlas of Lymph Node Anatomy*. Springer, New York, Heidelberg, Dordrecht, London; and ten Donkelaar HJ, Oostra R-J 2014 *Klinische Anatomie en Embryologie*, fourth ed. Springer Media, Houten). Levels IA and IB are similar to the submental and submandibular triangles, respectively. Levels II, III and IV include the sternocleidomastoid region and the carotid triangle. Levels IIA and IIB are divided by the course of the accessory nerve. The horizontal borders are formed by the hyoid bone (levels II-III) and the lower border of the cricoid cartilage (levels III-IV). The classical Virchow-Troisier node belongs to Level IV. Level V corresponds to the lateral cervical region and level VI roughly corresponds to the muscular or omotracheal triangle

astatizing to come to uniformity in therapeutic approach (Robbins 1998; Harisinghani 2013; Fig. 26.2). These levels partly overlap anatomical regions. Levels IA and IB are similar to the submental and submandibular triangles, respectively. Levels II, III and IV include the sternocleidomastoid region and the carotid triangle. Levels IIA and IIB are divided by the course of the accessory nerve. The horizontal borders are formed by the hyoid bone (levels II-III) and the lower border of the cricoid cartilage (levels III-IV). The classical Virchow-Troisier node belongs to Level IV. Level V corresponds to the lateral cervical region and level VI roughly corresponds to the muscular or omotracheal triangle.

26.3 Eponyms on the Lymphatic Drainage of the Breast

Studies on the lymphatic drainage of the **breast** date back to the studies of Sappey (1874–1885; *see* Sappey, Marie Phillibert Constant). His classical description of the breast emphasized the subareolar plexus (*eponym*: **subareolar plexus of Sappey**), which was said to receive most of the lymph vessels from the deeper tissue of the breast and to give rise to major efferent lymph vessels running subcutaneously towards the axilla (Fig. 26.3). Sappey injected mercury into the skin of a very thin adult cadaver. Poirier and Cuéno (1902) summarized Sappey's results and added their own data from infant studies using the method of Gerota (1896; *see* Gerota, Dumitru) with oil painting dye to stain the lymph vessels (Fig. 26.4a). Later studies showed that the density of the plexus has been overemphasized (Turner-Warwick 1959). All studies agree that the principal lymphatic drainage of all the segments of the breast is to the axilla with lesser drainage to the parasternal and supraclavicular lymph nodes (Fig. 26.4b). A **sentinel node** for the mammary gland (*eponym*: **gland** or **node of Sorgius**; Sorgius 1880; *see* Sorgius, Wilhelm) was identified as the caudalmost of the pectoral lymph nodes located on the second or third slip of the anterior serratus muscle, comparable to the **inferior thoracic nodes of Oelsner** (1901; *see* Oelsner, L). Oelsner described superior and inferior axillary lymph nodes in the 2nd and 3rd intercostal spaces and in the 4th and 5th intercostal spaces, respectively. Rotter described the interpectoral axillary lymph nodes between the pectoralis major and minor muscles (*eponym*: **nodes of Rotter**; Rotter 1896; *see* Rotter, Josef). Suami et al. (2008) studied the lymphatic anatomy of the breast and its implication for sentinel node biopsy. They found that in some cases one sentinel node in the axilla drained almost the entire breast but that in most cases more than one sentinel node were represented. From the pectoral and subscapular lymph nodes, lymphatic drainage of the breast goes to the central and apical lymph nodes, and from here to the venous angle to the venous system or to the supraclavicular lymph nodes.

26.4 Eponyms on the Lymphatic Drainage of the Lungs

The traditional scheme for the lymphatic drainage of the **lungs** from three parts, superior, middle and inferior, not restricted to the borders of the lung lobes, dates back to Rouvière (1932; Fig. 26.5). Lymphatic vessels of the lung drain to pulmonary and bronchopulmonary lymph nodes, the first within the pulmonary hilum, the second outside. These regional lymph nodes drain to the tracheobronchial lymph

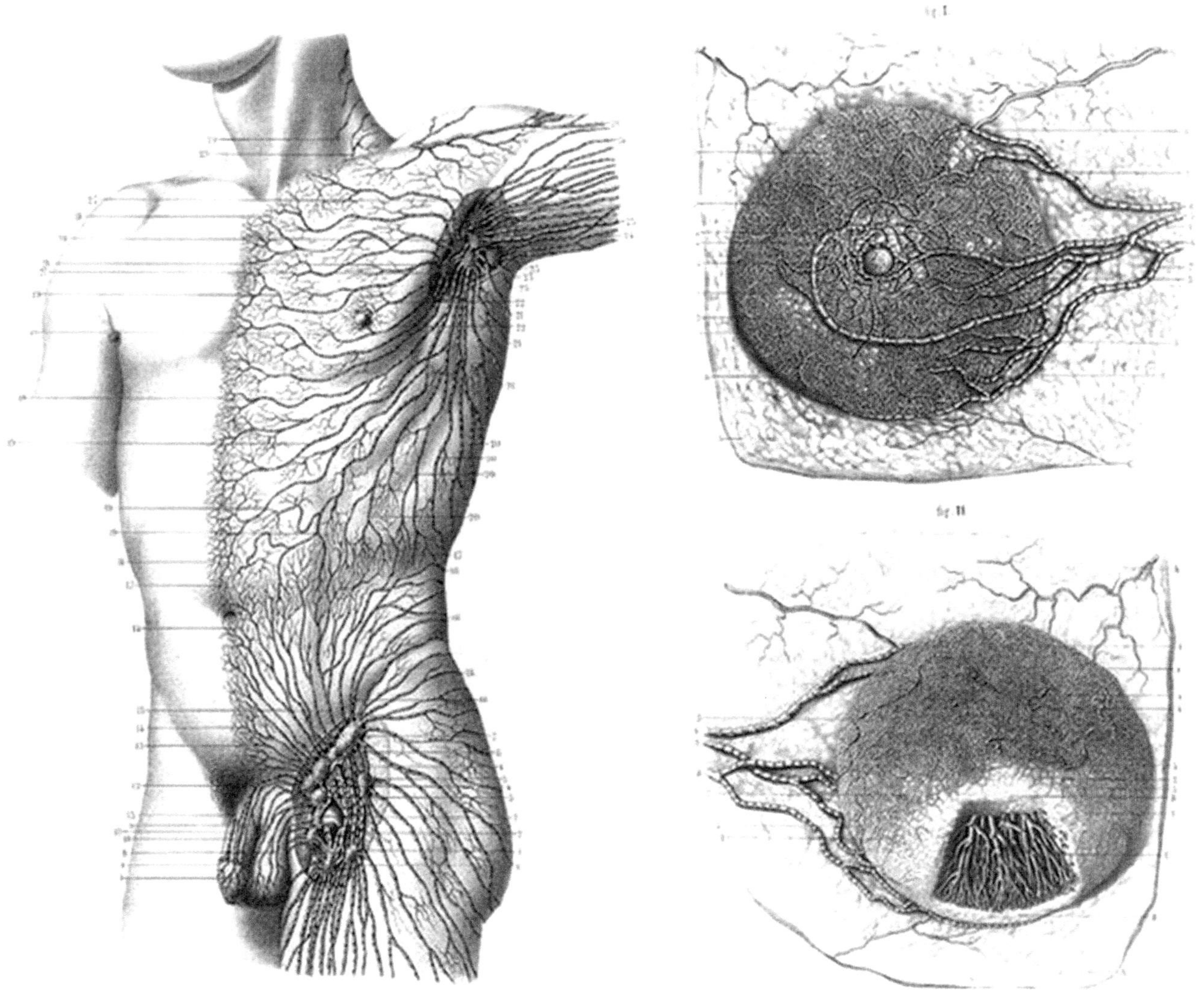

Fig. 26.3 Lymphatic subareolar plexus of Sappey. (From Suami H, Pan WR, Mann GB, Taylor GI *The lymphatic anatomy of the breast and its implications for sentinel lymph node biopsy: A human cadaver study.* Ann Surg Oncol 2007; 15:863–871; from Wikimedia Commons; in public domain)

nodes at the bifurcation of the trachea and paratracheal lymph nodes along the trachea, and via the bronchomediastinal trunks to the venous system. Modern studies have included further primary, stage N1 nodes ranging from hilar, interlobar, lobar, segmental to subsegmental (Harisinghani 2013; Fig. 26.6). Bronchial lymph nodes were already described by Vesalius (1543). Arnold (1880) described intrapleural lymph nodes, small lymph nodes along the suprapleural lymph vessels.

26.5 Eponyms on the Lymphatic Drainage of the Abdominal Viscera

In the **abdomen**, lymph nodes are found mainly along the arteries. Primary lymph nodes of the **stomach** include three major groups (Fig. 26.7a): (a) the left gastric lymph nodes along the left gastric artery draining the area of the lesser curvature, (b) the right gastric lymph nodes along the right gastric artery draining the area of the greater curvature, and (c) the splenic lymph nodes along the splenic artery draining the fundus of the stomach. A cystic lymph node found in the cystohepatic triangle of Calot was described by Mascagni (1787; *see* Mascagni, Paolo; *eponym*: **lymph node of Mascagni**) and Calori (1850–1855; *see* Calori, Luigi; *eponym*: **lymph node of Calori**), and more recently described as sentinel lymph node of the gallbladder by the American surgeon Fred Bates Lund (1865–1950; *eponym*: **lymph node of Lund**). The **intestines** drain to the coeliac, superior mesenteric and inferior mesenteric lymph nodes. These preaortic lymph nodes lie ventral to the abdominal aorta at the roots of the coeliac trunk and the superior and inferior mesenteric arteries (Fig. 26.7b). The superior mesenteric lymph nodes drain to the coeliac lymph nodes. From

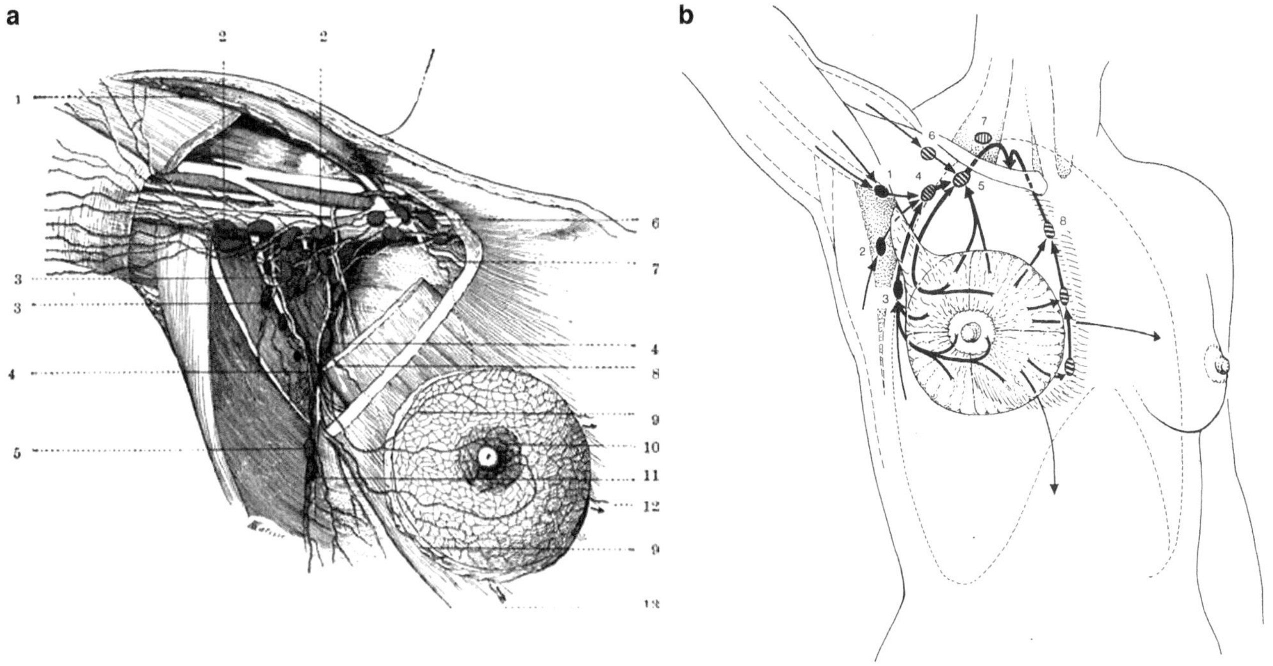

Fig. 26.4 (**a**) Lymph nodes of Poirier and Cuéno: Lymphatic drainage of the breast. (From Poirier PJ, Cuéno B *Système lymphatique.* In Poirier PJ, Charpy A, eds *Traité d'anatomie humaine,* Tome 2. Masson, Paris, 1902); (**b**) lymphatic drainage of the mamma. (After ten Donkelaar HJ, Oostra R-J 2014 *Klinische Anatomie en Embryologie*, 4th ed. Springer Media, Houten). The lateral axillary and deltoideopectoral lymph nodes lie along the axillary and proximal part of the cephalic vein, respectively

here, right and left intestinal trunks enter the cisterna chyli. Lumbar, clinically described as paraaortic lymph nodes drain the suprarenal glands, the kidneys, the ureters and the testes or ovaries, and via lumbar trunks to the cisterna chyli. A peculiar nodule was noted by Sister Mary Joseph Dempsey, the head nurse of William Mayo, at the umbilicus, indicating a malignant intraabdominal neoplasm, usually from the ovary or stomach, metastatic to the umbilicus (Mayo 1928). Aggregates of lymphatic nodules of the ileum were described by Pechlin (1672) and Peyer (1677), but are usually described with the *eponym* **plaques of Peyer**.

26.6 Eponyms on the Pelvic Viscera and the External Genital Organs

The lymphatic drainage of the **pelvic organs** and **external genital organs** predominantly passes by lymphatic vessels along arteries but also via the peritoneum. The lymphatic vessels along the arteries continue as lumbar trunks along the abdominal aorta (Fig. 26.8). Distinguished are superficial and deep inguinal lymph nodes, external, internal and common iliac lymph nodes along the respective iliac arteries, inferior mesenteric lymph nodes along the inferior mesenteric artery, lumbar lymph nodes, and sacral lymph nodes in the concavity of the sacrum. The **superficial** and **deep inguinal lymph nodes** are the regional lymph nodes for the lower extremity, the pelvic floor, the external genitalia (including the lower part of the vagina), the skin of the buttocks and the lower part of the back as well as the anterior abdominal wall beneath the level of the umbilicus, part of the uterus and part of the anal canal beneath the pectinate line. These lymph nodes drain to the external iliac lymph nodes. The **inferior mesenteric lymph nodes** are the regional lymph nodes for the upper part of the rectum and drain to the lumbar lymph nodes. The **lumbar lymph nodes** drain the retroperitoneal area and large part of the internal genital organs (the testis and epididymis in males, and the ovary, the uterine tube and the fundus of the uterus in females). The three groups of **iliac lymph nodes** and the **sacral lymph nodes** are the regional lymph nodes for the body and cervix of the uterus, the vagina, the urinary bladder and the lower part of the ureter and the urethra in females, and the ductus deferens, the prostate, the vesicular glands, the urinary bladder and the membranous and prostatic parts of the urethra in males. An inconstant parauterine lymph node at the crossing of the uterine artery and the ureter received the *eponym* **parauterine lymph node of Bayer** (*see* Bayer, Heinrich, and Eliška 2018 for discussion).

In the past, various names and eponyms were given to the inguinal lymph nodes. Cloquet (1825; *see* Cloquet, Jules Germain) described the deepest or proximal deep inguinal lymph node adjacent to the inguinal ligament (*eponym*: **gland** or **node of Cloquet**), earlier descibed by Rosenmüller

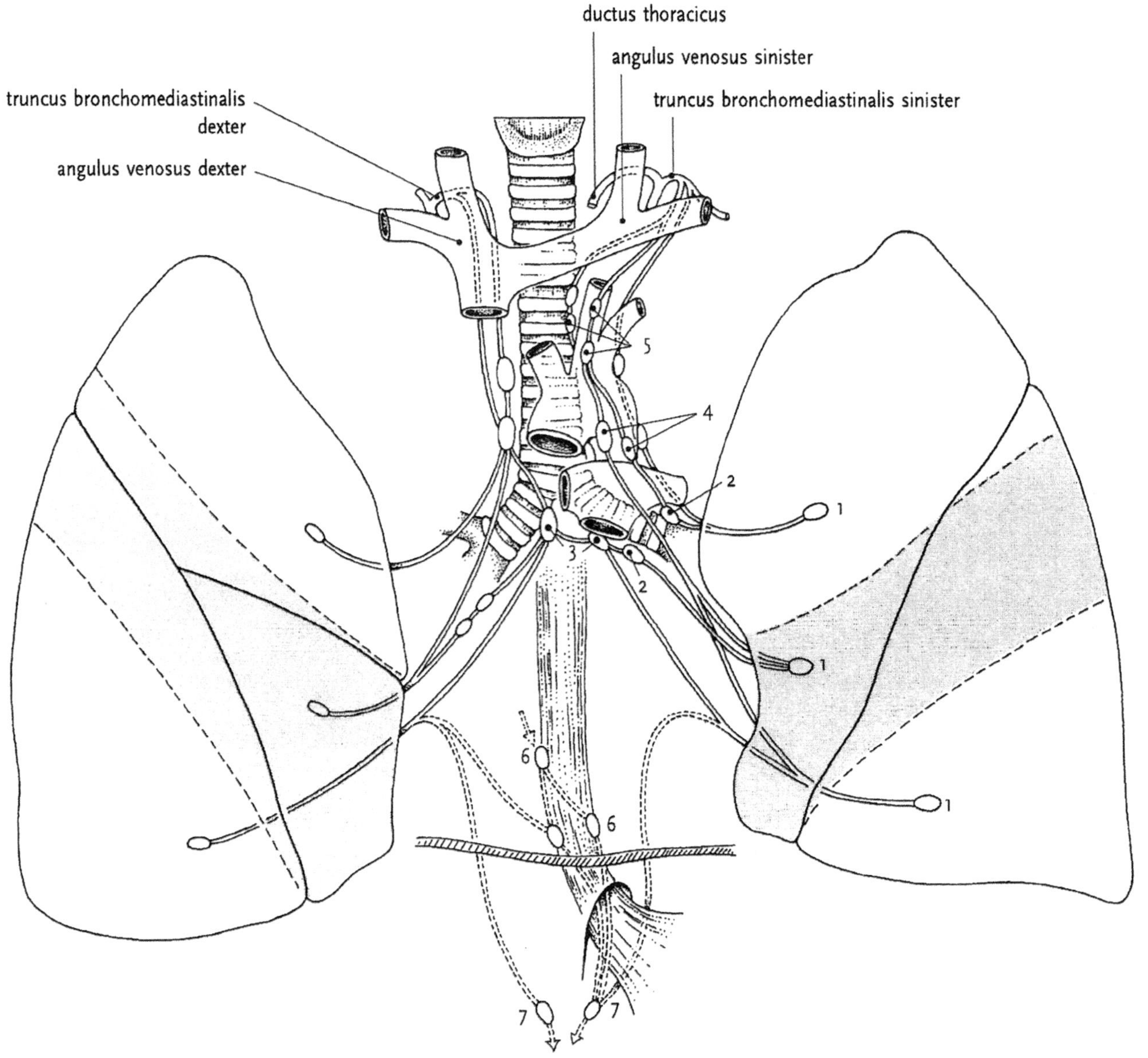

Fig. 26.5 Lymphatic drainage of the lungs. (After Rouvière H 1932 *Anatomie des lymphatiques de l'homme*. Masson, Paris; and Lohman AHM, ten Donkelaar HJ 1997 *Klinische Anatomie en Embryologie,* first ed. Bunge, Utrecht). *1* Pulmonary lymph nodes, *2* bronchopulmonary lymph nodes, *3* inferior tracheopulmonary lymph nodes; *4* superior tracheopulmonary lymph nodes, *5* paratracheal lymph nodes, *6* posterior mediastinal lymph nodes, *7* coeliac lymph nodes

(1805; *see* Rosenmüller, Johann Christian), so the *eponym* **node of Cloquet-Rosenmüller**. The deep inguinal nodes lie deep to the fascia lata located in the lacunar space adjacent to the inguinal canal, the superficial lymph nodes with horizontal and vertical tracts subcutaneously. Quénu (1893; *see* Quénu, Edouard) distinguished not only four groups of superficial inguinal nodes (*eponym*: **inguinal lymph nodes of Quénu**), 'supérior-interne' (superomedial), 'supérior externe' (superolateral), 'inférior-interne' (inferomedial) and 'inférior-externe' (inferolateral) but also a 'groupe présaphène', currently known as saphenofemoral lymph nodes. His subdivision has been used in modern studies on the lymphatic drainage of the penis and its sentinel node in penile cancer. Several *eponyms* were given to lymph nodes of the

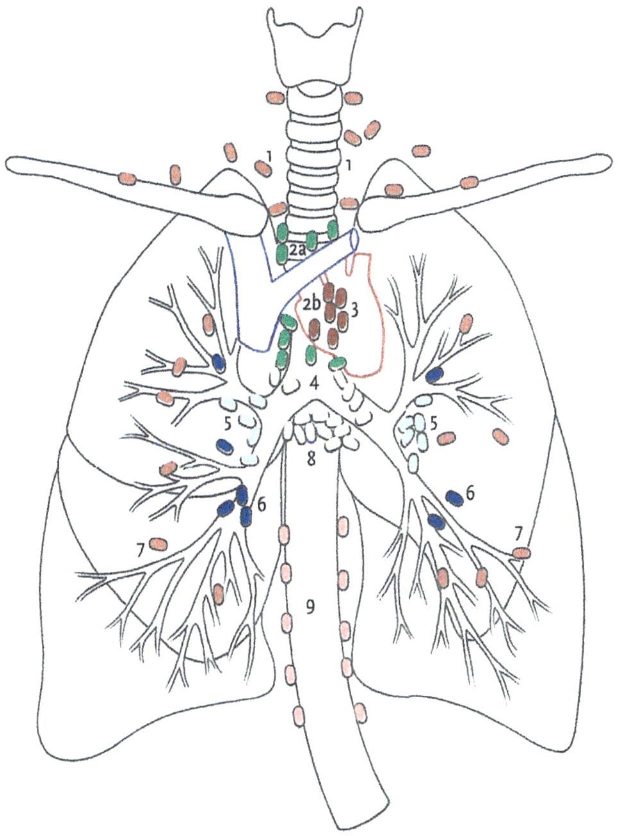

Fig. 26.6 Lymph node map based on a modern subdivision of the lymph nodes of the lungs by the International Association for the Study of Lung Cancer. (After Harisinghani MG, ed. 2013 *Atlas of Lymph Node Anatomy*. Springer, New York, Heidelberg, Dordrecht, London, and ten Donkelaar HJ, Oostra R-J 2014 *Klinische Anatomie en Embryologie*, 4th ed. Springer Media, Houten). *1* Supraclavicular zone with low cervical and supraclavicular lymph nodes, *2*, *3* lymph nodes of the superior mediastinum (*2a* upper and *2b* lower paratracheal; *3* retrotracheal lymph nodes), *4*–*7* N1 lymph nodes (*4* hilar, *5* interlobar, *6* lobar, *7* segmental lymph nodes), *8*, *9* lymph nodes of the inferior mediastinum (*8* subcarinal, *9* paraoesophageal lymph nodes)

pelvic organs: **anorectal lymph nodes of Gerota** (Gerota 1895; *see* Gerota, Dumitru), comparable to the **pararectal lymph nodes of Poirier and Cunéo** (1902). Poirier and Cuéno (1902) also described abdominoaortic lymph nodes, currently known as lumbar lymph nodes.

More recently, emphasis has been given to the **sentinel lymph node** for the **penis** (Cabanas 1977, 1992). The superomedial superficial inguinal lymph node appeared to be the sentinel lymph node for the penis (*eponym*: **lymph node of Cabanas**). The different lymphatic drainage of the penis and scrotum is shown in Fig. 26.9. The skin and preputium of the penis as well as the scrotum drain to the superficial inguinal lymph nodes, and the glans penis and the spongy part of the urethra to the deep inguinal and external iliac lymph nodes. The **plexus of Panizza** is an *eponym* for a plexus of lymph vessels in the lateral fossae of the prepuce (*see* Panizza, Bartolomeo).

26.7 Eponyms on Lymph Nodes of the Lower Limb

Various lymph nodes were described for the lower limb, among others popliteal and tibial lymph nodes (*eponyms*: **articular lymph node of Bardeleben**; von Bardeleben 1896; *see* Bardeleben, Karl von; **'ganglions poplitée'** of Poirier and Cuéno 1902), and an accessory lymph node at the tendinous arch of the soleus muscle (*eponym*: **tibiopopliteal lymph node of Bourgery**; Bourgery 1837–1840; *see* Bourgery, Marc Jean), but are hardly used.

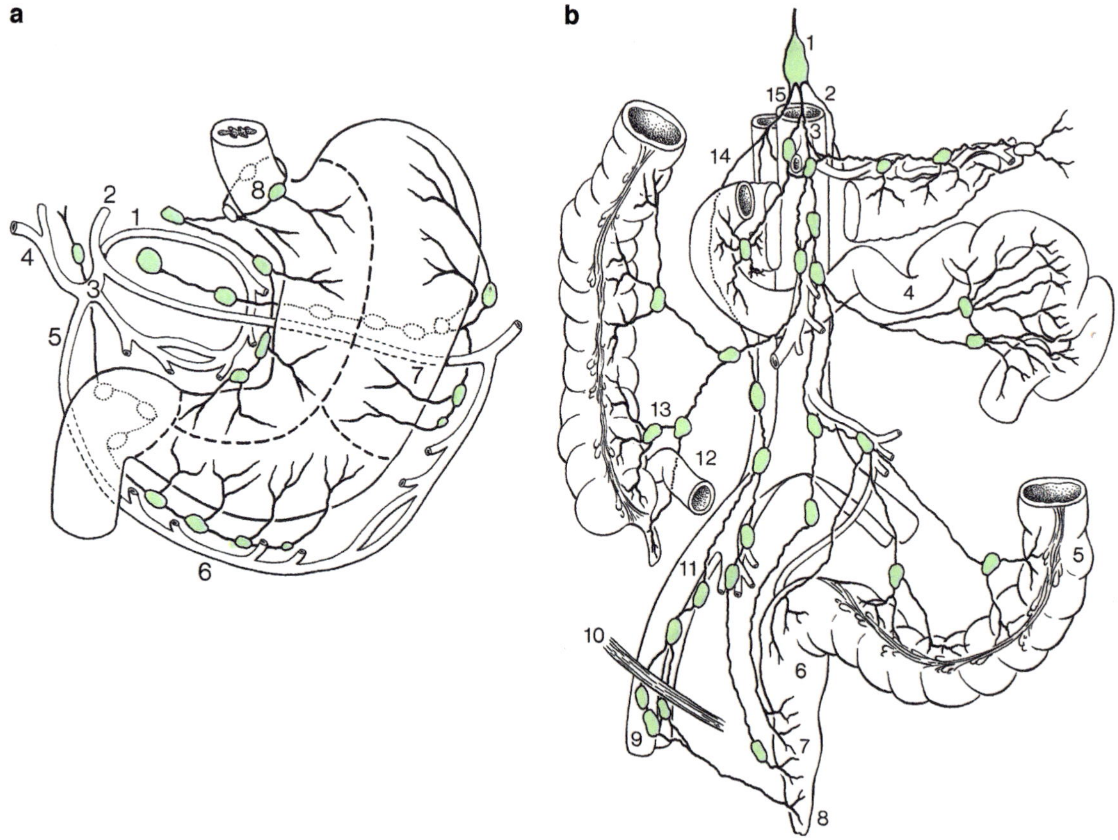

Fig. 26.7 (**a**) Lymphatic drainage of the stomach, *1* left gastric artery with left gastric lymph nodes, *2* coeliac trunk, *3* common hepatic artery, *4* proper hepatic artery, *5* gastroduodenal artery, *6* right gastroomental (gastroepiploic) artery with right gastric lymph nodes, *7* splenic artery with splenic lymph nodes, *8* cardiac lymph nodes; (**b**) lymphatic drainage of the intestines, *1* cysterna chyli, *2* left lumbar trunk, *3* intestinal trunk, *4* jejunum, *5* sigmoid colon, *6* rectum, *7* anal canal, *8* anus, *9* superficial inguinal lymph nodes, *10* inguinal ligament, *11* external and internal iliac lymph nodes, *12* ileum and caecum, *13* ileocaecal lymph nodes, *14* coeliac lymph nodes, *15* right lumbar trunk. (Both after ten Donkelaar HJ, Oostra R-J 2014 *Klinische Anatomie en Embryologie*, 4th ed. Springer Media, Houten)

Fig. 26.8 Lymphatic drainage of the pelvic organs in females. (After ten Donkelaar HJ, Oostra R-J 2014 *Klinische Anatomie en Embryologie*, 4th ed. Springer Media, Houten). *1* Superficial and deep inguinal lymph nodes, *2* external iliac lymph nodes, *3* internal iliac lymph nodes, *4* common iliac lymph nodes, *5* sacral lymph nodes, *6* inferior mesenteric lymph nodes, *7* lumbar lymph nodes

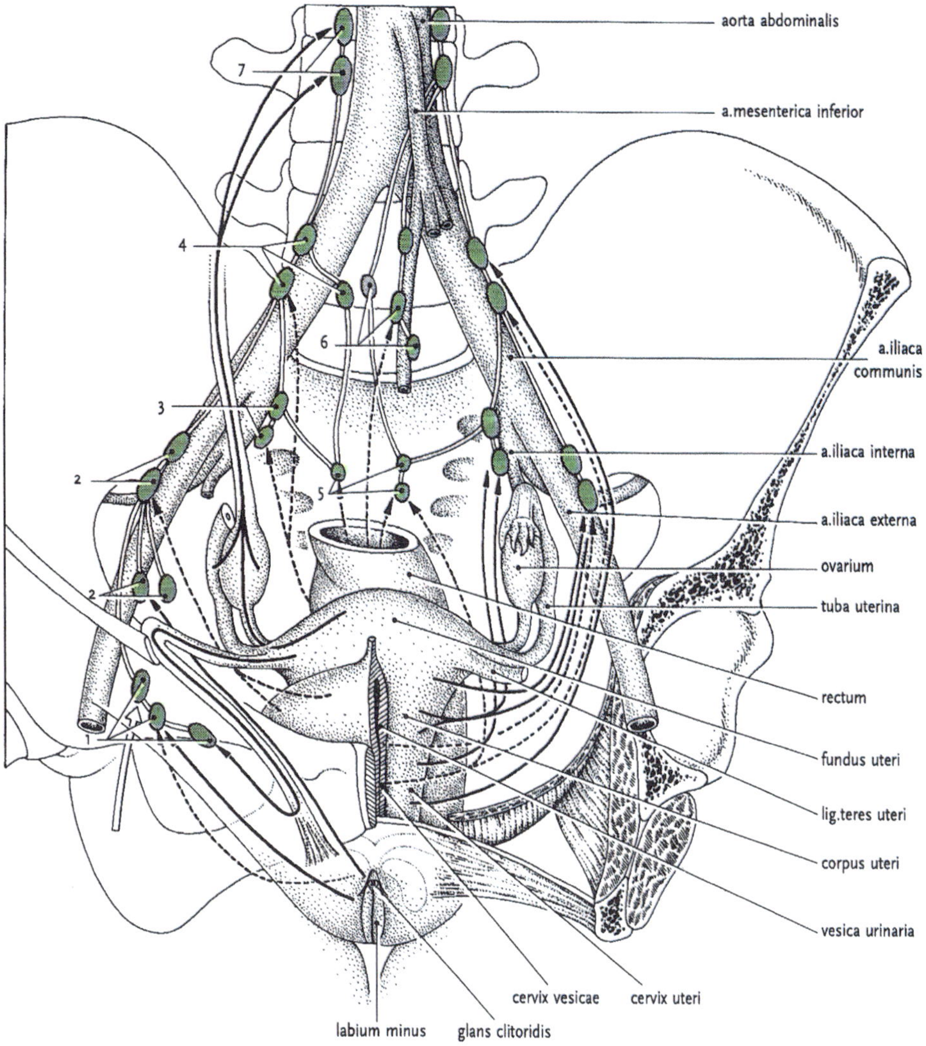

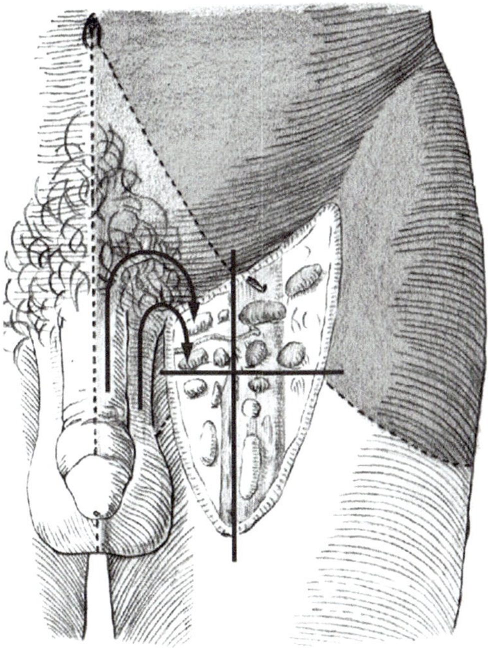

Fig. 26.9 Lymphatic drainage of the penis and scrotum to the superomedial superficial inguinal lymph nodes. (From Martins de Carvalho JP, Patricio BF, Medeiros J, Sampaio FJB, Favorito LA 2011 *Anatomic aspects of the inguinal lymph nodes applied to lymphadenectomy in penile cancer.* Adv Urology, Article ID 952532; Open Access, courtesy Luciano A Favorito, Rio de Janeiro)

References

Arnold F (1880) Ueber das Vorkommen lymphatischen Gewebes in den Lungen. Arch Pathol Anat Physiol Klin Med 80:315–336

Bourgery J-BM (1837–1840) Traité complet de l'anatomie de l'homme comprenant la médecine opératoire. Paris

Cabanas RM (1977) An approach for the treatment of penile carcinoma. Cancer 39:456–466

Cabanas RM (1992) Anatomy and biopsy of sentinel lymph nodes. Urol Clin North Am 19:267–276

Calori L (1850–1855) Tavole anatomiche rappresentanti la struttura del corpo umano. Tipografia Sassi nelle Spaderie, Bologna

Cloquet JG (1825) Manuel d'anatomie descriptive. Paris

Eliška O (2018) Lymfologie. Grada, Prague (in Czech)

Gerota D (1895) Der anorectale Lymphapparat. Sitz Preus Akad Wiss, pp 253–256

Gerota D (1896) Zur Technik der Lymphgefässinjektion. Eine neue Injektionsmasse für Lymphgefässe. Ann Anz 12:216

Harisinghani MG (ed) (2013) Atlas of lymph node anatomy. Springer, New York/Heidelberg/Dordrecht/London

Hollinshead WH (1971) Anatomy for surgeons, Vol 2: The Thorax, Abdomen and Pelvis, 2nd edn. Harper & Row, New York

Küttner A (1897) Über die Lymphgefässe der Zunge, mit Beziehung auf die Verbreitungsweise des Zungencarcinoms. Verh Deutsch Ges Chir Berlin 14:120–123

Mascagni P (1787) Vasorum lymphaticorum corporis humani historia et ichnographia. Pazzini Carli, Siena

Mayo WJ (1928) Metastasis in cancer. Proc Staff Meetings Mayo Clinic 3:227

Oelsner L (1901) Anatomische Untersuchungen über die Lymphwege der Brust mit Bezug auf die Ausbreitung des Mammacarcinoms (Thesis, Breslau). Arch Klin Chir 64:134–158

Pechlin JN (1672) De purgantium medicamentorum facultatibus exercitio nova. Leiden and Amsterdam

Peyer JC (1677) Exercitio anatomico-media de glandulis intestinorum carumque usu et affectionibus. Onophrinus and Waldkirch, Basel

Poirier P, Cuéno B (1902) Système lymphatique. In: Poirier P, Charpy A (eds) Traité d'anatomie humaine, Tome 2. Masson, Paris

Quénu EAVA (1893) Vaisseaux lymphatiques de l'anus. Bull Soc Anat Paris 7:399–409

Robbins KT (1998) Classification of neck dissection: current concepts and future considerations. Otolaryngol Clin N Am 31:639–655

Rosenmüller JC (1805) Chirurgisch-Anatomische Abbildungen für Aertze und Wundärzte. Weimar

Rotter J (1896) Dauerfolge durch ein verbessertes Operationsverfahren der Mammacarcinome. Berl Klin Wochenschr 33(69):99

Rouvière H (1932) Anatomie des lymphatiques de l'homme. Masson, Paris

Sappey MPC (1874–1885) Anatomie, physiologie et pathologie des vaisseaux lymphatiques considérés chez l'homme et les vertébrés. Delahaye and Lecrosnier, Paris

Sorgius W (1880) Ueber die Lymphgefässe der weiblichen Brustdrüße. Thesis, University of Strassburg

Stahr H (1899) Ueber den Lymphapparat des ausseren Ohres. Anat Anz 15:381–387

Suami H, Pan W-R, Mann GB, Taylor GI (2008) The lymphatic anatomy of the breast and its implications for sentinel lymph node biopsy: a human cadaver study. Ann Surg Oncol 15:863–871

Turner-Warwick RT (1959) The lymphatics of the breast. Br J Surg 46:574–582

Vesalius A (1543) De humani corporis fabrica. Oporinus, Basel

Virchow RLU (1856) Die Cellularpathologie in ihrer Begründung auf physiologische und pathologische Gewebelehre. Hirschwald, Berlin

von Bardeleben K (1896) Handbuch der Anatomie des Menschen. Fischer, Jena

27 Anatomical Eponyms in Ophthalmology

Hans J. ten Donkelaar, David Kachlík,
and Johannes R. M. Cruysberg

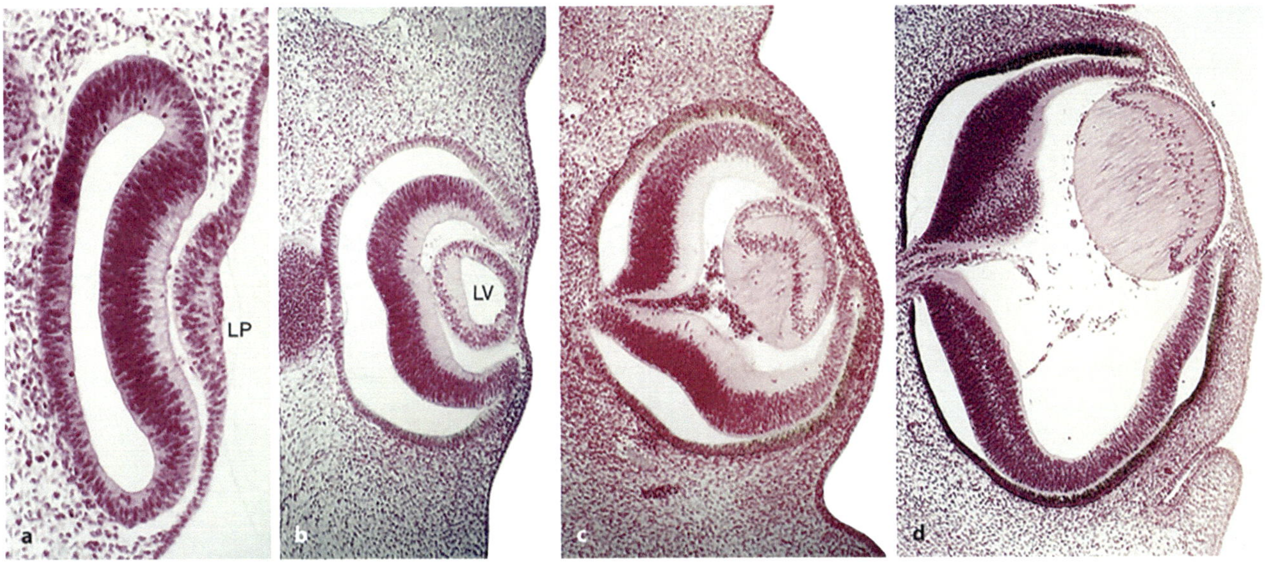

Vignette Series of photomicrographs showing the development of the human eye at Carnegie stages 14 **a**, 16 **b**, 18 **c** and 20 **d**. *LP* lens placode, *LV* lens vesicle. (From ten Donkelaar HJ, Lammens M, Hori A *Clinical Neuroembryology*, 3rd ed. Springer, Cham, 2023)

27.1 Introduction

The **eye** forms with the optic nerve the visual organ. The **vignette** shows the early development of the human eye. This consists of the eyeball and the accessory visual structures, including (Fig. 27.1): (1) the orbital fasciae; (2) the extraocular muscles; (3) the eyebrows; (4) the eyelids; (5) the conjunctiva; and (6) the lacrimal apparatus. For many structures of the eye, eponyms are still in use, especially for certain (extra)ocular muscles and the cornea. The four rectus muscles of the eyeball arise from the common tendinous ring (*eponym*: **anulus of Zinn**; Zinn 1755; *see* Zinn, Johann Friedrich). The deep layer of the levator palpebrae superioris muscle is also known as superior tarsal muscle (*eponym*: **muscle of Müller**; Müller 1856; *see* Müller, Heinrich). Many eponyms were given to the ciliary muscle and its fibres. New imaging techniques such as optical coherence imaging (OCT) allow excellent images of the retina. From the inside outwards, 18 alternating hyperreflective and hyporeflective zones were listed (Staurenghi et al. 2014), including the layers of the retina (zones 3–14) and the vascular layer (zones 15–18). For some of these zones eponyms were revitalized to honour their discoverers. The posterior cortical vitreous represents two zones: zone 1 and a preretinal space zone 2. After a few notes on the development of the eye, *eponyms* for the accessory visual structures, the eyeball and its walls, the retina and the optic disc and nerve will be discussed. For further data on the authors quoted, reference is made to Part II (*The Anatomical Eponyms from A to Z*) and figures in this part.

H. J. ten Donkelaar et al., *An Illustrated Guide to Anatomical Eponyms*, https://doi.org/10.1007/978-3-031-91664-9_27

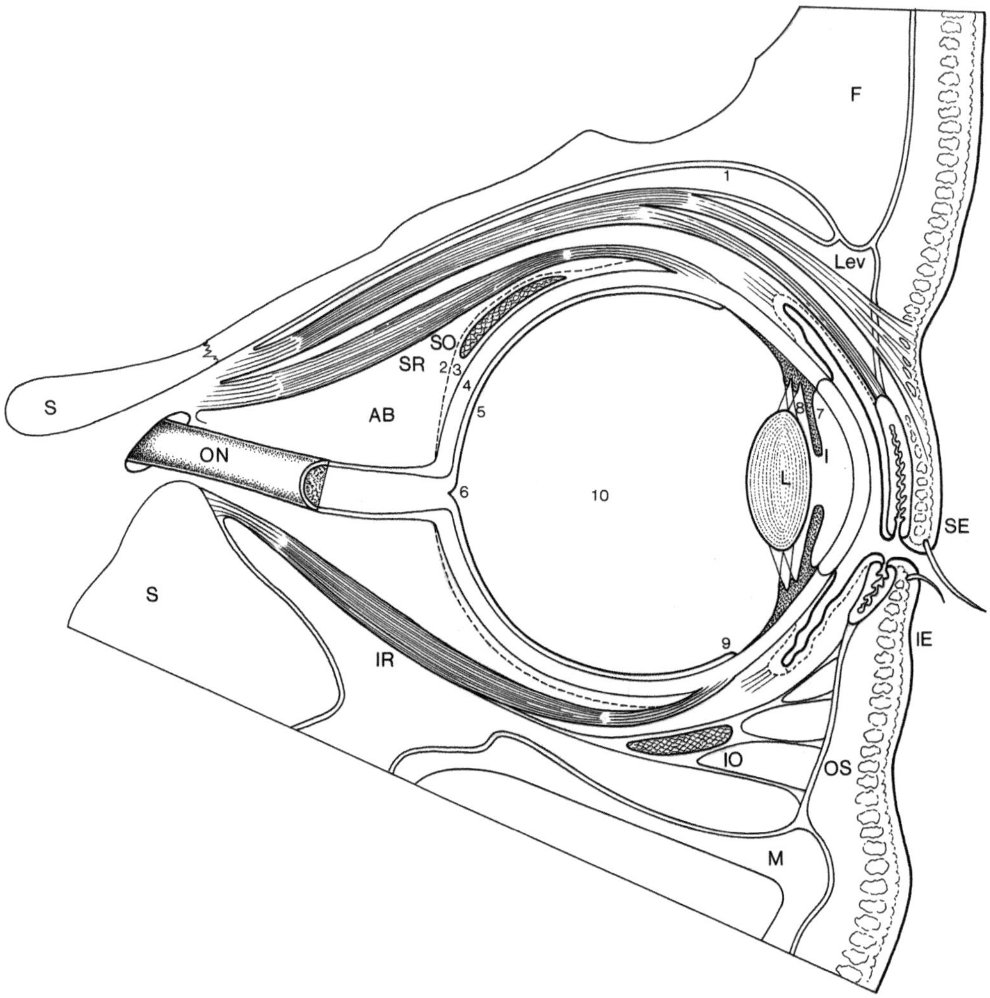

Fig. 27.1 Eye, overview. (From ten Donkelaar HJ, Kachlik D, Tubbs RS *An Illustrated Terminologia Neuroanatomica*. Springer Nature, Cham 2018). *AB* adipose body, *F* frontal bone, *I* iris, *IE* inferior eyelid, *IO, IR* inferior oblique and rectus muscles, *L* lens, *Lev* levator palpebrae superioris muscle, *M* maxilla, *ON* optic nerve, *OS* orbital septum, *S* sphenoid bone, *SE* superior eyelid, *SO* superior oblique muscle, *SR* superior rectus muscle, *1* periorbita, *2* vagina bulbi, *3* episcleral space, *4* sclera, *5* retina, *6* optic disc, *7* anterior chamber, *8* posterior chamber, *9* ora serrata, *10* corpus vitreum

27.2 A Few Notes on the Development of the Eye

In human embryos, the **optic primordium** becomes first visible as the **optic sulcus** appears in each neural fold at Carnegie stage (CS) 10, i.e. at about 4 weeks of development. The right and left primordia are connected by a ridge that will become the optic chiasm. The optic sulcus deepens and forms an evagination at CS 11, leading to the formation of the **optic vesicle** at CS 12 (Fig. 27.2). By 4.5 weeks, the optic vesicle and the adjacent surface ectoderm thicken and form at CS 13 the **retinal disc** and the **lens placode** or **disc**, respectively. The retinal disc becomes invaginated, resulting in a double-layered **optic cup**. This invagination extends partly onto the stalk of the optic cup to form the transient **retinal** or **choroid fissure**, that received the *eponym* **fissure of Schwalbe** (*see* Schwalbe, Gustav Albert). The inverted layer of the optic cup is comparable to the wall of the developing forebrain. The walls of the optic cup form the optic part of the retina and the epithelium of the ciliary body and the iris. The lens placode becomes indented as the **lens pit** at CS 14, which closes at CS 15 as the **lens vesicle**. At the same time, retinal pigment appears in the external layer of the optic cup. The cells of the deep wall of the lens vesicle become elongated and form the primary lens fibres. Anterior cells of the lens form a simple epithelium from which the secondary lens fibres arise that form the bulk of the mature lens. Moreover, the **primary vitreous body** and the **hyaloid artery** develop. The hyaloid artery enters the vitreous cavity through the retinal fissure and supplies the thickened inner layer, the lens vesicle and the intervening mesenchyme. Later, the hyaloid artery disappears within the eye. Its stem forms the central retinal artery. The optic cup is anchored to the forebrain by the **optic stalk** that will form the optic nerve. Axons of retinal ganglion cells grow via the optic stalk to the

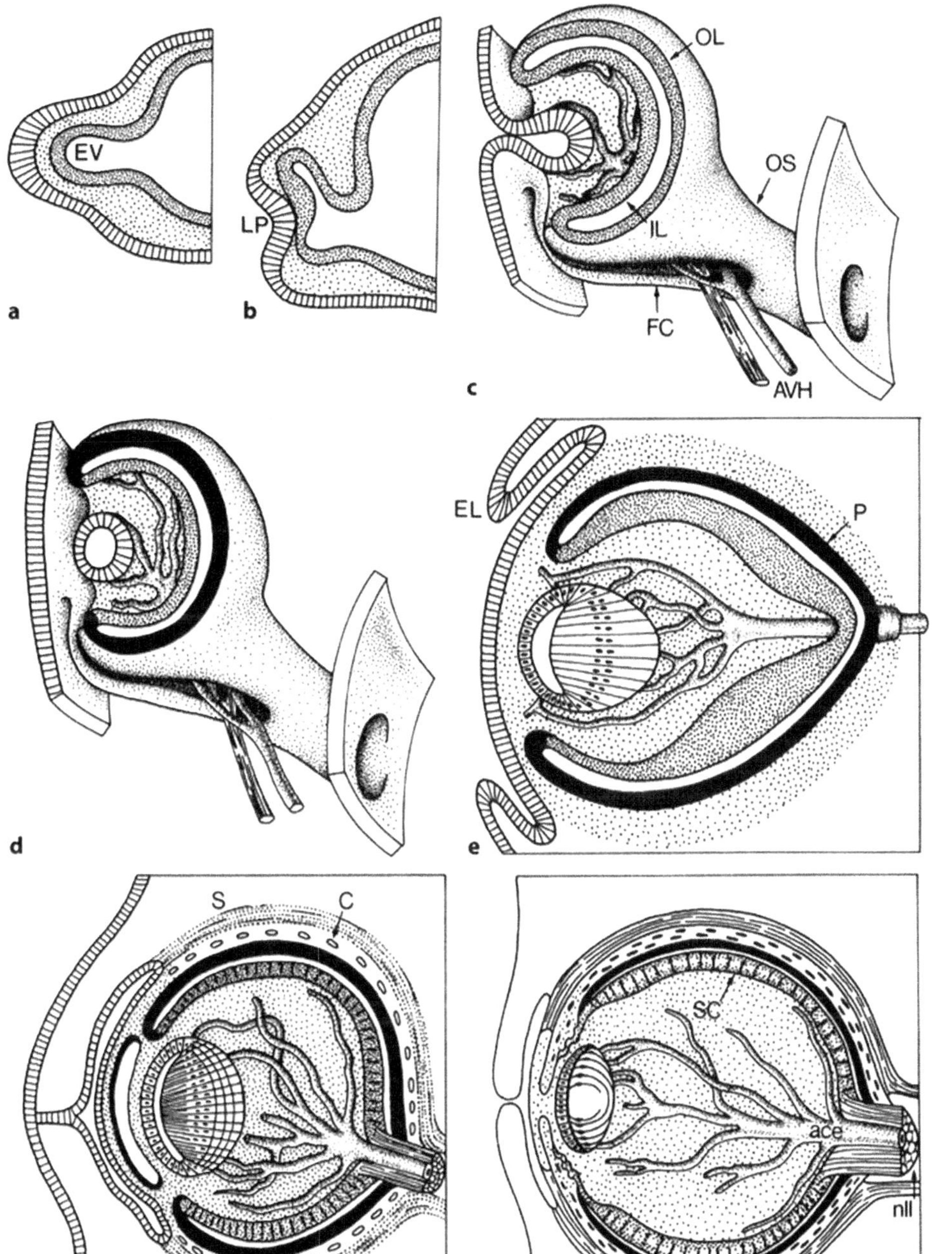

Fig. 27.2 Development of the human eye in the fourth (**a**), fifth (**b**), sixth (**c**), seventh (**d**), eighth (**e**) and ninth week (**f**) and in the seventh month (**g**) of development. *ace* a. centralis retinae, *AVH* hyaloid artery and vein, *C* choroid, *EL* eyelid, *EV* eye vesicle, *FC* fissura choroidea, *IL* inner layer of eye cup, *LP* lens placode, *nII* optic nerve, *OL* outer layer of eye cup, *OS* optic stalk, *P* pigment layer of retina, *S* sclera, *SC* stratum cerebrale of retina. (After Hamilton WJ, Mossman HW *Hamilton, Boyd and Mossman's Human Embryology*, fourth ed. Heffer, Cambridge, 1972, based on reconstructions by Mann IC *The Development of the Human Eye.* Cambridge University Press, London, 1928; from ten Donkelaar HJ, Lammens M, Hori A *Clinical Neuroembryology*, 3rd ed. Springer, Cham, 2023)

brain. The lumen of the optic stalk is gradually obliterated as axons of ganglion cells accumulate in the inner layer of the optic stalk, resulting in the formation of the optic nerve between the sixth and eighth weeks of development.

27.3 The Accessory Visual Structures and Their Eponyms

Eponyms are still in use for the eyelids and the lacrimal apparatus (Fig. 27.3). For the eyelids, eponyms were given to various glands: (1) ciliary glands (*eponym*: **glands of Zeis**; Zeis 1835; *see* Zeis, Eduard); (2) tarsal glands (*eponym*: **glands of Meibom**; Meibom 1666; *see* Meibom, Heinrich); and (3) ciliary or palpebral sweat glands (*eponyms*: **glands of Moll** and **of Sattler**; Moll 1857, *see* Moll, Jacob Antonius; Sattler 1877, *see* Sattler, Hubert). For the lacrimal apparatus, accessory lacrimal glands of the eyelids, mainly of the upper one, were originally described by Krause (*eponym*: **glands of Krause**; Krause 1833; *see* Krause, Karl Friedrich Theodor), but later also received the *eponyms* **glands of Ciaccio**, **Henle**, **WJF Krause**, **Wolfring** and **Terson** (Krause 1879, *see* Krause, Wilhelm Johann Friedrich; Ciaccio 1874, *see* Ciaccio, Giuseppe Vincenzo; Henle

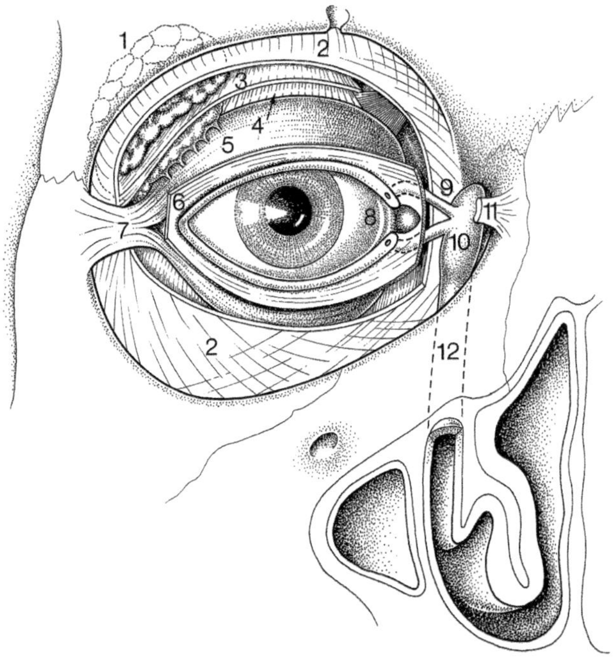

Fig. 27.3 The lacrimal apparatus. (After ten Donkelaar HJ, Oostra R-J *Klinische anatomie en embryologie*, 4th ed. Springer-Media, Houten, 2014). *1* Lacrimal gland, *2* orbital septum, *3* levator palpebrae superioris muscle, *4* superior tarsal muscle, *5* superior tarsus, *6* superior palpebra, *7* lateral palpebral ligament, *8* bulbar conjunctiva and lacus lacrimalis, *9* superior lacrimal canaliculus, *10* lacrimal sac, *11* medial palpebral ligament, *12* nasolacrimal duct

1855–1872, *see* Henle, Friedrich Gustav Jacob; von Wolfring 1872, *see* Wolfring, Ernst von; Terson 1892, *see* Terson, Albert).

Many *eponyms* were given to a lacrimal fold or valve in the past but most of them are not true mucosal valves but merely developmental irregularities in the wall of the nasolacrimal duct or may be based on different swelling states of the cavernous body of the duct (Paulsen 2003). Mucosal folds at the puncta lacrimalia of the lacrimal canaliculi were described as **valves of Foltz** and **of Bochdalek** (Foltz 1862; *see* Foltz, Jean Charles Eugène; Bochdalek 1866; *see* Bochdalek, Vincenz Alexander). The opening of the common lacrimal canaliculus into the lacrimal sac is marked by a prominent upper margin, bearing the *eponym* **valve of Rosenmüller** (Rosenmüller 1805; *see* Rosenmüller, Johann Christian), and also by a prominent lower margin, the **valve of Huschke** (Huschke 1854; *see* Huschke, Emil). A fold between the lacrimal sac and the nasolacrimal duct is known by the *eponyms* **valve of Béraud** (Béraud 1854; *see* Béraud, Bruno Jacques) and **valve of Krause** (Krause 1833). A fold of the lower part of the nasolacrimal duct is known under the *eponyms* **valve of Taillefer** (Taillefer 1826; *see* Taillefer, Louis) and **valve of Hyrtl** (Hyrtl 1847; *see* Hyrtl, Joseph). The slit-like opening of the nasolacrimal duct into the inferior nasal meatus is covered by a mucosal fold called the lacrimal fold or **valve of Bianchi** (1723; *see* Bianchi, Giovanni Battista), **of Hasner** (1850; *see* Hasner, Joseph), and **of Cruveilhier** (1871; *see* Cruveilhier, Jean Baptiste). Other *eponyms* for the lacrimal apparatus are **sinus of Maier** (Maier 1859; *see* Maier, Rudolf) for the fornix of the lacrimal sac, **sinus of Arlt** (1854–1855; *see* Arlt, Carl Ferdinand von) for a dilatation of the lacrimal sac, and **canal of Ferrein** (Ferrein 1749; *see* Ferrein, Antoine) for the lacus lacrimalis.

27.4 The Eyeball and Its Eponyms

The **eyeball** comprises a smaller anterior segment, surrounded by the cornea, and a larger, posterior segment, surrounded by the sclera (Fig. 27.4). Externally, the two segments are separated by a shallow, circular groove, the scleral sulcus, which is situated at the **corneoscleral junction** or **limbus**, the transition between the cornea and the sclera. The eyeball is located in the anterior part of the orbit. Behind the eyeball, the retrobulbar fat pad or orbital fat body is found, which is traversed by the optic nerve. The eyeball is separated from the retrobulbar fat pad by the fascial sheath of the eyeball (*eponym*: **capsule of Tenon**; Tenon 1816; *see* Tenon, Jacques René). The episcleral space between the fascial sheath of the eye and the sclera bears the *eponym* **space**

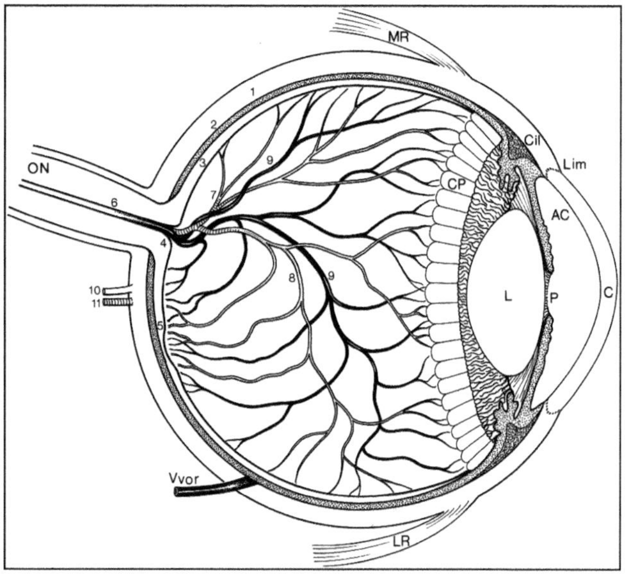

Fig. 27.4 Eye, overview. (From ten Donkelaar HJ, Kachlik D, Tubbs RS *An Illustrated Terminologia Neuroanatomica*. Springer, Cham, 2018). *AC* anterior chamber, *C* cornea, *Cil* ciliary muscle, *CP* ciliary processes, *L* lens, *Lim* corneoscleral junction or limbus corneae, *LR* lateral rectus muscle, *MR* medial rectus muscle, *ON* optic nerve, *P* pupil, *Vvor* vorticose vein; the following structures are indicated by number: *1* sclera, *2* choroid, *3* retina, *4* optic disc, *5* fovea centralis, *6* central retinal vessels, *7* nasal branch of central retinal artery, *8* temporal branch of central retinal artery, *9* temporal branch of central retinal vein, *10* long ciliary nerve, 11 posterior ciliary artery

of Tenon (*see* Fig. 27.1). A thickened portion of the bulbar sheath is continuous with the muscular fasciae of the inferior oblique and inferior rectus muscles, and is known as the suspensory ligament of the eyeball (*eponym*: **suspensory ligament of Lockwood**; Lockwood 1886; *see* Lockwood, Charles Barrett). A condensation of fascial sheaths around the levator palpebrae superioris muscle is known as the superior transverse ligament (*eponym*: **ligament of Whitnall**; Whitnall 1911; *see* Whitnall, Samuel Ernest).

The interior of the eyeball (Fig. 27.5) is formed by: (1) immediately behind the cornea: the anterior chamber of the eyeball, which contains aqueous humor; it is bounded by the iris, which separates it from the posterior chamber; (2) the lens suspended by means of a system of fibres, the ciliary zonule or suspensory ligament of the lens (*eponym*: **suspensory ligament** or **zonule of Zinn**; Zinn 1755; *see* Zinn, Johann Friedrich), which attaches it to the ciliary body; (3) the posterior chamber of the eyeball, which also contains aqueous humor; the two chambers are in continuity through a narrow slit between the lens and the iris (Fig. 27.4); shallow openings into the posterior chamber between the ciliary body and the ciliary zonules are known as the *eponym* **spaces of Kuhnt** (Kuhnt 1898; *see* Kuhnt, Hermann); and (4) the postremal or vitreous chamber, which contains the vitreous body. The *eponym* **ring of Soemmerring** is used for a mass of lenticular fibres between the anterior and posterior parts of the lens capsule, leaving the pupillary area relatively free (*see* Soemmerring, Samuel Thomas von).

The **vitreous body** has many anatomical features with *eponyms*. During development, the hyaloid artery passes through the hyaloid canal (*eponyms*: **canal of Cloquet** and **of Stilling**; Cloquet 1825, *see* Cloquet, Jules Germain; Stilling 1832, *see* Stilling, Benedict). A funnel-shaped dilatation of the optic disc marks the beginning of the hyaloid canal (*eponym*: **area** or **space of Martegiani**; Martegiani 1814; *see* Martegiani, Francesco). A funnel-shaped dilatation of the anterior end of the hyaloid canal behind the lens received the *eponym* **space of Erggelet** (Erggelet 1914; *see* Erggelet, Heinrich). A more lateral, retrolenticular space is known as the *eponym* **space of Berger** (Berger 1887; *see* Berger, Emil). Here, the **ligament of Wieger** is found, an *eponym* for the ring-shaped hyalocapsular ligament, a thickening of the vitreous body attached to the posterior surface of the lens (Wieger 1883; *see* Wieger, Germain). The **line of Egger** is an *eponym* for a circular line, the outer border of the ligament of Wieger (Egger 1888; *see* Egger, Fritz). A frequently seen anterior remnant of the hyaloid artery at the posterior surface of the lens is known as the *eponym* **dot of Mittendorf** (Mittendorf 1881; *see* Mittendorf, William Frederick). At the beginning of the hyaloid artery, the **papilla of Bergmeister** is found, an *eponym* for a small mass of glial tissue found in the hyaloid artery during foetal life, covering the optic disc area (Bergmeister 1877; *see* Bergmeister, Otto). It may persist as a prepapillary membrane. The optic disc is covered by the **meniscus of Kuhnt**, a fibrous plate, forming a sheath around the hyaloid artery where it leaves the optic disc (Kuhnt 1898; *see* Kuhnt, Hermann). In case of incomplete regression, the papilla of Bergmeister persists as a remnant of the fibrous sheath of the hyaloid artery. At the posterior aspect of the vitreous body, the **ring of Weiss** and **of Vogt** is found, an *eponym* for the anular glial tissue attachment of the posterior aspect of the vitreous body to the optic

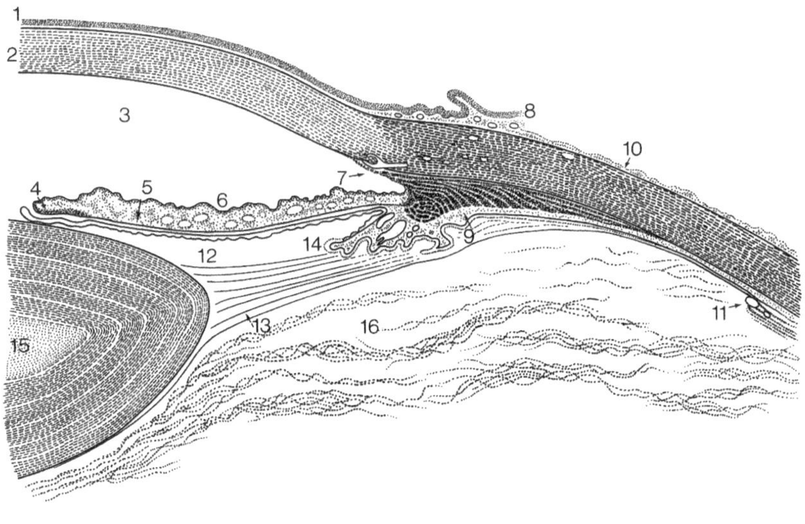

Fig. 27.5 Schematic drawing of part of a meridional section through the anterior part of the eyeball. (After Bloom W, Fawcett DW *A Textbook of Histology*, 10th ed. Saunders, Philadelphia, PA, 1975; from ten Donkelaar HJ, Kachlik D, Tubbs RS *An Illustrated Terminologia Neuroanatomica*. Springer, Cham, 2018). *1* anterior corneal epithelium, *2* cornea, *3* anterior chamber, *4* sphincter pupillae muscle, *5* dilator pupillae muscle, *6* iris, *7* scleral venous sinus, *8* bulbar conjunctiva, *9* ciliary muscle, *10* sclera, *11* ora serrata, *12* posterior chamber, *13* ciliary zonules, *14* ciliary process, *15* lens, *16* vitreous body

nerve, visible as a partial or complete ring after complete detachment of the vitreous body (Weiss 1897; *see* Weiss, Leopold; Vogt 1930; *see* Vogt, Alfred).

27.4.1 The Walls of the Eyeball

The wall of the eyeball consists of three tissue layers, all of which have a different structure in the anterior and posterior halves of the eyeball: (1) the fibrous layer of the eyeball; (2) the vascular layer of the eyeball or uvea; and (3) the inner layer of the eyeball. The eponyms on the retina will be described in Sect. 27.5.

27.4.2 The Fibrous Layer

The **fibrous layer** of the **eyeball** is the outer protective capsule, which entirely encloses the other components of the eye. It consists of the cornea and the sclera, which are connected by a transitional zone, termed the corneoscleral junction or limbus. The **cornea** is a transparent, avascular membrane and is resistant to mechanical deformation. Histologically, the cornea consists of five layers, from the outside inwards (Fig. 27.6a): (1) the anterior epithelium of the cornea or corneal epithelium; (2) the anterior limiting membrane of the cornea (*eponym*: **membrane of Bowman**; Bowman 1849; *see* Bowman, William); (3) the substantia propria or stroma of the cornea; (4) the posterior limiting membrane of the cornea (*eponym*: **membrane of Descemet**; Descemet 1758, *see* Descemet, Jean; an earlier description was given by Duddell in 1729 (*see* Duddell, Benedict) and one later by **Demours** in 1767; *see* Demours, Pierre); and (5) the posterior epithelium of the cornea or corneal endothelium. Recently, Dua et al. (2013) described a preposterior limiting layer, anterior to the posterior limiting membrane of Descemet as **layer of Dua** (*see* Dua, Harminder S). The **tendinous ring of Dollinger** is an *eponym* for an elastic ring around the periphery of the cornea formed by a thickening of the posterior limiting membrane (Dollinger 1817; *see* Dollinger, Johann Ignaz Josef). **Corneal corpuscles** between the fibrous layers of the cornea were described by Toynbee (*see* Toynbee, Joseph) and Virchow (*see* Virchow, Rudolf Ludwig Karl).

The **sclera** is a strong, tough membrane, which surrounds the posterior five sixths of the eyeball (Fig. 27.6b). It is pierced posteriorly by the optic nerve. The external two thirds of the sclera continue posteriorly into the dural sheath around the nerve. Its innermost part, the cribrosal layer of the sclera, contains numerous small holes through which the individual fascicles of the optic nerve pass. Histologically, although of rather uniform structure, three layers can be distinguished from the outside inwards: (1) the episcleral layer of the sclera; (2) the substantia propria or stroma of the sclera; and (3) the suprachoroid layer of the sclera.

The **corneoscleral junction** or **corneal limbus** forms a 1.5–2 mm wide transitional zone between the cornea and the sclera (Fig. 27.7). Its constituent structures are of utmost importance for the maintenance of intraocular pressure, since the removal of aqueous humor occurs here. Its anterior limit is known as the corneolimbal junction. Here, the less curved sclera meets the more convex cornea, resulting in the formation of an external shallow groove, the scleral sulcus. At the inner side of the limbus, an internal scleral sulcus is entirely filled by the trabecular meshwork and the scleral venous sinus (*eponyms*: **canal of Fontana**, **of Lauth** and **of Schlemm**; Fontana 1765, *see* Fontana, Felice Gaspard Ferdinand; Lauth 1829, *see* Lauth, Ernest Alexandre; Schlemm 1830, *see* Schlemm, Friedrich; mostly Schlemm gets the credit). Here, the sclera forms a circular projection, the scleral spur. The outflow of the aqueous humor occurs at the periphery of the anterior chamber of the eyeball in a narrow angle between the iris and the cornea, the iridocorneal angle. Here, the corneal limbus consists of a network of anastomosing lamellae and trabeculae, known as the trabecular meshwork (*eponym*: **spaces of Fontana**; Fontana 1765), also known as the pectinate ligament of the iridocorneal angle. It has two parts: (1) a uveal part; and (2) a corneoscleral part, the main part. A small venous plexus between the canal of Schlemm and the spaces of Fontana bears the *eponym* **plexus of Leber** (*see* Leber, Theodor Karl Gustav von). Aqueous veins (*eponym*: **veins of Ascher**; Ascher 1942; *see* Ascher, Karl) collect the aqueous humor from the scleral venous sinus and drain into the episcleral veins.

27.4.3 The Vascular Layer

The **vascular layer** of the **eyeball** or **uvea** is a highly vascularized, pigment-rich layer. Its posterior, larger part is the **choroid** (*choroidea*) with the following layers, from the outside inwards (Fig. 27.8a):

1. The suprachoroidal layer, which contains the choroidal melanocytes and corresponds to the OCT zone 18 at the outer choroid (Staurenghi et al. 2014); it is separated from the next layer by the perichoroidal space;
2. the vascular layer of the choroid or vascular choroidal layer (*eponym*: **vascular layer of Haller**; von Haller 1746–1752; *see* Haller, Albrecht von), the OCT zone 17, a thick layer of oval-shaped hyperreflective profiles with hyporeflective cores in the outer choroid;
3. the recently rediscovered intervascular layer (*eponym*: **layer of Sattler**; Sattler 1876; *see* Sattler, Hubert), the OCT zone 16, a thick layer of round or oval-shaped

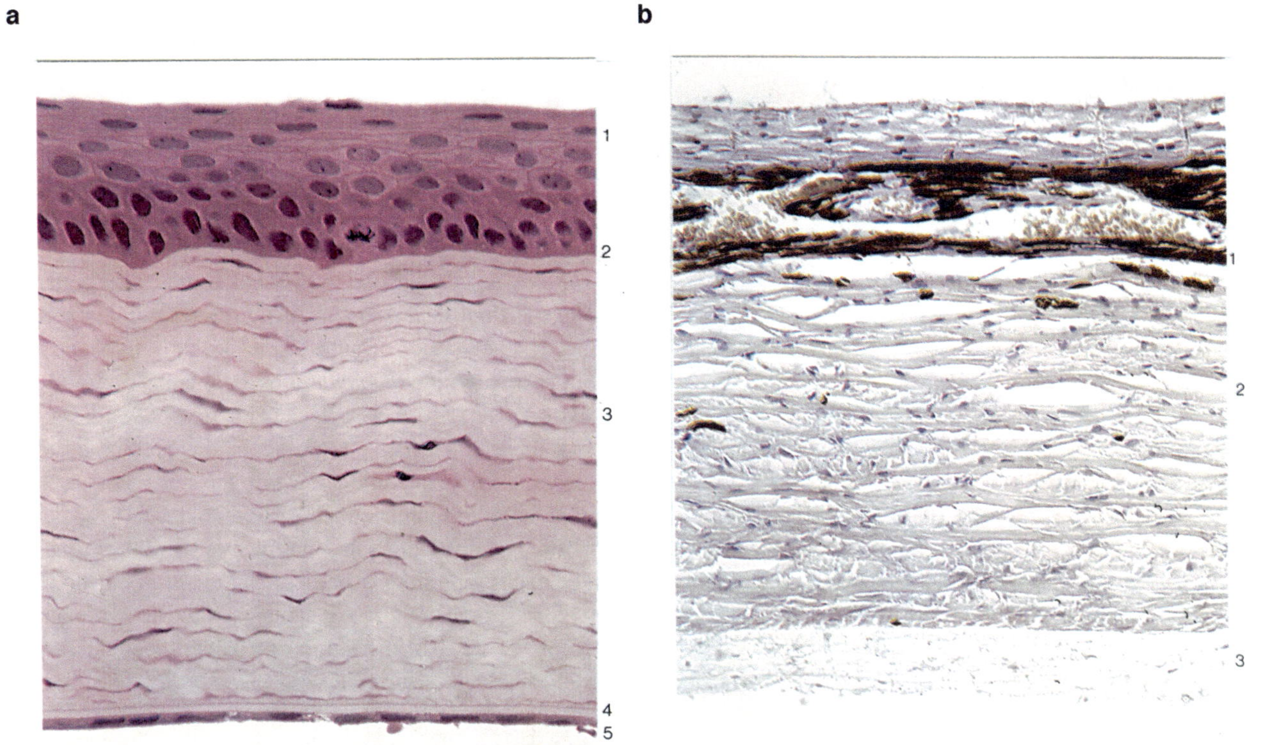

Fig. 27.6 Histology of (**a**) a hematoxylin-eosin-stained section of part of the cornea from a guinea pig. *1* anterior epithelium of cornea, *2* anterior limiting membrane of cornea (membrane of Bowman), *3* substantia propria of cornea, *4* posterior limiting membrane of cornea (membrane of Descemet), *5* posterior epithelium of cornea; and (**b**) a hematoxylin-eosin-stained section of part of a meridional section through the sclera. (Both from Brüel A, Christensen EI, Tranum-Jensen J, Qvortrup K, Geneser F 2012 *Geneser's Histologi*. Munksgaard, Copenhagen; with permission). *1* suprachoroidal layer, *2* substantia propria, *3* episcleral layer

hyperreflective profiles with hyporeflective cores in the mid-choroid;

4. the choroidocapillary layer (*eponym*: **membrane of Ruysch**; Ruysch 1744; *see* Ruysch, Frederik), the OCT zone 15, a thin layer of moderate reflectivity in the inner choroid; and
5. the basal complex (*eponym*: **membrane of Bruch**; Bruch 1844; *see* Bruch, Carl Wilhelm Ludwig), the OCT zone 14, a hyperreflective band, with the following sublayers: (a) an inner basal layer; (b) an inner fibrous layer; (c) an elastic layer; (d) an outer fibrous layer; and (e) the basal lamina of the capillaries of the choroid. In the OCT nomenclature, zone 14 is described as the RPE/Bruch membrane complex, RPE being the retinal pigment epithelium. Occasionally, it can be divided into more than one band (Staurenghi et al. 2014).

The anterior part of the vascular layer is specialized as the ciliary body and the iris. The **ciliary body** extends as a ring from the ora serrata anteriorly to the root of the iris. Its external surface faces the sclera and its internal surface the vitreous body. The ciliary body (Fig. 27.8b) consists of a posterior region, the **ciliary ring**, and an anterior region, the **corona ciliaris**. The corona ciliaris has 70–80 meridionally running folds, the ciliary processes, whose fibres form the zonule of Zinn. Between the processes are smaller, linear ridges, the ciliary folds. Between the ciliary body and the vitreous body, a potential space is known with the *eponyms* **canal of Hannover** (Hannover 1845; *see* Hannover, Adolph) or **space of Cloquet** and **of Petit** (Cloquet 1818, *see* Cloquet, Jules Germain; Petit 1726 *see* Petit, François Pourfour du). The ciliary body consists of the following layers:

1. The supraciliary lamina, which resembles the suprachoroidal layer of the sclera and forms a transitional zone between the ciliary body and the sclera; this layer is traversed by the long posterior ciliary arteries and the anterior ciliary arteries;
2. the ciliary muscle (*eponym*: **muscle of Riolan**; *see* Riolan, Jean), consisting of meridional or longitudinal fibres (*eponyms*: **fibres of Brücke**, **of Bowman** and **of Crampton**; Brücke 1847, *see* Brücke, Ernst Wilhelm von; Bowman 1849, *see* Bowman, William; Crampton 1805, *see* Crampton, Philip; here, usually Brücke gets the

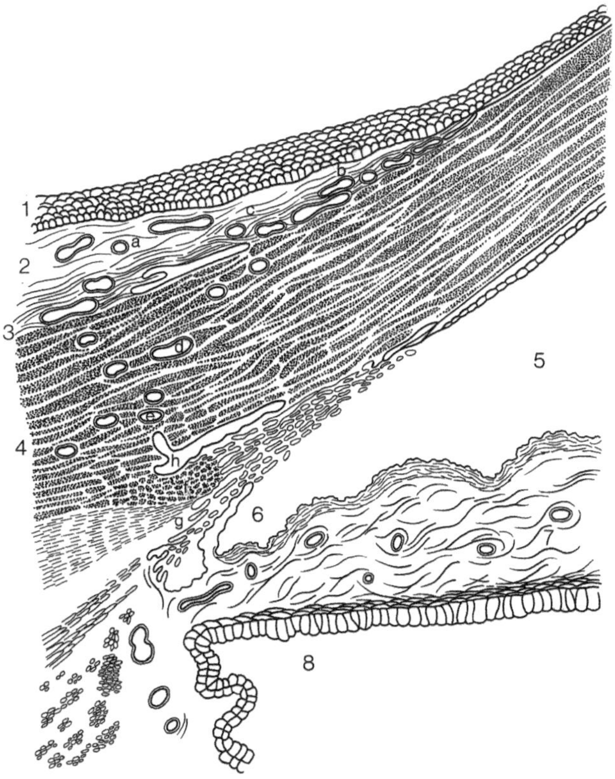

Fig. 27.7 Schematic drawing of the organization of the corneoscleral junction or limbus corneae and the iridocorneal angle. (After Hogan MJ, Alvarado JA, Weddell JE *Histology of the Human Eye*. Saunders, Philadelphia, 1971; from ten Donkelaar, HJ, Kachlik D, Tubbs RS *An Illustrated Terminologia Neuroanatomica*. Springer, Cham, 2018). *1* conjunctival epithelium, *2* conjunctival connective tissue with: *a* conjunctival vessels, *b* peripheral corneal arcades of vessels, *c* episcleral vessels, *d* intrascleral vessels, *e* deep scleral vessels, *f* scleral spur, *g* ciliary muscle, *h* scleral venous sinus (canal of Schlemm), *3* episcleral layer, *4* limbal (corneoscleral) stroma, *5* anterior chamber, *6* iridocorneal angle, *7* iris, *8* posterior chamber

credit), radial fibres (*eponym*: **fibres of Rouget**; Rouguet 1856; *see* Rouguet, Charles Marie Benjamin) and circular fibres (*eponym*: **fibres of Müller**; Müller 1857; *see* Müller, Heinrich);

3. the ciliary stroma, mainly containing blood vessels;
4. the basal layer, the continuation of the corresponding layer in the choroid; and
5. the ciliary epithelium or ciliary part of the retina, composed of an outer, pigmented epithelium and an inner, nonpigmented epithelium.

The **iris** is the most anterior part of the vascular layer of the eyeball. The pupil is a dark, round opening in the centre of the iris. The iris has a diameter of about 12 mm, a peripheral margin, the ciliary margin, a central margin, the pupillary margin, anterior and posterior surfaces and outer and inner borders. Its anterior layer received the *eponym* **membrane of Zinn** (*see* Zinn, Johann Gottfried). The anterior surface of the iris can be divided into two zones by an irregular, serrated, circular line, the collarette, about 1.5 mm from the pupillary margin. In the narrower inner zone, the pupillary zone, trabeculae are present, separated by depressions, the pupillary crypts. The outer zone, the ciliary zone, shows radial ridges, which are caused by radially oriented, underlying blood vessels in the stroma of the iris. Peripheral crypts of the iris (*eponym*: **crypts of Fuchs**; Fuchs 1885; *see* Fuchs, Ernst) are seen adjacent to the collarette. The collarette marks the line of attachment of the foetal pupillary membrane. In foetuses, the pupil is closed by a thin, vascular pupillary membrane (*eponym*: **membrane of Wachendorff**; Wachendorff 1740; *see* Wachendorff, Eberhard Jacob). About the sixth month of gestation, the membrane starts absorption from the centre towards the periphery; at birth only scattered fragments remain.

Histologically, the iris has the following layers (Fig. 27.9): (1) the anterior limiting or border layer of the iris, a modification of the stroma of the iris; it contains fibrocytes and melanocytes of the iris; (2) the stroma of the iris, composed of a nonvascular layer, which contains the sphincter pupillae muscle, and a vascular layer with the major and minor arterial circles of the iris; the clump cells (*eponym*: **cells of Koganei**; Koganei 1885; *see* Koganei, Yoshikiyo) are thought to represent histiocytes; (3) the epithelium of the iris, composed of an anterior, nonpigmented epithelium and a posterior, pigmented epithelium, which together form the anterior, light-insensible part of the retina, the iridial part of the retina. The anterior nonpigmented epithelium contains the dilator pupillae muscle (*eponym*: **muscle of Grynfeltt**; Grynfeltt 1899, *see* Grynfeltt, Joseph Casimir, but earlier described by von Kölliker 1852; see Kölliker, Rudolf Albert von), whereas the posterior epithelium consists of a single layer of columnar cells, the pigment cells.

27.5 Eponyms on the Retina

The **inner layer** of the **eyeball** is formed by the retina. Embryologically, the retina differentiates from the two layers of the optic vesicle (Fig. 27.2; ten Donkelaar et al. 2023). The thinner outer layer of the optic cup becomes the **pigmented layer** of the retina or **retinal pigment epithelium**. The inner layer can soon be divided into a thicker nine-layered posterior part, the **optic part of the retina**, which develops into the visual receptive part of the retina, and a thinner, one-layered part, the **nonvisual part**. This part does not develop photoreceptive cells and becomes subdivided into a posterior part, the **ciliary part**, and an anterior part firmly fused with the outer pigmented layer to form the **iridial part**. The cavity of the optic vesicle is soon occluded by the apposition of its inner layer to the pigmented layer. The **intraretinal space**, separating the two layers, is obliterated as the retina develops, but the two layers never fuse firmly.

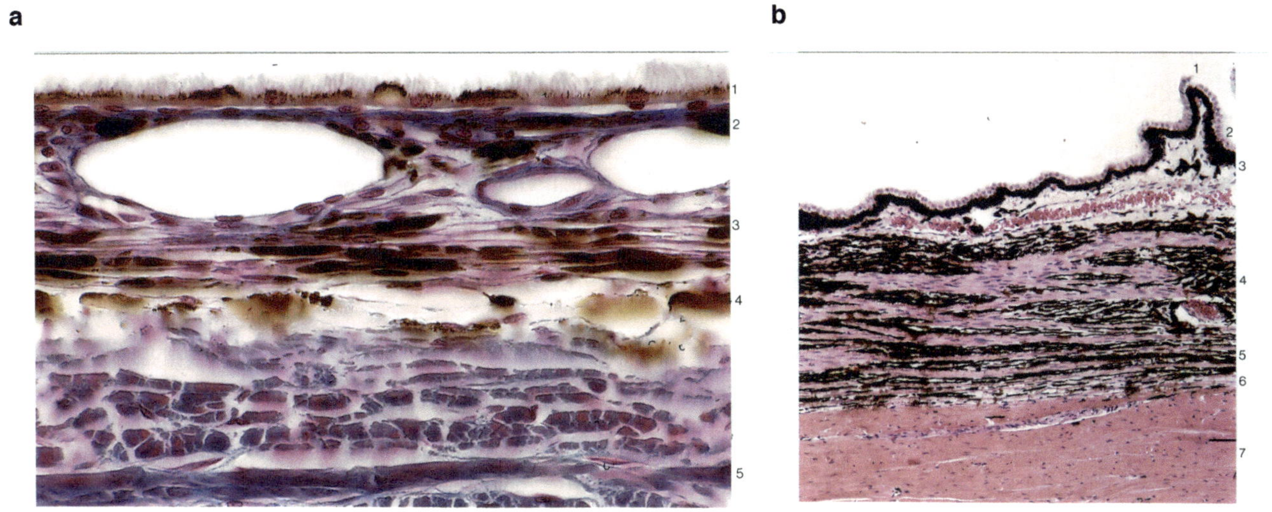

Fig. 27.8 Histology of (**a**) the choroid: *1* basal complex (membrane of Bruch), *2* intervascular layer of Sattler with above located choroidocapillary layer (membrane of Ruysch), and below the vascular layer of von Haller with its large venules, *3* substantia propria, *4* suprachoroidal layer, *5* sclera; and (**b**) the ciliary body. (Both from Brüel A, Christensen EI, Tranum-Jensen J, Qvortrup K, Geneser F 2012 *Geneser's Histologi*. Munksgaard, Copenhagen; with permission). *1* ciliary process, *2* ciliary epithelium, *3* basal complex (membrane of Bruch), *4* ciliary stroma, *5* ciliary muscle, *6* supraciliary layer, *7* sclera

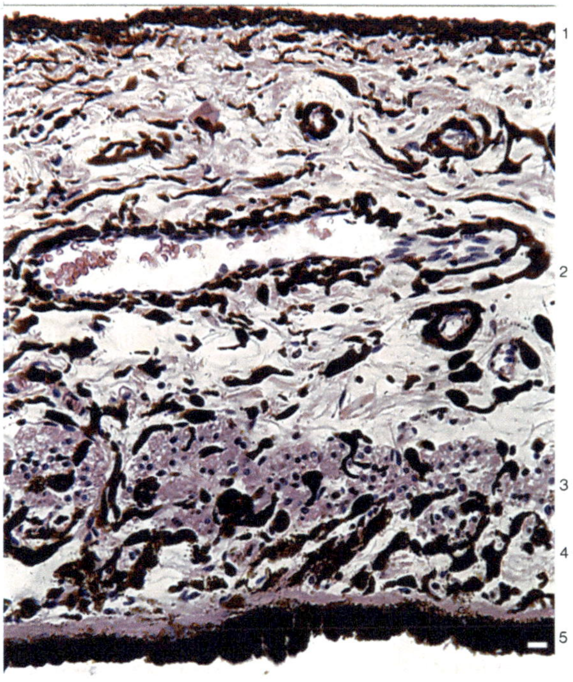

Fig. 27.9 Histology of the iris. (From Brüel A, Christensen EI, Tranum-Jensen J, Qvortrup K, Geneser F 2012 *Geneser's Histologi*. Munksgaard, Copenhagen; with permission). *1* anterior limiting layer, *2* stroma of iris, *3* sphincter pupillae muscle, *4* anterior, nonpigmented epithelium of iris with dilator pupillae muscle, *5* posterior, pigmented epithelium of iris

The differentiation of the neural retina takes place between the sixth week and the eighth month. By the fifth week, the first of a series of shifts in nuclear arrangement occurs, resulting in the formation of an inner layer devoid of cells, and an outer nuclear layer. About the middle of the sixth week, cells of the outer nuclear layer migrate centrally and form an inner cellular zone. Between the two layers the transient layer of Chievitz (*eponym*: **layer of Chievitz**; Chievitz 1887; *see* Chievitz, Johan Henrik) appears. In the third month, the ganglion cells arise from the inner neuroblastic layer. As additional cells migrate from the outer neuroblastic layer to the inner zone, the layer of Chievitz disappears. Now, three nuclear zones can be distinguished separated by fibrous layers. From within outwards, these nuclear zones are the ganglion cell layer, the inner nuclear layer and the outer nuclear layer. The majority of the cells of the inner nuclear layer give rise to the bipolar neurons, which relay impulses from the rod cells or rods and the cone cells or cones to the ganglion cells. The cells in the outer layer form the rods and cones. Two parts of the retina deserve to be emphasized: (1) the macula lutea (*eponym*: **spot of Soemmerring**; *see* Soemmerring, Samuel Thomas von), with its centre, the fovea centralis, the deepest part of which contains only cones; and (2) the optic disc or papilla nervi optici, the place where the optic nerve leaves the eyeball and the central retinal artery enters it (*see* Sect. 27.6).

Externally, the retina is firmly attached to the choroid since the basal complex of the choroid (the membrane of Bruch) forms also the basement membrane of the retinal pigment epithelium. In the OCT nomenclature, the term RPE/Bruch membrane complex is used, a hyperreflective band (zone 14). The neural layer is only loosely attached to the pigment epithelium, except at the ora serrata and the optic disc. It forms the optic part of the retina. In front of the ora serrata, the neural layer is reduced to a simple epithelium, which with the pigment layer covers the ciliary body and the posterior surface of the iris as the ciliary part of the retina and iridial part of the retina, together known as the nonvisual retina.

Histologically, ten well-defined layers of the retina can be distinguished (Fig. 27.10), described from the outside inwards as in the anatomical nomenclature or as with modern imaging techniques such as optical coherence tomography (OCT) from the inside of the retina outwards (Staurenghi et al. 2014). In the OCT nomenclature (Fig. 27.11), 18 zones are listed, progressing from the innermost to the outermost OCT-determined layers of the retina, but including the vascular layer (zones 18–15), and starting with the posterior cortical vitreous as the hyperreflective zone 1, and a preretinal space as the hyporeflective zone 2. The inner limiting layer is not numbered. In the TNA (2017, and digital version of 2022), the following layers are distinguished:

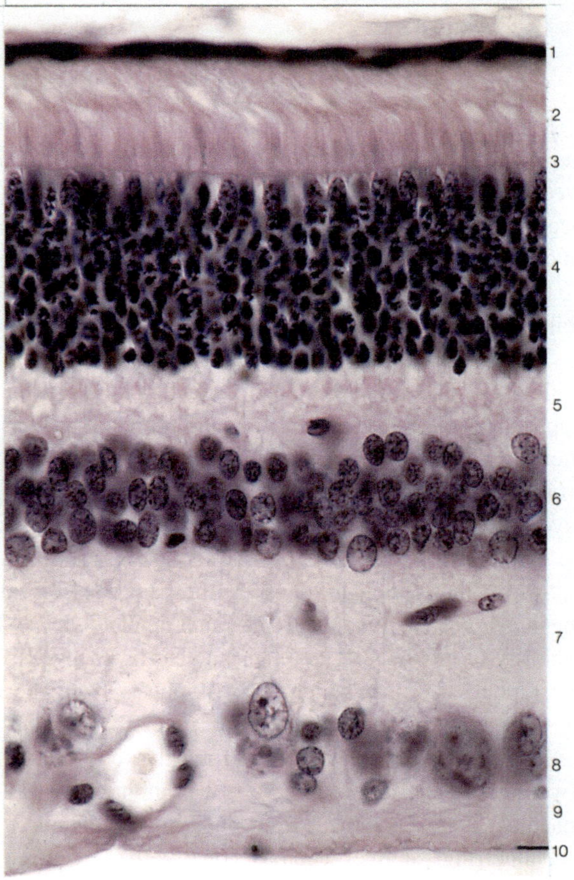

Fig. 27.10 The retina. (From Brüel A, Christensen EI, Tranum-Jensen J, Qvortrup K, Geneser F 2012 *Genesers histologi*. Munksgaard, Copenhagen; with permission). *1* pigment layer of retina, 2 layer of rods and cones, 3 outer limiting layer, 4 outer nuclear layer, 5 outer plexiform layer, 6 inner nuclear layer, 7 inner plexiform layer, 8 ganglionic layer, 9 layer of nerve fibres, 10 inner limiting layer

1. The pigmented layer or retinal pigment epithelial layer, clinically known as retinal pigment epithelium (RPE). In OCT, zone 14 is formed by the retinal pigment epithelial layer and the basal complex or membrane of Bruch, together described as the RPE/Bruch complex. Zone 13 is the interdigitation zone described with the *eponym* **membrane of Verhoeff** (Verhoeff 1903; *see* Verhoeff, Frederick) and as cone interdigitation with the RPE (Staurenghi et al. 2014).
2. The layer of inner and outer segments or layer of rods and cones; here, the outer and inner segments of the rod cell or rod and cone cell or cone are found. Both types of photoreceptor cells comprise a soma or cell body and a rod or cone, respectively. This layer includes the OCT hyperreflective zone 12 (outer segments of photoreceptors), the hyperreflective zone 11 (ellipsoid zone) and the hyporeflective zone 10 (myoid zone).
3. The outer or external limiting layer (OCT hyperreflective zone 9) is an exceedingly thin, fenestrated membrane, which delimits the rod and cone layer from the outer nuclear layer. It is not a true membrane but consists of contacts between the cell membranes of rods and cones and of the radial glial cells (*eponym*: **cells of Müller**; Müller 1856; *see* Müller, Heinrich). Through the fenestrations in the membrane, the most proximal part of the inner segments of the rods and cones pass and continue into the cell bodies of the rod and cone cells in the outer nuclear layer.
4. The outer or external nuclear layer (OCT zone 8.1) contains the cell bodies of the rod and cone cells. Their outer fibres connect the rods and cones, respectively with their somata. An inner fibre is the axonal process of the rod and cone cells, and has the character of an axon, which terminates in the outer plexiform layer. The axonal processes are also known as the nerve fibre layer of Henle (*eponym*: **fibre layer** or **fibres of Henle**; Henle 1855–1872, *see* Henle, Ferdinand Gustav Jacob; OCT zone 8.2).
5. In the outer or external plexiform layer (OCT hyporeflective zone 7), the rod and cone axons form synapses with bipolar and horizontal neurons. Histologically, three zones can be distinguished: an external zone receiving the rod and cone axons, a middle zone with the rod and cone axonal processes, the rod spherule and the

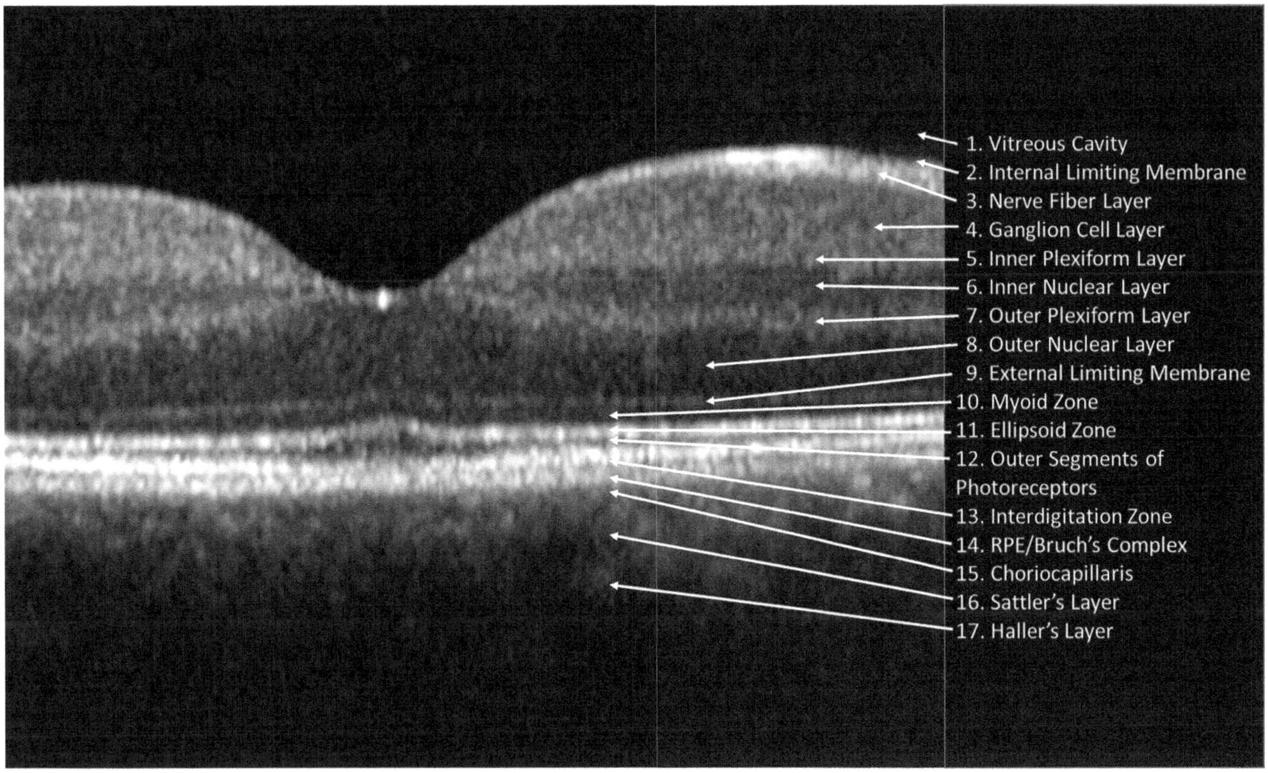

Fig. 27.11 OCT of a normal retina, showing its layers. (*See* text for explanation; courtesy Thomas Theelen, Nijmegen)

cone pedicle, and an internal zone containing the branched processes of bipolar, horizontal and radial glial cells.

6. The inner or internal nuclear layer (OCT zone 6) contains four cell types: the horizontal neurons, the bipolar neurons, the amacrine cells and the radial glial cells (*eponym*: **cells of Müller**; Müller 1856).
7. The inner or internal plexiform layer (OCT hyperreflective zone 5) is a dense network of the processes of amacrine cells, bipolar neuron axons and retinal ganglion cell dendrites.
8. The ganglionic layer (OCT hyporeflective zone 4) contains the cell bodies of the retinal ganglion cells.
9. The layer of nerve fibres (OCT hyperreflective zone 3) is formed by the axons of the retinal ganglion cells, which from all parts of the retina are heading for the optic disc.
10. The inner or internal limiting layer consists of the internal expanded termination of the radial glial cells and the associated basement membrane.

27.6 Eponyms on the Optic Disc and the Optic Nerve

At the **optic disc**, the optic nerve leaves the eyeball (Fig. 27.12). The optic disc is located about 3 mm nasally to the posterior pole of the eyeball. The old designation **papilla of the optic nerve** is misleading, since only the edge of the disc protrudes. The more central part is excavated as the **optic disc excavation** or **excavation of** the **optic nerve head**. Here, the central retinal artery and vein enter or leave, respectively. At the optic disc, the **intermediary tissue of Kuhnt** is found, an *eponym* for a glial spezialization at the optic disc. It surrounds the intraocular part of the optic nerve and acts as a barrier between the optic nerve and the retina (Kuhnt 1898; *see* Kuhnt, Hermann). The head of the optic nerve is covered by the **membrane of Elschnig**, an *eponym* for the internal limiting membrane, isolating the ganglion cells from the surface of the vitreous body (Elschnig 1901; *see* Elschnig, Anton). The **scleral ring of Elschnig** is an *eponym* for a white circular band visible on the fundus of the eye, formed by an extension of the sclera between the choroid and the optic nerve, separating the intrapapillary area of the optic disc from the peripapillary area. Two border tissues are found here (Barishak 2001): (1) the **border tissue of Jacoby**, present at the circumference of the scleral layer, separating the sclera from the axons traversing the optic nerve head (Jacoby 1905); and (2) the **border tissue of Elschnig**, an *eponym* for a ring of scleral collagenous tissue lying external to the glial sheaths, separating the cribrosal layer of the sclera from the peripapillary scleral flange (Elschnig 1901).

The **optic nerve** perforates the sclera and extends caudalwards to the optic chiasm. The **vascular circle** or **corona of**

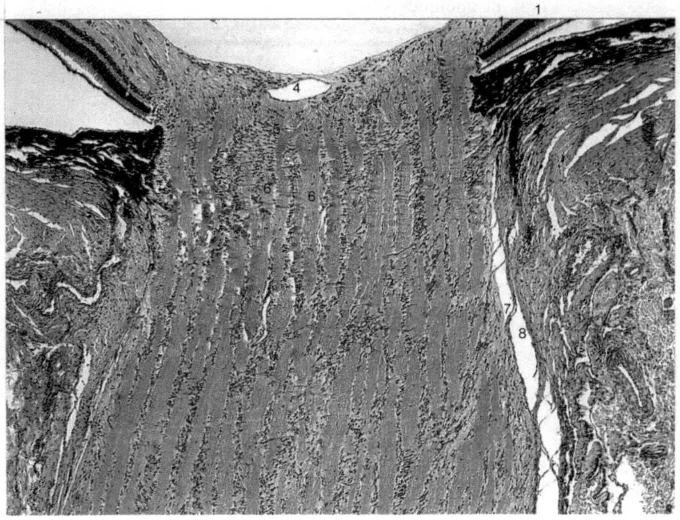

Fig. 27.12 The optic disc and nerve. (From Brüel A, Christensen EI, Tranum-Jensen J, Qvortrup K, Geneser F 2012 *Genesers histologi.* Munksgaard, Copenhagen; with permission). *1* retina, *2* choroid, *3* sclera, *4* central vein of retina, *5* lamina cribrosa of sclera, *6* fibre bundles, *7* pia sheath, *8* dura sheath (note subarachnoid space between *7* and *8*)

Zinn, also described as **circle of Haller**, is an *eponym* for the vascular circle of the optic nerve, a network of branches of the short ciliary arteries on the sclera around the exit of the optic nerve (Zinn 1755; *see* Zinn, Johann Gottfried, and Haller, Albrecht von). As a protruding part of the forebrain, the optic nerve is surrounded by a prolongation of the three meningeal membranes: an outer or external sheath, corresponding to the dura mater, and an inner or internal sheath, corresponding to the leptomeningeal arachnoid and pia mater, separated by the subarachnoid or leptomeningeal space. The spaces within the internal sheath of the optic nerve, filled with cerebrospinal fluid, are known as the *eponym* **space of Schwalbe** (*see* Schwalbe, Gustav Albert). The subdural and subarachnoid spaces of the brain, therefore, are continuous along the optic nerve and an increase in intracranial pressure is readily transmitted along the optic nerve to the optic disc, where it may manifest itself as ***papilloedema***. At the eyeball, the dura mater continues into the sclera, whereas the arachnoid and the pia mater are continuous with the choroid. The first part of the optic nerve, the intraocular part begins at the optic disc and extends posteriorly through the three layers of the eyeball. The passage through the sclera takes place through the cribrosal layer of the sclera, where the fibre bundles of the optic nerve pass through its holes. The intraocular part consists of a prelaminar part in which the axons of the optic nerve are still unmyelinated and surrounded by astroglial processes, an intralaminar part in the cribrosal layer of the sclera, where the fibres become myelinated and are surrounded by oligodendrocytes, and a postlaminar part, a transitional zone from the intralaminar part to the optic nerve proper.

References

Arlt CF (1854–1855) Ueber den Thränenschlauch, vol 2. Von Graefe's Arch Ophthalmol, pp 135–160

Ascher KW (1942) The aqueous veins. I. Physiologic importance of the visible elimination of intraocular fluid. Am J Ophthalmol 25:1174–1209

Barishak YR (2001) Embryology of the eye and its adnexa, 2nd edn. Karger, Basel

Béraud BJ (1854) Essai sur le cathéterisme du canal nasal, suivant le méthode de Laforest, nouveau procédé. Paris

Berger E (1887) Beiträge zur Anatomie des Auges in normalen und pathologischen Zustände. Bergmann, Wiesbaden

Bergmeister O (1877) Beiträge zur Entwicklungsgeschichte des Säugethierauges. Wien Embryol Inst Mitteilungen 1:63–84

Bianchi GB (1723) Ductuum lacrymalium novorum epistolaris dissertatio. Leiden

Bochdalek VA (1866) Beitrag zur Anatomie der Thränenorgane. Vjschr Prakt Heilkd 90:121

Bowman W (1849) Lectures on parts of the eye. Longmans, London

Bruch CWL (1844) Untersuchungen zur Kenntniss des körnigen Pigments der Wirbelthiere in physiologischer und pathologischer Hinsicht. Meyer und Zeller, Zürich

Brücke EW (1847) Anatomische Beschreibung des menschlichen Augapfels. Berlin

Chievitz JH (1887) Die Area und Fovea centralis retinae beim menschlichen Fetus. Int Monatsschr Anat Physiol IV:201

Ciaccio GV (1874) Osservazioni intorno alla struttura della congiuntiva umana. Bologna

Cloquet JG (1818) Mémoire sur la membrane pupillaire et sur la formation du petit cercle artériel de l'iris. Paris

Cloquet JG (1825) Manuel d'anatomie descriptive. Paris

Crampton P (1805) An essay on the Entropion or inversion of the eyelids. London

Cruveilhier JB (1871) Traité d'anatomie descriptive. Asselin, Paris

Demours P (1767) Lettre de Demours à M. Petit en response à sa critique d'un rapport sur une maladie de l'oeil survenu après inoculation de la petite vérole, contenant de nouvelles observations sur

la structure de l'oeil, et quelques remarques générales de pratique relatives aux maladies de cet organe. Paris
Descemet J (1758) Au sola lens crystalline cataractae sedes? Paris
Dollinger JIJ (1817) Illustratio iconographica fabricae oculi humani. Würzburg
Dua HS, Faraj LA, Said DG et al (2013) Human corneal anatomy redefined: a novel pre-Descemet layer (Dua' layer). Ophthalmology 120:1778–1785
Duddell B (1729) Treatise on the horny coat of the eye and the various kinds of cataracts; to which is prefixed a method entirely new of scarifying the eyes for several disorders. With remarks on the practice of some oculists both at home and abroad. Clark, London
Egger F (1888) Über Keratitis parenchymatosa circumscripta. Thesis, Basel
Elschnig A (1901) Der normale Sehnerveneintritt des menschlichen Auges. Denkschr Kais Akad Wiss Wien, Math-Naturw Kl 70:219–310
Erggelet H (1914) Klinische Befunde bei fokaler Beleuchtung mit den Gullstrandschen Nernst Spaltlampe. Klin Monatsbl Augenheilkd 53:449–470
Ferrein A (1749) Sur la structure des viscères glanduleux et en particulier sur celle des reins et de foie. Histoire de l'Academie Royale des Sciences, Paris
Foltz JCE (1862) Anatomie et physiologie des conduits lacrymeux. Ann Soc Méd Lyon
Fontana FGF (1765) Dei moti dell'iride. Lucca
Fuchs E (1885) Beiträge zur normalen Anatomie der menschlichen Iris. Von Graefe's Arch Ophthalmol 31:39–96
Grynfeltt JC (1899) Le muscle dilatateur de la pupille chez les mammifères. Firmin and Montaine, Montpellier
Hannover A (1845) On the anatomy, Physiology and Pathology of the Eye (in Danish). Copenhagen
Hasner J (1850) Physiologie und Pathologie des Thränenableitungsapparates, Prague
Henle FGJ (1855–1872) Handbuch der systematischen Anatomie des Menschen. Vieweg, Braunschweig
Huschke E (1854) Schädel, Hirn und Scala des Menschen und der Thiere nach Alter, Geschlecht und Race. Mauke, Jena
Hyrtl J (1847) Handbuch der topographischen Anatomie. Braumüller, Vienna
Jacoby E (1905) Über die Neuroglia des Sehnerven. Klin Monatsbl Augenheilkd 43:129–137
Koganei J (1885) Untersuchungen über den Bau der Iris des Menschen und der Wirbelthiere. Arch Mikr Anat 25:1–10
Krause KFT (1833) Handbuch der Anatomie des Menschen. Hannover
Krause WJF (1879) Handbuch der menschlichen Anatomie. Hannover
Kuhnt H (1898) Über die Verwertbarkeit der Bindehaut in der praktischen und in der operativen Augenheilkunde. Heidelberg
Lauth EA (1829) Nouveau manuel de l'anatomiste, comprenent le description succinte de toutes les parties du corps humain et la manière de les préparer survie de préceptes sur la confection des pièces de cabinet et sur leur conservation. Strassburg
Lockwood CB (1886) The anatomy of the muscles, ligaments and fasciae of the orbit, including an account of the capsule of Tenon, the check ligaments of the recti and the suspensory ligaments of the eye. J Anat (Lond) 20:1–25
Maier R (1859) Ueber den Bau der Thränenorgane. Freiburg
Martegiani F (1814) Novae observationes de oculo humano. Cajetani Eboli, Napoli
Meibom H (1666) De vasis palpebrarum novis epistola. Müller, Helmstädt
Mittendorf WF (1881) A manual on diseases of the eye and ear. Putnam, New York
Moll JA (1857) Dissertatio continens quaedam de anatomia et physiologia palpebrarum. Utrecht
Müller H (1856) Untersuchungen über den Bau der Retina des Menschen. Leipzig
Müller H (1857) Ueber einen ringformigen Muskel am Ciliar-Körper des Menschen und über den Mechanismus der Accommodation. Von Graefes Arch Ophthalmol 3:1–14
Paulsen F (2003) The human nasolacrimal ducts. Adv Anat Embryol Cell Biol 170:1–106
Petit F Pourfour du (1726) Mémoire sur plusiers découvertes faites dans les yeux de l'homme, des animaux à quatre pieds, des oiseaux et des poissons. Mém Acad Sci, Paris, pp 69–96
Rosenmüller JC (1805) Chirurgisch-anatomische Abbildungen für Aerzte und Wundärzte. Weimar
Rouguet C (1856) Note sur le structure de l'oeil et au particulier sur l'appareil iridochoroideum. C R Soc Biol Paris, pp 113–132
Ruysch F (1744) Alle de ontleed-, genees- en heelkundige werken. Amsterdam
Sattler H (1876) Ueber den feineren Bau der Choroidea des Menschen nebst Beiträgen zur pathologischen und vergleichenden Anatomie der Aderhaut. Von Graefes Arch Ophthalmol 22:1–99
Sattler H (1877) Beitrag zur Kenntniss der normalen Bindehaut des Menschen. Von Graefe's Arch Ophthalmol 23:1–28
Schlemm F (1830) Arteriarum capitis superficialum icon nova. Berlin
Staurenghi G, Sadda S, Chakravarthy U, Spaide RF, International Nomenclature for Optical Coherence Tomography (2014) Proposed lexicon for anatomic landmarks in normal posterior segment spectral-domain optical coherence tomography. The IN.OCT consensus. Ophthalmology 121:1572–1578
Stilling B (1832) De pupilla artificiali in sclerotica conformanda. Thesis
Taillefer LAHST (1826) Quelques remarques sur la disposition anatomique du canal nasal suivies de la description d'un nouveau procédé pour la cure de la fistula lacrymale. Arch Gen Med 11:438–444
ten Donkelaar HJ, Lammens M, Hori A (2023) Clinical Neuroembryology: development and developmental disorders of the human central nervous system, 3rd edn. Springer, Cham
Tenon JR (1816) Mémoires sur l'anatomie, la pathologie et la chirurgie et principalement sur l'organ de l'oeil. Paris
Terson A (1892) Les glandes lacrymales conjunctivales et orbito-palpébrales; l'ablation des glandes lacrymales palpébrales. Thesis, Paris
TNA (2017) Terminologia Neuroanatomica. http://FIPAT.library.dal.ca; Digital, updated version of 2022 available at https://ifaa.unifr.ch
Verhoeff FH (1903) A hitherto undescribed membrane of the eye and its significance. Boston Med Surg J 149:456–458. (quoted from Staurenghi et al. 2014)
Vogt A (1930) Lehrbuch und Atlas der Spaltlampenmikroskopie des lebenden Auges, vol XI. Springer, Berlin
Haller A von (1746–1752) Disputationes anatomicae sclerae. Göttingen
von Kölliker RA (1852) Handbuch der Gewebelehre für Ärzte und Studierende. Engelmann, Leipzig
von Wolfring E (1872) Untersuchungen über die Drüsen der Bindehaut des Auges. Zbl Med Wiss 10:852–854
Wachendorff EJ (1740) Commercii litterarii ad rei medicae et scientiae naturalis incrementum instituti. Nuremberg, Hebdomas 18:137–141
Weiss L (1897) Über das Vorkommen von scharfbegrenzten Ektasien im Augengrunde und über partielle Farbenblindheit bei hochgradiger Myopie. Wiesbaden
Whitnall SE (1911) On the tubercle of the malar bone and on the lateral attachments of the tarsal plates. J Anat (Lond) 45:426–432
Wieger, G (1883) Über den Canalis Petiti und ein Ligamentum hyaloideocapsularis. Thesis, University of Strassburg
Zeis E (1835) Anatomische Untersuchungen der Meibomischen Drüsen des Menschen und der Thiere mit besonderer Berücksichtigung ihres Verhaltens zum Tarsus. Zschr Ophthal 3:231–271
Zinn JG (1755) Descriptio anatomica oculi humani. Vandenhoeck, Göttingen

28 Anatomical Eponyms in Orthopaedics

Hans J. ten Donkelaar, Marina Quartu, Ondřej Naňka, and Michael Pretterklieber

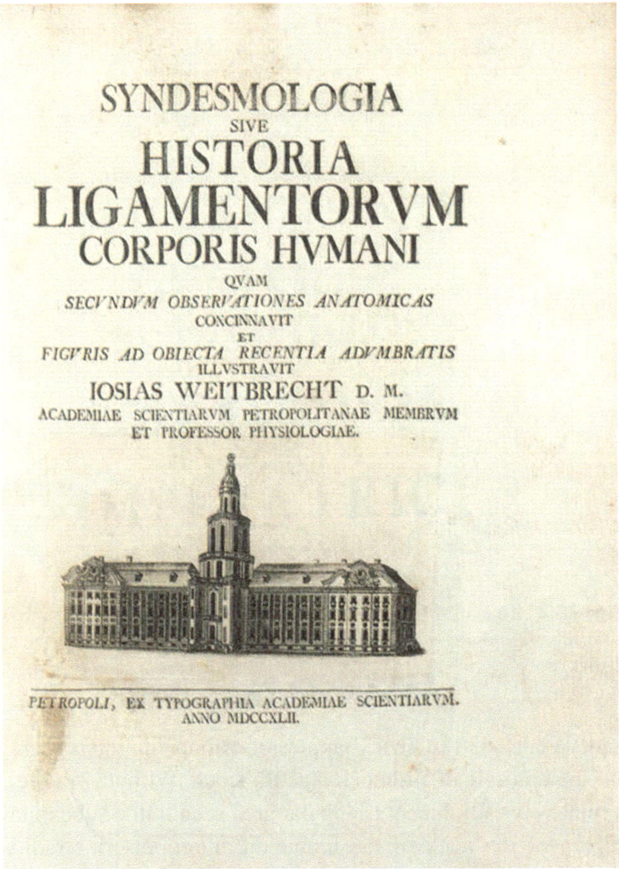
SYNDESMOLOGIA
SIVE
HISTORIA
LIGAMENTORVM
CORPORIS HVMANI
QVAM
SECVNDVM OBSERVATIONES ANATOMICAS
CONCINNAVIT
ET
FIGVRIS AD OBIECTA RECENTIA ADVMBRATIS
ILLVSTRAVIT
IOSIAS WEITBRECHT D. M.
ACADEMIAE SCIENTIARVM PETROPOLITANAE MEMBRVM
ET PROFESSOR PHYSIOLOGIAE.
PETROPOLI, EX TYPOGRAPHIA ACADEMIAE SCIENTIARVM.
ANNO MDCCXLII.

Vignette Title page of *Syndesmologia sive historia ligamentorum corporis humani quam secundum observationes anatomicas concinnavit* by Josias Weitbrecht, published 1742 in St. Petersburg

28.1 Introduction

Anatomical eponyms that may be in use in orthopaedics can be listed for lines, points, triangles, joints, tubercles and, especially for ligaments. Many of the ligaments currently in use date back to Weitbrecht (1742), who founded syndesmology (*see* **Vignette**), and Henle (1856). Certain pertinent aspects are illustrated in the Figs. 28.1, 28.2, 28.3, 28.4, 28.5, 28.6, 28.7, 28.8, 28.9, 28.10 and 28.11. Two joints even received an eponym (Fig. 28.1): the **joint of Chopart**, an *eponym* for the transverse tarsal joint (*see* Chopart, François), and the **joint of Lisfranc**, an *eponym* for the tarsometatarsal joint (Lisfranc 1815; *see* Lisfranc, Jacques;). For further data on the various authors quoted, reference is made to Part II (*The Anatomical Eponyms from A to Z*) and figures in this part.

28.2 Eponyms for Orthopedic Lines and Triangles

Various **lines** were described, especially for the lower limb. Jan von Mikulicz-Radecki described the **line of Mikulicz**, an *eponym* for a vertical line from the head of the femur to the middle of the ankle joint (Fig. 28.2). In the Anglo-Saxon literature it begins at the middle of the inguinal ligament. Wilhelm and Eduard Weber defined the **point of Weber**, an *eponym* for the centre of gravity of the body, about 1 cm below the sacral promontory (Weber and Weber 1836; *see* Weber, Wilhelm, and Weber, Eduard). Eponyms for terms marking the position of the trochanter major of the femur include the **line of Roser-Nélaton**, an *eponym* for a line between the anterior superior iliac spine and the ischial tuberosity (Roser 1844; Nélaton 1844–1860; Fig. 28.3a; *see* Nélaton, Auguste, and Roser, Wilhelm), the **line of Shoemaker** passing from the trochanter major to the umbilicus (Fig. 28.3b), and the **triangle of Bryant** (1828–1914), an *eponym* for the iliofemoral triangle, a surgical marking point for diagnosing fractures of the neck of the femur (Bryant 1861; Fig. 28.3c; *see* Bryant, Thomas). A comparable triangle is the **triangle of Langenbeck**, an *eponym* for a triangular area formed by lines drawn from the anterior superior iliac spine to the greater trochanter of the femur and to the surgical neck of the femur (von Langenbeck 1874; *see* Langenbeck, Bernhard Rudolf Konrad von). Other lines include the **line of Farabeuf**, an *eponym* for the iliotrochanteric line (Farabeuf

H. J. ten Donkelaar et al., *An Illustrated Guide to Anatomical Eponyms*, https://doi.org/10.1007/978-3-031-91664-9_28

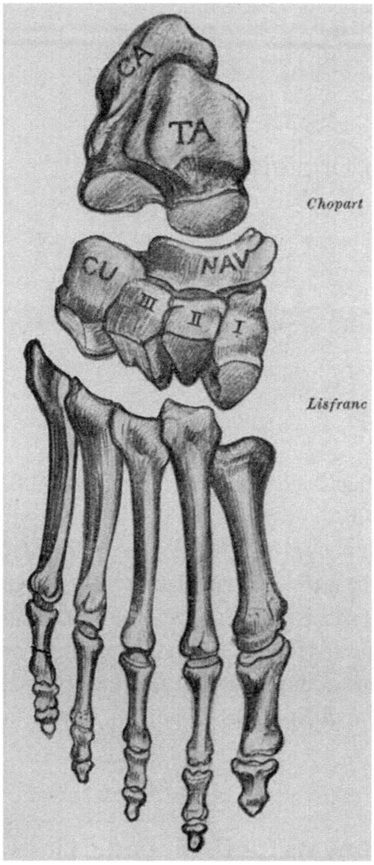

Fig. 28.1 The joints of Chopart and Lisfranc

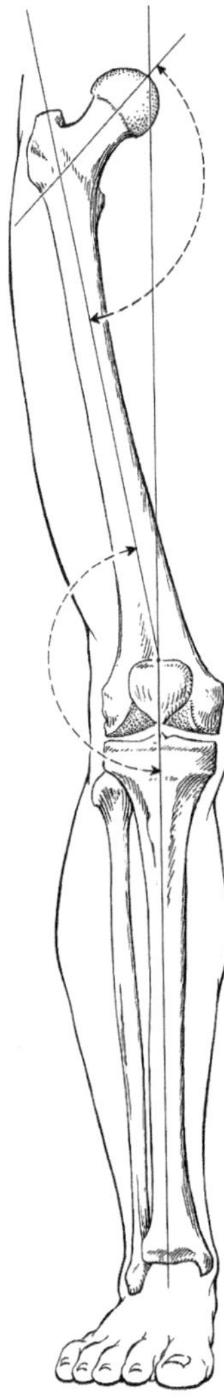

Fig. 28.2 Line of Mikulicz. (After ten Donkelaar HJ, Oostra R-J *Klinische Anatomie en Embryologie,* 4th ed. Springer Media, Houten, 2014)

1872; *see* Farabeuf, Louis Hubert), the **line of Hilgenreiter**, an *eponym* for a line drawn on pelvic radiographs to evaluate the angle of the acetabulum (*see* Hilgenreiter, Heinrich), the **line of Ogston**, an *eponym* for an imaginary line from the greater trochanter to the intercondylar fossa (Ogston 1876–1877; *see* Ogston, Alexander), the **line of Perkins**, an *eponym* for a perpendicular line drawn on anteroposterior radiographs of the pelvis, intersecting the lateral edge of the roof of the acetabulum (*see* Perkins, George), the **line of Shenton-Ménard**, an *eponym* for a curved line formed by the top of the obturator foramen and the inner side of the neck of the femur, seen on an anteroposterior X-ray of a normal hip joint (*see* Shenton, Edward Warren Hine, and Ménard, Maxime), and the **line of Blumensaat**, an *eponym* for the intercondylar line corresponding to the roof of the intercondylar fossa of the femur seen on a lateral radiograph of the knee joint, used to determine the position of the patella (Blumensaat 1938). On the medial side of the neck of the femur, the **arch of Adams** is found, an *eponym* for thick cortical bone continuous in the diaphysis of the femur (Adams 1836–1839; *see* Adams, Robert). The Webers also defined the **triangle of Weber**, an *eponym* for an area on the sole of the foot formed by the heads of the first and fifth metatarsals and the midpoint of the plantar surface of the heel (Weber and Weber 1836). Other triangles of orthopedic interest are: (a) the **triangle of Babcock** (*see* Babcock, William Wayne), a relatively radiolucent triangular area seen in the subcapital region of the neck of the femur on anteroposterior radiographs of the hip, caused by loosely arranged trabeculae; (b) the **triangle of Kager**, a sharply marginated radiolucent triangle seen posteriorly on lateral radiographs of the ankle, caused by the fat pad of Kager (Kager 1939; *see* Kager, Hans); and (c) the **triangle of Ward** (Ward 1838; *see* Ward, Frederick Oldfield), a radiolucent area between prinicipal compressive, secondary compressive and primary tensile trabeculae in the neck of the femur.

For the elbow, the **triangle of Hueter**, an *eponym* for an equilateral triangle formed by the medial and lateral epicondyles of the humerus and the olecranon of the ulna in a 90° flexed elbow has to be mentioned (Hueter 1870; *see* Hueter, Carl). For the wrist, the **lines** or **arcs of Gilula** form an

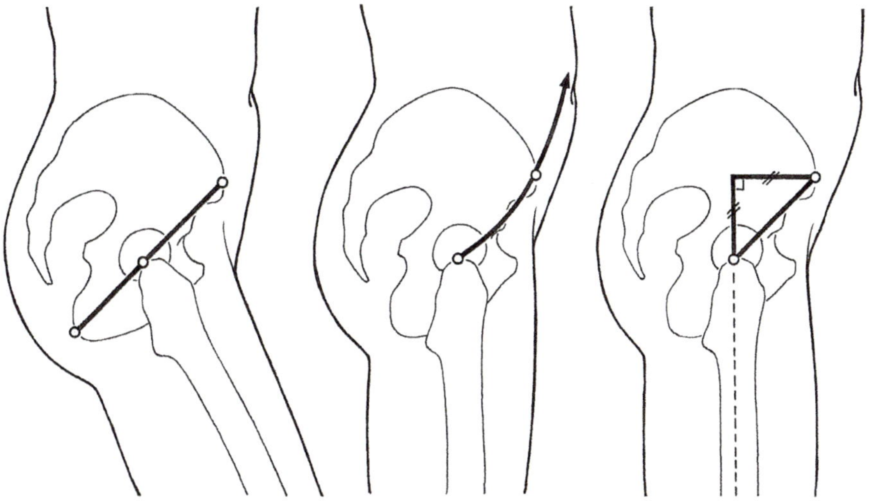

Fig. 28.3 Lines of (**a**) Roser-Nélaton, (**b**) Shoemaker and **c** the triangle of Bryant

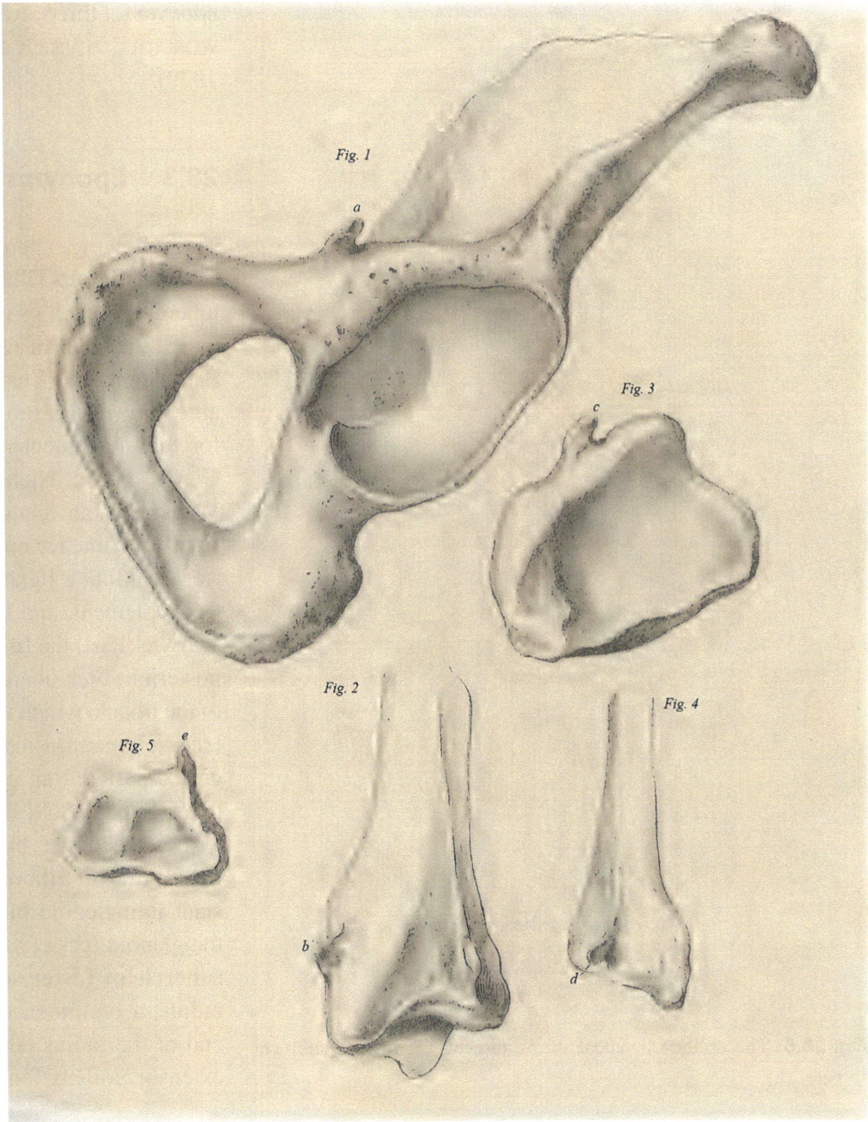

Fig. 28.4 Trochlear processes on the hip bone, the tibia and the talus. (From Hyrtl J 1860 *Über die Trochlearfortsätze des menschlichen Knochen.* Denkschr Kais Akad Wiss Wien 18:141–155). On the left hip bone (Fig. 1), a trochlear process (**a**) was found dorsal to the iliopectineal eminence. On the distal end of the right tibia (Fig. 2), a trochlear process (**b**) for the tendon of the right flexor hallucis longus muscle was found, whereas on the left tibia (Fig. 3), a very prominent trochlear process (**c**) for the tendon of the left flexor hallucis longus muscle was present. On the posterior aspect of the distal part of the right radius (Fig. 4), a trochlear process (**d**) for the tendon of the abductor pollicis longus muscle was found. On the carpal articular surface of the same radius (Fig. 5), a trochlear process (**e**) was present for the tendon of the abductor pollicis longus muscle. (Courtesy Michael Pretterklieber, Graz)

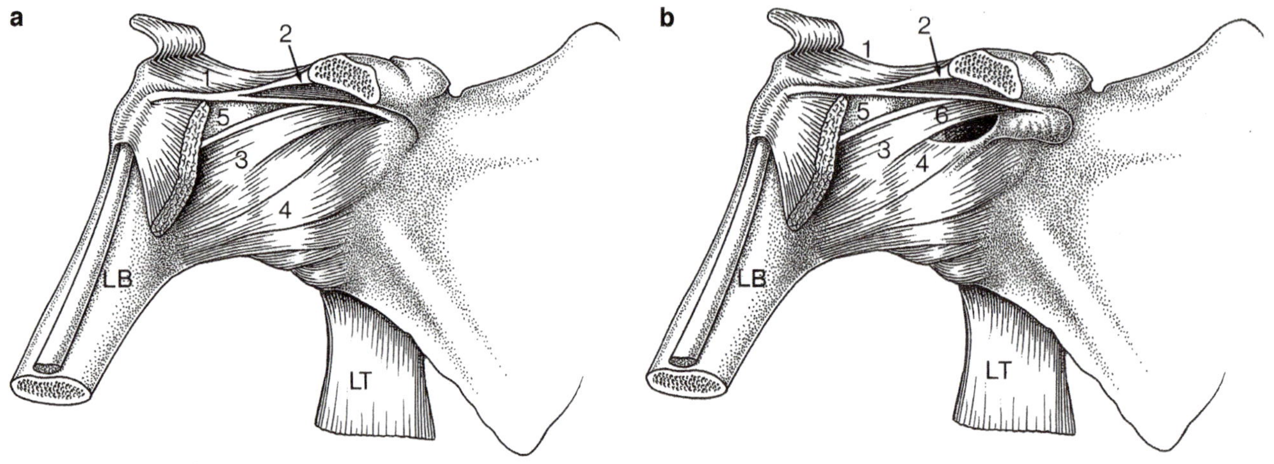

Fig. 28.5 The glenohumeral ligaments with (**a**) the foramen of Weitbrecht and (**b**) the foramen of Rouvière. (After Rouvière 1967)

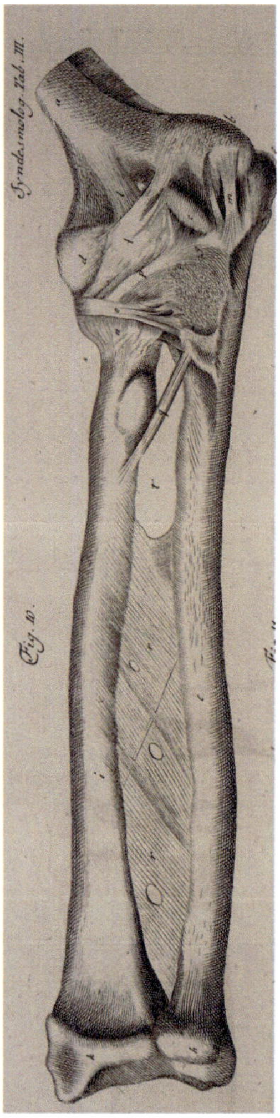

Fig. 28.6 The oblique ligament of Weitbrecht. (*q*; from Weitbrecht 1742)

eponym for three arcs on an anteroposterior radiograph of the wrist used to assess the alignment of the carpal bones (Gilula 1979).

28.3 Eponyms for Tubercles

The following tubercles deserve mentioning: (a) the **tubercle of Chaput-Tillaux**, an *eponym* for the anterior tibial tubercle, a small prominence on the distal end of the tibia to which the anterior tibiofibular ligament inserts (*see* Tillaux, Paul Henri, and Chaput, Henri); (b) the **tubercle of Gerdy**, an *eponym* for a tubercle on the lateral condyle of the tibia for the attachment of the iliotibial tract (Gerdy 1851; *see* Gerdy, Pierre Nicholas); (c) the **tubercle of Le Fort-Wagstaffe**, an *eponym* for the anterior fibular tubercle, a small prominence on the lateral malleolus to which the anterior tibiofibular ligament inserts (Le Fort 1866; *see* Le Fort, Léon Clément, and Wagstaffe 1875; *see* Wagstaffe, William Warwick); (d) the **tubercle of Volkmann**. an *eponym* for the posterior tibial tubercle, a small prominence on the distal end of the tibia to which the posterior tibiofibular ligament inserts (*see* Volkmann, Richard von); and (e) the **tubercle** or **process of Stieda**, an *eponym* for the posterior process of the talus (Stieda 1869; *see* Stieda, Christian Hermann Ludwig).

For the upper limb, the following tubercles should be noted: (a) the **tubercle of Assaki**, an *eponym* for an inconstant intraglenoid tubercle located just below the centre of the glenoid fossa (Assaki 1885; see Assaki, George); (b) the **tubercle of Lister**, an *eponym* for the dorsal tubercle of the radius, a prominence on the posterior surface of the distal end of the radius between the groove for the tendon of the extensor pollicis longus muscle medially, and the extensor

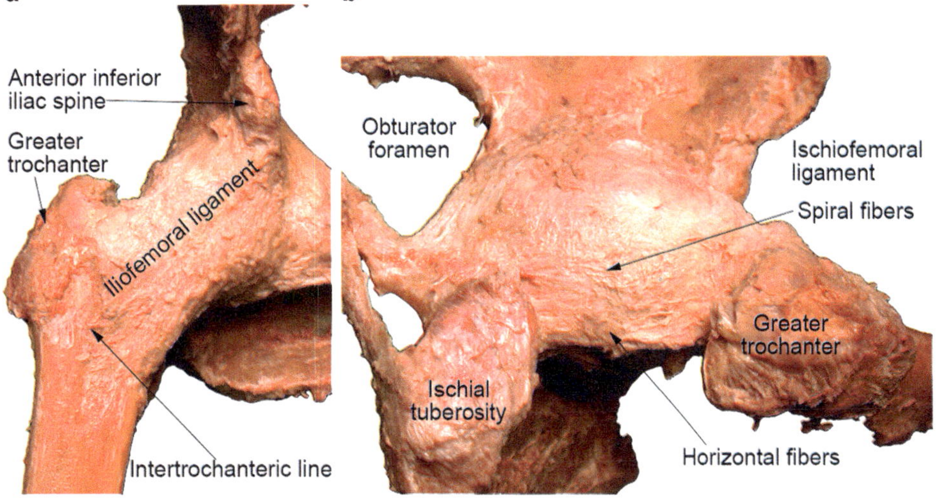

Fig. 28.7 (**a**) Ventral and (**b**) dorsal views of the ligaments of a right hip joint (photographs from dissected specimens; courtesy Michael Pretterklieber, Graz)

carpi radialis brevis muscle laterally (*see* Lister, Joseph); and (c) the **trochlear processes of Hyrtl** (Hyrtl 1860; *see* Hyrtl, Joseph), an *eponym* for various bony spurs on bones of the limbs, representing hypomochlia for tendons of muscles. In the upper limb, he found them only on the distal dorsal aspect of the radius to guide either the tendon of the extensor pollicis longus muscle or the abductor pollicis longus muscle. In the lower limb, Hyrtl found several trochlear processes on the tibia, calcaneus, navicular bone and talus for various tendons (Figs. 28.4 and 28.11).

28.4 Eponyms for Ligaments per Joint

28.4.1 Ligaments of the Shoulder Joint

The glenohumeral ligaments were described in detail by Farabeuf (1882) and received the *eponym* **ligaments of Farabeuf** (Fig. 28.5a; *see* Farabeuf, Louis Hubert). The superior glenohumeral ligament is also known as the **ligament of Flood** (Flood 1829–1830; *see* Flood, Valentine). A supernumerary ligament of the shoulder between the acromion and the greater tubercle of the humerus is known under the *eponym* **ligament of Malgaigne** (Malgaigne 1837; *see* Malgaigne, Joseph François). Between the three glenohumeral ligaments openings or dehiscences are found known as the **oval foramen of Weitbrecht**, an *eponym* for an opening between the superior and middle glenohumeral ligaments (Weitbrecht 1742; Fig. 28.5a; *see* Weitbrecht, Josias), and the **foramen of Rouvière**, an *eponym* for a dehiscence between the middle and the inferior glenohumeral ligaments (Rouvière 1967; Fig. 28.5b; *see* Rouvière, Henri). Both openings serve as communications between the articular cavity of the shoulder joint and the subcutaneous bursa of the subscapularis muscle.

28.4.2 Ligaments of the Elbow Joint

For the elbow joint, Weitbrecht (1742) described the oblique radioulnar ligament (Fig. 28.6), known with the *eponym* **ligament of Weitbrecht** and nowadays as chorda obliqua. Other ligaments are (a) the **ligament of Bardinet**, an *eponym* for the posterior part of the ulnar collateral ligament of the elbow joint connecting the olecranon and the lateral epicondyle of humerus (Bardinet 1869; *see* Bardinet, Bartholomy Alphonse); (b) the **ligaments of Barkow**, an *eponym* for the anterior and posterior parts of the fibrous capsule of the elbow joint (Barkow 1841; see Barkow, Hans Leopold Karl); (c) the **ligament of Brodie**, an *eponym* for the transverse humeral ligament (Brodie 1850; *see* Brodie, Charles Gordon); (d) the **ligament of Cooper**, an *eponym* for an oblique ligament from the olecranon to the coronoid process of the ulna, part of the ulnar collateral ligament of the elbow joint (Cooper 1825; *see* Cooper, Astley Paston); (e) the **ligament of Denucé**, an *eponym* for the quadrate ligament of the proximal radioulnar joint (Denucé 1854; *see* Denucé, Jean Louis Paul); and (f) the **ligament of Osborne**, an *eponym* for the triangular band of connective tissue forming the roof of the cubital tunnel. It connects the humeral and ulnar heads of the flexor carpi ulnaris muscle and invests the ulnar nerve (Osborne 1957, 1959; *see* Osborne, Geoffrey Vaughan). Related to the elbow joint are the arcade and ligament of Struthers (Struthers 1848, 1854; *see* Struthers, John). The

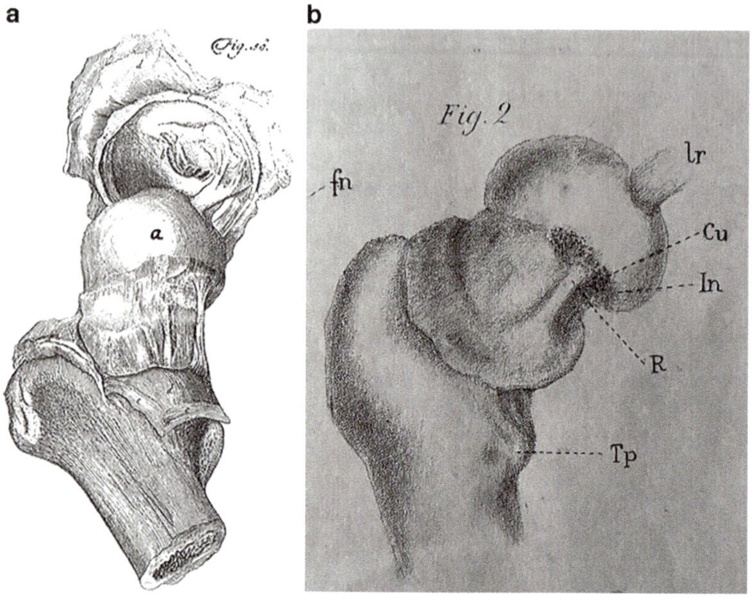

Fig. 28.8 (**a**) The retinacula of Weitbrecht. (From Bartoníček and Naňka 2019; courtesy Jan Bartoníček, Prague); (**b**) the prominent medial fold, known as pectineofoveal fold of Amantini (*R*; from Amantini 1899; courtesy Michael Pretterklieber, Graz)

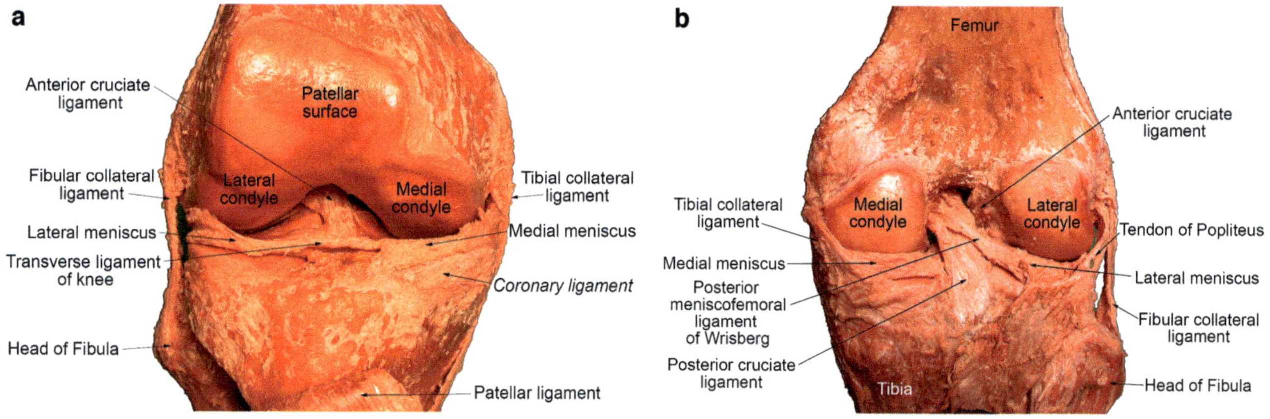

Fig. 28.9 (**a**) Ventral and (**b**) dorsal views of the ligaments of a right knee joint. (Photographs from dissected specimens; courtesy Michael Pretterklieber, Graz)

term **arcade of Struthers** is an *eponym* for an arcade in the distal third of the arm on its medial aspect. Here, the ulnar nerve passes from the anterior to the posterior compartment of the arm, which may lead to ulnar nerve compression (Spinner and Kaplan 1976). The **ligament of Struthers** is an *eponym* for an inconstant fibrous band located on the medial side of the distal part of the humerus, earlier described by Tiedemann (1818; *see* Tiedemann, Friedrich). It arises from a supracondylar process and forms a canal, through which the median nerve may pass, leading to median nerve compression (Lordan et al. 2008).

28.4.3 Ligaments of the Wrist and Hand Joints

For the wrist joint, the **space of Poirier** should be mentioned, an *eponym* for a palmar area of the wrist between the capitate and lunate with poor ligamentous support, a weak spot through which the lunate might dislocate (Poirier and Charpy 1911; *see* Poirier, Paul Julien, and Charpy, Adrien), and also the **canal of Guyon**, an *eponym* for the ulnar canal through which the ulnar nerve and vessels pass at the wrist (Guyon 1861; *see* Guyon, Jean Casimir Félix). Ligaments worth mentioning are (a) the **ligament of Barkow**, an *eponym* for the pisohamate ligament (Barkow 1841; *see* Barkow, Hans Karl Leopold); (b) the **ligament of Gerdy**, an *eponym* for the superficial transverse metacarpal ligament (Gerdy 1851; *see* Gerdy, Pierre Nicholas); (c) the **ligament of Testut**, an *eponym* for the radioscapholunate ligament (Testut 1891; *see* Testut, Jean Léon); and (d) another **ligament of Weitbrecht**, an *eponym* for the anterior ligament of the radiocarpal joint (Weitbrecht 1742). A peculiar ligament was described by Henle (Henle 1856; *see* Henle, Friedrich Gustav Jacob) as **ligamentum subcruentum of Henle**. It described the split

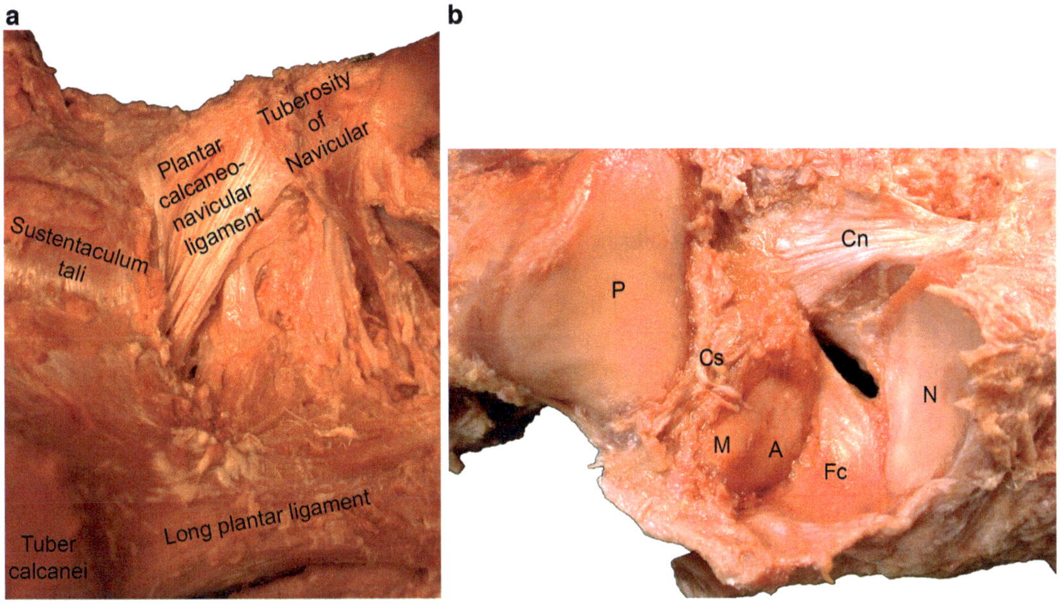

Fig. 28.10 (**a**) Plantar and (**b**) dorsal view of the plantar calcaneonavicular ligament of the foot (photographs from dissected specimens; courtesy Michael Pretterklieber, Graz). To expose the ligament from dorsal, the talus has been removed. *A* Anterior talar articular surface, *Cn* calcaneonavicular ligament (medial part of bifurcate ligament), *Cs* calcaneal sulcus with remnants of the interosseous talocalcaneal ligament, *M* middle talar articular surface, *N* articular surface of the navicular bone, *P* posterior talar articular surface

Fig. 28.11 Trochlear processes on the calcaneus. (From Hyrtl J 1860 *Über die Trochlearfortsätze des menschlichen Knochen.* Denkschr Kais Akad Wiss Wien 18:141–155). A dissection by his assistant Karl Gilewski (Fig. 1) shows the trochlear or inframalleolar process formed at the fibular (or peroneal) trochlea (**a**), serving as hypomochlion for the tendon of the fibularis (or peroneus) longus muscle (**b**); *c* tendon of the fibularis (or peroneus) brevis muscle, *d* abductor digiti minimi muscle. Fig. 2: a right calcaneus viewed from above; *g* trochlear or inframalleolar process, *f* posterior talar articular surface, *g* middle and anterior talar articular surface, *h* dorsal margin of articular surface for the cuboid bone. Fig. 3: another right calcaneus viewed from above; *i* trochlear or inframalleolar process, *j* posterior talar articular surface, *k* middle and anterior talar articular surfaces. Fig. 4: a right calcaneus from a foot with a pes varus deformity: the trochlear or inframalleolar process is of considerable size (courtesy Michael Pretterklieber, Graz)

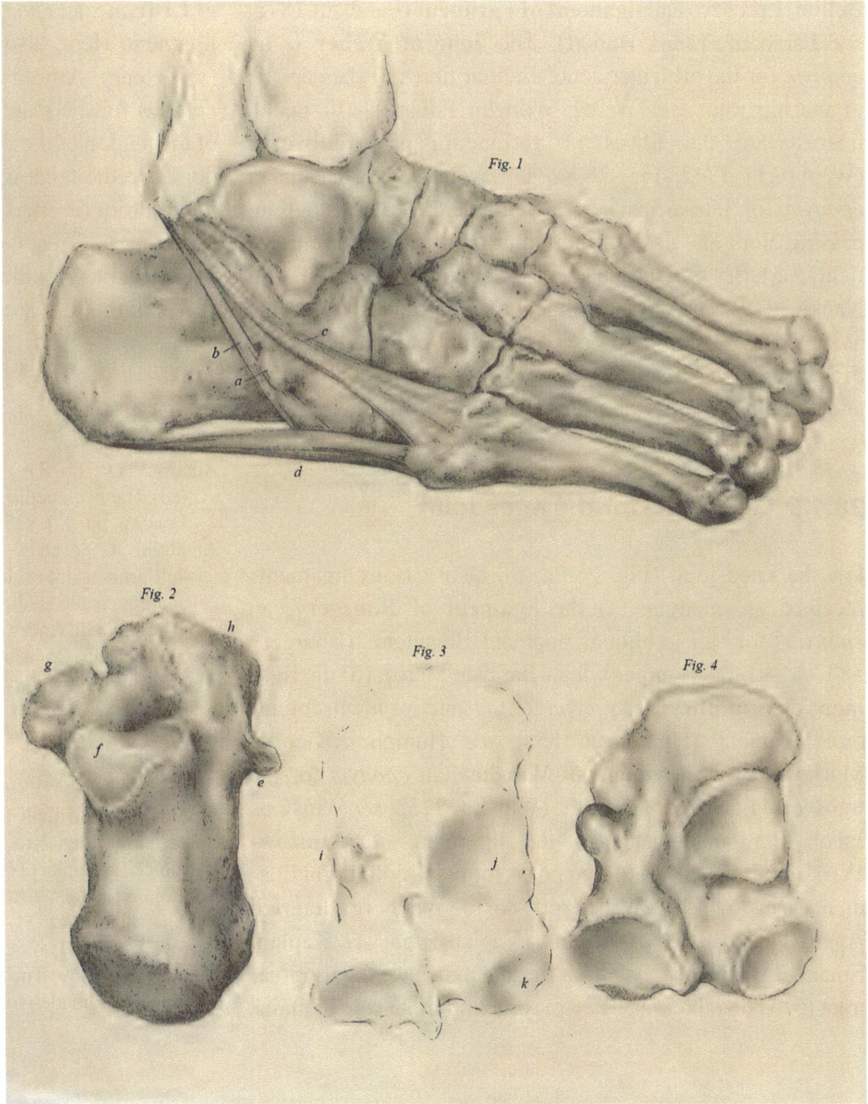

medial attachment of the articular disc. As vessels pass between the two structures, Henle used the term subcruentum, meaning 'under the stream of blood'. More recently, it has been interpreted as loose connective tissue radial to the middle of the styloid process of the ulna and interposed between the layers of the radioulnar ligaments attached to the styloid process (Palmer and Werner 1981). These authors viewed the ligament of Henle as part of their triangular fibrocartilaginous or ulnocarpal complex, which is formed by ligamentous and cartilaginous structures that suspend the distal end of the radius and ulnar part of the wrist from the distal part of the ulna (Palmer and Werner 1981).

28.4.4 Ligaments of the Hip Joint

For the hip joint, *eponyms* were given to the iliofemoral ligament (Fig. 28.7a): the **ligament of Bertin** (Bertin 1754; *see* Bertin, Exupère Joseph), also known as the **Y-ligament of the hip joint of Bigelow** (Bigelow 1869; *see* Bigelow, Henry Jacob). The ischiofemoral ligament of the hip joint (Fig. 28.7b) received the *eponyms* **ligament of Bellini** (*see* Bellini, Lorenzo) and **ligament of Farabeuf** (Farabeuf 1872; *see* Farabeuf, Louis Hubert). The **zone of Weber** is an *eponym* for the orbicular zone, circular fibres of the capsule of the hip joint (*see* Weber, Wilhelm Eduard). Of special interest are the **fibres** or **retinacula of Weitbrecht** (Weitbrecht 1742; Fig. 28.8a; *see* Weitbrecht, Josias), an *eponym* for intraarticular synovial plicae of the hip joint passing along the neck of the femur partly covering intraarticular arteries for the head of the femur (Bartoníček and Naňka 2019). The most prominent medial fold is also known as **pectineofoveal fold of Amantini** (Amantini 1889; Fig. 28.8b; *see* Amantini, Cesare, and St-Pierre et al. 2024), and contains the inferior artery of the neck of the femur.

28.4.5 Ligaments for the Knee Joint

For the knee joint (Fig. 28.9a, b), also various ligaments received an eponym: (a) the **ligament of Bourgery**, an *eponym* for the oblique popliteal ligament (Bourgery 1837–1840; *see* Bourgery, Jean-Baptiste Marc); (b) the **ligament of Humphry**, an *eponym* for the anterior meniscofemoral ligament (Humphry 1858; *see* Humphry, George Murray); (c) the **ligament of Winslow**, an *eponym* for the transverse ligament of the knee (Winslow 1732; *see* Winslow, Jacob Benignus); and (d) the **ligament of Winslow-Wrisberg**, an *eponym* for the posterior meniscofemoral ligament (Winslow 1732; Wrisberg 1786; *see* Wrisberg, Heinrich August). The **fibres of Kaplan** (Kaplan 1958; *see* Kaplan, Emanuel B) are an *eponym* for the attachment of the iliotibial tract to the distal femur. They play an important role in maintaining rotatory stability of the knee, particularly during greater degrees of knee flexion. Also worth mentioning is the infrapatellar **fat pad of Hoffa** (Hoffa 1898; *see* Hoffa, Albert).

28.4.6 Ligaments for the Ankle and Foot Joints

For the ankle and foot joints, various ligaments received an eponym: (a) the **ligament of Meckel**, an *eponym* for the interosseous tibiofibular ligament (Meckel and The Younger 1816; *see* Meckel, Johann Friedrich, The Younger); (b) the *eponym* **ligament of Barkow** is in use for the oblique tibiofibular ligament, located deep to the muscles of the anterior compartment of the leg, and for the plantar calcaneonavicular ligament (Barkow 1841; *see* Barkow, Hans Karl Leopold), also known as spring ligament or as part of the spring ligament complex (Fig. 28.10); (c) the **ligament of Chopart**, an *eponym* for the bifurcate ligament of the foot (*see* Chopart, François); (d) the **ligaments of Cruveilhier**, an *eponym* for the plantar metatarsophalangeal ligaments (Cruveilhier 1871; *see* Cruveilhier, Jean Baptiste); and (e) the **ligament of Lisfranc**, an *eponym* for the interosseous cuneometatarsal ligament. Here, also the **master knot** or **chiasm of Henry** (*see* Henry, Arnold Kirkpatrick) should be mentioned, an *eponym* for the plantar chiasm, the variable plantar crossing of the tendon of the flexor digitorum longus muscle obliquely dorsal to the tendon of the flexor hallucis longus muscle in the midfoot (Pretterklieber 2017). The inferior extensor retinaculum has received the *eponym* **cruciate ligament of Weitbrecht**. Trochlear processes found by Hyrtl (1860) on bones of the foot are shown in Fig. 28.11.

References

Adams R (1836–1839) Hip joint abnormal conditions. In: Todd RB (ed) The cyclopedia of anatomy and physiology, vol II. Longman, London, pp 780–825

Amantini C (1889) Di una men nota ripiegatura sinoviale dell'articolazione dell'anca (Piega pettineofoveale); contributo anatomico allo studio delle fratture intra-articolare del collo femore. Tipografia di Vincenzo Santucci, Perugia

Assaki G (1885) Contribution à l'anatomie et le physiologie de la cavité glénoide de l'omoplate. CR Hebdom Séc Mém Soc Biol, Paris, pp 371–373

Bardinet BA (1869) Cas intéressant de fracture de l'olécrâne. Gaz Méd Chir 6:110

Barkow HKL (1841) Syndesmologie oder die Lehre von den Bändern, durch welche die Knochen des menschlichen Körpers zum Gerippe vereint werden. Breslau

Bartoníček J, Naňka O (2019) Josias Weitbrecht the founder of syndesmology, and the history of the retinacula of Weitbrecht. Surg Radiol Anat 41:1103–1111

Bertin EJ (1754) Traité de ostéologie, 4 Vols. Paris

Bigelow HJ (1869) The mechanism of dislocation and fractures of the hip. Lea, Philadelphia

Blumensaat C (1938) Die Lageabweichungen und Verrenkungen der Kniescheibe. Ergebn Chir Orthop 31:149–223
Bourgery JBM (1837–1840) Traité complet de l'anatomie de l'homme comprenant la médecine opératoire. Paris
Brodie BC (1850) Pathological and surgical observations on the diseases of joints, 5th ed. London
Bryant T (1861) The diseases and injuries of the joints. Clinical and pathological observations. London
Cooper AP (1825) Treatise on dislocation and on fractures of the joints. London
Cruveilhier JB (1871) Traité d'anatomie descriptive. Asselin, Paris
Denucé JLP (1854) Mémoire sur les luxations du coude. Thesis No 69, Paris
Farabeuf LH (1872) Précis de manuel opératoire. Paris
Farabeuf LH (1882) Sur l'articulation scapulo-humerale. Bull Soc Chir
Flood V (1829–1830) Discovery of a new ligament of the shoulder joint, vol 13. Lancet, pp 672–673
Gerdy PN (1851) Chirurgie pratique complète. Breslau
Gilula LA (1979) Carpal injuries: analytic approach and case exercise. AJR Am J Roentgenol 133:503–517
Guyon JCF (1861) Note sur une disposition anatomique propre à la face antérieure de la région du poignet et non encore décrite. Bull Soc Anat Paris 6:184–186
Henle J (1856) Handbuch der Bänderlehre des Menschen. Vieweg, Braunschweig
Hoffa A (1898) Lehrbuch der orthopädischen Chirurgie, 3. Aufl, Stuttgart
Hueter C (1870) Klinik der Gelenkkrankheiten mit Einschluss der Orthopädie
Humphry GM (1858) A treatise on the human skeleton. Macmillan, Cambridge
Hyrtl J (1860) Über die Trochlearfortsätze des menschlichen Knochen. Denkschr Kais Akad Wiss Wien 18:141–155
Kager H (1939) Zur Klinik und Diagnostik des Achillessehnenrisses. Chirurg 11:691–695
Kaplan EB (1958) The iliotibial tract: clinical and morphological significance. J Bone Joint Surg Am 46A:817–832
Le Fort LC (1866) Note sur une variété non décrite de fracture verticale de la malléole externe par arrachment. Bull Thér Méd Chir 110:193–199
Lisfranc J (1815) Mémoire sur l'amputation du pied dans son articulation tarso-métatarsienne. Gabon, Paris
Lordan J, Rauh P, Spinner RJ (2008) The clinical anatomy of the supracondylar spur and the ligament of Struthers. Clin Anat 18:548–551
Malgaigne JF (1837) Traité d'anatomie chirurgicale. Paris
Meckel JF, The Younger (1816) Handbuch der menschlichen Anatomie, Vol II: Besondere Anatomie, Knochenlehre, Bänderlehre, Muskellehre. Buchhandlung des Hallischen Waisenhaus, Halle, Berlin
Nélaton A (1844–1860) Eléments de pathologie chirurgicale, 5 Vols. Baillière, Paris
Ogston A (1876–1877) The operative treatment in Ogston's operation. Edinb Med J 22:782–785
Osborne G (1957) The surgical treatment of tardy ulnar neuritis. J Bone Joint Surg 39B:782
Osborne G (1959) Ulnar neuritis. Postgrad Med J 35:392–396
Palmer AK, Werner FW (1981) The triangular fibrocartilage complex of the distal wrist – anatomy and formation. J Hand Surg 6:153–162
Poirier P, Charpy A (1911) Traité d'anatomie humaine. Tome I. Masson, Paris
Pretterklieber B (2017) The high variability of the chiasma plantare and the long flexor tendons: anatomical aspects of tendon transfer in foot surgery. Ann Anat 211:11–32
Roser W (1844) Handbuch der anatomischen Chirurgie. Tübingen
Rouvière H (1967) Anatomie humaine, descriptive et topographique, 3 Vols. Masson, Paris
Spinner M, Kaplan EB (1976) The relationship of the ulnar nerve to the medial intermuscular septum in the arm and its clinical significance. Hand 8:239–242
Stieda CHL (1869) Über secundäre Fusswurzelknochen. Anat Physiol Wissensch Med, pp 108–111
St-Pierre MO, Sobczak S, Olry R (2024) History of the hip joint pectineofoveal fold. Clin Anat 37:635–639
Struthers J (1848) On a peculiarity of the humers and humeral artery. Monthly J Med Sci, New Series XXVIII: 264–267
Struthers J (1854) Anatomical and physiological observations, Part I. Sutherland and Knox, Edinburgh
Testut JL (1891) Traité d'anatomie humaine. Doin, Paris
Tiedemann F (1818) Ueber einen am Oberarmgebiet bei mehreren geschwanzten Affen vorkommenden Kanal und eine damit in Verbindung stehende besondere Anordnung der Arterien und Nerven des Armes. Deutsch Arch Physiol 4:544–549
von Langenbeck BRK (1874) Ueber Schussverletzungen des Hüftgelenkes. Berlin
Wagstaffe W (1875) An unusual form of fracture of the fibula. St Thomas Hosp Rep 6:43–49
Ward O (1838) Outline of human osteology. Renshaw, London
Weber WE, Weber EF (1836) Mechanik der menschlichen Gehwerkzeuge; eine anatomisch-physiologische Untersuchung. Göttingen
Weitbrecht J (1742) Syndesmologia sive historia ligamentorum corporis humani quam secundum observationes anatomicae concinnavit. St. Petersburg
Winslow JD (1732) Exposition anatomique de la structure du corps humain. Desprez, Paris
Wrisberg HA (1786) Sylloge Commentationum Anatomicorum. Dieterich, Göttingen

Anatomical Eponyms in Otology

29

Hans J. ten Donkelaar, David Kachlík, and Vedat Topsakal

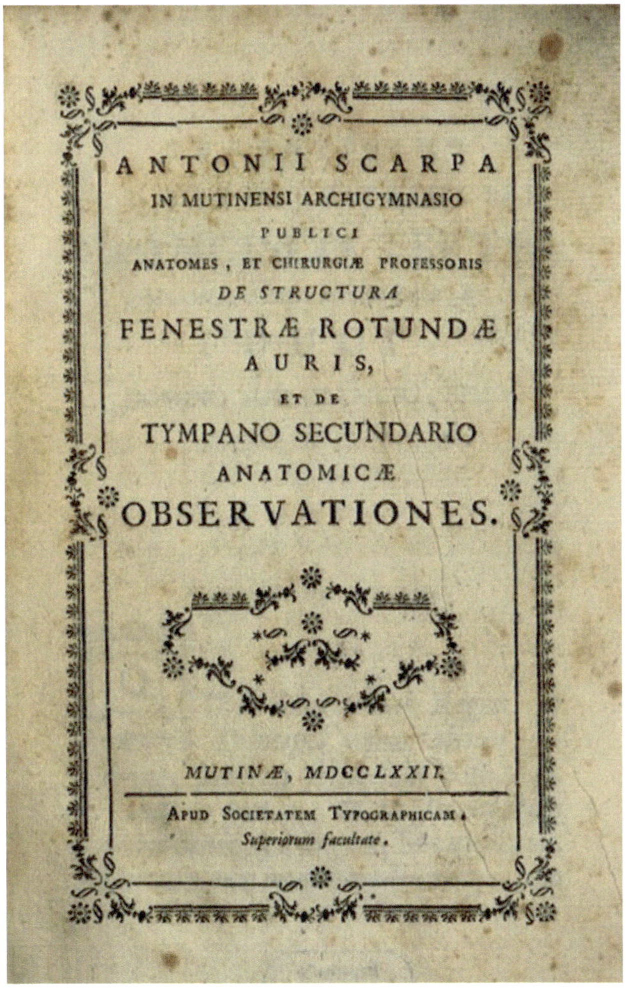
ANTONII SCARPA
IN MUTINENSI ARCHIGYMNASIO
PUBLICI
ANATOMES, ET CHIRURGIÆ PROFESSORIS
DE STRUCTURA
FENESTRÆ ROTUNDÆ
AURIS,
ET DE
TYMPANO SECUNDARIO
ANATOMICÆ
OBSERVATIONES.
MUTINÆ, MDCCLXXII.
Apud Societatem Typographicam.
Superiorum facultate.

Vignette Title page of *De structura fenestrae rotundae aurii et de tympano secundario anatomicae observationes* by Antonio Scarpa, published 1772 in Modena

29.1 Introduction

Studies on the anatomy of the ear go back to Duverney (1683), Cotugno (1760), Scarpa (1772, 1789, *see* **Vignette**), Soemmerring (1806) and later Retzius (1881, 1884). But their extensive studies only sparsely earned them an eponym. It was up to Corti and Reissner to start the long series of eponyms on the ear, particularly on the inner ear. Kölliker (1852, 1867) referred to Corti's description (1851) of the spiral organ as the **organ of Corti**. Reissner described the membrane separating the cochlear duct or scala media (*eponym*: **canal of Reissner**; Reissner 1851; see Reissner, Ernst; also described by Löwenberg 1864; *see* Löwenberg, Benjamin Benno) from the scala vestibuli (*eponym*: **membrane of Reissner**; Reissner 1851), identifying three distinct channels in the cochlear canal instead of two as previously thought. From now on, many new structures, cells in particular, have been identified with the rationale that names of first identifiers should be associated with the structure they identified. Almost all of these new eponyms date back to the second half of the nineteenth century. Only a few were added in the twentieth century, including some clinical eponyms. The official terminology for the ear is found in TNA (2017) and in a digital version of the TNA of 2022 at https://ifaa.unifr.ch. After an overview of the ear, eponyms for structures of the middle ear, the bony labyrinth, the cochlear labyrinth, the spiral organ, and on neural components of the ear will be discussed. For further data on the various authors quoted, reference is made to Part II (*The Anatomical Eponyms from A to Z*) and figures in this part.

H. J. ten Donkelaar et al., *An Illustrated Guide to Anatomical Eponyms*, https://doi.org/10.1007/978-3-031-91664-9_29

29.2 The Ear in Overview

The ear is composed of external, middle and internal parts (Fig. 29.1). The external ear consists of the auricle or pinna and the external acoustic meatus with the outer, cutaneous layer of the tympanic membrane. The helix of the auricle may show a small prominence known as the *eponym* **tubercle of Darwin** (*see* Darwin, Charles Robert). The middle ear is formed by the tympanic cavity, which contains the auditory ossicles with their muscles, the inner layers of the tympanic membrane, air-filled cavities in the mastoid part of the temporal bone, and the auditory tube (*eponym*: **Eustachian tube**; Eustachi 1563; *see* Eustachi, Bartolomeo).

The **external acoustic meatus** extends from the concha of the auricle to the tympanic membrane. Its lateral part is largely cartilagineous, whereas its longer medial part is bony. Fissures of the external acoustic meatus (*eponyms*: **fissures of Duverney** and **of Santorini**; Duverney 1683, *see* Duverney, Joseph Guichard; Santorini 1724, *see* Santorini, Giovanni Domenico) may be present in the hyaline cartilage of the external acoustic meatus. A variable deficiency in the floor of the bony part is known as the tympanic foramen (*eponym*: **foramen of Huschke**; Huschke 1844; *see* Huschke, Emil). Pirsig and Mudry (2015) found several earlier descriptions going back to a Renaissance portrait of Saint Jerome by Marinus Claeszoon van Reymerswaele (ca. 1490—after 1567). The tympanic ring forms the most medial part of the tympanic part of the temporal bone. It is C-shaped and forms the frame in which most of the tympanic membrane inserts. The superior part of the tympanic ring is deficient and forms the tympanic notch (*eponym*: **notch of Rivinus**; *see* Rivinus, Augustus Quirinus), bridged by the flaccid part of the tympanic membrane. A fibrocartilaginous anulus known with the *eponym* **anular tendon of Gerlach** (Gerlach 1848; *see* Gerlach, Joseph) maintains the insertion of the tympanic membrane in the tympanic sulcus but is absent at the level of the notch of Rivinus (Fig. 29.2). Of clinical importance is the **zone of Ramsay Hunt**, an *eponym* for a delimited skin area in the external acoustic meatus supplied by somatosensory fibres, travelling via the auricular branch of the vagus nerve and connected to the ganglion geniculi, relevant in otic herpes zoster (Ramsay Hunt 1907; *see* Ramsay Hunt, James).

The **auditory tube** shows two fat pads near its pharyngeal opening (*eponym*: **fat pads of Ostmann**; Ostmann 1892; *see* Ostmann, Friedrich Wilhelm Paul). The tubal cartilage consists of lateral and medial laminae. The voluminous medial lamina protrudes into the nasopharynx and forms the skeleton of the torus tubarius. The mucosa of the torus tubarius is rich in glands and lymphatic tissue and forms the (tubal) **tonsil of Gerlach** (Gerlach 1875). From the torus the **salpingopharyngeal fold of Zaufal** (Zaufal 1880; *see* Zaufal or Coufal, Emanuel) and the **salpingopalatine fold of Tourtual** (Tourtual 1846; *see* Tourtual, Kaspar Theobald) arise. Just medial and behind the torus, the pharyngeal **recess** or **fossa of Rosenmüller** (Rosenmüller 1805; *see* Rosenmüller, Joseph) is found. An inconstant lateral outgrowth of the pha-

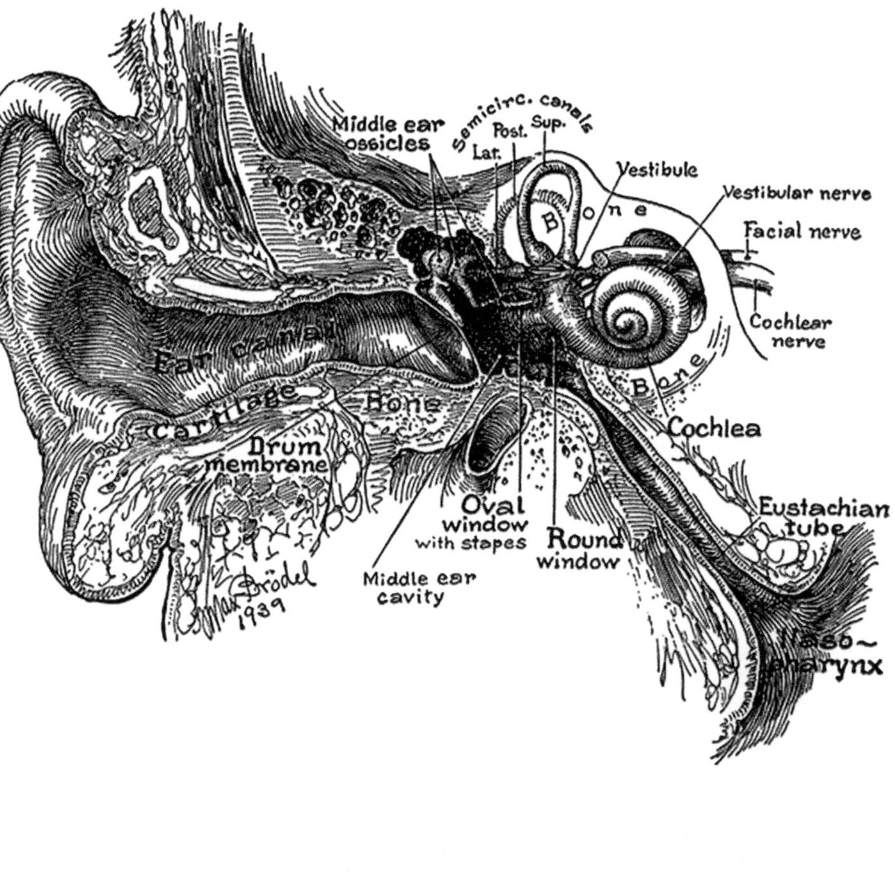

Fig. 29.1 Overview of the external, middle and internal ear. (From Brödel M *Three unpublished drawings of the anatomy of the ear.* Saunders, Philadelphia, PA, 1938; with permission)

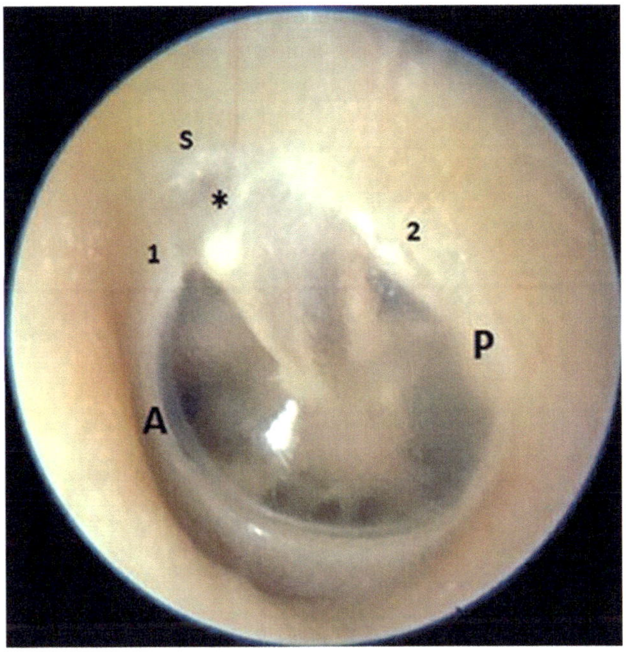

Fig. 29.2 The left tympanic membrane showing the notch of Rivinus, marked by an *asterisk*, limited by the anterior (*1*) and posterior (*2*) tympanic spines. The tympanic membrane is inserted directly to the scutum (*S*). The tympanic ring is absent in this zone. *A*, *P* anterior and posterior parts of tympanic ring. (From Mansour S, Magnan J, Haidar Ahmad H, Nicolas K, Louryan S *Comprehensive and Clinical Anatomy of the Middle Ear*, 2nd ed. Springer, Cham, 2019; with permission)

ryngeal recess received the *eponym* **diverticulum of Pertick** (Pertick 1883; *see* Pertick, Otto Johann Christian) and an inconstant diverticulum between the tensor and levator veli palatini muscles the *eponym* **diverticulum of Kirchner** (*see* Kirchner, Wilhelm). Another inconstant recess is known as the *eponym* **salpingopharyngeal recess of Zuckerkandl** (*see* Zuckerkandl, Emil). A gap between the superior constrictor muscle of the pharynx and the skull base, within the pharyngobasilar fascia, transmitting the auditory tube and the levator veli palatini muscle, bears the *eponym* **sinus of Morgagni** (Morgagni 1761; *see* Morgagni, Giovanni Battista). The fibrous part of the fibrocartilaginous auditory tube bears the *eponym* **salpingopharyngeal fascia of Troeltsch** (Troeltsch 1860; *see* Troeltsch, Anton Friedrich von), which is laterally inserted to the skull base at the sphenopetrous suture as the **ligament of Proctor** (Proctor 1967; *see* Proctor, Bruce).

29.3 Eponyms for the Middle Ear

In contrast to the inner ear, anatomical eponyms for the middle ear are rather restricted. Apart from the flaccid part of the tympanic membrane (*eponym*: **membrane of Shrapnell**; Shrapnell 1832; *see* Shrapnell, Henry Jones; also described by Rivinus), eponyms are still in use for the anterior and posterior recesses (*eponyms*: **anterior and posterior spaces of Troeltsch**; Troeltsch 1860; *see* Troeltsch, Anton Friedrich von), located between the tympanic membrane and the mallear folds, and for the superior recess (*eponym*: **space of Prussak**; Prussak 1867; *see* Prussak, Alexander) between the lateral mallear fold and the fold of the incus. The middle ear has been compared to a box with six walls (Fig. 29.3), to which many new terms have been added without eponyms (Lang 1992; Marchioni et al. 2016; Mansour et al. 2019; Topsakal et al. 2021) such as:

1. An osseous ridge projecting downwards from the tegmental wall, the tegmen tympani, for which Lang (1992) used the term inferior tegmental process, but which in ear surgery became known as the **cog**, separating the epitympanum into anterior and posterior parts, useful for defining the position of the geniculate ganglion of the facial nerve, found medial and inferior to it.
2. The finiculus or sustentaculum of the promontory, a variable osseous ridge that may contain the tympanic nerve of Jacobson.
3. As shown by Bolobas (1972), the promontory overhangs the round window. This bony overhang, comprises an anterior pillar, a tegmen and a posterior pillar. Endoscopy of the middle ear clearly showed the bony overhang, for which the term **canopy** has been suggested that partially covers and protects the round window (Topsakal et al. 2020, 2021; Fig. 29.4).
4. Several surgical landmarks for the mastoid antrum have been described: (a) the suprameatal spine of the squamous part of the temporal bone (*eponym*: **spine of Henle**; Henle 1855; *see* Henle, Friedrich Gustav Jacob), a variable superficial landmark; and (b) the suprameatal triangle or fossa on the squamous part of the temporal bone (*eponym*: **triangle of Macewen**; Macewen 1893; *see* Macewen, William), which marks the lateral wall of the mastoid antrum. The petrosquamous septum (*eponym*: **septum of Körner**; Körner 1927; *see* Körner, Otto) separates central and peripheral parts of the mastoid cavity.

The **facial nerve** is related closely to the tympanic cavity and consists of two roots (**Figs. 29.3** and **29.5**), the proper facial nerve and the intermediate nerve (*eponym*: **nerve of Wrisberg**; Wrisberg 1777; *see* Wrisberg, Heinrich August; also described by Sapolini; *see* Sapolini, Giuseppe). Via the **hiatus of Falloppio**, an *eponym* for the opening in the cranial base leading to the facial canal, they pass through the facial canal (*eponym*: **aqueduct** or **canal of Falloppio**; Falloppio 1561; *see* Falloppio, Gabriele). Intracranial branches of the facial nerve include (Fig. 29.5): (a) the greater petrosal nerve; (b) the nerve to the stapedius muscle; and (c) the chorda tympani. The **chorda tympani** passes from the facial canal through the posterior canaliculus of the

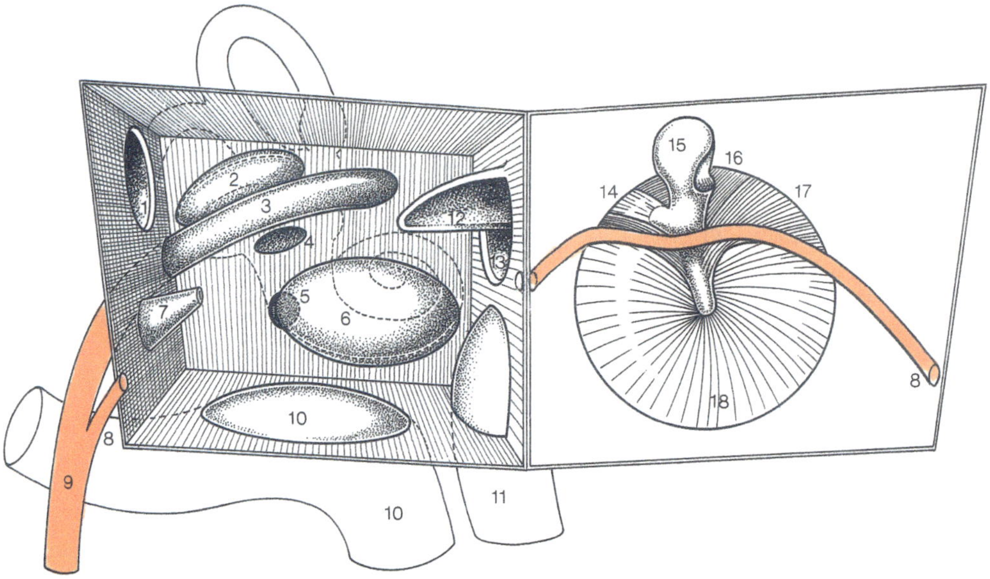

Fig. 29.3 The tympanic cavity (based on O'Rahilly R *Gardner-Gray-O'Rahilly Anatomy,* 5th ed. Saunders, Philadelphia, PA, 1986; and ten Donkelaar, HJ, Oostra, R-J *Klinische Anatomie en Embryologie,* 4th ed. Springer Media, Houten, 2014). The following structures are indicated by numbers: *1* aditus to mastoid antrum, *2* lateral semicircular canal, *3* facial canal, *4* vestibular or oval window, 5 cochlear or round window, *6* promontory, *7* pyramidal eminence, *8* chorda tympani, *9* facial nerve, *10* internal jugular vein, *11* internal carotid artery, *12* semicanal for tensor tympani muscle, *13* semicanal for auditory tube, *14* anterior mallear fold, *15* malleus, *16* flaccid part of tympanic membrane, *17* posterior mallear fold, *18* tense part of tympanic membrane

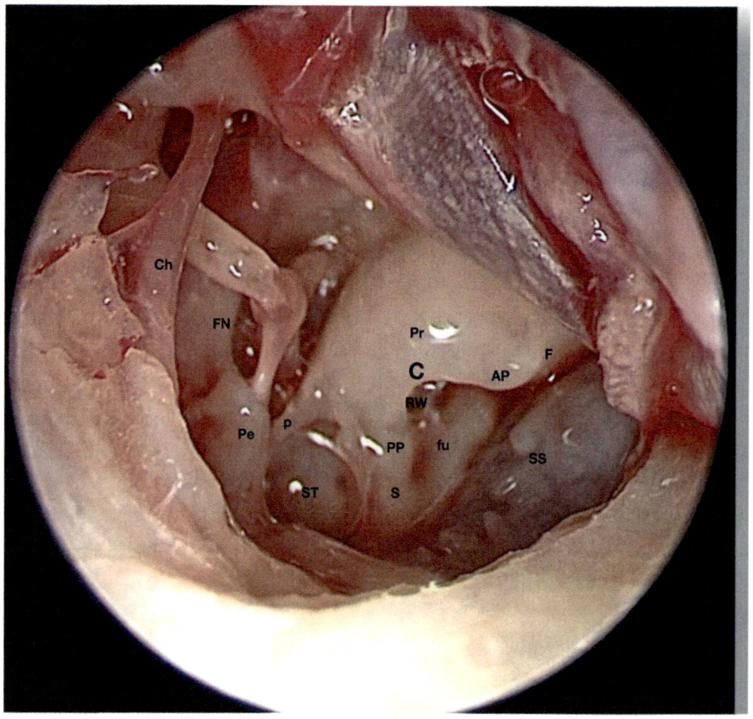

Fig. 29.4 Endoscopic picture of a fresh postmortem body with eardrum and osteomeatal flap opened (courtesy Vedat Topsakal, Brussels). *AP* anterior pillar, *C* canopy, *Ch* chorda tympani, *F* finiculus, *FN* facial nerve, *fu* fustis, *P* ponticulus, *Pe* pyramidal eminence, *PP* posterior pillar, *Pr* promontory, *RW* round window, *S* subiculum, *SS* subtympanic sinus, *ST* tympanic sinus

chorda tympani to enter the tympanic cavity via the tympanic opening of this posterior canaliculus. Inside the tympanic cavity, the chorda tympani passes within a mucosal fold between the malleus and the incus (Fig. 29.3). Via the tympanic opening of the anterior canaliculus of the chorda tympani, the chorda tympani enters the anterior canaliculus of the chorda tympani (*eponyms*: **canaliculus of Civinini** and **of Huguier**; Civinini 1830, *see* Civinini, Filippo; Huguier 1834, *see* Huguier, Pierre Charles) on its way to the petrotympanic fissure (*eponym*: **fissure of Glaser**; Glaser

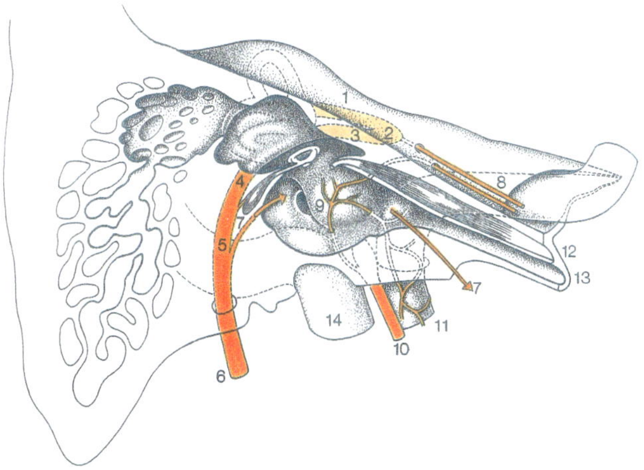

Fig. 29.5 The facial nerve (based on ten Donkelaar, HJ, Oostra, R-J *Klinische Anatomie en Embryologie,* 4th ed. Springer Media, Houten, 2014). The following structures are indicated by numbers: *1* medial segment, *2* geniculum, *3* labyrinthine segment, *4* pyramidal segment, *5* mastoid segment, *6* extracranial segment, *7* chorda tympani, *8* greater petrosal nerve (and below it the lesser petrosal nerve), *9* tympanic plexus, *10* glossopharyngeal nerve, *11* carotid plexus, *12* semicanal for tensor tympani muscle, *13* semicanal for auditory tube, *14* internal carotid artery

1680, *see* Glaser, Johann Heinrich). Mudry (2015) rightfully questioned the correctness of the use of these eponyms. He showed that Huguier and Civinini were not the first to describe the anterior canaliculus, and that Glaser did not describe the petrotympanic fissure in 1680.

29.4 Eponyms for the Bony Labyrinth

Outer and inner views of the bony labyrinth are shown in Fig. 29.6. The bony labyrinth contains the perilymph in the perilymphatic space, surrounding the membranous labyrinth. At places, the walls of the vestibule are perforated by several minute foramina, the maculae cribrosae. One of these, the macula cribrosa superior, became known in otologic surgery as **Mike's dot** (after Michael E Glasscock III, MD; Glasscock and Dickins 1982). It is an important landmark for translabyrinthine resection of an acoustic neurinoma.

29.5 Eponyms for the Cochlear Labyrinth

The **cochlear labyrinth** is formed by the **cochlear duct** or **scala media** (*eponym*: **canal of Reissner**; Reissner 1851, also known as **canal of Löwenberg**; Löwenberg 1864), which contains the spiral organ of Corti, resting on the basal lamina or basilar membrane (*eponym*: **spiral membrane of Duverney**; Duverney 1683; Fig. 29.7). It is filled with endolymph. The vestibular surface of the cochlear duct, located superiorly, consists of the vestibular membrane (*eponym*: **membrane of Reissner**; Reissner 1851) and angles downwards from lateral to medial, making the cochlear duct wedge-shaped on cross section. The external surface is formed by the spiral ligament, the thickened periosteum of the cochlea at the level of the cochlear duct. Its main part is specialized as the stria vascularis, which produces the endolymph. The spiral prominence is a highly vascularized thickening of the connective tissue of the spiral ligament, containing the vas prominens. The outer spiral sulcus separates the stria vascularis from the spiral organ. It is covered by the root cells of the cochlea (Fritzsch and Elliott 2018; ten Donkelaar et al. 2020).

The **osseous spiral lamina** consists of vestibular and tympanic lamellae with the habenula perforata in between. The habenulata perforata, a perforated rein, contains about 4000 openings, known as the foramina nervosa (*eponym*: **foramina of Morgagni**; Morgagni 1761; *see* Morgagni, Giovanni Battista), through which the fibres of the cochlear nerve pass. The osseous spiral lamina extends only halfway into the lumen of the cochlea and thereby incompletely divides the cochlea into an upper or lateral scala vestibuli and the lower or medial scala tympani. The two scalae communicate at the apex of the cochlea through an opening, the helicotrema (*eponym*: **orifice of Scarpa**; Scarpa 1772, 1789, *see* Scarpa, Antonio; also described by Breschet 1834; *see* Breschet, Gilbert). The scala vestibuli begins in the vestibule in the basal part of the cochlea, whereas the scala tympani begins at the cochlear or round window. The separation of the two scalae is completed by a connective tissue membrane, the basal lamina or basilar membrane (*eponym*: **spiral membrane of Duverney**; Duverney 1683). At its attachment to the external wall of the cochlea, the periosteum is thickened as the basal or spiral crest.

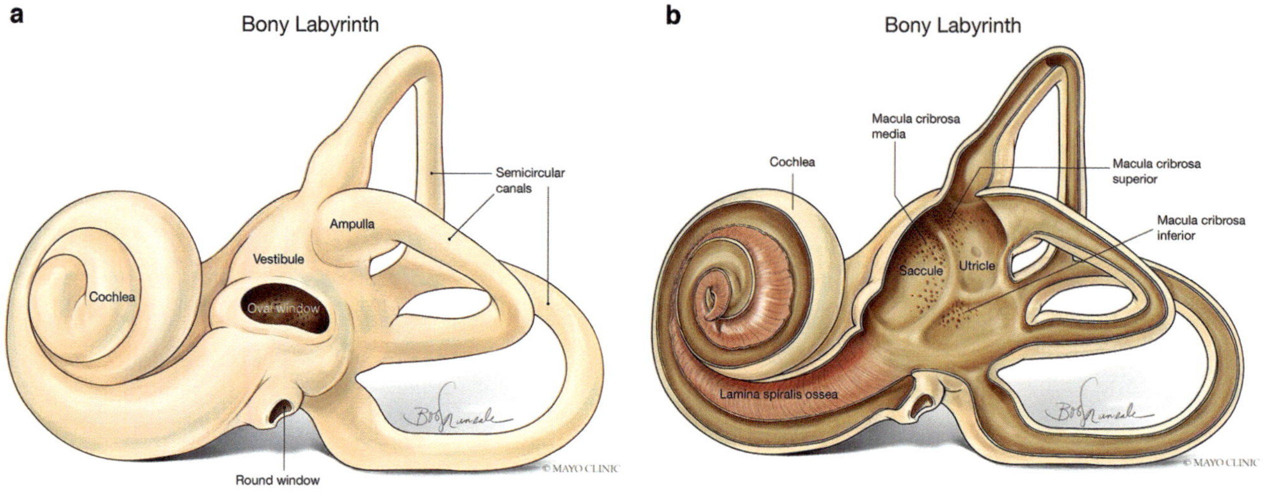

Fig. 29.6 The bony labyrinth: (**a**) outer view, (**b**) inner view. (From ten Donkelaar HJ, Kachlik D, Tubbs RS *An Illustrated Terminologia Neuroanatomica*. Springer, Cham, 2018; courtesy Bob Morreale, Mayo Clinic)

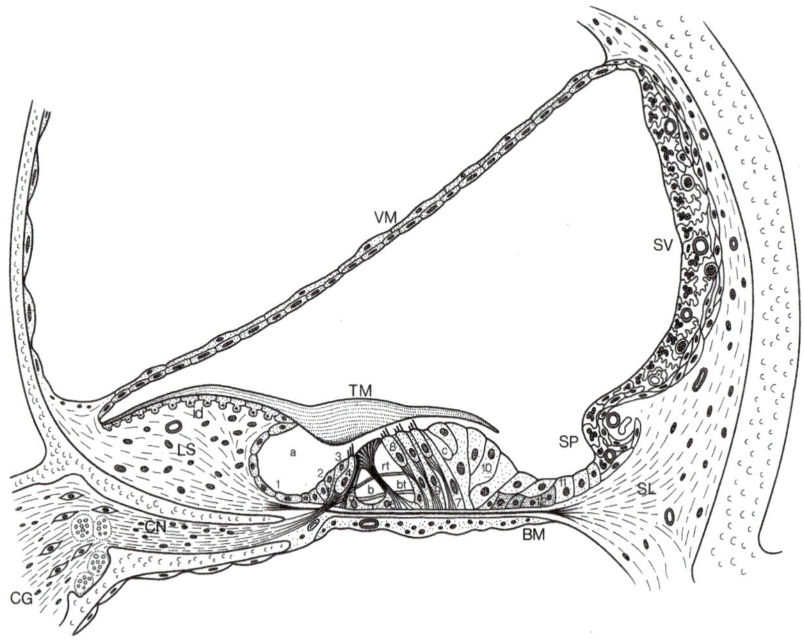

Fig. 29.7 The spiral organ of Corti (after Schuknecht HF *Pathology of the Ear,* 2nd ed. Lea and Febiger, Philadelphia, PA, 1993; from ten Donkelaar HJ, Kaga K *The auditory system.* In: ten Donkelaar HJ *Clinical Neuroanatomy,* 2nd ed. Springer, Cham, 2020, pp 373–407). *a* inner spiral tunnel, *b* inner tunnel of Corti, *BM* basal lamina, *bt* basal tunnel fiber (afferent), *c* outer tunnel, *CG* cochlear ganglion, *CN* cochlear nerve, *id* interdental cells, *LS* spiral limbus, *rt* radial tunnel fiber (efferent), *SL* spiral ligament, *SP* spiral prominence, *SV* stria vascularis, *TM* tectorial membrane, *VM* vestibular membrane, *1* inner sulcus cell, *2* inner border cell of Held, *3* inner phalangeal cell, *4* inner hair cell, *5* inner pillar cell, *6* outer pillar cell, *7*, *9* outer phalangeal cells of Deiters, *8* outer hair cell, *10* outer border cell of Hensen, *11* outer sulcus cell of Claudius, *12* basal external glandular cell of Boettcher

29.6 Eponyms for the Spiral Organ

The **spiral organ** (*eponym*: organ **of Corti**; Corti 1851; *see* Corti, Alfonso) is located on the tympanic surface, the inferior or medial wall of the cochlear duct (Fritzsch and Elliott 2018; ten Donkelaar et al. 2020). It contains hair cells, supporting cells and border cells (Fig. 29.7). The **cochlear hair cells** (*eponym*: **hair cells of Corti**; Corti 1851) are the sensory receptor cells of which there is a single row of inner hair cells and three rows of outer hair cells. They exhibit a tonotopic distribution: receptors sensitive to high frequencies are located near the cochlear base and those sensitive to low frequencies near the cochlear apex. The **supporting cells** include the inner and outer pillar cells, also known with the *eponym* **rods of Corti**, which are separated by the inner tunnel (*eponym*: **tunnel of Corti**; Corti 1851) extending along the whole length of the cochlea, and the inner and outer phalangeal cells. The **border cells** include inner and outer sulcus and border cells. The cochlear hair cells and part of the supporting cells are overlaid by the gelatinous tectorial membrane (*eponym*: **membrane of Corti**; Corti 1851). The connective tissue of the limbus forms teeth-like protrusions, the auditory teeth (*eponym*: **teeth of Huschke**; Huschke 1844), which are separated by deep furrows. Towards the cochlear duct, the auditory teeth are covered by a simple epithelium, formed by the interdental cells. The tectorial membrane is anchored to the interdental cells of the vestibular lip of the spiral limbus. At the site of the inner hair cells, the inner stripe of the tectorial membrane (*eponym*: **stripe of Hensen**; Hensen 1863; *see* Hensen, Viktor) is found, a thickening of the tectorial membrane. Here, the tectorial membrane is anchored to the inner phalangeal and border cells. More externally, the outer stripe of the tectorial membrane (*eponyms*: **membrane of Hardesty**; Hardesty 1915; *see* Hardesty, Irving; **stripe of Kimura**; Kimura 1975; *see* Kimura, Robert) is found. Here, the tallest rows of the outer hair cells are embedded in the tectorial membrane (Kimura 1966, 1975).

The sulcus, border and supporting cells of the spiral organ are arranged from the inner spiral sulcus in an external direction (Fig. 29.7): (1) the inner or cuboidal inner sulcus cell (no eponym); (2) the inner or internal border cell (*eponym*: **inner border cell of Held**; Held 1902, 1926; *see* Held, Hans); (3) the inner or internal phalangeal cell (no eponym); (4) the inner or internal pillar cell (*eponym*: **rod of Corti**; Corti 1851); (5) the outer or external pillar cell (*eponym*: **rod of Corti**; Corti 1851); (6) the outer or external phalangeal cell (*eponym*: **epithelial cell of Deiters**; Deiters 1860; *see* Deiters, Otto Friedrich Karl); (7) the outer or external border cell (*eponym*: **border cell of Hensen**; Hensen 1863; *see* Hensen, Viktor); (8) the basal external glandular cell (*eponym*: **glandular cell of Boettcher**; Boettcher 1859; *see* Boettcher, Arthur), restricted to the basal turn; and (9) the outer or cuboidal sulcus cell (*eponym*: **supporting cell of Claudius**; Claudius 1856; *see* Claudius, Friedrich Matthias).

The supporting cells all rest on the basal lamina of the epithelium that separates the epithelial cells of the spiral organ from the connective tissue of the epithelium. At their free surface, the cochlear hair, pillar and phalangeal cells are interconnected by junctional complexes, light-microscopically with the appearance of a continuous thin membrane, the reticular membrane (*eponym*: **reticular membrane of Kölliker**; Kölliker 1852; *see* Kölliker, Rudolf Albert von). In the epithelium of the spiral organ, several clefts or tunnels exist between the lateral cell surfaces: (1) the inner or internal tunnel (*eponym*: **tunnel of Corti**; Corti 1851); (2) the intermediate tunnel (*eponym*: **tunnel** or **space of Nuel**; Nuel 1872; *see* Nuel, Jean-Pierre); (3) the outer or external tunnel (*eponym*: **tunnel of Held**; Held 1902, 1926; *see* Held, Hans).

29.7 Eponyms on Neural Components of the Ear

The neural components of the inner ear are the cochlear ganglion, the vestibular ganglion and the vestibulocochlear nerve (Fig. 29.8). The cochlear ganglion, known as the *eponym* **ganglion of Corti** (Corti 1851; *see* Corti, Alfonso), extends only halfway from the base of the cochlea to the apex. The vestibular ganglion, known as the *eponym* **ganglion of Scarpa** (Scarpa 1789; *see* Scarpa, Antonio), lies outside the semicircular canals. Its superior and inferior parts continue into the superior and inferior parts of the vestibular nerve. The vestibulocochlear nerve enters the internal acoustic meatus through the internal acoustic opening. The **internal acoustic meatus** ends blindly as the **fundus of the internal acoustic opening**, a thin cribriform plate of bone

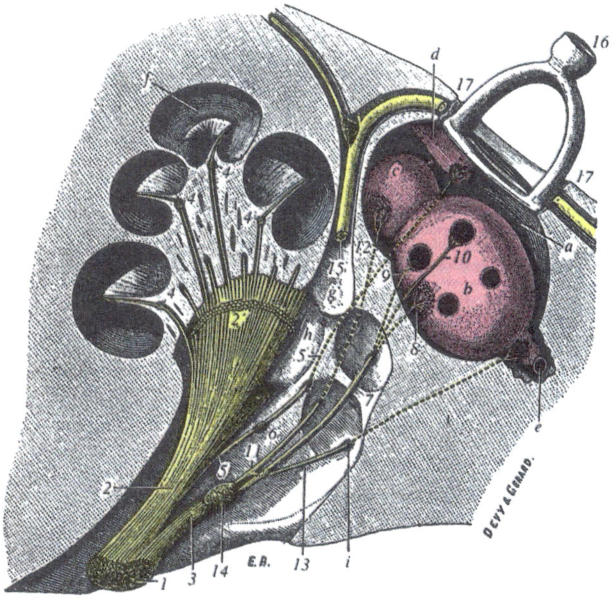

Fig. 29.8 The distribution of the eight cranial nerve (from Testut L, Latarjet A *Traité d'anatomie humaine*, Tome Troisième, Doin, Paris, 1930). *a* vestibular part with *b* utriculus, *c* sacculus, and *d* initial part of cochlear canal, *e* ampulla of posterior semicircular canal, *f* cochlea, *g* aqueduct of Falloppio (facial canal), *h* fundus of internal acoustic meatus, *i* singular foramen, *1* trunk of acoustic nerve, *2* cochlear branch, *3* vestibular branch, *4* spiral ganglion of Corti, *5* small branch for vestibular part in cochlear canal, *6* ganglion of Boettcher, **7** superior vestibular nerve, *8* utricular nerve, *9* superior ampullary nerve, *10* external ampullary nerve, *11* inferior vestibular nerve, *12* saccular nerve, *13* posterior ampullary nerve, *14* ganglion of Scarpa, *15* facial nerve, *16* stapes, *17* oval window

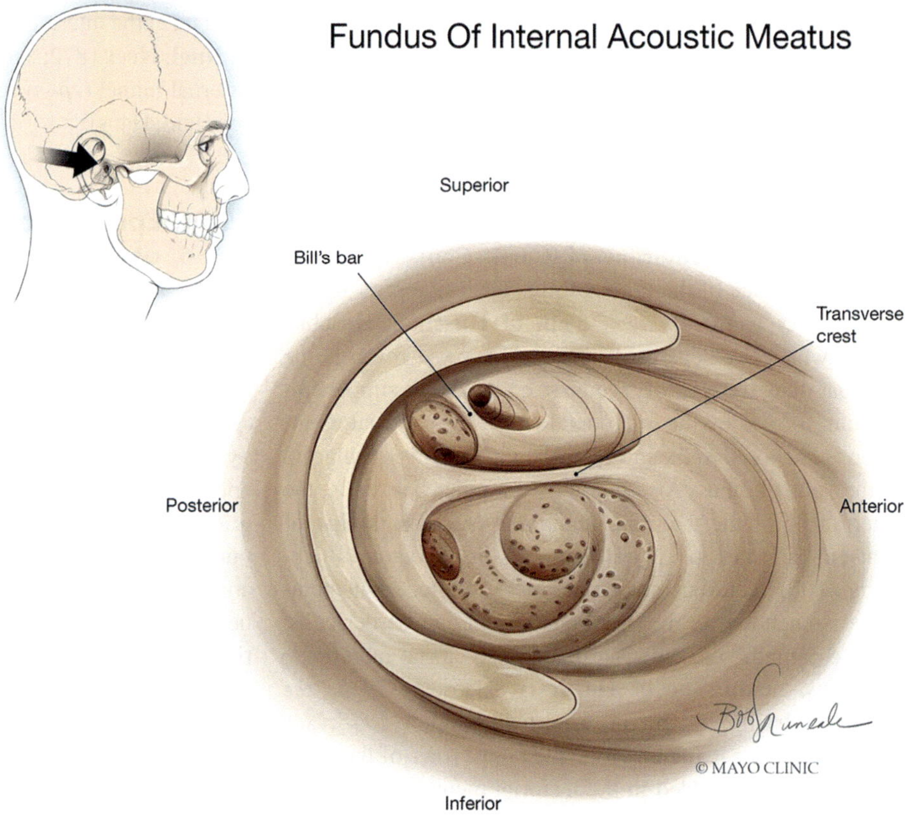

Fig. 29.9 The fundus of the internal acoustic meatus. (From ten Donkelaar HJ, Kachlik D, Tubbs RS *An Illustrated Terminologia Neuroanatomica*. Springer, Cham, 2018; courtesy Bob Morreale, Mayo Clinic)

separating the cochlea and the vestibule from the internal acoustic meatus (Fig. 29.9). The **transverse crest** divides the fundus into superior and inferior regions. The superior region is divided by a **vertical crest**, named **Bill's bar** after William F. House, D.D.S., M.D., into an anterior, facial area for the facial nerve, and a posterior or superior vestibular area for the superior part of the vestibular nerve. The inferior region consists of the inferior vestibular area and the cochlear area. Two nerves received eponyms: (a) the **nerve of Oort** (Oort 1918; *see* Oort, Henricus), an *eponym* for efferent fibres of the cochlear nerve, which leave the brain stem with the vestibular nerve and join the cochlear nerve to end in the spiral organ of Corti; and (b) the superior saccular branch, known as the *eponym* **nerve of Voit** (Voit 1907; *see* Voit, Max). From the cochlear nerve, inside the internal acoustic meatus, a vestibular branch comes off with its own ganglion, the **ganglion of Boettcher** (Fig. 29.8), which innervates the portion of the cochlear duct located in the cavity of the vestibule (Boettcher 1856; *see* Boettcher, Arthur).

References

Boettcher A (1856) Observationes microscopiae de ratione, qua nervus cochleae mammalium terminatur. Schünmann and Mathiesen, Dorpat

Boettcher A (1859) Weitere Beiträge zur Anatomie der Schnecke. Arch Pathol Anat Physiol Klin Med 17:243–281

Bolobas B (1972) A hallószery mikrochirurgial anatomiaja. Könykiadó, Budapest (in Hungarian)

Breschet G (1834) Etudes sur l'ouie, Paris

Civinini F (1830) Linee anatomiche. Fasc. II: In osteologia, sulla scissura di Glaser nel temporale. Bracali, Pistoja

Claudius M (1856) Bemerkungen über den Bau der hautigen Spiralleiste der Schnecke. Z Wiss Zool 7:154–161

Corti A (1851) Recherches sur l'organe de l'ouie des mammifères. Z Wiss Zool 3:109–172

Cotugno D (1760) De acquaeductus auris humanae internae (Dissertation). Typographia Sanctae Thomas Aquinatis, Napoli

Deiters O (1860) Beiträge zur Kenntniss der Lamina spiralis membranacea der Schnecke. Z Wiss Zool 10:1–14

Duverney G (1683) Traité de l'organe de l'ouie. Michallet, Paris. (also published by Langerak, Leyden, in Latin, 1730, and a new French edition in 1731; and an English version by Baker, London, 1737, translated by John Marshall)

Eustachi B (1563) Epistola de auditus organis. In: Opercula Anatomica. Luchinus, Venice

Falloppio G (1561) Observationes anatomicae. Ulmo and Perchachino, Venice

Fritzsch B, Elliott KL (2018) Auditory nomenclature: combining name recognition with anatomical description. Front Neuroanat 12:99

Gerlach J (1848) Handbuch der allgemeinen und spezielle Gewebelehre, Mainz

Gerlach J (1875) Zur Morphologie der Tuba Eustachii. Beilage zur Mschr Ohrenheilk 9:49

Glaser JH (1680) Tractatus posthumus de cerebro in quo ujus non fabrica tantum. Bertschi, Basel

Glasscock ME, Dickins JRE (1982) Complications of acoustic tumor surgery. Otolaryngol Clin N Am 15:883–895

Hardesty I (1915) On the nature of the tectorial membrane and its probable role in the anatomy of hearing. Am J Anat 8: 109–179

Held H (1902) Untersuchungen über den feineren Bau des Ohrlabyrinthes der Wirbelthiere. Teubner, Leipzig

Held H (1926) Die Cochlea der Säuger und der Vögel, ihre Entwicklung und ihr Bau, Receptionsorgane 1. Springer, Berlin

Henle J (1855) Handbuch der systematischen Anatomie des Menschen, Vol I: Knochenlehre. Vieweg, Braunschweig

Hensen V (1863) Zur Morphologie der Schnecke des Menschen und der Säugethiere. Z Wiss Zool 13:481–512

Huguier PC (1834) Dissertation sur quelques points d'anatomie, de physiologie et de pathologie. Didot, Paris

Huschke E (1844) Sömmerring Vom Baue des menschlichen Körper, Vol. V. Voss, Leipzig, pp 822–823

Kimura RS (1966) Hairs of the cochlear sensory cells and their attachments to the tectorial membrane. Acta Otolaryngol 61:55–72

Kimura RS (1975) The ultrastructure of the organ of Corti. Int Rev Cytol 42:173–222, Elsevier

Körner O (1927) Das Septum petrosquamosum (mastoideum) und seine klinische Bedeutung. Fortschr Hals Nasen Ohrenheilkd 17:137–140

Lang J (1992) Klinische Anatomie des Ohres. Springer, Wien/New York

Löwenberg BB (1864) La lame spirale du limaçon de l'oreille et des mammifères, Paris

Macewen W (1893) Pyogenic infections of the brain. Maclehose, Glasgow

Mansour S, Magnan J, Haidar Ahmad H, Nicolas K, Louryan S (2019) Comprehensive and clinical anatomy of the middle ear, 2nd edn. Springer Nature, Cham

Marchioni D, Soloporto D, Colleselli E, Totti M, Patel N, Jufas N (2016) Round window chamber and fustis: endoscopic anatomy and surgical implications. Surg Radiol Anat 38:1013–1019

Morgagni GB (1761) De sedibus et causismorborum per anatomen indigeris. Lib V, Venice

Mudry A (2015) Glaser fissure, Huguier canal, and Civinini canal. A confused eponymical imbroglio. Otol Neurotol 36:1115–1120

Nuel JP (1872) Beitrag zur Kenntniss der Säugethierschnecke. Arch Mikr Anat 8:200–215

Oort H (1918) Über die Verästelung des Nervus octavus bei Säugetiere. Anat Anz 51:272–280

Ostmann L (1892) Druck und Drucksteigerung im Labyrinth. Arch Ohrenheilkd 34:35–53

Pertick O (1883) Neues Divertikel des Nasenrachenraums. Virchow's Arch 94:1–12

Pirsig W, Mudry A (2015) Huschke's anterior external auditory canal foramen: art before medicinae? Otol Neurotol 36:555–560

Proctor B (1967) Embryology and anatomy of the Eustachian tube. Arch Otolaryngol 86:503–526

Prussak A (1867) Zur Anatomie des menschlichen Trommelfells. Arch Ohrenheilkd 3:255–280

Ramsay Hunt J (1907) On herpetic inflammations of the geniculate ganglion. A new syndrome and its complications. J Nerv Ment Dis 34:73–96

Reissner E (1851) De auris internae formatione (Dissertation). Laakmann, Dorpat

Retzius G (1881) Das Gehörorgan der Wirbelthiere, Vol. I. Samson and Wallin, Stockholm

Retzius G (1884) Das Gehörorgan der Wirbelthiere, Vol. II. Samson and Wallin, Stockholm

Rosenmüller JC (1805) Chirurgisch-anatomische Abbildungen für Aerzte und Wundärzte, Weimar

Santorini GD (1724) Observationes anatomicae. Recurti, Venice

Scarpa A (1772) De structura fenestrae rotundae aurii et de tympano secundario anatomicae observationes, Modena

Scarpa A (1789) Anatomicae disquisitiones de auditu et olfactu. Galeati, Pavia

Shrapnell HJ (1832) On the form and structure of the membrana tympani. Lond Med Gaz 10:120–124

ten Donkelaar HJ, Elliott KJ, Fritzsch B, Kachlik D, Carlson M, Isaacson B, Topsakal V, Broman J, Tubbs RS, Baud R (2020) An updated terminology for the internal ear with combined anatomical and clinical terms. J Phonet Audiol 6:1000147

TNA (2017) Terminologia Neuroanatomica. http://FIPAT.library.dal.ca; Digital, updated version of 2022 available at https://ifaa.unifr.ch

Topsakal V, Matulic M, Assadi MZ, Mertens G, Assadi MZ, Mertens G, Van Rompaey V, Van de Heyning P (2020) Comparison of the surgical techniques and robotic techniques for cochlear implantation in terms of the trajectories toward the inner ear. J Int Adv Otol 16:3–7

Topsakal V, Kachlik D, Bahşi I, Carlson M, Isaacson B, Broman J, Tubbs RS, Baud R, ten Donkelaar HJ (2021) Relevant temporal bone anatomy for robotic cochlear implantation: an updated terminology combined with anatomical and clinical terms. Transl Res Anat 25:100138

Tourtual KT (1846) Neue Untersuchungen ueber den Bau des menschlichen Schlund- und vergleichend-anatomische Bemerkungen, Leipzig

Voit M (1907) Zur Frage der Verästelung des Nervus acusticus bei den Säugethieren. Anat Anz 31:635–655

von Kölliker A (1852) Mikroskopische Anatomie oder Gewebelehre des Menschen. Engelmann, Leipzig

von Kölliker A (1867) Handbuch der Gewebelehre des Menschen für Ärzte und Studierende. Engelmann, Leipzig

von Soemmerring T (1806) Abbildungen des menschlichen Hörorgans. Varrentrapp & Wenner, Frankfurt am Main

von Troeltsch AF (1860) Die Anatomie des Ohres in ihrer Anwendung auf die Praxis und der Krankheiten des Gehörorgans. Stahel, Würzburg

Wrisberg HA (1777) Observationes anatomicae de quinto pare nervorum encephali. Johann Christian Dietrich, Göttingen

Zaufal E (1880) Die Plica salpingo-pharyngea (Wulstfalte). Arch Ohr-Heilkunde 15:96–144

30 Anatomical Eponyms in Phlebology

David Kachlík, Alberto Caggiati,
and Hans J. ten Donkelaar

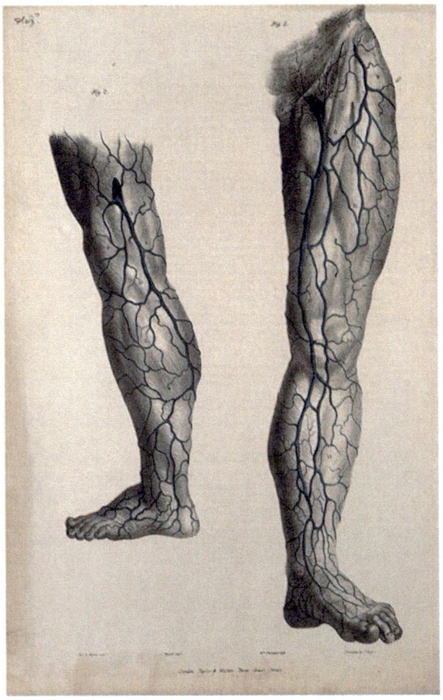

Vignette Veins of the lower limb shown in a coloured lithograph by William Fairland (1837; from the Wellcome Collection; in public domain)

30.1 Introduction

In phlebology, many eponyms are in use for perforating veins. The **vignette** shows a coloured lithograph of the veins of the leg by William Fairland from 1837. Perforating veins or perforators in the lower limb are numerous and quite variable (Cockett 1953; Dodd and Cockett 1956; van Limborgh et al. 1962; Caggiati et al. 2002, 2014; Kachlik et al. 2019). For the upper limb, only one perforating vein has received clinical relevance for applying an arteriovenous fistula for maintenance of hemodialysis, and is known as the *eponym* **cubital vein of Gracz**, after Kenneth Charles Gracz, an American surgeon in Chicago (Gracz et al. 1977). In clinical practice, many perforators of the lower limb have been associated with authorities in the field including Bassi, Boyd, Cockett, Dodd, Hach, Hunter, Kuster, May and Sherman, sometimes misleading. Therefore, descriptive terms are preferred (Caggiati et al. 2014; Kachlík et al. 2019). Van Limborgh et al. (1962) listed already 49 perforators (Fig. 30.1). When perforating veins pass through the muscular fascia, they are mostly accompanied by a small artery and a cutaneous nerve, together known as the *eponym* **triad of Staubesand** (Staubesand 1994; *see* Staubesand, Jochen). A lateral marginal vein may persist from an embryonic vein of the limb and is known as the **vein of Servelle** (Servelle 1985). Here, we concentrate on the perforating veins of the lower limb. For further data on the authors quoted, reference is made to Part II (*The Anatomical Eponyms from A to Z*) and figures in this part.

30.2 Perforators of the Lower Limb

A more recent schematic representation of the topography of the main groups of perforating veins is shown in Fig. 30.2. Grouped on the basis of their topography, Caggiati et al. (2002, 2014) distinguished from the foot to the thigh upwards:

1. **Perforators of the foot:** dorsal, medial, lateral and plantar foot perforators, according to their location. Kuster et al. (1968) gave a detailed subdivision of them.
2. **Ankle perforators:** medial, anterior and lateral ankle perforators, according to their location (*eponym*: **perforating veins of Kuster**; Kuster et al. 1968; *see* Kuster, G; Fig. 30.3a). The medial ankle perforators are found fairly constantly at 4 cm from the ground immediately behind the medial malleolus at the beginning of the posterior accessory great saphenous vein, also known as the *eponym* **vein of Leonardo** (Caggiati et al. 2014).
3. **Perforators of the leg:** (a) **Medial leg perforators**: paratibial perforators, corresponding to the Sherman per-

H. J. ten Donkelaar et al., *An Illustrated Guide to Anatomical Eponyms*, https://doi.org/10.1007/978-3-031-91664-9_30

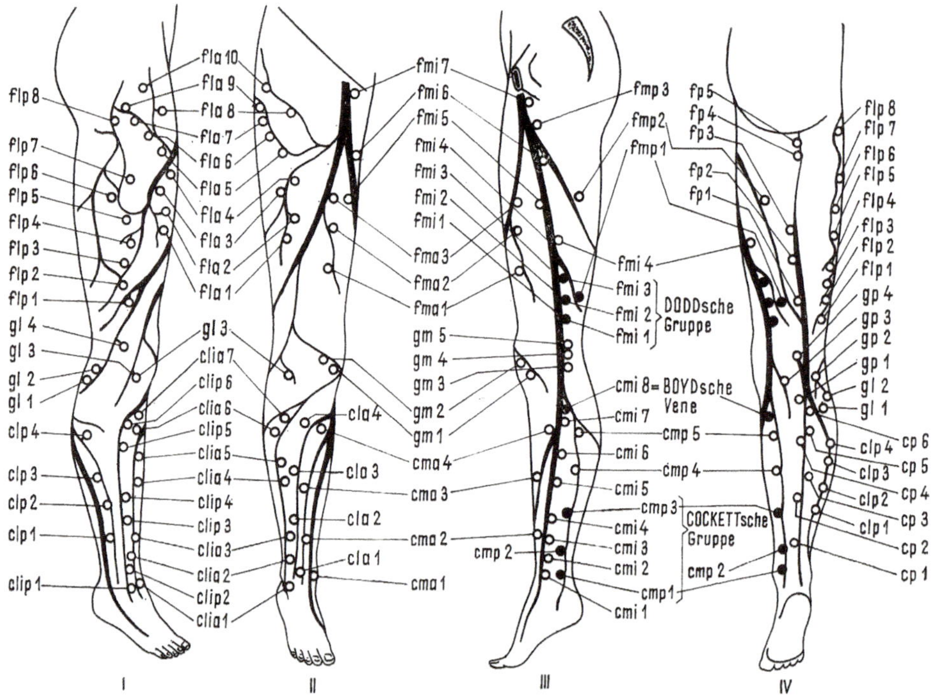

Fig. 30.1 Summary of the communicating or perforating veins of the lower limb (after van Limborgh J, Banga DA, Meijerink CJH, Luigiesi HH *De venae communicantes van het been*. Ned T Geneeskd 1962; 106:415–420). *cla, clia, clip, clp, cma, cmi, cmp, cp* anterior lateral, anterior intermediate lateral, posterior intermediate lateral, posterior lateral, anterior medial, intermediate medial, posterior medial and posterior crural communicating veins; *fla, flp, fma, fmi, fmp, fp* anterior lateral, posterior lateral, anterior medial, intermediate medial, posterior medial and posterior femoral communicating veins; *gl, gm, gp* lateral, medial and posterior genual communicating veins

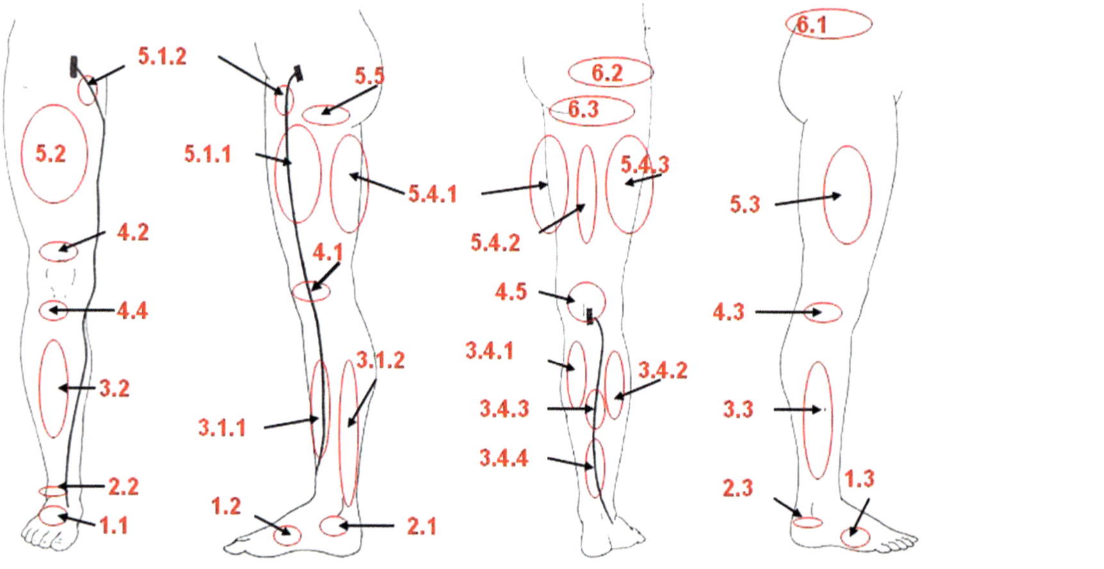

Fig. 30.2 Schematic representation of the topography of the main groups of perforating veins (after Caggiati A, Bergan JJ, Gloviczki P, Jantet G, Wendell-Smith CP, Partsch H, International Interdisciplinary Consensus Committee on Venous Anatomical Terminology *Nomenclature of the veins of the lower limbs: An international interdisciplinary consensus statement*. J Vasc Surg 2002; 36:416–422). *1.1, 1.2, 1.3* dorsal, medial and lateral foot perforators; *2.1, 2.2, 2.3* medial, anterior and lateral ankle perforating veins; *3.1.1, 3.1.2, 3.2, 3.3, 3.4.1, 3.4.2, 3.4.3, 3.4.4* paratibial, posterior tibial, anterior leg, lateral leg, medial gastrocnemius, lateral gastrocnemius, intergemellar and para-Achillean perforators; *4.1, 4.2, 4.3, 4.4, 4.5* medial knee, suprapatellar, lateral knee, inrapatellar and popliteal fossa perforators; *5.1.1, 5.1.2, 5.2, 5.3, 5.4.1, 5.4.2, 5.4.3, 5.5* femoral canal, inguinal, anterior thigh, lateral thigh, posteromedial thigh, sciatic, posterolateral and pudendal perforators; *6.1, 6.2, 6.3* superior, middle and lower gluteal perforators

forators (lower and mid leg; *eponym*: **perforating veins of Sherman**; Sherman 1944; *see* Sherman, R Shanton; Fig. 30.3a), the Boyd perforator (upper leg; *eponym*: **perforating vein of Boyd**; Boyd et al. 1953; *see* Boyd, Alexander Michael; Fig. 30.4), found fairly constantly immediately below the knee, and the posterior tibial perforators, corresponding to the Cockett perforators (*eponym*: **perforating veins of Cockett**; Cockett 1953, *see* Cockett, Frank Bernard; Dodd and Cockett 1956; Fig. 30.3a); The lower, middle and superior Cockett per-

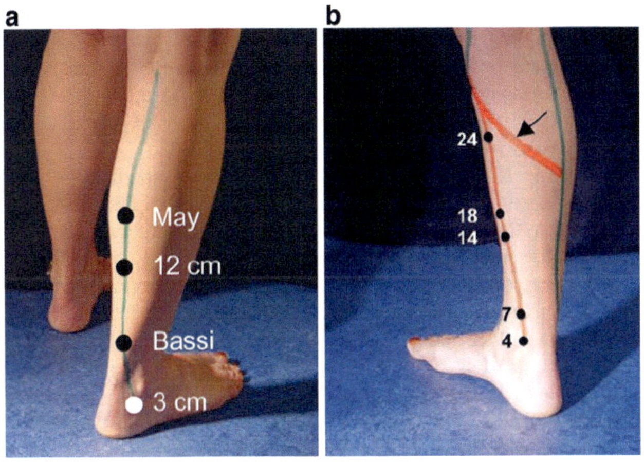

Fig. 30.3 (**a**) Location of the soleal perforating vein of May and the para-Achillean perforating vein of Bassi; (**b**) Posteromedial view of the leg demonstrating the intersaphenous vein (*arrow*) between the small and great saphenous veins. The *black dots* represent the locations of perforating veins. The *numbers* by the perforating veins show their distance in cm from the ground in a standing patient. *4* medial ankle perforating vein of Kuster, *7*, *14*, *18* lower, middle and upper posterior tibial perforating veins of Cockett, *24* medial paratibial perforating vein of Sherman. (Both from Caggiati A, Mendoza E, Murena-Schmidt R, Lattimer CR *Anatomy of the superficial veins*. In: Mendoza E et al. (eds) *Duplex Ultrasound of Superficial Leg Veins*. Springer, Berlin, Heidelberg, New York, 2014, pp 19–47; courtesy Alberto Caggiati, Rome)

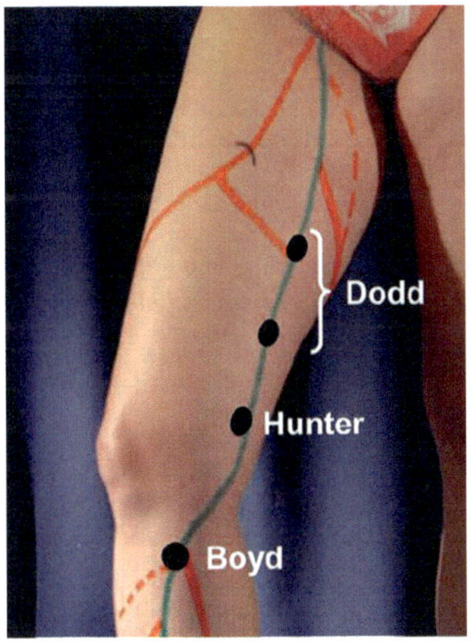

Fig. 30.4 Location of the perforating veins of Dodd and Hunter in the adductor canal, and the paratibial perforating vein of Boyd. (From Caggiati A, Mendoza E, Murena-Schmidt R, Lattimer CR *Anatomy of the superficial veins*. In: Mendoza E et al. (eds) *Duplex Ultrasound of Superficial Leg Veins*. Springer, Berlin, Heidelberg, New York, 2014, pp 19–47; courtesy Alberto Caggiati, Rome)

forating veins (Cockett I, II and III) are located generally at 7, 14 and 18 cm from the ground, but this may vary, depending on the length of the subject; (b) **anterior leg perforators**; (c) **lateral leg perforators**, and (d) **posterior leg perforators**, divided into medial (*eponym*: **perforating vein of Gillot**; Gillot 1994; *see* Gillot, Claude) and lateral gastrocnemius perforators, intergemellar perforators between the heads of the gastrocnemius muscle (*eponym*: **midcalf perforating vein of May**; May et al. 1980; see May, Robert; Fig. 30.3b), and para-Achillean perforators (*eponym*: **perforating vein of Bassi**; Bassi 1962; *see* Bassi, Glauco; Fig. 30.3b). Gillot et al. (2019) also described bony perforating veins of the knee.

4. **Perforators of the knee:** medial knee, suprapatellar, lateral knee, infrapatellar and popliteal (*eponym*: **perforating vein of Thierry**; Thierry 1986; *see* Thierry, L) perforators.
5. **Perforators of the thigh:** (a) **Medial thigh perforators:** femoral (*eponym*: **perforating veins of Dodd**; Dodd 1959; *see* Dodd, William Harold Alfred; Fig. 30.4), and inguinal perforators. A perforator in the adductor canal is usually ascribed to John Hunter (*eponym*: **perforating vein of Hunter**; Fig. 30.4), but is rarely present within the adductor canal itself; (b) **anterior thigh perforators**, piercing the quadriceps femoris muscle; (c) **lateral thigh perforators**, piercing the lateral muscles of the thigh, and (d) **posterior thigh perforators**, posteromedial, sciatic, posterolateral (*eponym*: **perforating vein of Hach**; *see* Hach, Wolfgang), breaking through the biceps femoris and semitendinosus muscles, and pudendal.
6. **Perforators of the gluteal muscles**, superior, middle and inferior.

30.3 A Few Notes on Eponyms for Superficial Veins of the Lower Limb

The terminology of the superficial veins of the limbs is not entirely clear, especially due to their high variability. A few *eponyms* exist for superficial veins of the lower limb. The membranous layer of the subcutaneous tissue of the lower limb, known as saphenous fascia, received the *eponym* **fascia of Sherman** (Sherman 1944; *see* Sherman, R Shanton). Following the direction of the blood flow, the *eponym* **arcade of Braune** (Braune 1890; *see* Braune, Christian Wilhelm) is used for a superficial vein at the root of the toes, opening into the superficial plantar venous plexus by intercapitular veins. This arcade also drains the anterior part of this plexus and interconnects it with the veins of the dorsum of the foot. The *eponym* **plexus of Lejars** (Lejars 1890; *see* Lejars, Félix-Marie Louis) is used for the superficial plantar venous plexus, composed of fine venules within the dermis and subcutaneous tissue of the sole, earlier described by Bourceret (1885). The tributaries of the saphenous veins on the leg are known as the **veins of Braune** (Braune 1890), comprising the anterior and posterior saphenotibial veins and the saphenofibular veins. The cranial extension of the small saphenous vein was first described as the femoropopliteal vein by Hyrtl (*eponym*: **vein of Hyrtl**; Hyrtl 1864; *see* Hyrtl, Joseph). One of its potential terminations, the intersaphenous vein of the thigh, mainly known as the **vein of Giacomini** (Giacomini 1873; *see* Giacomini, Carlo), connects it to the great saphenous vein, thus corresponding to the posterior femoral circumflex vein. For an independent third superficial venous system on the lateral aspect of the knee, the *eponym* **vein** or **system of Albanese** is used (Albanese et al. 1969; *see* Albanese, Alfonso). One of the main contributors to the great saphenous vein, the accessory great saphenous vein, was probably first described by Cruveilhier (*see* Cruveilhier, Jean Baptiste), but the *eponym* **vein of Cruveilhier** is unprecisely applied to different kinds of the accessory saphenous vein.

References

Albanese AR, Albanese AM, Albanese EF (1969) Lateral subdermic varicose vein system of the legs. Its surgical treatment by the chiseling tube method. Vasc Surg 3:81–89

Bassi G (1962) The varices of the lower limbs. Edizione Minerva Medica, Torino. (French translation: Les varices des membres inférieurs. Doin, Paris, 1967)

Bourceret H (1885) Circulation locale. Procédé d injection des veines. Note Acad Sciences

Boyd AM, Catchpole BN, Jepson RP, Rose SS (1953) Major venous ligation in the treatment of postphlebotic sequelae. Lancet 265:113–116

Braune CW (1890) Die Venen des usses und Unterschenkels, Leipzig

Caggiati A, Bergan JJ, Gloviczki P, Jantet G, Wendell-Smith CP, Partsch H, International Interdisciplinary Consensus Committee on Venous Anatomical Terminology (2002) Nomenclature of the veins of the lower limbs: an international interdisciplinary consensus statement. J Vasc Surg 36:416–422

Caggiati A, Mendoza E, Murena-Schmidt R, Lattimer CR (2014) Anatomy of the superficial veins. In: Mendoza E et al (eds) Duplex ultrasound of superficial leg veins. Springer, Berlin/Heidelberg/New York, pp 19–47

Cockett FB (1953) The ankle blow-out syndrome: a new approach to the varicose ulcer problem. Lancet 1(6749):17–23

Dodd WHA (1959) The varicose tributaries of the superficial femoral vein passing into Hunter's canal. Postgraduate Med J 35:18–23

Dodd WHA, Cockett FB (1956) The pathology and surgery of the veins of the lower limb, 2nd edn. Churchill Livingstone, London. 1976

Giacomini C (1873) Osservazioni anatomiche per servire allo studio della circolazione venosa della estremita inferiori fatte all'isituto anatomica di Torino. Vercellino, Torino

Gillot C (1994) Les veines perforantes inférieures de la jambe, de la cheville et du pied. Phlebologie 47:76–104

Gillot C, Uhl JF, Ovelar J, Merino J (2019) Anatomy of the bony perforator veins of the knee. Ann Med 51(Suppl 1):60

Gracz KC, Ing TS, Soung LS, Armbruster KFW, Seim SK, Merkel FH (1977) Proximal forearm fistula for maintenance hemodialysis. Kidney Int 11:71–75

Hyrtl J (1864) Über normale und abnormale Verhältnisse der Schlagadern des Unterschenkels. Sitz Math-Naturwiss Classe Kais Akad Wiss

Kachlik D, Pechacek V, Hnatkova G, Hnatek L, Musil V, Baca B (2019) The venous perforators of the lower limb – a new terminology. Phlebology 34:650–668

Kuster G, Lofgren EP, Hollinshead WH (1968) Anatomy of the veins of the foot. Surg Gynecol Obstet 127:817–823

Lejars FML (1890) Les veines de la plante du pied chez l'homme et les grands animaux. Ach Physiol, 5e Série; II:89

May R, Partsch H, Staubesand J (1980) Perforating veins. Urban & Schwarzenberg, Munich

Servelle M (1985) Klippel and Trenaunay's syndrome. Ann Surg 201:365–373

Sherman RS (1944) Varicose veins. Anatomic findings and operative procedure based upon them. Ann Surg 120:772–784

Staubesand J (1994) Die Perforanten-Trias: Ein funktionelles System. Vasomed 6:447–450

Thierry L (1986) Anatomie chirurgicale de la fosse poplitée. Phlebologie 39:57–66

van Limborgh J, Banga DA, Meijerink CJH, Luigiesi HH (1962) De venae communicantes van het been. Ned T Geneeskd 106:415–420

31 Anatomical Eponyms in Radiology of the Digestive System

Marina Quartu and Hans J. ten Donkelaar

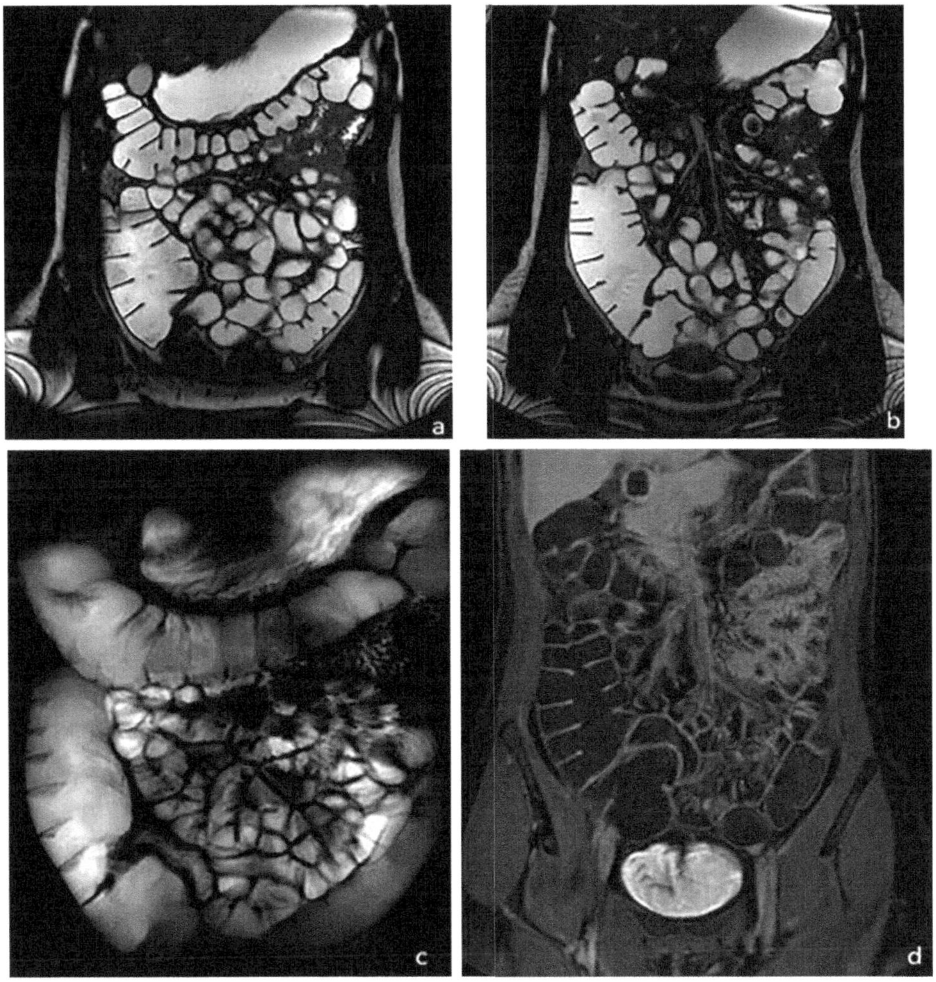

Vignette MR enterography in a healthy woman: (**a, b**) T2-balanced coronal images, (**c**) T2-weighted fat-suppressed single shot coronal image, and (**d**) a coronal T1-weighted image after intravenous gadolinium-chelate injection. (From Maccioni et al. 2023 Diagnostics 13:241; Open Access; courtesy Francesca Maccioni, Rome)

31.1 Introduction

Eponyms used in radiology of the digestive system are numerous and classically refer to the imaging appearance of the alimentary canal in physiological and pathological conditions from the pharynx to the anorectal canal (Kanne et al. 2006a, b). Some examples on the pharynx and oesophagus, the stomach, the intestines, anorectal defects, the liver and extrahepatic biliary ducts, the gallbladder and the pancreas are presented with relevant illustrations. For further data on

H. J. ten Donkelaar et al., *An Illustrated Guide to Anatomical Eponyms*, https://doi.org/10.1007/978-3-031-91664-9_31

the various authors quoted, reference is made to Part II (*The Anatomical Eponyms from A to Z*) and figures in this part. The **vignette** shows the possibilities of wide-field enterography of a healthy adult (Maccioni et al. 2023) through T2-balanced magnetic resonance imaging (MRI) in coronal images (**a**, **b**), in a T2-weighted fat-suppressed single shot coronal image (**c**), and in a coronal T1-weighted image after intravenous gadolinium-chelate injection (**d**). Both imaging sequences are essential for satisfactory bowel MRI examination, allowing radiologists to make accurate diagnoses based on the characteristics of the examined tissues.

31.2 The Pharynx and the Oesophagus

Pulsion diverticula of the border region of the pharynx and the oesophagus can be divided into diverticula of Zenker, Killian-Jamieson and Laimer. They involve herniations of the mucosa through defects in the muscular wall of that region. As shown in Fig. 31.1a, the more common Zenker diverticulum arises between the inferior pharyngeal constrictor muscle and the cricopharyngeal muscle (*eponym*: **diverticulum of Zenker**; Zenker and Ziemssen 1877; *see* Zenker, Friedrich Albert von). The Laimer diverticulum is a common diverticle of the posterior upper oesophagus, arising from the triangle of Laimer, inferior to the transverse fibres of the cricopharyngeal muscle (*eponym*: **diverticulum of Laimer**; Laimer 1883; *see* Laimer, Edouard). The less known diverticulum of Killian-Jamieson arises from the triangle of Killian-Jamieson, delimited by the oblique fibres of the thyropharyngeal muscle and the transverse fibres of the cricopharyngeal muscle (*eponym*: **diverticulum of Killian-Jamieson**; Killian 1908; *see* Killian, Gustav). Figure 31.1b, c shows oesophagograms of a Zenker diverticulum (**b**) and a Killian-Jamieson diverticulum (**c**). For further discussion, *see* Rubesin and Levine (2001) and Watanabe et al. (2023).

An 'inferior oesophageal sphincter' is physiologically independent of the rest of the oesophagus, but anatomically indistinguishable when relaxed (Gray and Skandalakis 1972). Apaydin et al. (2008), however, demonstrated a clear thickening of the circular muscle layer at the gastroesophageal junction. A few clinical findings on pathology of the oesophagus will be briefly discussed: (a) the Mallory-Weiss tear; (b) the Boerhaave syndrome; (c) the Schatzki ring; and (d) the Barrett oesophagus. Longitudinal lacerations of the distal oesophageal mucosa, often extending into the gastric cardia and usually occurring after any event that provokes a sudden rise in the intragastric pressure or gastric prolapse into the oesophagus (Fig. 31.2a), are known with the *eponym* **Mallory-Weiss tear**, after George Kenneth Mallory (1900–1986), an American pathologist born in Boston and working at Boston University, and Soma Weiss (1898–1942), an Hungarian pathologist, who emigrated to the United States during World War I, and working at Peter Bent Brigham Hospital in Boston (Mallory and Weiss 1929). 'Effort rupture' or 'spontaneous rupture' of the oesophagus may result from profound stress such as forceful vomiting or excessive straining, and received the *eponym* **Boerhaave syndrome** (Boerhaave 1724; *see* Boerhaave, Herman). The **Schatzki ring** (Fig. 31.2b) is an *eponym* for a symptomatic mucosal fold found in the distal oesophagus above the gastrooesophageal junction (**A ring**) or at the squamocolumnar junction of the lower oesophagus (**B ring**; *see* Fig. 31.3), becoming symptomatic when the diameter of the ring narrows to 11 mm. It was described by Richard Schatzki (1901–1992), a German diagnostic radiologist, born in Clafeld and working in Berlin and, later, at Harvard Medical School, Cambridge, Massachusetts (Schatzki and Gary 1953). The *eponym* **Barrett oesophagus** (Fig. 31.2c) is used

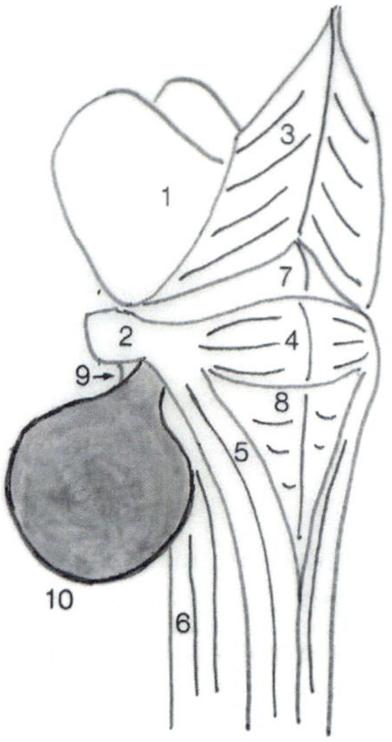

Fig. 31.1 (**a**) Anatomy of the pharyngeal-oesophageal border; (**b**, **c**) comparison of (**b**) a Zenker diverticulum versus (**c**) a Killian-Jamieson diverticulum. (From Watanabe et al. 2023 Surg Case Reports 9:21, Open Access; courtesy Yuki Watanabe, Miyagi, Japan)

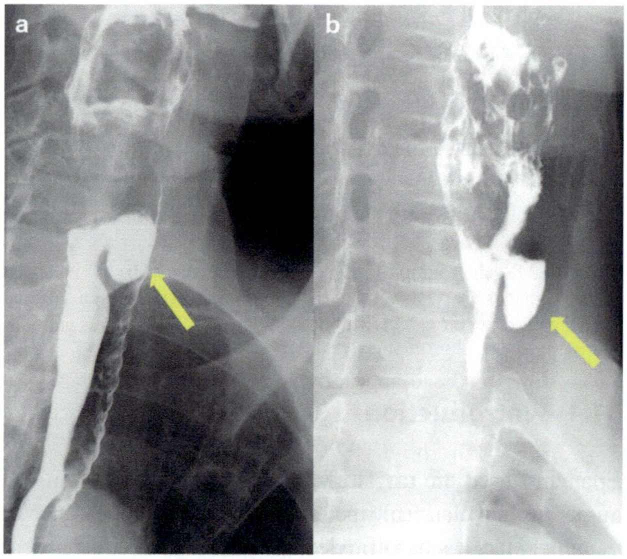

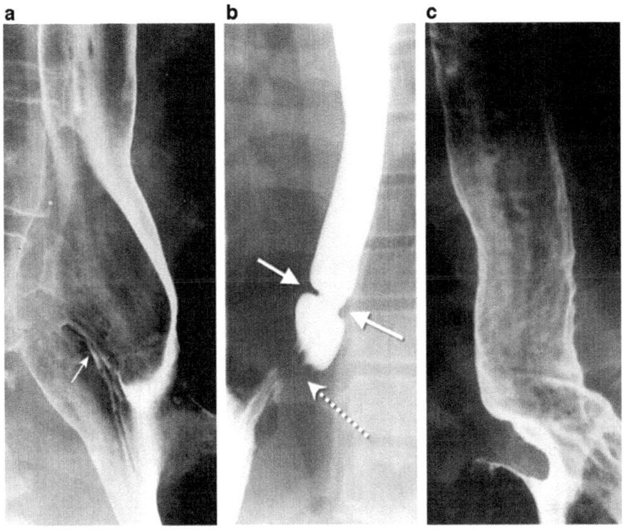

Fig. 31.2 Oesophagograms of (**a**) Mallory-Weiss tear, (**b**) Schatzki ring, and (**c**) Barrett oesophagus. ((**a**) From Wikipedia, courtesy Harvey M. Goldstein, M.D., San Antonio, TX; (**b**) from Towbin AJ, Diniz LO 2012 *Schatzki ring in pediatric and young adult patients.* Pediatr Radiol 42:1437–1440, with permission, courtesy Alexander J. Towbin, Cincinnati, OH; (**c**) from Levine MS 2005 *Barrett esophagus: Update for radiologists.* Abdom Imaging 30:133–141, with permission, courtesy Marc S Levine, Philadelphia, PA)

for intestinal metaplasia of the oesophageal mucosa, thought to arise from chronic reflux oesophagitis. It was described by Norman Rupert Barrett (1903–1979), a British surgeon born in Adelaide, Australia, and working at the St Thomas Hospital in London, where he established one of the first thoracic surgery and intensive care units (Barrett 1950).

31.3 The Stomach

The Gastrooesophageal Junction: At the gastrooesophageal junction, the circular muscular layer of the lower part of the oesophagus is continuous with that of the stomach (Fig. 31.3). This junction is usually known as a functional sphincter but morphological data point at a true anatomical sphincter (Apaydin et al. 2008). During swallowing, the oesophagus and stomach are drawn into the thorax for about 1 cm. The elasticity of the fibrous tissue with which the oesophagus and stomach are anchored to the diaphragm, redraws the stomach downwards again. This fibrous tissue bears the *eponym* **membrane of Bertelli and Laimer** (*see* Bertelli, Dante, and Laimer, Edouard; Laimer 1883) for the phrenicooesophageal membrane or ligament, connecting the oesophagus to the diaphragm, extensions of the diaphragmatic fascia to the oesophagus and stomach (Eliška 1973; Friedland 1978). Other *eponyms* for the gastrooesophageal junction are (a) the **angle of His** for the cardiac notch (*see* His, Wilhelm Jr); (b) the **semilunar fold** or **valve of Gubaroff** (*see* Gubaroff, Alexander Petrowitsch von) for a perpendicular fold at the cardiac ostium of the stomach, described as cardiac sphincter between the oesophagus and stomach; this fold contains the **sling fibres**, these are known as (c) the *eponym* **loop of Willis** (*see* Willis, Thomas) that became also known as the **collar of Helvetius** (Helvetius 1719; *see* Helvetius, Claude-Adrien), an *eponym* for the transition of the circular muscle fibres of the oesophagus to the oblique muscle fibres of the stomach at the cardiac notch of the stomach (*see* Zifan et al. 2017 for discussion); (d) the **ligament of Helvetius** for lateral muscle bands of the stomach; and (e) the **fibres of Rouget-Juvara**, an *eponym* for phrenicooesophageal muscle fibres; usually, ascending fibres in the thorax are called **fibres of Juvara** (Juvara 1899; *see* Juvara, Ernest) and the descending fibres in the abdomen **fibres of Rouget** (*see* Rouget, Charles Marie Benjamin).

Infantile hypertrophic pyloric stenosis, a most common gastric surgical disorder in neonates, is due to a narrowed and elongated pyloric canal. It causes projectile vomiting and can lead to severe dehydration in babies. The disease was described by Harald Hirschsprung (1830–1916), a Danish pediatrician born and working in Copenhagen, as 'Angeborener Pylorusstenose' (*eponym*: **congenital pyloric stenosis of Hirschsprung**; Hirschsprung 1888; *see* Hirschsprung, Harald). Its treatment involves pylorotomy as promoted by Conrad Ramstedt (1867–1963), a German surgeon born in Hamersleben, Saxony, and working in Halle and Münster (*eponym*: **pylorotomy of Ramstedt**; Ramstedt 1912). This technique was developed in France by Pierre Fredet (Dufour and Fredet 1908) and in Germany by Wilhelm Weber (1910).

31.4 The Intestines

For the small and large intestines, it is important to start with their development. In the fifth week of development, the future stomach is enlarged, and from the midgut the primary intestinal loop is formed with the superior mesenteric artery as the axis. From its proximal limb the jejunum and part of the ileum are formed, and from the distal limb the distal part of the ileum, the caecum, the ascending colon and the transverse colon. At the point between the proximal and distal limbs, the primary intestinal loop is continuous with the omphalomesenteric or vitelline duct in the umbilical cord. During the sixth week of development, the **first stage** of intestinal rotation and development takes place, due to the large growth of the liver, there is less space for the midgut, resulting in the **physiological umbilical hernia**. The proximal limb enlarges extensively to form the jejunum and the proximal part of the ileum. The distal part shows the beginning of the caecum and the vermiform appendix. At the 10th week of development, the midgut returns to the abdominal cavity, the **second phase**, and rotation of the midgut occurs,

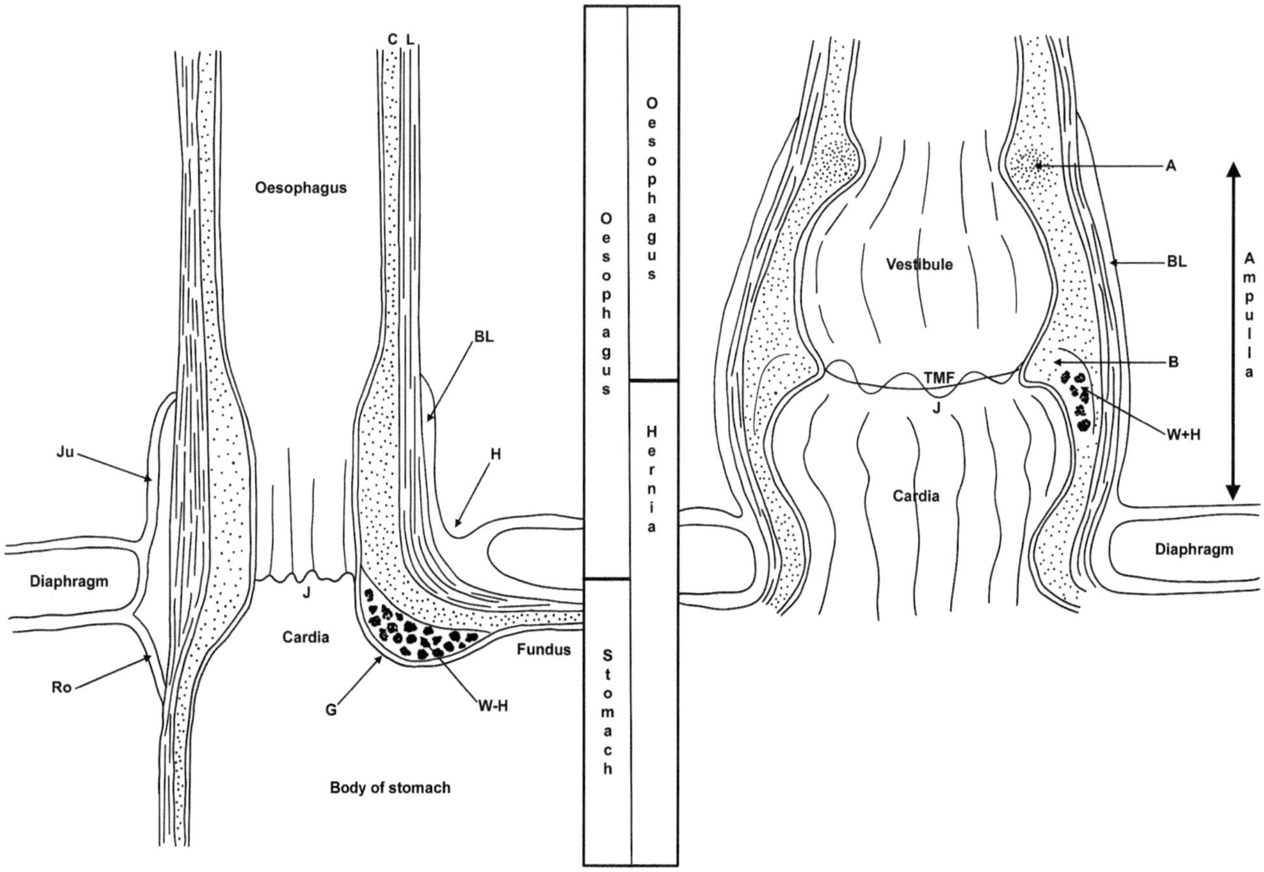

Fig. 31.3 The gastrooesophageal junction during rest (at the *left*) and during swallowing (at the *right*; based on Friedland 1978; and O'Rahilly R *Gardner-Gray-O'Rahilly Anatomy. A regional study of human structure,* Fifth ed. Saunders, Philadelphia, 1986; courtesy David Kachlík and Anhelina Khadanovich, Prague). *A* A ring (inferior oesophageal sphincter), *B* B ring (mucosal fold present in 15% of cases, as Schatzki ring if symptomatic), *BL* phrenicooesophageal membrane of Bertelli and Laimer, *C* circular muscle layer, *G* semilunar fold or valve of Gubaroff, *H* cardiac notch or angle of His, *J* gastrooesophageal junction (Z-line), *Ju* ascending phrenicooesophageal muscle fibres of Juvara, *L* longitudinal muscle layer, *Ro* descending phrenicooesophageal muscle fibres of Rouget, *TMF* transverse muscular fold, *W-H* loop of Willis or collar of Helvetius (containing the sling fibres (*black dots*)

resulting normally in the fixation of the caecum in the right lower part of the abdominal cavity, the **third stage**.

The omphalomesenteric or vitelline duct is normally cut off from the primary intestinal loop, but may persist as the **diverticulum of Meckel** (Meckel 1817; *see* Meckel, Johann Friedrich, the Younger). When present, it usually occurs within 60 cm of the ileocaecal valve on the antimesenteric side of the bowel, and occurs in 2–3% of the population. **Anomalies of intestinal rotation and fixation** form a spectrum of conditions that occur during the normal embryological process of intestinal rotation. The range of abnormalities arises from perturbations in the sequence of herniation, rotation, and fixation of the midgut. Most of the schematic descriptions and illustrations have been presented by several embryologists and anatomists starting with Meckel (1817), Mall (1889), Schwalbe (1906) and by many others later, but their observations did not earn them an eponym. Abnormalities may be asymptomatic and symptomatic and include nonrotation, mixed rotation, reversed rotation and hyperrotation (Estrada 1958; Snyder and Chaffin 1954; Gray and Skandalakis 1972; Steward et al. 1976; Langer 2017), summarized in Fig. 31.4. During the returning of the intestines, it appears important that the distal, caecocolic loop does not return to the abdomen before the return of the proximal loop so that the duodenum and jejunum are not pushed superolaterally and undergo only 180° of rotation. This condition is termed **nonrotation**, and since it is associated with a wide-based mesentery, does not cause a midgut volvulus. Classical **malrotation** occurs as a result of failed extracoelomic rotation and results in a narrowed mesenteric base, which predisposes the child to potentially fatal midgut volvulus. In typical malrotation, the duodenojejunal junction is located in the right upper quadrant and the caecum located in the middle to upper abdomen (Fig. 31.5), and fixed in place by the **bands of Ladd** (Ladd 1932), an *eponym* for fibrous shoots of peritoneal tissue to the gallbladder, duodenum, and right-sided abdominal wall, as described by William Edward Ladd (1880–1967), an American paediatric surgeon, born in

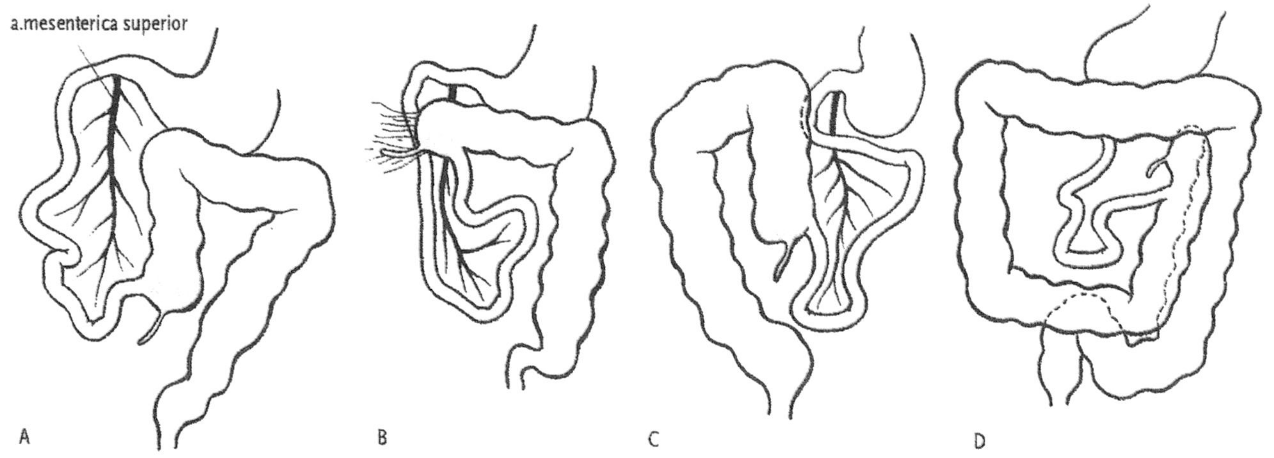

Fig. 31.4 Schematic overview of malrotations of the intestines: (**a**) 'nonrotation', (**b**) mixed rotation, (**c**) reverse rotation, and (**d**) hyperrotation. (After Gray and Skandalakis 1972; after ten Donkelaar HJ and Oostra R-J 2014 *Klinische Anatomie en Embryologie*, 4th ed. Springer Media, Houten)

Milton, Massachussets, and working at the Boston Children's Hospital. A less common condition is **reverse rotation**, in which the duodenum is placed anterior to the superior mesenteric artery and the colon lies posteriorly, producing a retroarterial tunnel. If the mesoderm does not fuse to the retroperitoneum during the fourth and fifth months of gestation, paraduodenal or paracolic hernias may form. In **hyperrotation**, the caecum can be found in the left upper abdominal cavity. **Congenital internal herniae** may occur through peritoneal fossae and apertures (Kelahan et al. 2021) and may lead to small bowel obstruction into adulthood. They include left and right paraduodenal, pericaecal, sigmoid mesocolic, omental foramen, transomental, small intestine mesenteric and others (Kelahan et al. 2021). A left paraduodenal hernia occurs at the left paraduodenal fossa (Fig. 31.6), known with the *eponym* **fossa of Gruber-Lanzert** (*see* **Fig. G25**, Gruber, Wenzel (Václav) Leopold, and Landzert, Theodor). A right paraduodenal hernia occurs at the **mesentericoparietal fossa of Waldeyer** (*see* **Fig. W3** and Waldeyer, Heinrich Wilhelm Gottfried).

Structural arrangement of groups of circular fibres in the wall of the large intestine have been noted (Fig. 31.7): (a) from the ileum to the caecum (ascribed to Varolio); (b) from the caecum to the ascending colon (the **sphincter of Busi**; Busi 1933; *see* Busi, Aristide); (c) in the proximal part of the ascending colon (the **sphincter of Hirsch**; Hirsch 1924; *see* Hirsch, I Seth); (d) in the transverse colon at the junction of the proximal and middle thirds (the **sphincter of Cannon-Böhm** or **ring of Cannon** (Cannon 1902, 1911), described by Walter Bradford Cannon, an American physiologist, 1871–1945 (*see* Cannon, Walter Bradford), and Gottfried Böhm, a German radiologist, 1880–1952); (e) at the splenic colic flexure (the **sphincter of Payr-Strauss**), described by Erwin Payr, an Austrian surgeon (1871–1946) and Ludwig Georg Strauss, a German radiologist (1949–2013); (f) between the descending colon and the sigmoid colon (the **sphincter of Balli**; Balli 1939; *see* Balli, Ruggero); (g) in the sigmoid colon (the **sphincter of Moultier**); and (h) between the sigmoid colon and the rectum (the **sphincter of O'Beirne** and **of Hyrtl**; O'Beirne 1833, *see* O'Beirne, James; Hyrtl 1846, *see* Hyrtl, Josef). A functional sphincter in this area received the *eponym* **sphincter of Rossi**. Gagliardi et al. (1994) showed that among these sphincters those of Balli and Rossi are involved in nerve reflexes and are functional sphincters, but that those of Hirsch, Moultier and Busi are true sphincters due to thickenings of circular muscle fibres. Cannon's ring is the site of overlap between the innervation of the superior and inferior mesenteric plexuses.

31.5 Anorectal Defects

Functional colonic obstructions may result from congenital or infection-derived absence of the myenteric plexus. They include **Hirschsprung disease**, described by Hirschsprung in (1887), an *eponym* for a functional colonic obstruction resulting from the failure of neural crest cells to migrate the entire length of the gastrointestinal tract (Fig. 31.8), and **Chagas disease**, described by Carlos Justitiano Ribeiro Chagas (1879–1934), a Brazilian physician born in Oliveira (Chagas 1909), a digestive tract disorder deriving from infection by *Trypanosoma cruzi*, causing megaviscera, primarily megaoesophagus and megacolon and resulting from destruction of the myenteric plexus by the protozoan parasite.

Anorectal malformation is a collective term for a variety of congenital anomalies also referred to as imperforate anus. It is often linked to genetic syndromes such as the Currarino triad and trisomy 21. In these patients, the opening of the anus is narrow, covered with skin, or missing; often, a fistulous tract opens onto the perineum anterior to the anal mus-

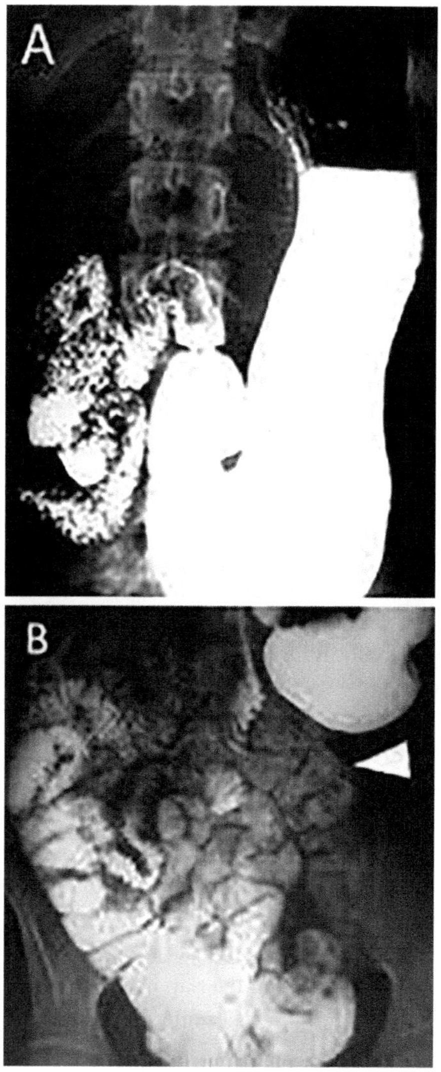

Fig. 31.5 Malrotation of the intestines with the duodenum, the proximal jejunum and the duodenojejunal angle to the right of the midline (**a**) and the caecum to the left of the midline ((**b**), arrowhead; from Del Toro C, Cabrera-Aguirre A, l Casillas J, Ivanovic A, Scortegagno E, Estango I, Alessandrino F 2023 *Imaging spectrum of non-neoplastic and neoplastic conditions of the duodenum*. Abdom Radiol 48:2237–2257, with permission, courtesy Francesco Alessandrino, Miami)

cle complex or into adjacent anatomical structures. The fistulous tract associated with anal and rectal obstructions can connect the rectum with the urinary bladder, urethra, vagina, navicular fossa of the urethra, or the perineum. The current classification is based upon the seminal work of William E Ladd and Robert E Gross (1934) and illustrated in Fig. 31.9. The farther the fistulous tract opens from the normal anatomic location, the more likely it is that there are additional associated issues such as underdeveloped musculature of the anal muscle complex. Correctly classifying the anorectal malformation has significance regarding the patient's prognosis, and it is a crucial component of determining the patients' long-term potential for bowel control. MRIs of some examples are shown in Fig. 31.10.

31.6 The Liver

Macroscopically, the liver is subdivided into four lobes, right, left, caudate and quadrate. Its internal anatomy is different, however. Already Rex and Cantlie suggested a subdivision based on the distribution of the portal vein. Hugo Rex can be seen as a pioneer (Rex 1888; *see* Rex, Hugo). In 1897, James Cantlie (1851–1926) described his eponymous line (*eponym*: **line of Cantlie**; Cantlie 1897; *see* Cantlie, James), which divides the two halves of the liver, extending from the fossa of the gallbladder anteriorly to the inferior vena cava posteriorly, which he based on autoptic findings, in which the right side of the liver was reduced to a mass of fibrotic tissue whereas the left side of the liver showed a marked hypertrophy (Fig. 31.11). The line of Cantlie or **line of Cantlie-Rex**, also known as the **principal plane** of the liver, is nowadays used by hepatic surgeons to check for the portal blood supply and look upon the preresection use of portal vein ligation or embolization to induce hypertrophy of the part of the liver the surgeons intend to preserve. Further work on the human liver, such as that from Healey and Schroy (1953), paved the way for the studies of Claude Maurice Couinaud (1922–2008), a French surgeon and anatomist who proposed the concept of segmental anatomy (*see* **Fig. C24**; Couinaud 1957; *see* Couinaud, Claude Maurice), that remains the standard and established the nomenclature by which surgeons classify liver resections (*see* Strasberg 2005).

Accessory liver lobules are relatively uncommon and are characterized by the presence of a supernumerary lobe of normal hepatic parenchyma in continuity with the liver, directly or through stalks (for review *see* Medeiros de Araújo et al. 2022). They are thought to represent a developmental error during the third week of gestation, marked by an erroneous development of the endodermal caudal foregut and segmentation of the hepatic bud. The very first review of the various anatomical variants of the liver, including an extensive study on accessory liver lobes, was presented by Thomas Cullen (1925). The first description of accessory lobules dates back to Morgagni. Cullen provided a comprehensive review of accessory liver lobules connected to the rest of the liver by a liver pedicle or mesentery. Most of these structures are usually asymptomatic and without clinical interest. However, there have been case reports of resection because of a concurrent focal nodular hyperplasia and adenomas, hepatic carcinoma, or due to symptoms following torsion. Indeed, most descriptions of this alteration originate from autopsy and laparoscopy reports as an incidental finding. Epidemiologic studies point out an incidence, found by CT/

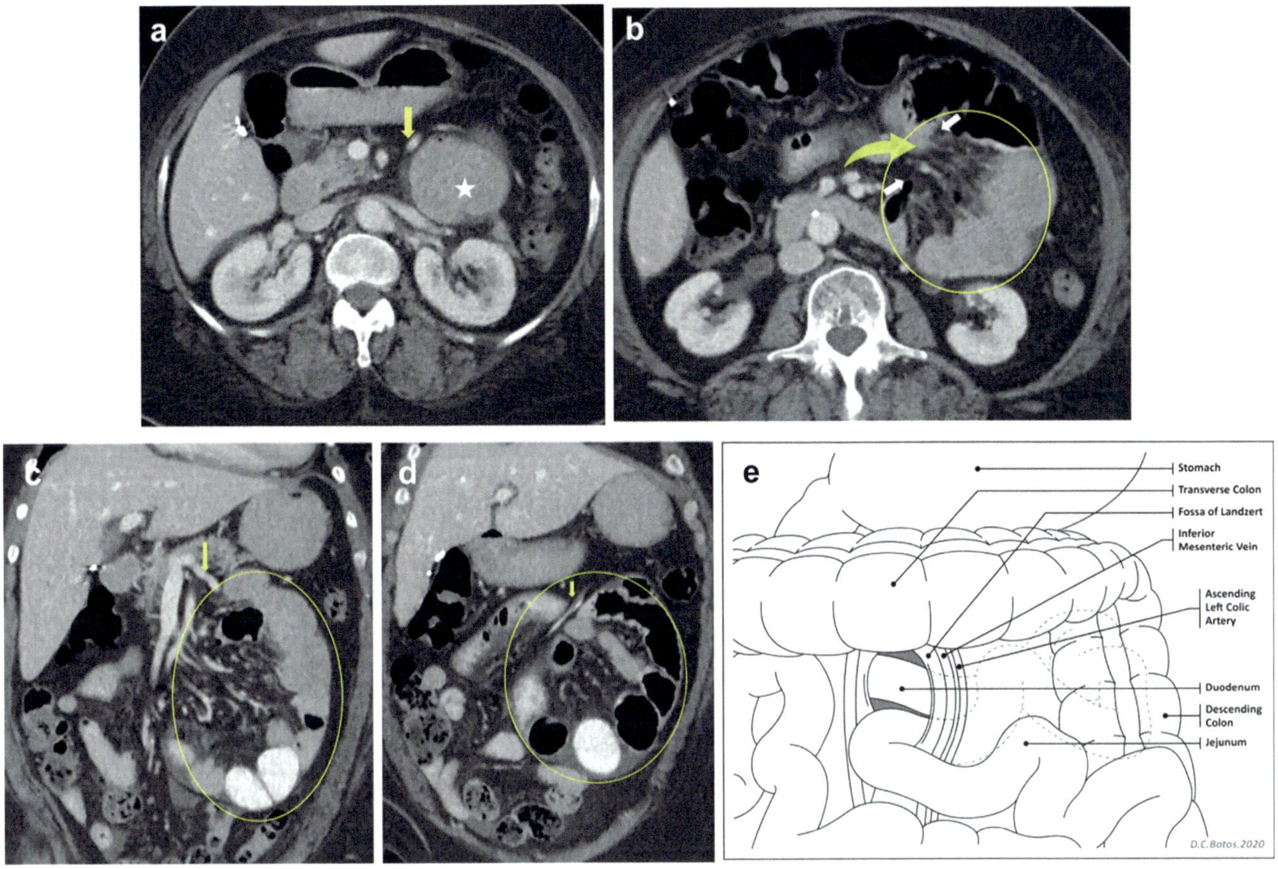

Fig. 31.6 A left paraduodenal hernia. (From Kelahan L, Menias CO and Chow L 2021 Abdom Radiol 46:1825–1836; with permission courtesy Linda Kelahan, Chicago)

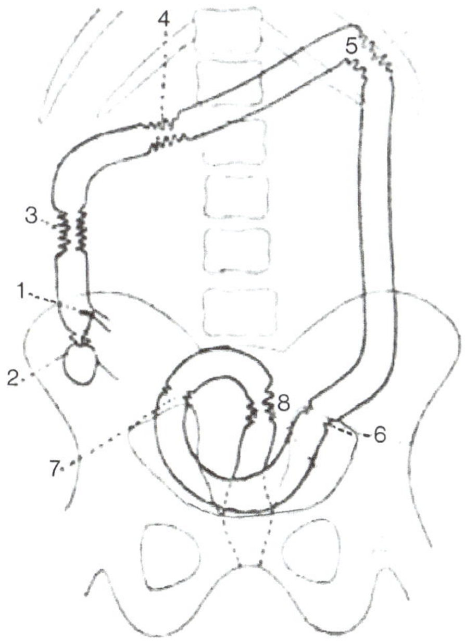

Fig. 31.7 Sphincters described for the colon, starting from the caecum (after Balli 1939): *1* sphincter of Varolio, *2* sphincter of Busi, *3* sphincter of Hirsch, *4* sphincter of Cannon-Böhm, *5* sphincter of Payr-Strauss, *6* sphincter of Balli, *7* sphincter of Moultier, *8* sphincter of Rossi

MRI, of 0.7–1% (Kapoor et al. 2017) A more recent classification by Collan et al. (1978) defines four main types of accessory lobules: (a) an ectopic liver, not connected to the hepatic parenchyma and, usually, attached to the gallbladder or intra-abdominal ligaments; (b) a microscopic ectopic liver that may be occasionally found in the gallbladder wall; (c) a large, pedunculated accessory liver lobule, attached to the main liver by a stalk; and (d) a small accessory liver lobule attached to the main liver without a stalk (sessile). The **lobe of Riedel** is the most well-known type of accessory liver lobule (*see* **Fig. R21**; Riedel 1888; *see* Riedel, Bernhard Moritz Carl Gustav).

31.7 The Extrahepatic Biliary Ducts, the Gallbladder and the Pancreas

The intrahepatic and extrahepatic biliary ducts are rather variable (*see* Hollinshead 1971). Large intrahepatic bile ducts may be affected in **Caroli disease**, an *eponym* for a congenital anomaly of the biliary tree, characterized by saccular dilatation of the intrahepatic bile ducts (Fig. 31.12a), with a

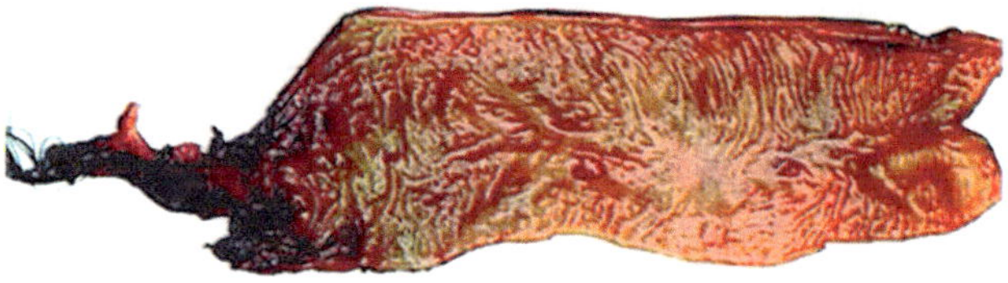

Fig. 31.8 Postmortem preparation of the distal colon in Hirschsprung disease. (From ten Donkelaar HJ, Lammens M and Hori A 2023 *Clinical Neuroembryology*, 3rd ed. Springer Nature, Cham; courtesy Martin Lammens, Antwerp)

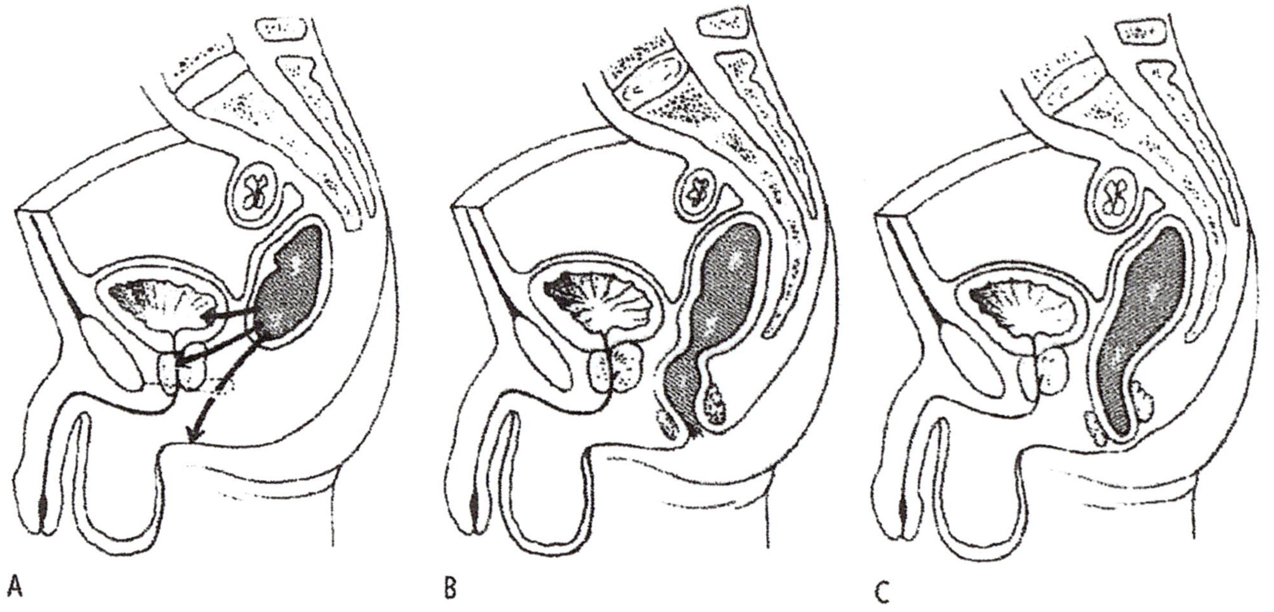

Fig. 31.9 Schematic overview of anorectal defects: (**a**) ectopic fistulae of the anorectal canal into the urogenital tract and perineum, (**b**) stenosis of the anorectal canal, and (**c**) atresia of the anorectal canal. (After Ladd WE and Gross RE 1934 *Congenital malformations of anus and rectum.* Am J Surgery 23:167–183; and ten Donkelaar HJ and Oostra R-J 2014 *Klinische Anatomie en Embryologie*, 4th ed. Springer Media, Houten)

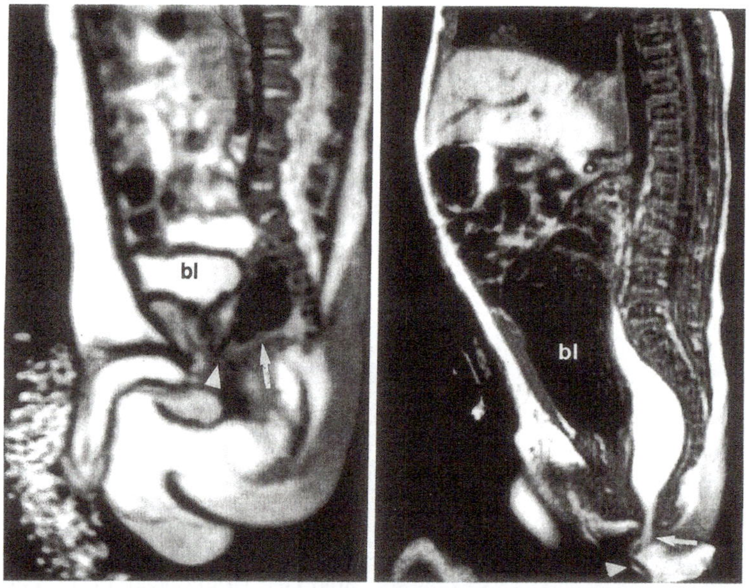

Fig. 31.10 MRIs of (**a**) high anorectal malformation with a rectourethral fistula (*white arrowhead*) in a 2.5 months-old boy; the rectum ends above the levator ani muscle (*white arrow*), and (**b**) stenosis of the anorectal canal with an anoscrotal fistula (*white arrowhead*) in a three-days-old boy. (After Nievelstein RAJ, Vos A, Valk J, Vermeij-Keers C *Magnetic resonance imaging in children with anorectal malformations: Embryological implications*. J Pediatr Surg 2002; 37:1138–1145; and ten Donkelaar HJ, Oostra R-J 2014 *Klinische Anatomie en* Embryologie, 4th ed. Springer Media, Houten; courtesy Rutger Jan Nievelstein, Utrecht)

possible autosomal recessive trait in some individuals. It is named after Jacques Caroli (1902–1979), a French physician, born near Versailles and working in Paris (Caroli et al. 1958).

The cystic duct (approximately 2–3 cm long and 2–3 mm in diameter) comprises multiple spiral arranged endoluminal folds known as the spiral mucosal folds or **spiral valves of Heister** (*see* Heister, Lorenz, and Fig. 31.12b). The cystic mucosal folds contain neurohormonally responsive muscle fibres and their unique shape is thought to support the patency of the cystic duct. Close to the spiral valves of Heister, the **spiral sphincter of Lütkens** is found, an *eponym* for a circular bundle of smooth muscle fibres in the neck of the gallbladder (Lütkens 1926; *see* Lütjens, Ulrich). A physiological sphincter at the confluence of the right and left hepatic ducts is known as the *eponym* **sphincter of Mirizzi** (Mirizzi 1940; *see* Mirizzi, Pablo Luis).

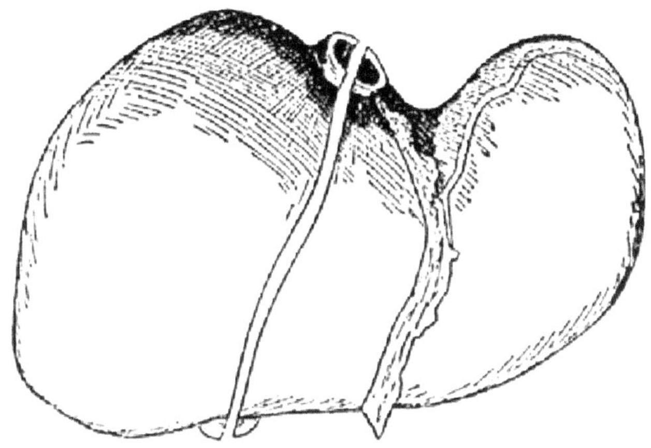

Fig. 31.11 The line of Cantlie. (From Cantlie J 1897 *On a new arrangement of the right and left lobes of the liver.* Proc Anat Soc Great Britain & Ireland. J Anat Physiol 32:4–9)

Variant anatomy of the **gallbladder** can be found in its morphology, number (either multiple gallbladders or agenesis), and location (intrahepatic, retrohepatic, transverse, retroperitoneal, left-sided; *see* Meilstrup et al. 1991). Among the morphological variants can be recognized: (a) the commonly found folding of the fundus upon itself, a feature known as *Phrygian cap* deformity, the appearance being reminiscent of a head garment worn by inhabitants of Phrygia (modern Turkey) 1200–700 BC (Meilstrup et al. 1991; Boyden 1935; Nayak et al. 2018); (b) the *sigmoid* gallbladder (Ryan et al. 2011); (c) stricture of the gallbladder lumen due to mural thickening, termed as *'hourglass'* appearance (Wong and Lee 2017); (d) septation, such as the *'honeycomb'* gallbladder, caused by a thickened gallbladder wall with numerous nonshadowing echogenic septa traversing the lumen (Robinson et al. 2019; Fig. 31.12c); (e) the **pouch of Hartmann**, an *eponym* for a dilatation of the neck of the gallbladder (*see* Hartmann, Henri Albert Charles Antoine); (f) a floating gallbladder, due to the presence of a gallbladder peritoneal mesentery; and (g) diverticula, containing all layers of the gallbladder wall. Moreover, the mucous crypts in the wall of the gallbladder may undergo mucosal and smooth muscle hyperplasia, first described by both Rokitansky (*see* Rokitansky, Carl Joseph Wenzel Prokop von) and Aschoff (*see* Aschoff, Karl Albert Ludwig), and known as the *eponym* **sinuses of Rokitansky-Aschoff**.

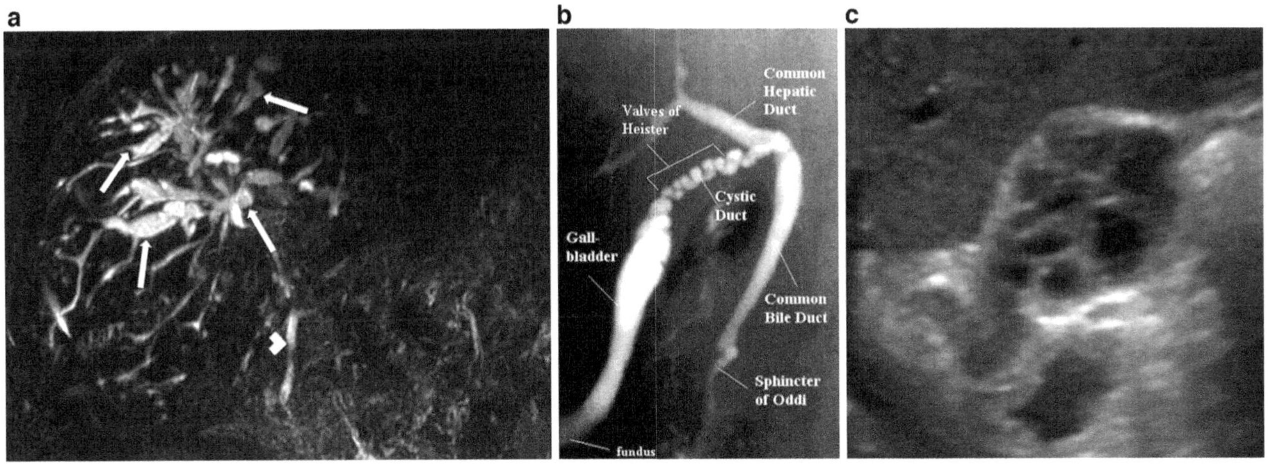

Fig. 31.12 (**a**) Calori disease, (**b**) valves of Heister, and (**c**) a 'honeycomb' gallbladder. ((**a**) From Karaosmanoglu AD, Ardslan S, Akata D, Ozmen M, Haliloglu M, Oguz B, Karcaaltincaba M 2020 *Congenital and hereditary cystic disease of the abdomen.* Insights in Imaging 11:90; Open Access, courtesy Ali Devrim Karaosmanoglu, Ankara; (**b**) from Al-Atabi M, Chin CB, Luo XY 2006 *Cystic duct visual-based evaluation of gallstones formation risk factors.* J Engin Sci Technol 1:1–9, courtesy Mushtak Al-Atabi, Selangor, Malaysia, and (**c**) from Robinson DW, Oliphant M and Dyer RB 2019 *The 'honeycomb' gallbladder.* Abdom Radiol 44:1627–1628; courtesy David W Robinson, Winston-Salem; with permission)

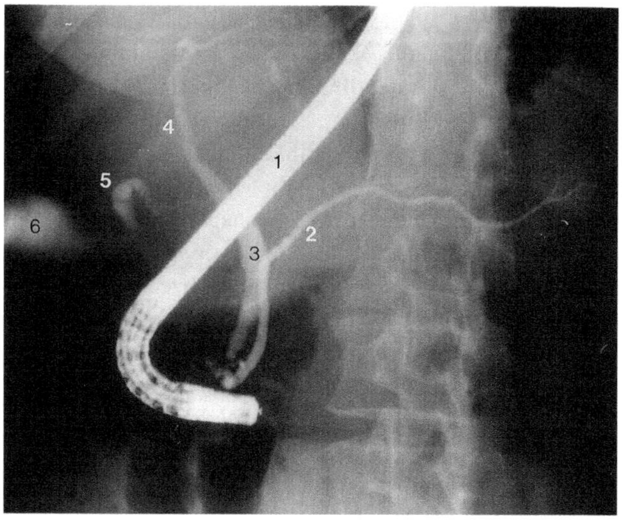

Fig. 31.13 Endoscopic retrograde cholangiopancreaticography: *1* endoscope, *2* pancreatic duct, *3* bile duct, *4* common hepatic duct, *5* cystic duct, *6* gallbladder. (After ten Donkelaar HJ and Oostra R-J 2014 *Klinische Anatomie en Embryologie*, 4th ed. Springer Media, Houten; courtesy Ton van der Vliet, Groningen)

Boyden (1941a, b, 1957) described a **four-part** sphincter system for the bile duct, the pancreatic duct and the hepatopancreatic duct: (a) superior and (b) inferior (or preampullary) sphincters of the bile duct, (c) the sphincter of the pancreatic duct, and (d) the sphincter of the hepatopancreatic or biliaropancreatic ampulla (*see* **Fig. B31**). The **sphincter of Oddi** is the sphincter of the hepatopancreatic ampulla (*see* Oddi, Ruggero). The biliary duct and the pancreatic **duct of Wirsung** (*see* Wirsung, Johann Georg) open into the duodenum at the **ampulla of Vater** (*see* Vater, Abraham), also known as **major caruncle of Santorini** (Fig. 31.13). The accessory pancreatic **duct of Santorini** (*see* Santorini, Giovanni Domenico) opens on the minor duodenal papilla or **minor caruncle of Santorini**.

References

Apaydin N, Uz A, Elhan A, Loukas M, Tubbs RS (2008) Does an anatomical sphincter exists in the distal esophagus? Surg Radiol Anat 30:11–16

Balli R (1939) The sphincters of the colon. Radiology 33:372–376

Barrett NR (1950) Chronic peptic ulcer of the oesophagus and 'oesophagitis'. Br J Surg 38:175–182

Boerhaave H (1724) Atrocis, nec descripti prius, morbii historia: secundum medicae artis leges conscripta, Leiden

Boyden EA (1935) The phrygian cap in cholecystography: a congenital anomaly of the gallbladder. Am J Radiol 33:589

Boyden EA (1941a) The sphincter of Oddi. Surgery 9:443–445

Boyden EA (1941b) Hypertrophy of the sphincter choledochus: a cause of internal ciliary fistula. Surgery 10:567–571

Boyden EA (1957) The anatomy of the choledochoduodenal junction in man. Surg Gynecol Obst 104:1641–1652

Busi A (1933) Tecnica e diagnostica radiologica delle malattie chirurgiche. Utet, Torino

Cannon WB (1902) The movements of the intestines studied by means of Röntgen rays. Am J Phys 6:251–277

Cannon WB (1911) The mechanical factors of digestion. Edward Arnold, London

Cantlie J (1897) On a new arrangement of the right and left lobes of the liver. Proc Anat Soc Great Britain and Ireland 32:4–9

Caroli J, Soupault R, Kossakowski J, Plocker L, Paradowska M (1958) La dilatation polykystigul congénitale devoies biliares intrahépatigues: Essai de classification. Sém Hop 34:488–495

Chagas C (1909) Nova trypanozomiase humana: Estudos sobre a morfolojia e o cíclo evolutivo do schizotripanum cruzi n.gen., n.sp., ajente etiolojico de nova entidade morbida do homem. Mem Inst Oswaldo Cruz 1:159–218

Collan Y, Hakkiluoto A, Hästbacka J (1978) Ectopic liver. Ann Chir Gynaecol 67:27–29

Couinaud C (1957) Le Foie. Etudes anatomiques et chirurgicales. Masson, Paris

Cullen TS (1925) Accessory lobes of the liver: an accessory hepatic lobe springing from the surface of the gallbladder. Arch Surg 11:718–764

Dufour H, Fredet P (1908) La sténose hypertrophique du pylore chez le nourisson et son traitement chirurgical. Rev Chir 37:208

Eliška O (1973) Phreno-oesophageal membrane and its role in the development of hiatal hernia. Acta Anat 86:137–150

Estrada RL (1958) Anomalies of intestinal rotation and fixation. Charles C Thomas, Springfield

Friedland GW (1978) Historical review of the changing concepts of lower esophageal anatomy: 430 B.C.-1977. Am J Roentgenol 131:373–388

Gagliardi JA, Radvany MG, Kilkenny TE, Russo RD Jr (1994) Colonic sphincters revisited: simulation of organic disease. Hawaii Med J 53:278–282

Gray SW, Skandalakis JE (1972) Embryology for surgeons. Saunders, Philadelphia

Healey JE Jr, Schroy PC (1953) Anatomy of the biliary ducts within the human liver; analysis of the prevailing pattern of branchings and the major variations of the biliary ducts. Arch Surg 66:599–616

Helvetius JCA (1719) Observations anatomiques sur l'estomac de l'homme, avec des reflexions sur le système nerveux, qui regarde la trituration dans l'estomac, comme la cause de la digestion des ailments. Hist Acad Roy Sci, Paris, pp 336–349

Hirsch I (1924) The cecocolic sphinteric tract. Med J Rec 6:541–549

Hirschsprung H (1887) Stuhlträgheit Neugeborener in Folge von Dilatation und Hypertrophie des Colons. Jahrb Kinderheilk physische Erziehung 27:1–7

Hirschsprung H (1888) Fälle von angeborener Pylorusstenose beobachtet bei Säuglingen. Jahrb Kinderh 28:61–68

Hollinshead WH (1971) Anatomy for surgeons, Vol 2: the thorax, abdomen and pelvis. Harper & Row, New York

Hyrtl J (1846) Lehrbuch der Anatomie des Menschen mit Rücksicht auf physiologische Begründung und praktische Anwendung. Friedrich Ehrlich, Prague

Juvara E (1899) Leitfaden für die chirurgische Anatomie. Hirschwald, Berlin

Kanne JP, Rohrmann CA, Lichtenstein JE (2006a) Eponyms in radiology of the digestive tract: historical perspectives and imaging appearances. Part 1. Pharynx, esophagus, stomach and intestine. Radiographics 26:129–142

Kanne JP, Rohrmann CA, Lichtenstein JE (2006b) Eponyms in radiology of the digestive tract: historical perspectives and imaging appearances. Part 2. Liver, biliary system, pancreas, peritoneum, and systemic disease. Radiographics 26:465–480

Kapoor A, Harshavardhan KR, Mutnuru PC, Narayanan R (2017) Accessory hepatic lobe – a 'not so rare' entity. Int J Anat Radiol Surg 6:26–30

Kelahan L, Menias CO, Chow L (2021) A review of internal hernias related to congenital peritoneal fossae and apertures. Abdom Radiol 46:1825–1836

Killian G (1908) La bouche de l'oesophage. Ann Mal Oreil Larynx 34:1–52

Ladd WE (1932) Congenital obstruction of the duodenum in children. New Engl J Med 206:277–283

Ladd WE, Gross RE (1934) Congenital malformations of anus and rectum. Am J Surg 23:167–183

Laimer E (1883) Beiträge zur Anatomie des Oesophagus. Med Jahrb Wien:333–388

Langer JC (2017) Intestinal rotation abnormalities and midgut volvulus. Surg Clin North Am 97:147–159

Lütkens U (1926) Aufbau und Funktion der extrahepatischen Gallenwege. Vogel

Maccioni F, Busato L, Valenti A, Cardaccio S, Longhi A, Catalano C (2023) Magnetic resonance imaging of the gastrointestinal tract: current role, recent advancements and future prospectives. Diagnostics 13:241

Mall FP (1889) Development of the human intestine and its position in the adult. Bull Johns Hopkins Hosp 9:197–208

Mallory GK, Weiss S (1929) Hemorrhage from laceration of the cardiac orifice of the stomach due to vomiting. Am J Med Sci 178:506–514

Meckel JF (1817) Bildungsgeschichte des Darmkanals der Saugtiere und namentlich des Menschen. Deutsch Arch Physiol 3:1–84

Medeiros de Araújo EM, Torres US, Pria HD, Torres LR, Pedroso MHNI, Racy DJ, D'Ippolito G (2022) Anatomy and imaging of accessory liver lobes: what radiologists should know. Semin Ultrasound CT MR 43:476–489

Meilstrup J, Hopper K, Thieme G (1991) Imaging of gallbladder variants. AJR Am J Roentgenol 157:1205–1208

Mirizzi PB (1940) Physiologic sphincter o hepatic bile duct. Arch Surg 41:1325–1335

Nayak S, Aithal A, Padavinangadi A, Prabhu G (2018) Double pouched, sigmoid gallbladder that can cause a diagnostic dilemma to radiologists: a case report. Anat Cell Biol 51:209–211

O'Beirne J (1833) New views of the process of defecation, and their application to the pathology and treatment of diseases of the stomach, bowels, and other organs. Together with an analytical correction of Sir Charles Bell's views respecting the nerves of the face. Dublin

Ramstedt C (1912) Zur Operation der angeborenen Pylorusstenose. Med Klin 8:1702

Rex H (1888) Beiträge zur Morphologie der Säugerleber. Morphol Jahrb 14:517–615

Riedel BMCL (1888) Ueber den zungenformigen Fortsatz des rechten Leberlappens und seine pathognostische Bedeutung für die Erkrankung der Gallenblase nebst Bemerkungen ueber Gallensteinoperationen. Berlin Klin Wschr 577–581:602–607

Robinson DW, Oliphant M, Dyer RB (2019) The "honeycomb" gallbladder. Abdom Radiol 44:1627–1628

Rubesin SE, Levine MS (2001) Killian-Jamieson diverticula: radiographic findings in 16 patients. AJR Am J Roentgenol 172:85–89

Ryan S, McNicholas M, Eustace SJ (2011) Anatomy for diagnostic imaging. ISBN: 9780702029714 – Google Books

Schatzki R, Gary JE (1953) Dysphagia due to a diaphragm-like localized narrowing in the lower esophagus ("lower esophageal ring"). Am J Roentgenol Radium Therapy, Nucl Med 70:911–922

Schwalbe E (1906) Die Morphologie der Missbildungen des Menschen und der Tiere, Teil 1 Allgemeine Missbildungslehre (Teratologie). Fischer, Jena

Snyder WH Jr, Chaffin L (1954) Embryology and pathology of the intestinal tract: presentation of 40 cases of malrotation. Ann Surg 140:368–379

Steward DR, Colodny AL, Dagget WC (1976) Malrotation of the bowel in infants and children: a 15 year review. Surgery 79:716–720

Strasberg SM (2005) Nomenclature of hepatic anatomy and resections: a review of the Brisbane 2000 system. J Hepatobiliary Pancreatic Surg 12:351–355

Watanabe Y, Taniyama Y, Koseki K, Ishid H, Ozawa Y, Okamoto H, Sato C et al (2023) Distinguishing Killian-Jamieson diverticulum from Zenker's diverticulum. Surg Case Rep 9:21

Weber W (1910) Ueber eine technische Neuerung bei der Operation der Pylorusstenose des Säuglings. Berl Klin Wschr 47:763

Wong H, Lee K (2017) The hourglass gallbladder. Abdom Radiol 43:1268–1269

Zenker FA, von Ziemssen H (1877) Krankheiten des Oesophagus. In: von Ziemssen H, Hrsg Handbuch der speciellen Pathologie und Therapie. Siebenter Band, Erste Hälfte. Vogel, Leipzig, pp 1–87

Zifan A, Kumar D, Cheng LK, Mittal RK (2017) Three-dimensional myoarchitecture of the lower esophageal sphincter and esophageal hiatus using optical sectioning microscopy. Sci Rep 7:13188

GPSR Compliance

The European Union's (EU) General Product Safety Regulation (GPSR) is a set of rules that requires consumer products to be safe and our obligations to ensure this.

If you have any concerns about our products, you can contact us on ProductSafety@springernature.com

In case Publisher is established outside the EU, the EU authorized representative is:

Springer Nature Customer Service Center GmbH
Europaplatz 3
69115 Heidelberg, Germany

Batch number: 10755966

Printed by Printforce, the Netherlands